Geophysical Monograph Series

Including

IUGG Volumes
Maurice Ewing Volumes
Mineral Physics Volumes

Geophysical Monograph Series

Geophysical Monograph 152

Sea Salt Aerosol Production: Mechanisms, Methods, Measurements and Models—A Critical Review

Ernie R. Lewis
Stephen E. Schwartz
Authors

American Geophysical Union
Washington, DC 2004

Library of Congress Cataloging-in-Publication Data

Sea salt aerosol production : mechanisms, methods, measurements and models : a critical review/Ernie R. Lewis, Stephen E. Schwartz.
   p. cm. – (Geophysical monograph ; 152)
 Includes bibliographical references and index.
 1. Sea salt aerosols. 2. Sea salt aerosols–Measurement. 3. Ocean-atmosphere interaction. 4. Marine meteorology. I. Schwartz, Stephen E. (Stephen Eugene) II. Title. III. Series.

 QC929.S43L49 2004
 551.5'246–dc22

                   2004027707

ISBN 087590-417-3
ISSN 0065-8448

CONTENTS

Preface

Sea salt aerosol (SSA) exerts a major influence over a broad reach of geophysics. It is important to the physics and chemistry of the marine atmosphere and to marine geochemistry and biogeochemistry generally. It affects visibility, remote sensing, atmospheric chemistry, and air quality. Sea salt aerosol particles interact with other atmospheric gaseous and aerosol constituents by acting as sinks for condensable gases and suppressing new particle formation, thus influencing the size distribution of these other aerosols and more broadly influencing the geochemical cycles of substances with which they interact. As the key aerosol constituent over much of Earth's surface at present, and all the more so in pre-industrial times, SSA is central to description of Earth's aerosol burden.

There is, however, a further compelling reason for current interest in SSA—its influence on climate. Sea salt aerosol particles scatter solar radiation and, by serving as seed particles for cloud drops, greatly affect the microphysics and reflectivity of marine clouds and development of precipitation. Understanding these influences of SSA is essential to understanding and quantifying such influences of anthropogenic aerosols. These effects of anthropogenic aerosols have been identified by the Intergovernmental Panel on Climate Change as the greatest uncertainty in radiative forcing of climate change over the industrial period [*IPCC*, 2001, p. 8]. Understanding SSA and representing it in models is thus central to decreasing this key uncertainty in climate change. Unfortunately the requisite understanding is not yet at hand. There is much of importance concerning SSA for which uncertainties reside not at the 10% level or even at the factor-of-two level, but rather at the order-of-magnitude level.

This book focuses on the production of SSA particles and their subsequent behavior, transport, and removal—all as a function of particle size—and on the factors that control these processes. It is a thorough review of SSA, treating the topics that are required to understand and evaluate SSA properties and production and examining data from a large number of sources, and a critical one, carefully examining what is demonstrated in measurements as opposed to merely quoting conclusions stated by other investigators. While this book is a self-contained reference on SSA and its production, with little need for reliance on other works but with numerous internal cross-references between sections, it also serves as an introduction and guide to the literature on SSA, with over 1800 references.

This review had its genesis in an attempt to develop a representation of the size-dependent production flux of SSA particles and its dependence on controlling factors for use in chemical transport models suitable for examining the climate influence of aerosols. An initial examination of the literature revealed that several expressions had been developed to describe the size-dependent SSA production flux and its dependence on wind speed. These expressions, which have been widely used in chemical transport and climate models, led to an initial impression, based on their rather detailed dependencies on particle size and wind speed and on the numbers of significant digits in these expressions, that quantitative description of the SSA production flux was well in hand. However, the more we looked into the origin of these expressions and the extent to which they were supported by measurements, the more we came to question the confidence that could be placed in them. We were ultimately led to the conclusion that SSA production is far less well understood than is indicated by much of the modeling community and indeed than appears to be appreciated by some investigators, who may be familiar with only a portion of the relevant literature.

These considerations motivated a more thorough review of the literature pertinent to SSA and its production. In conducting this review, we examined the great majority of the literature that is readily available (and much that is not). This literature is quite voluminous and diverse, touching on many aspects of physics and chemistry as well as meteorology, geophysics, and geochemistry. Inevitably in such a broad literature there are many data on numerous aspects of SSA and its production, and good measurements of SSA concentrations extend back some fifty years. However there has been little synthesis of this literature, and little critical examination. It was the need for such a synthesis and critical evaluation that led to the review presented here.

In denoting this review as a critical evaluation we use the term in the sense described by *Brewer* [1988], who observed that "Critical evaluation is not a simple job; it requires a thorough study of all information related to the system being reviewed." He elaborates:

> Published results are often in error...One must often reject most or all of the reported results. One must review carefully the experimental procedures used and examine the publication or previous publications by the same authors to determine whether they have been taking the effort to pin down systematic errors. It is necessary to utilize any reliable predictive models to determine if the experimental data are reasonable even within an order of magnitude. It is most important to

give the reader some feeling for the possible uncertainty of the final accepted values. The ± values assigned should not be just the statistical uncertainty (which is a measure of the scatter around the average). The uncertainty cited should include an estimate of possible systematic errors... It is particularly important to not arrive at the recommended value by just averaging all the reported results; that would just mix the reasonably correct results with those in serious error.

In this spirit our intent has been not simply to compile what has been reported in the literature, but to critically analyze and examine the various investigations and their results, and thus to assess the extent to which they can provide meaningful quantitative information pertaining to the size-dependent SSA production flux and the factors upon which it depends. In our efforts we have found encouragement in an exhortation articulated by the eleventh century scholar al-Hasan ibn al-Haytham [*Sabra*, 2003]:

The seeker after the truth is not one who studies the writings of the ancients and, following his natural disposition, puts his trust in them, but rather the one who suspects his faith in them and questions what he gathers from them... The duty of the man who investigates the writings of scientists, if learning the truth is his goal, is to make himself an enemy of all that he reads, and applying his mind to the core and margins of its content, attack it from every side. He should also suspect himself as he performs his critical examination of it, so that he may avoid falling into either prejudice or leniency.

We have endeavored to follow these precepts and hope thereby to have achieved the objective of bounding confidence in the knowledge of sea salt aerosol and its production.

A guiding principle in preparing this review has been to carry out our assessments wherever possible by examining original sources rather than relying on secondary literature. In doing so we have found that not infrequently substantial errors, some of them so-called "illegitimate errors"—those of the sort "which originate from mistakes or blunders in computation or measurement" [*Bevington*, 1969, p. 2]—had been made and in some instances these have been propagated by subsequent investigators. While mistakes and blunders are perhaps inevitable in scientific research (as elsewhere), it is essential that when they are found they be identified and corrected so that their consequences can be eliminated.

Another guiding principle has been to examine and compare data from many different sources. Comparison of large numbers of data sets provides insight into the variability that can be expected for measurements taken using different techniques and at different locations and times for nominally the same conditions, thus yielding some measure of the extent to which any one set of measurements is representative of others and can be said to accurately represent a given quantity. For many quantities of interest pertinent to SSA, examination of multiple data sets results in much greater variability than is apparent from just one or a few such data sets, and comparison of data reported by numerous investigators, some of which individually suggest rather narrow uncertainty ranges, shows differences of an order of magnitude or more among studies using the same method, and often of several orders of magnitude among studies using different methods.

Throughout this review we have also sought to compare multiple determinations of the same quantity by plotting them on the same graph. In this we have been stimulated by the work of *Tufte*, who writes in his book *The Visual Display of Quantitative Information* [1983, p. 9]:

Modern data graphics can do much more than simply substitute for small statistical tables. At their best, graphics are instruments for reasoning about quantitative information. Often the most effective way to describe, explore, and summarize a set of numbers— even a very large set—is to look at pictures of those numbers. Furthermore, of all methods for analyzing and communicating statistical information, well-designed data graphics are usually the simplest and at the same time the most powerful.

We rely heavily on such graphical comparisons in this review, and have applied this principle to many quantities pertinent to sea salt aerosol and its production, such as total concentrations of SSA mass and number, size distributions of SSA concentration, size distributions of bubble concentration, and fractional whitecap coverage. For each of these quantities numerous data sets have been digitized, converted to common units and means of representation, and presented graphically, resulting in more than twenty color plates. The compilation and presentation of this large and diverse body of measurements is necessary to evaluate the formulations that have been presented for the SSA production flux, and we hope that the graphics we present meet the above objectives and that the reader will concur in their utility. Additionally, these graphs provide a framework against which any new data can be compared, and it is hoped that investigators presenting such data will use these graphs for this purpose. The "inertia" of the large number of data sets presented here for various quantities of interest should temper claims that rest on a single data set.

In our quest to determine the size-dependent SSA production flux and its dependence on controlling factors we have identified several independent approaches. We compare the results obtained with these approaches by converting them to the same representation and presenting them on the same graph. This comparison likewise reveals large discrepancies.

Of course meaningful comparison requires consideration of uncertainties due to natural variability, as manifested in the spread of the data within individual studies and between multiple studies, and also consideration of systematic errors or biases in the approaches, the consequences of which must be recognized and dealt with. Results have frequently been reported with a precision that cannot be justified based on the reproducibility and the scatter of the data within a given study—all the more so when results from multiple studies are compared.

A further benefit of comparing multiple data sets for a given quantity is that such a comparison provides insight into the accuracy and utility of analytic expressions that have been presented to parameterize this quantity and its dependence on controlling factors, thus allowing assessment of the confidence that can be placed in these parameterizations. Measured data have commonly been fitted to analytic expressions, often rather complex ones with high implied precision. Here we find ourselves in agreement with *Mason* [2001], who lamented the complexity of proposed functions "with up to nine numerical coefficients...some quoted to four significant figures!" Such fits often obscure the findings and fail to convey the variability of the quantities under examination and the uncertainties of the measurements.

In turn these fits and parameterizations have been used in models or for estimating other quantities of interest, again implying that these quantities are known to greater accuracy than can be justified from the available data, which can lead to conclusions not substantiated by these data. Unfortunately, several of these expressions appear to have become accepted by a rather large community, perhaps more through repeated use than by objective standard.

In the course of this work we found it necessary to apply many fundamental principles and properties of sea salt aerosol particles and their dynamics. As description of these phenomena draws on quite a diverse literature, and as many of these key fundamentals were often not readily accessible or easily applied, we found it helpful to provide numerical expressions for these properties so that we could effectively use and refer to them. In doing so we have generally relied on simple models and approximations for processes and quantities of interest, the objective being to obtain expressions that both reflect the underlying physics of the processes under consideration and provide realistic numerical estimates. The goal was to keep the physical sense of the subject foremost, allowing identification of what is important and what is secondary. Numerous formulas and expressions that can readily be used to estimate important quantities are presented with an accuracy typically of a few percent; almost always the uncertainties in the various aspects of the production of SSA and its behavior in the marine atmosphere are such that little is to be gained from the use of more detailed

models, and more precise numerical results can rarely be justified. We hope that this compilation will be useful to future investigators.

Although this review is concerned primarily with SSA production, in view of its wide-ranging scope much of it will be of broader interest. For instance, the discussion of fundamental principles and properties will be useful to investigators interested in atmospheric aerosols in general and marine aerosols in particular, and to others who desire a reference on the behavior of aerosol particles with respect to relative humidity, the kinematics and dynamics of aerosol particles, and like topics. Additionally, the compilation and presentation of the large and diverse body of measurements will be of value to atmospheric scientists and oceanographers concerned with air-sea interaction, biogeochemical cycles, atmospheric radiation and its transmission in the marine atmosphere (especially with regard to Earth's albedo and radiation budget), and remote sensing of the ocean, especially by satellite.

To call attention to the rich body of research pertinent to SSA production, we occasionally present rather long lists of references in the hope that these citations will be of value to future investigators. While we have tried to maintain the convention of citing references in the text by author and year, when such references become so numerous as to impede the readability of the text, we have placed them in footnotes.

We hope this book will be useful to investigators with much experience with SSA as well as to those with relatively little experience. Some of the former may wish to skip some of the fundamental material, but we would hope that even they would find much here that is valuable. While some classical "textbook" material is included, such as the description of representation of aerosol size distributions, we have included this material deliberately to ensure, by means of definitions and equations, that a term such as "size distribution of number concentration" is unambiguously defined. Definitions of such quantities differ substantially among different references, so that without explicit definitions the same words can mean different things to different readers. We have seen, and call attention in the text to, all too many instances of ambiguity on the part of authors and erroneous interpretation on the part readers when quantities are not explicitly defined. This applies even to the "well-understood" lognormal distribution, for which different references give different variants and ascribe different meanings to the same symbol or named quantity.

While some readers may wish to skip certain sections on first reading, we would encourage all readers to read §2.1.1 (description of particle sizes), §2.1.2 (classification into size ranges), §2.1.3 (description of size distributions), §2.1.4 (description of concentrations), and §2.1.6 (description of fluxes), which introduce important terminology that is used throughout the book. Sections that might be skipped on first reading include §2.1.5 (cloud drop activation, uptake and reaction of gases and particles, and light scattering), §2.1.7 (fluxes of SSA properties) and §2.6 (kinematics and dynamics of SSA particles), but we expect that these sections will be useful to many investigators concerned with these geophysical processes involving SSA. We expect that much of the book might be useful as supplemental reading for graduate or advanced undergraduate courses in atmospheric aerosols.

So, at the end of the day, where are we? If nothing else we hope that this book leaves the reader with the impression that much work remains to be done. While it is possible to present a picture of the production of sea salt aerosol, this picture is still painted with a rather broad brush. One is thus hard put to use with confidence any single model or to recommend any single formulation for the size-dependent production flux of SSA and its dependence on controlling factors with anywhere near the precision that is required, for example, to quantify aerosol influences on climate change. However, at least the uncertainty in SSA production is now much better quantified, so that the implications of this uncertainty can be recognized. Additionally, many reasons for the uncertainties in present estimates are identified. We thus hope that this book suggests fruitful directions for future research. New, more direct approaches to measurement of the size-dependent production flux of sea salt aerosol are now at hand, and it is hoped that with concerted effort rapid progress can be made in determining this important but elusive quantity and its dependence on controlling factors.

We thank Duncan Blanchard, Donald Lenschow, Peter Minnett, and Harry ten Brink, and our Brookhaven National Laboratory (BNL) colleagues Mark Miller, Tim Onasch, and Gunnar Senum for valuable discussions. We thank the research librarians at BNL for their great help in tracking down obscure publications, Judith Williams for her expertise in preparation of the copy, and Peter Daum and Leonard Newman for their support. We especially thank Antony Clarke, Edward Monahan, and Jeffrey Reid, who provided anonymous reviews of the manuscript to the American Geophysical Union and who subsequently revealed their identities to us, for their expertise, the time and effort they gave to their reviews, their challenges to arguments contained within the book, their referring us to additional works, and their overall contribution to substantially enhancing this book. We, the authors, take full responsibility for the viewpoints expressed within the book (and for any remaining errors). This review has benefitted from the continued support of the Atmospheric Sciences Program of the United States Department of Energy, for which we thank the program officers. We dedicate this book to our parents, Ray and Jennie Lewis, and Alfred and Ellen Jane Schwartz, for their love and encouragement.

Ernie R. Lewis
Stephen E. Schwartz
July 2004

1. Introduction

Sea salt aerosol (SSA) particles[1] are formed predominantly by the action of the wind on the ocean. The wind stress on the ocean surface forms waves, some of which break and entrain air to various depths. The bubbles thus formed rise to the surface, creating whitecaps, and burst, injecting seawater drops into the atmosphere. Seawater drops are also produced from bubbles produced by other mechanisms (§4.7). Additionally, under sufficiently strong winds, drops are torn directly off wave crests. The rate of production of seawater drops, per unit area of ocean surface, is denoted the interfacial SSA production flux (§2.1.6).

Once produced, some of the drops, depending on their size and on meteorological conditions (primarily wind speed), are entrained upward by turbulent eddies—small parcels of air with fluctuating vertical velocity components which characterize the small-scale motion of the marine atmosphere—and by large-scale convection, while others fall back to the sea. The rate of production of those drops that remain in the atmosphere sufficiently long to contribute appreciably to the aerosol in the marine (atmospheric) boundary layer, per unit area of ocean surface, is denoted the effective SSA production flux (§2.1.6). It is important to distinguish this effective SSA production flux from the interfacial SSA production flux. Depending mainly on drop size the effective SSA production flux may be nearly equal to the interfacial SSA production flux (small drops) or it may be substantially less because of rapid removal by gravity (large drops).

Sea salt aerosol particles are important for a variety of reasons (§1.1): they act as cloud condensation nuclei to form cloud drops, exchange gases with the atmosphere and engage in chemical reactions, scatter light, exchange moisture with the atmosphere, and participate in geochemical cycles of elements (§2.1.5; §2.1.7). The importance of SSA particles

of a given size to a particular atmospheric process depends on the size, the concentration of these particles, their mean residence time in the atmosphere, and the extent to which they are mixed vertically. This importance is thus controlled by meteorological and environmental factors that affect formation and breaking of ocean waves; properties and areal coverage of whitecaps; production, dynamics, and bursting of bubbles; the formation of drops, their upward entrainment, behavior in the atmosphere, and transport; and the processes that act to remove these drops. These topics define the focus of this review: the production of SSA particles and their subsequent behavior, transport, and removal—all as a function of particle size—and the factors that control these processes.

1.1. SEA SALT AEROSOL AND ITS IMPORTANCE

Marine aerosol—the aerosol in the marine atmospheric boundary layer—is comprised of a variety of components, both natural and anthropogenic: liquid seawater drops and dry sea salt particles, dusts and minerals from continental regions and from volcanos, pollen and other biological particles, sulfates, nitrates, and organic compounds, including soot and other carbonaceous material [*Prospero*, 2002]. The various aerosol components are generated from the ocean surface; formed by gas-to-particle conversion mechanisms such as homogeneous nucleation, heterogeneous nucleation, and condensation; transported from land; produced by anthropogenic activities at sea (e.g., ship exhaust); injected by volcanic eruptions; and entrained from the free troposphere. Marine aerosol may be internally mixed (individual particles having nonuniform composition), externally mixed (individual particles having uniform composition, which may differ between particles), or both, and the radii of the aerosol particles range from nanometers to millimeters. Because approximately two-thirds of Earth's surface is covered by oceans, marine aerosol exerts a large impact on various geochemical and geophysical processes and on Earth's climate.

Sea salt aerosol is defined as that aerosol component consisting of seawater drops and dry sea salt particles. The radii of SSA particles range from less than 0.1 μm to greater than 1000 μm. Sea salt aerosol is an important component of the marine aerosol, and in the undisturbed marine environment, far from continental and anthropogenic sources, it is often the

---

[1] We distinguish here and throughout this review between sea salt aerosol particles (which can refer to liquid, solid, or mixed-phase material) and sea salt aerosol, which consists of the suspension of these particles in the atmosphere.

Sea Salt Aerosol Production: Mechanisms, Methods, Measurements, and Models
Geophysical Monograph Series 152
Copyright 2004 by the American Geophysical Union.
10.1029/152GM01

Table 1. Estimates of the Annual Global Sea Salt Aerosol Mass Production

Reference	Annual Production/ (10^{12} kg)	Basis of Estimate
Köhler, 1925	1.96985	Budget of cloud condensation nuclei
Eriksson, 1959	1	Mass balance using estimated dry and wet deposition and measured SSA concentrations
Eriksson, 1960	1.70	Estimate from continental run-off of sulfur
Eriksson, 1963	1.75	Probably rounded off from estimate in *Eriksson* [1960], which was stated in grams of sulfur per year
Blanchard, 1963 (p. 187)	10	Mass balance using estimated dry and wet deposition and measured SSA concentrations at 600-800 m
Hidy & Brock, 1971[a]	1.095	Average of estimates of *Eriksson* [1960] and of *Junge* [1963, p. 159], which in turn was based on *Eriksson* [1959]
SMIC, 1971 (Table 8.1)	0.3	Based on estimate of *Eriksson* [1959] of SSA dry deposition over land
Hsu & Whelan, 1976	1.1	Model of whitecaps as line sources, using estimated distance between whitecaps, wind speed, and SSA mass concentration
Petrenchuk, 1980	1.3-1.9	Mass balance using estimated wet deposition and measured chloride concentration in precipitation
Kritz & Rancher, 1980	0.537	Mass balance using measured sodium concentrations and estimated mean residence time
Mészáros, 1982	> 6.31	Mass balance using estimated sulfate dry and wet deposition
Várhelyi & Gravenhorst, 1983	5-10	Mass balance using estimated sulfate dry and wet deposition
Weisel et al., 1984	3	Geometric mean of range of estimates presented by *Eriksson* [1959], *Blanchard* [1963], *Ahmed* [1978], and *Petrenchuk* [1980]
Möller, 1984	6.81 ± 3.41	Based on estimates of SSA sulfur emission rates by *Eriksson* [1959, 1960], *Mészáros* [1982], and *Várhelyi and Gravenhorst* [1983]
Blanchard, 1985a	1-10	Estimates of *Eriksson* [1959] and *Blanchard* [1963]
Monahan, 1986	2.89[b]	SSA production flux formulation of *Monahan et al.* [1986] multiplied by whitecap ratios determined from global maps of wind speeds
Spillane et al., 1986	3.50[c]	SSA production flux formulation of *Monahan et al.* [1986] multiplied by whitecap ratios determined from global maps of wind speed
Woolf et al., 1988	20[d]	SSA production flux formulation of *Woolf et al.* [1988] assuming average whitecap ratio of 0.01
Erickson & Duce, 1988	10-30	Mass balance using global maps of estimated dry and wet deposition and SSA concentrations
Möller, 1990	10	Mass balance using estimated dry deposition velocity, mixing height, and SSA concentration
Genthon, 1992b	7	Global transport model using SSA concentrations estimated from wind speeds
Andreae, 1995	1.3	Based on estimates of SSA emission rates from *SMIC* [1971, Table 8.1] (which was based on the estimate of *Eriksson* [1959]), *Blanchard* [1983], and *Petrenchuk* [1980]
Tegen et al., 1997	5.9	Global transport model using SSA concentrations estimated from wind speeds
Gong et al., 1997b	11.7	Based on SSA production flux formulation of *Monahan et al.* [1986] and wind speed history at several locations
Gong et al., 1998	3.33	Global transport model using SSA production flux formulation of *Monahan et al.* [1986]
Erickson et al., 1999	3.244	Based on chlorine input rate from global transport model of *Gong et al.* [1997a,b] using SSA production flux formulation of *Monahan et al.* [1986]
Takemura et al., 2000	3.529	Global transport model using SSA concentrations estimated from wind speeds
Penner et al., 2001[e]	3.77	Based on sodium input rate from global transport model of *Gong et al.* [1997a,b, 1998] using SSA production flux formulation of *Monahan et al.* [1986]
Grini et al., 2002	6.5	Based on global transport model using SSA production flux formulation of *Monahan et al.* [1986] for $r_{80} < 7$ μm and that of *Smith et al.* [1993] for $r_{80} > 7$ μm

Table 1. (Continued)

Reference	Annual Production/ (10^{12} kg)	Basis of Estimate
Chin et al., 2002	5.81[f] 7.56[g] 7.48[h]	Based on global transport model using SSA production flux formulation of *Monahan et al.* [1986]
Schulz et al., 2004	19.8	Based on comparison of SSA production flux formulations by *Guelle et al.* [2001]
This review (§5.11.1)	5 ⊠ 4[i]	Based on application of dry deposition method (§5.2.1.5) to canonical SSA size distribution of concentration (§4.1.4.4)

Numbers of significant digits have been retained from original references including entries for which units have been converted.

For original values stated in tons there is ambiguity of up to 10% if "ton" was not specified as metric ton, long ton, or short ton.

Some estimates implicitly refer to effective production fluxes and some (e.g., those based on the SSA production flux formulation of *Monahan et al.* [1986]) to interfacial production fluxes (§2.1.6).

Despite the appearance of independence, many other estimates in the literature (especially those in compilations of global sources) are ultimately based on those presented by *Eriksson* [1959, 1960], *Blanchard* [1963], or *Petrenchuk* [1980], although this may not be apparent without extensive data archeology. For comparison, an annual global production of $10 \cdot 10^{12}$ kg corresponds to a total effective SSA mass production flux of approximately $F_{eff}^M = 1$ μg m^{-2} s^{-1} (§2.1.7).

[a] *Junge* [1963, p. 159] estimated a global mean SSA particle production flux of 1 cm^{-2} s^{-1}, and *Hidy and Brock* [1971] assumed all of the particles had a (presumably dry) radius of 1 μm, although this yields a sea salt mass flux of $3.2 \cdot 10^6$ tons d^{-1} as opposed to the $2 \cdot 10^6$ tons d^{-1} they stated.

[b] For bubble-ejected drops with radii (presumably at 80% RH) in the range 0.8-10 μm between latitudes 70°N and 70°S.

[c] For bubble-ejected drops with 0.8 μm $< r_{80} <$ 10 μm between latitudes 70°N and 70°S.

[d] For bubble-produced drops with 0.8 μm $< r_{80} <$ 10 μm.

[e] From Table 5.7 of *Penner et al.* [2001]; the value listed in Table 5.3 of that reference is $0.334 \cdot 10^{16}$ g (citing *Gong et al.* [1998]). *Penner et al.* [2001, p. 298] stated that this value is within the range of previous estimates, but the value attributed there to *Erickson and Duce* [1988] is a factor of 10 low.

[f] Estimate for 1990.

[g] Estimate for 1996.

[h] Estimate for 1997.

[i] The symbol ⊠ ("times or divided by") denotes multiplicative uncertainty (§4.1.2.2).

dominant component (§4.1.1.1). Additionally, sea salt is one of the major contributors to the mass of particulate matter injected into the atmosphere globally [*Peterson and Junge*, 1971; *Hidy and Brock*, 1971; *SMIC*, 1971, Table 8.1; *Andreae*, 1995; *Tegen et al.*, 1997; *Raes et al.*, 2000; *Penner et al.*, 2001, Table 5.3], with estimates of the annual contribution varying from $0.3 \cdot 10^{12}$ kg to $30 \cdot 10^{12}$ kg (Table 1), corresponding to a sea salt mass flux over the oceans of $0.03 \cdot 10^{-6}$ g m^{-2} s^{-1} to $3 \cdot 10^{-6}$ g m^{-2} s^{-1}. Equivalently, this flux corresponds to a daily removal of a layer of water 0.07 μm to 7 μm thick, the evaporation of which would result in a latent heat flux of 0.002 W m^{-2} to 0.2 W m^{-2}. By contrast the global annual mean water evaporation over the ocean is 100 cm or greater [*Sverdrup*, 1951; *Defant*, 1961, Chap. 7; *Kraus and Morrison*, 1966; *Penman*, 1970; *Baumgartner and Reichel*, 1975; *Bunker*, 1976; *Weare et al.*, 1981; *Hsiung*, 1986], corresponding to a mean daily removal of a layer of water at least 0.3 cm thick with an accompanying latent heat flux of

80 W m^{-2} or more. Thus, the production of sea salt aerosol, though it contributes greatly to the mass of particulate matter injected into the atmosphere, does not contribute appreciably to the global exchange of water or latent heat between the ocean and the atmosphere, nor can considerations of the global water budget constrain estimates of SSA production.

Sea salt aerosol is of fundamental importance in air-sea interaction, and plays a dominant role in many aspects of atmospheric chemistry (recent reviews are presented by *Finlayson-Pitts and Hemminger* [2000], *Finlayson-Pitts* [2003], *Rossi* [2003], and *von Glasow and Crutzen*, [2004]), atmospheric radiation, geochemistry (including the geochemical cycles of various elements), meteorology, cloud physics, climate, oceanography, and coastal ecology. Additionally, SSA can have consequences of economic and technological importance and may play a role in human health. Many investigations have examined the roles of SSA in the atmospheric cycles of chlorine and other halogens[1] and in the atmospheric

---

[1] *Cauer* [1951]; *Robbins et al.* [1959]; *Eriksson* [1959, 1960]; *Bruyevich and Kulik* [1967]; *Duce* [1969]; *Finlayson-Pitts et al.* [1989]; *Keene et al.* [1990]; *Möller* [1990]; *Vogt and Findlayson-Pitts* [1994]; *Graedel and Keene* [1995]; *Vogt et al.* [1996]; *Sander and Crutzen* [1996]; *Andreae and Crutzen* [1997]; *de Haan and Finlayson-Pitts* [1997]; *Langer et al.* [1997]; *ten Harkel* [1997]; *Pio and Lopes* [1998]; *Kerminen et al.* [1998]; *Gard et al.* [1998]; *Keene and Savoie* [1998, 1999]; *Erickson et al.* [1999]; *Toyota et al.* [2001]; *Moldanová and Ljungström* [2001]; *von Glasow and Sander* [2001]; *Sellegri et al.* [2001]; *Caffrey et al.* [2001]; *Hoppel et al.* [2001]; *Knipping and Dabdub* [2003].

nitrogen cycle[1]. Some of these investigations and other recent studies have focused on reactions occurring inside seawater drops and their effect on the atmospheric sulfur cycle[2]. Under certain circumstances, uptake on SSA particles may act as a sink for condensable atmospheric gases, affecting the deposition rate of nitrogen to the ocean and possibly inhibiting nucleation of new particles[3]. Sea salt aerosol affects the transmission of electromagnetic radiation in the atmosphere by scattering light, thus affecting visibility, the performance of electro-optical systems, and remote sensing[4]. Sea salt aerosol exerts a major influence on the clear-sky radiation balance over oceans through changes to system albedo, thus impacting Earth's radiation budget[5]. Sea spray can contribute to the local fluxes of water, momentum, and sensible and latent heat in the ocean-atmosphere system, and recent investigations have investigated possible effects of sea spray on hurricane strength and development[6]. Many investigations have dealt with the influence of SSA on the formation and microphysical properties of fog and rain and its role as an important component of cloud condensation nuclei[7]. Sea salt aerosol is a primary contributor to the ocean-atmosphere fluxes of organic substances[8], electric charge[9], radioactivity[10], microorganisms and viruses[11], and pollen [*Valenci*, 1967]. SSA also plays a role in coastal ecology[12], corrosion and fouling of sensors[13], and weathering of historical stone and marble monuments[14]. Additionally, sea salt concentrations in glacial ice cores have

---

[1] *Robbins et al.* [1959]; *Savoie and Prospero* [1982]; *Finlayson-Pitts* [1983]; *Harrison and Pio* [1983]; *Mamane and Gottlieb* [1992]; *Karlsson and Ljungström* [1995]; *Pakkanen* [1996]; *ten Brink* [1998]; *Gard et al.* [1998]; *Karlsson and Ljungström* [1998]; *Pryor and Sørenson* [2000]; *Spokes et al.* [2000]; *Song and Carmichael* [2001b].

[2] *Clarke and Radojevic* [1984]; *Miller et al.* [1987]; *Twohy et al.* [1989]; *Sievering et al.* [1991, 1992]; *Chameides and Stelson* [1992]; *McInnes et al.* [1994]; *Sievering et al.* [1995]; *Pósfai et al.* [1995]; *Mouri et al.* [1997]; *McInnes et al.* [1997]; *O'Dowd et al.* [1997a]; *Clegg and Toumi* [1997, 1998]; *Keene et al.* [1998]; *Mouri et al.* [1999]; *Sievering et al.* [1999]; *Song and Carmichael* [1999]; *O'Dowd et al.* [1999a,c]; *Gebel et al.* [2000]; *Kerminen et al.* [2000]; *Krischke et al.* [2000]; *van den Berg et al.* [2000]; *Song and Carmichael* [2001a].

[3] *Savoie and Prospero* [1982]; *Huebert et al.* [1993]; *Gras* [1993b]; *O'Dowd et al.* [1997a]; *Geernaert et al.* [1998]; *O'Dowd et al.* [1999a]; *Pryor and Sørenson* [2000]; *Pirjola et al.* [2000]; *O'Dowd et al.* [2001].

[4] *Wright* [1939, 1940c]; *Valenzuela and Laing* [1970]; *Lai and Shemden* [1974]; *Tang* [1974]; *Wells et al.* [1977]; *Hughes* [1980]; *Hughes and Richter* [1980]; *Gathman* [1983a]; *Gathman and Ulfers* [1983]; *Hughes* [1987]; *de Leeuw* [1989b]; *Gerber* [1991]; *Barber and Wu* [1997]; *Flamant et al.* [1998]; *Murayama et al.* [1999]; *Sutton et al.* [2004].

[5] *Quinn et al.* [1995, 1996]; *Winter and Chýlek* [1997]; *Carrico et al.* [1998]; *McInnes et al.* [1998]; *Berg et al.* [1998]; *Quinn et al.* [1998]; *Murphy et al.* [1998a]; *Quinn and Coffman* [1998, 1999]; *Haywood et al.* [1999]; *Jacobson* [2001]; *Satheesh* [2002]; *Penner et al.* [2002]; *Dobbie et al.* [2003]; *Satheesh and Lubin* [2003].

[6] *Woodcock* [1950b, 1951]; *McDonald* [1951]; *Munk* [1955]; *Woodcock* [1958]; *Okuda and Hayami* [1959]; *Woodcock et al.* [1963]; *Kraus* [1967]; *Mangarella et al.* [1973]; *Bortkovskii* [1973]; *Donelan and Miyake* [1973]; *Wu* [1973, 1974]; *Lai and Shemdin* [1974]; *Ling and Kao* [1976]; *Borisenkov and Kuznetsov* [1978]; *Wang and Street* [1978a,b]; *Wu* [1979a]; *Ling et al.* [1980]; *Borisenkov and Kuznetsov* [1985]; *Bortkovskii* [1987]; *Stramska* [1987]; *Mestayer and Lefauconnier* [1988]; *de Leeuw* [1989a]; *Vonorov and Gavrilov* [1989]; *Wu* [1990d]; *Andreas* [1990b]; *Korolev et al.* [1990]; *Rouault et al.* [1991]; *Pielke and Lee* [1991]; *Andreas* [1992]; *Ling* [1993]; *Smith et al.* [1993]; *Edson and Fairall* [1994]; *Fairall et al.* [1994]; *Lighthill et al.* [1994]; *Andreas et al.* [1995]; *Edson et al.* [1996]; *Makin* [1998]; *Pattison and Belcher* [1999]; *Bao et al.* [2000]; *Andreas and Emanuel* [2001]; *Meirink and Makin* [2001]; *Wang et al.* [2001]; *Van Eijk et al.* [2001]; *Emanuel* [2003].

[7] *Houghton* [1932]; *Bennett* [1934]; *Köhler* [1936]; *Jacobs* [1936, 1937]; *Houghton* [1938]; *Wright* [1939, 1940a,b]; *Owens* [1940]; *Simpson* [1941a,b]; *Neiburger and Wurtele* [1949]; *Woodcock and Gifford* [1949]; *Woodcock* [1950a]; *Mason and Ludlam* [1951]; *Ludlam* [1951]; *Moore* [1952]; *Woodcock* [1952, 1953]; *Reitan and Braham* [1954]; *Turner* [1955]; *Woodcock and Blanchard* [1955]; *Kuroiwa* [1956]; *Woodcock and Spencer* [1957]; *Isono* [1957]; *Twomey* [1960]; *Woodcock et al.* [1963]; *Blanchard* [1969]; *Woodcock et al.* [1971]; *Woodcock and Duce* [1972]; *Woodcock* [1978]; *Woodcock et al.* [1981]; *Johnson* [1982]; *Ghan et al.* [1998]; *Murphy et al.* [1998a]; *Covert et al.* [1998]; *Hegg* [1999]; *Feingold et al.* [1999]; *O'Dowd et al.* [1999a,b,c]; *Mason* [2001]; *Yoon and Brimblecombe* [2002]; *Rosenfeld et al.* [2002].

[8] *Wilson* [1959a,b]; *Blanchard* [1963, 1964]; *Barger and Garrett* [1970]; *Bezdek and Carlucci* [1974]; *Pueschel and van Valin* [1974]; *Blanchard* [1978]; *Tusseau et al.* [1980]; *Barbier et al.* [1981]; *Gershey* [1983a,b]; *Tseng et al.* [1992]; *Oppo et al.* [1999].

[9] *Blanchard* [1955]; *Schaefer* [1956]; *Blanchard* [1958, 1963, 1966]; *Gathman and Trent*, 1968; *Strong*, 1969; *Gathman and Trent*, 1969; *Gathman and Hoppel* [1970]; *Blanchard* [1985b].

[10] *Martin et al.* [1981]; *Cambray and Eakins* [1982]; *Walker et al.* [1986]; *McKay and Pattenden* [1990]; *Nelis et al.* [1994]; *McKay et al.* [1994].

[11] *ZoBell and Mathews* [1936]; *Jacobs* [1939a]; *Woodcock* [1948, 1955]; *Aubert* [1974]; *Baylor et al.* [1977b].

[12] *Beck* [1819]; *Wells and Shunk* [1937]; *Oosting and Billings* [1942]; *Boyce* [1954]; *Wilson* [1959a,b]; *Edwards and Claxton* [1964]; *Stong* [1966]; *Clayton* [1972]; *Art et al.* [1974]; *Williams and Moser* [1976]; *Blanchard* [1977].

[13] *Ambler and Bain* [1955]; *Preobrazhenskii* [1973]; *Schmitt et al.* [1978]; *Davidson et al.* [1978]; *Fairall et al.* [1979]; *Kulkarni et al.* [1982]; *Ohba et al.* [1990]; *Dean* [1993]; *Morcillo et al.* [1999, 2000]; *Cole et al.* [2003a,b,c, 2004].

[14] *Fassina* [1978]; *Moropoulou et al.* [1995]; *Chabas et al.* [2000]; *Chabas and Lefèvre* [2000].

been used in paleoclimate studies to infer changes in wind speed associated with climate change under the assumption that SSA production and deposition increase with increasing wind speed[1].

1.2. METHODS OF DETERMINING SIZE-DEPENDENT SSA PRODUCTION FLUXES

The interfacial and effective production fluxes of SSA particles are difficult to determine by direct measurement and therefore determination of these fluxes has most often been by indirect methods based on laboratory and/or field measurements. Specification of the quantities that are required to evaluate the SSA production fluxes by these methods has motivated much of this review. These methods, examined in more detail in §3, are briefly described as follows:

- **The steady state dry deposition method** (§3.1) calculates the effective SSA production flux from field measurements of the number concentration of SSA particles and modeled dry deposition velocities, under assumption of local balance between production and removal by dry deposition.
- **The whitecap method** (§3.2) calculates the interfacial SSA production flux from laboratory measurements of drop production and the decay of whitecap area from simulated whitecaps, and field measurements of the fraction of the ocean surface covered with whitecaps.
- **The concentration buildup method** (§3.3) calculates the effective SSA production flux from field measurements of the increase in the vertical integral of SSA particle number concentration as a function of the distance in the along-wind direction, such as the distance from the coast during conditions of offshore winds (i.e., blowing from land to the sea).
- **The bubble method** (§3.4) calculates the interfacial SSA production flux from laboratory measurements of the number and sizes of drops produced by individual bubbles as a function of bubble size, field measurements of size-dependent oceanic bubble concentrations, and modeled or measured bubble rise velocities.
- **Micrometeorological methods** (§3.5) determine the effective SSA production flux from field measurements of the turbulent vertical flux of particles using techniques such as the eddy correlation method, which simultaneously measures fluctuations in number concentrations and the vertical component of the wind velocity; the eddy accumulation method (or relaxed eddy accumulation method), which measures the difference in the

average number concentrations in upward-moving and downward-moving eddies; and the profile (gradient) method, in which the difference of the measured number concentrations at two heights is multiplied by the ratio of the flux of another quantity (such as momentum, heat, or water vapor) to the difference in the concentrations of that quantity at those same heights.

- **The along-wind flux method** (§3.6) calculates the interfacial SSA production flux from laboratory or field measurements of the near-surface (< 1 m) flux of SSA particles that pass through a vertical area perpendicular to the mean flow or of the near-surface SSA concentration, along with certain assumptions concerning the mean distances traveled by drops and/or their mean velocity components in the direction of the wind. Typically this method has been employed to obtain the relative size dependence of the interfacial SSA production flux, with the absolute magnitude being obtained by equating this flux at a given drop size to that determined by some other method.
- **The direct observation method** (§3.7) determines the interfacial SSA production flux from time-resolved photographic observation or video imaging of individual drops leaving the surface or in the immediate vicinity of the surface, in the laboratory or in the field.
- **The vertical impaction method** (§3.8) determines the interfacial SSA production flux directly from measurements of the rate at which SSA particles impact the underside of a horizontal collection device near (< 1 m above) the ocean surface.
- **The statistical wet deposition method** (§3.9) estimates the effective SSA production flux from field measurements of SSA number concentration and estimates of the height of the marine atmospheric boundary layer and of the mean time between appreciable rainfall events, on the assumption that the vast majority of SSA particles present at the time of measurement had been produced since the last rain event.

1.3. SCOPE OF THIS WORK

This work is a comprehensive review of size-dependent SSA production over the open ocean: the mechanisms responsible for this production, the rate at which it occurs, and the factors upon which it depends. Methods that have been employed, measurements that have been made, and models that have been developed to understand, describe, and parameterize this production (characterized by both the interfacial SSA production flux at the ocean surface and by the effective

---

[1] *Legrand et al.* [1998]; *Genthon* [1992a,b]; *Mayewski et al.* [1993, 1994, 1997]; *Woolf et al.* [2003]; *Reader and McFarlane* [2003].

SSA production flux, consisting of those particles that contribute to the SSA in the marine atmosphere) are critically examined and compared. Because of the wide range of approaches that have been applied to estimate these SSA production fluxes, many other pertinent aspects of SSA and its behavior that are required to evaluate the formulations that have been presented for these fluxes are also examined.

Because of its geophysical and geochemical importance, SSA has been the subject of much research over many years, and numerous reviews have examined many facets of its production and behavior[1]. Despite this extensive literature, the current review, especially in view of its scope and magnitude, can nonetheless be justified on several grounds. Some of these earlier reviews are quite dated, and the primary literature on which they are based even more so, with much subsequent work now available. More substantively, some of these reviews have presented the topic of SSA production as if it were well understood, leading to an impression that SSA production fluxes and the corresponding production mechanisms are accurately known and represented in models. However, recent exchanges in the literature make it evident that this is not the situation, and that many fundamental questions remain unanswered[2]. Furthermore, as noted by *Andreas* and coworkers [*Andreas et al.*, 1995; *Andreas*, 1998], estimates of the SSA production flux for a given particle size vary over many orders of magnitude (§5). Additionally, some of the concepts and arguments presented here do not appear to be broadly appreciated by the research community and bear repeating, such as the distinction between interfacial fluxes and effective fluxes (§2.1.6) and the extent to which SSA particles of a given size equilibrate to the ambient relative humidity (§2.9.1), whereas others are new, such as the classification of SSA particles by size based on the factors controlling their atmospheric residence times (§2.1.2). Finally, many of the analyses in the literature of formulations of SSA production fluxes and of the measurements upon which they are based are cursory and of insufficient depth to provide a critical evaluation of their accuracy and limitations.

This review provides a framework for the whole of SSA production and behavior. It is organized around the physics, chemistry, and meteorology necessary to evaluate formulations of SSA production fluxes resulting from the methods described in §1.2. It is fundamentally based on measurements, which are stressed more than theory, as data form the underlying foundation against which models and theories must be tested. Data from a large number of sources are included, as reported values of quantities of interest from different investigators often vary greatly. Only by presenting data from a number of sources is it possible to estimate the variability in a given quantity under nominally the same conditions and to ascertain the extent to which a given measurement or set of measurements is representative of others taken at different locations perhaps by different investigators using different instrumentation. Emphasis is given to field measurements over laboratory investigations where available; similarly, laboratory results from experiments using seawater are stressed more than those involving freshwater.

This review is divided into the following major sections, which are briefly outlined here:

Fundamental background information necessary to understand properties and behavior of SSA and SSA particles pertinent to their fluxes (Section 2). Key topics examined in this section are:

- Means of describing SSA particle sizes (§2.1.1) and the specification of size ranges into which SSA particles naturally fit based on their behavior in the atmosphere (§2.1.2), the size dependence of quantities pertinent to SSA (§2.1.3), SSA concentrations (§2.1.4) and their properties (§2.1.5), and SSA fluxes (§2.1.6) and their properties (§2.1.7);
- Major SSA production mechanisms (§2.2);
- Factors affecting SSA production, entrainment, transport, and removal (§2.3);
- The structure of the atmosphere over the ocean as it pertains to SSA and its behavior (§2.4);
- Composition and chemical and physical properties of seawater (§2.5.1), composition of SSA drops (§2.5.2), the equilibrium and non-equilibrium behavior of these drops with respect to relative humidity (§2.5.3 and §2.5.4), and phase transitions of SSA particles (§2.5.5);
- Key mechanisms that control the motion of SSA particles, and the kinematics and dynamics of individual particles in the atmosphere (§2.6);
- The main removal mechanisms of SSA particles from the atmosphere: wet deposition (§2.7.1) and dry deposition (§2.7.2), and their relative importance (§2.7.3);

---

[1] *Eriksson* [1959, 1960]; *Blanchard* [1963]; *Toba* [1965a,b]; *Junge* [1972]; *Wu* [1979a]; *Blanchard* and *Woodcock* [1980]; *Podzimek* [1980]; *Monahan et al.* [1983a]; *Blanchard* [1983]; *Buat-Ménard* [1983]; *Monahan* [1986]; *Wu* [1986b]; *Miller* [1987]; *Bortkovskii* [1987]; *Fitzgerald* [1991]; *de Leeuw* [1993]; *Andreas et al.* [1995]; *O'Dowd et al.* [1997a]; *Gong et al.* [1997b]; *Heintzenberg et al.* [2000]; *Hoppel et al.* [2002]; *Schulz et al.* [2004]; *von Glasow* and *Crutzen* [2004].

[2] *Blanchard* and *Syzdek* [1988], *Wu* [1990b], and *Blanchard* and *Syzdek* [1990a]; *Wu* [1988b], *Monahan* and *Woolf* [1989], and *Wu* [1989b]; *Wu* [1989a], *Woolf* [1990], and *Wu* [1990a]; *Wu* [1990c], and *de Leeuw* [1990b]; *Hasse*, 1992, *Andreas* [1994b], and *Hasse* [1994]; *Andreas* [1992], *Katsaros* and *de Leeuw* [1994], and *Andreas* [1994a]; *Wu* [2000a], *Monahan* [2001], and *Wu* [2002a].

- Transport of SSA in the marine atmospheric boundary layer, resulting in a mathematical basis for several of the methods that have been used to determine the effective SSA production flux (§2.8); and
- Vertical distribution of SSA particles in the marine atmosphere and implications of the various time scales pertaining to the behavior and lifetime of SSA particles of a given size, leading to an upper limit of the size of SSA particles that need be considered for the effective SSA production flux (§2.9).

In examining the various processes relating to SSA production discussed in this section, simple models and typical conditions are employed, and numerous simple formulae and expressions that can readily be used to estimate important quantities are presented, the objectives being to reflect the underlying physics of the processes under consideration and to provide realistic numerical estimates without introducing undue complexity.

Methods of determining size-dependent interfacial and effective SSA production fluxes (Section 3). The several methods listed in §1.2 are examined in detail in this section. The equations that describe these fluxes and that readily identify the information that must be known from measurements or models to evaluate them according to each method, are presented. The governing physical processes underlying each method are discussed and the size range of SSA particles for which each method is best suited is identified. Additionally, the conditions that must be satisfied for the successful implementation of each method, and the difficulties and concerns with each, are examined.

Measurements and models of quantities required to evaluate SSA production fluxes (Section 4). The main topics examined in this section are:

- Field measurements of total SSA mass and number concentrations and of size distributions of SSA concentration in the marine atmosphere; knowledge of these quantities allows identification of the abundance of particles in the size ranges that are important for various processes of interest and numerical estimates for quantities pertinent to these processes, and provides a consistency check on SSA production flux formulations (§4.1);
- Field measurements of near-surface concentration and fluxes of SSA particles (§4.2);
- Laboratory measurements of the numbers, sizes, and heights of drops ejected from individual bubbles and their dependence on bubble size and other factors; SSA production from laboratory simulated whitecaps; and SSA production in wind/wave tunnels (§4.3);
- Field measurements of size distributions of bubble concentration in the oceans, and modeled rise velocities

and gas exchange properties of individual bubbles (§4.4);
- Measured oceanic fractional whitecap coverage and its dependence on wind speed, air and sea temperatures, and other controlling factors, and properties of individual whitecaps (§4.5);
- Modeled dry deposition velocities of SSA particles over the ocean and their dependence on various quantities, and measurements that would yield information on these velocities (§4.6); and
- Minor SSA production mechanisms, including wave breaking at coasts, direct and indirect production by rainfall, miscellaneous mechanisms such as volcanoes and merging of capillary waves, and the possible shattering of dry sea salt particles (§4.7); the latter, although not a surface production mechanism, has been hypothesized to be a source of smaller particles, and a sink of larger ones (thus modifying the shape of the SSA concentration size distribution), and therefore is pertinent to this review.

As the several quantities of interest have been reported in a multiplicity of ways, for each such quantity a single unambiguous way of reporting data has been adopted, and the original data have been converted into these consistent units and expressions. The results of numerous investigations are compared in tables and graphs. Expressions that have been presented to represent these quantities and their dependencies on controlling factors are also examined and compared with the observations themselves.

Estimates of SSA production fluxes according to the several methods (Section 5). Formulations proposed to describe interfacial and effective SSA production fluxes according to the several methods listed in §1.2 are evaluated and compared, and the methodologies that have been used to estimate these fluxes are critically examined to determine the extent to which the assumptions and conditions necessary for their successful implementations are satisfied, and hence the confidence that can be placed in determination of SSA production fluxes by each of the methods. Those flux estimates that can be considered most credible are identified and the extent to which these estimates provide meaningful descriptions of the interfacial and effective SSA production fluxes is examined. Finally, the basic conclusions that can be reached concerning production fluxes in each of several size ranges of SSA particles are summarized.

Applications, Implications, and Future Directions (Section 6). The implications of present knowledge of SSA production on determining the climatic influence of anthropogenic aerosols are examined. Representation of SSA production in chemical transport models and implications on calculated SSA properties are also examined. Finally, possible future research directions are noted.

2. Fundamentals

The investigation and description of size-dependent production, entrainment, transport, and removal of SSA particles, and the determination and understanding of the factors that control these processes, require consideration of the entire life cycle of SSA particles. Key topics that are important in this regard are size-dependent SSA concentrations, fluxes, and related properties; the main production mechanisms of SSA particles; factors that affect these properties and mechanisms; the structure of the marine atmospheric boundary layer, in which these particles mainly reside; chemical and physical properties of seawater and SSA particles; kinematics and dynamics of SSA particles; mechanisms by which SSA particles are removed from the atmosphere; transport of SSA particles in the marine atmosphere; and the vertical distribution of SSA particles in the marine atmosphere. These topics, which constitute the fundamental background information that is necessary to describe and quantify SSA concentrations, production, and behavior, are examined in this section.

In considering these topics several simplifying assumptions are made, the intent being to retain the essential physics of the processes affecting SSA production and behavior, while at the same time avoiding unnecessary details and complications. Sea salt aerosol particles are considered as non-interacting particles whose behavior is determined by meteorological and environmental conditions while not affecting these conditions in return [*Bortkovskii*, 1987, discusses these and related topics]; this is justified because of the generally dilute concentrations of SSA particles in the marine atmosphere (§4.1.4.5). Additionally, all SSA particles are assumed to have the same relative composition of dissolved substances (§2.5.2), implying that their size can be described by a single quantity, e.g., dry mass, or radius at a given relative humidity (§2.1.1). Implications of this assumption are that for a given set of conditions the concentrations of SSA dry mass and volume are directly proportional to each other (§2.5.3) and that properties of individual SSA particles such as density and index of refraction are independent of particle size (§2.1.5.3). For most considerations, the sea

surface is assumed to be a horizontal plane with production occurring uniformly over this surface. This assumption implies spatial and/or temporal averaging over times and distances sufficiently large to account for the sporadic nature of breaking waves and inhomogeneities caused by wave intermittency, and sufficiently small to be characterized by uniform meteorological conditions (§2.1.6). In considering the marine atmospheric boundary layer, a simplified model is presented (§2.4) in which various sublayers are defined; though highly idealized, this is thought to represent the essential features pertinent to SSA production and behavior.

Many formulas and expressions are presented that can readily be used to estimate important quantities to accuracies of several percent; this is all that is necessary in many instances, and more than is warranted by the data in many others. The wide range of environmental and meteorological conditions occurring in the marine environment necessitates the use of typical values of quantities for illustrative purposes; the values chosen in this review yield realistic numerical estimates, and conclusions that are reached using them are expected to be valid in most circumstances. Almost always the uncertainties of the various aspects of the production of SSA and its behavior in the marine atmosphere are such that little is to be gained from the use of more detailed models, and more precise numerical results can rarely be justified.

2.1. DESCRIPTION OF SEA SALT AEROSOL PARTICLE SIZES, CONCENTRATIONS, AND FLUXES

The size of an SSA particle is central to all aspects of its behavior, as SSA particles of different sizes have different concentrations, production rates, intrinsic properties (e.g., mass, surface area), residence times, and mixing heights. The relative importance of SSA particles of a given size to atmospheric, meteorological, and geophysical processes is determined by this size and by the relative abundance and rate of production of SSA particles of this size. These properties in turn are described by size-dependent concentrations and fluxes, and by properties of these distributions. Thus it is necessary to discuss the descriptions of SSA particle size, and of SSA concentrations and fluxes and related quantities and their dependencies on SSA particle size. As generally the desired SSA concentrations and fluxes are average values over temporal and spatial scales longer than those characterizing

Sea Salt Aerosol Production: Mechanisms, Methods, Measurements, and Models
Geophysical Monograph Series 152
Copyright 2004 by the American Geophysical Union.
10.1029/152GM02

production, that is, those characterizing breaking waves and whitecaps (for which temporal scales are of order seconds to minutes, and spatial scales are of order tens of meters), quantities refer to average values unless otherwise stated; this takes into account the inhomogeneities in space and time due to the sporadic nature of breaking waves.

2.1.1. Description of SSA Particle Sizes

Throughout this review the assumption is made that the only feature distinguishing one SSA particle from another, and necessary to describe the behavior of an SSA particle for a given set of conditions, is m_{dry}, the mass of solute it contains (§2.5.3); hence an SSA particle can be described solely by any quantity uniquely determined by its solute mass. One such quantity is the equivalent dry radius r_{dry}, defined by

$$m_{dry} = \frac{4\pi}{3}\rho_{ss} r_{dry}^3 \qquad (2.1\text{-}1)$$

where ρ_{ss} is the density of dry sea salt (the definition of r_{dry} thus depends on the value specified for ρ_{ss}; consequences of this choice are examined in §2.5.1). However, it is important to note that this equivalent dry radius is not a physical quantity, as dry SSA particles can assume various shapes other than spheres and may have densities different from that of bulk sea salt (§2.5.5). Seawater drops are hygroscopic, meaning that they readily exchange moisture with their surroundings, and therefore they change their equilibrium water content and radii under different atmospheric conditions. These equilibrium radii are determined almost entirely by the ambient relative humidity RH (§2.5.3), defined to be 100% times the fractional relative humidity rh, the ratio of the vapor pressure of water in the atmosphere to the saturation vapor pressure of bulk water at the ambient temperature. Thus, the mass of solute in an SSA particle can also be uniquely represented by the radius of the particle in equilibrium with the atmosphere at a given RH. Throughout this review this standard relative humidity is taken to be 80%, and the corresponding radius denoted by r_{80} (§2.5.3). For a given SSA particle, r_{80} is almost exactly twice r_{dry} (§2.5.3). Likewise, r_{98} refers to the radius of an SSA particle in equilibrium with the atmosphere at 98% RH, and r_{form} denotes its radius at formation. For most of the world's oceans, formation conditions corresponds to 98% RH, and $r_{form} \approx r_{98}$ (§2.5.3). This formation radius r_{98} is almost exactly twice r_{80} (§2.5.3). For consistency, results of measurements on dry particles other than SSA particles are reported in this review in terms of r_{80} evaluated as $2r_{dry}$, despite the fact that for these substances the radius at 80% RH may be different from $2r_{dry}$. Likewise, results of experiments using liquids other than seawater, including freshwater and pure water, are reported in this review in terms of r_{80} evaluated as $0.5r_{form}$, despite the fact that for these liquids r_{80} may

not be meaningful or may not be equal to $0.5r_{form}$. Radii of SSA particles in this review are presented in micrometers, μm.

Labeling SSA particles by r_{80} facilitates the description of their behavior; it provides a unambiguous description the amount of solute present in the drop that is independent of local conditions, and furthermore it describes a physically relevant size, 80% RH being typical over the oceans (§2.4). Implicit in this parameterization are the assumptions that seawater drops are spherical—a good approximation under nearly all situations considered in this review (§2.6.3), and that the relative composition of the dissolved substances in SSA particles is the same and independent of the amount of solute in the particle. Although fractionation resulting in enrichment of some substances may occur during the production of seawater drops, and after some time in the atmosphere SSA particle composition may be modified by uptake of and reaction with other gases and particles, these effects are thought to be unimportant for the consideration of SSA production fluxes (§2.5.2). Over a range of RH (~45% to ~75%) an SSA particle can exist as either a seawater drop or a dry salt particle (§2.5.5), but because such a particle starts as a liquid drop, and because the RH in the lowest tens of meters over the ocean is typically above this range, SSA particles are present almost always as liquid drops. Consequently, they are treated as such in this review. At 80% RH, SSA particles are unambiguously liquid drops; thus labeling particles by their r_{80} values does not introduce any difficulties or ambiguities arising from this complication.

The fact that an SSA particle of a given r_{80} does not instantaneously respond to changes in RH (§2.5.4) implies that the geometric radius of the particle r, which largely determines its behavior in the atmosphere, is not uniquely determined by local conditions, and thus not necessarily equal to its equilibrium radius with respect to the ambient RH r_{amb}, unless further assumptions are made (§2.9.1). These assumptions are that the time required for an SSA particle with $r_{80} \lesssim 25$ μm to attain its equilibrium radius with respect to the ambient RH is sufficiently short that under most circumstances its geometric radius r is very nearly equal to r_{amb} (i.e., the particle has equilibrated to the ambient RH), and that the time required for a larger SSA particle to equilibrate to the ambient RH is so long that it retains a radius near its formation value during its lifetime in the atmosphere (and thus $r \neq r_{amb}$). These assumptions allow the behavior of an SSA particle to be completely described by the value of r_{80} and the local conditions, and the geometric radius of an SSA particle can thus be written as $r = r(r_{80}, rh)$. For an SSA particle with $r_{80} \lesssim 25$ μm, for which the geometric radius r and the ambient radius r_{amb} are equal under the above assumptions, this relation can be further simplified, as under most circumstances the ratio r/r_{80} is independent of r_{80} and depends only on rh (§2.5.3, §2.9.1);

i.e., $r/r_{80} = \zeta(rh)$. The quantity r_{80} does not represent a physically meaningful size for a larger SSA particle, although it can still be unambiguously used to describe the size of such a particle. However, SSA particles with $r_{80} \gtrsim 25$ μm typically occur in low concentrations at heights more than a few meters above the sea surface (§4.1.4.3), spend little time in the atmosphere (§2.9.1), and are not entrained to appreciable heights (§2.9.5), and they are generally of minimal importance in considerations of cloud formation, atmospheric chemistry, and atmospheric radiation (although they may play a large role in other processes such as air-sea momentum and moisture exchange). Thus there is a rather sharp demarcation in particle size that provides a practical upper limit for the sizes of particles that need be considered in reference to participation in atmospheric processes (§2.1.2).

As the size of an SSA particle controls many aspects of its behavior, the range of sizes of SSA particles present in the marine atmosphere determines the topics that must be treated and the parameterizations that must be employed to describe this behavior. Results from field investigations have indicated the presence of SSA particles with r_{80} at least as small as 0.01 μm, although often no SSA particles with $r_{80} \lesssim 0.1$ μm are detected, and when present they typically comprise a very small fraction of the number of marine aerosol particles in this size range (§4.1.1.1). Sea salt aerosol particles with $r_{80} \gtrsim 0.1$ μm occur in appreciable concentrations and play important roles in a variety of atmospheric phenomena. Therefore a practical lower limit for the range of SSA particle sizes may generally be taken as $r_{80} \approx 0.1$ μm (§4.1.1.1; §4.1.4.5). If in a given situation it is determined that SSA particles smaller than this are present in sufficient numbers to be important then minor modifications will be required, but the basic conclusions obtained in this review will not be changed. At the other extreme, although SSA particles with r_{80} up to several hundred micrometers have been reported over the ocean, and larger drops certainly occur (as is well known by anyone who has encountered spray from breaking waves at sea), a practical upper limit for this review is several millimeters. These considerations allow several simplifications to be made concerning the behavior of SSA particles (such as the neglect of the Kelvin effect on the relationship between r_{80} and RH; §2.5.3) and delimit the range of sizes that must be treated when examining the physical properties and kinematics and dynamics of SSA particles (§2.6).

2.1.2. Specification of SSA Particle Size Ranges

From considerations of several factors, including dry deposition, residence times in the atmosphere, concentration and flux gradients near the sea surface, and response times with respect to RH and wind speed, it proves convenient to classify SSA particles into three size ranges based on their behavior in the atmosphere and on their importance with regard to effective and interfacial fluxes (§2.1.6). These ranges are: $r_{80} \lesssim 1$ μm, denoted small SSA particles; 1 μm $\lesssim r_{80} \lesssim 25$ μm, denoted medium SSA particles; and 25 μm $\lesssim r_{80}$, denoted large SSA particles. The limits of these ranges should not be viewed as rigidly specified—for instance, the upper limit of the intermediate range could arguably be from 25 μm to 40 μm (although these values comprise only a factor of 1.6, or a difference in $\log_{10} r_{80}$ of 0.2). However, these limits, while dependent to some extent on meteorological conditions, nonetheless possess rather distinct values for most conditions typically encountered. A multitude of names have been used to classify aerosol particles into size ranges, including nuclei, CN (condensation nuclei), CCN (cloud condensation nuclei), accumulation, Aitken, fine, coarse, ultrafine, very very small, giant, and ultragiant, several of which have been applied to the size range of interest for SSA particles. *Junge* [1956], for instance, stated that the generally accepted nomenclature is that particles with 0.1 μm $< r < 1$ μm are "large nuclei" and those with $r > 1$ μm are "giant nuclei" [see also *Junge*, 1955, and *Mason*, 1957b] and *Toba* [1965a,b, 1966] titled his papers "On the giant sea-salt particles in the atmosphere," referring to particles with $r_{80} \gtrsim 1$ μm, whereas *Whitby* [1978] discussed the distinction between "fine particles," for which $r < 1$ μm, and "coarse particles," with $r > 1$ μm. Thus it is with some hesitancy that new names are proposed here, and to avoid confusion it is essential that the designations "small", "medium", and "large" should always be followed by "sea salt aerosol particles" (or "SSA particles") rather than merely "particles", as these designations apply here only to SSA particles. However, the convenience of such a classification, and the observation that fluxes and other properties of SSA particles, such as their characteristic times for various processes and their removal mechanisms, fit naturally into these three classes (the demarcation at the upper size limit with respect to the time of equilibration of size with regard to change in RH was noted in §2.1.1), argue for its utility. It is important to note that this classification is not based upon the mechanism of production of SSA particles (as large drops can result from either bubble bursting or wave tearing, for instance), but rather on the behavior of particles once formed and the importance of various mechanisms upon this behavior. Aspects of this behavior are summarized in Table 2.

Small SSA particles—those with $r_{80} \lesssim 1$ μm—dominate the concentrations and fluxes of SSA number and radius in most circumstances, but their contribution to the concentrations and fluxes of SSA surface area and volume (or mass) is typically much less important. The size dependencies of the concentrations and fluxes of these particles are not well known, but size is not an important consideration for some of the processes for which these particles exert the most

Table 2. Properties of Sea Salt Aerosol Particles by Size Range

Property	SSA Particle Size Class			Section
	Small	Medium	Large	
r_{80} μm/range	~0.1 - ~1	~1 - ~25	$\gtrsim 25$	2.1.2
Atmospheric importance	CCN, atmospheric chemistry	Atmospheric chemistry, light scattering	Fluxes of sensible and latent heat	2.1.5, 2.1.7
Provides dominant contribution to:	Concentrations and fluxes of number and radius	Concentrations of surface area, volume, and mass	Interfacial fluxes of surface area, volume, and mass	2.1.5, 2.1.7, 4.1.4.5
Pertinent flux	Effective	Effective	Interfacial	2.1.6
Ψ_f, ratio of effective to interfacial SSA production flux	≈ 1	≈ 1 to < 1	$\ll 1$	2.1.6, 2.9.4
Predominant drop type	Film	Jet	Jet, spume	2.2, 4.3.7
Response to RH	$\lesssim 0.1$ s	~0.1 s to ~50 s	$\gtrsim 50$ s	2.5.4
Role of gravity	Negligible	Important	Dominant	2.6.1, 2.9.3
Stokes' Law valid?	Yes, but unimportant	Yes	No	2.6.2, 2.6.3
Response to wind speed	$\lesssim 3 \cdot 10^{-5}$ s	$\sim 3 \cdot 10^{-5}$ s to ~0.02 s	$\gtrsim 0.1$ s	2.6.2, 2.6.4
Attainment of horizontal component of wind speed?	Yes	Yes	No	2.6.2, 2.6.4
Primary removal mechanism	Wet deposition	Dry deposition	Gravitational fallout	2.7.3
Attainment of equilibrium with ambient RH before measurement at 10 m?	Yes	Yes	No	2.9.1
Probability of being entrained upward from near the sea surface	High	High to low	Low	2.9.2
Vertical concentration gradient near 10 m	Negligible	Small	Very large	2.9.4
Mixing heights	Entire marine boundary layer	Varies greatly with r_{80}	At most a few meters above sea surface	2.9.5
Steady state with respect to dry deposition attained?	No	Depends on r_{80}	Not pertinent	2.9.6
Atmospheric residence times	Weeks to days	Days to minutes	Seconds	2.9.6
Methods of production flux determination	Micrometeorological, statistical wet deposition	Steady state dry deposition, whitecap, concentration buildup, bubble, vertical impaction	Along-wind flux, direct observation, vertical impaction	3
Abundance compared to non-SSA particles	Low to high	High	High	4.1.1.1

influence. For instance, because of the relatively large sizes and low concentrations of small SSA particles compared to other marine aerosol particles, nearly all SSA particles present can function as cloud condensation nuclei under most circumstances (§2.1.5.1). The interfacial and effective production fluxes of these particles (§2.1.6) are expected to be nearly the same, meaning that virtually all of the particles in this size range that are produced at the surface are entrained upward to at least 10 m before they return to the surface. Gravity plays little role in the motion of these particles; for example, an SSA particle with radius 1 μm requires nearly a day to fall 10 m in still air (§2.6.2). Thus, small SSA particles are expected to be nearly uniformly mixed over the marine boundary layer (§2.9.5). Consequently (§2.9.5), a measurement of the number concentration of small SSA particles over a wide range of heights will provide an accurate representation of this quantity. Likewise, the size-dependent vertical integral of number concentration (§2.1.5) may be approximated by $n(r_{80})H_{mbl}$, where H_{mbl} is the height of the marine boundary layer (§2.4). The mean atmospheric residence times of these particles are long, often several days or more (§2.9.6), and their principal mechanism of removal from the atmosphere is through wet deposition (§2.7.1).

Medium SSA particles—those with 1 μm $\lesssim r_{80} \lesssim 25$ μm—provide the dominant contribution to the concentrations and effective fluxes of SSA surface area and volume; thus these particles are the ones most important for most applications involving light scattering, for instance (§2.1.5.3).

With increasing r_{80}, gravity plays an increasingly greater role in the motion and lifetime of these particles, and the effective SSA production flux becomes increasingly less than the interfacial SSA production flux. The extent to which medium SSA particles are vertically mixed throughout the marine boundary layer depends strongly on their size (§2.9.5), as does their mean atmospheric residence time, which varies from days to hours. Both wet and dry deposition contribute to the removal of SSA particles in this size range.

Large SSA particles—those with $r_{80} \gtrsim 25$ μm—provide the dominant contribution to the SSA fluxes of moisture, momentum, and latent and sensible heat between the ocean and the atmosphere. Gravity plays a large role in controlling the motion of these particles, resulting in strong vertical gradients of concentration, limited vertical mixing, and extremely short atmospheric residence times, often of order seconds. Because of the large gravitational sedimentation velocities of these particles (§2.6.3) and their long characteristic times for equilibration with respect to RH (§2.5.4), only a small fraction of particles of this size produced at the sea surface attain heights of 10 m or more (§2.9.4); thus under most circumstances the effective production flux of these particles is quite low, and as discussed in §2.1.6 and §2.9.5, for all practical purposes it can be considered equal to zero.

2.1.3. Description of Size Distributions

Many quantities of interest pertaining to SSA particles, such as numbers, concentrations, and fluxes, depend on the sizes of these particles. To describe these dependencies, these quantities are treated as size distributions. These size distributions can be (and have been) represented in numerous different forms (similar considerations apply to bubbles). For instance, size distributions of SSA particle concentration can be represented as the number, area, volume, or mass concentration (i.e., amount per unit volume of air) per unit interval of some quantity describing the size of the particle, which may be the radius or diameter, the mass, or the logarithm (natural or common) of one of these (a number of different representations are presented in *Slinn* [1975], and similar considerations for size distributions of SSA fluxes are presented in *Andreas et al.* [2001]). Additionally, radii and diameters can refer to geometric, aerodynamic, or mobility (§2.6.2) values. Size distributions have also commonly been reported as cumulative distributions. Unfortunately no one formulation is standard. An additional complicating factor for quantities pertaining directly to SSA is the dependence of particle size on RH (§2.5.3), and size distributions are often ambiguous unless RH is specified, as the radius or diameter (or the mass) can refer to that of the particle at formation (98% RH; §2.5.3), at given reference RH such as 80%, at ambient RH, or while dry.

To prevent ambiguity it is necessary to precisely specify the form used to represent the size distribution of a quantity; the phrase "the number concentration of SSA," for instance, can refer to several different representations, and not infrequently ambiguities in defining size distributions have resulted in errors when different measurements have been compared (e.g., §4.1.4.2; §4.2; §5.2.1.1). Throughout this review, the size dependence of a quantity pertaining to SSA concentrations or fluxes is expressed as the amount of that quantity per unit logarithmic interval (base 10) of r_{80} (the size dependence of a quantity pertaining to concentrations or fluxes of bubbles is analogous, with R_{bub} being used instead of r_{80}). Hence if $Q(r_{80})$ denotes the cumulative distribution of some quantity, that is, the contribution to the quantity from all SSA particles with r_{80} less than the value of the argument, then the size distribution $q(r_{80})$ of this quantity is defined such that $dQ(r_{80}) = q(r_{80}) \, d\log r_{80}$ is the contribution to it by all SSA particles in the size interval $(\log r_{80}, \log r_{80} + d\log r_{80})$. Throughout this review, "log" refers to the common (Napierian) logarithm, i.e., $\log_{10}$, and natural logarithms are expressed as "ln"; the choice of common logarithms rather than natural logarithms is arbitrary but convenient. From this definition it follows that

$$q(r_{80}) = \frac{dQ(r_{80})}{d\log r_{80}}$$

or equivalently,

$$Q(r_{80}) = \int_{0}^{r_{80}} q(\tilde{r}_{80}) \, d\log \tilde{r}_{80}$$

where $\tilde{r}_{80}$ denotes a variable of integration. Lower case letters are used to represent the size distributions of quantities, and the corresponding capital letters are used to represent their cumulative distributions; a capital letter without an argument refers to the total integral quantity, i.e, Q is equal to $Q(r_{80} = \infty) \equiv \int q(r_{80}) d\log r_{80}$ (where no limits of integration are specified the integral is assumed to be over the range of r_{80} sufficient to encompass the bulk of the integral). The partial integral quantity δQ is used to denote the contribution to Q within a given range $\delta \log r_{80}$. For small intervals of $\delta \log r_{80}$, $\delta Q \approx (dQ/d\log r_{80})\delta \log r_{80} \equiv q(r_{80})\delta \log r_{80}$. As the equilibrium size of an SSA particle depends on the ambient RH (§2.5.3), so do many quantities describing SSA, such as the concentrations of radius, surface area, and volume. It is often convenient to specify these quantities at a given RH, such as 80%; in this case the size-dependent and total values of the quantity are labeled as $q_{80}(r_{80})$ and Q_{80}, respectively. When no RH is specified, it is to be understood that the quantities refer to their values at the ambient RH.

This method of representing size distributions (that is, the choice of $\log r_{80}$ to describe the size interval of SSA particles) has several advantages: it allows the size distribution of a quantity to retain the same units as the quantity itself, and it facilitates reporting this size distribution over a wide range of sizes. Size distributions represented by $dQ(r_{80})/d\log r_{80}$ implicitly assume equal logarithmic intervals of r_{80}, as opposed to size distributions represented by $dQ(r_{80})/dr_{80}$, in which equal weight is given to the r_{80} ranges 0-10 μm and 100-110 μm, for instance. Although little difference is expected among the behavior of SSA particles with r_{80} between 100 and 110 μm, extremely large differences in behavior are expected among those with r_{80} less than 10 μm. Another advantage is that when a size distribution of a quantity in this form is plotted on a linear scale against $\log r_{80}$ a so-called "equal-area" graph [*Whitby et al.*, 1972] results, in which the contribution to the quantity from SSA particles between any two radii is directly proportional to the area under the curve between these two radii, readily allowing the relative contributions from SSA particles of different sizes to be discerned. In principle $dN(r_{80})/dr_{80}$ vs. r_{80} yields an equal-area plot when both quantities are plotted on linear scales, but such a graph typically results in too much compression at low r_{80} of quantities that contain useful information and exhibit much variability, and thus this approach is not commonly used. Additionally, another advantage to the choice of $\log r_{80}$ to describe size intervals is that size distributions of concentrations of atmospheric aerosol particles are commonly fitted by a lognormal distribution or a sum of lognormal distributions, for which $\log r_{80}$ is the most natural argument.

Additional information on size distributions (with specific reference to atmospheric aerosols) and on differential and integral properties (§2.1.5) derived from them can be found in *Davies* [1974], *Slinn* [1975], *Willeke and Whitby* [1975], *Jaenicke and Davies* [1976], and *Jaenicke* [1978a].

2.1.4. Description of Concentrations

Of the many representations that have been used to describe the size dependence of SSA concentrations, perhaps the most fundamental is the number concentration $n(r_{80})$, the number of SSA particles per unit volume at a given location and time in a unit logarithmic interval of r_{80}, defined by

$$n(r_{80}) \equiv \frac{dN(r_{80})}{d \log r_{80}}, \qquad (2.1\text{-}2)$$

where $N(r_{80})$ denotes the number concentration of all SSA particles with r_{80} less than the value of the argument. The total number concentration N can be determined from $n(r_{80})$ by

$$N = \int n(r_{80}) d \log r_{80}, \qquad (2.1\text{-}3)$$

and likewise the partial number concentration δN for a given range of r_{80} can be determined. The representation of size distributions of SSA concentration in terms of $n(r_{80})$ reflects the way in which these distributions are often obtained: by counting the number of particles in a known volume of air in a given size range (the advantages of the choice of $\log r_{80}$ to describe the size of an SSA particle were discussed in §2.1.3), and knowledge of $n(r_{80})$ permits determination of properties of SSA concentration that are important for various processes of interest (§1.1).

2.1.4.1. Representations of size distributions of concentration. Other methods of representing size distributions of concentration have commonly been employed and are often useful when examining particular properties of SSA concentration. These other methods describe the relative importance of SSA particles of a given size to different concentration properties of interest. As comparison of SSA size distributions of concentration presented in different representations requires conversion to a unique representation, it proves useful to discuss relationships between $n(r_{80})$ and some of these other representations, which differ in the independent variable (which describes particle size), the dependent variable (which describes the amount of SSA present), or both. Particle sizes were represented by several early investigators, such as *Woodcock* [1950b, 1952, 1953], *Eriksson* [1959], *Toba* [1965a,b, 1966], and *Chaen* [1973], by the dry particle mass m_{dry}, which is related to r_{80} from (2.1-1) by

$$m_{dry} = \frac{4\pi}{3}\rho_{ss}\left(\frac{r_{dry}}{r_{80}}\right)^3 r_{80}^3 \approx \frac{1}{8}\frac{4\pi}{3}\rho_{ss} r_{80}^3, \qquad (2.1\text{-}4)$$

where the factor 8 arises because $r_{80}/r_{dry} \approx 2$ (§2.5.3). The number concentration in terms of m_{dry} is thus related to $n(r_{80})$ by

$$\frac{dN(m_{dry})}{dm_{dry}} = \left(\frac{r_{80}}{r_{dry}}\right)^3 \frac{n(r_{80})}{4\pi\rho_{ss} r_{80}^3 \ln 10}$$
$$\approx \frac{8 n(r_{80})}{4\pi\rho_{ss} r_{80}^3 \ln 10}, \qquad (2.1\text{-}5)$$

or equivalently,

$$n(r_{80}) = 3 \ln 10 \, m_{dry} \frac{dN(m_{dry})}{dm_{dry}}. \qquad (2.1\text{-}6)$$

Often the radius at a given RH, here denoted by r, is used to describe SSA particle size, in which situation the number concentration $dN(r)/dr$—also commonly referred to by some investigators as $n(r)$—is related to $n(r_{80})$ by

$$\frac{dN(r)}{dr} = \frac{n(r_{80})}{r \ln 10}, \qquad (2.1\text{-}7)$$

or equivalently,

$$n(r_{80}) = r \ln 10 \frac{dN(r)}{dr}. \qquad (2.1\text{-}8)$$

The size distributions of the concentrations of SSA radius, surface area, volume, and dry mass are important representations and have also been commonly used. With the choice of $\log r_{80}$ as the independent variable, these quantities can be defined analogously to $n(r_{80})$ and are related to it by

$$\frac{dR(r_{80})}{d \log r_{80}} = r\, n(r_{80}),$$

(to avoid confusion $dR(r_{80})/d\log r_{80}$ is not abbreviated),

$$a(r_{80}) \equiv \frac{dA(r_{80})}{d \log r_{80}} = 4\pi r^2 n(r_{80}),$$

$$v(r_{80}) \equiv \frac{dV(r_{80})}{d \log r_{80}} = \frac{4\pi}{3} r^3 n(r_{80}),$$

and

$$m(r_{80}) \equiv \frac{dM(r_{80})}{d \log r_{80}} = \rho_{ss} \left(\frac{r_{dry}}{r_{80}}\right)^3 \left(\frac{r_{80}}{r}\right)^3 \frac{4\pi}{3} r^3 n(r_{80}),$$

respectively, where r is the radius of the SSA particle at a given RH, $r_{80}/r_{dry} \approx 2$, and r/r_{80} is a known function of RH (§2.5.3). The cumulative distributions of these quantities are denoted by $N(r_{80})$, $R(r_{80})$, $A(r_{80})$, $V(r_{80})$, and $M(r_{80})$ and the total concentrations by N, R, A, V, and M, respectively. Likewise, the partial concentrations of these quantities (the contributions to these quantities from a given range of r_{80}) are denoted by δN, δR, δA, δV, and δM, respectively. The values of the size-dependent number concentration $n(r_{80})$ and dry mass concentration $m(r_{80})$, and those of $N(r_{80})$ and N, and of $M(r_{80})$ and M, do not vary with RH, whereas the concentrations of SSA radius, surface area, and volume depend on RH, which must therefore be specified to unambiguously determine these quantities. This dependence on RH can be eliminated by describing the size-dependent concentrations of these other quantities in terms of the values they would possess at a reference RH, again taken as 80%; the known relation between the radius of an SSA particle at a given RH and that at 80% RH (§2.5.3) allows calculation of these quantities at other RH values. The size-dependent concentrations of radius, surface area, and volume at 80%

RH are labeled by $dR_{80}/d\log r_{80}$, $a_{80}(r_{80})$, and $v_{80}(r_{80})$, respectively (similarly for the cumulative distributions $R_{80}(r_{80})$, $A_{80}(r_{80})$, and $V_{80}(r_{80})$, and the total concentrations R_{80}, A_{80}, and V_{80}). In some instances partial concentrations are determined by measurements over an interval of r_{amb} rather than r_{80}; in these situations δN and δM may depend on RH because of the dependence of the limits of the interval on RH.

The shape of a graph of a size distribution depends strongly on the representation used to describe it. This is illustrated in Fig. 1, in which the representations $n(r)$, $dN(r)/dr$, $a(r)$, and $v(r)$ are plotted against $\log r$ for an artificial (but not unrealistic) size distribution of concentration (the vertical scale is arbitrary, and the magnitudes of the various representations have been scaled so that the peak values occur at the same height). The equal-area representations $n(r)$, $a(r)$, and $v(r)$ readily show the size ranges that provide the majority of the contribution to each of these concentrations. As size-dependent concentrations typically range over many orders of magnitude, using a linear ordinate does not allow the full extent of the information available to be presented, and in practice concentration size distributions are often presented with a logarithmic ordinate. Plots on a

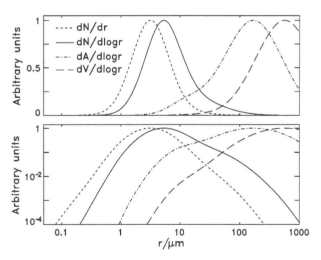

Figure 1. Size distribution of number concentration of a hypothetical but not unrealistic aerosol, and alternative representations calculated from it, normalized to maximum values, as a function of radius r on linear (upper panel) and logarithmic (lower panel) ordinate scales. For differential number concentration $n(r) \equiv dN(r)/d\log r$, surface area concentration $a(r) \equiv dA(r)/d\log r$ and volume concentration $v(r) \equiv dV(r)/d\log r$, upper panel presents "equal-area" plots in which the area under the curve for any radius range is proportional to the contribution of that radius range to the total concentrations of aerosol number N, area A, and volume V. The logarithmic ordinate in the lower panel allows examination of size distributions at values well below their peak values, but the curves are no longer equal-area plots. Also shown for comparison is $dN(r)/dr$, which is not an equal-area representation in either plot.

logarithmic scale also allow inflections indicative of multiple modes to be more readily discerned, e.g., in the graphs of $n(r)$ and $a(r)$ in Fig. 1. It is often useful to present size distributions of concentration in more than one representation, such as $n(r_{80})$ and $v_{80}(r_{80})$ [*Whitby*, 1978; *Reid et al.*, 2001]; while seemingly providing no new information, this practice can make the particle size range contributing to an SSA concentration property of interest more apparent and can lead to discerning properties of the size distribution that might otherwise not be apparent.

A consequence of multiple methods of representing size distributions is that the locations of any maxima and/or minima in the size distribution may differ greatly with the choice of representations; as seen in Fig. 1 the maxima of $n(r)$, $dN(r)/dr$, $a(r)$, and $v(r)$ occur near $r = 5$, 3, 170, and 560 μm, respectively. There is thus inherent ambiguity, often by orders of magnitude, in what is meant by a statement that the "peak of the size distribution" of a given aerosol concentration occurs at a particular radius value without specification of which representation is being considered ("number distribution" is obviously insufficient, as noted above), and it must be concluded that no strong significance should be attached to the exact locations of maxima or minima. More importantly, the fact that in the several representations in Fig. 1 the greatest contributions to the size distribution of concentration occur in wholly different radius ranges has major implications concerning fitting analytical functions to measured size distributions of concentration, and concerning using such fits. In the example shown in Fig. 1, the maximum value of $v(r)$ is four orders of magnitude greater than its value at the radius at which $n(r)$ takes its maximum value, and vice versa. Fits to one or another of these representations will be weighted by entirely different parts of the distribution, and an analytical expression describing a given representation of a size distribution that is obtained by a fit that is most heavily weighted in one radius range may completely misrepresent the size distribution outside that radius range. Of particular relevance to SSA is the fact that the volume distribution $v(r)$, for instance, may be dominated by a relatively small number of particles in the large-radius "tail" of $n(r)$. Any function that is fitted to the measured number distribution would be determined mainly by the bulk of the measurements in the number distribution, with the consequence that the fitted function transformed to the representation $v(r)$ might depart substantially from the data in the region of the dominant contribution to $v(r)$ (this concern is in addition to any concerns with the accuracy of the number concentrations at the high end of the distribution associated with the sampling of large particles, RH equilibration, or counting statistics for low numbers of counts; §4.1.4.3). Finally, while extrapolation of fits beyond the range of the data, especially where there is no theoretical justification for the fitting function, is a dubious practice in

any event, it is even more dubious for a fit based on a number distribution that has been transformed to a volume distribution, for which the portion of the distribution yielding the greatest contribution to the volume concentration may even lie outside the range of the measurements (as would likely occur, for instance, for the size distribution shown in Fig. 1, for which the maximum of $v(r)$ occurs at $r \approx 560$ μm).

2.1.4.2. Parameterizations of size distributions of concentration. Several different functional forms have been used to parameterize size distributions of SSA concentration. Criteria proposed by *Whitby* [1978] for such a form are that it should fit the distribution over the entire size range, fit size distributions in terms of number, area, or volume equally well, and have some physical basis, or, barring that, be as simple as possible [*Jaenicke and Davies*, 1976]. Three functional forms that have been widely used are power laws, gamma or modified gamma distributions, and lognormal distributions [these and other forms are discussed in *Liu and Liu*, 1994, for example]. These forms are examined briefly here in reference to size distributions of SSA concentration; similar considerations apply to size distributions of concentrations of bubbles and other quantities, and to fluxes of SSA particles and of bubbles.

Power law fits are of the form $n(r_{80}) = ar_{80}^{-b}$, equivalent to $dN(r_{80})/dr_{80} = (a/\ln 10)r_{80}^{-(b+1)}$, with $b > 0$ and a and b independent of r_{80}. They are also known as Jungian fits after *Junge* [1953], who proposed $b = 3$ for atmospheric aerosols (this choice yields equal contributions to the volume concentration from equal intervals of $\log r_{80}$). Power law fits are represented by straight lines on a log-log plot of $n(r_{80})$ vs. r_{80} with slope $-b$, and they require two quantities for their complete specification: the exponent of r_{80} and the magnitude of the concentration at a given r_{80}. If $n(r_{80})$ is of this form, then $dN(r_{80})/dr_{80}$, $dR_{80}(r_{80})/d\log r_{80}$, $a_{80}(r_{80})$, and $v_{80}(r_{80})$ will also be of this form, although with different exponents and magnitudes. These fits, which yield a monotonic decrease in $n(r_{80})$ with increasing r_{80}, cannot describe the size distribution for all particle sizes and thus they are necessarily restricted in their radius range of applicability. Additionally, as noted by *Whitby* [1978], apparently minor deviations from the power law fit may actually represent different aerosol populations (modes). Despite much effort to determine a physical basis for fits of this form, little theoretical justification for their use can be made.

Gamma distributions and modified gamma distributions represent size distributions of concentration as $n(r_{80}) = ar_{80}^{b} \exp\{-cr_{80}^{-\gamma}\}$, with $a > 0$, $c > 0$, and $\gamma > 0$, and where a, b, c, and γ are independent of r_{80} but may depend on wind speed and other meteorological quantities (for $\gamma = 1$ the distribution is the ordinary gamma distribution). If $n(r_{80})$ is represented by a gamma or modified gamma distribution, then $dN(r_{80})/dr_{80}$, $dR_{80}(r_{80})/d\log r_{80}$, $a_{80}(r_{80})$, and $v_{80}(r_{80})$ are also of this form, although with different values of the coefficient and the power of r_{80} (characterized by a and b for $n(r_{80})$). Some widely used expressions, including those of *Rosin and Rammler* [1933], *Nukiyama and Tanasawa* [1938], *Weibull* [1951], and *Deirmendjian* [1964] are of this form. These formulations often provide a good fit to the data, and for a size distribution represented by a gamma distribution or a modified gamma distribution all moments (integrals of powers of radius over the size distribution) exist; that is, the integrals converge. Although the use of these distributions was formerly common, they have not gained widespread acceptance for representing SSA concentration size distributions, and lognormal distributions or sums of lognormal distributions are now typically used.

Lognormal distributions [*Galton*, 1879; *McAlister*, 1879] have long been used for a wide variety of applications[1]. Although lognormal distributions had previously been used to represent cloud drop size distributions [*East and Marshall*, 1954] and atmospheric aerosol concentration size distributions (see above references and *Sinclair* [1946]; *Junge* [1957b, 1969]; *Kerker et al.* [1964]; and *Quenzel* [1970], for example), *Blifford and Gillette* [1971] speculated that the size distribution of the number concentration of the global aerosol could be represented by the sum of three lognormal distributions, and *Whitby et al.* [1972] reported that measured size distributions of aerosol volume concentration were bimodal and suggested that the mode comprised of the smaller particles could be represented by a lognormal distribution. Since that time it has become common to represent size distributions of aerosol number concentration as the sum of lognormal "modes" [*Whitby*, 1978]. In some circumstances arguments can be made for a theoretical basis for fits of this form, but although a sum of lognormal modes often provides a reasonable representation of the data and has become a standard way to represent concentration size distributions, it should be stressed that this does not necessarily imply that the data actually behave in such a fashion, or that there is a physical basis for this representation. In most situations lognormal distributions merely provide a convenient and commonly accepted method of fitting such distributions. As lognormal modes are widely used, several of their properties are discussed.

Each lognormal mode can be written in the form

$$n(r_{80}) = n_0 \exp\left\{-\frac{1}{2}\left[\frac{\ln(r_{80}/r_{80}')}{\ln \sigma}\right]^2\right\}, \qquad (2.1\text{-}9)$$

and thus is parameterized by three quantities: the maximum amplitude of $n(r_{80})$, n_0; the geometric mean r_{80}', the value of r_{80} at which the maximum amplitude of $n(r_{80})$ occurs (hence $n_0 = n(r_{80}')$); and the geometric standard deviation σ, a dimensionless number that characterizes the width of the mode ("log" can also be substituted for "ln" in this expression without changing the values of the parameters). As a number of variants of the lognormal distribution have been employed by various investigators (for instance, the symbol n_0 has sometimes been used to denote the total number concentration of particles in a given mode, and likewise the symbol σ has sometimes been used differently from how it is here), a statement of values of parameters of a lognormal distribution without specification of the expression may lead to error. The mean, median, and mode values of r_{80} for a size distribution $n(r_{80})$ described by (2.1-9) are $r_{80}'\exp\{(\ln \sigma)^2/2\}$, r_{80}', and $r_{80}'\exp\{-(\ln\sigma)^2\}$, respectively. The full width at half maximum of $n(r_{80})$ is $\Delta_{1/2}\log r_{80} = 2(2\ln2)^{1/2}\log \sigma$. When plotted as $n(r_{80})$ vs. $\log r_{80}$, a lognormal mode is a gaussian; when plotted as $\log n(r_{80})$ vs. $\log r_{80}$ it is a downward-facing parabola. The size-dependent number concentration at a given value of r_{80} can readily be determined from the three quantities n_0, r_{80}', and σ; using relations such as $n(r_{80}'/\sigma) = n(r_{80}'\sigma) \approx 0.61n(r_{80}')$ and $n(r_{80}'/\sigma^2) = n(r_{80}'\sigma^2) \approx 0.14n(r_{80}')$. Approximately 40% of the contribution to the total number concentration is from particles with r_{80} between $r_{80}'/\sigma^{1/2}$ and $r_{80}'\sigma^{1/2}$ and nearly 70% of the contribution is from particles with r_{80} between r_{80}'/σ and $r_{80}'\sigma$.

If $n(r_{80})$ is a lognormal distribution representing a single mode of a concentration size distribution described by (2.1-9), then $dN(r_{80})/dr_{80}$, $dR_{80}(r_{80})/d\log r_{80}$, $a_{80}(r_{80})$, $v_{80}(r_{80})$, and $m(r_{80})$ are also lognormal distributions. The parameters describing the maximum amplitude and its location differ for these other representations (Table 3), although the value of σ remains unchanged. Numerical results are generally more sensitive to the value of σ than to n_0 or r_{80}', and thus it may be justified to give this quantity to higher precision. As with gamma and modified gamma distributions, all moments exist for a size distribution expressed as a lognormal (or as a sum of lognormal modes). Expressions for the integral quantities N, R, A, V, and M for a lognormal size distribution of SSA number concentration of the form (2.1-9) are presented in Table 4. For such a lognormal distribution each integral is well approximated (to within 6%) as the product of the maximum value of the integrand and the full width at half maximum of the distribution $\Delta_{1/2}\log r_{80}$, with the range of integration between the half-maximum points contributing approximately three-fourths of the value of the integral.

---

[1] Information on the lognormal distribution and its applications, fitting data to this form, and relations involving this distribution, are discussed in *Gaddum* [1945]; *Howell* [1949]; *Kottler* [1950, 1952]; *Aitchison and Brown* [1957]; *Shockley* [1957]; *Herdan* [1960, pp. 81-85]; *Espenscheid et al.* [1964]; *Mitchell* [1968]; *Raabe* [1971]; *Heintzenberg* [1994]; *Limpert et al.* [2001], and references therein.

Table 3. Alternative Representations of Lognormal Size Distributions of Sea Salt Aerosol Concentration

Differential Property $q_i(r_{80})$	i	C_i
$\dfrac{dN(r_{80})}{dr_{80}}$	-1	$\dfrac{1}{\ln 10}$
$n(r_{80}) \equiv \dfrac{dN(r_{80})}{d\log r_{80}}$	0	1
$\dfrac{dR_{80}(r_{80})}{d\log r_{80}}$	1	1
$a_{80}(r_{80}) \equiv \dfrac{dA_{80}(r_{80})}{d\log r_{80}}$	2	4π
$v_{80}(r_{80}) \equiv \dfrac{dV_{80}(r_{80})}{d\log r_{80}}$	3	$\dfrac{4\pi}{3}$
$m(r_{80}) \equiv \dfrac{dM(r_{80})}{d\log r_{80}}$	3	$\dfrac{4\pi}{3}\rho_{ss}\left(\dfrac{r_{dry}}{r_{80}}\right)^3$

For the lognormal size distribution of SSA number concentration given by (2.1-9),

$$n(r_{80}) = n_0 \exp\left\{-\frac{1}{2}\left[\frac{\ln(r_{80}/r'_{80})}{\ln\sigma}\right]^2\right\},$$

alternative representations for the size distribution are of the form

$$q_i(r_{80}) = n_0 C_i (r'_{80})^i \exp\left[\frac{i^2}{2}(\ln\sigma)^2\right]\exp\left\{-\frac{1}{2}\left[\frac{\ln(r_{80}/r^*_{80})}{\ln\sigma}\right]^2\right\},$$

where r^*_{80}, the value of r_{80} at the peak of the distribution in the given representation, is given by

$$r^*_{80} = r'_{80}\exp\{i(\ln\sigma)\}^2.$$

All representations are characterized by the same geometric standard deviation σ and thus have the same full width at half maximum $\Delta_{1/2}\log r_{80} = 2(2\ln 2)^{1/2}\log\sigma$.
In the expression for $m(r_{80})$ the density of dry sea salt $\rho_{ss} \approx 2.2$ g cm^{-3} (§2.5.1) and the ratio r_{dry}/r_{80} is almost exactly 0.5 (§2.5.3).

Table 4. Integral Properties of Lognormal Size Distributions of Sea Salt Aerosol Concentration

Integral Property Q_i	i	C_i	$\left(\dfrac{\sqrt{2\pi}}{\ln 10}C_i\right)$
$\dfrac{N}{\text{cm}^{-3}}$	0	1	1.1
$\dfrac{R_{80}}{\mu\text{m cm}^{-3}}$	1	1	1.1
$\dfrac{A_{80}}{\mu\text{m}^2\,\text{cm}^{-3}}$	2	4π	1.4
$\dfrac{V_{80}}{\mu\text{m}^3\,\text{cm}^{-3}}$	3	$\dfrac{4\pi}{3}$	4.6
$\dfrac{M}{\mu\text{g m}^{-3}}$	3	$\dfrac{4\pi}{3}\left(\dfrac{\rho_{ss}}{\text{g cm}^{-3}}\right)\left(\dfrac{r_{dry}}{r_{80}}\right)^3$	1.3

For the lognormal size distribution of SSA number concentration given by (2.1-9),

$$n(r_{80}) = n_0 \exp\left\{-\frac{1}{2}\left[\frac{\ln(r_{80}/r'_{80})}{\ln\sigma}\right]^2\right\},$$

integral properties Q_i are evaluated as

$$Q_i = \left(\frac{\sqrt{2\pi}}{\ln 10}C_i\right)\left(\frac{n_0}{\text{cm}^{-3}}\right)\left(\frac{r'_{80}}{\mu\text{m}}\right)^i(\ln\sigma)\exp\left\{\frac{i^2}{2}(\ln\sigma)^2\right\}.$$

The integral $N = n_0(\pi)^{1/2}\log\sigma \approx 2.5 n_0\log\sigma$ compares closely to its approximation as the product of the maximum value of the integrand (i.e., n_0) and the full width at half maximum of the distribution, $N \approx n_0(\Delta_{1/2}\log r_{80}) \approx n_0 2(2\ln 2)^{1/2}\log\sigma \approx 2.4 n_0\log\sigma$, and similarly for the other tabulated integral properties.
In the expression for M the density of dry sea salt $\rho_{ss} \approx 2.2$ g cm^{-3} (§2.5.1) and the ratio r_{dry}/r_{80} is almost exactly 0.5 (§2.5.3).

The cautions noted above regarding transformation of fits based on one representation of a size distribution of concentration to other representations are especially pertinent to lognormal distributions (for which these transformations are readily determined; Table 3), as peak values for $a_{80}(r_{80})$ and $v_{80}(r_{80})$ typically occur beyond the size range that played the dominant role in determining the parameters of the fit describing $n(r_{80})$, and often beyond the size range for which data exist. Additionally, the effects of any uncertainties in the parameters describing the lognormal mode are amplified when the representations $a_{80}(r_{80})$ and especially $v_{80}(r_{80})$ are obtained from $n(r_{80})$, and the accuracies of the parameters describing these resulting distributions are lower [*Heintzenberg*, 1994], often to the point that little confidence can be placed in their values (§4.1.4.4).

2.1.5. Concentrations of SSA Properties

Many properties of SSA that characterize its role in processes of geophysical importance can be evaluated from size distributions of concentration. Some of these properties can be categorized into local differential (i.e., size distributions) and integral (cumulative distributions and partial and total values) properties, and column burdens (integrals over height) of these quantities. Such properties, which scale linearly with $n(r_{80})$, are denoted extensive properties of the aerosol [*Ogren*, 1995]. Similar considerations apply to fluxes. Local differential properties of size distributions of concentration can be written in the form $W(r)n(r_{80})$, where r is the geometric radius of the particle and the quantity $W(r)$ is the size-dependent weighting factor for the property of interest; for example, $W(r) = 4\pi r^2$ for $a(r_{80})$, and $W(r) = 4\pi r_{80}^2 = 4\pi r^2/[\zeta(rh)]^2$ for $a_{80}(r_{80})$, where $\zeta(rh) = r/r_{80}$ is known as a function of RH (§2.1.1; §2.5.3). Local integral properties of size distributions of concentration can be written in the form $I = \int W(r)n(r_{80})d \log r_{80}$; for partial integral quantities (such as δA) the integral is over the specified size range and for total integral quantities (such as A) the integral is over the size range necessary to include the vast majority of the contribution, whereas for cumulative distributions (such as $A(r_{80})$) the upper limit of the integral is a given value of r_{80} or r. Size-dependent column burdens can be written in the form $\int W(r)n(r_{80},z)dz$ and total column burdens in the form $I_{col} = \int[\int W(r)n(r_{80},z)d \log r_{80}]dz = \int I dz$, where again the integral is over the range of heights necessary to provide the vast majority of the contribution.

The contribution of SSA particles of a given r_{80} to a property of the concentration is determined by $W(r)$, which describes the relative importance of a single SSA particle with geometric radius r that corresponds to r_{80} at the given conditions (i.e., RH), and by $n(r_{80})$, which determines the relative abundance of SSA particles with the given r_{80}. As in nearly all situations of interest the geometric radius of a particle depends only on r_{80} and the ambient RH (§2.1.1; §2.9.1), $W(r)$ can often be written as $W(r_{80}, rh)$. In these situations total integral quantities, for a given $n(r_{80})$, depend only on RH, and total column burdens, for a given $n(r_{80}, z)$, depend only on the vertical distribution of RH. For example, the column burden of the concentration of SSA surface area, $A_{col} = \int[\int 4\pi r^2 n(r_{80},z)d \log r_{80}]dz$, can also be written $A_{col} = \int[\int 4\pi r_{80}^2(\zeta(rh))^2 n(r_{80},z)d \log r_{80}]dz$, which under change of order of integration can be expressed as $A_{col} = \int 4\pi r_{80}^2[\int(\zeta(rh))^2 n(r_{80},z)dz]d \log r_{80}$.

Many quantities of interest, such as the total concentrations of SSA number, radius, surface area, volume, and mass, are directly proportional to one of the lower-order moments of radius (or r_{80}) over the size distribution of the number concentration, the kth moment being defined as

$$\mu_k = \int r^k n(r_{80})d \log r_{80};$$

thus the weighting factor for any of these quantities is directly proportional to a power of the radius (i.e., $W(r) \propto r^k$). For instance, the zeroth moment μ_0 is equal to N, the total SSA number concentration; the first moment μ_1 is one-half the length that would be obtained if all the SSA particles in a given volume were placed side-by-side; the second moment μ_2 is directly proportional to A, the concentration of SSA surface area; and the third moment μ_3 is directly proportional

to the V, the volume concentration of SSA particles. For a given RH, the third moment is also directly proportional to the concentration of liquid water in the SSA particles and to the SSA dry mass concentration M, which describes the amount of dissolved material in SSA particles that can be exchanged with the atmosphere and that is available to participate in chemical reactions with other substances. The quantities V, the volume fraction of SSA particles in the atmosphere, and M, which is directly proportional to the dry mass fraction of SSA particles in the atmosphere, determine the extent to which the SSA affects the physical properties of the atmosphere (§4.1.2). Additionally, many other integral properties can be well represented as sums of the low-order moments, and a size distribution that accurately estimates the several lowest order moments will accurately estimate most total aerosol properties [e.g., McGraw et al., 1995]. The integral moments of the radius over the size distribution of SSA number concentration are thus important properties in their own right in addition to serving as approximations or bounds for other properties of interest.

For properties that are directly proportional to the kth moment of the radius, $W(r) \equiv W(r_{80}, rh)$ can be written as $[\zeta(rh)]^k W(r_{80})$, with the consequence that the differential and total integral properties of the quantity under consideration for a given RH are related to those at 80% RH by the known factor $[\zeta(rh)]^k$, which depends only on RH and the moment of interest k. Thus, the quantities represented by powers of r_{80} are important; not only are they necessary to determine the corresponding values at RH other than 80%, additionally they provide physically meaningful values, as 80% is typical of RH in the marine boundary layer.

Other important SSA properties, which are independent of the amount of SSA present, are denoted intensive properties of the aerosol [Ogren, 1995]. These properties, which are characteristic of the shape of the size distribution of SSA concentration, are frequently evaluated as ratios of extensive SSA properties, and in many instances are ratios of low-order moments of radius over the size distribution of SSA number concentration [e.g., Mugele and Evans, 1951]. Often a size distribution of SSA concentration is characterized by a single radius typifying, in some way, its importance to some property of interest. However, as with the representation in which a size distribution is presented, these characteristic radii are often not clearly defined and their use differs among different investigators, sometimes leading to confusion and error (e.g., §4.1.4.4). Several such characteristic radii are defined here.

The median radius (i.e., number-median radius) r_{nm} is the radius at which one-half of the SSA number concentration is provided by radii less than this value; thus this quantity is defined by $N(r_{nm}) = 0.5N$. Likewise, the area-median radius r_{am} is the radius for which $A(r_{am}) = 0.5A$, and the mass-median (or volume-median) radius r_{mm} is the radius for which

$M(r_{\text{mm}}) = 0.5M$. The arithmetic mean radius is the ratio of the first moment of the radius with respect to the size distribution of SSA number concentration to the zeroth moment, μ_1/μ_0, whereas the geometric mean radius is e raised to the power of the mean value of $\ln r$. The root-mean-square radius, or mean area radius, is the square root of the ratio of the second moment of the radius with respect to the size distribution of SSA number concentration to the zeroth moment, $(\mu_2/\mu_0)^{1/2}$, which is also equal to $[A/(4\pi N)]^{1/2}$. Likewise the mean volume radius, or mean mass radius, is the cube root of the ratio of the third moment of the radius with respect to the size distribution of SSA number concentration to the zeroth moment, $(\mu_3/\mu_0)^{1/3}$, which is also equal to $[3V/(4\pi N)]^{1/3}$, or equivalently $\zeta(rh) \times [6M/(\pi\rho_{\text{ss}}N)]^{1/3}$, where $\zeta(rh) = r/r_{80}$ (§2.1.1; §2.5.3). This mean mass radius is to be distinguished from the mass-mean radius (or volume-mean radius), which is the ratio of the fourth moment of the radius over the size distribution of SSA number concentration to the third moment, μ_4/μ_3. Likewise, the area-mean radius, also called the mean volume-surface radius, the Sauter radius, or the effective radius r_{eff}, is the ratio of the third moment of the radius over the size distribution of SSA number concentration to the second moment, μ_3/μ_2, also equal to $3V/A$.

For a lognormal distribution consisting of a single mode described by (2.1-9), these characteristic radii can be evaluated from Tables 3 and 4 and are presented here for 80% RH (values for each of these characteristic radii increase with increasing RH and can be determined at other RH by the multiplication by the factor $\zeta(rh)$). The (number-) median radius is r_{80}', the area-median radius is $r_{80}' \exp\{2(\ln\sigma)^2\}$, and the mass-median radius is $r_{80}' \exp\{3(\ln\sigma)^2\}$. The geometric mean radius is r_{80}'. The mean radius is $r_{80}' \exp\{(1/2)(\ln\sigma)^2\}$ the root-mean-square radius is $r_{80}' \exp\{(\ln\sigma)^2\}$ and the mean volume, or mean mass, radius is $r_{80}' \exp\{(3/2)(\ln\sigma)^2\}$. The area-mean, or effective, radius is $r_{\text{eff}} = r_{80}' \exp\{(5/2)(\ln\sigma)^2\}$, and the mass-mean, or volume-mean, radius is given by $r_{80}' \exp\{(7/2)(\ln\sigma)^2\}$.

A quantity commonly used in radiative transfer models [*Hansen and Travis*, 1974] describing atmospheric transmission and reflection of radiation and often also called the effective radius is the ratio of the column burden of the third moment of radius with respect to the size distribution of SSA number concentration to the column burden of the second moment. If RH and $n(r_{80})$ are independent of height then this quantity is the same as the value of r_{eff} defined above, but otherwise it may be different.

Three key roles played by SSA particles are their ability to act as cloud condensation nuclei to form cloud drops; to exchange gases with the atmosphere, serve as a medium for reaction with these gases and engage in chemical reactions, and act as a sink for condensable gases and for particles; and to scatter light and thus directly affect visibility and the albedo of Earth (see references in §1.1). These roles are examined in the following sections. Weighting factors $W(r)$ are presented which characterize the relative importance of different particle sizes to the processes of interest.

For scoping purposes, a lognormal distribution is used to parameterize the size-dependent SSA number concentration here and throughout this review. This distribution is of the form (2.1-9) with amplitude $n_0/\text{cm}^{-3} = 0.07[U_{10}/(\text{m s}^{-1})]^2$, where U_{10} is the mean wind speed at 10 m above the sea surface (§2.3.1), geometric mean $r_{80}' = 0.3$ μm, and geometric standard deviation $\sigma = 2.8$. This distribution, which is purely empirical, is based on size distributions of SSA concentration presented in §4.1.4. It is denoted the canonical size distribution of SSA concentration. For this distribution, the maximum contributions to $n(r_{80})$, dN/dr_{80}, $dR_{80}/d\log r_{80}$, $a_{80}(r_{80})$, and $v_{80}(r_{80})$ (and hence $m(r_{80})$) occur at r_{80} values of approximately 0.3, 0.1, 0.9, 2.5, and 7 μm, respectively (Table 3). At 80% RH the effective radius is $r_{\text{eff},80} \approx 4.2$ μm, the mean mass radius is approximately 1.5 μm, and the mass-mean radius is approximately 12 μm. The large spread of these several characteristic radii, two orders of magnitude, underlies the importance of the different weighting factors associated with different physical phenomena. The full width at half maximum (§2.1.4.2) for this parameterization is given by $\Delta_{1/2}\log r_{80} \approx 1.05$, independent of wind speed. The fact that this quantity is nearly unity allows integral properties of quantities that are directly proportional to the moments of radius over the size distribution of SSA concentration to be approximated as equal to the maximum value of the quantity (§2.1.4.2). The shape (i.e., size dependence) of the SSA number concentration according to this distribution is shown in Fig. 2 (§2.1.5.1), as is the shape of the corresponding size distribution of SSA mass concentration. The canonical size distribution of SSA concentration illustrates the size range of SSA particles that play the dominant role in various processes discussed in the following sections and allows estimates to be made of the accuracies of approximations that are used.

2.1.5.1. Cloud drop activation. Sea salt aerosol particles exert an important influence on clouds in the marine atmosphere by serving as cloud condensation nuclei (CCN), upon which cloud drops form. The size distribution of SSA number concentration influences the size distribution of number concentration of the drops comprising maritime clouds, in turn affecting their reflectance [*Twomey*, 1977b, 1991], coalescence, drizzle and rain formation, and persistence [*Albrecht*, 1989], and possibly playing a role in cleansing the air of anthropogenic constituents [e.g., *Rosenfeld et al.*, 2002].

The ability of an SSA particle to form a cloud drop is determined by its size, by local meteorological conditions—specifically the updraft velocity influencing the ambient supersaturation (the amount by which the ambient RH is greater than 100%), and by the compositions and size

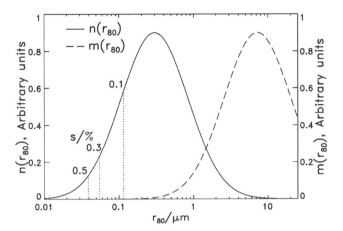

Figure 2. Equal-area plot of SSA number concentration described by the canonical SSA size distribution of concentration: a log-normal distribution in the representation $n(r_{80}) \equiv dN(r_{80})/d\log r_{80}$ having peak value $r'_{80} = 0.3\,\mu\text{m}$ and geometric standard deviation $\sigma = 2.8$ (§2.1.5), as a function of the equilibrium radius at 80% RH, r_{80}. Vertical lines show r_{80} for the smallest SSA particles that can activate to form cloud drops at supersaturation s equal to 0.5, 0.3, and 0.1%. Dashed curve shows size distribution of SSA mass concentration $m(r_{80}) \equiv dM(r_{80})/d\log r_{80}$.

distributions of concentration of other aerosol particles present. When an air parcel rises in the atmosphere it does work on its surroundings and is cooled adiabatically at a rate that depends mainly on the updraft velocity. The decrease in temperature results in an increase in ambient RH given by the Clausius-Clapeyron relation [*Clapeyron*, 1834; *Clausius*, 1850]. As the ambient RH increases, water vapor condenses on existing aerosol particles (i.e., drops) at rates that are controlled by the ambient temperature and by the differences between the ambient RH and the equilibrium RH values of the drops. The equilibrium RH of a given drop is equal to the ratio of its equilibrium water vapor pressure to the saturation vapor pressure of pure bulk water at the ambient temperature, and depends on drop size and composition (§2.5.3). This equilibrium RH generally differs from that of pure bulk water (i.e., 100%) both because of the presence of dissolved substances (Raoult effect, §2.5.3), which decreases this equilibrium RH, and because of the curvature of the drop in conjunction with surface tension (Kelvin effect, §2.5.3), which increases this equilibrium RH. These two effects result in a maximum equilibrium RH for an aerosol particle containing a given amount and composition of solute [*Köhler*, 1936], often expressed as a critical supersaturation s_c equal to the maximum equilibrium RH minus

100%. For a particular dissolved solute (i.e., particle composition), the critical supersaturation decreases with increasing solute mass, or for SSA particles, with increasing r_{80}. Thus water vapor condenses preferentially on particles containing greater concentrations of dissolved salts because of Raoult vapor-pressure reduction and on larger particles because of their larger radius, with resultant lesser enhancement of vapor pressure due to the Kelvin effect (§2.5.3).

With increasing ambient RH above 100% those particles for which the ambient supersaturation s exceeds the critical supersaturation s_c will accrete water and grow to a radius that exceeds the critical radius (the radius at which the equilibrium supersaturation is equal to the critical supersaturation s_c) and thereafter continue to accrete water, as the equilibrium supersaturation decreases as the radius increases above the critical radius. Such drops are said to be activated. This critical, or activation, radius r_{act} for an SSA particle with given r_{80} is given approximately by

$$\frac{r_{act}}{\mu\text{m}} \approx \left(\frac{r_{80}}{0.13\,\mu\text{m}}\right)^{\frac{3}{2}} \qquad (2.1\text{-}10)$$

at 0 °C, and is approximately 6% greater at 20 °C ([1]).

At a given supersaturation s all SSA particles with r_{80} greater than a value $r_{80,s}$ can activate; this value is given approximately by

$$\frac{r_{80,s}}{\mu\text{m}} \approx \left(\frac{0.0039\%}{s}\right)^{\frac{2}{3}} \qquad (2.1\text{-}11)$$

at 0 °C, and is approximately 10% less at 20 °C. Equivalently, the critical supersaturation s_c corresponding to a given r_{80} (that is, the supersaturation at which all SSA particles larger than the given r_{80} can activate) is given approximately by

$$\frac{s_c}{\%} \approx \left(\frac{0.023\,\mu\text{m}}{r_{80}}\right)^{\frac{3}{2}} \qquad (2.1\text{-}12)$$

at 0 °C, and is approximately 10% less at 20 °C. Several phenomena prevent this growth from continuing indefinitely: the ambient supersaturation decreases because of uptake of water by SSA and other aerosol particles, and some SSA particles having $r_{80} > r_{80,s}$ may not activate because of their relatively large response times to changes in RH (§2.5.4), although such drops can nonetheless effectively function as cloud drops [*Hänel*, 1987]. For example, an SSA particle with radius

---

[1] Here and throughout this review parameterizations are often presented in a form similar to that of Eq. 2.1-10, as such expressions readily permit mental calculation of estimates accurate to a few percent thereby providing immediate insight into the quantity being thus parameterized.

(0.5, 1, 2) μm at 80% RH instantaneously exposed to a supersaturation of 0.1% at 0° C would require a time of approximately (0.12, 1, 8) h to grow to its activation radius $r_{act} \approx$ (7, 20, 60) μm, during which time it would fall in still air under the influence of gravity approximately (2, 110, 5500) m.

An important property of SSA governing its role in cloud formation is the number concentration of SSA particles that can activate and function as CCN at a given supersaturation s. This number concentration $N_{CCN,s}$ is given by

$$N_{CCN,s} = \int_{r_{80,s}}^{\infty} n(r_{80}) \, d \log r_{80}, \qquad (2.1\text{-}13)$$

where $r_{80,s}$ is the value of r_{80} for the smallest SSA particle that can activate to form a cloud drop at the given ambient supersaturation s given by (2.1-11). For maximum supersaturations 0.1% to 0.5%, thought to be typical of marine clouds [*Baker and Charlson*, 1990; *Hudson*, 1993; *Martin et al.*, 1994; *Twohy and Hudson*, 1995; *Leaitch et al.*, 1996; *Yum et al.*, 1998; *Cantrell et al.*, 1999; *Yum and Hudson*, 2002; *Hudson and Yum*, 2002], values of $r_{80,s}$ range from near 0.1 μm to near 0.04 μm, respectively, according to (2.1-11). Values of $r_{80,s}$ corresponding to supersaturations of 0.1, 0.3, and 0.5% are indicated in Fig. 2 for the canonical SSA size distribution of concentration introduced above (§2.1.5). For this size distribution, the great majority of SSA particles (81, 94, and 97% of the number concentration, respectively, for these supersaturations), containing virtually all of the SSA mass concentration (more than 99.995% in each case), can function as CCN and activate to form cloud drops. Thus, to a good approximation the property characterizing the ability of SSA particles to act as CCN is their total number concentration, N, and the weighting factor $W(r)$ is nearly equal to unity. Measurements of N and thus the extent to which SSA can provide CCN are discussed in §4.1.3 and §4.1.4.5.

2.1.5.2. Uptake and reaction of gases and particles. Sea salt aerosol particles serve as a sink for condensable trace atmospheric gases and for smaller aerosol particles (those having radii from several nanometers to several tens of nanometers), such as result from gas-to-particle conversion. Sea salt aerosol particles also serve as a medium for aqueous-phase reaction of reactive gases. The uptake of a gas by an aqueous SSA particle (drop) may be reversible or, for highly soluble or highly reactive gases, may be essentially irreversible and complete, depending on the solubility of the gas and/or the reaction kinetics.

The steady-state reactive (irreversible) uptake of trace gases by an aqueous SSA drop may be viewed as occurring by a sequence of processes [*Schwartz and Freiberg*, 1981;

Schwartz, 1986]: reversible mass transport of the reactive trace gas through the atmosphere to the surface of the drop by molecular diffusion, reversible interfacial mass transfer across the surface of the drop by free-molecular transport, dissolution in the aqueous phase, and simultaneous diffusion and irreversible reaction within the drop. Similar considerations apply to the uptake of smaller particles by SSA, except that the sequence of processes consists only of diffusion of the smaller particles in the gas phase and collisional uptake at the interface, the latter process being irreversible.

The rate of uptake of a trace reactive gas by an individual SSA drop depends on the geometric radius r of the drop, which for a given r_{80} depends only on the RH (§2.5.3), and on properties of the trace species, specifically its gas- and aqueous-phase diffusion coefficients D_g and D_{aq}, respectively, its mass accommodation coefficient upon an SSA drop α_M (the fraction of those molecules striking the SSA drop that are captured and dissolve in it), and its solubility and reactive properties, all of which may be temperature-dependent. Some controlling properties may also depend on pH or other properties of the SSA drop such as ionic strength, which are controlled by the relative concentrations of water and of dissolved solids in the drop and hence on RH. Depending on solubility, kinetics, and drop size, often one or another of the several processes controls the overall rate at which uptake occurs. The role of SSA as a sink for smaller particles is analogous to its role in the condensation of highly reactive gases, but with some important differences. As the diffusion coefficient of such a smaller particle depends strongly on its radius, the rate at which smaller particles of a given size coagulate onto a larger SSA particle depends strongly on the radii of the particles being scavenged, which also may depend on RH.

The steady-state rate of reactive (irreversible) uptake of a trace gas by SSA is expressed as an integral over the SSA size distribution of number concentration,

$$-\frac{dC_g}{dt} = \int \frac{4\pi}{3} r^3 \left(\frac{dC_B(r)}{dt} \right) n(r_{80}) \, d \log r_{80},$$

where C_g denotes the bulk concentration of the reactive trace gas (that is, the concentration in the gas phase not in the immediate vicinity of an SSA drop) and $C_B(r)$ denotes the average concentration of the reaction product B in an SSA drop having geometric radius r. The characteristic time of this uptake τ_u is generally defined by its inverse, the effective first-order rate constant for this process τ_u^{-1}, the fractional rate of decrease of the bulk gas-phase concentration C_g due to uptake by SSA,

$$\tau_u^{-1} = -\frac{1}{C_g} \frac{dC_g}{dt}. \qquad (2.1\text{-}14)$$

Hence τ_u^{-1} is an extensive property of SSA concentration (§2.1.5). In this section, which follows *Schwartz* [1986], expressions for this steady-state rate constant τ_u^{-1} are developed, and it is demonstrated that in the limit of one or another process controlling the overall rate, τ_u^{-1} becomes directly proportional to one or another of the lower-order moments of the radius over $n(r_{80})$.

In order to develop an expression for this rate of uptake it is useful to treat the sequence of processes by which a trace gas is taken up by and reacts with an SSA particle (drop) with radius r as a series of chemical reactions:

$$G_\infty \underset{k_{-d}(r)}{\overset{k_d(r)}{\rightleftharpoons}} G_i \underset{k_{-i}(r)}{\overset{k_i(r)}{\rightleftharpoons}} A_i \overset{k_{rxn}Q(r)}{\longrightarrow} B.$$

Here G_∞ denotes the trace substance in the bulk gas phase (with concentration C_g), G_i denotes the trace substance in the gas phase at the surface of the drop, with concentration $C_{g,i}$, and A_i denotes the aqueous species at the interface, with concentration $C_{aq,i}$. The quantities $k_d(r)$ and $k_{-d}(r)$ denote rate constants for diffusive mass transport from the bulk gas phase to the surface of the drop and from the surface of the drop to the bulk gas phase, respectively. The rates of mass-transport processes are expressed as averages over the volume of the drop and are given as the product of the rate constant and the pertinent concentration, here $k_d(r)C_g$ and $k_{-d}(r)C_{g,i}$. Similarly, $k_i(r)$ and $k_{-i}(r)$ denote rate constants for the reversible mass transfer of the trace species across the drop interface. The quantity $k_{rxn}Q(r)$ denotes an effective first-order rate constant for diffusion and irreversible reaction occurring simultaneously within the drop, averaged over the volume of the drop; these processes occur in parallel and thus are not separable. The rate constant k_{rxn} is the first-order reaction rate constant (or the effective first-order rate constant if the reaction is not first order) and $Q(r)$ is a factor that accounts for the possible spatial non-uniformity of the concentration of the reacting dissolved gas in the drop.

This treatment differs from the conventional treatment in which mass-transport processes are represented by fluxes. These fluxes can be expressed as rates averaged over the volume of the drop by noting that the rate at which the average concentration of a substance in the drop increases because of a given flux into the drop is equal to the product of this flux and the surface area of the drop, divided by its volume. This is equivalent to multiplying the flux by $3/r$, the ratio of the surface area of the drop to the volume. Thus for a given flux F the corresponding uptake rate is $3F/r$ and the uptake rate constant is the quantity $3F/r$ divided by the pertinent concentration.

The net inward diffusion flux of a trace gas through the atmosphere at the surface of a SSA drop of radius r (that is, from the bulk gas phase to the interface of the drop) is

$F_d = (C_g - C_{g,i})D_g/r$ [*Maxwell*, 1877], which can be viewed as the difference between an inward diffusion flux $C_g D_g/r$ and an outward diffusion flux $C_{g,i}D_g/r$. The corresponding rates of these processes are $3C_g D_g/r^2$ and $3C_{g,i}D_g/r^2$, respectively, from which the corresponding rate constants are seen to be equal: $k_d(r) = k_{-d}(r) = 3D_g/r^2$.

Similarly, mass transfer of the trace gas across the interface of the drop from the gas phase to the aqueous phase occurs by molecular collisions at this interface, and the rate of transfer depends on the mass accommodation coefficient of the trace gas upon an SSA particle α_M and on the mean molecular speed of the trace gas c_g, given by $c_g = [8RT/(\pi M_g)]^{1/2}$, where M_g is its molecular weight, R is the gas constant, and T is the absolute temperature [*Maxwell*, 1860]. From gas kinetic theory, the flux of the trace gas from the gas phase across the interface is given by $F_i = C_{g,i}(\alpha_M c_g/4)$ [*Hertz*, 1882; *Langmuir*, 1913; *Marcelin*, 1914b; *Knudsen*, 1915], from which the interfacial mass-transfer rate is given by $3C_{g,i}(\alpha_M c_g/4r)$ and the interfacial mass-transfer rate constant is $k_i(r) = 3\alpha_M c_g/4r$. The rate constant for the reverse process $k_{-i}(r)$, is obtained from detailed-balance considerations. At equilibrium the interfacial transfer rates in each direction are equal:

$$k_i(r)\, C_{g,i}^{eq} = k_{-i}(r)\, C_{aq,i}^{eq},$$

from which the ratio of the forward to the reverse rate constants is

$$\frac{k_i(r)}{k_{-i}(r)} = \frac{C_{aq,i}^{eq}}{C_{g,i}^{eq}} = H,$$

where H is the (dimensionless) Henry's Law solubility constant [*Henry*, 1803a,b] for the gas in the sea salt solution at the ambient relative humidity (this solubility coefficient will generally be substantially less than that for the gas in pure water because of the high salt concentration characteristic of SSA particles at ambient RH; §2.5.3), or the effective Henry constant, which takes into account rapid equilibria on the water side (such as hydrolysis of SO_2 to HSO_3^-) and may be pH-dependent. Hence, $k_{-i}(r) = k_i(r)/H$.

Within the steady-state approximation, for which the net rate of change of the concentration of the intermediate species $C_{g,i}$ is much less than the rates of production and loss of this species, and under the assumption that diffusional transport of the trace gas to the drop interface and its kinetic transfer across this interface act in series (as assumed throughout this discussion), a mass-transport rate constant for combined diffusional and interfacial transfer $k_{d,i}(r)$ can be obtained from $k_d(r)$ and $k_i(r)$ as

$$\frac{1}{k_{d,i}(r)} = \frac{1}{k_d(r)} + \frac{1}{k_i(r)}. \tag{2.1-15}$$

This so-called "flux matching" expression for $k_{d,i}$ results from equating the net inward diffusion flux at the surface of the drop and the net flux from the gas phase across the interface [*Fuchs*, 1934 (cited in *Fuchs*, 1959, p. 8); *Schwartz*, 1986]. Equivalently, from the above expressions for $k_d(r)$ and $k_i(r)$, $k_{d,i}(r)$ can be expressed as

$$k_{d,i}(r) = \frac{3D_g}{r^2}\left(\frac{1}{1+\dfrac{r_{tr}}{r}}\right), \qquad (2.1\text{-}16)$$

where $r_{tr} = 4D_g/(\alpha_M c_g)$ defines a critical transition radius. For most gases in air at temperatures typical of those in the marine boundary layer, D_g is in the range 0.05-0.2 cm^2 s^{-1}; choosing $D_g = 0.1$ cm^2 s^{-1} yields $r_{tr} \approx (0.1 \, \mu m)/\alpha_M$ for $M_g = 50$ g mol^{-1} (as c_g and for the most part D_g decrease with increasing M_g, r_{tr} is within at most a factor of 1.5 of this value for most gases of interest).

For $r \succ r_{tr}$ ([1]), or equivalently, $k_i(r) \succ k_d(r)$, $k_{d,i}(r) \approx k_d(r)$ and mass transport is controlled by diffusion through the atmosphere (continuum, or diffusion, limit), whereas for $r \prec r_{tr}$, $k_{d,i}(r) \approx k_i(r)$ and mass transport is controlled by kinetic transfer across the drop interface (free-molecular, or kinetic, limit); a similar discussion pertinent to water vapor is presented in §2.5.4. In the transition regime between these two limits, the rate of mass transport is controlled by both diffusion and interfacial transfer. Alternative expressions for the flux in the transition regime, and hence for the combined mass-transport rate constant $k_{d,i}(r)$, that also yield the correct limits for both large and small values of r_{tr}/r have been employed [*Fuchs and Sutugin*, 1971; *Smirnov*, 1971; *Dahneke*, 1983; *Seinfeld and Pandis*, 1998, pp. 600-607 provides further discussion]. However, even in the transition regime these expressions differ little from (2.1-15).

The spatial-average rate of reaction within the drop is given by the product of the interfacial aqueous concentration $C_{aq,i}$, the first-order rate constant for reaction (or the effective first-order rate constant for a reaction that proceeds by a more complex chemical mechanism) k_{rxn}, and a factor $Q(r) \leq 1$ that accounts for the fact that the reaction rate averaged over the drop may be less than the rate near the interface owing to depletion by reaction and the inability of aqueous-phase

diffusion to replenish the dissolved reactant gas:

$$\frac{dC_B}{dt} = k_{rxn}Q(r)C_{aq,i}(r).$$

The quantity $Q(r)$ is a function of the ratio $q = r/r_{aq}$, where $r_{aq} = (D_{aq}/k_{rxn})^{1/2}$ defines a penetration distance characterizing the competing effects of diffusion and reaction; for a reaction that is first order in the concentration of the dissolved reagent, this quantity is given by

$$Q(q) = 3\left(\frac{\coth q}{q} - \frac{1}{q^2}\right)$$

[*Danckwerts*, 1951; *Cadle and Robbins*, 1960; *Schwartz and Freiberg*, 1981]. For $q \prec 1$, corresponding to $r \prec r_{aq}$, the reaction occurs slowly compared to diffusion. In this limit the concentration of the trace substance is nearly uniform throughout the drop, and $Q \approx 1$. Alternatively, for $q \succ 1$, corresponding to $r \succ r_{aq}$, the reaction occurs before the trace substance can diffuse very far into the drop, and most of the reaction takes place within the penetration layer thickness r_{aq}, which is the same for all SSA particles, independent of radius. In this limit the average rate of reaction over the drop is given by the ratio of the volume over which the reaction occurs—the product of the surface area $4\pi r^2$ and the thickness r_{aq}—and the volume of the drop; $Q \approx (4\pi r^2)r_{aq}/[(4\pi/3)r^3] = 3r_{aq}/r = 3/q$.

Steady-state treatment of the concentration of the intermediate species $C_{aq,i}$ similar to that for $C_{g,i}$ yields an expression for the size-dependent uptake rate, which accounts for the kinetics of mass transport and reaction, in terms of the bulk concentration of the trace gas C_g:

$$\frac{dC_B(r)}{dt} = k_{tot}(r)C_g,$$

where

$$\frac{1}{k_{tot}(r)} = \frac{1}{k_{d,i}(r)} + \frac{1}{Hk_{rxn}Q(r)}$$
$$= \frac{1}{k_d(r)} + \frac{1}{k_i(r)} + \frac{1}{Hk_{rxn}Q(r)}. \qquad (2.1\text{-}17)$$

---

[1] In order to express a relation between quantities that must be satisfied for some condition to hold it is useful to introduce a symbol to denote a relation that is intermediate between > (simply greater than, e.g., 1 > 0.99) and >> (much greater than, typically by at least an order of magnitude, e.g., 1 >> 0.1). The symbol $\succ$ (considerably greater than, typically by at least about half an order of magnitude, e.g., 1 $\succ$ 0.3) is thus introduced to denote such a relation. The corresponding symbol $\prec$ is likewise used to denote considerably less than, again by at least about half an order of magnitude. Thus for quantities a_1 and a_2 representing effects of mechanisms that are additive in magnitude, the statement $a_1 \succ a_2$, roughly equivalent to $a_1 > 3a_2$, implies that the dominant contribution (more than 75%) to the overall effect is from the mechanism characterized by a_1.

Thus the overall rate constant characterizing the uptake of the reactive trace gas (Eq. 2.1-14) can be expressed as

$$\tau_{u}^{-1} = \int \frac{4\pi}{3} r^3 n(r_{80}) \times \left[(k_d(r))^{-1} + (k_i(r))^{-1} \right.$$
$$\left. + (Hk_{rxn}Q(r))^{-1} \right]^{-1} d\log r_{80}. \qquad (2.1\text{-}18)$$

Evaluation of τ_u^{-1} for a specific gas (for which H and k_{rxn} are known) requires knowledge of $Q(r)$, $k_d(r)$, and $k_i(r)$, all of which depend on the gas considered and also on temperature.

One important consequence of (2.1-17) is that the rate of uptake of a trace reactive gas on an SSA particle per unit volume of particle decreases with increasing SSA particle size, implying that the effect of reactive uptake on particle composition will be greater for smaller SSA particles than for larger ones (§2.5.2); the same result applies for uptake of smaller particles by larger SSA particles discussed below.

Several limiting cases are considered for which simple expressions are obtained for τ_u^{-1}, as shown in Table 5. Here the inequalities, which represent conditions that must be fulfilled for the system to be in one or another limit, must hold over the range of r_{80} that provides the dominant contribution to the integral. Depending on the trace substance and its properties (its Henry's Law solubility constant H; gas-phase and aqueous-phase diffusion coefficients D_g and D_{aq}, respectively; mass accommodation coefficient α_M; and mean molecular speed c_g), the reaction (characterized by k_{rxn}), and the size distribution of the SSA concentration $n(r_{80})$, the uptake rate may be governed by the concentration of SSA radius R, surface area A, or volume V. If the solubility of the trace gas is high and the reaction kinetics are fast, then the rate of uptake is controlled by diffusion in the gas phase and/or mass transfer at the interface. If further, as is frequently the situation, the rate of interfacial mass transfer is high (mass accommodation coefficient near unity), then for particles in the size range of SSA the overall rate of uptake is controlled by gas-phase diffusion to the drop (continuum limit). In this limit the weighting factor in the integral for uptake rate is given by $W(r) = 4\pi D_g r$ and the uptake rate constant is directly proportional to R (Table 5). Uptake of substances with sufficiently low values of the mass accommodation coefficient is controlled by the collision rate at the interface (free-molecular limit), and the weighting factor is given by $W(r) = \pi r^2 \alpha_M c_g$—the uptake rate constant is directly proportional to A (Table 5). In contrast to these situations, if the solubility of the reacting gas is low (H small) and the kinetics of reaction are slow (k_{rxn} small) compared to the rates of mass transport in the gas phase and of mass transfer across the interface, then the rate of uptake is limited by the absorption of the gas and/or its rates of diffusion and reaction in the aqueous phase. If the trace substance reacts before appreciable diffusion occurs ($q \succ 1$), then the weighting factor $W(r) \approx 4\pi r^2 (D_{aq}k_{rxn})^{1/2}$ and the uptake rate constant is again nearly directly proportional to A (Table 5). Finally if the reaction rate is sufficiently slow that the dissolved reacting gas is nearly uniformly distributed through the drop ($q \prec 1$), then $W(r) \approx (4\pi/3) r^3 Hk_{rxn}$ and the uptake rate constant is nearly directly proportional to V (Table 5). In all instances the uptake rate constant depends on RH through the dependencies of R, A, and V (§2.1.4.1), and it may exhibit further dependence on RH because of RH influences on H and/or k_{rxn}.

Table 5. Limiting Cases for Reactive Uptake of Trace Substances by Sea Salt Aerosol Particles

Condition	τ_u^{-1}	Condition	τ_u^{-1}	Limiting Factor
		$r \succ \dfrac{4D_g}{\alpha_M c_g}$	$4\pi D_g R$	Gas-phase diffusion
$Hk_{rxn}Q(r) \succ k_{d,i}(r)$	$4\pi D_g \int \left(\dfrac{1}{r+r_{tr}}\right) r^2 n(r_{80})\, d\log r_{80}$	$r \prec \dfrac{4D_g}{\alpha_M c_g}$	$\dfrac{\alpha_M c_g A}{4}$	Interfacial mass transfer
		$r \succ \left(\dfrac{D_{aq}}{k_{rxn}}\right)^{1/2}$	$H(D_{aq}k_{rxn})^{1/2}A$	Aqueous-phase diffusion
$Hk_{rxn}Q(r) \prec k_{d,i}(r)$	$\dfrac{4\pi}{3} Hk_{rxn}\int Q(r) r^3 n(r_{80})\, d\log r_{80}$	$r \prec \left(\dfrac{D_{aq}}{k_{rxn}}\right)^{1/2}$	$Hk_{rxn}V$	Aqueous-phase reaction

The conditions denoted by inequalities (columns 1 and 3) must be satisfied for the expression for uptake rate constant τ_u^{-1} in the succeeding column to apply. For conditions in both columns 1 and 3 satisfied, the uptake is controlled by a single process (column 5) and the uptake rate constant is proportional to the indicated integral property of the SSA (column 4): the radius concentration R, surface area concentration A, or volume concentration V. The symbols $\succ$ and $\prec$ denote "considerably greater than" and "considerably less than," respectively.

For many gases of atmospheric importance the uptake rate is controlled by gas-phase diffusion and/or interfacial mass transfer. In these situations $Hk_{rxn}Q(r) \succ k_{d,i}(r)$ over most or all of the size range of SSA particles; the uptake rate constant is therefore determined primarily by $k_{d,i}(r)$ (Table 5) and the contribution of SSA particles of a given r_{80} to τ_u^{-1} is given by the differential quantity $d\tau_u^{-1}/d\log r_{80} = (4\pi/3)r^3 n(r_{80})k_{d,i}(r)$. This quantity, evaluated at 80% RH for several values of α_M (with the assumptions $D_g = 0.1$ cm^2 s^{-1} and $M_g = 50$ g mol^{-1}, characterizing a typical gas of interest), is shown in Fig. 3 as a function of r_{80} for the canonical SSA size distribution of concentration presented in Fig. 2. Also shown are the limits for uptake controlled by gas-phase diffusion (continuum limit) and by interfacial collision (free-molecular limit), which additionally represent the size dependence of the concentrations of SSA radius and surface area at 80% RH, $dR_{80}(r_{80})/d\log r_{80}$ and $a_{80}(r_{80})$, respectively, for the canonical lognormal distribution. For a given α_M (i.e., a given r_{tr}) the continuum approximation improves with increasing r_{80}. Likewise, for a given r_{80} the value of $d(\tau_{u,80})^{-1}/d\log r_{80}$ increases with increasing α_M, and the

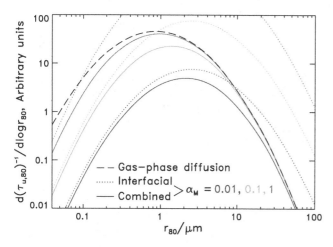

Figure 3. Logarithmic plot of the size distribution of uptake rate constant for reactive gases by SSA particles at 80% RH, $d(\tau_{u,80})^{-1}/d\log r_{80} = (4\pi/3)r_{80}^3 n(r_{80})k_{d,i}(r_{80})$ (Eq. 2.1-18), as a function of r_{80}. Uptake rate constant (solid curves) was evaluated (Eq. 2.1-16) taking into account both diffusional and interfacial mass transport, for the canonical SSA size distribution of concentration $n(r_{80})$ shown in Fig. 2, for mass accommodation coefficients $\alpha_M = 0.01, 0.1,$ and 1, and for diffusion coefficient $D_g = 0.1$ cm^2 s^{-1} and molecular weight $M_g = 50$ g mol^{-1} at 20 °C, typical for trace atmospheric gases. Approximations are also shown for the diffusion (continuum) and interfacial mass transfer (free-molecular) limits; the uptake rate constants in these limits are proportional to the concentrations of SSA radius and surface area at 80% RH, $dR_{80}(r_{80})/d\log r_{80}$ and $a_{80}(r_{80})$, respectively, and thus these curves also represent those concentrations.

transition from the free-molecular limit to the continuum limit occurs as α_M increases from 0.01 to 1. For $\alpha_M = 0.01$ the free-molecular limit provides a fairly accurate approximation over the range of importance of r_{80}, whereas for $\alpha_M = 1$, corresponding to $r_{tr} \approx 0.1$ μm (the practical lower limit of SSA particle size; §2.1.1; §4.1.1.1; §4.1.4.5) for the values used for D_g and c_g, the continuum limit is a good approximation over the range of r_{80} that provides most of the contribution to $(\tau_{u,80})^{-1}$.

This situation applies to uptake by SSA of sulfuric acid (H$_2$SO$_4$), for which the aqueous solubility and dissociation rates are sufficiently high that they are not controlling and for which a value of α_M near unity has been reported [*Pöschl et al.*, 1998]. The uptake rate is therefore limited by gas-phase diffusion throughout most of the range of integration (i.e., continuum limit) and the uptake rate constant is nearly directly proportional to R, the SSA radius concentration. For a diffusion-controlled process with $D_g = 0.1$ cm^2 s^{-1}, the uptake rate constant is given by $\tau_u^{-1}/s^{-1} \approx 1.3 \cdot 10^{-4}[R/(\mu m \ cm^{-3})]$, yielding a characteristic uptake time of

$$\frac{\tau_u}{h} = \frac{2}{\left(\dfrac{R}{\mu m \, cm^{-3}}\right)}. \qquad (2.1\text{-}19)$$

Uptake of gas-phase H$_2$SO$_4$ by SSA particles has important atmospheric implications. By providing a sink for H$_2$SO$_4$ (nucleation of which is thought to be a major source of new particles in the atmosphere), SSA can suppress nucleation, and thereby affect the size distribution of other components of the marine aerosol. This topic is examined further in §4.1.4.5.

It has often been assumed (explicitly or tacitly) that the relevant parameter for nucleation suppression (or reactive uptake of trace gases) is the concentration of aerosol surface area [e.g., *Savoie and Prospero*, 1982; *Hegg*, 1991; *Hegg et al.*, 1992; *Covert et al.*, 1992; *Clarke*, 1992; *Gras*, 1993b; *Huebert et al.*, 1993; *Clarke and Porter*, 1993; *Bigg et al.*, 1995; *Clarke et al.*, 1998; *Pryor and Sørenson*, 2000]. *Covert et al.* [1992], for instance, reported a critical value of the concentration of marine aerosol surface area above which nucleation was not observed. The surface area concentration would indeed be the controlling variable if the vast majority of the uptake of gas molecules by aerosol particles were limited by molecular transfer at the interface (i.e., the free-molecular limit). This might occur in situations for which the aerosol is dominated by particles with $r \prec r_{tr}$ (that is, for particles other than sea salt) or for very small values of the accommodation coefficient, i.e., $\alpha_M \ll 1$. However, in general the relevant parameter for nucleation suppression depends on the size distribution of the concentration of all aerosol particles, and for SSA particles, and all the more for cloud drops (to which some of these investigations specifically referred), the pertinent quantity is not surface area concentration but instead the radius concentration.

Reported values of the mass accommodation coefficient upon water for nitric acid (HNO_3), which is also highly soluble, range from ~0.1 to ~0.2 [*van Doren et al.*, 1990; *Ponche et al.*, 1993; *Abbatt and Waschewsky*, 1998] to ~0.5 [*Guimbaud et al.*, 2002] to near unity [*Rudolf et al.*, 2001], also indicative of uptake at or near the diffusion-controlled limit.

In some instances of importance the rates of aqueous-phase reaction are sufficiently slow compared to rates of gas-phase and interfacial mass transport that the uptake is controlled largely by aqueous-phase processes. In these situations $Hk_{rxn}Q(r) \prec k_{d,i}(r)$ over most or all of the range of sizes of SSA particles and the uptake rate constant is therefore determined primarily by the product $Hk_{rxn}Q(r)$ (Table 5). Two such examples are reaction of sulfur dioxide (SO_2) with ozone (O_3) and reaction of sulfur dioxide with hydrogen peroxide (H_2O_2), both of which involve simultaneous uptake and reaction of two gases, so that the uptake of both species must be considered. Further, for both examples the reaction is not strictly first order in either reagent. However, it is often the situation that at least one of the reacting gases is depleted sufficiently slowly that its aqueous-phase concentration profile is nearly uniform, permitting treatment as an effective first-order reaction, with k_{rxn} equal to the second-order rate constant times the concentration of the "other" reagent. Evaluations of the rates of these reactions, following *Schwartz* [1988a] and using the solubilities, mass accommodation coefficients, and rate expressions given there, but for pH 8, indicate that for the O_3-SO_2 reaction aqueous sulfur-IV establishes a nearly uniform concentration profile in drops throughout the size range of SSA drops but that over much of this size range O_3 reacts preferentially near the drop surface ($q \succ 1$; $Q \prec 1$), at least at the initial high pH before the drop is acidified by the sulfuric acid reaction product and the reaction slows down. In this limit the uptake rate constant is directly proportional to SSA surface area concentration A (Table 5). In contrast aqueous-phase reaction of SO_2 and H_2O_2 in SSA drops is sufficiently slow compared to aqueous diffusion that the concentrations of both dissolved gases are nearly uniform throughout the drop, and hence the uptake rate constant is nearly directly proportional to SSA volume concentration V (Table 5).

As noted above similar formalism applies to scavenging of smaller particles by SSA drops. This process, for which there is no resistance to uptake by aqueous-phase reaction, is limited by gas-phase diffusion and/or interfacial mass transfer. For such a smaller particle with radius r_s up to a few tens of nanometers, the diffusion coefficient D_s is roughly proportional to $(r_s)^{-2}$ [*Friedlander*, 2000, pp. 33-34] and can be expressed approximately as $D_s/(cm^2 \ s^{-1}) \approx [(0.11 \ nm)/r_s]^2$.

Additionally, as smaller particles are thought to remain attached to SSA drops when they collide, α_M is taken as unity, resulting in a value of r_{tr} that is given approximately by $r_{tr}/\mu m \approx [(0.011 \ nm)/r_s]^{1/2}$ for particles of density 1 g cm^{-3} (this value is relatively insensitive to density, being only about 25% greater for particles with density 2 g cm^{-3}). For smaller particles with $r_s > 1$ nm the value of r_{tr} is less than 0.1 μm, and as the vast majority of SSA particles have radii greater than this value (with the vast majority of the contribution to the concentration of SSA radius deriving from particles with radii greater than this value; §4.1.4.3; §4.1.4.5), uptake of these smaller particles by SSA is limited by gas-phase diffusion. Therefore, the uptake rate constant τ_u^{-1} for this process is given by an expression that is identical to that for gases in this limit (Table 5), $\tau_u^{-1} = 4\pi D_s R$, yielding a characteristic uptake time

$$\frac{\tau_u}{h} \approx 18 \frac{\left(\dfrac{r_s}{nm}\right)^2}{\left(\dfrac{R}{\mu m \, cm^{-3}}\right)}. \qquad (2.1\text{-}20)$$

This expression together with (2.1-19) and the several limiting cases in Table 5 define the moments of the radius over the size distribution of SSA concentration pertinent to uptake of gases and smaller particles by SSA and permit identification of the SSA particle size ranges that predominantly contribute to the uptake for the various processes. These expressions also allow the rates of uptake of gases and smaller particles to be readily estimated for known SSA concentration size distributions (§4.1.4.5).

2.1.5.3. Light scattering. By scattering light, SSA particles affect visibility and influence Earth's radiation budget. Sea salt aerosol is widely viewed as providing a major contribution, and often the dominant contribution, to light scattering by aerosols over the oceans [*Quinn et al.*, 1996; *Li et al.*, 1996; *Murphy et al.*, 1998a; *Li-Jones et al.*, 1998; *Carrico et al.*, 1998; *Quinn et al.*, 1998; *Haywood et al.*, 1999; *Quinn and Coffman*, 1999; *Jacobson*, 2001; *Penner et al.*, 2002; *Dobbie et al.*, 2003]. Additionally, scattering by SSA is pertinent to many remote sensing applications, both ground-based and satellite-based. Thus knowledge of the size-dependent optical properties of SSA particles is important to understanding and quantifying the role of SSA in light scattering.

The light-scattering efficiency $Q_s(r/\lambda, m)$ of a spherical particle with radius r and index of refraction m for monochromatic light of wavelength λ is the ratio of the energy of the light scattered by the particle to the energy incident upon its geometric cross section, πr^2. The light-scattering efficiency can be calculated for a given radius as a function of λ and m [*Mie*, 1908; *Debye*, 1909; *van de Hulst*, 1957]. The light-scattering cross section of an SSA particle is the product of this light-scattering efficiency and the particle's geometric cross section, i.e., $\pi r^2 Q_s (r/\lambda, m)$. The key property of SSA that characterizes its role in radiation transfer is $\sigma_{sp}(\lambda)$, the total concentration of light-scattering cross section due to SSA particles (commonly denoted the SSA light-scattering coefficient) for light of wavelength λ. This quantity is given by the integral of the light-scattering cross section over the size distribution of SSA number concentration:

$$\sigma_{sp}(\lambda) = \int \pi r^2 Q_s(r/\lambda, m) n(r_{80}) \, d\log r_{80}, \quad (2.1\text{-}21)$$

and is equal to the fractional rate of decrease in intensity of a monochromatic beam of light of wavelength λ per unit length along the direction of the beam due to scattering by SSA particles. The numerical value of the concentration of light-scattering cross section $\sigma_{sp}(\lambda)$ in units $\mu m^2 \, cm^{-3}$ is coincidentally equal to its value in the units commonly employed for the light-scattering coefficient, Mm^{-1} ($= 10^{-6} \, m^{-1}$). As there is virtually no absorption of visible light by SSA particles, the scattering coefficient is nearly identical to the extinction coefficient; consequently, the single scattering albedo, defined as the ratio of these two quantities, is near unity. The wavelength dependence of $\sigma_{sp}(\lambda)$ is characterized by the Ångström exponent α [Ångström, 1929],

$$\alpha = -\left(\frac{d \ln \sigma_{sp}(\lambda)}{d \ln \lambda} \right). \quad (2.1\text{-}22)$$

This quantity, which can also be expressed as the ratio of concentrations of SSA properties, is an intensive property of SSA concentration (§2.1.5). Other important intensive optical properties of SSA concentration are the phase function, which describes the angular distribution of light scattering; the asymmetry parameter, which is the mean cosine of the phase function; the backscatter fraction, which is the fraction of the scattering phase function in the backward hemisphere; and the mass scattering efficiency, which is the ratio of the concentration of light-scattering cross section to the concentration of SSA dry mass. All of these properties are also wavelength dependent.

For a given size distribution of SSA number concentration $n(r_{80})$, the SSA light-scattering coefficient at a particular wavelength, $\sigma_{sp}(\lambda)$, increases strongly with increasing RH because of the factor r^2 in the integrand of (2.1-21). There is also an RH dependence of the index of refraction m of a seawater drop with fixed amount of solute, shown in Fig. 4, but this dependence is weak, with m increasing from near 1.340 at 98% RH (formation) to near 1.375 at 80% RH [Tang et al., 1997]. The index of refraction m is also expected to be nearly independent of temperature over the range 0 °C to

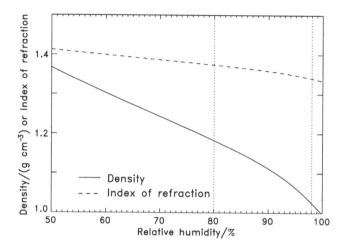

Figure 4. Density and index of refraction (at visible wavelengths) for a liquid drop of sea salt solution in equilibrium with water vapor as a function of relative humidity RH at temperature 25 °C, from *Tang et al.* [1997]. Vertical dotted lines denote 98% RH, characteristic of the atmosphere in equilibrium with seawater of salinity 35 (corresponding to a drop at formation; §2.5.3) and 80% RH, typical of marine boundary layer and used as a reference RH throughout this review.

25 °C and nearly independent of wavelength over the visible spectrum, based on measurements using seawater in concentrations near those typically found in the oceans [*Millard and Seaver*, 1990]. Values of Q_s for $m = 1.33$ (and for $m = 1.5$) are presented by *van de Hulst* [1957, Fig. 32], from which the behavior of this quantity for seawater drops can be inferred. For a given wavelength and for $r \gtrsim 0.6\,\lambda$ (corresponding to $r \gtrsim 0.24$-0.4 μm for the range of wavelengths of visible light), the value of Q_s, for increasing radius, oscillates with decreasing amplitude between 1.5 and a value slightly greater than 4, asymptotically approaching the value 2. Consequently, the light-scattering cross section of an SSA particle, $\pi r^2 Q_s$, is a non-monotonically increasing function of radius, which can result in ambiguity in measurements that utilize light-scattering properties to infer the geometric size of an SSA particle (§4.1.4.1). As most of the contribution to the concentration of SSA surface area is from particles with radius greater than several tenths of a micrometer (§4.1.4.3), to first approximation the value of Q_s can be taken as constant, equal to 2, for purposes of estimating the role of SSA in light scattering, corresponding to a weighting factor for $\sigma_{sp}(\lambda)$ given by $W(r) \approx 2\pi r^2$. Within this approximation, the relations $d\sigma_{sp}(r_{80})/d\log r_{80} = a(r_{80})/2$ and $\sigma_{sp} = A/2$ hold (a and A pertain to the surface area, which for a spherical particle is 4 times greater than the geometric cross-sectional area). A further consequence of the assumption $Q_s = 2$ is that the Ångström exponent α given by (2.1-22) is equal to zero (§2.1.5.3).

The size-dependent SSA light-scattering coefficient $d\sigma_{\mathrm{sp},80}/d\log r_{80} \equiv \pi r_{80}^2 Q_{\mathrm{s}}(r/\lambda,m) \times n(r_{80})$, which describes the contribution of SSA particles of a given r_{80} to the light-scattering coefficient at 80% RH, is shown in Fig. 5 for the canonical SSA size distribution of concentration, for wavelength $\lambda = 0.532$ µm, as an average over the solar spectrum (wavelengths 0.3 µm to 1.0 µm), and using the approximation $Q_{\mathrm{s}} = 2$ (which also represents the size dependence of the concentration of SSA surface area at 80% RH, $a_{80}(r_{80})$, for this canonical distribution). The ratio $Q_{\mathrm{s}}/2$ for wavelength $\lambda = 0.532$ µm and for the solar spectrum is also shown (lower panel). The large oscillations in the scattering coefficient for monochromatic light are somewhat smoothed out for the solar spectrum. Although $d\sigma_{\mathrm{sp},80}/d\log r_{80}$ varies with wavelength for a given r_{80} (especially for $r_{80} \lesssim 1$ µm), over the visible spectrum the differences are minimal for $r_{80} \gtrsim 3$ µm. The light-scattering coefficient σ_{sp} is nearly independent of wavelength for the canonical SSA size distribution of concentration, varying by less than a few percent over the range of visible wavelengths; the corresponding value of the Ångström exponent α is approximately -0.09 to within a few percent over the range of visible wavelengths (0.4-0.7 µm), very close to zero as

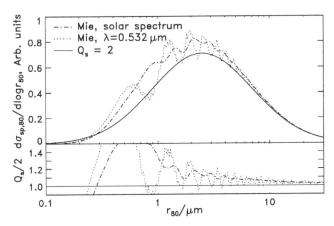

Figure 5. Upper panel: Equal-area plot of the size distribution of coefficient of light scattering by SSA particles at 80% RH, $d\sigma_{\mathrm{sp},80}/d\log r_{80} \equiv \pi r_{80}^2 Q_{\mathrm{s}}(r_{80}) n(r_{80})$, as a function of r_{80}, where $Q_{\mathrm{s}}(r)$ is the Mie scattering kernel for radius r and index of refraction $m = 1.375$. Scattering coefficients were evaluated by Mie calculations using a code based on that of *Bohren and Huffman* [1983, pp. 479-481] for the canonical SSA size distribution of concentration $n(r_{80})$ shown in Fig. 2, as a weighted average of 7 wavelengths over the visible portion of the solar spectrum [*Coakley et al.*, 1983] and for wavelength $\lambda = 0.532$ µm. Also shown is the distribution evaluated using the approximation $Q_{\mathrm{s}} = 2$, which represents the size dependence of the concentration of SSA surface area at 80% RH $a_{80}(r_{80})$ for this distribution. Lower panel: Ratio $Q_{\mathrm{s}}/2$ for weighted average of 7 wavelengths over the visible portion of the solar spectrum and for wavelength $\lambda = 0.532$ µm.

expected. For this SSA concentration size distribution, the major contribution to the light scattering over the solar spectrum is from SSA particles with r_{80} from ~0.5 µm to ~10 µm; over this range the approximation $Q_{\mathrm{s}} = 2$ is fairly accurate, becoming increasingly so as r_{80} increases. This approximation underestimates the SSA light-scattering coefficient σ_{sp} by about 15% for any visible wavelength and hence for the solar spectrum.

Another important quantity parameterizing the light-scattering ability of SSA is the mass scattering efficiency, the ratio of the light-scattering coefficient to the dry mass concentration σ_{sp}/M (also an intensive property of the SSA concentration). Under the approximation $Q_{\mathrm{s}} = 2$ (i.e., $\sigma_{\mathrm{sp}} = A/2$) this quantity is independent of wavelength, and in terms of the effective radius r_{eff}, which was defined above as $3V/A$, the mass scattering efficiency is given by

$$\frac{\sigma_{\mathrm{sp}}}{M} = \frac{\left\{ \dfrac{3}{2}\left(\dfrac{r_{80}}{r_{\mathrm{dry}}}\right)^3 \dfrac{[\zeta(rh)]^3}{\rho_{\mathrm{ss}}} \right\}}{r_{\mathrm{eff}}} \approx \frac{\left\{ \dfrac{12[\zeta(rh)]^3}{\rho_{\mathrm{ss}}} \right\}}{r_{\mathrm{eff}}}, \quad (2.1\text{-}23)$$

where $\zeta(rh) = r/r_{80}$ (§2.1.1; §2.5.3) and $r_{80}/r_{\mathrm{dry}} \approx 2$ (§2.5.3). At 80% RH the SSA mass scattering efficiency $\sigma_{\mathrm{sp},80}/M$ is approximately equal to 5.5 m^2 g^{-1}/($r_{\mathrm{eff},80}$/µm). For the canonical SSA size distribution of concentration at 80% RH, for which $r_{\mathrm{eff},80} \approx 4.2$ µm, the mass scattering efficiency $\sigma_{\mathrm{sp},80}/M \approx 1.3$ m^2 g^{-1} (§4.1.4.4; §4.1.4.5).

The total SSA optical thickness $\tau_{\mathrm{sp}}(\lambda)$ is the vertical integral of the light-scattering coefficient:

$$\tau_{\mathrm{sp}}(\lambda) = \int \sigma_{\mathrm{sp}}(\lambda, z)\, dz. \quad (2.1\text{-}24)$$

It is a measure of the fraction of the incident solar radiation of wavelength λ that is scattered by SSA and is important to considerations of remote sensing and of the energy balance in the marine atmosphere. The wavelength dependence of $\tau_{\mathrm{sp}}(\lambda)$ can also be characterized by an Ångström exponent, similar to (2.1-22), but within the approximation $Q_{\mathrm{s}} = 2$, τ_{sp} is also independent of wavelength and this Ångström exponent is equal to zero. Retrievals of the Ångström exponent over the ocean (which include contributions from aerosols other than SSA) from ground-based or satellite-based remote sensing frequently result in low values (i.e., less than ~1) indicative of SSA (or dust) particles, contrasting with higher values characteristic of smaller particles such as those arising from gas-to-particle conversion [e.g., *Tomasi and Prodi*, 1982; *Volgin et al.*, 1988; *Yershov et al.*, 1990; *Hoppel et al.*, 1990; *Villevalde et al.*, 1994; *Quinn et al.*, 1998; *Carrico et al.*, 1998; *Deuzé et al.*, 1999; *Kaufmann et al.*, 2001; *Wilson and Forgan*, 2002; *Smirnov et al.*, 2002; *Smirnov et al.*, 2003b].

As particles with r_{80} in the range that provides the dominant contribution to σ_{sp} (~0.5 µm to ~10 µm) are expected to be well mixed over the marine boundary layer of height H_{mbl} (and are confined mainly to this layer; §2.9.5), for situations in which RH is nearly independent of height, σ_{sp} also will be nearly independent of height. Under these circumstances τ_{sp} can be roughly approximated by $\tau_{sp} \approx \sigma_{sp}H_{mbl}$ (this is examined further in §4.1.4.5), allowing τ_{sp} to be estimated from measurements of σ_{sp} (or equivalently from measurements of $n(r_{80})$) taken at a height near 10 m. In contrast, larger SSA particles are confined mainly to the lowest portion of the marine boundary layer (§2.9). Their contribution to light scattering is often evident in horizontal viewing as a layer of intense scattering within the lowest few meters above the sea surface as a consequence of their high concentration there, but not so much in vertical viewing (i.e., with regard to optical depth). The fact that the scattered light appears whitish is consistent with wavelength-independent scattering ($Q_s = 2$).

2.1.6. Description of Fluxes

Fluxes of SSA particles and their properties are of fundamental importance to air-sea interaction. Such fluxes can be categorized in several different ways: production vs. deposition, net vs. gross, interfacial vs. effective, or horizontal vs. vertical. Recognition of the distinctions between these several fluxes is fundamental to understanding and describing the role played by SSA particles in a number of applications. These several fluxes are introduced and defined in this section. Although many different representations have been used to describe SSA fluxes, in this review the most fundamental representation is considered to be the size-dependent number flux of SSA particles through a given surface, $j(r_{80})$, from which fluxes of other quantities of interest can readily be determined. This SSA number flux is defined analogously to the SSA number concentration $n(r_{80})$ (§2.1.4) as the number of SSA particles in the size interval ($\log r_{80}$, $\log r_{80} + d\log r_{80}$) that pass through a unit area of a given reference plane per unit time. Likewise, $dJ(r_{80}) = j(r_{80})d\log r_{80}$ defines the quantity $J(r_{80})$, the number flux of SSA particles with values of r_{80} less than the argument (from which it follows that $j(r_{80}) = dJ(r_{80})/d\log r_{80}$), and the total number flux of SSA particles is given by $J = \int j(r_{80})d\log r_{80}$.

The size-dependent interfacial SSA production flux $f_{int}(r_{80})$ is defined to be the rate of upward transport of SSA particles per unit area from the ocean to the atmosphere, averaged over temporal and spatial scales much greater than those characterizing individual breaking waves and whitecaps; that is, for averaging periods of several minutes to several tens of minutes and length scales several tens to several hundreds of meters. Knowledge of $f_{int}(r_{80})$ permits

calculation of other properties of the interfacial SSA production flux, such as the size-dependent and total interfacial production fluxes of SSA radius, surface area, volume, and mass. Therefore, a fundamental research objective is determination of this size-dependent flux $f_{int}(r_{80})$, and it is this quantity that is the subject of this book. A further research objective is determination of the dependence of f_{int} (the dependence on r_{80} is hereinafter suppressed) on controlling factors. Knowledge of this dependence permits calculation of f_{int} for a given set of meteorological and environmental conditions. Such an approach implies that these controlling factors are known and can be obtained by measurements or models; lack of knowledge of the dependence of f_{int} on its controlling factors results in a spread of values of f_{int} for a given set of values of the factors thought to be controlling.

The quantity f_{int} denotes the gross, or one-way, flux of particles from the ocean to the atmosphere (throughout this review it is assumed that all production occurs at the surface of the ocean, described by a horizontal plane $z = 0$, with z being positive upward). This production flux differs from $j_z(0)$ (the argument refers to $z = 0$), the net vertical flux of SSA particles at the sea surface, by the flux of SSA particles deposited to the surface, $f_d(0)$; these quantities also are averages over suitable length and time scales. Thus, $j_z(0) = f_{int} - f_d(0)$, or equivalently,

$$f_{int} = j_z(0) + f_d(0) \qquad (2.1\text{-}25)$$

(here and throughout this review f_{int} and j_z are taken as positive upward, and f_d is taken as positive downward, as is conventional). The quantity f_d is referred to as the SSA dry deposition flux (§2.7.2), and describes removal by gravitational sedimentation, turbulent (eddy diffusion), Brownian diffusion, impaction on the ocean surface, and the like, but not by wet deposition (§2.7.1), which refers to removal through precipitation. Because the processes responsible for the formation of SSA particles and the processes responsible for the removal of SSA particles to the sea surface through dry deposition are fundamentally different and are controlled by different mechanisms, it is essential to distinguish between the gross flux f_{int} and the net flux $j_z(0)$ and to treat them separately.

Some arbitrariness is necessarily attached to the definition of the interfacial SSA production flux, especially concerning drops that are ejected to extremely low heights (less than a millimeter or so) above the sea surface and are not further entrained upward (§4.3.2.3). This concern must be recognized in attempts to determine the interfacial production flux by methods that employ measurements at a nonzero height above the surface, such as the whitecap method (§3.2), the along-wind flux method (§3.6), and the vertical impaction method (§3.8), and by methods that employ laboratory results based on the number of drops produced that can be detected above a still water surface, such as the bubble method (§3.4). Additionally, this concern must be addressed with regard to the application for which the interfacial production flux is desired. For instance, all drops that enter the atmosphere contribute to the exchange of water vapor, momentum, and sensible and latent heat (§2.1.7), but in amounts that depend on their atmospheric residence times and thus to a large extent on the heights they attain (in addition to their size and meteorological conditions), and estimates of the interfacial SSA production flux based on measurements at any given height must make allowances for SSA particles that do not attain this height.

For many applications, it is not the flux of SSA particles produced at the sea surface, f_{int}, that is of interest, but rather only a subset of that flux, specifically the flux of SSA particles that attain a given height above the surface. Of particular interest is the flux of particles that attain a height of 10 m (discussed further below), a height at which measurements are commonly made, as it is these particles that will remain in the atmosphere long enough to contribute to the sea salt burden, to be transported for appreciable horizontal distances, to interact with gases and other particles, and therefore to be important for representing the SSA production flux in atmospheric chemical transport models. Thus, it is necessary to define and discuss the effective SSA fluxes through a horizontal reference plane at an arbitrary given height, z_1, and to distinguish these effective fluxes from interfacial fluxes.

These effective fluxes (the size dependence of which again is suppressed but should be understood) are the effective net SSA flux upward through this height $j_z(z_1)$, the effective SSA production flux $f_{eff}(z_1)$, and the effective SSA dry deposition flux $f_d(z_1)$. Care is necessary in the definition of these quantities, as the distinctions among them are not so immediately evident as those for the corresponding interfacial SSA fluxes. The effective SSA dry deposition flux $f_d(z_1)$ is defined to be the net downward flux through a horizontal reference plane at height z_1 in the absence of a surface source; that is, the net downward flux that would occur for a given vertical distribution of SSA concentration with production hypothetically turned off. It is important that $f_d(z_1)$ be defined as the *net* downward flux at this height (in the absence of surface sources) rather than the *gross* downward flux, because a particle repeatedly passing through a horizontal plane at this height will contribute multiple times to the gross upward and downward fluxes through this plane, but not to the net fluxes. For this same reason, the effective SSA production flux $f_{eff}(z_1)$ is not the gross flux upward through this height; by analogy with the interfacial SSA production flux (Eq. 2.1-25), it is defined as the sum of the effective net SSA flux and the effective SSA dry deposition flux:

$$f_{eff}(z_1) = j_z(z_1) + f_d(z_1). \qquad (2.1-26)$$

That is, the effective SSA production flux at height z_1, $f_{eff}(z_1)$, is equal to the net vertical flux $j_z(z_1)$ at this height augmented by the dry deposition flux $f_d(z_1)$, which decreases the net flux from what it would otherwise be in the absence of dry deposition. Thus the effective SSA production flux through the height z_1, $f_{eff}(z_1)$, is the number of SSA particles per unit interval of $\log r_{80}$ produced at the sea surface that attain height z_1 above mean sea level, per unit area of the sea surface per unit time, likewise averaged over time and length scales much greater than those characterizing production.

It is important to distinguish conceptually between the effective net SSA flux $j_z(z_1)$ and the effective SSA production flux $f_{eff}(z_1)$. For many purposes it might be argued that it is the effective net flux rather than effective SSA production flux that is desired; $j_z(z_1)$ is the measurable quantity, and it describes the rate, per unit area of the sea surface, at which the number of SSA particles of a given size above the height of measurement is changing. However, as with interfacial SSA fluxes, because the processes governing the production of SSA particles and their subsequent transport to height z_1 differ from those governing their deposition, and because aerosol transport and chemistry models often include terms describing deposition, the distinction between the effective SSA production flux and the effective net SSA flux is nevertheless important.

The distinction between the interfacial SSA production flux and the effective SSA production flux is likewise important, and although noted by some investigators [e.g., *Blanchard*, 1963, p. 127; *Toba*, 1965a,b, 1966; *Chaen*, 1973; *Fairall and Larsen*, 1984; *de Leeuw*, 1990a; *Iida et al.*, 1992; *Reid et al.*, 2001; *Hoppel et al.*, 2002], it does not seem to be fully appreciated (possibly because it does not exist for most other applications; e.g., for gases in most situations). It is the effective SSA production flux and not the interfacial SSA production flux that constitutes a source of SSA particles to the atmosphere in any practical sense, and which is necessary for models that represent the effects of SSA on climate change and atmospheric chemistry, for example. The effective SSA production flux at a given height is less than the interfacial SSA production flux because of those particles produced at the sea surface that do not attain this height and that thus contribute to the interfacial SSA production flux but not to the effective SSA production flux. Because in general there are no source or sink terms for SSA particles below this height other than the sea surface at $z = 0$, where production occurs, these same particles will likewise contribute to the interfacial dry deposition flux but not to the effective dry deposition flux, and the difference between these fluxes will be the same as the difference between the interfacial SSA production flux and the effective SSA production flux. In effect, the particles that do not attain the given height provide a size-dependent sink of

SSA production flux, and a corresponding source of SSA dry deposition flux, at intermediate heights where particles reach their maximum elevations and begin their descents back to the sea. This situation might be described, in principle, by a volumetric loss term of upward (production) flux and a corresponding volumetric gain term of downward (dry deposition) flux (cf., §4.6.5) whose magnitudes depend on the height above the surface and on particle size. Other approaches have also been suggested (§4.6.5; §5.1.2).

The difference between the interfacial SSA production flux and the effective SSA production flux depends strongly on particle size. It is small for particles with r_{80} up to several micrometers, for which gravity is not an important factor in removal (§2.6.1) and which to a good approximation are well mixed to heights greater than the height of measurement (§2.9.5). However, the difference becomes much more pronounced for larger particles, for which gravity becomes increasingly important, because of the decrease with increasing particle size in the fraction of particles produced at the ocean surface that attain a given height. For the same reason, the effective SSA production flux, for a given particle size, is also a decreasing function of height. Because of the distinction between the interfacial SSA production flux and the effective SSA production flux, attempts to determine the effective SSA production flux using laboratory studies are severely limited for a wide range of particle sizes, primarily because of the inability of such studies to accurately simulate the transport of particles produced at the surface to heights typical of these at which SSA particles are measured (~10 m) under realistic conditions. Hence, determinations of the effective SSA production flux for these particle sizes must be based on field measurements of SSA concentration (§3).

Specification of the effective SSA production flux (and the effective SSA dry deposition flux) requires specification of z_1, the height to which these fluxes pertain. As SSA concentration generally decreases with increasing height at a rate that greatly increases with increasing particle size (§2.9), any specification of the height z_1 implies an upper size limit on SSA particles that are included in these fluxes in any practical sense. Lack of specification of the height particles must attain to be considered as contributing to the effective SSA production flux, and the resultant lack of specification of the size of particles that are included in this flux, contributes greatly to the large variation in the estimates of the global annual SSA mass production as summarized in Table 1; several investigators [e.g., *Petrenchuk*, 1980; *Blanchard*, 1983; *Blanchard et al.*, 1984; *Blanchard*, 1985a; *Andreae*, 1986b; *Jaenicke and Matthias-Maser*, 1992; *Jacob*, 1995] have examined difficulties in interpreting such data in light of the decrease in concentration with increasing height as a function of particle size. For consideration of the effective SSA production flux, the choice $z_1 \approx 10$ m is made

here and is used throughout this review (this value should be understood for any reference to f_{eff} without specification of a reference height). This is a typical height at which measurements are made, and it can be argued that SSA particles that do not attain this height do not contribute appreciably to phenomena of interest for the effective SSA production flux, whereas SSA particles that do attain this height will remain in the atmosphere for some time and hence can contribute appreciably to these phenomena. Implicit in this assumption is that the exact value of z_1 is not of great importance so long as it is near 10 m; that is, the dependencies of the various SSA effective fluxes on this height are slight. Equivalently, concentrations of SSA particles for the sizes for which the effective SSA production flux is desired must not exhibit strong vertical gradients near this height. For a wide range of particle sizes this assumption is valid (§2.9.4). However, as particle size increases SSA concentration decreases more rapidly with height, and this assumption becomes increasingly less likely to be satisfied. There is a rather strong upper cutoff at $r_{80} \approx 25$ μm in the size of SSA particles that have appreciable probability to attain a height of 10 m and thus to contribute to the effective SSA production flux (§2.9); this size marks the upper limit of the size range of medium SSA particles and the lower limit of the size range of large SSA particles (§2.1.2). As SSA particles with $r_{80} \gtrsim 25$ μm are generally not present in appreciable concentrations at heights of 10 m or greater, and thus do not play an important role in many atmospheric processes such as cloud formation, chemistry, and radiation transfer, it is reasonable to exclude these particles with regard to the effective SSA production flux (Table 2).

The quantity

$$\Psi_f (r_{80}) \equiv \frac{f_{eff} (r_{80})}{f_{int} (r_{80})} \qquad (2.1\text{-}27)$$

is introduced here to denote the ratio of the size-dependent effective and interfacial SSA production fluxes, where the effective SSA production flux is defined to be that for a height near 10 m, as discussed above. The relation $0 \leq \Psi_f(r_{80}) \leq 1$ follows readily from this definition. The quantity $\Psi_f(r_{80})$ is also equal to the probability that an SSA particle of a given r_{80} produced at the sea surface attains a height of 10 m, as those particles that attain this height contribute to the effective SSA production flux. For small SSA particles ($r_{80} \lesssim 1$ μm), Ψ_f is near unity, meaning that nearly all of these particles that are formed at the sea surface attain a height of 10 m or more. With increasing r_{80} over the size range comprising medium SSA particles (1 μm $\lesssim r_{80} \lesssim 25$ μm), Ψ_f decreases as an increasingly large fraction of the particles produced at the surface do not attain this height. For large SSA particles ($r_{80} \gtrsim 25$ μm), Ψ_f is much less than unity; because of their large mass and hence large gravitational fall velocity, only a small fraction of large SSA particles produced at the sea surface attain heights very far above the sea surface (§2.9.2; Table 2). Other estimates of the size dependence of $\Psi_f(r_{80})$ are presented in §2.9.4 and §4.6.5.

Exchange of gases (including water vapor), momentum, and heat between the ocean and the atmosphere under nearly all circumstances are characterized by effective fluxes. As there are no appreciable sinks for any of these substances in the lowest 10 m or so above the sea surface (as was provided by gravity for SSA particles), and as the marine atmosphere is generally well mixed (except perhaps in situations of extreme atmospheric stability), the vast majority of any of these substances transferred from the ocean to the atmosphere will be further transferred upward above 10 m.

Although the fluxes introduced so far have all been vertical fluxes, it is also important to consider horizontal fluxes. The most pertinent horizontal flux for SSA is the along-wind number flux of SSA particles at a given height, $j_x(r_{80}, z)$, defined to be the size-dependent flux of SSA particles through a reference plane normal to the mean flow (which is assumed to in the x-direction) at height z (§3.6). This along-wind flux $j_x(r_{80}, z)$ is equal to the product of the number concentration of SSA particles of a given size at that height, $n(r_{80}, z)$, and the average velocity component in the direction of the mean flow of all SSA particles of that size at the height z, $\bar{u}_m(r_{80}, z)$

$$j_x(r_{80}, z) = n(r_{80}, z)\, \bar{u}_m(r_{80}, z). \qquad (2.1\text{-}28)$$

As discussed in §4.3.6.2, certain measurements that are presumed to measure $n(r_{80}, z)$ actually measure $j_x(r_{80}, z)$ and use (2.1-28) to infer the number concentration on the assumption that particles have no horizontal component of relative velocity with respect to the mean flow. This is often a valid assumption for small and medium SSA particles, which are rapidly entrained into the mean flow (§2.6.2), and the along-wind flux of these particles at a given height is to good approximation equal to the product of their number concentration at that height and the mean wind speed of the surrounding air at that height $U(z)$:

$$j_x(r_{80}, z) = n(r_{80}, z)\, U(z). \qquad (2.1\text{-}29)$$

From (2.1-29) it follows that for these particles $j_x(r_{80}, z)$ and $n(r_{80}, z)$ exhibit the same dependence on r_{80}. However, this assumption is not valid for sufficiently large SSA particles, which often do not attain a mean horizontal velocity component equal to that of the surrounding flow during their time in the atmosphere (§2.6.4). The along-wind flux and the number concentration for these particles will therefore not have the same size dependence (§3.6).

2.1.7. Fluxes of SSA Properties

Fluxes of SSA properties of geophysical interest can be evaluated from knowledge of the size-dependent interfacial and effective SSA production fluxes. Key among these are the flux of sea salt mass and the fluxes of moisture (water substance or latent heat), sensible heat, and momentum associated with SSA production. As with SSA concentration (§2.1.5), certain properties of the SSA production fluxes can also be categorized into extensive and intensive properties. Similarly, extensive properties of the interfacial SSA production flux can be categorized as differential properties, which can be expressed as $W(r)f_{int}(r_{80})$, and integral properties, which can be expressed as $\int W(r)f_{int}(r_{80})\, d\log r_{80}$, where $W(r)$ is the pertinent weighting factor (as noted in §2.1.1, the variable r_{80} uniquely identifies the drop size whether or not the drop has equilibrated to ambient RH, which may or may not be 80%). Differential and integral properties of the effective SSA production flux can likewise be expressed as $W(r)f_{eff}(r_{80})$ and $\int W(r)f_{eff}(r_{80})\, d\log r_{80}$, respectively. Upon substitution of $\Psi_f(r_{80})$, the ratio of the effective SSA production flux to the interfacial SSA production flux (Eq. 2.1-27), these differential and integral properties of the effective SSA production flux can also be expressed in terms of the interfacial SSA production flux as $W(r)\Psi_f(r_{80})f_{int}(r_{80})$ and $\int W(r)\Psi_f(r_{80})f_{int}(r_{80})\, d\log r_{80}$, respectively. For example, the weighting factor for dry sea salt mass from (2.1-4) is $W(r) = (4\pi/3)\rho_{ss}(r_{dry}/r_{80})^3 r_{80}^3$, where the ratio r_{dry}/r_{80} is very nearly 0.5 (§2.5.3). Thus, the total interfacial flux of dry sea salt mass is

$$F_{int}^M = \frac{4\pi}{3}\rho_{ss}\left(\frac{r_{dry}}{r_{80}}\right)^3 \int r_{80}^3 f_{int}(r_{80})\, d\log r_{80}. \qquad (2.1\text{-}30)$$

The flux of SSA mass that remains in the atmosphere for an appreciable time—the effective SSA mass flux, to which most estimates in Table 1 refer (§1.1, §2.1.6)—can be evaluated as an integral over the effective SSA production flux:

$$F_{eff}^M = \frac{4\pi}{3}\rho_{ss}\left(\frac{r_{dry}}{r_{80}}\right)^3 \int r_{80}^3 f_{eff}(r_{80})\, d\log r_{80}. \qquad (2.1\text{-}31)$$

This flux could in principle also be written as an integral over the interfacial SSA production flux:

$$F_{eff}^M = \frac{4\pi}{3}\rho_{ss}\left(\frac{r_{dry}}{r_{80}}\right)^3$$
$$\times \int r_{80}^3 \Psi_f(r_{80})f_{int}(r_{80})\, d\log r_{80}, \qquad (2.1\text{-}32)$$

although little would be gained by such an approach. For comparison with estimates in Table 1, a total effective SSA mass flux $F_{\text{eff}}^{\text{M}} = 1\ \mu\text{g m}^{-2}\text{s}^{-1}$ yields an annual global effective SSA production of $10 \cdot 10^{12}$ kg. Similar integral expressions can be written for the total fluxes of moisture (latent heat), momentum, and sensible heat associated with SSA production, and by evaluating these expressions and comparing them with the contributions of these fluxes from direct mechanisms—evaporation, wind stress (§2.3.2), and conduction and convection, respectively—the geophysical importance of SSA production to these fluxes can be ascertained.

The situation concerning the fluxes of water is more complicated because water can be exchanged between the liquid phase and the gas phase, and it is therefore necessary to distinguish between the effective and interfacial fluxes of liquid water and of water vapor. The interfacial flux of water associated with SSA production is entirely in the liquid phase and is directly proportional to $F_{\text{int}}^{\text{M}}$, being greater by a factor of nearly 30, the ratio of the mass of water to the mass of sea salt for most of the ocean surface waters (§2.5.3). Of this flux, some of the water evaporates (if the ambient RH is less than the equilibrium RH of seawater, 98%, §2.5.3, as is typical, §2.4), some remains in drops that contribute to the effective SSA production flux (i.e., those with $r_{80} \lesssim 25\ \mu\text{m}$; §2.1.6), and some is returned to the sea surface in drops that do not contribute to the effective SSA production flux (i.e., mainly those with $r_{80} \gtrsim 25\ \mu\text{m}$; §2.1.6). As under nearly all circumstances the water vapor resulting from evaporation from drops remains in the atmosphere and is mixed to heights greater than 10 m, the sum of the first two contributions (that is., the evaporated water and the water associated with drops that contribute to the effective SSA production flux) constitutes the total effective production flux of water vapor associated with SSA production (likewise, the fluxes of sensible heat and momentum associated with SSA production are also effective fluxes; §2.1.6). The liquid water associated with drops that contribute to the effective SSA production flux is generally small compared to the amount that has evaporated from these drops; for example, a drop that has equilibrated to a typical ambient RH of 80% retains only ~10% of its initial water (§2.5.3). Therefore, to good approximation, the effective production flux of water associated with SSA production is entirely in the vapor phase, consisting of water evaporated from drops. If for a given set of meteorological conditions (i.e., RH, temperature, and wind speed), $m_{\text{w,e}}(r_{80})$ denotes the mass of water evaporated from a drop of given r_{80} during its residence in the atmosphere, then the total flux of water mass evaporated from SSA particles is given by

$$F_{\text{eff}}^{\text{W}} = \int m_{\text{w,e}}(r_{80}) f_{\text{int}}(r_{80})\, d\log r_{80}. \qquad (2.1\text{-}33)$$

This flux consists of two contributions: the water evaporated from those drops that contribute to the effective SSA production flux (i.e., small and medium SSA particles), and the water evaporated from those drops that do not contribute to the effective SSA production flux (large SSA particles). For small and medium SSA particles, which yield the former contribution, $m_{\text{w,e}}(r_{80})$ is nearly equal to the entire initial water content, and this contribution is directly proportional to $F_{\text{eff}}^{\text{M}}$. For large drops, which generally do not attain their equilibrium radii with respect to the ambient RH before they fall back into the sea (§2.9.1) and thus transfer only some of their water to the atmosphere, $m_{\text{w,e}}(r_{80})$ depends on the residence time of the drop in the atmosphere and on the rate at which it evaporates, both of which are size-dependent. The expression for the rate of evaporation of large SSA particles is more complicated than that of small or medium SSA particles, and depends also on the relative velocity of the drop in air; these so-called ventilation effects becomes increasingly important with increasing particle size (§2.5.4). For these reasons the quantity $m_{\text{w,e}}(r_{80})$ would seem difficult to measure or model under realistic conditions.

Associated with the flux of water mass evaporated from SSA particles is a flux of heat to the seawater drops, equal in magnitude to the product of $F_{\text{eff}}^{\text{W}}$ and the latent heat of vaporization of water from seawater $L_{\text{w,sw}}$. The latent heat of vaporization of water from seawater changes as the concentration of sea salt in the drop changes, but it is expected to be comparable to the latent heat of vaporization of bulk water. As this heat is supplied by the atmosphere, this flux must be viewed as an effective flux, as noted above. Additionally, there is an exchange of sensible heat between the atmosphere and the SSA particles, the direction and magnitude of which depend on the temperature difference between the ocean surface water (from which SSA particles are formed) and the overlying atmosphere, and on the extent to which drops equilibrate to the temperature of the surrounding air during their time in the atmosphere. In contrast to the time required for large SSA drops to evaporate and thus equilibrate to the ambient RH (typically minutes; §2.5.4), they equilibrate to the ambient temperature extremely rapidly (typically much less than a few seconds for drops up to more than a millimeter in radius; §2.5.4); thus nearly all SSA drops equilibrate with respect to the temperature of the surrounding air and exchange nearly all of their available sensible heat before falling back into the sea. The sensible heat flux, also an effective flux, is therefore nearly proportional to the interfacial SSA mass flux $F_{\text{int}}^{\text{M}}$ and is given by

$$F_{\text{eff}}^{\text{SH}} = \frac{4\pi}{3} \rho_{\text{sw}} \left(\frac{r_{\text{form}}}{r_{80}} \right)^{3} C_{\text{p,sw}} \Delta T$$
$$\times \int r_{80}^{3} f_{\text{int}}(r_{80})\, d\log r_{80}, \qquad (2.1\text{-}34)$$

where $C_{p,sw}$ is the specific heat of seawater at constant pressure (approximately 5% less than that of pure water at the same temperature [*Millero et al.*, 1973]), ΔT is the temperature difference between the air and the ocean surface waters, and the ratio $r_{form}/r_{80} \approx 2$ (§2.5.3). Alternatively, F_{eff}^{SH} can be expressed as

$$F_{eff}^{SH} = \frac{\rho_{sw}}{\rho_{ss}} \left(\frac{r_{form}}{r_{dry}} \right)^3 (C_{p,sw} \Delta T) F_{int}^M, \qquad (2.1\text{-}35)$$

where ρ_{sw} is the density of seawater; the ratio ρ_{sw}/ρ_{ss} is approximately 0.45 (§2.5.1), and the ratio r_{form}/r_{dry} is very nearly 4 (§2.5.3).

The momentum flux, describing the transfer of momentum from the atmosphere to SSA particles (which contributes to the total wind stress), can likewise be obtained as an integral over the interfacial SSA production flux. As with the fluxes of water vapor and heat, this flux is an effective flux. Under the assumptions that drops are initially produced at the sea surface with no horizontal velocity component (this might not be true for drops torn from wave crests) and that they do not appreciably change their mass during their residence in the atmosphere (this assumption will hold for large SSA particles which are expected to contribute most of this flux), the momentum transferred to an SSA particle with given r_{80} from the surrounding atmosphere is given by $\Delta P(r_{80}) \approx (4\pi/3)\rho_{sw}r_{80}^3(r_{form}/r_{80})^3 U_f(r_{80})$, where $U_f(r_{80})$ is the final horizontal velocity acquired by the drop. Large drops may not attain horizontal velocity components equal to that of the mean flow before falling back into the sea (§2.6.4), thus $U_f(r_{80})$ cannot necessarily be taken equal to the mean wind speed U. However, an upper limit of the momentum flux can be obtained by assuming that all drops attain a horizontal velocity equal to U, leading to the following inequality:

$$F_{eff}^P \leq \frac{4\pi}{3}\rho_{sw} \left(\frac{r_{form}}{r_{80}} \right)^3 U \int r_{80}^3\, f_{int}(r_{80})\, d\log r_{80}, \quad (2.1\text{-}36)$$

which can also be written as

$$F_{eff}^P \leq \frac{\rho_{sw}}{\rho_{ss}} \left(\frac{r_{form}}{r_{dry}} \right)^3 U \times F_{int}^M. \qquad (2.1\text{-}37)$$

Evaluation of the effective fluxes of water (or latent heat), sensible heat, and momentum associated with SSA production requires knowledge not only of quantities such as $m_{w,e}(r_{80})$, the mass of water evaporated from a drop of given r_{80} during its residence in the atmosphere, which would seem difficult to obtain for large SSA particles, but also of the size-dependent interfacial SSA production flux, $f_{int}(r_{80})$. However, as this latter quantity is poorly known, both with regard to the size dependence and to the magnitude (§5.11.2), little confidence can be placed in any of the estimates that have been proposed for it. Therefore, at the present time it is not possible to provide even rough estimates of the effective fluxes of water, sensible heat, or momentum associated with SSA production.

2.2. MAJOR SSA PRODUCTION MECHANISMS

The presence of sea salt particles in the atmosphere has long been known [*Beck*, 1819; *Sigerson*, 1870; *Pošepný*, 1877 (cited in *Rankama and Sahama*, 1950, p. 317); *Aitken*, 1881; *Kinch*, 1887; *Gautier*, 1899], and early investigators demonstrated that SSA particles could be produced by spraying seawater or by bubbling air through seawater [*Coste and Wright*, 1935; *Aliverti and Lovera*, 1939 (cited in *Blanchard and Woodcock*, 1957); *Köhler*, 1941 (cited in *Moore and Mason*, 1954); *Aliverti and Lovera*, 1950; *Köhler and Båth*, 1953], but the exact processes by which SSA particles are introduced into the atmosphere from the ocean have only more recently been established. Several mechanisms have been considered (§4.7), but currently only two are believed to be major sources of SSA for the oceans as a whole: bursting of bubbles formed primarily by breaking waves, and mechanical disruption of wave crests by the wind. These mechanisms have been denoted by some investigators as indirect and direct, respectively. Additionally, it has been suggested that the size distribution of SSA particles in the atmosphere can change by processes subsequent to particle formation at the surface, such as the fracture of an SSA particle upon drying, but evidence for these processes in the atmosphere is scant, and they are thought to be unimportant (§4.7.4).

Although it had long been known that bursting bubbles could produce drops [*Foulk*, 1927 (cited in *Foulk*, 1932); *Craven and Stuhlman*, 1931; *Foulk*, 1932; *Stuhlman*, 1932, *Davis*, 1940], and that bubbles play a role in SSA production [*Jacobs*, 1936, 1937; *Owens*, 1940; *Simpson*, 1941b], it was *Woodcock* [1948] who demonstrated that drops ejected from bursting bubbles in the oceans were a major source of SSA particles. When the wind speed is sufficiently high, typically ~5 m s^{-1} or greater at 10 m above the ocean surface (as conventionally reported), the stress of the wind on the ocean causes the water near the surface to move faster than the underlying water, forming a wave that breaks under the influence of gravity. This breaking wave entrains bubbles into the ocean, and these bubbles subsequently rise, forming a whitecap, and burst at the surface. *Boyce* [1951, 1954], by holding filter paper sensitive to chlorine above the sea surface, demonstrated that it was the bubbles created *after* the breaking of a wave rather than the wave itself that produced most of the drops. Investigations utilizing high-speed photography [*Woodcock et al.*, 1953; *Knelman et al.*, 1954; *Kientzler et al.*, 1954; *Newitt et al.*, 1954; *Mason*, 1954] established that bursting bubbles can produce two types of drops, denoted by *Blanchard* [1963] as "film" drops, of which there seem to be at least two production mechanisms (§4.3.1), and "jet" drops (although both of these types of drops had been known much earlier; *Foulk* [1932]). The process by which a bubble bursts and produces drops is illustrated schematically in Fig. 6 (and discussed in more detail in §4.3.1). Film drops (§4.3.3) are produced from bursting of the film, or cap, of the bubble (Fig. 6*d*). These drops can number up to several hundred per bubble, depending on bubble size (§4.3.3.2). Some film drops, presumably formed by the shatter of the film itself, are ejected vertically, and others, perhaps formed by the collision of the collapsing film with the water surface, are ejected at a low angle to the horizon. The radii of film drops vary greatly, from less than 0.1 μm to several hundred micrometers (§4.3.3.1). Jet drops (§4.3.2) are ejected vertically from the breakup of a jet (i.e., a vertical column of liquid) formed when the cavity left by the bubble fills with liquid (Fig. 6*f*). The number of these jet drops generally decreases with increasing bubble size from around 6 for bubbles with radii up to 0.35 mm to as few as one or none for bubbles with radii greater than 1.5 mm (§4.3.2.1). Even though the number of jet drops produced per bubble may be less than the number of film drops, jet drops occur for bubble sizes that are more numerous in the oceans. The radius of a jet drop at formation is typically from 5% to 15% of the radius of the parent bubble (§4.3.2.2). These two types of drops are thought to comprise the majority of SSA particles of all sizes at low wind speeds and to comprise the majority of small and medium SSA particles at higher wind speeds (Table 2).

When the wind stress is sufficiently great, drops are torn directly from wave crests, in contrast to production of film and jet drops associated with bubble bursting, which occurs away from the crest, after the wave has broken [*Zubov*, 1938 (cited in *Bruyevich and Kulik*, 1967); *Owens*, 1940; *Köhler*, 1941 (cited in *Woodcock et al.*, 1953; *Boyce*, 1951, 1954; *Moore and Mason*, 1954]. These directly produced drops, denoted "spume" drops by *Monahan et al.* [1983b], which are projected nearly horizontally, are thought to be larger than most jet and film drops, with radii at formation ranging up to several hundred micrometers. Several explanations of the spume drop mechanism have been proposed [*Phillips*, 1966; *Kraus*, 1967; *Koga*, 1981; *Bortkovskii*; 1987, pp. 1-5], but the formation process of these drops has been little studied and does not appear to be well understood nor well quantified. The spume drop formation process in freshwater has been investigated in the laboratory using multiple-color photography by *Koga* [1981], who reported that during high winds the surface becomes irregular, and small, isolated, bell-shaped projections occur (even in the absence of breaking, bubble entrainment, or splashing). These projections, which are ~1 cm across at the base and roughly half as high, are typically stationary or moving slowly forward relative to the carrier waves; some of them grow "in the manner of stretched millet jelly" and become unstable, forming drops with r_{80} up to 750 μm. *Koga* reported that most of the direct production of drops occurred on the leading slope of the carrier wave near the crest, and that the drops had initial speeds comparable to or somewhat greater than the speed of the projection. He attributed the formation and growth of these projections to a Kelvin-Helmholtz instability in which aerodynamic suction created by air flow over the crest of the wave exceeds the restoring force of gravity and surface tension. *Bortkovskii* [1987, p. 5] argued that because the laboratory conditions

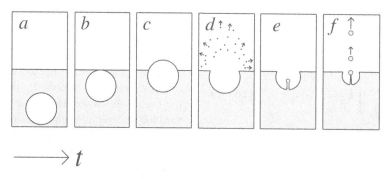

Figure 6. Drop formation from bubble bursting. *a-c*) Rise of bubble to the surface and formation of the bubble film or cap. *d*) Production of film drops from the bursting of the bubble film. *e*) Shrinkage of the cavity left by the bubble and formation of a jet rising from the center of this cavity. *f*) Further decrease in size of the cavity as it is filled by the surrounding liquid, and formation of several jet drops from breakup of the jet.

under which these experiments were performed did not reflect those typically occurring over the oceans, the role of this mechanism in drop production is relatively minor under natural conditions. *Anguelova et al.* [1999] argued that these drops were not technically spume drops but rather splash drops (according to *Andreas et al.* [1995]) or chop drops (according to *Monahan et al.* [1983, 1986]; *Monahan* [1986]; and *Woolf et al.* [1987]); however, as the distinction between spume drops and splash (or chop) drops has not been clearly made and is of little importance for the present concern, both of these types of drops will be referred to as spume drops.

Most information on spume drop production in the oceans, specifically the size range of drops produced and wind speeds for which this mechanism operates, has been inferred from measurements of SSA concentration near the ocean surface, it having been assumed that the large increase in the number concentrations of SSA particles, especially for large sizes, with increasing wind speed is due to spume drops. Some investigators, based on size-dependent SSA number concentration measurements at heights near 10 m (§4.1.4.3), have suggested the possibility of spume drop production of drops with $r_{80} \gtrsim 7$ μm when the mean 10 m wind speed U_{10} approaches 15 m s^{-1} (different investigators have suggested different size ranges and different wind speeds), but overall the data do not support this contention (§4.1.4.3).

Most small SSA particles ($r_{80} \lesssim 1$ μm) in the atmosphere are film drops (§4.3.7). Over most of the size range characterizing medium SSA particles (1 μm $\lesssim r_{80} \lesssim 25$ μm) jet drops predominate, with the relative contribution of jet drops increasing with increasing particle size. Some particles in this size range may be spume drops, but knowledge of the size distribution of spume drop production is insufficient to conclude this with any confidence. Most large SSA particles are jet drops or spume drops, and nearly all drops with r_{80} greater than a few hundred micrometers are spume drops (Table 2).

2.3. FACTORS AFFECTING SSA PRODUCTION, ENTRAINMENT, TRANSPORT, AND REMOVAL

From consideration of the processes leading to formation of SSA particles at the sea surface and their subsequent upward entrainment into the atmosphere, it can be concluded that production fluxes of SSA particles (both interfacial and effective) are affected by any meteorological or environmental factor that affects the surface properties of the ocean, wave breaking, or the turbulence level in the ocean surface waters; the formation, size distribution, entrainment, rise velocity, gas exchange characteristics, or bursting of bubbles; or the formation, entrainment, or behavior of drops. Additionally, any factor affecting the production, entrainment, transport, or removal of SSA particles also affects SSA concentrations and their properties.

Meteorological and environmental factors that have been proposed in this regard include the wind speed at a given height, atmospheric stability, present and prior rain or snow, amount and nature of surface-active materials in the near-surface ocean waters, sea and air temperatures, RH, sea state (i.e., the size distribution of waves and their temporal development), height of the marine boundary layer, fetch (the distance over water that the wind has blown prior to the measurement), saturation of the ocean surface waters with respect to the major atmospheric gases, salinity of the surface waters, and sea bottom depth and topography. Many of these factors may also affect air-sea interaction more generally and the exchange of gases and other substances between the ocean and the atmosphere. Although it is easy to understand how these factors might affect the production, entrainment, transport, and removal of SSA particles, systematic examination of the effects of these factors has proven difficult, and there are few quantitative results in which the effects of one factor are unambiguously separated from other confounding effects. Furthermore, some of these factors themselves are difficult to quantitatively characterize. In this section, some of these factors and their possible effects on interfacial and effective production fluxes, and concentrations, of SSA particles are examined.

2.3.1. Wind Speed

The key meteorological factor governing the production and life cycle of SSA particles is the wind. It is wind that causes waves to break, forming whitecaps and resulting in bubble-produced film and jet drops, and it is wind that tears spume drops from wave crests. Wind also affects the entrainment of SSA particles upward into the atmosphere and controls their transport. Thus wind is an important factor controlling both interfacial and effective SSA production, and it is also an important factor controlling SSA concentration and its properties. Additionally, wind speed is a commonly measured and modeled quantity, the latter making it useful in estimating SSA production when it cannot be directly measured (for instance, in calculating SSA production fluxes for use in large-scale chemical transport models). Finally, wind speed is the primary meteorological variable by which measurements pertinent to all aspects of SSA production and concentrations have been classified by virtually all investigators.

Typically the reported wind speed is the mean (i.e., time-average) wind speed at a height of 10 m above the sea surface, U_{10} (similarly, U_5 refers to the mean wind speed at a height of 5 m, for example), although several investigators have reported mean wind speed at a height of 19.5 m, and confusion has sometimes resulted when height is not specified, as noted by *Wu* [1995b]. However, under neutrally stable atmospheric conditions the mean wind speed is only weakly dependent on height (§2.4), and over a wide range of heights this dependence can usually be neglected in light of uncertainties (discussed below) in determining the mean wind speed. In reference to wind-wave interactions, such as whitecap formation, the pertinent wind speed is the relative wind speed of the air (at a given height) to the water. However, the motion of the water has rarely been taken into account despite the fact that for a water current of 2 m s^{-1} and a wind speed of 7 m s^{-1}, for example, quantities that vary as the cube of the wind speed differ by nearly a factor of six depending on whether the wind is along or against the direction of the water current.

The wind speed over the ocean exhibits great variability, both spatially and temporally on a variety of scales. Average wind speeds over the ocean vary from ~5 m s^{-1} to ~10 m s^{-1}, depending on latitude and season, with the vast majority of individual values in the range 3-12 m s^{-1}, and with values greater than 20 m s^{-1} occurring only a small fraction of the time at most locations[1]. Although these high wind speed events are likely very important for the injection of SSA particles into the atmosphere, they often reflect isolated occurrences not representative of steady state conditions; wind speeds greater than 20 m s^{-1} are typically transient occurrences of duration less than several hours [*Bortkovskii*, 1987, pp. 152-153]. As the sea state is typically not fully developed in such situations, the energy flux from the air to the oceans may differ from that under steady state conditions, as may wave breaking and hence SSA production. Thus the extent to which any individual high wind speed event is representative of other occurrences at that wind speed is questionable. Although the majority of wind speeds over the ocean fall in the range 3-12 m s^{-1}, most quantities pertinent to SSA and its production increase with wind speed greater than linearly, and the most important range of wind speeds for considerations of SSA is 5-15 m s^{-1}. In this review, $U_{10} = 10$ m s^{-1} is taken as a typical value for estimation of representative quantities.

The mean wind speed is generally assumed to be a well-known and easily measured quantity, but its determination from towers, buoys, ships, or aircraft (the latter two requiring corrections for relative motion) is not free from difficulties, and errors and uncertainties in reported values occur because of sensor inaccuracy and calibration difficulties; location of the sensor on the platform, flow distortion effects, and motion of the measurement platform (e.g., ship roll); ambiguities resulting from different conventions of reporting data (including different reference heights); physical separation between the location at which the wind speed is desired and the sensor (if on a buoy or a tower, for instance); roundoff (that is, wind speeds recorded to the nearest integer value of m s^{-1} or knot); and incorrect conversion from apparent wind to true wind and failure to report data necessary to perform this conversion accurately [*Telford et al.*, 1977; *Weller et al.*, 1983; *Gilhousen*, 1987; *Rahmstorf*, 1989; *Wilkerson and Earle*, 1990; *Pierson*, 1990; *Cardone et al.*, 1990; *Wu*, 1995c; *Freilich and Dunbar*, 1999; *Smith et al.*, 1999]. For example, during one study the mean wind speed over a two day period (corrected to 10 m height and neutrally stable atmospheric conditions) measured from a ship was 10.4 m s^{-1}, which was 0.8 m s^{-1} greater than that measured from a nearby drifting buoy and 2.6 m s^{-1} greater than that measured from a nearby fixed meteorological buoy [*Wanninkhof et al.*, 1993]. For the entire study period (approximately two weeks) the mean wind speed measured from the ship (again corrected to 10 m height and neutrally stable atmospheric conditions) was 8.5 m s^{-1}, which was 1.5 m s^{-1} higher than that measured by the fixed buoy.

Many measurements have been and continue to be reported on the Beaufort scale, a subjective scale (actually several different scales are in use) in which visual observations of the sea state are used to specify a wind force value, which corresponds to a narrow range of U_{10} values [*Alcock and Morgan*, 1978; *Quayle*, 1980; *Forrester*, 1986; *Peterson and Hasse*, 1987; *Cardone et al.*, 1990; *Isemer and Hasse*, 1991; *Kent and Taylor*, 1997]. For many purposes, including modeling, 10 m wind speeds over ocean regions for which direct measurements are not readily available are estimated using algorithms (typically *ad hoc* empirical fits) relating the wind speed to the short-wavelength roughness of the ocean surface characterized by its mean square slope [*Cox and Munk*, 1954; *Wu*, 1972; *Swift*, 1980], which is typically detected from backscatter radar cross section from the sea surface by remote sensing, including measurements from satellites. These algorithms require calibration and validation, usually by comparison with measurements of wind speeds from buoys, and tend to be less accurate for low ($U_{10} \lesssim 3$ m s^{-1}) and high ($U_{10} \gtrsim 15$ m s^{-1}) wind speeds; they suffer as well from any inaccuracies in the in-situ wind speeds used for calibration. However, it has been noted that the backscattered radar signal depends also on quantities other than wind speed such as sea state, atmospheric stability, and rainfall; additionally, remote sensing suffers from limited sampling coverage and from spatial and temporal separation between sampled regions and locations where

---

[1] *Kraus and Morrison* [1966]; *Quayle* [1980]; *Gathman* [1982]; *Chamberlain* [1983]; *Sandwell and Agreen* [1984]; *Chelton and McCabe* [1985]; *Erickson et al.* [1986]; *Bortkovskii* [1987, p. 153]; *Freilich and Challenor* [1994]; *Kent and Taylor* [1997]; *Wentz and Smith* [1999]; *Ebuchi* [1999]; *Meissner et al.* [2001]; *Ebuchi et al.* [2002].

wind speeds are desired[1]. On the basis of these considerations and the intercalibrations that have been performed, an estimate of the relative uncertainty contained in any wind speed determination or reported wind speed value is 10-20% for wind speeds 5-15 m s^{-1} and somewhat greater at wind speeds outside this range; this corresponds to roughly 1-2 m s^{-1} at typical wind speeds. Quantities dependent on U_{10} will likewise contain some associated uncertainty (for a quantity that varies as $U_{10}{}^3$, for example, this may result in a relative uncertainty exceeding 50%); this should be kept in mind with regard to the precision reported for measurements and results that are classified by wind speed.

Although the wind may be one of the most important factors affecting processes pertinent to SSA production, and although many measurements have been classified by the mean wind speed at the time and location of measurement, it is clear that mean wind speed alone is not sufficient to characterize SSA production, and that other factors can and do influence SSA production and contribute to variation in SSA production at a given wind speed. This dependence on factors other than wind speed at the time and location of measurement holds all the more for concentrations of SSA particles, which depend not only on factors affecting SSA production, but additionally on factors affecting entrainment, advection, and removal of SSA particles, not just at the time and location of measurement, but at prior times and locations as governed by the (size-dependent) atmospheric residence times of SSA particles and by wind velocities encountered by particles during their lifetimes. Additionally, the correlation between local wind speed at the time of measurement and the wind speed at the time measured SSA particles were produced generally decreases as the lifetimes of the particles increase and as the distances and times between production and measurement increase. A further consideration is that a single value of wind speed, typically the value of U_{10} averaged over some time interval, cannot provide information about the gustiness or steadiness of the wind. Moreover, it cannot provide information concerning the extent to which steady state conditions apply, which depends on the characteristic times for the processes of interest— specifically, the mean residence time of SSA particles of a given size, the response time of the concentration of SSA particles of this size in the atmosphere to factors affecting this concentration (§2.9.6), and the response time of the marine atmospheric boundary layer and the ocean to changing meteorological conditions. These characteristic times are difficult to define and estimate, and the number of possibilities of changing conditions suggests that it will be difficult to parameterize and describe meteorological conditions affecting SSA production at a given time and location by a single quantity or even a few quantities. In view of all these considerations, use of a single parameter such as U_{10} at the time and location of measurement to parameterize quantities pertinent to SSA production should not be expected to fully capture the range of conditions that may occur for a given value of this parameter, resulting inevitably in scatter in the results of measurements taken under what might be viewed as nominally the same conditions.

2.3.2. Atmospheric Stability and Wind Friction Velocity

Atmospheric stability is the tendency of the atmosphere to enhance or suppress vertical motion. Under conditions of neutral stability further motion of a small vertical displacement of a parcel of air is neither enhanced nor suppressed. In a stable atmosphere the vertical displacement of a parcel of air is inhibited by buoyancy, whereas under unstable conditions there is a driving force whereby an air parcel that is displaced vertically continues to move in the same direction, resulting in increased vertical mixing. Thus a neutrally stable atmosphere neither enhances nor suppresses turbulence, an unstable atmosphere enhances turbulence, and a stable atmosphere suppresses turbulence.

Atmospheric stability is determined by the vertical profiles of wind speed, temperature, and RH near the sea surface but is often inferred or parameterized simply by the difference between the temperature of the surface water and that of the atmosphere at a height of 10 m, or alternatively by the Monin-Obukhov length, which is related to the vertical flux of heat (§2.4). Atmospheric stability affects the SSA production process itself, and thus the interfacial SSA production flux f_{int}, by affecting the sea state, wave breaking, and whitecap formation. Additionally, atmospheric stability affects the vertical entrainment of SSA particles once they are formed (§2.9.5) and thus Ψ_f, the ratio of the effective SSA production flux at 10 m f_{eff} to the interfacial SSA production flux f_{int}, and also concentrations of SSA particles and their vertical distribution (§2.9.3, §2.9.5).

A key variable for the description of all atmosphere-ocean interactions (and thus f_{int}) is the wind friction velocity u_*, defined as $u_* = (\tau/\rho_a)^{1/2}$, where τ is the stress (horizontal force per unit area) exerted on the ocean by the wind and ρ_a is the density of air. The wind stress, and thus the wind friction velocity u_*, depend on the wind speed, atmospheric

---

[1] *Brown* [1979]; *Wentz et al.* [1982]; *Wentz et al.* [1984]; *Chelton and McCabe* [1985]; *Chelton and Wentz* [1986]; *Wentz and Mattox* [1986]; *Dobson* [1987]; *Monaldo* [1988]; *Glazman and Pilorz* [1990]; *Boutin and Etcheto* [1990]; *Witter and Chelton* [1991]; *Guillaume and Mognard* [1992]; *Carter et al.* [1992]; *Echeto and Banege* [1992]; *Freilich and Challenor* [1994]; *Wu* [1995c]; *Stoffelen* [1998]; *Hwang et al.* [1998]; *Freilich and Dunbar* [1999]; *Weissman et al.* [2002]; *Ebuchi et al.* [2002].

stability, and sea state, and in turn the wind stress affects the sea state and thus the formation of breaking waves. For a given stability condition, the quantities u_* and U_{10} are nearly proportional over a wide range of wind speeds, being related by the dimensionless wind stress coefficient C_D, defined by

$$C_D = \left(\frac{u_*}{U_{10}} \right)^2 \qquad (2.3\text{-}1)$$

and having typical values 0.0010-0.0015 for neutrally stable conditions. The wind stress coefficient is generally considered to be a function of wind speed and atmospheric stability, allowing various processes to be described wholly in terms of macroscopic, measurable quantities such as U_{10} and the air-sea temperature difference. A variety of formulations relating C_D and U_{10} have been proposed (some of which involve other parameters too), most of which are linear relations and yield a slight increase in C_D with increasing U_{10} [reviews and compilations are presented in *Wu*, 1969a,b; *Bunker*, 1976; *Garratt*, 1977; *Wu*, 1980; *Charnock*, 1981; *Brutsaert*, 1982; *Wu*, 1982c; *Blanc*, 1985; *Huang et al.*, 1986a; *Geernaert*, 1990; and *Rieder*, 1997, for example], but the scatter of the data preclude choosing one relation over another, and the differences between the formulations are not important for purposes of this review (and are certainly less than the uncertainties in the pertinent data). In any event the dependence on U_{10} is rather weak. A value of 0.0013 is used for C_D in this review, yielding $u_* \approx 36$ cm s^{-1} for $U_{10} = 10$ m s^{-1}.

The wind friction velocity has often been used to extrapolate results of laboratory (wind/wave tunnel) measurements to oceanic conditions, typically by determining the oceanic value of U_{10} corresponding to the laboratory value of u_* for conditions of neutral stability, the presumption being that wind stress is the principal controlling factor for SSA production and other phenomena. However, there are concerns with this approach (§4.3.6.2), and determinations of f_{int} using laboratory methods are subject to large uncertainty; at best, such methods can be expected to yield only qualitative insight into the dependence of SSA production at the sea surface on factors such as water temperature. Furthermore, because of the inability of laboratory conditions to accurately simulate the entrainment of drops to 10 m, little if any confidence can be placed in laboratory determinations of the effective SSA production flux f_{eff}.

2.3.3. Rainfall

Rainfall affects numerous aspects of SSA production (much of this discussion also applies to snowfall). Raindrops falling on the sea surface can, under certain conditions, produce SSA particles directly as a result of their impact (§4.7.2). Raindrops can also produce SSA particles indirectly through the entrainment of bubbles that later rise to the sea surface and burst (§4.7.2). Rain lowers the salinity of the sea surface layer [*Katsaros and Buettner*, 1969; *Marks*, 1990], resulting in SSA particles that contain less dissolved material (and thus have smaller r_{80}) for a given formation radius than those produced from seawater that has not been influenced by rain. Rain may also change the sea surface temperature. The calming effect of rain on the sea surface has long been known, and rainfall can also influence the conditions and properties of the sea surface[1] by removing organic surface films and by affecting the stress on the ocean surface, turbulence levels in the upper ocean (and thus the transport of bubbles to the surface), the roughness of the sea, the damping and breaking of waves, and possibly the wind profile over the sea (although this latter effect is thought to be slight). These effects will influence the interfacial SSA production flux. Falling raindrops can also function as efficient scavengers of particles in the atmosphere, and thus rainfall can influence the effective SSA production flux and SSA concentrations as well (§2.7.1)

The effects of rainfall on SSA production are not well quantified (e.g., with respect to the rainfall rate and the size distribution of the raindrops) and do not lend themselves to accurate parameterization. Additionally, measurements that would yield insight into SSA production have typically not been made during rain conditions. Rainfall is extremely variable in its extent (both temporal and spatial), nature, and intensity, complicating understanding of its effects on SSA production and hindering progress in investigating these effects through laboratory experiments. Furthermore, these effects depend on other factors besides those associated with rain; for instance, the production of SSA particles by raindrops differs greatly between conditions of no wind and conditions of very light ($U_{10} = 1.5$ m s^{-1}) winds (§4.7.2). Laboratory experiments afford the potential of providing some insight into the effects of rainfall on SSA production, but the difficulties in matching realistic conditions over the ocean, and the wide variety of possible conditions, diminish the likelihood that results of such experiments can be accurately extrapolated to ocean conditions. Because the effects of rain on both f_{int} and f_{eff} are poorly known, most methods for determining these fluxes implicitly or explicitly assume absence of rain.

Rainfall can have a large impact on SSA concentration. Sea salt aerosol particles are efficiently removed from the atmosphere through wet deposition (§2.7.1). They serve as

---

[1] *Reynolds* [1874 (in *Reynolds*, 1900, pp. 86-88)]; *Barnaby* [1949]; *Sainsbury and Cheeseman* [1950]; *van Dorn* [1953]; *Boys* [1959, pp. 36-39, Fig. 15]; *Caldwell and Elliott* [1971, 1972]; *Manton* [1973]; *Houk and Green* [1976]; *Green and Houk* [1979]; *Thorpe and Hall* [1983]; *Tsimplis and Thorpe* [1989]; *Nystuen* [1990]; *Le Méhauté and Khangaonkar* [1990]; *Pielke and Lee* [1991]; *Tsimplis* [1992]; *Poon et al.* [1992]; *Yang et al.* [1997]; *Schlüssel et al.* [1997]; *Craeye and Schlüssel* [1998].

nuclei for the formation of cloud drops (§2.1.5.1), which may subsequently be removed in precipitation either through gravitational deposition or by interception and accretion by falling hydrometeors. Sea salt aerosol particles themselves are also removed by precipitation below clouds by interception and accretion by falling hydrometeors. Therefore, after a precipitation event of appreciable intensity and duration, the concentration of SSA particles of a given size will be much lower than it was before (§2.7.1). The concentration subsequently increases, thus providing an opportunity for employing the concentration buildup method of determining the effective SSA production flux (§3.3). The rate of increase of concentration of SSA particles of a given size depends on the effective production flux of SSA particles of that size, the extent to which such particles are mixed vertically, and their rate of removal through dry deposition (§2.7.2). The concentration increases until a steady state value is approached, at which production is balanced by removal through dry deposition (thus providing the basis for the steady state dry deposition method of determining the effective SSA production flux; §3.1), or until another precipitation event occurs and the SSA particles are removed by wet deposition. The latter scenario is expected to be the usual one for sufficiently small SSA particles, i.e., $r_{80} \lesssim 1$ μm, in nearly all circumstances (providing the basis for the statistical wet deposition method of determining the effective SSA production flux, §3.9), and for larger SSA particles in some circumstances. Thus, the time since the sea salt aerosol experienced its last major precipitation event (in a Lagrangian sense) is expected to be a major factor controlling concentrations of SSA particles and hence must be considered when applying the steady state dry deposition method (§3.1).

2.3.4. Surface-Active Materials

Many aspects of SSA production and behavior involve interfacial phenomena: wind-wave interaction, which determines sea state, sea surface roughness, and whitecap formation; bubble dynamics and gas exchange, which influence the size distribution of bubbles bursting at the sea surface; bubble residence time on the surface, which affects bursting behavior and whitecap persistence; drop formation; and the exchange of moisture between drops and the atmosphere. Hence such phenomena are inherently subject to the properties of the air-water interface, and any factor that affects these properties can potentially affect SSA production and behavior.

Surface-active substances are materials that appreciably affect interfacial properties when present in small amounts. Of interest to the present discussion are those substances that affect properties of the air-water interface. These substances, of which surfactants comprise a large fraction, consist mainly of organic chemical substances and can be of either natural or anthropogenic origin. The cause of their surface activity is usually that individual molecules contain both hydrophobic and hydrophilic moeities, causing these substances to concentrate at an air-water interface—at the surface of the ocean, bubbles, or drops—and to spread spontaneously over this interface, lowering the surface tension and possibly affecting interfacial mass transport and other surface properties. Surface-active substances are scavenged by rising bubbles, influencing both bubble dynamics and gas exchange, and these substances are transported thereby to the surface of the ocean, where they can become increasingly concentrated. They can likewise, as a consequence, occur in enhanced concentrations in drops formed from these bubbles (§2.5.2). For all these reasons the effects of surface-active materials on surface properties and in turn on SSA can be substantial despite low bulk concentrations.

Numerous investigations have dealt with the effects of surface-active substances on these several processes: bubble coalescence [Anderson and Quinn, 1970]; bubble motion [Haberman and Morton, 1953; Detwiler and Blanchard, 1978; and numerous other references discussed in §4.4.2.1]; bubble residence time on the surface before bursting[1]; bubble bursting and drop formation[2]; drop coalescence [Ueno, 1976]; the behavior of a drop with respect to changes in RH and its rate of evaporation, potentially affecting visibility, cloud formation, and albedo[3]; evaporation from and gas

---

[1] Hardy [1925]; Blanchard [1963]; Garrett [1967a]; Blanchard and Syzdek [1972a]; Detwiler and Blanchard [1978]; Blanchard and Hoffman [1978]; Blanchard [1990]; §4.3.2.3; §4.3.3.3.

[2] Blanchard [1963]; Blanchard [1967, pp. 126-127]; Garrett [1968]; Paterson and Spillane [1969]; Blanchard and Syzdek [1972a]; Morelli et al. [1974]; Blanchard and Syzdek [1975, 1978a,b]; Blanchard and Hoffman [1978]; Detwiler and Blanchard [1978]; Blanchard [1990]; §4.3.1; §4.3.2; §4.3.3.

[3] Bradley [1955]; Fuchs [1959, p. 10]; Eisner et al. [1960]; Derjaguin et al. [1966a]; Mihara [1966]; Snead and Zung [1968]; Bigg et al. [1969]; Warner and Warne [1970]; Hoffer and Mallen [1970]; Ueno and Sano [1971b]; Garrett [1971]; Hughes and Stampfer [1971]; Kocmond et al. [1972]; May [1972]; Garrett [1972]; Pueschel and van Valin [1974]; Podzimek and Saad [1975]; Giddings and Baker [1977]; Bullrich and Hänel [1978]; Corradini and Tonna [1979]; Chang and Hill [1980]; Otani and Wang [1984]; Rubel and Gentry [1984]; Frey and King [1986]; Bortkovskii [1987, p. 140]; Andrews and Larson [1993]; Saxena et al. [1995]; Shulman et al. [1996]; Shulman et al. [1997]; Gorbunov et al. [1998]; Li et al. [1998]; Facchini et al. [1999, 2001]; Ming and Russell [2001]; Rood and Williams [2001]; Feingold and Chuang [2002].

exchange with bulk water surfaces, possibly changing the surface temperature[1]; the generation of streaks on the sea surface by wind [*Welander*, 1963]; and wave damping and surface roughness on water bodies, affecting remote sensing of the sea surface, ocean noise, wave breaking, and possibly the wind speed over the oceans[2]; further discussion and examples are presented in *La Mer* [1962], *Hühnerfuss and Garrett* [1981], *Gill et al.* [1983], and *Bortkovskii* [1987, pp. 79-87, 135-142].

The effects of surface-active substances, which depend on their amount and nature, are difficult to determine or characterize in any systematic way, and as results from different investigators often vary greatly, it is difficult to draw convincing conclusions regarding these effects. Additionally, such substances are ubiquitous in the ocean but exhibit large spatial and temporal variability in both composition and abundance. Although many arguments have been made as to the relative importance of the effects of these substances under ocean conditions, little is actually known about this importance. These substances are present also in laboratory experiments and in fact are extremely difficult to eliminate, making the production of clean (i.e., surfactant-free) water, and especially a clean water surface, challenging, to say the least. *Wangersky* [1965] reported that organic substances persisted in water that had been triply distilled and then distilled with alkaline permanganate. *MacIntyre* [1974a] stated that only once had he been able to keep a surface clean for 24 hours, and that was after a year of preparation and mostly because of good luck, and *MacIntyre* [1974c] stated that only once in the previous decade had he produced a solution surface that was not detectably dirty by bubble-experiment criteria. *Detwiler and Blanchard* [1978] stated, "No water sample [including tap water, distilled water, and salt solutions] was ever encountered that did not contain at least traces of surface-active material." Furthermore, the influence of such substances can be far out of proportion to their bulk concentration, with only a slight amount being necessary to produce a large effect. For instance, *Kitchener and Cooper* [1959] reported that a 0.0002 M solution of sodium cetyl sulfate was sufficient to produce foaming in water. *Manley* [1960] calculated that the volume fraction of an organic impurity in water had to be less than one part in 10^7 in order not to affect gas dissolution rates, and *MacIntyre*

[1974a] stated that impurities as low as one part in 10^9 can change the surface properties of water. *Broecker et al.* [1978] reported that monolayer films resulted in extreme damping of waves and consequent large reduction in air-water gas transfer in laboratory experiments. Additionally, *Rohr and Updegraff* [1993] reported that the presence of monolayer films resulted in a reduction in surface-related ambient acoustic noise in the ocean by up to 8 db (nearly an order of magnitude in power).

In view of these considerations, the influences of surface-active substances must be recognized in the conduct and interpretation of laboratory experiments, and may account for the non-reproducibility of results from different investigations and for some reported anomalies [*Anderson and Quinn*, 1970]. For instance, *Blanchard* [1963, pp. 139-140] determined that the non-reproducibility of jet drop ejection heights as a function of bubble rise time (§4.3.2.3), which occurred for both distilled water and artificial seawater, was related to the time since the water had been in the experimental apparatus and was ultimately traced to the presence of a rubber stopper. In the discussion that follows, attention is called to the influences of these substances where they have been determined, but it must be borne in mind that these influences can pervade the results of all studies of drop formation, both in the laboratory and in the ocean, leading to effects that are difficult to quantify, reproduce, and explain, and are therefore manifested as a spread of results under nominally identical conditions.

2.3.5. Other Factors

Additional factors identified as potentially affecting SSA production, entrainment, transport, and removal, and thus SSA concentration, are examined here.

The sea temperature determines the kinematic viscosity of seawater (§2.5.1) and thus affects bubble rise velocities; it affects the rate of gas exchange between a bubble and the surrounding fluid (§4.4.2) and thus the number and size distribution of bubbles arriving at the surface (§4.4.2); it influences bubble bursting behavior and thus the drop formation process (§4.3); and it appears to be a factor affecting the oceanic whitecap coverage (§4.5.3). Thus, the sea temperature may play a large role in the interfacial production flux

---

[1] *Rideal* [1925]; *Ramdas* [1927]; *Langmuir and Schaefer* [1943]; *Archer and La Mer* [1954]; *Mansfield* [1955]; *Jarvis* [1962]; *J. Davies* [1966]; *MacRitchie* [1968, 1969]; *Cadenhead* [1969]; *Grossman et al.* [1969]; *Springer and Pigford* [1970]; *Wu* [1971d]; *Mansfield* [1972]; *Dragčević et al.* [1979]; *Lee et al.* [1980]; *McCoy* [1982]; *Goldman et al.* [1988]; *Frew et al.* [1990].

[2] *Franklin* [1762 (in *Franklin*, 1987, pp. 794-795), 1773 (in *Franklin*, 1987, pp. 889-898), and 1774 (cited in *Giles*, 1969)]; *Reynolds* [1880 (in *Reynolds*, 1900)]; *Aitken* [1883]; *Rayleigh* [1890]; *Pockels* [1891]; *Ewing* [1950]; *van Dorn* [1953]; *Cox and Munk* [1954]; *Munk* [1955]; *Garrett and Bultman* [1963]; *Garrett* [1967c]; *Barger et al.* [1970]; *Scott* [1972]; *Mallinger and Mickelson* [1973]; *Broecker et al.* [1978]; *Hühnerfuss et al.* [1981]; *Lucassen* [1982]; *Rohr et al.* [1989]; *Alpers and Hühnerfuss* [1989]; *Wei and Wu* [1992]; *Rohr and Detsch* [1992]; *Tang and Wu* [1992]; *Rohr and Updegraff* [1993].

of SSA particles of a given size. Atmospheric stability (§2.5.1) and the relative importance of convective and mechanical mixing of the atmosphere above the ocean (§2.4) are determined primarily by the difference between the air temperature and the sea temperature. Hence these temperatures play a role in the upward entrainment of SSA particles into the atmosphere (§2.9.3; §2.9.5). Additionally, the wind friction velocity (§2.3.2) is often calculated from wind speed and the air-sea temperature difference. As these temperatures are easily measured (or readily available), they are potentially well suited to parameterize SSA flux formulations, especially for use in models for which forecast or archived data are available.

Relative humidity (RH), the ratio of the water vapor pressure to the saturation vapor pressure of bulk water at the ambient temperature (§2.1.1), affects the physical size, density (Fig. 4), and mass of a seawater drop containing a given mass of solute (§2.5.3), and thus affects the concentration of these dissolved substances in seawater drops and the behavior of these drops in the atmosphere. The dependencies of these quantities on RH has important consequences for measurements that resolve particles by physical size, such as those utilizing optical detectors and impactors. Additionally, the terminal velocity of an SSA particle with a given r_{80} depends on RH—the terminal velocity of an SSA particle in equilibrium at 95% RH is nearly twice the terminal velocity of the same particle in equilibrium at 80% RH (§2.5.3). The importance of this effect depends on the sizes of particles that contribute most to the quantity of interest. For instance, RH has little effect on N, the total SSA number concentration, as gravity plays a negligible role in the fate of small SSA particles, which contribute the most to this quantity (however, RH can greatly affect the concentration of particles in a certain physical size range as smaller particles grow into this size range [Exton et al., 1985, 1986; Kowalski, 2001]; §3.5.5; §4.1.3.3). The use of a reference RH such as 80% to describe SSA particle sizes, concentrations, and fluxes, along with knowledge of the dependence of equilibrium SSA particle size on RH (§2.5.3), allows the influence of RH on SSA properties to be easily taken into account. Relative humidity has also been reported [Burger and Blanchard, 1983] as affecting the time a bubble spends on the surface before bursting, implying its possible influence on drop production and whitecap decay rates.

The sea state, which is quantitatively characterized by the wave spectrum [Phillips, 1966], plays a large role in determining the frequency and nature of wave breaking. A few studies have parameterized SSA concentration and production in terms of properties of the wave spectrum such as the

phase velocity c_s, the frequency σ_s, or the period T_s ($\propto 1/\sigma_s$) of the significant (or dominant) wave—roughly, the wave with frequency near the peak of the energy spectrum. These quantities, along with the acceleration due to gravity g and the kinematic viscosity of air v_a, have been combined to form dimensionless groups, such as u_*/c_s, used by several investigators; $gT_s^2u_*/v_a$, used by Toba and Chaen [1973] to fit the number concentration of SSA particles with $r_{80} > 8$ µm based on the assumption that the concentration of these larger particles would reflect local production and by Chaen [1973] to classify measurements of size-dependent SSA concentrations; and $u_*^2/(v_a\sigma_s)$, or equivalently, $u_*^2T_s/v_a$, proposed by Toba and Koga [1986], who refitted data of Toba and Chaen [1973] using this parameter, and later used by Iida et al. [1992], who refitted data of Chaen [1973] to this parameter and used it to classify SSA production, and by Zhao and Toba [2001] to parameterize whitecap coverage. Similar attempts to classify the areal coverage of whitecaps and its dependence on meteorological factors are discussed in §4.5.3. Most of these investigations were based on the measurements of Toba and Chaen [1973], who subjectively measured T_s, from which c_s and σ_s were calculated. These quantities have generally not been reported by other investigators making measurements pertinent to SSA production; it is thus difficult to evaluate their utility for parameterizing SSA concentrations and fluxes (and areal whitecap coverage). Furthermore, v_a varies little with air temperature, and c_s and T_s are often highly correlated with U_{10} (or u_*); the linear correlation coefficient between T_s and U_{10} for the 45 data pairs presented by Toba and Chaen [1973] is $r_{corr} = 0.855$, and when a single outlier is removed the correlation coefficient increases to $r_{corr} = 0.94$, with 75% of the estimated values being within 25% of the actual values[1]. Therefore, as these quantities provide little independent information, there is nothing to be gained by using them to further classify SSA production beyond the classification by U_{10}.

The height of the marine boundary layer (§2.4), over which small SSA particles are generally well mixed (§2.9.5), determines the concentration of these particles for given production and removal fluxes (§2.9.6). However, ascertaining the influence on concentrations of small SSA particles of factors that affect both SSA production and the marine boundary layer height is difficult, as the confounding effects of this height on SSA particle concentration will obscure such influences. Fetch, the distance over the ocean along which SSA particles may have been produced and transported to the location of interest, is also a controlling factor for SSA concentration. Additionally fetch may affect SSA production directly by its influence on sea state.

---

[1] Although the data pairs of T_s and U_{10} presented in Chaen [1973] are from these same measurements, they differ from those in Toba and Chaen [1973] in a few instances—specifically runs 23, 24, 25, 26, and 40; however, such differences do not appreciably alter these results.

The saturation of the ocean surface waters (§2.5.1) with respect to the major atmospheric gases (i.e., nitrogen and oxygen) can potentially affect SSA production through its influence on gas exchange between the ocean and bubbles that are injected by breaking waves (§4.4.2). Situations in which the gas pressure inside a bubble is greater than that in the surrounding water will result in gas transfer from the bubble to the ocean, leading to a decrease in the bubble size. Conversely, situations in which the gas pressure inside a bubble is less than that in the surrounding water will result in gas transfer from the ocean to the bubble, and the bubble will increase in size. As the size of a bubble is the dominant factor affecting its rise speed (§4.4.2) and the spectrum of drops it produces (§4.3), saturation of the major atmospheric gases in seawater is potentially an important parameter. However, because this saturation typically varies over such a narrow range of values, its effect on SSA production is probably unimportant in nearly all situations (§2.5.1, §4.4.2).

Salinity, a measure of the concentration of dissolved substances in seawater, has been suggested as being important in various aspects of SSA production because of the strong influence of salinity on many seawater properties (§2.5.1). However, over the narrow range of salinities encompassing most of the ocean surface waters, the variation of these properties is so slight as to be negligible (an exception may be SSA production in the Baltic Sea, because of its low salinity; §2.5.1; §2.5.3; §4.4.1.4). In contrast, there exist large differences between seawater and freshwater in the magnitude and shape of bubble spectra, coalescence properties of bubbles, the behavior of individual bubbles, and various aspects of bubble bursting and drop formation (§4.4.1.4). Additionally, the behavior of freshwater drops and seawater drops subsequent to formation differs greatly because of the absence of dissolved salts in freshwater (§2.5.3). Thus the ability of laboratory simulations involving whitecap formation, bubble bursting, drop formation, and particle behavior in freshwater to accurately represent these processes in the ocean must be questioned.

Other factors have been suggested as playing a role in wave breaking and thus in SSA production, such as the depth and topography of the sea floor. However, little systematic quantitative support has been adduced for these factors, and although these factors may be important in coastal regions or in special circumstances, they are not important for SSA production over the open ocean.

2.4. STRUCTURE OF THE ATMOSPHERE OVER THE OCEAN

In examining the life cycle of SSA particles in the marine atmosphere, it is necessary to consider the structure of the marine atmospheric boundary layer as it pertains to SSA production, entrainment, transport, and removal. The marine atmospheric boundary layer (hereinafter denoted the marine boundary layer) is defined to be that portion of the atmosphere above the oceans that is appreciably affected by the surface of the ocean and its properties. Sea salt aerosol is confined mainly to this layer. Turbulence is an intrinsic feature of the motion of the air and of transport of SSA in the marine boundary layer, and both mechanical turbulence caused by wind shear and convective turbulence associated with the thermodynamic stability of the atmosphere in this layer are important. In contrast to the land, the ocean possesses a wet, movable surface that can strongly interact with, and affect, the wind in a much more dynamic sense [e.g., *Hristov et al.*, 2003]. Additionally, the surface of the ocean is much more uniform and has a smaller roughness length (defined below) than the land. Consequently, the properties of the atmosphere over the ocean differ considerably from those over land [*Lilly*, 1968; *Smith and Carson*, 1977; *Nicholls*, 1984; *Stull*, 1988; *Garratt*, 1992; *Albrecht et al.*, 1995b; *Miller et al.*, 1998]. Over the ocean RH is typically greater, and as the ocean provides a nearly infinite reservoir of heat and moisture, the diurnal variations in surface temperature and resultant forcing of the marine boundary layer are much less, than over land. Turbulent mixing in the marine boundary layer is typically less energetic than that observed over land, and mechanical mixing (i.e., wind-shear driven) is generally less important [*Lilly*, 1968]. Additionally, the atmosphere above the oceans is much more characterized by stability conditions that are near neutral or conditionally unstable, although substantial local perturbations in the typical thermodynamic structure may be observed, especially when a continental air mass is advected over a coast or when there is a strong ocean thermal front.

The depth and structure of the marine boundary layer vary as a function of latitude and sea-surface temperature. Outside the tropics and away from frontal zones the marine boundary layer is generally well mixed, as evidenced by near-uniform concentrations of various trace substances. Sea salt aerosol production is expected to be greater in middle and high latitudes because of higher average wind speeds near the ocean surface. Therefore this review and calculations presented below will focus on this middle and high latitude region, where the marine boundary layer is generally well mixed, cloudy, and capped by a thermal inversion.

The marine boundary layer in this region is conceptually divided into several sublayers which, though not rigidly demarcated, serve to identify the important processes that control the transport and evolution of SSA [similar arguments were presented in *Davidson et al.*, 1982; *Fairall et al.*, 1982; *Fairall and Larsen*, 1984; *Fairall and Davidson*, 1986; and *Garratt*, 1992]. The several sublayers are characterized by differences in wind speed and its controlling factors, specifically the stresses (momentum fluxes) that act to determine the air flow and the mechanisms that control these stresses: viscosity, form drag (resulting from pressure gradients) from waves and additional stresses experienced by flow over breaking waves, turbulence, Coriolis forces, buoyancy, and large-scale pressure gradients. These several sublayers are: the viscous (laminar) sublayer of thickness δ, typically taken as ~1 mm or less; the surface layer, of thickness several tens of meters; and the Ekman (also called mixing, or outer) layer, which extends to the top of the marine boundary layer at height H_{mbl}, which is typically between several hundred meters and a few kilometers. At the top of the marine boundary layer there is typically a strong temperature inversion resulting in an abrupt change in the density and relative humidity of the air between this layer and the free troposphere above. This conceptual model and the several sublayers, their properties, and the transitions between them, are briefly described here.

According to this conceptual picture, immediately adjacent to the sea surface is a laminar sublayer, assumed to be minimally affected by turbulent or convective motion on account of viscous coupling to the relatively stationary surface [Rossby, 1936 (cited in Sverdrup, 1937); Sverdrup, 1937]. The RH in this layer is near 98% on account of the proximity to the ocean surface together with the vapor pressure lowering of water above seawater due to its salt content (Raoult's Law; §2.5.3). In this layer transport of gases and momentum is provided mainly by molecular processes; hence the flow is controlled by molecular viscosity. However, the flow is characterized also by the existence of intermittent turbulent streaks or bursts which erupt abruptly from near the surface [Kline et al., 1967] and which may contribute to impaction of SSA particles on the sea surface [Owen, 1969; Slinn, 1974]. The existence of such a laminar layer is usually assumed based on experience with respect to flow over a flat, rigid surface, but the validity of this analogy is questionable for the ocean surface which is neither rigid or planar, and which is not uniform on a variety of scales because of waves, ripples, and bubbles. Additionally, near the surface, there are interactions between the wind and the waves, and there are stresses due to form drag on moving waves, and over breaking waves, which have little correspondence on land.

The thickness of the viscous sublayer δ, although not rigorously defined, is typically considered to be around 1 mm or less for wind speeds encountered over the ocean. Some estimates for δ are based on dimensional arguments, such as $\delta \approx v_a/u_*$, where v_a is the kinematic viscosity of air [e.g., Slinn, 1983a], which yields $\delta \approx 0.04$ mm for $U_{10} = 10$ m s^{-1}. Giorgi [1986] assumed $\delta = 30v_a/u_*$, which would correspond to $\delta \approx 1$ mm at $U_{10} = 10$ m s^{-1}. Another estimate can be obtained by considering the evaporative flux of water from the ocean. Although this flux is affected by wind speed and

other factors, for the purposes of this argument it is assumed that it is due solely to molecular diffusion through the viscous sublayer. The mass evaporative flux of water is given by $j_w = D_w(dC_w/dz)$, where D_w is the diffusion coefficient of water vapor in air and C_w is the mass concentration of water vapor. The quantity C_w is given in terms of the molecular weight of water M_w, the saturation vapor pressure of water $p_{w,sat}$, and the fractional relative humidity rh by $C_w = M_w(p_{w,sat}/RT)rh$. Approximating the derivative by a difference across the viscous sublayer yields $j_w \approx D_w M_w(p_{w,sat}/RT)(\Delta rh)/\delta$, where Δrh is the difference in fractional RH across the laminar sublayer, taken to be 0.2 (corresponding to the approximate difference between 98% RH at the water surface and a typical value of 80% RH in the surface layer above the laminar sublayer, as discussed below). Substituting the values of D_w and $p_{w,sat}$ at 20 °C, the mean sea surface temperature (§2.5.1), and equating this flux to the value $j_w \approx 0.03$ g m^{-2} s^{-1} (equivalent to $3 \cdot 10^{-6}$ cm s^{-1}), corresponding to the annual mean water evaporation over the ocean (~100 cm; §1.1), yields $\delta \approx 3$ mm. These estimates for δ, which span two orders of magnitude, serve to bound this quantity.

Above this viscous sublayer is the surface layer, which generally extends to heights of a few tens of meters (and thus includes heights at which shipboard measurements are typically made). In this layer, the dominant stresses are due to turbulence and are nearly independent of height; this layer is often referred to as the constant stress, or constant flux, layer. The RH of this layer varies spatially and temporally, but 80% is a typical value [e.g. Eriksson, 1959; Gathman, 1982; Liu et al., 1991]; this value is used for scoping purposes throughout this review. For neutrally stable conditions the vertical dependence of the mean wind speed in this layer is directly proportional to the logarithm of the height above the sea surface z [Ruggles, 1970; Tennekes, 1973]:

$$U(z) = \frac{u_*}{\kappa} \ln\left(\frac{z}{z_0}\right), \qquad (2.4\text{-}1)$$

where u_* is the wind friction velocity, $\kappa \approx 0.40$ is the von Karman constant [compilations of various determinations of κ are presented in Hicks, 1976; Garratt, 1992, p. 289; Garratt et al., 1996; and Högström, 1996, for example], and z_0 is the so-called roughness length, which for a wind stress coefficient $C_D = 0.0013$ (§2.3.2) is given approximately by $z_0 \approx 0.15$ mm. With increasing height above the surface the dependence of the mean wind speed on height rapidly decreases, and the mean wind speed approaches a value that changes little with increasing height. For example, if the above value of C_D (and hence z_0) is used, the mean wind speed at heights between 3.5 m and 30 m, at which most field measurements of SSA concentrations are made, differs from the value at 10 m by less than 10%. As this difference is less than the typical uncertainty in determinations of the mean wind speed (§2.3.1), the exact height at which the mean wind speed is reported is not of great consequence provided it is in this range.

Mixing of heat, moisture, and momentum in this layer is controlled by turbulent eddies and depends on height, wind speed, and atmospheric stability. This mixing is generally parameterized by an eddy (turbulent) diffusion coefficient, or eddy diffusivity, D_{eddy}, which for neutrally stable atmospheres is given by

$$D_{\text{eddy}} = \kappa u_* z \qquad (2.4\text{-}2)$$

[*Tennekes*, 1973]. Particle transfer in this layer is also controlled by these turbulent eddies (as well as by gravitational sedimentation, which depends on particle size), and it is therefore generally assumed that this eddy diffusivity also describes particle transfer (§2.8.2) and thus can be used to calculate the steady state SSA concentration profile in this layer (§2.9.3).

For stable or unstable atmospheric conditions the relations expressing the dependencies of velocity and eddy diffusivity on height contain additional terms that depend on the ratio of the height z to the Monin-Obukhov length L [*Monin and Obukhov*, 1954 (cited in *Phillips*, 1966)]. The Monin-Obukhov length characterizes the extent to which stability conditions differ from neutral, with $L > 0$ for stable conditions and $L < 0$ for unstable conditions; when positive L defines the height at which the mechanical (i.e., shear) production and the buoyancy production of turbulent kinetic energy are equal. With increasingly unstable conditions the effects of the vertical fluxes of moisture and heat, which result in convective mixing, become relatively more important compared to the effects of wind shear. Under unstable conditions turbulent (eddy) diffusion is greater, and particles are entrained upward much more efficiently, than under neutral conditions; conversely, under stable conditions turbulent (eddy) diffusion is less, and particles are entrained upward much less efficiently, than under neutral conditions.

The strength of turbulent eddies and their ability to transport SSA particles upward from near the sea surface are characterized by σ_w, the root-mean-squared value of the vertical component of the wind speed. At heights of several meters above the surface σ_w is slightly (~30%) greater than u_* [*Miyake et al.*, 1970a,b; *Smith*, 1970; *Pond et al.*, 1971; *Donelan and Miyake*, 1973; *Pennell and LeMone*, 1974; *Smith and Banke*, 1975; *Leavitt and Paulson*, 1975; *Merry and Panofsky*, 1976; *Geernaert et al.*, 1987; *Garratt*, 1992], and thus is approximately equal to $0.05U_{10}$ for moderate wind speeds, resulting in $\sigma_w \approx 50$ cm s^{-1} for $U_{10} = 10$ m s^{-1}. Under neutrally stable atmospheric conditions and for heights sufficiently great that effects of wave motion and of viscous

stresses very near the surface can be neglected, σ_w is expected to be independent of height [*Tennekes*, 1982], although changes in the wind flow over breaking waves and changes in surface roughness caused by bubbles may alter this relationship somewhat closer to the surface. Arguments involving σ_w to determine an upper limit for the size of drops that have an appreciable probability of being entrained upward near the surface are presented in §2.9.2 and §4.3.1.3.

With increasing height the surface layer gradually gives way to what is known as the Ekman layer [*Ekman*, 1905], which extends to the top of the marine boundary layer. In the Ekman layer, Coriolis effects are increasingly important in the motion of the atmosphere, and as a consequence, the mean wind speed changes direction with height. The flow in this layer is still turbulent, but the turbulent stresses, and thus the eddy diffusivity, do not continually increase with height and in fact decrease to near zero at the top of this layer because of inhibition of vertical transport due to the inversion at the top of the marine boundary layer. In the marine boundary layer, vertical mixing of particles is effected by convection in addition to mechanical turbulence. The rate of this mixing is described by the convective mixing velocity v_{conv} [*Kaimal et al.*, 1976], for which a typical measured value over the ocean is ~0.6 m s^{-1} [*Pennell and LeMone*, 1974; *Markson et al.*, 1981; *Fairall et al.*, 1982, 1983]; this convective mixing velocity can be thought of as a mean value of the updraft velocity of convective eddies. A characteristic time for convective mixing can thus be defined by

$$\tau_{\text{conv}} = \frac{H_{\text{mbl}}}{v_{\text{conv}}}; \qquad (2.4\text{-}3)$$

for $v_{\text{conv}} = 0.6$ m s^{-1} and $H_{\text{mbl}} = 0.5$ km (as discussed below), $\tau_{\text{conv}} \approx 8 \cdot 10^2$ s (~15 min). An alternative estimate for the time characterizing the mixing of the marine boundary layer τ_{mix} can be obtained from the relation

$$\tau_{\text{mix}} \approx \frac{z^2}{2D_{\text{eddy}}}, \qquad (2.4\text{-}4)$$

where z is taken to be the height of the marine boundary layer H_{mbl}, ~0.5 km, and D_{eddy} is evaluated at half this height (although this height may exceed that for which the expression used for D_{eddy}, Eq. 2.4-2, is valid, this estimate should nevertheless yield values that are accurate to within a factor of three or so). At $U_{10} = 10$ m s^{-1} these quantities yield an estimate $\tau_{\text{mix}} \approx H_{\text{mbl}}/(\kappa u_*) \approx 3 \cdot 10^3$ s (~1 h). Thus the time characterizing the mixing of the marine boundary layer is of order 10^3 s, and typically not more than an hour. This time should also reflect the time required for small SSA particles, for which gravity is unimportant, to become mixed throughout the marine boundary layer.

Because of the temperature inversion typically present at the top of the marine boundary layer, vertical motion is inhibited, and hence there is an abrupt change in concentrations of trace substances, including SSA. The height at which this occurs, H_{mbl}, is for the most part governed by the balance between fluxes of heat and moisture at the surface, radiative cooling at the top of the layer, large-scale subsidence, and cloud microphysical processes. As these factors depend on location and time on several scales, variations in H_{mbl} are to be expected, and over the ocean H_{mbl} may vary from a few hundred meters (under stable conditions such as a cold ocean and warm atmosphere) to slightly more than a kilometer. A typical value for H_{mbl} is 0.5 km, and this choice is made in this review as a representative value for scoping purposes, although a more suitable value should be used in specific situations where appropriate (and when known).

In the free troposphere above the marine boundary layer, the flow is controlled by large-scale pressure gradients and Coriolis forces. In this region the flow, denoted the geostrophic flow, is characterized by low turbulence. Because of the inversion at the top of the marine boundary layer and the resultant low mixing of material across the inversion, concentrations of SSA particles and other materials whose origin is within the marine boundary layer are generally much less in the free troposphere than in the marine boundary layer below.

Entrainment of air from the free troposphere into the marine boundary layer results in an increase in the height of the marine boundary layer at a rate described by the entrainment velocity v_{ent}, defined by

$$v_{ent} = \frac{dH_{mbl}}{dt}.$$
(2.4-5)

Values of v_{ent} over the oceans in the range 0.3-0.8 cm s^{-1} have been reported, based on measurements of the rate of change of the height of the marine boundary layer or on measurements of fluxes or rates of change of conserved quantities or concentrations of trace species [*Lenschow et al.*, 1982; *Davidson et al.*, 1982; *Kritz*, 1983; *Fairall et al.*, 1983, 1984; *Davidson et al.*, 1984; *Boers and Betts*, 1988; *Bretherton et al.*, 1995; *Clarke et al.*, 1996; *Huebert et al.*, 1996; *Boers et al.*, 1998; *Lenschow et al.*, 1999; *Sollazzo et al.*, 2000]. For a wind speed of 10 m s^{-1}, this range of entrainment velocities is equivalent to a gradient in the height of the marine boundary layer of 0.3-0.8 m per km in the direction of the wind. A characteristic time associated with entrainment useful for comparison with characteristic times of other processes can be defined by

$$\tau_{ent} = \frac{H_{mbl}}{v_{ent}},$$
(2.4-6)

or equivalently,

$$\tau_{ent} = \left(\frac{d \ln H_{mbl}}{dt} \right)^{-1},$$

for which the above range of values of v_{ent} results in values of τ_{ent} of ~0.5 d to ~2 d for $H_{mbl} = 0.5$ km. Entrainment of the free troposphere and the accompanying growth of the marine boundary layer results in a decrease (due to dilution) in the concentration of SSA particles that are mixed over the height of the marine boundary layer and have much lower concentrations in the free troposphere above (§2.8.3); that is, particles with $r_{80} \lesssim 10$-15 μm (§2.9.5). In contrast, however, growth of the marine boundary layer should have little or no effect on the concentration of larger SSA particles that are not mixed over this height because of the greater role played by gravity in their behavior (§2.9.5).

Examination of the behavior of SSA and its concentration is facilitated by use of an even more simplified model of the marine boundary layer introduced here. This model is used in the discussion of the mechanisms that affect SSA particles (§2.6), in the treatment of dry deposition (§2.7, §4.6), in the investigation of the continuity equation of SSA concentration and the specification of the vertical flux that result in expressions for the determination of the effective SSA production flux by various methods (§2.8), and in considerations of the vertical distribution of SSA (§2.9). According to this model, the sea surface is a horizontal plane on which SSA production occurs uniformly and continuously at a constant rate. The marine boundary layer, in which all SSA is confined, is separated into two distinct layers: a thin (~0.1-1 mm) viscous sublayer with RH 98% and a surface layer with RH 80% that extends to the top of the marine boundary layer of height 0.5 km. Coriolis effects and the change in the direction of the wind speed with height are neglected, and conditions of neutral atmospheric stability are assumed to hold. This latter assumption is less restrictive than it might seem because typically situations in which atmospheric stability is appreciably different from neutral do not represent steady state and are not well described solely by U_{10}; they are often accompanied by rainfall and do not last long. Although this model is highly idealized and cannot be expected to represent all conditions present over the ocean, it is expected to capture most of the essential features that affect SSA behavior and lifetime, and further that the basic results obtained from this model are qualitatively valid.

2.5. CHEMICAL AND PHYSICAL PROPERTIES OF SEAWATER AND SSA PARTICLES

Seawater is the medium in which waves break, whitecaps form, and bubbles rise to the surface and burst, producing drops which are likewise composed of seawater. The transport, removal, and behavior of these drops in the atmosphere are affected by their size, density, and composition. These drops can exchange moisture with the atmosphere, in turn changing their size and density, and the rate of this exchange is determined by their composition and size. Thus the composition of seawater and its physical properties, the composition of seawater drops, and the behavior of these drops under different conditions of RH, are important in understanding the production of these drops and their subsequent evolution.

2.5.1. Composition and Properties of Seawater

Seawater contains most elements and a wide variety of organic compounds, many of which are biological in origin. However, most of the dissolved material in seawater consists of only a few ionic species; only Cl^-, Na^+, Mg^{2+}, SO_4^{2-}, Ca^{2+}, K^+, and HCO_3^- normally occur in concentrations greater than 0.001 mol per kg of seawater. To a good approximation, these substances, except for HCO_3^-, are fully dissociated in the ocean. The ratios of these major species, which comprise virtually all the mass of the dissolved substances in seawater (>99.5%), are quite constant throughout the oceans [Marcet, 1819 (cited in Culkin, 1965); Culkin, 1965; Wilson, 1975;

Millero, 1982; DOE, 1994]. The concentration of dissolved material is described by the salinity, which to a good approximation is equal to the mass ratio of dissolved material (i.e., sea salt) in grams per kilogram of seawater. By definition [UNESCO, 1981], salinity is a dimensionless number based on conductivity. The salinity of the vast majority of the world oceans, and of the majority of ocean surface water, is between 33 and 37, with 35 being a typical value [Sverdrup et al., 1942, pp. 124-127 and Chart VI; Montgomery, 1958; Levitus and Oort, 1977; Worthington, 1981; Emery and Meincke, 1986; Millero and Sohn, 1992, p. 21], although in some locations the salinity of the surface water differs appreciably from 35. For example, in the Red Sea the salinity can approach 40, in the Mediterranean it is around 38.5, in Hudson Bay it is in the range 20-30, in the Black Sea it is around 17, and in most of the Baltic Sea it is around 7.

The composition of seawater of salinity 35 is well approximated by values presented in Table 6, and the composition of seawater of other salinities can be determined from these values. The mean molecular weight of sea salt is 62.8 and the mean equivalent weight is 58.1 (these values differ because of the presence of both singly and doubly valent ions in seawater; Table 6); thus 1 kg of seawater of salinity 35 contains 0.56 mol, or 0.60 equiv, of sea salt and has an

Table 6. Composition of Seawater of Salinity 35

Species	Mass Fraction	Mass Ratio g (kg-sw)$^{-1}$	g (kg-H$_2$O)$^{-1}$	Mole Fraction	Molarity mol (kg-sw)$^{-1}$	mol (kg-H$_2$O)$^{-1}$	equiv (kg-sw)$^{-1}$	Ionic Strength mol (kg-sw)$^{-1}$
H$_2$O	0.9648	964.83	1000.00	0.9795	53.558	55.510	—	—
Cl$^-$	0.0194	19.35	20.06	0.0100	0.546	0.566	0.546	0.273
SO$_4^{2-}$	0.0027	2.71	2.81	0.0005	0.028	0.029	0.056	0.056
HCO$_3^-$	0.0001	0.11	0.11	< 0.0001	0.002	0.002	0.002	0.001
Br$^-$	0.0001	0.07	0.07	< 0.0001	0.001	0.001	0.001	< 0.001
Na$^+$	0.0108	10.78	11.18	0.0086	0.469	0.486	0.469	0.235
Mg^{2+}	0.0013	1.28	1.33	0.0010	0.053	0.055	0.106	0.106
Ca^{2+}	0.0004	0.41	0.43	0.0002	0.010	0.011	0.020	0.020
K$^+$	0.0004	0.40	0.41	0.0002	0.010	0.011	0.010	0.005
minor species	< 0.0001	0.06	0.05	< 0.0001	0.001	0.001	0.001	0.001
Σ salts	0.0352	35.17[a]	36.45	0.0205	1.120	1.162	1.211	0.698
Total	**1.0000**	**1000.00**	**1036.45**	**1.0000**	**54.678**	**56.672**	**54.769**	**0.698**

Based on DOE [1994, Table 6.2].

[a] The discrepancy between the mass ratio of the dissolved solids in g per kg of seawater (35.17) and the salinity (35) results from artifacts of the history of the salinity concept and the several definitions, both official and practical, that have been used over the years (see, for instance, Wallace [1974]). Salinity was originally defined to be "the total amount of solid in grams contained in one kilogram of sea water when all the carbonate has been converted to oxide, the bromine and iodine replaced by chlorine and all organic material completely oxidized" [Forch et al., 1902 (translated in Morris and Riley, 1964)] and was calculated from the chlorinity (the mass ratio of halides in seawater, calculated as chloride, in grams per kilogram of seawater), which was determined by titration with silver nitrate. Different definitions of salinity that have been used over the years, necessitated partly by changes in the accepted values of the atomic weights, and partly by the desire to make the definition of chlorinity (and hence the values of salinity) independent of changes in accepted atomic weights and numerically identical to earlier definitions, have resulted in a 0.5% difference between the salinity and the mass ratio of dissolved solids [Reeburgh, 1966]. The current definition is based on electrical conductivity [UNESCO, 1981].

ionic strength of approximately 0.70 mol (kg-sw)$^{-1}$, or equivalently 0.72 mol (kg-H$_2$O)$^{-1}$ [*Millero*, 1974; *Millero and Leung*, 1976; *Millero*, 1982].

In addition to the substances listed in Table 6, a wide variety of dissolved organic substances may be present in seawater [*Garrett*, 1967b; *Wagner*, 1969; *Marty et al.*, 1979; *Barbier et al.*, 1981; *Benner et al.*, 1992; *Lepri et al.*, 1995; *Cincinelli et al.*, 2001]. Typical concentrations are 0.5-2 mg of carbon per kg of seawater [*Williams*, 1975; *Martin and Fitzwater*, 1992; *Fry et al.*, 1996], but the concentration varies both spatially and temporally, and during phytoplankton blooms it can be much greater than this range. Although the concentrations of these dissolved organic substances in bulk seawater may be very low (the above values are more than a factor of 10^4 lower than the total concentration of dissolved salts), these substances can affect various processes in a manner far out of proportion to their low concentrations because of surface activity that preferentially concentrates some of them at the surface (§2.3.4, §2.5.2).

Depending on the application, many different recipes for artificial seawater can be constructed, the main difference being how minor elements are treated. For biological applications, trace elements and nutrients might need to be included, but for laboratory experiments simulating SSA particles as seawater drops, it is reasonable to assume that only the species with the greatest concentrations are required (except for considerations of effects of surface-active substances). One recipe for 1 kg of seawater of salinity 35, containing the species listed in Table 6 in the concentrations presented there but with Cl$^-$ replacing Br$^-$ and HCO$_3^-$, and Na$^+$ replacing Ca^{2+} and K$^+$ on an equivalent basis, is presented in Table 7; a similar recipe was used by *Tang et al.* [1997]. This recipe, under the assumption of volume additivity, yields $\rho_{ss} \approx 2.2$ g cm^{-3} for the density of dry sea salt, equal (or very nearly equal) to the value used by several earlier investigators [e.g., *Woodcock*, 1952; *Twomey*, 1954; *Hänel*, 1976]. Various alternative recipes may yield

Table 7. Recipe for Artificial Seawater of Salinity 35

Species	Grams	Moles
NaCl	25.9	0.443
MgCl$_2$	5.0 (10.8^a)	0.053
Na$_2$SO$_4$	4.0	0.028
H$_2$O	965.1 (959.3^a)	
Total	1000.0	

This recipe contains the species listed in Table 5 and results in a solution having the same concentrations, except with Cl$^-$ replacing Br$^-$ and HCO$_3^-$, and Na$^+$ replacing Ca^{2+} and K$^+$, on an equivalent basis.

^{a} For MgCl$_2$ in the form MgCl$_2 \cdot$ 6H$_2$O.

slightly different values, and other values have been stated (typically without discussion), but because the radius of a seawater drop containing a given mass of solute is proportional to the cube root of the density (§2.5.3), differences among these recipes are generally unimportant for applications pertinent to SSA.

Physical properties of bulk seawater pertinent to SSA production are the vapor pressure of water in equilibrium with seawater (discussed below in §2.5.3), the density, surface tension, and kinematic viscosity, the solubilities of the major atmospheric gases, and the dependencies of these quantities on temperature and salinity. As noted above, for most of the ocean surface waters the salinity varies little from 35. In contrast, the temperature varies substantially, from −2 °C to 30 °C, with a global mean value of 20 °C [*Sverdrup et al.*, 1942, pp. 127-128 and Charts II and III; *Alexander and Mobley*, 1976; *Levitus*, 1982; *Shea et al.*, 1992]. The density of seawater [*Millero and Poisson*, 1981, 1982] of salinity 35 is about 3% greater than that of pure water at the same temperature, varying by only ~0.3% over the salinity range 33-37 at a given temperature. The density of seawater in this salinity range decreases slightly with increasing temperature, by less than 1% over the above temperature range. The surface tension of seawater [*Krümmel*, 1907 (cited in *Fleming and Revelle*, 1939)] of salinity 35, approximately 0.075 N m^{-1}, is about 1% greater than that of pure water at the same temperature, varying by only ~0.1% over the salinity range 33-37 at a given temperature. The surface tension of seawater decreases with increasing temperature, by around 6% over this same temperature range. The kinematic viscosity of seawater [*Krümmel and Ruppin*, 1906 (cited in *Dorsey*, 1940); *Miyake and Koizumi*, 1948; *Dexter*, 1968 (cited in *Millero*, 1974)] is roughly 4% greater than that of pure water at the same temperature, varying by only ~0.7% over the salinity range 33-37 at a given temperature. The kinematic viscosity of seawater exhibits a strong dependence on temperature, decreasing with increasing temperature by slightly over a factor of 2 over the above temperature range, from near 1.8·10^{-6} m^2 s^{-1} at 0 °C to near 0.8·10^{-6} m^2 s^{-1} at 30 °C.

The solubilities of the major atmospheric gases (nitrogen, oxygen, and argon) in seawater are defined as their equilibrium concentrations with respect to standard atmospheric composition including equilibrium water vapor at the given temperature at 1 atm (101325 Pa) total pressure. At 20 °C the solubilities of nitrogen, oxygen, and argon in seawater of salinity 35 are 414, 225, and 11 µmol per kg of seawater, respectively [*Weiss*, 1970]. At a given temperature, the solubilities of these gases in seawater of salinity 35 are 20-30% less than their solubilities in pure water, varying by only around 3% over the salinity range 33 to 37 [*Weiss*, 1970]. Each of these solubilities exhibits roughly the same temperature dependence, decreasing by a factor of approximately two as the temperature increases from $-2\,°C$ to $30\,°C$ at a given salinity [*Weiss*, 1970].

The saturations of nitrogen, oxygen, and argon in seawater are defined as the ratios of their concentrations to their solubilities at the temperature of the seawater. For the vast majority of the ocean surface waters the saturations of these gases are near 100%. Saturation anomalies (deviations from 100%) of these gases in surface waters due to deviations of barometric pressure from 1 atm, heating of surface waters, upwelling, mixing of water masses, or other effects such as those from sea ice or subsurface melting of glacial ice are generally negligible, but the injection of bubbles by wave action can modify the saturation of the ocean surface waters with respect to each of these gases. A bubble exchanges mass with the surrounding water at a rate that depends on the difference in the partial pressures of the gases inside the bubble and the saturation pressures of these gases in the surrounding water. Because of the increase in pressure inside the bubble due to hydrostatic pressure (10% for every 1 m depth) and surface tension (3% for a bubble with radius $R_{bub} = 50\ \mu m$, varying inversely as R_{bub}), a bubble can dissolve even in water that is supersaturated, and the ocean surface waters are often somewhat supersaturated with respect to these gases (that is, saturations are typically greater than 100%). However, based on measurements and on results from models[1] supersaturations are typically only a few percent, and rarely more than 10%.

Biological activity can affect the oxygen saturation but it has virtually no effect on the saturations of nitrogen and argon. Oxygen saturation anomalies of up to 50% or more have been observed in the oceans [*Ramsey*, 1962a,b; *Simpson*, 1984], but these situations are rare and local in character, and for most of the ocean surface waters the oxygen saturation anomaly is less than 10% [*Najjar and Keeling*, 1997]. Because the equilibrium concentration of dissolved nitrogen in seawater is roughly twice that of oxygen, the change in total gas saturation due to biological activity is approximately a factor of three lower than the change in oxygen saturation (the argon concentration is too small to appreciably affect this calculation). For example, even a relatively large oxygen saturation anomaly of 15% (i.e., an oxygen saturation of 115%) corresponds to a total gas saturation anomaly of only 5%. Thus it can be concluded that in most situations likely to be encountered in the ocean, the saturation anomalies are quite small—a few percent at most. The effects of the saturation anomalies on SSA production are examined in §4.3.4.3.

Because of the small variations in the density, surface tension, and kinematic viscosity of seawater and in its solubility with respect to any of the major atmospheric gases over the range of salinities normally encountered in ocean surface waters (approximately 0.3%, 0.1%, 0.7%, and 3%, respectively, at a given temperature), it can confidently be concluded that variations of salinity in the ocean can account for no substantial differences in factors and processes that affect oceanic SSA production, such as the frequency and extent of wave breaking, the occurrence and lifetimes of whitecaps, the formation and dynamics of bubbles, and the formation of drops. This conclusion applies also to locations where the salinity differs appreciably from 35, e.g., the Red Sea, Mediterranean, Hudson Bay, and Black Sea (although the difference in the vapor pressure of water in equilibrium with seawater does affect the relation between the formation radius and r_{80} for these locations; §2.5.3). Furthermore, the differences in the densities, surface tensions, viscosities, and solubilities between pure water and seawater are so small that it is unlikely that they contribute to differences in behavior of drops, bubbles, or whitecaps between these two media. Differences in bubble spectra and coalescence properties between freshwater and seawater that have been reported are attributed to differences in the electrolytic natures of these media (and not to differences in density, surface tension, or viscosity; §4.4.1); such differences limit the ability of experiments using freshwater to accurately simulate oceanic processes pertinent to SSA production. Surface-active materials can appreciably lower the surface tension, and are probably more important in affecting SSA production and behavior than temperature or salinity, but the effects of these substances are extremely variable and are difficult to predict and to quantify under oceanic conditions (§2.3.4).

2.5.2. Composition of SSA Particles

The assumption is made throughout this review that the only quantity required to characterize an SSA particle and its behavior for a given set of conditions is its value of r_{80} (§2.1.1). This assumption greatly simplifies description of SSA particles and their behavior. It also implies that properties of these particles are independent of their age (time since formation). It is further assumed that the composition of SSA particles, at least with respect to the major dissolved components (Table 6), is the same as that of bulk seawater, permitting properties of these particles to be inferred from those of seawater (§2.1.5.3; §2.5.1; §2.5.3). However,

---

[1] *Wyman et al.* [1952]; *Blanchard and Woodcock* [1957]; *Benson and Parker* [1961]; *Kanwisher* [1963]; *Atkinson* [1973]; *Thorpe* [1982]; *Merlivat and Memery* [1983]; *Thorpe* [1984a]; *Memery and Merlivat* [1985]; *Craig and Hayward* [1987]; *Musgrave et al.* [1988]; *Spitzer and Jenkins* [1989]; *Woolf and Thorpe* [1991]; *Wallace and Wirick* [1992]; *Farmer et al.* [1993]; *Woolf* [1993]; *Keeling* [1993]; *Emerson et al.* [1995]; *Schudlich and Emerson* [1996]; *Asher et al.* [1996]; *Najjar and Keeling* [1997].

fractionation—a change in the concentration of one substance relative to another (usually taken to be sodium for SSA particles [*Duce et al.*, 1972a]), may occur during formation of SSA particles or during their residence in the atmosphere. Fractionation results in enrichment, defined to be amount by which the ratio of the concentration of a substance to the concentration of sodium in the particle divided by the ratio of these concentrations in seawater is greater than unity [*Duce et al.*, 1972a]; thus enrichment can be positive or negative. As properties of an aerosol particle depend on its composition, enrichment of an SSA particle can affect its hygroscopic growth (§2.5.4), phase transitions (§2.5.5), and optical (§2.1.5.3) and other properties compared to those of a pure SSA particle, requiring a more complex means of description of SSA than that employed here and affecting measurements that infer sizes of SSA particles based on their physical (including optical) properties. Hence it is necessary to examine the composition of SSA particles in the atmosphere to ascertain the extent to which these compositions differ from that of bulk seawater and to examine the consequences of any such differences, to assess the validity of the assumption that an SSA particle can be characterized solely by its value of r_{80}.

Early investigations, based on the composition of rain water or marine aerosol particulate matter, concluded that SSA particles were enriched (compared to sodium) with respect to some of the major components of seawater. Experimental results appeared to demonstrate that this enrichment occurred during formation by bubble bursting [*Köhler and Båth*, 1953; *Sugawara*, 1959; *Bloch et al.*, 1966; *Bloch and Luecke*, 1972; *Chesselet et al.*, 1972; *Wilkness and Bressan*, 1972; *Glass and Matteson*, 1973; *Morelli et al.*, 1974], but concern has been raised that the apparent fractionation resulted because the experiments did not accurately simulate the bubble-formation and bursting processes in the oceans. Other investigations [*Hoffman and Duce*, 1972; *Paterson and Scorer*, 1975; *Hoffman and Duce*, 1977a; *Hoffman et al.*, 1977; *Gravenhorst*, 1978; *Hoffman et al.*, 1980; *Savoie and Prospero*, 1980; *Weisel et al.*, 1984; *Keene et al.*, 1986] and reviews by *Junge* [1972], *MacIntyre* [1974b], and *Duce and Hoffman* [1976] convincingly concluded that there is little, if any, fractionation (and hence little enrichment) of the major components of seawater during the bubble-bursting process. These reviews attributed the earlier conclusions to poor experimental techniques, failure

to consider non-SSA (especially crustal) sources of some components, exchange of gases, and chemical reactions occurring on and in SSA particles, particularly those involving Cl^- and SO_4^{2-}. Enrichment of sulfate (referenced to sodium) of 10-30% in SSA particles was reported by *Garland* [1981] to have occurred during bubble bursting in laboratory experiments, but this result was challenged by *Duce et al.* [1982; see also *Garland*, 1982] and by *Keene et al.* [1986]. Additionally, no significant enrichment of sulfate (or of calcium or magnesium) was reported in precipitation near the coast of Norway by *Skarveit* [1982]. Although enrichment of sulfate in SSA particles can result from uptake from the atmosphere, as discussed below, little enrichment is expected during drop formation at the sea surface.

In contrast to the situation for the major ions, it has been established that fractionation can occur during bubble bursting for minor constituents such as iodide [*Seto and Duce*, 1972; *Korzh*, 1984], potassium [*Morelli et al.*, 1974], and phosphate [*Sutcliffe et al.*, 1963; *MacIntyre and Winchester*, 1969; *MacIntyre*, 1970; *Graham et al.*, 1979]. Enrichment can be extreme for trace metals (including radioactive elements), with reported enrichments up to 10^4 [*Van Grieken et al.*, 1974; *Cattell and Scott*, 1978; *Piotrowicz et al.*, 1979; *Martin et al.*, 1981; *Pattenden et al.*, 1981; *Belot et al.*, 1982; *Cambray and Eakins*, 1982; *Weisel et al.*, 1984; *Walker et al.*, 1986]. Fractionation also occurs for particulate matter, bacteria and viruses, and organic substances (see references in §4.2). This fractionation is due partly to the difference between the composition of bulk seawater and that of the thin layer of the ocean surface, termed the sea surface microlayer [reviews of various aspects of this layer are presented by *Parker and Barsom*, 1970; *MacIntyre*, 1974a,b; *Baier et al.*, 1974; *Liss*, 1975; *Wangersky*, 1976; *Lion and Leckie*, 1981; *Hardy*, 1982]. Several investigations have demonstrated that concentrations of trace elements in this layer differ greatly from those of bulk seawater [*Duce et al.*, 1972b; *Barker and Zeitlin*, 1972; *Szekielda et al.*, 1972; *Hunter*, 1980]; some of these differences may be due to atmospheric deposition of trace substances, but they are probably caused mostly by the accumulation and transport of substances to the sea surface by rising bubbles. During their ascent bubbles scavenge surface-active materials, in the process incorporating bacteria and viruses, particulate organic matter, iodine, potassium, and inorganic phosphate[1]. These substances accumulate at the surface in concentrations

---

[1] *McBain and DuBois* [1929]; *Baylor et al.* [1962]; *Blanchard* [1963]; *Sutcliffe et al.* [1963]; *Carlucci and Williams* [1965]; *Jarvis* [1967]; *MacIntyre and Winchester* [1969]; *MacIntyre* [1970]; *Wallace et al.* [1972]; *MacIntyre* [1972]; *Lemlich* [1972]; *Seto and Duce* [1972]; *Morelli et al.* [1974]; *Blanchard and Syzdek* [1974a,b]; *Wallace and Duce* [1975]; *Quinn et al.* [1975]; *Hoffman and Duce* [1976]; *Baylor et al.* [1977b]; *Wallace and Duce* [1978]; *Hunter* [1980]; *Blanchard et al.* [1981]; *Gershey* [1983a,b]; *Weber et al.* [1983]; *Blanchard* [1989b]; *Cloke et al.* [1991]; *Tseng et al.* [1992]; *Skop et al.* [1994]; *Stefan and Szeri* [1999]; *Cincinelli et al.* [1999]; *Saint-Louis et al.* [2004].

far exceeding those in bulk seawater. As bubbles do not directly rise to the surface but are transported by turbulence and wave motion, their scavenging length may be substantially greater than their maximum depth, depending on their size, the amount of turbulence in the ocean near the surface, and other factors.

As suggested early on [*Rossby*, 1959; *Wilson*, 1959a; *Eriksson*, 1959, 1960; *Oddie*, 1960], fractionation during bubble bursting at the sea surface is due to the presence of organic substances in the sea surface microlayer. As an SSA particle (i.e., a seawater drop) is composed of water and other substances from this layer in amounts that depend on the size and type of the drop (i.e., film or jet; §2.2), the amount of fractionation likewise depends on drop size and type, in addition to the amount and nature of the surface-active materials in the sea surface microlayer. Despite this recognition, quantitative characterization of the amount of fractionation is extremely difficult, and drops of the same size and type may experience different fractionation because of other factors such as the time after the breaking of the wave from which they are ultimately formed. Immediately following the breaking of a wave, large number of bubbles that are injected into the ocean rise to the surface, scavenging surface-active material during their ascent. Depending on the concentration of these bubbles and the amount of surface-active material in the ocean surface waters, much of the surface-active material in the uppermost meter or so of the ocean may be removed by the bubbles that reach the surface first. Furthermore, the ocean surface near the center of the breaking wave, which is transported radially outward by the upwelling of a large number of bubbles, may be depleted of surface-active substances. In such a situation, although fractionation of minor components might occur during bursting of the first bubbles that arrive at the surface (because of the organic substances adhering to them), later bubbles would be relatively free of surface-active materials and would burst on a relatively clean surface; thus the composition of the drops formed by these bursts would more closely reflect that of bulk seawater.

Laboratory experiments by *Seto and Duce* [1972] and *Korzh* [1984] demonstrated a substantial enrichment of organically-bound iodine for all particle sizes, but no enrichment of inorganic iodine, during bubble bursting in seawater. A general trend of the highest enrichment on the smallest particles was noted by *Seto and Duce* [1972], although this was not observed by *Korzh* [1984]. Iodine in marine aerosol particles has been shown to be correlated with the presence of organic material [*Murphy et al.*, 1997]. Enrichment of

potassium relative to sodium in precipitation at the Norwegian coast was reported by *Skartveit* [1978], who attributed it to attachment to organic substances, concluding that this enrichment reflected the high biological production in the North Sea and the Norwegian Sea.

The ratio of iodine to chlorine in SSA particles produced in the surf zone reported by *Duce and Woodcock* [1971] was up to two orders of magnitude greater than the ratio in seawater, increasing greatly with decreasing particle size, with the greatest values occurring for SSA particles with $r_{80} < 1$ μm. The ratio of iodine to chlorine in SSA particles produced over the open ocean was greater than that from SSA particles produced in the surf zone, up to three orders of magnitude greater than the seawater ratio, with similar increase with decreasing particle size. The total iodine content in SSA particles produced in the surf zone was roughly twice as great as that in SSA particles produced over the open ocean. In contrast, the ratio of bromine to chlorine was either the same as the seawater ratio or somewhat lower, with no apparent trend with particle size, for both SSA particles produced over the open ocean and those produced in the surf zone. These investigators hypothesized that several factors may have played a role in these results: the difference in formation mechanisms (bubble bursting vs. mechanical tearing of waves striking the rocky coast), organic films of surface-active materials of both natural and anthropogenic origin, and differences in the chemical composition of near-surface waters because of river runoff, ship operations, and the like.

The composition of SSA particles can be modified after formation during their residence in the atmosphere by exchange of and reaction with other gases and particles [*Martens et al.*, 1973; *Bonsang et al.*, 1980; *Andreae et al.*, 1986; *McInnes et al.*, 1994; *Anderson et al.*, 1996; *Buseck and Pósfai*, 1999; *Pryor et al.*, 1999; *Zhang and Iwasaka*, 2001; *Sellegri et al.*, 2001; *Tervahattu et al.*, 2002] as discussed in §2.1.5.2. The increase in sulfate content and the decrease in chloride content of sea salt particles with time in the atmosphere is shown clearly by *Buseck and Pósfai* [1999, Fig. 6], for example [and also in *Mouri et al.*, 1995; *Mouri et al.*, 1999], and the anticorrelation between the Cl:Na ratio and the excess S (the amount in excess of what would derive from sea salt) in aerosol samples collected at a coastal site is illustrated in *Chabas and Lefèvre* [2000, Fig. 5]. In some instances complete chloride loss from individual particles has been reported [e.g., *Pakkanen*, 1996; *Roth and Okada*, 1998; *Kerminen et al.*, 1998]. These effects depend on particle size and are more pronounced for smaller particles[1], for

---

[1] *Martens et al.* [1973]; *Sadasivan* [1978]; *Bonsang et al.* [1980]; *Khemani et al.* [1985]; *Raemdonck et al.* [1986]; *Ikegami et al.* [1994]; *Pakkanen* [1996]; *Pio and Lopes* [1998]; *Kerminen et al.* [1998]; *Roth and Odada* [1998]; *Kerminen et al.* [2000]; *Caffrey et al.* [2001]; *Sellegri et al.* [2001]; *Yao et al.* [2003].

which uptake of gaseous or particulate substances from the atmosphere is favored by their longer atmospheric residence times and higher radius-to-volume or surface-area-to-volume ratios (§2.1.5.2) compared to larger SSA particles, resulting in relatively greater uptake of gaseous or particulate substances per unit volume. The possibility of metamorphosis of a pure SSA particle into one that has been transformed by clouds and chemical reactions results in ambiguity in the definition of what constitutes an SSA particle, especially for small SSA particles.

The presence of organic substances associated with SSA particles has been reported by several investigators, with the proportion of the mass of organic substances varying greatly and depending also on particle size [*Blanchard*, 1964; *Barger and Garrett*, 1970; *Woodcock et al.*, 1971; *Hoffman and Duce*, 1974; *Pueschel and van Valin*, 1974; *Hoffman and Duce*, 1977b; *Lowe et al.*, 1996; *Middlebrook et al.*, 1998; *Uematsu et al.*, 2001; *Tervahattu et al.*, 2002]. Analysis of individual particles in clean marine conditions at Cape Grim, Tasmania [*Middlebrook et al.*, 1998] revealed the presence of organic species in more than half of the SSA particles, consisting on average of an estimated 10% of the sea salt mass. These organic species appeared to have two distinct sources, depending on the extent of particle age: incorporation during drop formation and scavenging of organic substances (both particulate and gaseous) during the atmospheric residence of the SSA particles. Likewise, based on the similar temporal behavior and the high correlation between the mass concentration of organic aerosols and mass concentrations of SSA particles in various size ranges, *Uematsu et al.* [2001] concluded that organic substances were associated with SSA particles and that this association probably occurred during bubble bursting because of organic enrichment of the sea surface microlayer. These investigators reported measured mass concentrations of organic carbon that were roughly 40% of those of SSA mass for $r_{80} \lesssim 1$ μm.

The presence of surface-active material in SSA particles, whether incorporated at the time of formation or during their residence in the atmosphere, can have a pronounced effect on their hygroscopic behavior, as noted above and in §2.3.4. As a result, uncertainty may arise in determinations of SSA particle size by methods that rely on physical size or other properties related to hygroscopic properties, and conversions of geometric radii from those at ambient RH to a given reference value may be in error. Because surface-active materials are ubiquitous in the atmosphere and in the ocean, they have the potential to exert large influences on SSA and its properties, although the extent of these influences is difficult to evaluate. It is implicitly assumed here and by most investigators that these influences are small, and thus that surface-active materials do not appreciably alter the behavior of the majority of SSA particles. It might be argued that this assumption is questionable, although it appears necessary given the lack of knowledge concerning these factors, and without such an assumption the description of SSA would encounter enormous complications.

As no appreciable fractionation of the major components of seawater occurs during drop formation, the composition of all drops immediately subsequent to formation is for the most part very nearly the same as that of bulk seawater with respect to these major components. Properties of SSA particles can thus be determined from those of concentrated seawater and are, at least for some considerable time, independent of the age of the particles. This justifies the assumption made throughout this review that the only quantity required to characterize an SSA particle and its behavior for a given set of conditions is its value of r_{80} (§2.1.1).

2.5.3. Equilibrium Drop Behavior With Relative Humidity

Because sea salt is hygroscopic (§2.1.1), a seawater drop with a given mass of solute, by exchanging water with the atmosphere, changes its equilibrium size, mass, water content, solute concentration, and pH, influencing its role in chemical reactions and the rate at which it uptakes gases, its light-scattering properties, and its kinematic behavior. For a given mass of solute, the equilibrium radius (and thus the mass, water content, solute concentration, and pH) of an SSA particle depends primarily on RH and is nearly independent of temperature [*Arons and Kientzler*, 1954; *Salhotra et al.*, 1987]. Henceforth in this review RH is considered the only quantity affecting this equilibrium radius. Knowledge of the dependencies of density (Fig. 4) and of equilibrium water vapor pressure of seawater solutions on concentration [*Tang et al.*, 1997] permits calculation of the equilibrium radius of a drop containing a specified mass of sea salt as a function of RH.

The equilibrium vapor pressure of water above a solution drop is less than that of bulk water at the same temperature by an amount that depends on the concentration of the solute and which for small concentrations is nearly proportional to this concentration [*Raoult*, 1887]. As the mole fraction of water in seawater of salinity 35 is very nearly 0.98 (Table 6), the vapor pressure of water in equilibrium with seawater of salinity 35 is therefore expected to be 98% of the vapor pressure of water at the same temperature; experimentally this has been found to hold to good accuracy [*Robinson*, 1954]. Thus, at formation, a drop of seawater of salinity 35 ejected into the atmosphere has a water vapor pressure that corresponds to 98% RH in air at the temperature of the drop. As this value of RH varies by only about 0.1% over the range of salinities 33 to 37, 98% can be considered as the equilibrium

RH of a newly formed SSA particle over the majority of the world's oceans with little loss in accuracy (this is discussed further below). The ratio of the radius of a seawater drop at formation r_{98} to the equivalent dry radius r_{dry} (Eq. 2.1-1) can be evaluated as $r_{98}/r_{dry} = [(\rho_{ss}/\rho_{sw})(m_{98}/m_{dry})]^{1/3}$, where ρ_{ss}, the density of sea salt, is approximately 2.2 g cm^{-3} (§2.5.1), ρ_{sw}, the density of seawater of salinity 35, is near 1.0 g cm^{-3} (§2.5.1), and the ratio of the mass of a seawater drop at 98% RH to its dry mass m_{98}/m_{dry} is equal to 1000/35 (from the definition of salinity; §2.5.1); thus $r_{98}/r_{dry} \approx 4.0$.

Knowledge of the dependence of the radius of a drop with a given mass of solute on relative humidity permits scaling to a reference RH. This reference RH has been taken by several investigators to be 80% [e.g., *Toba*, 1965a; *Davidson et al.*, 1982; *Fairall et al.*, 1982, 1983; *Monahan et al.*, 1983a; *Fairall et al.*, 1984; *Burk*, 1984], typical of the marine boundary layer (§2.4). As the use of r_{80} (the equilibrium radius at 80% RH; §2.1.1) not only provides an unambiguous means of describing the mass of solute in an SSA particle but also represents a physically relevant size (§2.1.1), this convention is adopted here and used throughout this review. At 80% RH the mass fraction of sea salt in a seawater drop is about 23% (i.e., salinity 230), the density is near 1.2 g cm^{-3} (Fig. 4), and the ratio r_{80}/r_{dry} is nearly equal to 2.0 [*Tang et al.*, 1997]. (The relation $r_{80} = 2r_{dry}$ is used in this review for particles other than sea salt that are measured dry to allow comparison with SSA particles; e.g., §4.1.1.1). Thus, for a drop of seawater of salinity 35, the radius at formation is twice that at 80% RH, which in turn is twice the equivalent dry radius (§2.1.1); $r_{98}:r_{80}:r_{dry} \approx 4:2:1$. Likewise, the ratios of the surface areas of a drop at formation (98% RH), the same drop at 80% RH, and that computed from the equivalent dry radius under the assumption that the dry particle is spherical are approximately 16:4:1 and the ratios of the volumes are approximately 64:8:1, all from geometry. In contrast to the above properties, the masses are roughly in ratio 30:4.5:1 as a consequence of the changes in density with RH. Thus for a seawater drop of salinity 35 at formation (98% RH) the mass of water in the drop exceeds that of the sea salt by a factor of ~29, and at 80% RH the mass of water remaining in the drop is only about 10% of that at formation. The mass of solute m_{dry} (i.e., the dry sea salt mass) in a seawater drop of given r_{80} is given approximately by

$$\frac{m_{dry}}{10^{-12}\,g} \approx 1.15\left(\frac{r_{80}}{\mu m}\right)^3. \qquad (2.5\text{-}1)$$

This expression can be inverted to yield

$$\frac{r_{80}}{\mu m} \approx 0.95\left(\frac{m_{dry}}{10^{-12}\,g}\right)^{\frac{1}{3}}. \qquad (2.5\text{-}2)$$

Thus a drop with $r_{80} = 1$ μm contains just over 10^{-12} g of sea salt. The SSA dry mass concentration M and the concentration of SSA volume at 80% RH, V_{80}, are related by $M = \rho_{ss}(r_{dry}/r_{80})^3 V_{80}$ (§2.1.4); thus

$$\frac{M}{\mu g\,m^{-3}} \approx 0.28\,\frac{V_{80}}{\mu m^3\,cm^{-3}} \qquad (2.5\text{-}3)$$

Because of the variety of different methods of describing particle size that are in common use, great ambiguity results if the RH to which the size pertains is not specified, and of course whether "size" refers to radius or diameter. Classification of SSA particles as "submicrometer" or "supermicrometer" and statements such as "the size of the particle is 1 μm" are likewise ambiguous, as the values of the dry radius and the diameter at formation that correspond to a given r_{80} span nearly an order of magnitude.

The dependence of the equilibrium radius of an SSA particle with a given solute mass on RH is described by the ratio $r/r_{80} = \zeta(rh)$ (§2.1.1), shown in Fig. 7 as a function of RH for both SSA and NaCl particles. On a mass fraction basis (but not on a mole fraction basis), the growth of a sea salt drop for RH above 70% is very close to that of a drop consisting of NaCl, as would be expected, because Na and Cl comprise the vast majority of the mass of sea salt (Table 6). The value of the equivalent dry radius r_{dry} (§2.1.1) is seen to be very nearly one half that of r_{80}, and the value of the radius at 98% RH, r_{98}, almost exactly twice that of r_{80}, as discussed above. The radius of a seawater drop varies slowly with RH between ~70% and ~90% RH; r_{70}/r_{80} and r_{90}/r_{80} are approximately 0.91 and 1.19, respectively. A large number of expressions have been proposed to describe the dependence of the ratio r/r_{80} on RH for SSA (or NaCl or marine aerosol) particles[1], with varying degrees of accuracy and ranges of validity. The expressions $\zeta(rh) = 0.54/(1-rh)^{1/3}$ for RH greater than 93% (i.e., $rh > 0.93$) and $\zeta(rh) = 0.67/(1-rh)^{1/4}$ for RH less than 93% yield results that agree to within ~1% with measurements of *Tang et al.* [1997]. The expression $\zeta(rh) = 0.54[1 + 1/(1-rh)]^{1/3}$ (similar to ones proposed earlier by *Fitzgerald* [1978] and *Hughes* [1987]) agrees with the measurements of *Tang et al.* [1997] to within ~2.5% for RH greater than or equal to 50%.

---

[1] *Keith and Arons* [1954]; *Kasten* [1969]; *Barnhardt and Streete* [1970]; *Fitzgerald* [1975]; *Wells et al.* [1977]; *Fitzgerald* [1978]; *Hughes and Richter* [1980]; *Davidson et al.* [1984]; *Gerber* [1985]; *Hughes* [1987]; *Gathman* [1989]; *Fitzgerald et al.* [1998a].

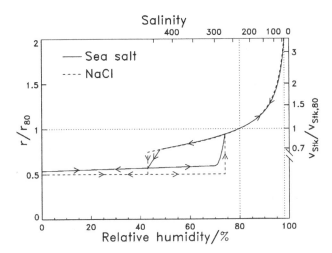

Figure 7. Dependence of the equilibrium radii and Stokes velocity of SSA and NaCl particles on relative humidity RH, after *Tang et al.* [1997]. The ratio r/r_{80} denotes the ratio of r, the equilibrium radius at RH, to r_{80}, the equilibrium radius at 80% RH; Kelvin effects are neglected. The ratio $v_{Stk}/v_{Stk,80}$ (right axis) denotes the similar ratio for the gravitational sedimentation velocity of SSA particles that are liquid solution drops, calculated under the Stokes approximation, (2.6-7) with the Cunningham slip factor set to unity (§2.6.2). Phase transitions from dry particle to solution (deliquescence) and solution to dry particle (efflorescence) are also shown. As relative humidity is increased an initially dry particle remains near its dry radius (evaluated as the radius of a sphere containing the same mass of substance at the bulk density) until the deliquescence point is reached. The curve for sea salt near and below the deliquescence transition is somewhat schematic as values presented by *Tang et al.* were in terms of the mass ratio, and conversion to the radius ratio requires density, which has not been reported for dry SSA particles just below the deliquescence transition. As RH is reduced below the deliquescence point, the particle remains as a supersaturated (with respect to the solute) solution until the efflorescence point is reached. The efflorescence phase transition for SSA particles may be more distinct than that shown, as *Tang et al.* did not present measurements that would allow determination of this. In the case of sea salt some residual water remains in the particle at all relative humidities, as illustrated schematically by the fact that the radius of an SSA particle at 0% RH is slightly greater than r_{dry} ($= 0.5r_{80}$), the definition of which was based on the density of dry sea salt (i.e., containing no residual water). Upper axis shows the salinity of SSA particles corresponding to the given RH. Vertical dotted lines denote RH 80%, typical of the marine boundary layer and used as a reference RH throughout this review, and RH 98%, characteristic of the atmosphere in equilibrium with seawater of salinity 35.

The appreciable difference in salinity from 35 that occurs in surface waters over some regions of the ocean (§2.5.1) results in different amounts of sea salt mass in a drop of a given formation radius, different equilibrium relative

humidities for drops at formation (i.e., $r_{form} \neq r_{98}$), and thus different relations between the formation radii and r_{80} values; r_{80}/r_{dry} remains unchanged, of course. However, as r_{form}/r_{80} is inversely proportional to the cube root of the salinity, this ratio differs appreciably from 2 only for extreme situations—for salinity 40 and 38.5 (typical of surface waters of the Red Sea and the Mediterranean, respectively) the ratio is near 1.9; for salinity 25 (in the range of salinities of surface waters in Hudson Bay), near 2.2; and for salinity 17 (typical of surface waters of the Black Sea), approximately 2.5. In contrast, for salinity 7 (typical of surface waters of the Baltic Sea) the ratio is around 3.4. The equilibrium RH values for drops at formation for these salinities are near 97.7%, 97.8%, 98.6%, 99.0%, and 99.6%, respectively. As r_{form}/r_{80} is near 2 over a wide range of salinities, this value, and the relation $r_{form} = r_{98}$, are assumed throughout this review, although for considerations of SSA production in the Baltic Sea the inaccuracy of these assumptions may lead to error. Of course, for drops formed from freshwater the quantity r_{80} is not defined, and for drops formed from solutions other than seawater the equilibrium radius at 80% RH may not be equal to $0.5r_{form}$. For comparison of results of laboratory studies using such liquids with those using seawater, the sizes of drops formed from such liquids (§4.3) are described by $0.5r_{form}$, consistent with use of r_{80} for the description of sizes of seawater drops.

The equilibrium vapor pressure of water in equilibrium with a liquid drop, on account of its curvature, is greater than that above a plane surface of the same solution by an amount that increases with decreasing radius. This effect, known as the Kelvin effect [*Thompson*, 1871], results in the radius of a drop with a given mass of solute, for a given RH, being less than that calculated without taking this effect into account. Consequently, the ratio r/r_{80} depends on r_{80} in addition to RH. This effect is important in situations when RH approaches 100% (i.e., fog and cloud formation; §2.1.5.1), but for RH less than or equal to 98% it is not important except possibly for very small drops. For an SSA particle with $r_{80} = 0.01$ μm, the ratio $r_{98}/r_{80} \approx 1.5$, but this ratio rapidly approaches 2 as drop size increases; for $r_{80} = 0.03$ μm and 0.05 μm, the ratio r_{98}/r_{80} is approximately equal to 1.8 and 1.9, respectively. Neglect of the Kelvin effect results in an overestimation of the radius of a SSA particle by an amount that is approximately equal to (0.0004 μm)/(1−rh), independent of the radius. For a SSA drop with $r_{80} = 1$ μm, this amount corresponds to an overestimation of the radius of only about 0.2% at 80% RH and 1% at 98% RH (for which the radius has increased to 2 μm). Even for an SSA drop with $r_{80} = 0.01$ μm neglect of the Kelvin effect results in overestimation of the radius by only a factor of 1.2 at 80% RH. Another way to look at the Kelvin effect is that it increases the equilibrium RH above the drop by 0.1% for a seawater drop with radius 1 μm at 80% RH, and 0.06% for this same drop at 98% RH (formation), compared to a seawater solution of the same concentration with a flat surface; for drops with greater solute mass the effect is less. Thus under most circumstances other than those involving fog or cloud formation, to good approximation the Kelvin effect for SSA particles can be neglected (as is done hereinafter in this review), with the result that r/r_{80}, where r is the radius of the SSA particle in equilibrium at a given RH, is independent of r_{80} and can be determined as a function only of RH; i.e., $r/r_{80} = \zeta(rh)$ (§2.1.1).

2.5.4. Non-Equilibrium Drop Behavior With Relative Humidity

Seawater drops that are not in equilibrium with respect to RH exchange water substance with the surrounding air (discussions pertinent to this topic are presented also in *Fuchs* [1959] and *Sedunov* [1974]). The driving force for this exchange is the difference between the concentration of water vapor in the surrounding air, which depends on the RH and temperature of the surrounding air, and the concentration of water vapor in air in equilibrium with the aqueous solution in the drop, which depends on the salt content, size, and temperature of the drop. Key determinants that control the rate of this exchange are the transfer of mass and heat in the gas phase (described by the molecular diffusivity of water vapor in air and the thermal conductivity of air, respectively) and kinetic transfer of water substance and heat at the surface of the drop (described by the mass and thermal accommodation coefficients, respectively), analogous to uptake of gases by seawater drops examined in §2.1.5.2. The relative motion of sufficiently large particles with respect to the surrounding atmosphere results in their rate of exchange of water substance being several times greater than that calculated without taking into account this motion, but as discussed below, this effect is generally not important for SSA drops with $r_{80} \lesssim 25$ μm.

The rate of water exchange (and hence the rate of approach to equilibrium) for a wide range of sizes of SSA particles (i.e., drops) is controlled by heat and mass transfer in the gas phase; the limit of validity of this condition is examined below. As an SSA drop loses (gains) water, the concentration of the salt in the drop increases (decreases), the vapor pressure at the surface of the drop decreases (increases), and the radius of the drop decreases (increases). At the same time, the drop exchanges heat with the surrounding air as it approaches equilibrium with respect to temperature. Along with the loss (gain) of water there is an associated loss (gain) of latent heat, which causes the temperature of the drop to be lower (higher) than that of the surrounding air (this latent heat effect is generally not important

for the uptake of gases other than water vapor and was omitted as unnecessary in §2.1.5.2 because of the low concentrations of these gases and consequently low latent heat deposition). For a seawater drop at 98% RH ejected into an atmosphere of lower RH, this temperature difference can initially be several degrees, independent of the radius of the drop, as readily shown following *Maxwell* [1877]. The fluxes of water substance and heat are coupled, but in nearly all situations of atmospheric interest simplifying assumptions can be made [*Howell*, 1949], resulting in a single equation describing the exchange of water in which the diffusion coefficient, or diffusivity, of water vapor in air D_w is replaced by an "effective" diffusivity $D_{w,eff}$ that accounts for the effects of latent heat and of thermal conduction of air in addition to the usual gaseous diffusion. This effective diffusivity is given by

$$D_{w,eff} = \left[\frac{1}{D_w} + \frac{L_w^2 M_w^2 p_{w,sat}}{k_a R^2 T^3} \right]^{-1}, \qquad (2.5\text{-}4)$$

where L_w is the latent heat of vaporization of water, $p_{w,sat}$ is the saturation vapor pressure of water, and k_a is the thermal conductivity of air, all evaluated at T, the absolute temperature of the surrounding air (M_w is the molecular weight of water and R is the gas constant). According to this approach the driving force for the exchange of water is proportional to the difference in RH of the surrounding air and the equilibrium RH of the drop based on its salt content and size, but evaluated at the temperature of the surrounding air. For an air temperature of 20 °C, $D_{w,eff} \approx 0.3 D_w$ and for an air temperature of 0 °C, $D_{w,eff} \approx 0.6 D_w$. For air temperatures greater than around 5 °C, $D_{w,eff} < 0.5 D_w$ (that is, the second term in the expression on the right hand side of Eq. 2.5-4 is greater than the first), implying that the effect of latent heat is more important than diffusion of water vapor in determining $D_{w,eff}$. This effect, which has been neglected by some investigators, amounts to a factor of 2 or more in the rate of exchange of water at temperatures greater than ~5 °C.

Of primary interest is the rate of change in radius that accompanies the exchange of water substance. The rate at which the radius of a seawater drop, initially in equilibrium at a given RH, approaches its new equilibrium radius when exposed to a different RH depends on its value of r_{80} and on the initial and final equilibrium values of RH; this rate also depends on temperature mainly because of the rather strong temperature dependence of the saturation vapor pressure of water (as described by the Clausius-Clareyron relation [*Clapeyron*, 1834; *Clausius*, 1850]), which increases from near 600 Pa (~0.006 atm) at 0 °C to more than 2300 Pa (~0.023 atm) at 20 °C. Over a wide range of initial and final

RH values, the rate of decrease of the radius of a seawater drop initially at one RH exposed to a considerably lower RH is approximately independent of time, and for 20 °C is given approximately by $(dr/dt)/(\mu m \ s^{-1}) \approx -[(60 \ \mu m)/r_{80}] \times \Delta rh$, where Δrh is the difference between the initial and final fractional relative humidites; similar relations hold for drops composed of other substances. This relation allows evaluation of rate of transfer of moisture and of latent heat from an SSA drop to the atmosphere (§2.1.7).

The rate of change of the radius of a seawater drop initially at 98% RH (i.e., formation) when suddenly exposed to a lower ambient RH is important to the dynamics of SSA particles in the marine atmosphere. The approach of the radius of a seawater drop to its equilibrium value at 80% RH (typical of the marine boundary layer; §2.4), r_{80}, is shown in Fig. 8 and Fig. 9. To good approximation, the fractional departure of the radius from its equilibrium value decreases nearly linearly with time until it approaches ~10%, i.e., until $r = 1.1r_{80}$ (the equilibrium RH for this size is near 87%), after which this rate of decrease is less (Figs. 8 and 9). A characteristic time $\tau_{98,80}$ (which depends on r_{80}) can be defined as the time at which this fractional departure equals

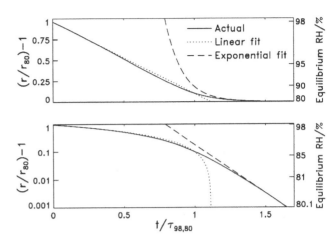

Figure 9. Time dependence of fractional departure of drop radius r from its equilibrium value at 80% RH, r_{80}, following abrupt decrease of RH from initial value of 98%, for which initial value of radius $r_{98} = 2r_{80}$, on linear (upper panel) and logarithmic (lower panel) scale. Time is shown as ratio to characteristic time $\tau_{98,80}$, given by $\tau_{98,80}/s \approx [r_{80}/(3.5 \ \mu m)]^2$ at 20 °C (Eq. 2.5-6). Dotted line shows linear approximation (Eq. 2.5-5); dashed line shows long-time exponential approximation. Right axis gives equilibrium RH of drop with value of r indicated on left axis.

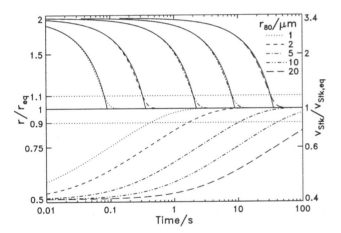

Figure 8. Time dependence of approach of the radius of an SSA particle to its new equilibrium radius following an abrupt change in relative humidity RH. Top panel shows the response of SSA particles with given r_{80} initially at 98% RH, after RH is instantaneously decreased to 80%, as the ratio of the time-dependent radius $r(t)$ to the equilibrium radius r_{80}; lower panel shows the response of SSA particles of given r_{80} initially at 80% RH, after RH is instantaneously increased to 98%, as the ratio $r(t)$ to the equilibrium radius r_{98}. Horizontal lines denote approach of the radius to within 10% of its equilibrium value. Solid lines in upper panel denote linear approximation, (2.5-5). Right axis shows the similar ratio for the gravitational sedimentation velocity under the Stokes approximation, (2.6-7). Calculations are for 20 °C.

10%. For $t \lesssim \tau_{98,80}$, the radius of such a drop is accurately described by

$$r(t) \approx 2r_{80}\left[1 - 0.45\left(\frac{t}{\tau_{98,80}}\right)\right] \qquad (2.5\text{-}5)$$

(Figs. 8 and 9). As the drop radius approaches its equilibrium radius, the fractional departure of the radius from its equilibrium value decreases exponentially with time [*Andreas*, 1992; *Andreas et al.*, 1995], but this dependence is accurate only when the radius is within ~1% of its equilibrium value and the difference in RH from its equilibrium value is also less than 1% (Figs. 8 and 9). Under these circumstances, the e-folding time describing the approach of the radius to its equilibrium value, applicable to $t \gtrsim 1.4\tau_{98,80}$, is approximately equal to $\tau_{98,80}/8$.

The characteristic time $\tau_{98,80}$ varies with particle size as

$$\frac{\tau_{98,80}}{s} \approx \left(\frac{r_{80}}{3.5 \ \mu m}\right)^2 \qquad (2.5\text{-}6)$$

at 20 °C; it is nearly twice as large at 0 °C. This time characterizes the rate at which the radius of a drop formed at the ocean surface approaches its equilibrium radius and thus identifies how long the physical (geometric) radius, which governs important properties such as the gravitational sedimentation velocity (§2.6.2), remains close to the formation radius versus being characterized by the equilibrium radius (§2.9.1). For comparison, $\tau_{98,80}$ is only about 10% greater than the time required for a drop of pure water (i.e., containing no dissolved substances) of the same formation radius to evaporate completely when exposed to 80% RH. Although it might be argued that the requirement that a drop attain a radius within 10% of its equilibrium value is too stringent, the time for a drop initially at 98% RH to attain a radius within 50% of its equilibrium value at 80% RH is only about a factor of 2 less than $\tau_{98,80}$ (Figs. 8 and 9). This characteristic time is not strongly dependent on the final RH; the characteristic time for a drop of a given r_{80}, initially at 98% RH, to attain a radius within 10% of its equilibrium value at a final RH of 90% is only about a factor of two greater than $\tau_{98,80}$, and that for a final RH of 70% is roughly 30% less than $\tau_{98,80}$. Thus it can be concluded that under most conditions small and medium SSA particles ($r_{80} \lesssim 25\ \mu m$) attain radii near their equilibrium values with respect to RH within a minute or so after formation (Table 2).

Similarly, a characteristic time for equilibration of a drop initially at 80% RH exposed to 98% RH, $\tau_{80,98}$, is defined as the time for the radius of such a drop to approach within 10% of its new equilibrium value. At time $\tau_{80,98}$, the radius is equal to $1.8r_{80}$ and the corresponding equilibrium RH is slightly greater than 97%. At 20 °C this characteristic time is given by

$$\frac{\tau_{80,98}}{s} \approx \left(\frac{r_{80}}{1.6\ \mu m}\right)^2, \qquad (2.5\text{-}7)$$

and at 0 °C it is also nearly twice as long. The approach of the ratio of the radius of a seawater drop to its equilibrium value for this situation is shown in Fig. 8 as a function of time.

The characteristic times defined by (2.5-6) and (2.5-7) are proportional to $(r^2/D_{w,\text{eff}})(\rho_w/\rho_{wv})$, where ρ_w is the density of liquid water and ρ_{wv} is the density of water vapor in air (that is, the mass of water vapor per volume of air at the ambient RH). The factor $r^2/D_{w,\text{eff}}$ describes the diffusion of water vapor through the atmosphere to the surface of the drop, and the ratio ρ_w/ρ_{wv}, near $6 \cdot 10^4$ in water-saturated air (i.e., 100% RH) at 20 °C and near $2 \cdot 10^5$ at 0 °C, describes the factor by which the volume of water vapor decreases when it condenses on the drop. For the relations given by (2.5-6) and (2.5-7) to be valid, the change in radius due to exchange of moisture with the surrounding air must be sufficiently slower than other processes that affect this exchange that these other processes need not be considered. These processes are (1) the response of the water vapor density field in the surrounding air, (2) the response of the temperature field in the surrounding air, (3) the equilibration of temperature within the drop to that of the surrounding air, and (4) the equilibration of solute concentration within the drop

[discussions of characteristic times for these and other processes are presented in *Sedunov*, 1974, Chap. 2]. Characteristic times for these processes for sufficiently small drops ($r \lesssim 50\ \mu m$), for which the principal mass transport processes are diffusional, are likewise proportional to the square of the drop radius divided by the pertinent diffusion coefficient. For these four processes, the respective diffusion coefficients are the diffusivity of water vapor in air, the thermal diffusivity of air, the thermal diffusivity of the seawater solution in the drop, and the diffusivity of solute (sea salt) in the drop. At 20 °C these characteristic times, in seconds, are approximately $[r/(8000\ \mu m)]^2$, $[r/(8000\ \mu m)]^2$, $[r/(600\ \mu m)]^2$, and $[r/(60\ \mu m)]^2$, respectively. There is some variation among definitions of the characteristic times for these processes because they are not described by pure exponential behavior; also, some investigators have given incorrect expressions or values (for example, the characteristic time of equilibration of the water vapor field given by *Pruppacher and Klett* [1997, p. 504], is too small by a factor of 10^2, and the equation for the characteristic time for the change of radius of a particle with respect to RH given by *Pilinis et al.* [1989] results in a value that is too small by a factor of nearly 10^6). These several characteristic times vary slightly with temperature; at 0 °C the characteristic time for the concentration of the solute to become uniform throughout the drop is nearly twice the value at 20 °C, whereas the characteristic times for equilibration of the water vapor field, for the response of the drop to the temperature of the surrounding air, and for the temperature inside the drop to become uniform, are only 10-15% greater than the values at 20 °C. Thus, as the characteristic times describing these other processes are considerably less than (and vary as the same power of the radius as) the characteristic time for the radius to become within 10% of its equilibrium value, it can be concluded that all seawater drops respond to the temperature of the surrounding air, and attain uniform temperature and solute concentration, sufficiently rapidly compared to the time for the radius to change by any appreciable extent that these other processes can be neglected and need not be considered further here (Table 2).

These considerations permit the conclusion that even large SSA drops, which do not remain long in the atmosphere and fall rapidly back to the sea surface under the influence of gravity, thermally equilibrate with respect to the temperature of the surrounding air during their brief atmospheric sojourn, thereby transferring much or most of their available sensible heat to the atmosphere (warm ocean, cold air), or extracting sensible heat from the atmosphere (cold ocean, warm air). In contrast, sufficiently large SSA particles, which retain radii near their formation values during their atmospheric residence (§2.9.1), transfer only some of their available latent heat to the atmosphere (§2.1.7).

For sufficiently small drops, the rate of growth of the radius is controlled by kinetic transfer at the surface of the drop rather than by molecular diffusion of water vapor and heat in the surrounding atmosphere, resulting in characteristic times for these drops being greater than those described by (2.5-6) and (2.5-7) by an amount that depends on α_M, the mass accommodation coefficient [*Langmuir*, 1913; *Marcelin*, 1914b] of water upon seawater—the fraction of water molecules impacting the surface of the drop that are exchanged across the interface, and on α_T, the thermal accommodation coefficient [*Knudsen*, 1911] of water upon seawater—the ratio of the mean energy transferred by molecules during collisions with the drop to the amount that would be transferred if the colliding molecules left thermally equilibrated at the temperature of the drop (a related discussion was presented in §2.1.5.2). It is therefore important to examine the consequences of these effects to determine the extent, if any, to which they are important for SSA particles in the atmosphere.

The relative importance of mass accommodation at the drop surface to gaseous and thermal diffusion through the atmosphere for the evaporational shrinkage or condensational growth of a drop with radius r is determined by the ratio r_M/r, where $r_M = 4D_{w,eff}/(\alpha_M c_w)$ and c_w is the mean speed of the water vapor molecules in air, given (§2.1.5.2) by $c_w = [8RT/(\pi M_w)]^{1/2}$. If $r_M \prec r$ ([1]) then the rate of transfer of moisture to the drop is controlled by diffusion of water vapor through the atmosphere, whereas if $r_M \succ r$ then the rate of transfer is limited by mass accommodation (kinetic transport) near the drop surface. At 20 °C, $r_M \approx (0.05\ \mu m)/\alpha_M$, and at 0 °C it is nearly twice this value. No measurements of the mass accommodation coefficient of water upon seawater have been reported, but it can reasonably be assumed to be near that of water upon water, also known as the evaporation coefficient or condensation coefficient, for which values ranging from less than 0.01 to 1 have been reported [compilations are presented in *Nabavian and Bromley*, 1963; *Mills and Seban*, 1967; *Cammenga*, 1980; *Wagner*, 1982; *Mozurkewich*, 1986; and *Pruppacher and Klett*, 1997, Table 5.4]. Older values are typically at the low end of this range whereas more recent determinations have resulted in values near unity [*Wagner and Pohl*, 1979; *Richardson et al.*, 1986; *Fung et al.*, 1987; *Rudolf et al.*, 2001; *Fladerer and Strey*, 2003]; several reviews have also concluded that α_M is near unity [*Mills and Seban*, 1967; *Cammenga*, 1980; *Wagner*, 1982; *Mozurkewich*, 1986]. Because of high latent heat of water, any experimental determination of mass transfer of water is inevitably also a problem in heat transfer, and the failure to account for lack of thermal equilibrium, either in experiment or interpretation, results in apparent low values of α_M. As numerical results for growth rates of

particles depend on the value chosen for α_M (especially for particles with $r \lesssim 1\ \mu m$), the growth rates described here differ from some in the literature; several investigations reported calculations using α_M near 0.04 [*Fukuta and Walter*, 1970; *Ferron*, 1977; *Andreas*, 1989, 1990a; *Ferron and Soderholm*, 1990; *Andreas*, 1995; *Kepert*, 1996; *Andreas*, 1996; *Pruppacher and Klett*, 1997, pp. 507, 511], from sources as far back as *Alty and MacKay* [1935], who concluded that $\alpha_M = 0.036$. For $\alpha_M = 0.036$, $r_M \approx 1.3\ \mu m$ at 20 °C, and thus mass accommodation will be a limiting factor for moisture transfer for particles with radii up to about 1 μm; the choice $\alpha_M = 1$ yields $r_M \approx 0.05\ \mu m$ at 20 °C. However, the time characterizing the rate of approach of the radius to its equilibrium value for a drop in this size range is so small—the value of $\tau_{98,80}$ from (2.5-6) is less than 0.1 s for $r_{80} < 1\ \mu m$—that such drops rapidly attain their equilibrium radii in nearly all situations, and thus these effects are unimportant for SSA particles in the atmosphere.

Analogous to the role of mass accommodation in mass transfer, thermal accommodation effectively reduces the thermal diffusivity and thermal conductivity of air and thus the effective diffusivity of water vapor in air, also slowing the rate of water exchange between a drop and the surrounding air. The relative importance of thermal accommodation at the drop surface to diffusion of water vapor and heat through the atmosphere for evaporational shrinkage or condensational growth is determined by the ratio r_T/r, where $r_T = [4\kappa_a/(\alpha_T c_a)] \times (1 - D_{w,eff}/D_w)$, where κ_a is the thermal diffusivity of air and c_a is the mean molecular speed of the air molecules, given (§2.1.5.2) by $c_a = [8RT/(\pi M_a)]^{1/2}$, M_a being the mean molecular weight of air. If $r_T \prec r$ then diffusion of water vapor and heat through the atmosphere is more important than kinetic transport of heat near the drop surface in limiting the rate of moisture transfer between the atmosphere and the drop, whereas if $r_T \succ r$ the opposite is true. At 20 °C, $r_T \approx (0.13\ \mu m)/\alpha_T$, and at 0 °C it is roughly half this value. The thermal accommodation coefficient α_T is thought to be near unity, a value supported by the few measurements that have been reported [*Alty and McKay*, 1935; *Sageev et al.*, 1986; *Shaw and Lamb*, 1999; *Li et al.*, 2001]. Thus, $r_T \approx 0.13\ \mu m$ at 20 °C and the effects of thermal accommodation are unimportant for all except possibly the smallest seawater drops, for which, again, the process is so rapid as to be unimportant relative to other processes involving these drops in the atmosphere.

Other factors that can affect the exchange of moisture of a drop are internal circulation, radiative cooling, and ventilation (increased transport due to the relative motion of the drop and the surrounding air), but the effects of these factors are minor for drops with $r_{80} \lesssim 25\ \mu m$ at RH not greater than

---

[1] The symbols $\prec$ and $\succ$, introduced in §2.1.5.2, denote "considerably less than" and "considerably greater than", respectively.

98% and thus are not considered here in relation to the effective SSA production flux. Although these factors affect the behavior of large SSA particles and become increasingly more pronounced as drop size increases, the atmospheric lifetimes of these particles also decrease rapidly with increasing drop size (§2.9.1), and these effects are not expected to contribute to the overall behavior of large drops to any appreciable extent during the short time they remain airborne before falling back into the sea; hence these effects also need not be considered in relation to the interfacial SSA production flux. Surface-active materials also can affect the rate of moisture exchange with the atmosphere (§2.3.4); by forming a film on the surface of a drop they can hinder mass transport by reducing α_M and increasing the characteristic times of equilibration. The importance of these effects to drop behavior in the atmosphere, and their dependencies on drop size, are not well known, but it is assumed that under most circumstances these effects will not appreciably alter the results obtained above.

From the above considerations some general conclusions can be drawn concerning the behavior of SSA particles with respect to changes in RH and the dependence of this behavior on particle size (Table 2). In the marine environment small SSA particles (1 μm $\lesssim r_{80}$) typically equilibrate to the ambient RH in much less than 1 s. Non-continuum effects (i.e., mass and thermal accommodation) are important if at all only for the smallest seawater drops, but equilibration of these drops is so rapid that these effects can be neglected for considerations of their behavior in the marine environment (although they might be important in measuring devices). Medium SSA particles (1 μm $\lesssim r_{80} \lesssim$ 25 μm), although they require some time ($\lesssim$ 1 min) to equilibrate with respect to the ambient RH, will typically have attained their equilibrium radii by the time of measurement at any height greater than a few meters above the sea surface (§2.9.1). Large SSA particles ($r_{80} \gtrsim$ 25 μm) require extremely long times ($\gtrsim$ 1 min) to attain their equilibrium radii—in many situations much greater than the time they spend in the atmosphere, and to good approximation these particles can be assumed to retain their formation radii during their atmospheric lifetimes (§2.9.1). Consequently, RH has little effect on the behavior of these particles during their residence in the atmosphere. Measured radii of these particles are unlikely to reflect their equilibrium radii, and the assumption that these particles have attained their equilibrium radii leads to incorrect inferences concerning their size distributions—a conclusion that has not been widely appreciated (§2.9.1).

2.5.5. Efflorescence and Deliquescence of SSA Particles

When a dry particle of a single salt such as NaCl is exposed to increasing RH it remains dry and experiences no change in size until at a critical RH the particle undergoes a phase transition known as deliquescence to a saturated solution drop. This deliquescence RH is near 75% for NaCl[1]; it increases very slightly with decreasing temperature [*Wexler and Seinfeld*, 1991; *Cziczo and Abbatt*, 2000] and increases with decreasing particle size [*Hämeri et al.*, 2001], although this latter dependence (probably due to the Kelvin effect; §2.5.3) is negligible for particles with $r_{80} > 0.1$ μm. With further increase in RH, the equilibrium radius of the drop increases continuously as water accretes on the particle, and the concentration of the solute decreases (Fig. 7; §2.5.3). There is some evidence of uptake of water below the deliquescence RH [*Dai et al.*, 1995; *Vogt and Finlayson-Pitts*, 1994; *Allen et al.*, 1996; *Neubauer et al.*, 1998; *Finlayson-Pitts and Hemminger*, 2000; *Gysel et al.*, 2002], possibly due to surface irregularities, but the effect of this uptake on particle size is negligible.

When a liquid drop consisting of an aqueous solution of a single solute that is a solid material in the dry state is exposed to decreasing RH, evaporation occurs, the radius decreases, and the concentration of the solute increases until the deliquescence RH is reached, at which the solution becomes saturated with respect to the solute. With further decrease in RH a drop that is not in contact with a solid surface can become supersaturated with respect to the solute and remain in a thermodynamically unstable state until eventually a phase transition known as efflorescence to a dry particle occurs abruptly by spontaneous homogeneous nucleation [*Orr et al.*, 1958a,b; *Junge*, 1958; *Martin*, 2000]. The mean lifetime of a drop in this unstable state, which is inversely proportional to the volume of the drop, may be quite long (of order weeks or more), depending on the temperature and RH. As the RH approaches a value known as the efflorescence RH, the transition rate increases greatly with decreasing RH [*Tang*, 1980; *Tang and Munkelwitz*, 1984; *Cohen et al.*, 1987c]. This efflorescence RH, unlike the deliquescence RH, cannot be predicted by equilibrium thermodynamics, as efflorescence is not an equilibrium phenomenon—the drop being in a thermodynamically unstable state when efflorescence occurs. The efflorescence RH for NaCl is near 45% at 20 °C [*Orr* et al., 1958a,b; *Junge*, 1958; *Tang et al.*, 1977; *Tang*, 1980; *Cohen et al.*, 1987a,c; *Richardson and Snyder*, 1994; *Cziczo et al.*, 1997; *Cziczo*

---

[1] *Owens* [1926]; *Twomey* [1953]; *Orr et al.* [1958a,b]; *Junge* [1958]; *Tang et al.* [1977]; *Cohen et al.* [1987a]; *Tang and Munkelwitz* [1993]; *Richardson and Snyder* [1994]; *Weingartner et al.* [1995]; *Cziczo et al.* [1997]; *Neubauer et al.* [1998]; *Cziczo and Abbatt* [2000]; *Lee and Hsu* [2000]; *Krämer et al.* [2000]; *Koop et al.* [2000]; *Hämeri et al.* [2001]; *Braun and Krieger* [2001]; *Krieger and Braun* [2001]; *Gysel et al.* [2002].

and Abbatt, 2000; *Lee and Hsu*, 2000; *Hämeri et al.*, 2001] and appears to increase slightly with increasing temperature [*Koop et al.*, 2000; *Cziczo and Abbatt*, 2000], consistent with the value of 37% at −10 °C reported by *Braun and Krieger* [2001]. The value of 25% RH for NaCl assumed by *Quinn and Coffman* [1998] is certainly incorrect. Bulk solutions and drops that are mechanically suspended or otherwise attached to solid surfaces typically do not remain in liquid form at RH values below the deliquescence RH, as the container or point of attachment provides a site for nucleation to commence.

For NaCl and many other salts there exists a range of RH between the efflorescence RH and the deliquescence RH over which a particle may exist either as a dry particle or as a liquid solution drop supersaturated with respect to the solute [*Dessens*, 1946c, 1949; *Orr et al.*, 1958a,b]. The state of the drop/particle thus depends on previous relative humidities it has encountered (shown for NaCl by the upper and lower curves in Fig. 7). Laboratory experiments involving single particles that were electrostatically suspended have demonstrated that an individual particle can be repeatedly cycled through this hysteresis loop [e.g., *Tang and Munkelwitz*, 1984; *Krieger and Braun*, 2001].

The processes that occur during evaporation of drops of a single-salt solution, and the final shapes and structures of the resulting particles, have been described by several studies [*Charlesworth and Marshall*, 1960; *Leong*, 1981; *Beard et al.*, 1983; *Leong*, 1987a,b; *Baumgartner et al.*, 1989; *Mitra et al.*, 1992; *Weis and Ewing*, 1999; *Braun and Krieger*, 2001]. Depending on the rate of evaporation (which in turn depends on RH and temperature), the total mass of the solute, and other factors, a variety of shapes and structures may result for a dried salt particle, including solid particles, multiple crystals, or hollow spherical shells sometimes containing holes (which result from the growth of the crystal and not from ejection of mass). Such hollow particles may have an effective density well below that of the bulk material; mean densities of hollow NaCl particles formed by rapidly drying solution drops reported from one laboratory study [*Baumgartner et al.*, 1989] decreased from 1.6 g cm^{-3} for dry radii near 0.2 μm to near 1.0 g cm^{-3} for dry radii of 1.5 μm. Additionally, several investigations have demonstrated that small amounts of water are trapped in dried NaCl particles even at RH values well below the efflorescence RH [*Cziczo et al.*, 1997; *Weis and Ewing*, 1999; *Cziczo and Abbatt*, 2000].

The foregoing picture for solutions of single salts applies to SSA particles, but with important differences. Sea salt particles can undergo both efflorescence and deliquescence, and although these transitions are not so sharp as for NaCl, they are still distinct (Fig. 7). Mixed salt solutions exhibit lower deliquescence (and efflorescence) RH values than do saturated solutions of any of the individual salts [*Prideaux*, 1920; *Owens*, 1926; *Tang*, 1976, 1980; *Wexler and Seinfeld*, 1991], but because sea salt is composed largely of NaCl (Table 6), the differences in the deliquescence and efflorescence relative humidities of sea salt and those of NaCl are small (Fig. 7). For sea salt, deliquescence occurs between 70% and 75% [*Owens*, 1926; *Twomey*, 1953, 1959; *Junge*, 1972; *Winkler*, 1988; *Tang et al.*, 1997; *Koop et al.*, 2000; *Lee and Hsu*, 2000], and efflorescence occurs between 40% and 45% [*Tang et al.*, 1997; *Koop et al.*, 2000]. The deliquescence RH of sea salt particles appears to be nearly independent of temperature, while the efflorescence RH increases slightly with increasing temperature [*Koop et al.*, 2000]; the dependence of these RH values on particle size (caused by the Kelvin effect; §2.5.3) has not been reported but is probably negligible in most situations. Sea salt aerosol particles take up water at RH values well below the deliquescence RH (as schematically illustrated in Fig. 7) and do not fully dry even at very low RH [*Hänel and Zankl*, 1979; *Tang et al.*, 1997; *Cziczo et al.*, 1997] unless subjected to temperatures of nearly 650 °C [*Morris and Riley*, 1964; *Winkler and Junge*, 1971], primarily because of the presence of substances that form stable hydrates, most importantly MgCl$_2$, which forms MgCl$_2$·6(H$_2$O). This water of hydration contributes roughly 15% of the mass of dry sea salt (based on the amount of Mg in seawater; Table 6), consistent with measurements of *Winkler and Junge* [1971], and its presence may reduce the density of a "dry" sea salt particle to near 2.0 g cm^{-3} (as calculated from Table 7 under the assumption of volume additivity, similarly to the value of ρ_{ss} presented in §2.5.1). However, dry sea salt particles cannot be considered as homogeneous drops of solution below their efflorescence RH; this can also be seen from their light-scattering behavior [*D. Imre, personal communication*, 1999]. The conclusion reached by *McInnes et al.* [1996] that deliquescence of sea salt occurs in the range 9-20% RH, based on observed water uptake of sea salt particles on filters, was probably an artifact of the experimental technique (that is, the use of filters or the uptake of water by particles that had been dried by heating); in any event, it does not apply to SSA particles in the atmosphere.

As with single-solute solutions, there is also a range of RH values between the efflorescence RH and the deliquescence RH over which an SSA particle can exist either as a dry particle or as a liquid solution drop, depending on the RH it has previously experienced (shown by the upper and lower curves in Fig. 7) [*Tang et al.*, 1997]. Despite recognition of these properties of single-solute drops/particles and their demonstration for drops/particles of mixed-salt solutions [*Spann and Richardson*, 1985; *Cohen et al.*, 1987b; *Tang and Munkelwitz*, 1993, 1994; *Richardson and Snyder*, 1994], many investigators have stated or implied that a dry SSA particle is formed as soon as the RH becomes lower than the deliquescence RH[1]. However, SSA particles remain liquid drops until they experience an RH below the efflorescence RH (~45%), not the deliquescence RH (~75%). As the RH in the marine atmosphere, at least within tens of meters of the surface, is typically greater than 75% and almost always greater than 45%, virtually all SSA particles, and especially those at heights within ~10 m above the surface (where most measurements are made), will thus exist as liquid solution drops and not as dry or mixed-phase particles. Only those particles that have experienced RH values below the efflorescence RH, and that have remained at RH values below the deliquescence RH, will be present as dry particles. Based on these considerations, the hypothesis of *Murayama et al.* [1999; see also *Sugimoto et al.*, 2000] that the observed lidar depolarization ratios in the marine atmosphere can be explained by the presence of crystallized SSA particles seems unlikely, as the relative humidities were above the efflorescence RH for sea salt, thus requiring generation of SSA particles in the solid phase (as also suggested by these investigators, though without any proposed mechanism).

The processes that occur during evaporation and liquid-solid phase transition of drops containing more than one solute, such as seawater, differ from those for a single-solute solution. *Twomey* [1954; 1955] argued that as RH was decreased, various substances, upon reaching saturation, would crystallize out at their respective deliquescence values, and he listed various compounds in the order in which this would occur, based on studies using drops attached to spider webs. The sequence of minerals deposited from the evaporation of seawater has also been the focus of several other investigations [e.g., *Phillips*, 1947; *Eugster*, 1971; *Harvie et al.*, 1980; *Eugster et al.*, 1980; *Greenberg and Møller*, 1989; *Spencer et al.*, 1990; *Marion and Farren*, 1999], which concluded that the situation is much more complex than that described by *Twomey*, as back-reactions can occur between compounds previously crystallized and the remaining solution. However, such results may not be sufficient to describe processes in seawater drops in the atmosphere, not only because of the relative scarcity of data covering a sufficient temperature range to account for the temperature dependencies of the solubilities and reaction constants, but also because these results were based on equilibrium considerations and thus do not apply to drops below the deliquescence RH. Photomicrographs of the residues of dried seawater drops show various shapes and structures,

including hollow spheres and loosely bound aggregates, sometimes with crystals of $CaSO_4$ protruding from the surface of a larger NaCl crystal [*Facy*, 1951a; *Woodcock*, 1962; *Cheng et al.*, 1988; *Cheng*, 1988; *Mitra et al.*, 1992], often similar to sketches and photomicrographs of dried marine aerosol particles [e.g., *Sigerson*, 1870, Plate IV, especially his description on pp. 15-16; *Parungo et al.*, 1986a; *Harvey et al.*, 1991; *Artaxo et al.*, 1992; *Pósfai et al.*, 1995; *Buseck and Pósfai*, 1999]. As for single-salt solutions, evaporation rate and mass of the solute are expected to play major roles in determining the final structure of dry SSA particles.

Understanding of phase transitions of SSA particles has major implications for understanding of behavior of these particles in the atmosphere and in interpretations of observations. Dry particles that assume shapes other than spheres require a shape factor to account for the effect of their deviation from sphericity on their motion (§2.6.2), but this requires knowledge of the particle shape. As noted above, the densities of these dry particles can differ from that of the bulk material. These effects may lead to inaccurate determination of dry particle sizes in instruments such as impactors and mobility analyzers. Additionally, differences in size, shapes, and index of refraction between dry particles and solution drops will likely lead to inaccurate determinations of the amount of solute in the particle (and hence its value of r_{80}) when sized by optical methods [*Pinnick et al.*, 1976; *Perry et al.*, 1978; *Coletti*, 1984; *Bohren and Singham*, 1991; *Quinby-Hunt et al.*, 1997], and particles that contain crystalline solids are not isotropic (i.e., their light-scattering properties depend on their orientation). The assumption that SSA particles that have been exposed to RH well below the efflorescence value of ~40% RH can be treated as homogeneous liquid spheres [*Quinn et al.*, 1995; *Quinn et al.*, 1996; *Quinn et al.*, 1998; *Quinn and Coffman*, 1998; *Kleefeld et al.*, 2002], would not seem to be justified (some of these investigations assumed that the efflorescence RH of NaCl and of sea salt was 25%), as light-scattering properties of these particles (i.e., index of refraction, light-scattering cross section, and angular distribution of scatted light) are certainly different from those of liquid drops, as noted above. The use of a reference RH at which all measurements are made (or to which they are converted) can lead to ambiguous results if it is in the range where hysteresis can occur. Choice of a reference RH to which the aerosol is conditioned before being measured that is greater than the deliquescence RH for sea salt would result in particles that are

---

[1] *Woodcock and Gifford* [1949]; *Twomey* [1954]; *Ambler and Bain* [1955]; *Twomey and McMaster* [1955]; *Blanchard* [1958]; *Metnieks* [1958]; *Eriksson* [1959, 1960]; *Woodcock et al.* [1963]; *Duce et al.* [1967]; *Hoffman and Duce* [1977a]; *Blanchard and Woodcock* [1980]; *Kulkarni et al.* [1982]; *Blanchard* [1983]; *Cheng et al.* [1988]; *Dulac et al.* [1989]; *Andreas* [1989]; *Smirnov and Shifrin* [1989]; *Andreas* [1990a]; *Pósfai et al.* [1995]; *Andreas* [1996]; *Winter and Chýlek* [1997]; *Monahan and Dam* [2001]; *Andreas and Emanuel* [2001]; *Cole et al.* [2003a].

liquid drops and thus eliminate these ambiguities (in addition to eliminating difficulties associated with shape factors and porous structures), and would allow light-scattering measurements to be described by known theories. Additionally, such a reference RH would allow sizes of SSA particles to be unambiguously defined. The choice of 80% RH made in this review satisfies this requirement.

2.6 KINEMATICS AND DYNAMICS OF SSA PARTICLES

The motion of individual SSA particles is central to their vertical distribution and residence times in the atmosphere. This motion is controlled by several mechanisms, the relative importances of which are determined largely by particle size and location in the atmosphere. In this section these mechanisms and the resulting motion of SSA particles are examined, and velocities and characteristic times and distances that describe this motion are presented as a function of particle size. As the motion of an SSA particle depends on its physical (i.e., geometric) radius r as opposed to r_{80}, a key question is how long it takes a particle to attain its equilibrium radius at the ambient RH (§2.5.4). The comparison of this time and the time that the particles spends in the atmosphere, as governed by considerations discussed here, determines the geometric radius of a particle during its lifetime and hence the size range of interest with regard to atmospheric processes. This comparison is made in §2.9.1, with the result that small and medium SSA particles equilibrate sufficiently rapidly that their kinematics and dynamics are characteristic of their equilibrium radii at ambient RH, and further that they can be assumed to have attained their equilibrium radii by the time of measurement. However, this does not hold for large SSA particles, whose radii often remain near their formation values for a considerable portion of the time they spend in the atmosphere.

2.6.1. Mechanisms Controlling SSA Particle Motion

Key mechanisms that control the motions of SSA particles in the marine atmosphere are turbulent (eddy) diffusion (§2.4), gravity, and drag (Table 2). Turbulent diffusion controls the entrainment of small and medium SSA particles into the atmosphere, but it is generally insufficient to entrain large SSA particles very far above the sea surface. Gravity plays a dominant role in the behavior of medium and large SSA particles, but its effect on the motion of small SSA particles is generally insignificant compared to other mechanisms. Drag, the resistive force of the air to the motion of a particle, plays a role in the behavior of drops of nearly all sizes. The electrical force resulting from the charge on a drop, which depends on the type of the drop (§4.3.1), various properties of the bubble producing the drop [Blanchard, 1955, 1958, 1963, 1966], and the electric field over the ocean, can exceed the gravitational force on drops with r_{80} as great as several micrometers under certain conditions (such as during electrical storms), although typically these forces can be neglected, and they are not considered further here (they are discussed in Blanchard [1963, p. 127] and Bortkovskii [1987, pp. 93-96]). Impaction, which is important only at the sea surface for a restricted range of particle sizes, is discussed elsewhere (§2.7.2.2, §2.8.2, §4.6.1). Other mechanisms that may be important in certain circumstances are Stefan flow and Brownian diffusion. These mechanisms are considered with respect to the dry deposition of small particles over the ocean in §4.6, but they are thought to be important only for very small SSA particles ($r_{80} < 0.1$ μm), and then only in certain circumstances.

Phoretic effects resulting from gradients of water vapor and temperature [Goldsmith et al., 1963; Waldmann and Schmitt, 1966; Goldsmith and May, 1966] may act to transport particles through diffusiophoresis, thermophoresis, and Stefan flow, but only the latter is thought to be possibly important for SSA particles over the ocean [Slinn et al., 1978]. Stefan flow, which is caused by evaporation of water from the ocean surface, results in upward motion of particles. The velocity of this motion, v_{Stefan}, which is nearly independent of particle size (for most of the size range of SSA particles), is approximately equal to the evaporative mass flux of water j_w (§2.4) divided by the density of air ρ_a [Slinn, 1974; Slinn et al., 1978]; numerically

$$\frac{v_{Stefan}}{cm\ s^{-1}} \approx 0.1 \frac{j_w}{g\ m^{-2}\ s^{-1}}. \qquad (2.6-1)$$

As noted earlier (§1.1; §2.4), the evaporative mass flux of water corresponding to the global annual mean evaporation flux over the ocean is $j_w \approx 0.03$ g m^{-2} s^{-1} (equivalent to an evaporation rate of $3 \cdot 10^{-6}$ cm s^{-1}), corresponding to $v_{Stefan} \approx 0.003$ cm s^{-1}, although local instantaneous rates may be considerably greater. An evaporation rate of 10^{-5} cm s^{-1} yields a Stefan flow away from the surface of nearly 0.01 cm s^{-1}, in the range of model-derived values (0.001-0.015 cm s^{-1}) reported by Giorgi [1988] and equal in magnitude to the gravitational sedimentation velocity (§2.6.2) of an SSA particle with $r \approx 1$ μm. As this effect is typically much less than the motion induced by turbulence or that due to gravitational sedimentation (for SSA particles with $r \gtrsim 1$ μm), it is important, if at all, only for small SSA particles and then only in the viscous sublayer, and it can be neglected for virtually all considerations of SSA particles elsewhere in the marine atmosphere. Stefan flow is also discussed in §4.6.3 with regard to the dry deposition velocity.

Brownian (molecular) diffusion of particles is characterized by the Brownian diffusivity D_B, given [*Einstein*, 1905; *Friedlander*, 2000, pp. 32-34] by

$$D_B = \frac{kT}{6\pi \, \rho_a \, \nu_a \, r} \times C; \qquad (2.6\text{-}2)$$

where k is the Boltzmann constant, T is the absolute temperature, ν_a is the kinematic viscosity of air, and C is the Cunningham slip factor, also called the Cunningham correction. This Cunningham correction [*Cunningham*, 1910; *Knudsen and Weber*, 1911; *Millikan*, 1923] accounts for the non-continuum nature of air and depends on the ratio λ/r, where the mean free path of air λ depends primarily on the density of the air and is around 0.06 μm under conditions normally encountered near the sea surface[1]. The Cunningham correction is nearly proportional to r^{-1} for particles with $r \ll \lambda$ (with resulting diffusion coefficient proportional to r^{-2}; §2.1.5.2), but C approaches unity as r increases; $C \approx 1.85$ for $r = 0.1$ μm and $C \approx 1.08$ for $r = 1$ μm.

Where there are spatial inhomogeneities in the SSA number concentration, Brownian diffusion results in a size-dependent flux of SSA particles equal to

$$\vec{j}_B = -D_B(\nabla n). \qquad (2.6\text{-}3)$$

In the marine atmosphere, Brownian diffusion is important, if at all, only for small SSA particles in the viscous sublayer adjacent to the sea surface, as it is greatly exceeded by eddy diffusion in the turbulent air away from the surface (as shown below). As RH in the viscous sublayer is near 98% (§2.4) because of proximity to the surface together with vapor pressure lowering due to the salt content of seawater (§2.5.3), the geometric radius of an SSA particle in this layer will be close to that at formation r_{98} ($= 2r_{80}$), and the Brownian diffusion coefficient is given approximately by

$$\frac{D_B}{\text{m}^2 \, \text{s}^{-1}} \approx \frac{6 \cdot 10^{-12} C}{\left(\dfrac{r_{80}}{\mu\text{m}} \right)}. \qquad (2.6\text{-}4)$$

For SSA particles with $r_{80} \gtrsim 0.1$ μm (having geometric radii $r > 0.2$ μm in the viscous sublayer), corresponding to the majority of the size range comprising SSA particles in the marine atmosphere (§4.1.1; §4.1.4); $C \lesssim 1.5$ and $D_B \lesssim 10^{-10}$ m^2 s^{-1}. Using this value of the diffusion coefficient, a lower limit for the time required for a particle with

$r = 0.2$ μm ($r_{80} = 0.1$ μm) to diffuse through a viscous sublayer of thickness $\delta = 0.1$ (1) mm (§2.4) can be estimated from the expression $\delta^2/(2D_B)$ as 50 (5000) s; by contrast the time required for an SSA particle of this size to fall through this distance because of gravity is approximately 10 (100) s (§2.6.2), and the time required for its transport through this layer from Stefan flow (using $\nu_{\text{Stefan}} = 0.01$ cm s^{-1}) is approximately 1 (10) s. Thus, Brownian diffusion can be neglected for virtually all considerations of SSA particles in the atmosphere.

Turbulent, or eddy, diffusion characterizes the dispersion of particles in the surface layer, that part of the marine boundary layer directly above the viscous sublayer up to heights of at least several tens of meters (§2.4). For the one-dimensional situation in which the size-dependent SSA number concentration varies only with the height above the ocean surface z, the *ansatz* is typically made that the size-dependent flux of particles in the vertical direction due to turbulent diffusion, $j_{z,\text{turb}}$, is proportional to the vertical gradient of the size-dependent number concentration, and thus is given by

$$j_{z,\text{turb}} = -D_{\text{eddy}} \frac{\partial n}{\partial z}; \qquad (2.6\text{-}5)$$

for neutrally stable atmospheres the magnitude of this turbulent diffusion coefficient is given by (2.4-2). It is assumed that this value of the eddy diffusivity applies to SSA particles independent of size. Although this assumption is not valid for sufficiently large particles, which will not be entrained upward by the smallest eddies because of their large inertia and their greater response times to changes in the flow (§2.6.4), few of these particles remain in the atmosphere sufficiently long for this to be an issue. For $U_{10} = 10$ m s^{-1}, corresponding to $u_* = 36$ cm s^{-1} (§2.3.2), the eddy diffusion coefficient at 10 m above the surface is given by (2.4-2) as $D_{\text{eddy}} = 1.4$ m^2 s^{-1}—more than 11 orders of magnitude greater than the Brownian diffusion coefficient for a particle with $r_{80} = 1$ μm.

Gravitational sedimentation can be important throughout the marine boundary layer. The motion of a particle in the atmosphere under the influence of gravity is determined by its mass and by the drag force of the air on the particle, both of which depend primarily on particle size. A drag force occurs whenever there is a relative velocity between a particle and the surrounding air, and the magnitude of this force will determine the terminal velocity with respect to

---

[1] Some references use a value of λ based on an older definition; for instance, *Twomey* [1977a, p. 55] implicitly uses the value 0.093 μm. Additionally, the expression for C in Eq. 3.11 of that reference is incorrect, as a set of parentheses is missing. Likewise, the value stated for λ in Table 11.1 of *Pruppacher and Klett* [1997, p. 451] is too high by approximately 15%, as are the values of diffusivity presented in that table.

gravitational sedimentation of a particle of a given size, how rapidly this particle approaches this velocity, and how rapidly it attains the horizontal velocity component of the surrounding air. Drops with $r \lesssim 500$ μm falling under the influence of gravity behave similarly to solid rigid spheres, and most results for the motion of drops in this size range are based on the uniform motion of a solid sphere in still air. Effects such as internal circulation, deformation, oscillation, and break-up into smaller drops become increasingly important as drop size increases, and most results for drops with $r \gtrsim 500$ μm are based on the behavior of raindrops moving at their gravitational terminal velocities [more information on the motion of spheres and drops can be found in *Torobin and Gauvin*, 1959a,b,c, 1960a,b, 1961; *Wallis*, 1974; *Clift et al.*, 1978; and *Pruppacher and Klett*, 1997, Chap. 10]. For a particle undergoing acceleration other effects such as additive mass (due to the entrained air) and the time-history of the motion can contribute to the drag force, but it is generally assumed that the force on such a particle is determined only by its instantaneous relative velocity with respect to the surrounding air; this is typically a valid assumption [*Fuchs*, 1964, pp. 70-80; *Sartor and Abbott*, 1975; *Wang and Pruppacher*, 1977; *Beard*, 1977; *Clift et al.*, 1978, Chap. 11].

The following sections examine the motion of SSA particles under the influences of only gravity and drag. The key quantity parameterizing this motion is the gravitational terminal velocity of a particle falling in still air v_{term}, shown in Fig. 10 for a liquid drop with geometric radius r and density 1 g cm^{-3} under conditions typical of those near the sea surface. Expressions that readily allow calculation of v_{term} to an accuracy of ~10% or better are presented solely in terms of r for several size ranges; because of the dependencies of v_{term} on the atmospheric pressure, temperature, and RH (through their effects on the density and kinematic viscosity of air and on the density of the particle), greater accuracy requires more complicated expressions which include these dependencies. In most situations experienced by SSA particles the flow is turbulent, and especially within a few meters of the sea surface (where large SSA particles are almost exclusively present) the flow field is difficult to model accurately or determine quantitatively. It is therefore useful to characterize the motion of SSA particles by suitable times and lengths describing two idealized situations: (1) a drop initially at rest falling in still air under the influence of gravity, and (2) a drop initially at rest placed into a uniform horizontal flow. The dependencies of these times and distances on particle size aid in determining the important mechanisms that control the motion of SSA particles, and allow approximations to be made that greatly simplify the description of the behavior of SSA particles in the atmosphere with little sacrifice in accuracy (§2.9.1).

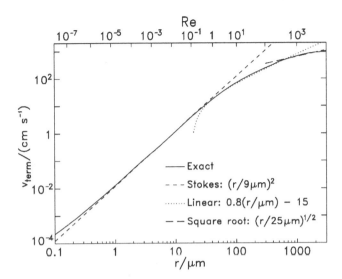

Figure 10. Terminal velocity, v_{term} of a particle with density 1 g cm^{-3} falling under the influence of gravity for conditions typical of those in the marine boundary layer as a function of physical radius r (lower axis) or Reynolds number Re (upper axis), evaluated following *Beard* [1976]. Also shown are the Stokes law approximation, (2.6-7), linear approximation given by (2.6-11), and square root approximation given by (2.6-12).

2.6.2. Stokesian Behavior

A good approximation to the drag force for small and medium SSA particles in the atmosphere under most circumstances is given by Stokes' Law [*Stokes*, 1850], which states that the magnitude of the drag force on a solid sphere of given radius r moving at a constant relative velocity v_{rel} with respect to a fluid (air in this case) is directly proportional to both r and v_{rel}:

$$F_{\text{drag}} = \frac{6\pi \rho_a v_a r v_{\text{rel}}}{C}, \tag{2.6-6}$$

where ρ_a and v_a are the density and kinematic viscosity of air, respectively, and C is the Cunningham slip factor (§2.6.1). This drag force is composed of friction drag (caused by viscosity), which accounts for two-thirds of the total drag, and form drag (caused by the imbalance of pressure on the sphere), which accounts for the remaining one-third. In situations for which Stokes' Law applies (termed "Stokesian" behavior), the motions of the particle in the horizontal and vertical directions are independent of each other (as a consequence of the drag force being directly proportional to the relative velocity); hence v_{term} is independent of the horizontal component of the relative velocity of the particle and the surrounding air. This gravitational terminal velocity v_{term} can be obtained by equating the drag force with the gravitational force on the particle, and is given by the Stokes velocity v_{Stk}:

$$v_{Stk} = \frac{2}{9} \frac{(\rho_p - \rho_a)}{\rho_a} \frac{g}{v_a} r^2 C, \qquad (2.6\text{-}7)$$

where ρ_p is the density of the particle and g is the acceleration due to gravity. As $\rho_p \gg \rho_a$, the quantity $(\rho_p - \rho_a)$ can be replaced by ρ_p with virtually no loss in accuracy. Likewise, the Cunningham slip factor C differs appreciably from unity only for the smallest particles (§2.6.1), for which gravity plays little role in their fate; its effect on v_{term} for small r is shown in Fig. 10. Thus in many situations it can be set to unity with little loss in accuracy.

The dependence on RH of the Stokes velocity of an SSA particle (under the assumption that the particle is a solution drop sufficiently large that the Cunningham slip factor can be set to unity) with a given mass of solute is shown in Fig. 7. Most of this dependence is through the effect of RH on radius, although there is a slight effect on density; an increase in RH from 80% to 98% results in a doubling of the radius but a decrease in the density of only ~15% (§2.5.3; Fig. 7). At high RH, v_{Stk} for an SSA particle with a given r_{80} exhibits a strong dependence on RH—increasing by a factor of 2.5 between 90% and 98% RH. At lower RH the dependence is much weaker—for a given r_{80}, v_{Stk} increases by a factor of 1.4 between 50% and 80% RH (again, under the assumption that the SSA particle remains a liquid drop), and by a factor of 1.3 between 80% and 90% RH (Fig. 7). For an SSA particle sufficiently large that the Cunningham correction can be set to unity, the Stokes velocity at formation (98% RH), at 80% RH, and for the dry particle (assuming it is cubical with density ρ_{ss} and with a shape factor, discussed below, of 0.92), are in the ratio 8:2.4:1. Under conditions typical of the marine boundary layer, the Stokes velocity of an SSA particle with geometric radius r (not necessarily equal to r_{80}), again under situations for which the Cunningham correction can be set to unity, can be approximated by

$$\frac{v_{Stk}}{cm\ s^{-1}} \approx \left(\frac{r}{8.5\ \mu m}\right)^2 \qquad (2.6\text{-}8)$$

(Fig. 10). The exact numerical value varies slightly with temperature and with particle density, and thus depends slightly on RH, but under most typical situations it is within ~10% of that given by (2.6-8). From (2.6-8) it follows that the distances fallen due to gravity in an hour by SSA particles with radii at their r_{80} values of 0.1 μm, 1 μm, and 10 μm are 0.5 cm, 50 cm, and 50 m, respectively (the distance fallen by the 0.1 μm radius particle is nearly doubled if the Cunningham slip correction is taken into account). Alternatively, the times required for SSA particles with radii at their r_{80} values of

0.1 μm, 1 μm, and 10 μm to fall 100 m in still air are of 2 y, 8 d, and 2 h, respectively (likewise, the time required for the 0.1 μm radius particle is reduced to nearly 1 y if the Cunningham slip correction is taken into account). Thus, the role of gravity can be neglected in the motion of small SSA particles, whereas for medium SSA particles, depending on size, gravity may be important (Table 2).

Particles that are not spherical (dry particles, for instance) require different parameterizations to describe their size and their motion. Their size is typically described by the radius of a sphere of equal volume (for dry particles this is the equivalent dry radius; Eq. 2.1-1), and a shape factor is used to describe the effects of their shape on their motion. For sufficiently small particles, this shape factor is defined here as the ratio of the drag given by Stokes' Law evaluated using the volume-equivalent radius (i.e., the radius of a sphere with equal volume) to the actual drag on the particle, or equivalently, the ratio of the terminal velocity of the particle to that given by Stokes' Law (Eq. 2.6-7) based on this equivalent-volume radius (in some references, e.g., *Hinds* [1999, p. 51], the shape factor is defined as the inverse of this). This shape factor is around 0.92 for cubes with edges less than ~20 μm [*Pettyjohn and Christiansen*, 1948; *McNown and Malaika*, 1950; *Heiss and Coull*, 1952; *Gurel et al.*, 1955; *Chowdhury and Fritz*, 1959; *Johnson et al.*, 1987]; dry SSA particles larger than this are unlikely to be found in the marine atmosphere (§2.9.5, §4.1.4.3). Shape factors for larger particles and in situations for which Stokes' Law does not apply are discussed in *Stringham et al.* [1969]; *Kasper* [1982]; and *Cheng et al.* [1990].

Measurements that classify SSA particles based on their aerodynamic behavior, such as those made with impactors and aerodynamic particle sizers, determine their aerodynamic radius, defined to be the radius of a spherical particle of density 1 g cm^{-3} that has the same aerodynamic properties (often referring to the gravitational terminal velocity) as the given particle. In situations for which Stokes' Law is valid, the aerodynamic radius of a (spherical) particle is greater than its geometric radius by the square root of the ratio of its density in g cm^{-3}. Thus under most conditions experienced by an SSA particle that exists as a liquid drop, its aerodynamic radius is roughly equal to its geometric radius (at 80% RH the aerodynamic radius is approximately 10% greater than its geometric radius). However, for a dry SSA particle that is cubical and has the same density as dry sea salt, the aerodynamic radius would be approximately 60% greater than its equivalent dry radius (§2.5.3).

The rate at which a particle responds to an external force is determined primarily by its radius. In situations for which Stokes' Law applies, the velocity of a particle asymptotically approaches its gravitational terminal (i.e., Stokes) velocity, or the velocity of the surrounding flow, by an exponential function of time (this is a consequence of the direct proportionality of the drag force and the relative velocity). Whether or not a particle can be considered to fall at its Stokes velocity, or to move with the velocity of the surrounding air, depends on how rapidly the particle would, in the absence of other forces, attain a substantial fraction of one or another of these velocities. Expressions are presented for the time required for an SSA particle to attain a given fraction of either of these velocities (in the absence of other forces), and for the distance it travels during this time, for conditions typical of the lower marine boundary layer, under the assumptions (discussed above) that SSA particles behave like rigid spheres and that the drag force on an accelerating particle is determined by its instantaneous velocity relative to the surrounding air. For small and medium SSA particles these assumptions are valid under nearly all circumstances of interest.

An SSA particle of radius r initially at rest falling in still air under the influence of gravity, under the assumption that the drag force is given by (2.6-6), attains 90% of its terminal velocity in a time $t_{90}/\text{s} \approx [r/(175 \ \mu\text{m})]^2$, during which it will have fallen a distance $z_{90}/\text{cm} \approx [r/(45 \ \mu\text{m})]^4$. For small and medium SSA particles falling in still air Stokes' Law applies (as discussed below), and these times and distances are very small: an SSA particle with radius equal to its r_{80} value of 25 μm (near the upper size range of medium SSA particles; §2.1.2), which at its gravitational terminal velocity of nearly 9 cm s^{-1} (Eq. 2.6-8) requires almost two minutes to fall 10 m in still air, attains 90% of this terminal velocity in 0.02 s, during which time it will have fallen only 0.1 cm. Thus, it can be concluded that small and medium SSA particles in the atmosphere attain their gravitational terminal velocities nearly instantaneously (Table 2).

Likewise, small and medium SSA particles can be considered to be transported at the horizontal velocity of the surrounding flow (that is, have zero horizontal component of the relative velocity between the particle and the air) if they attain the horizontal velocity component of the mean flow sufficiently rapidly. For an idealized situation in which an SSA particle is instantaneously introduced at rest into a horizontal flow of 10 m s^{-1}, the time required (according to Stokes' Law) for this particle to attain 90% of this velocity (that is, for the relative velocity to be reduced to 10% of its original value) is the same as the time given above for a particle initially at rest to reach 90% of its gravitational terminal velocity, $t_{90}/\text{s} \approx [r/(175 \ \mu\text{m})]^2$. During this time the particle will have been transported a horizontal distance $x_{90}/\text{cm} \approx [r/(5.5 \ \mu\text{m})]^2$. For an SSA particle with radius equal to its r_{80} value of 25 μm, these expressions yield a time of approximately 0.02 s and a distance of approximately 20 cm. However, for these conditions the drag force is greater than that given by Stokes' Law, initially by a factor of nearly 3 (this is discussed below), and the actual times and distances are roughly half of the values given. Thus,

small and medium SSA particles attain the horizontal velocity of the mean flow nearly instantaneously and can be considered to be transported at this velocity (Table 2).

2.6.3. Ultra-Stokesian Behavior

Although Stokes' Law (2.6-6) provides a good approximation for the drag force on small and medium SSA particles in most circumstances, there are some situations of interest, termed "ultra-Stokesian," in which it does not apply. Examples are the ejection of jet drops with large initial velocities (§4.3.2), the motion of large SSA particles (§4.3.6), drops that experience large relative velocities with respect to the air (especially spume drops torn off of wave crests), and falling raindrops (§4.7.2). Also aerodynamic particle sizers, impactors, and other instruments often operate in the ultra-Stokesian regime, depending on particle size for given operating conditions. In these situations the drag force on a particle of given radius increases at a rate greater than the first power of the relative velocity, and the description of the motion of such a particle requires a different expression for the drag force as a function of particle radius and relative velocity.

The key parameter that describes the motion in ultra-Stokesian situations, and which determines the range of validity of Stokes' Law for a spherical particle in air, is the Reynolds number Re [Reynolds, 1883], defined here as

$$Re = \frac{2v_{\text{rel}}r}{v_{\text{a}}} \tag{2.6-9}$$

(this definition of Re is not universal and it is often defined without the factor of 2). For $Re \lesssim 1$, Stokes' Law applies (in the example above, $Re \approx 0.3$ for an SSA particle with radius 25 μm falling at its gravitational terminal velocity of ~9 cm s^{-1}), but as Re increases above unity, Stokes' Law increasingly underestimates the drag force and thus overestimates both the gravitational terminal velocity (Fig. 10) and the time for a particle to attain a given fraction of this velocity or of the flow into which it is introduced. At sea level, the Reynolds numbers of particles of density 1 g cm^{-3} with radii 40 μm and 60 μm falling at their terminal velocities are near 1 and 3, respectively, and Stokes' Law overestimates the gravitational terminal velocities of these particles by around 10% and 25%, respectively (Fig. 10). However, radii of SSA particles measured as 40 μm and 60 μm probably reflect values close to formation radii (§2.9.1) and thus correspond to r_{80} of 20 μm and 30 μm, respectively. As these are near the upper limit of the size range of medium SSA particles, Stokes' Law can therefore be used with little loss of accuracy for small and medium SSA particles in the atmosphere, although for the motion of larger drops different expressions are required (Table 2).

In ultra-Stokesian situations the drag force on a sphere in air is generally parameterized by the dimensionless drag coefficient C_d (also referred to as the friction factor), defined here as

$$C_d = \frac{F_{drag}}{\left(\dfrac{\rho_a v_{rel}^2}{2}\right) \times (\pi r^2)}. \qquad (2.6\text{-}10)$$

The drag coefficient can be related to v_{term} by equating the drag force on the particle and its weight, resulting in

$$C_d = \frac{8}{3}\frac{\rho_p}{\rho_a}\frac{rg}{v_{term}^2}.$$

As with Re, there is no standard convention for the definition of C_d, and the drag coefficient defined in (2.6-10) is a factor of 2 greater than that used by many investigators (the definition used by van Dyke [1975, p. 5] is a factor of $2/\pi$ lower). The lack of standardization in the definitions of both Re and C_d can result in confusion not only in specifying ranges of validity of certain formulae and phenomena, but also when comparing results from different investigators and when using relations for C_d given in terms of Re [e.g., von Kármán, 1954, p. 87; §4.4.2].

For a rigid sphere moving at constant velocity in a still medium of much lower density, C_d is a function only of Re [Rayleigh, 1909-1910]; as discussed above, the assumption is generally made that the same relation between C_d and Re applies for spheres undergoing acceleration. Expressions have been proposed for C_d based on series expansions in terms of Re or $\ln Re$ derived from the equations describing the motion of a rigid sphere in a fluid [Oseen, 1913; Goldstein, 1929b (with a correction provided by Shanks, 1955, p. 36); Proudman and Pearson, 1957; Chester and Breach, 1969; Proudman, 1969], but these have not proven very useful, as their accuracy diminishes rapidly as Re increases above unity. For example, for $r \gtrsim 50\ \mu m$ the difference between the gravitational terminal velocity obtained from any of these expressions differs from the actual value by more than 10%, and for $r \gtrsim 125\ \mu m$ they differ by more than a factor of two (except for that calculated from the Oseen approximation, for which the difference is a factor of 1.5 at $r = 125\ \mu m$). An exception is the semi-theoretical expression $C_d = (24/Re)(1 + 0.08Re)$ derived by Carrier [1953], which provides a very accurate approximation for $Re \lesssim 15$, yielding gravitational terminal velocities that are within 5% of their correct values for particles with $r \lesssim 125\ \mu m$. In lieu of such theoretical expressions, empirical relations for C_d in terms of Re are typically used. Numerous such expressions have been proposed [compilations are presented in Graf, 1971, pp. 42-43, Fig. 4.2, and Clift et al., 1978, pp. 111-113, for example]; one such relation that is widely used is that due to Schiller and Nauman [1933]: $C_d = (24/Re)(1 + 0.15Re^{0.687})$. Other similar expressions are those due to Klyachko [1934 (cited in Fuchs, 1964, p. 33, with an obvious typographic error)]: $C_d = (24/Re) \times [1 + (1/6)Re^{2/3}]$, Serafini [1954]: $C_d = (24/Re)(1 + 0.158Re^{2/3})$, Olson [1961, Chap. 11]: $C_d = (24/Re)[1 + (3/16)Re]^{1/2}$, and the semi-theoretical expression of Abraham [1970]: $C_d = (24/Re)(1 + Re^{1/2}/a)^2$, where a is near 9. All of these expressions reduce to Stokes' Law (equivalent to $C_d = 24/Re$) at low Re, and all provide accurate representations for C_d (to within ~10%) over the entire range of Re up to near 1000, corresponding to SSA particles with $r \lesssim 1000\ \mu m$ falling in still air under the influence of gravity near the ocean surface. For a sphere of given radius and density, the fractional change in the gravitational terminal velocity is equal to -0.5 times the fractional change in C_d; thus the terminal velocities obtained from these relations for a given particle radius are accurate to within ~5%. Often a power law relation between C_d and Re is assumed; the expression $C_d = 12/Re^{1/2}$ is accurate to within ~10% for $10 \lesssim Re \lesssim 800$, corresponding roughly to SSA particles with $7.5\ \mu m \lesssim r \lesssim 600\ \mu m$ moving in air with a relative velocity of $10\ m\ s^{-1}$, or to SSA particles with $100\ \mu m \lesssim r \lesssim 900\ \mu m$ falling at their gravitational terminal velocities in still air. Other power laws have been used for other ranges of Re.

The gravitational terminal velocity of a spherical particle can be determined from the dimensionless quantity $C_d Re^2$, which is independent of the relative velocity and depends only on the particle radius (and density). This quantity has been termed the Davies number [after Davies, 1945] and also the Best number [after Best, 1950], although it was used earlier by Castleman [1926], Burke and Plummer [1928], Schiller and Naumann [1933], Lapple and Shepherd [1940], Krumbein [1942], and Langmuir [1943-1944 (in Langmuir, 1961)]. For a given radius (and density), the equation

$$C_d Re^2 = \frac{32}{3}\frac{\rho_p}{\rho_a}\frac{r^3 g}{v_a^2}$$

can be solved for Re, from which v_{term} can be obtained from the definition of Re (Eq. 2.6.9). However, as the solution of this equation from an expression giving C_d in terms of Re usually requires an iterative procedure, relations giving Re in terms of $C_d Re^2$ (which allow direct calculation of Re and thus v_{term}) have been presented [Langmuir 1948; Fuchs, 1964, Table 5 (p. 32); Cornford, 1965; Beard and Pruppacher, 1969; Berry and Pranger, 1974; Gay et al., 1974; Beard, 1976; Clift et al., 1978, p. 114]. For conditions typical of the marine atmosphere, the expression

$$\frac{v_{term}}{cm\ s^{-1}} \approx 0.8\left(\frac{r}{\mu m}\right) - 15 \qquad (2.6\text{-}11)$$

is accurate to ~10% for SSA particles in the size range 40 μm $\lesssim r \lesssim$ 800 μm (Fig. 10); as noted above it is not possible to determine simple expressions for v_{term} solely in terms of r that are more accurate because of the effects of temperature and RH. The relation $C_{\text{d}} = 12/Re^{1/2}$, which implies the drag force on a spherical particle of given radius is proportional to the three-halves power of its relative velocity, yields $v_{\text{term}}/(\text{cm s}^{-1}) \approx 0.76(r/\mu\text{m})$, also accurate to ~10% for SSA particles with 100 μm $\lesssim r \lesssim$ 800 μm.

Several effects that influence the free fall motion of drops start to become important at $r \approx 500$ μm and become increasingly important as drop size increases further. These effects must be taken into account when discussing the motion of sufficiently large SSA particles and are likewise important for raindrops (§4.7.2). The detachment of the vortex ring behind a falling drop results in unsteady motion of the drop, causing it to oscillate. Additionally, the weight of the drop results in a deviation from sphericity. For freely falling water drops with $r < 500$ μm the ratio of the length of the axis in the direction of motion to the length of the axis perpendicular to this motion is nearly unity, but for larger drops this ratio decreases nearly linearly with increasing equivalent radius (i.e., that of a sphere of equal volume) until it is approximately 0.5 for equivalent radius $r = 4000$ μm [Pruppacher and Beard, 1970; Andsager et al., 1999]. This deformation in turn changes the flow field around the drop and hence the drag force on it, resulting in a greater drag force (and hence greater drag coefficient) than that for a solid rigid sphere of the same density and equivalent radius, and leading to different expressions for gravitational terminal velocities and to different characteristic times and distances. For conditions in the lower portion of the marine boundary layer, with increasing drop size (and hence Re) the drag coefficient C_{d} reaches a broad minimum of near 0.5 for drops with radii from ~1000 to ~1700 μm falling at their gravitational terminal velocities (Re from ~900 to ~1900). With further increasing (equivalent) radius, C_{d} slowly increases (in contrast to the situation for a rigid sphere, in which C_{d} continues to decreases over this range of radii or Re), attaining a value of near 0.75 for a drop with $r \approx 3000$ μm. The gravitational terminal velocity of a drop of this size in still air increases less than linearly with (volume-equivalent) radius, from nearly 400 cm s^{-1} for $r = 500$ μm to 650 cm s^{-1} for $r = 1000$ μm to nearly 900 cm s^{-1} for a radius of 2000 μm (Fig. 10). For larger values of r the gravitational terminal velocity varies little (Fig. 10), but few raindrops with radii greater than a few thousand micrometers (i.e., a few millimeters) occur in the atmosphere, as they break up into smaller drops. An approximation for the gravitational terminal velocity for drops in the size range 500 μm $\lesssim r \lesssim$ 2500 μm accurate to ~10% for conditions typical of the marine boundary layer

is given by

$$\frac{v_{\text{term}}}{\text{m s}^{-1}} \approx \left(\frac{r}{25\,\mu\text{m}}\right)^{\frac{1}{2}}, \qquad (2.6\text{-}12)$$

which follows from the assumption of a constant drag coefficient $C_{\text{d}} = 0.54$ (implying that the drag on a particle of a given radius is proportional to the square of the velocity); a similar expression proposed by Kessler [1969, p. 20] is equivalent to $C_{\text{d}} = 0.64$.

2.6.4. Characteristic Times and Distances for Particles in Ultra-Stokesian Situations

The equations describing the two-dimensional motion of a particle with horizontal velocity component u and vertical velocity component w (positive upward) moving under the influence of gravity in a flow with constant horizontal velocity U follows immediately from Newton's second law and the definition of C_{d} given by (26-10):

$$\frac{du}{dt} = \left(\frac{3}{8}\frac{\rho_{\text{a}}}{\rho_{\text{p}}}\frac{C_{\text{d}}}{r}\right)\left(\sqrt{(U-u)^2 + w^2}\right)(U-u) \qquad (2.6\text{-}13\text{a})$$

and

$$\frac{dw}{dt} = -\left(\frac{3}{8}\frac{\rho_{\text{a}}}{\rho_{\text{p}}}\frac{C_{\text{d}}}{r}\right)\left(\sqrt{(U-u)^2 + w^2}\right)w - g, \qquad (2.6\text{-}13\text{b})$$

where C_{d} is assumed to be a function only of Re defined in terms of the relative velocity $v_{\text{rel}} = [(U-u)^2 + w^2]^{1/2}$. These equations can also be combined into a single second-order nonlinear differential equation for Re. Not infrequently these equations have been presented and applied incorrectly [Wu, 1979a; Koga and Toba, 1981; Nystuen and Farmer, 1987; de Leeuw, 1989a; Edson and Fairall, 1994; Mestayer et al., 1996]; they are given correctly in Fuchs [1964, p. 107] and Bortkovskii [1987, p. 97] (who used a definition of C_{d} different from that given by Eq. 2.6-10). The dependence of Re, and thus C_{d}, on both u and w implies that except for Stokesian situations the horizontal and vertical motions of a particle are not independent, and analytic solutions have been found for (2.6-13a) and (2.6-13b) only for the case in which the drag force is linear in the relative velocity, i.e., Stokes' Law holds (for which the equations are independent), and for the case in which the drag force is proportional to the square of the relative velocity, i.e., C_{d} is a constant [Kármán and Biot, 1940, pp. 139-143]. As a consequence of this interdependence, the drag force on a particle in one direction is increased by the presence of a relative velocity in a perpendicular direction. In the present context this conclusion is pertinent to drops ejected from the sea surface into the marine atmosphere, in which there is nearly always a horizontal component of the mean air flow. Consequently, these drops do not attain heights as great as those in laboratory conditions that are ejected into still air (§4.3.1.3). Likewise raindrops falling through a horizontal wind that changes magnitude with height have lower terminal velocities than they would falling through still air (4.7.2).

The one-dimensional motion of a particle falling in still air under the influence of gravity can be described by

$$\frac{d(Re)}{dt} = \frac{2rg}{v_a} - \frac{3}{16}\frac{\rho_a}{\rho_p}\frac{v_a}{r^2}(C_d Re^2), \qquad (2.6\text{-}14)$$

which follows from (2.6-9) and (2.6-13b). Likewise, the one-dimensional motion of a particle in the absence of external forces can be described by

$$\frac{d(Re)}{dt} = \frac{3}{16}\frac{\rho_a}{\rho_p}\frac{v_a}{r^2}(C_d Re^2), \qquad (2.6\text{-}15)$$

which follows from (2.6-9) and (2.6-13a). Equations 2.6-14 and 2.6-15 can be integrated to yield Re (and hence the velocity) as a function of time if C_d is known as a function of Re [Lapple and Shepherd, 1940]. An analytic solution of (2.6-14) can be obtained if the expression for C_d presented by Olson [1961, Chap. 11] or Abraham [1970] is used (this does not appear to have been previously reported), but in general the equation must be integrated numerically. Similarly, an analytic solution of (2.6-15) can be obtained using the expression for C_d presented by Klyachko [1934 (cited in Fuchs, 1964, p. 33)], Serafini [1954 (in which the correct solution is given; that quoted in Fuchs, 1964, Eq. 18.10, contains a typographic error)], Olson [1961, Chap. 11], or Abraham [1970].

As opposed to small and medium SSA particles, for which Stokes' Law is valid, large SSA particles do not approach their gravitational terminal velocities, or the velocity of the surrounding flow, via an exponential function of time. An estimate of the time required for a particle initially at rest falling in still air under the influence of gravity to attain 90% of its terminal velocity can be determined from (2.6-14). For an SSA particle with $100~\mu m \lesssim r \lesssim 500~\mu m$ this time is given approximately by $t_{90}/s \approx r/(700~\mu m)$, during which the particle will have fallen a distance $z_{90}/cm \approx [r/(40~\mu m)]^2$. For SSA particles with radii $100~\mu m$ and $500~\mu m$ these times are near 0.1 s and 0.7 s and the distances are near 6 cm and 150 cm, respectively. Similarly, a rough estimate of the time required for a particle initially at rest placed in a horizontal flow of 10 m s^{-1} to attain 90% of this velocity can be determined from (2.6-15). For an SSA particle in the size range $100~\mu m \lesssim r \lesssim 500~\mu m$ this time is given approximately by

$t_{90}/s \approx r/(400~\mu m) - 0.15$, during which the particle will have traveled a horizontal distance $x_{90}/cm \approx r/(1.25~\mu m) - 50$ (these expressions should serve only to provide approximate estimates; as noted above, for two-dimensional motion the equations for the horizontal and vertical components—Eqs. 2.6-13a and 2.6-13b, respectively—are coupled and cannot be solved separately, thus the effect of gravity cannot be neglected in calculations for which high accuracy is required). For a particle with $r = 100~\mu m$, this time is around 0.1 s, during which the particle will have traveled 30 cm (during this same time, this particle would have fallen a distance of around 6 cm in still air and would have attained roughly 90% of its gravitational terminal velocity), whereas for a particle with $r = 500~\mu m$ this time is around 1.1 s, during which the particle will have traveled 350 cm (during this the same time this particle, if released from rest and acted upon by gravity, would have fallen over 300 cm in still air and would have attained a velocity within 2% of its gravitational terminal velocity).

These results imply that with increasing particle size the assumption that a particle rapidly attains its gravitational terminal velocity or that of the surrounding flow becomes increasingly less valid. Sufficiently large SSA particles will fall back to the sea surface before they attain the horizontal velocity component of the surrounding flow and before they attain velocities close to their gravitational terminal velocities; this has important consequences concerning the along-wind flux of large particles near the sea surface, especially for spume drops torn from wave crests (§4.3.6, §5.7). Additionally, large SSA particles might not respond to sudden changes in the velocity of the surrounding air sufficiently rapidly to follow the flow; consequently larger SSA particles "miss" some of the smaller, higher frequency eddies [Yudine, 1959; Csanady, 1963], and the eddy diffusivity would not be independent of particle size (as assumed in §2.4) but would decrease with increasing particle size.

The estimates presented thus far are for drops with $r \lesssim 500~\mu m$; characteristic times and distances for larger drops are not accurately described by these relationships because of effects such as deformation that influence their motion (§2.6.3). Consideration of these effects is generally not necessary for the motion of extremely large SSA particles, as very few SSA particles with $r > 500~\mu m$ will be present more than a very short distance from the sea surface, and these particles will neither attain their gravitational terminal velocities nor the horizontal velocity component of the surrounding flow before falling back to the sea. However, consideration of the distance required for a liquid drop with $r > 500~\mu m$ falling from rest in still air to attain a given fraction of its gravitational terminal velocity is important to ascertain the extent to which laboratory experiments can accurately simulate processes of interest such as SSA production from the impact of raindrops on the sea surface (§4.7.2). As this mechanism is extremely sensitive to the size of the raindrop and its velocity at impact (§4.7.2), it is necessary to consider the distance required for such a drop to attain a large fraction of its gravitational terminal velocity. Estimates for this distance for values of 95% and 99% are presented here, based on experiment results of *Laws* [1941], *Gunn and Kinzer* [1949], and *Wang and Pruppacher* [1977]. The distance required for a drop to attain 95% of its gravitational terminal velocity when falling from rest in still air increases roughly linearly from ~2.5 m for $r = 500$ µm to ~7 m for $r = 1500$ µm, remaining near 7-8 m for drops with $r \gtrsim 1500$ µm. Likewise, the distance required for a drop to attain 99% of its gravitational terminal velocity under similar conditions increases roughly linearly from ~4 m for $r = 500$ µm to ~12 m for $r = 1500$ µm, remaining near 12-13 m (and requiring approximately 2 s) for larger drops. Because changes in the drop shape occur at different times than those at which they reach a given fraction of their terminal velocities, the distance required for a drop to attain a given fraction of its terminal velocity actually decreases with increasing radius in some size ranges. However, for drops with $r \gtrsim 1500$ µm the times and distances required to attain a given fraction of the gravitational terminal velocity are nearly independent of drop size (which can be shown to be a consequence of the near constancy of terminal velocity of drops in this size range; §2.6.3). Such large distances raise concerns about the ability of many laboratory experiments to accurately model drop impaction in relation to SSA production by rainfall (§4.7.2).

2.7. MECHANISMS FOR REMOVAL OF SSA PARTICLES FROM THE ATMOSPHERE

The mechanisms that act to transport SSA particles to the surface of Earth (in this context the ocean), and thus remove them from the atmosphere, are categorized into wet deposition and dry deposition. These two processes and their rates as a function of particle size are important in the context of estimating the size-dependent SSA production flux from knowledge of the size-dependent SSA number concentration (i.e., the steady state dry deposition method, §3.1, and the statistical wet deposition method, §3.9), and also in ascertaining the applicability of methods that rely on the SSA concentration being in, or not being in, steady state. In this section, the principal mechanisms for each process are examined, and the times characterizing the lifetimes of SSA

particles of a given size against removal by these processes are presented. The relative importance of wet deposition to dry deposition is also examined.

2.7.1. Wet Deposition

Wet deposition of particles refers to their transfer from the atmosphere to the surface by precipitation, either by in-cloud or below-cloud scavenging. In-cloud scavenging consists of particles forming cloud drops (§2.1.5.1) that are subsequently removed in precipitation either through gravitational deposition or by interception and accretion by falling hydrometeors. Below-cloud scavenging consists of particles being intercepted by falling drops and thereby removed from the atmosphere. Key factors affecting the rate of wet deposition of particles are their size, shape, and hygroscopicity; the concentrations of other particles and their sizes and hygroscopicities; RH, updraft velocity, and temperature; the frequency of rainfall, and the raindrop size distribution and rainfall rate.

Despite numerous attempts to parameterize the rates of both in-cloud and below-cloud scavenging of aerosol particles and the dependence of these rates on particle size, raindrop size distributions, rainfall rates, and the like [e.g., *Dana and Hales*, 1976; *Slinn*, 1977a,b; *Slinn et al.*, 1978; *Slinn*, 1983a, 1984], these rates remain highly uncertain[1]. Fortunately, the details of these dependencies are relatively unimportant for considerations pertinent to SSA production, as a general conclusion that can be drawn from these studies is that several features of the wet deposition process operate to make it an efficient removal mechanism for SSA particles of all sizes. Because of their low concentrations, high hygroscopicity, and large sizes compared to other particles typically present in the marine atmosphere (§4.1.1.1), SSA particles are highly effective cloud condensation nuclei, and even the low supersaturations associated with marine stratus clouds are sufficient to exceed their activation thresholds (§2.1.5.1). In-cloud scavenging is thought to be especially important for removal of small SSA particles, which are mixed throughout the marine boundary layer (§2.9.5). Additionally, interception by falling raindrops (below-cloud scavenging) is believed to be an efficient removal mechanism for larger SSA particles [*Twomey*, 1977a, p. 146]. For these reasons wet deposition is expected to remove a large fraction of SSA particles of all sizes from the atmosphere during precipitation events of any appreciable intensity and duration [*Slinn*, 1977b, 1983a, 1984].

---

[1] Despite the admonition of *Slinn* [1977b] that users be critical of the expression he proposed for the scavenging efficiency of aerosol particles by falling raindrops and consider it a formulation for researchers to test and not for modelers to use, this expression has, with a few modifications, become a standard formulation for this quantity.

The expected efficient removal of SSA particles in precipitation has been confirmed by measurements. For instance, *Twomey* [1955], based on measurements of SSA particles over land during conditions of on-shore flow (i.e., from the ocean to the land), reported that precipitation was very efficient in removing SSA particles, and that in air in which rain had fallen the size distribution of SSA particles was reduced by approximately the same ratio throughout the size range sampled (which was not explicitly stated but was probably $r_{80} \gtrsim 1$ μm). A decrease by nearly an order of magnitude in the number concentration of SSA particles with $r_{80} > 3$ μm after a brief (~10 min) shower was reported by *Lodge* [1955]. The total aerosol particle concentration reported by *Parungo et al.* [1986a] in a marine environment from shipboard measurements approximately 3 h after 10 h of intermittent rain, corresponding to ~0.5 cm of precipitation, was about one-fourth that before the rain, and when the ship had sailed out of the zone in which it had rained, the particle concentration, size distribution, and chemical composition were similar to those of pre-rain samples.

The efficiency of wet deposition as a removal mechanism for SSA particles of all sizes ensures that the occurrence of precipitation will under most circumstances greatly reduce SSA concentration and thus will affect the relation between SSA production and SSA concentration. Even if wet deposition does not decrease the SSA concentration to zero, it diminishes the abundance of SSA in the atmosphere sufficiently that the subsequent rate of increase of concentration can still be tracked to determine the production flux. The key quantity describing the effect of wet deposition on SSA concentration is the time since the aerosol last experienced a major precipitation event. In principle, if this time were known, then it might be possible to determine the production flux from the measured concentration of SSA particles for which the main removal mechanism is wet deposition (*cf.*, §3.3, §3.6, and §3.9). Knowledge of the elapsed time since wet removal is important also in identifying the applicability of the steady state dry deposition method for determining the SSA production flux (§3.1), which assumes that SSA production is in local steady state with dry deposition and thus that SSA concentration is controlled by dry deposition. The validity of this assumption depends on the time required for the concentration of SSA particles of a given size to attain steady state under the influence of dry deposition, which depends strongly on r_{80} (§2.9.6), compared to the time subsequent to the prior wet removal event. In view of the efficiency of wet deposition as a removal mechanism for SSA particles, it is clear that this assumption is not satisfied during a precipitation event, and also after such an event for a time that likewise depends to a large extent on r_{80} (§2.7.3).

As precipitation is episodic in its occurrence, the time since the last precipitation event experienced by an air parcel is generally difficult to determine, and certainly this elapsed time has rarely if ever been reported in conjunction with measurements of SSA concentration. Additionally, there are numerous complicating factors. For example, if precipitation commences above the marine boundary layer then although most larger SSA particles may be removed by below-cloud scavenging, smaller SSA particles will not necessarily be activated and removed by in-cloud scavenging. An alternative approach to estimating the time since the last appreciable precipitation event is to resort to the statistics of the time between precipitation events (leading to the statistical wet deposition method; §3.9). To this end, a characteristic time $\tau_{\mathrm{wet}}(r_{80})$ is defined as the mean residence time of SSA particles with given r_{80} against removal through wet deposition (on the assumption that no other processes operate to remove particles from the atmosphere). Under the assumption that wet deposition is a highly efficient removal process for SSA particles of all sizes, this quantity is roughly independent of r_{80}. This characteristic time τ_{wet} is related to the distribution of the frequency of air parcels encountering precipitation events, or, alternatively, to the distribution of the time between these events. An estimate for τ_{wet} is the mean time between rainfall events, which in marine environments is usually several days to a week [*Rodhe and Grandell*, 1972; *Hamrud et al.*, 1981; *Giorgi and Chameides*, 1986; *Hamrud and Rodhe*, 1986; *Pruppacher and Jaenicke*, 1995]. Thus, $\tau_{\mathrm{wet}} = 3$ d ($\approx 3 \cdot 10^5$ s) is assumed here as typical value. This time is comparable to values assumed by *Junge and Gustafson* [1957], *Giorgi and Chaemeides* [1986], and *Capaldo et al.* [1999], and under most circumstances this value is expected to be accurate to within a factor of two or three.

2.7.2. Dry Deposition

Dry deposition of gaseous and particulate matter refers to transfer of this material to Earth's surface (which may be ocean or land, including vegetated surfaces) by mechanisms not involving precipitation. Key factors affecting the rate of dry deposition of particles are their size, shape, density, and hygroscopicity; RH, wind speed, and atmospheric stability and turbulence; and properties of the surface to which the transfer occurs [*Sehmel*, 1980]. Over the ocean the principal mechanisms that have been considered for dry deposition of particles are gravitational sedimentation, turbulent transfer, Brownian diffusion, impaction, interception by waves, and scavenging by spray. For SSA particles, the rate of dry deposition is strongly dependent on particle size and wind speed (§4.6.2).

Dry deposition results in a downward flux of SSA particles. The interfacial dry deposition flux of SSA particles of a given r_{80} was defined in §2.1.6 as the flux of particles deposited to the ocean surface; it is equal to the difference between the interfacial production flux and the net vertical flux of SSA particles through the ocean surface. This definition is unambiguous, as the ejection of drops from the sea surface and their deposition to the surface are distinctly different processes. The situation with the effective dry deposition flux of SSA particles is not so unambiguous, and the presence of a surface source requires care in the definition of this quantity (§2.1.6). Although it may be going too far to conclude, as did *Wesely and Hicks* [2000], that the concept of a (dry) deposition velocity (and by extension the effective dry deposition flux) is not applicable under conditions where there is a surface source, the situation is certainly more complicated under these conditions. The effective SSA dry deposition flux in the presence of a surface source is taken (§2.1.6) as the net downward flux of SSA particles through a horizontal plane at $z = z_1$ (where z_1 is typically near 10 m) that would occur for a given concentration field in the absence of a surface source, but with all other processes operating.

Much of the early work on dry deposition involved the deposition of radioactive particles, and much subsequent work has focused on the transfer of gases to a surface. For these situations, as well as for the deposition of SSA particles over land, there is generally not a surface source (notable but important exceptions being water vapor and CO_2), and the mean net vertical flux is downward and independent of height. However, dry deposition of SSA particles over the oceans differs from most other situations involving dry deposition because of the presence of a source at the surface, resulting in the net flux of particles from the atmosphere to the surface, which is the measurable quantity, being less than the gross (one-way) flux of particles to the surface, the quantity that is referred to in (2.1-25) as the dry deposition flux and the quantity that is calculated in dry deposition models. Many investigators have taken the dry deposition flux to be the negative of the net flux and likewise have taken the dry deposition velocity as this dry deposition flux divided by the number concentration at the reference height (§2.7.2.1). However in the presence of the surface source this flux is less than that calculated by dry deposition models—indeed this flux can be negative in situations for which the net vertical flux is upward and it is often reported as such; negative dry deposition velocities have likewise been reported as the ratio of the net upward flux to the atmospheric concentration. As the production flux is entirely unrelated to the ambient concentration, expressing a production flux in this way is wholly without meaning. In this review the dry deposition flux is defined (§2.1.6) as the mean net downward flux through a horizontal reference plane at a given height that would occur in the absence of surface sources; the mean net flux is employed because particles can be repeatedly transported through this height by turbulent

motion of the air, so employing a gross downward flux would lead to multiple counting. This definition corresponds to the dry deposition flux calculated by models.

Implicit in most discussions of dry deposition are the assumptions that there is a large reservoir of the substance being deposited (in this context, SSA particles of a given r_{80}) at heights much greater than the height of measurement (a related discussion is presented in *Slinn* [1983a]), and that dry deposition acts sufficiently slowly compared to other processes of interest that it does not appreciably decrease the amount (or concentration) of the substance in the atmosphere over times characterizing these other processes. As a consequence of these assumptions, the system can be considered to be near steady state in the sense that removal through dry deposition acts on a longer time scale than that characterizing other mechanisms that act to change the concentration, and the substance that is experiencing dry deposition remains well mixed vertically over heights much greater than the height of measurement. For most applications of dry deposition other than for SSA particles these conditions are satisfied, and the substance of interest is sufficiently well mixed vertically that there is a net downward flux at the height of measurement that does not depend strongly on this height, in principle allowing the dry deposition flux to be determined from concentration measurements at a single height. However, for SSA particles, depending on their size, these conditions might not be satisfied. Although sufficiently small SSA particles are readily transported upward and become nearly uniformly mixed over great heights (§2.9.5), with increasing particle size an increasing fraction of particles that are produced at the ocean surface does not attain a height greater than the reference height. When SSA particles attain their maximum heights and begin their descents to the surface, a volumetric source of downward flux is created, the magnitude of which depends strongly on height. A similar situation occurs where height-dependent vertical fluxes result from chemical reactions, for example the loss of particulate ammonium nitrate and the production of gaseous NH_3 by dissociation of ammonium nitrate [*Kramm and Dlugi*, 1994] or production of nitric acid by reaction of OH with NO_2 [*Geernaert et al.*, 1998]. Such a volume source of downward flux violates the constant flux condition (discussed below) which allows flux matching and which is required to derive dry deposition velocities [*Slinn*, 1983a]. Although this condition could be restricted to account for the downward flux of only those SSA particles which have attained the height of measurement, the utility of this approach is diminished with increasing particle size as the concentration becomes more strongly dependent on the height. These considerations lead to an upper limit on the size of SSA particles of around $r_{80} = 25 \ \mu m$ for which dry deposition methods can be meaningfully applied (§2.9).

2.7.2.1. Dry deposition velocity. The *ansatz* is generally made that the dry deposition flux (positive downward) of SSA particles of a specific r_{80} (the dependence on which is suppressed but should be understood) through a horizontal plane $z = z_1$ is proportional to the mean number concentration of SSA particles of this size at some reference height z_{ref}:

$$f_d(z_1) = n(z_{ref}) \times v_d(z_1, z_{ref}), \qquad (2.7\text{-}2)$$

where the constant of proportionality $v_d(z_1, z_{ref})$ is denoted the dry deposition velocity of SSA particles of the given size [*Johnstone et al.*, 1944; *Gregory*, 1945; *Chamberlain and Chadwick*, 1953]. Dry deposition velocities of SSA particles are usually determined from models (§2.7.2.2). Knowledge of the size-dependent SSA dry deposition velocity, which depends on ambient conditions, permits calculation of the size-dependent SSA dry deposition flux from measurements of the size-dependent SSA concentration at a given reference height (typically near 10 m). As the dry deposition velocity defined by (2.7-2) is based on the downward flux and not the net vertical flux, it is necessarily positive.

The dependence of v_d on z_1 and z_{ref}, although explicitly noted by several investigators [*Hicks*, 1976; *Slinn and Slinn*, 1981; *Slinn*, 1983a], does not appear to be widely appreciated and is usually left implicit, but it is essential to the definition of this quantity. Difficulties arise if $z_1 \neq z_{ref}$ as both the concentration and the dry deposition flux of SSA particles of a given size can depend on height. In such situations it is not possible to apply the definition of the dry deposition velocity given by (2.7-2) unless assumptions (typically involving steady state) are made regarding the vertical profile of number concentration. For $z_1 > z_{ref}$, measurements of SSA concentrations at z_{ref} cannot provide any information on the maximum heights these particles attain, and thus cannot determine whether or not they contribute to the dry deposition flux at z_1. Likewise, for $z_1 < z_{ref}$, measurements of concentrations at height z_{ref} will not include those SSA particles produced at the sea surface that attain a maximum height is greater than z_1 but less than z_{ref}, and which therefore do not contribute to the SSA dry deposition flux at z_{ref} (§2.1.6). Thus, it is not possible to determine the interfacial SSA dry deposition flux (or the interfacial SSA production flux) from concentration measurements at any appreciable distance from the surface without additional assumptions. For these reasons, z_1 and z_{ref} are taken to be equal, typically 10 m, the height at which concentrations are often measured. With such a choice, (2.7-2) yields the effective SSA dry deposition flux (§2.1.6). For the dry deposition models and velocities discussed in this review, it should be understood, unless otherwise specified, that $z_1 = z_{ref}$; hence the dependence of v_d on z_{ref} and z_1 is typically omitted. As noted by *Slinn* [1983a], v_d generally depends only weakly on height, and the neglect of this dependence is often minor compared to other uncertainties.

2.7.2.2. Dry deposition models. Dry deposition velocities of SSA particles are usually derived from one-dimensional models. Many such models have been proposed, but most are based ultimately on that of *Slinn and Slinn* [1980, 1981]. According to this model, the atmosphere over the ocean below a reference height, z_{ref} (typically near 10 m), is conceptually divided into two layers, in a manner originally proposed by *Sverdrup* [1937] as described in §2.4. The lowest layer consists of a viscous sublayer adjacent to the ocean surface. This layer is also known as the diffusion, laminar, or quasi-laminar sublayer, the interfacial layer, or the deposition layer (*Slinn and Slinn* [1980] distinguished between the diffusive sublayer and the viscous sublayer). The thickness of this layer, δ, is usually given as 0.1-1 mm; estimates of ~0.04 mm, ~1 mm, and ~3 mm were presented in §2.4. Above this is the so-called surface layer, which generally extends to and above the reference height. Each layer is characterized by a (different) uniform RH (§2.4, §4.6.1) with which SSA particles equilibrate instantaneously (this assumption is discussed in §4.6.3); as SSA particles are hygroscopic, the equilibrium radius of a given particle differs between the two layers (§2.5.3). These models implicitly assume that SSA particles are sufficiently well mixed to heights greater than the height of measurement that their downward flux does not depend strongly on this height. Additionally, these models implicitly assume that dry deposition is sufficiently slow, and that the heights to which SSA particles of a given size are mixed are sufficiently greater than the reference height, that the reservoir of SSA particles of this size above the reference height is not appreciably depleted on a time scale characterizing the response of the vertical concentration profile to changing conditions. These assumptions eliminate some of the complications arising from the existence of a surface source and make it possible to discuss the downward (i.e., dry deposition) flux *through* a layer instead of the downward flux *through the top of the layer*, allowing the assumption to be made that the downward flux at the bottom of the surface layer is equal to the downward flux at the top of the viscous sublayer. This flux matching immediately yields an expression for the dry deposition velocity.

In determination of the size-dependent dry deposition velocity, mechanisms that operate to transport SSA particles (gravitational sedimentation, turbulent diffusion, Brownian diffusion, and impaction on the sea surface) are parameterized by transport velocities that depend strongly on particle size and on meteorological conditions. The downward flux through each layer is obtained in terms of these transport velocities (specific formulations for these velocities are presented in §4.6), and the overall dry deposition velocity is obtained by matching the downward fluxes through the two layers. Although these transport velocities, and the resultant dry deposition velocity, are often formulated in terms of the wind friction velocity u_*, they can equivalently be expressed in terms of macroscopic quantities such as U_{10} by using the wind stress coefficient C_D (§2.3.2), and thus can be parameterized (§4.6) in terms of r_{80}, U_{10}, and RH.

In the viscous sublayer, the important contributions to SSA particle transfer are Brownian diffusion, impaction, and gravitational sedimentation (§2.6.1); transport by diffusiophoresis and by thermophoresis, while often mentioned, are typically not included in dry deposition models but may be important in some circumstances (§4.6.2). The flux of SSA particles with given r_{80} through this layer due to Brownian diffusion is assumed to be proportional to the difference between the number concentration of particles of this size at the top of this layer n_δ, and that at the ocean surface n_0, the constant of proportionality being defined as the Brownian diffusion velocity v_{diff}. As impaction acts only to remove particles, the flux of particles of this size through this layer due to this mechanism is assumed [*Fairall and Larsen*, 1984] to be proportional to the number concentration of these particles at the top of the viscous sublayer n_δ (and not to the difference $n_\delta - n_0$ as assumed in earlier formulations, such as that of *Slinn and Slinn* [1980]), with the constant of proportionality denoted the impaction velocity v_{imp}. The flux of SSA particles of the given size through this layer due to gravitational sedimentation is given by the product of their number concentration at the top of this layer n_δ and the gravitational terminal velocity corresponding to their equilibrium radius with respect to the RH of this layer (i.e., 98%), $v_{V,Stk}$. For the range of particle sizes considered (i.e., small and medium SSA particles), v_{term} is equal to the Stokes velocity v_{Stk} given by (2.6-7). As these three fluxes act in parallel, the resultant size-dependent deposition flux of SSA particles through the viscous layer is given by their sum:

$$f_{d,V} = v_{diff}(n_\delta - n_0) + (v_{imp} + v_{v,Stk})n_\delta. \quad (2.7\text{-}3)$$

The ocean is treated as a perfect sink in that particles, once reaching the surface, do not rebound but remain in the water (although the ocean is also a source of SSA particles, the flux of these particles is treated separately; §2.1.6; §2.7.2.1). Thus n_0 is set to zero, and the transfer velocity for the viscous layer is therefore given by

$$v_V = v_{imp} + v_{diff} + v_{V,Stk}. \quad (2.7\text{-}4)$$

In the surface layer SSA particles are transported primarily by turbulent diffusion (which greatly exceeds Brownian diffusion; §2.6.1) and by gravitational sedimentation (§2.6). The flux of SSA particles of a given size due to turbulent diffusion under neutrally stable conditions is assumed to be

proportional to the derivative of the number concentration with respect to height (Eq. 2.6-5). Typically this flux is taken to be proportional to the difference between the number concentration of SSA particles of this size at the reference height and that at the lower surface of this layer, $n_{ref} - n_\delta$, with the constant of proportionality defining the turbulent diffusion velocity v_{turb} [*Slinn et al.*, 1978]. The flux of these particles through the surface layer due to gravitational sedimentation is given by the product of n_{ref}, the number concentration of SSA particles of the given size at the reference height z_{ref}, and the gravitational terminal velocity (also given by the Stokes velocity) corresponding to their equilibrium radius with respect to the RH of this layer (i.e., 80%), $v_{S,Stk}$. Because the radius of a given SSA particle in the surface layer is less than that of the same particle in the viscous sublayer as a consequence of the lower RH in the surface layer, $v_{S,Stk} < v_{V,Stk}$. As the gravitational and turbulent diffusion fluxes also act in parallel, the size-dependent SSA deposition flux through the surface layer is given by

$$f_{d,S} = v_{turb}(n_{ref} - n_\delta) + v_{S,Stk}\, n_{ref}, \quad (2.7\text{-}5)$$

from which a transfer velocity for the surface layer may be defined by

$$v_S = v_{turb} + v_{S,Stk}. \quad (2.7\text{-}6)$$

Equating the fluxes through the two layers (and assuming $n_0 = 0$) yields the overall size-dependent SSA dry deposition velocity,

$$\begin{aligned} v_d &= \frac{(v_{V,Stk} + v_{diff} + v_{imp}) \times (v_{S,Stk} + v_{turb})}{(v_{V,Stk} + v_{diff} + v_{imp}) + v_{turb}} \\ &= \frac{v_V v_S}{v_V + (v_S - v_{S,Stk})}, \end{aligned} \quad (2.7\text{-}7)$$

which can be calculated from the individual transfer velocities (usually determined from models; §4.6.1, §4.6.3). The ratio of the concentration at the reference height to that at the top of the viscous sublayer is given by

$$\begin{aligned} \frac{n_{ref}}{n_\delta} &= \frac{(v_{V,Stk} + v_{diff} + v_{imp}) + v_{turb}}{v_{S,Stk} + v_{turb}} \\ &= \frac{v_V + v_{turb}}{v_S}, \end{aligned} \quad (2.7\text{-}8)$$

implications of which are discussed in §4.6.5 and §5.1.2.

The expression for the dry deposition velocity, (2.7-7), essentially follows from a discretized version of the transport equation with three points: the surface, height δ, and height z_{ref} (a formulation for the dry deposition velocity in which transfer through the surface layer is modeled by a differential equation instead of a difference equation was presented by *Hoppel et al.* [2002], but the resulting dry deposition velocities differ little from those given by Eq. 2.7-7; §4.6.3). Because of the complexity of the processes occurring over the oceans, the idealized nature of dry deposition models and their treatment of the marine atmosphere and of dry deposition of SSA particles limit the confidence that can be placed in numerical results obtained from such models (§4.6.5).

The dry deposition velocity v_d is expected to increase with increasing r_{80} over nearly all of the size range comprising SSA particles primarily because of the contribution from gravitational sedimentation. Additionally, v_d is expected to increase with increasing U_{10} because of increased vertical mixing in the surface layer, increased diffusion through the viscous sublayer due to decreased thickness of that layer, and increased impact velocity. Because gravitational sedimentation does not depend on a difference in concentrations, a simple resistance analogy, commonly made for dry deposition of gases, no longer holds [*Slinn*, 1983a], although a more complex resistance analogy can be used [*Seinfeld and Pandis*, 1998, p. 961].

The dry deposition velocity given by (2.7-7), if used with the concentration at z_{ref} (typically near 10 m), yields the downward flux of SSA particles at this height. Under steady state conditions, this downward flux is equal to the upward flux of SSA particles at this height. This effective SSA production flux, $f_{eff}(r_{80})$, is the quantity that is desired for atmospheric chemistry considerations and for inputs into large-scale transport and chemistry models, as it comprises only those particles that are expected to remain in the atmosphere for an appreciable length of time. It is assumed that particles that attain this height remain in the atmosphere long enough to participate in the various processes of interest, whereas those that do not attain this height are not important in this regard (§2.1.6). However, comparison with other methods of determining SSA production fluxes that yield the production flux at the surface of the ocean, $f_{int}(r_{80})$, such as that using laboratory whitecaps extrapolated to ocean conditions (§3.2) or that involving bubble populations and the number of drops per bubble (§3.4), requires some method of relating the effective SSA production flux $f_{eff}(r_{80})$ and the interfacial SSA production flux $f_{int}(r_{80})$. As noted above, there is no general means of relating these quantities, although several methods have been proposed that are expected to yield results that are qualitatively valid (§2.9.4; §4.6.5; §5.1).

2.7.2.3. Mean atmospheric residence time against dry deposition. The mean atmospheric residence time of SSA particles of a given r_{80} with respect to dry deposition is denoted by $\tau_{dry}(r_{80})$, analogous to the quantity τ_{wet} defined in §2.7.1. The quantity τ_{dry} depends strongly on particle size and on meteorological conditions such as wind speed. Furthermore, it is expected to be inversely proportional to v_d and directly proportional to the height over which particles are mixed (§2.9.6). At $U_{10} = 10$ m s^{-1} and for a marine boundary layer height of 0.5 km (§2.4), estimated values of τ_{dry} for SSA particles with $r_{80} = 1, 5, 15,$ and 25 μm, based on arguments presented in §2.9.6 and on modeled dry deposition velocities presented in §4.6.2, are approximately $1 \cdot 10^6$ s (~1.5 wks), $3.3 \cdot 10^4$ s (~10 h), $5 \cdot 10^3$ s (~1.5 h), and 300 s (5 min), respectively (Table 8). Implications of these residence times and their dependence on particle size are examined in the next section.

Table 8. Characteristic Times and Distances for Removal of Sea Salt Aerosol Particles by Dry Deposition as a Function of r_{80} for Wind Speed $U_{10} = 10$ m s^{-1}

$r_{80}/\mu m$[a]	1	2	5	10	15	20	25
Dry deposition velocity, $v_d/(cm\ s^{-1})$[b]	0.05	0.25	1.5	3	5	7	10
Mixing height, H_{mix}/m[c]	500	500	500	500	230	60	30
Dry deposition residence time, τ_{dry}/s[d]	$1 \cdot 10^6$ (1.5 wks)	$2 \cdot 10^5$ (2.3 d)	$3.3 \cdot 10^4$ (10 h)	$1.7 \cdot 10^4$ (5 h)	5000 (1.5 h)	850 (15 min)	300 (5 min)
Transport distance, X/km[e]	$1 \cdot 10^4$	2000	330	170	50	8.5	3

[a] Assumed to have equilibrated at 80% RH.

[b] From §4.6.2.

[c] Taken to be the lesser of the height of the marine boundary layer height (assumed to be 0.5 km) and z_{50}, the height at which the steady state concentration is 50% of its value at 10 m (§2.9.5).

[d] Defined by H_{mix}/v_d.

[e] Defined by $H_{mix}U_{10}/v_d$.

2.7.3. Relative Importance of Wet Deposition and Dry Deposition

A wide range of estimates has been presented for the relative importance of wet deposition and dry deposition as removal processes for SSA particles [*Eriksson*, 1959; *Blanchard*, 1963, pp.153-155; *Junge*, 1972; *Slinn et al.*, 1978; *Slinn*, 1983a; *Blanchard*, 1985a; *Erickson and Duce*, 1988]. However, the assumptions inherent in such estimates and the large uncertainties of scavenging coefficients and other quantities upon which they are based diminish the confidence that can be placed in them for a wide range of particle sizes. Additionally, as mass-weighted deposition is dominated by larger particles, comparison of total dry deposition and total wet deposition fluxes of SSA mass does not provide information on the size dependence of the relative importance of these two mechanisms, nor does it lead to any indication of the size range of SSA particles for which one or the other mechanism is the dominant one.

The relative importance of wet and dry deposition of SSA particles depends strongly on particle size, and additionally on the meteorological conditions experienced by the particles subsequent to their production. Consequently, this relative importance can vary greatly with season and location. The strong dependence on particle size is a consequence of the vastly different size dependencies of the mean atmospheric residence times against removal through wet deposition and through dry deposition. The mean atmospheric residence time against removal through wet deposition, τ_{wet}, was estimated (§2.7.1) to be ~3 d (to within a factor of two or three), and it is assumed to be nearly independent of particle size. In contrast, the estimates for τ_{dry} presented in §2.7.2.3 exhibit a strong dependence on particle size—for $U_{10} = 10$ m s^{-1}, τ_{dry} decreases from approximately 1.5 weeks to 5 min as r_{80} increases from 1 μm to 25 μm. For small SSA particles ($r_{80} \lesssim 1$ μm), τ_{wet} is much less than τ_{dry}, and it can be concluded that wet deposition is more important than dry deposition as a removal mechanism in nearly all situations. An important consequence of this conclusion is that the size distribution of the number concentration of small SSA particles in the atmosphere should be nearly the same as the size distribution of their production flux (§3.9). With increasing particle size τ_{dry} decreases rapidly, and dry deposition becomes increasingly important compared to wet deposition. For SSA particles with $r_{80} \gtrsim 3$ μm, τ_{dry} is much less than τ_{wet}, and it can be concluded that these particles will likely be removed by dry deposition before they are removed by wet deposition. Comparison of τ_{wet} and $\tau_{dry}(r_{80})$ also provides a rough estimate for the lower bound on the size of SSA particles for which the production flux can be determined from the steady state dry deposition method, which requires that τ_{wet} be considerably greater than τ_{dry} (§3.1), and it can be concluded that

in general this method is restricted to SSA particles with r_{80} greater than several micrometers (§3.1).

2.8. TRANSPORT OF SSA IN THE MARINE BOUNDARY LAYER

Treating SSA particles collectively and characterizing them by their size-dependent number concentration $n(r_{80})$ allows derivation of a continuity equation that describes the temporal evolution of the concentration of SSA particles of a given size and yields insight into the mean lifetime of these particles and the extent to which they are mixed vertically in the marine atmosphere. In this section, this equation is considered in some detail, allowing derivation of expressions for the effective production flux of small and medium SSA particles and specification of the conditions required by the various methods used to determine this flux.

Several assumptions are made which greatly simplify this analysis and which are expected to be valid in most circumstances. Sea salt aerosol particles are assumed to attain their equilibrium radii with respect to the ambient RH instantaneously; this generally takes less than one minute for small and medium SSA particles (§2.5.4), much shorter than the times characterizing changes in the number concentration of SSA particles in this size range. These particles are assumed to attain the horizontal velocity component of the surrounding flow and also their gravitational terminal velocities instantaneously; this generally takes less than a few hundredths of a second for small and medium SSA particles (§2.6.2). For SSA particles of these sizes the horizontal and vertical motions are independent of each other (§2.6.2). It is assumed that SSA particles are present in sufficiently low concentrations that they do not appreciably interact with each other or affect the mean flow; that is, SSA concentration is treated as a passive scalar [§4.1.2.1; §4.1.3.1; *Bortkovskii*, 1987]. Finally, the simplified model of the marine boundary layer presented in (§2.4) is employed. This model treats the sea surface as a horizontal plane at which production occurs and the marine boundary layer as consisting of only two layers: a viscous sublayer, and a surface layer which extends to the top of the marine boundary layer, below which SSA particles are confined. Within this surface layer the dependence of the mean wind speed on height can be neglected with little loss in accuracy in many situations, and it to this layer that this analysis mainly pertains.

This analysis is restricted to determination of the effective production flux of small and medium SSA particles. Although in principle it might also be applied to determine the interfacial production flux of SSA particles of these sizes, the inability of determining the interfacial SSA dry deposition flux from concentration measurements at heights of several meters or more above the surface would hinder such an application. Likewise, this analysis might be applied to determination of the effective production flux of large SSA particles, but this quantity is near zero (§2.1.2), and there are several additional factors that would complicate such an attempt. Large SSA particles are confined mainly to the lowest few meters above the sea surface (§2.9), where waves make accurate determination of the velocity field difficult. The geometric radius of a large SSA particle with a given r_{80} depends on the time subsequent to its formation because of the long times required for equilibration with respect to RH (§2.5.4), and it is unlikely that measurements of the radii of these particles reflect their equilibrium values (§2.9.1). Additionally, large SSA particles often fall back to the sea surface before they attain the horizontal velocity component of the surrounding air or their gravitational terminal velocities, and their horizontal and vertical motions are not independent (§2.6.4).

2.8.1. Continuity Equation for the Size-Dependent SSA Number Concentration

The continuity equation for the size-dependent SSA number concentration n (the dependence of this and other quantities on r_{80} is often suppressed in the following discussion, but should be understood), which represents changes in the concentration of SSA particles of a given size due to transport and to gain and loss processes, is given by

$$\frac{\partial n}{\partial t} + \nabla \cdot (n\vec{U} + n\vec{v}_s - D_B \nabla n) = g_V - l_V, \qquad (2.8\text{-}1)$$

where $\vec{U}$ is the three-dimensional wind velocity, $\vec{v}_s$ is the slip velocity, D_B is the Brownian diffusion coefficient of the particles (§2.6.1), and g_V and l_V are the volumetric gain and loss terms, respectively (similar equations are presented in *Slinn et al.* [1978]; *Slinn and Slinn* [1981]; *Slinn* [1983a]; *Fairall and Larsen* [1984]; *Businger* [1986]; and *Fairall and Davidson* [1986]). The quantities $\vec{v}_s$, D_B, g_V, and l_V generally depend strongly on particle size. The slip velocity describes the motion of particles relative to the mean flow arising from mechanisms such as gravitational or electrical forces, diffusiophoresis and Stefan flow, thermophoresis, and the failure of particles due to their inertia to exactly follow changing flow conditions (which may result in impaction at the sea surface; *Slinn* [1983a]). The gain and loss terms describe the change in the number concentration resulting from processes that produce or remove SSA particles in a given size range. As particle sizes are described by r_{80}, there are no contributions to these terms from particles moving into and out of size bins because of condensation or evaporation of water, and the only mechanisms that might contribute are wet deposition (i.e., removal by precipitation;

§2.7.1), coagulation (which would result in a decrease in the number concentration of smaller particles and an increase in the number concentration of larger ones), and formation of new particles by fracture of larger ones (which would result in a decrease in the number concentration of larger particles and an increase in the number concentration of smaller ones). Wet deposition is typically not important as a loss term, as measurements of SSA concentration or flux are rarely made in conditions for which precipitation is occurring (however, in many situations the significance of wet removal is determined not by the length of the time of measurement, but rather by the lifetime of the particles in the atmosphere, which is typically much greater; §2.7.1; §3.1). Coagulation is not important for SSA particles in the marine atmosphere because of their relatively low concentrations and large sizes (§4.1.3.1; §4.1.4.5), and is also neglected here. There is little convincing observational support for SSA particle fracture in the atmosphere (§4.7.4) and it is likewise omitted from consideration here. Thus, the resulting equation is linear in n; for situations in which the concentrations are so low that they cannot be treated in a continuous fashion, this linearity allows the equation to be interpreted in a statistical sense.

With z being defined as the height above the sea surface (positive upward), x and y the along-wind and cross-wind directions, respectively, parallel to the plane of the mean sea surface, x being positive downwind (implicit in this definition is that a mean wind direction can be defined and is, to some extent, constant), and u, v, and w denoting the components of the wind velocity in the x-, y-, and z-directions, respectively, (2.8-1) can be written

$$\frac{\partial n}{\partial t} + \frac{\partial j_x}{\partial x} + \frac{\partial j_y}{\partial y} + \frac{\partial j_z}{\partial z} = g_V - l_V, \qquad (2.8\text{-}2)$$

where j_z, the net flux of SSA particles of a given size in the z-direction, is given by

$$j_z = nw + nv_{s,z} - D_B \frac{\partial n}{\partial z}, \qquad (2.8\text{-}3)$$

with similar expressions for j_x and j_y.

2.8.2. Decomposition Into Mean and Fluctuating Components

As the air flow in the marine boundary layer is nearly always turbulent, the several quantities in (2.8-1), (2.8-2), and (2.8-3)—the concentration n of SSA particles in a given range of r_{80}; the wind speed components u, v, and w; and, because of their dependence on the physical particle size and on air properties, the Brownian diffusivity D_B and the components of the slip velocity $v_{s,x}$, $v_{s,y}$, and $v_{s,z}$—all contain high-frequency fluctuations. For many purposes, the quantities of interest are averages over time scales longer than those characterizing these high-frequency fluctuations; these mean quantities are the ones obtained from measurements of SSA concentration or fluxes taking place over periods of order seconds to minutes. Mathematically, this situation is treated by the customary Reynolds method [*Reynolds*, 1894] in which a quantity q is decomposed into the sum of its mean component (denoted by an overbar) and its fluctuating component (denoted by a prime): $q = \bar{q} + q'$. Implicit in this decomposition is the occurrence of two separate time scales: one characterizing the fluctuations, and a much longer one characterizing the time variation of the mean quantities (this topic is discussed in *Bernstein* [1966] and *Robinson* [1967]). All relevant quantities are decomposed in this manner and are substituted into the pertinent equation, which is then averaged over a time intermediate to these two time scales; sufficiently long to smooth out fluctuations but sufficiently short that the mean quantities do not vary appreciably. As most equations of interest are nonlinear, cross terms from the products of fluctuating components generally result.

Applying this procedure to (2.8-2) yields

$$\frac{\partial \bar{n}}{\partial t} + \frac{\partial \bar{j}_x}{\partial x} + \frac{\partial \bar{j}_y}{\partial y} + \frac{\partial \bar{j}_z}{\partial z} = \bar{g}_V - \bar{l}_V, \qquad (2.8\text{-}4)$$

where the mean along-wind flux $\bar{j}_x$ positive downwind (§2.1.6), is given by

$$\bar{j}_x = \bar{n}\,\bar{u} + \overline{n'u'} + \bar{n}\,\bar{v}_{s,x} + \overline{n'v'_{s,x}} \\ - \bar{D}_B \frac{\partial \bar{n}}{\partial x} - \overline{D'_B \frac{\partial n'}{\partial x}}, \qquad (2.8\text{-}5)$$

with similar equations for $\bar{j}_y$ and $\bar{j}_z$. The first term on the right hand side of (2.8-5) represents the horizontal transport of particles by the mean flow and is generally the dominant term. The second term on the right hand side represents the time-average transport resulting from fluctuating components of the wind in the along-wind direction and is expected to be negligible compared to the first term. The next two terms represent the contributions from slip flows. The gravitational force does not have a component in the x-direction, and slip flows besides that due to gravity often result from gradients, which are predominantly in the vertical direction; thus these terms are expected to be negligible. The final two terms on the right hand side of (2.8-5) represent contributions from Brownian diffusion and are generally negligible. Because all of these terms appear in (2.8-4) as $\partial \bar{j}_x / \partial x$, it is their derivatives in the along-wind direction rather than their magnitudes that are important, although in practice only the first term in (2.8-5) is generally kept.

The mean flux in the cross-wind direction $\bar{j}_y$ is given by

$$\bar{j}_y = \bar{n}\,\bar{v} + \overline{n'v'} + \bar{n}\,\bar{v}_{s,y} + \overline{n'v'_{s,y}} \\ - \bar{D}_B \frac{\partial \bar{n}}{\partial y} - \overline{D'_B \frac{\partial n'}{\partial y}}. \qquad (2.8\text{-}6)$$

The first term on the right hand side can be set to zero because $\bar{v} = 0$ (which follows from the assumption made above that a mean wind direction can be defined and does not change over the time of averaging), and under the generally made assumption of horizontal homogeneity in the y-direction the other terms are likewise equal to 0; hence $\bar{j}_y = 0$.

The mean net vertical flux $\bar{j}_z$ (i.e., the effective net SSA flux through the height under consideration) is given by

$$\bar{j}_z = \bar{n}\,\bar{w} + \overline{n'w'} + \bar{n}\,\bar{v}_{s,z} + \overline{n'v'_{s,z}} \\ - \bar{D}_B \frac{\partial \bar{n}}{\partial z} - \overline{D'_B \frac{\partial n'}{\partial z}}. \qquad (2.8\text{-}7)$$

The first term on the right hand side of (2.8-7), which represents vertical transport of particles by mean vertical flow, is usually set to zero on account of there being no mean wind velocity in the vertical direction ($\bar{w} = 0$). However, this is not always the situation, and this term must be included if there is large-scale convective motion or if there are heat and water vapor fluxes resulting in Stefan flow (§2.6.1). The second term on the right hand side of (2.8-7) represents the vertical flux due to turbulent diffusion, which is assumed to be proportional to the vertical gradient of the mean number concentration (2.6-5):

$$\overline{n'w'} = -D_{eddy} \frac{\partial \bar{n}}{\partial z}, \qquad (2.8\text{-}8)$$

where D_{eddy}, which characterizes the turbulent diffusion of SSA particles, is generally assumed to be the same as that given by (2.4-2), which describes the turbulent transport of momentum. The next two terms represent the transport of particles by slip flow: the first is the transport of the mean concentration of particles of a given size, primarily through gravitational sedimentation, and the second (which has typically been omitted in previous treatments of this topic) accounts for contributions from correlations between the fluctuating components of the vertical slip flow and of the SSA number concentration. Gravity itself does not fluctuate, but the (gravitational) terminal velocity, given by (2.6-7), contains a fluctuating component because of its dependence on the density and kinematic viscosity of the air (which vary with temperature) and on the RH because of its influence on particle geometric radius and density, although these contributions are probably unimportant. Other mechanisms resulting in a slip velocity, such as diffusiophoresis, can also contribute to both of these terms. Impaction (§2.7.2.2, §4.6.1), caused by the inability of a particle to follow rapidly changing flow conditions in the region very near the surface, is included in the latter term only [Slinn et al., 1978]. The last two terms on the right hand side of (2.8-7) represent contributions from Brownian diffusion (§2.6.1); the first is the diffusion of the mean SSA concentration, and the second (also typically omitted in previous treatments) accounts for correlations between the fluctuating component of the Brownian diffusivity (caused by its dependence on temperature, density, and kinematic viscosity of the air, and on the geometric radius of the particle, through fluctuations in RH) and the vertical gradient of the number concentration. These contributions are unimportant in the surface layer.

The relative importance of the various terms in (2.8-7) for SSA particles of a given size depends primarily on the height above the ocean. Brownian diffusion, Stefan flow, slip flow from diffusiophoresis and thermophoresis, and impaction are generally important only in the viscous sublayer (§2.6.1), if at all. Turbulent diffusion is important only in the surface layer, where it provides a major, and often dominant, contribution to the vertical flux. The contribution to the slip flow from gravity, $\bar{v}_{\text{term}}$ (defined to be positive downward), can be important at any height, depending on particle size. Thus in the surface layer the mean net vertical flux $\bar{j}_z$ (for $\bar{w} = 0$) at any given height is given by

$$\bar{j}_z = \overline{n'w'} - \bar{n}\,\bar{v}_{\text{term}}; \qquad (2.8\text{-}9)$$

that is, for a given particle size, the effective net vertical SSA flux (positive upward) is equal to the turbulent vertical flux (also positive upward) minus the flux due to gravitational sedimentation.

2.8.3. Equations for the Effective SSA Production Flux

An expression for $\bar{j}_z(z_1)$, the mean net vertical flux of SSA particles of a given r_{80} through a horizontal plane at height z_1, can also be obtained by integrating (2.8-4) over the height z between heights z_1 and z_2 and rearranging:

$$\bar{j}_z(z_1) = \bar{j}_z(z_2) + \int_{z_1}^{z_2} \frac{\partial \bar{n}}{\partial t}\,dz$$

$$+ \int_{z_1}^{z_2} \frac{\partial \bar{j}_x}{\partial x}\,dz + \int_{z_1}^{z_2} \frac{\partial \bar{j}_y}{\partial y}\,dz - \int_{z_1}^{z_2} (\bar{g}_V - \bar{l}_V)\,dz \quad (2.8\text{-}10)$$

where $\bar{j}_z(z_2)$ is the mean net vertical flux through the upper surface of integration at height $z = z_2$. From (2.8-10) it

follows that if the mean SSA number concentration is independent of time (or nearly so over a sufficiently long time scale that the second term on the right hand side can be neglected), if horizontally homogeneity exists (so that the third and fourth terms on the right hand side can be neglected), and if the volumetric gain and loss terms can be neglected, then the mean net vertical flux is independent of height, i.e. $\bar{j}_z(z_1) = \bar{j}_z(z_2)$ for any z_1 and z_2. Thus, for example, the net interfacial SSA flux $\bar{j}_z(0)$—the mean net flux of SSA particles from the ocean to the atmosphere—would be the same as $\bar{j}_z(z)$—the mean net vertical flux of SSA particles through a horizontal plane at any height z—under situations for which these conditions were satisfied. However, (2.8-10) also demonstrates that time variations of the SSA number concentration and advective effects due to horizontal non-uniformity of the upwind concentration, as can occur during gusts or after breaking waves, can be manifested as vertical fluxes [Slinn, 1983b; Fairall, 1984; Wesely et al., 1985; Wesely, 1986; Sievering, 1987; Lamaud et al., 1994], and likewise any horizontal non-uniformity can result in differences between the net fluxes at any two heights.

In the present context the primary interest is the determination of the SSA production flux, not the net vertical SSA flux, yet it is the latter quantity that is represented in the continuity equation and that is measured; hence means of relating these fluxes are necessary. From §2.1.6 (especially Eq. 2.1-25 and Eq. 2.1-26), $\bar{j}_z(z_1)$ the mean net SSA vertical flux through height z_1, can be expressed as the difference between $f(z_1)$, the SSA production flux at this height (which depending on z_1 could be the interfacial SSA production flux or the effective SSA production flux), and $f_d(z_1)$, the SSA dry deposition flux at this height:

$$\bar{j}_z(z_1) = f(z_1) - f_d(z_1) \qquad (2.8\text{-}11)$$

(the overbar denoting time average is omitted on f and $f_d(z_1)$, but should be understood as implicit in the definition of these quantities). The effective SSA dry deposition flux is given by (2.7-2) as the product of the mean SSA number concentration at a given reference height z_{ref} and the SSA dry deposition velocity v_d, which depends on both the height z_1 for which the downward flux is defined and on the height z_{ref}: $f_d(z_1) = \bar{n}(z_{\text{ref}}) \times v_d(z_1, z_{\text{ref}})$. The quantity $\bar{j}_z(z_2)$ in (2.8-10) is the mean net vertical flux of SSA particles through a horizontal plane at height z_2. If this height is taken to be the height of the marine boundary layer H_{mbl}, then $\bar{j}_z(H_{\text{mbl}})$ describes the effects of an increase in this height with time and represents the SSA flux due to entrainment and mixing with the free troposphere (§2.4). This flux is given by

$$\bar{j}_z(H_{\text{mbl}}) = \bar{n}(H_{\text{mbl}}) \times v_{\text{ent}}, \qquad (2.8\text{-}12)$$

where the entrainment velocity v_{ent} is given by (2.4-5). For SSA particles that are not mixed to the height of the marine boundary layer to any appreciable extent, as will be the situation for sufficiently SSA large particles (§2.9.5), $n(H_{mbl}) \approx 0$ and this term does not provide an important contribution to the vertical flux of SSA particles in (2.8-10). With these substitutions, (2.8-10) becomes

$$f(z_1) = \bar{n}(z_{ref}) \times v_d(z_1, z_{ref}) + \bar{n}(H_{mbl}) \times v_{ent}$$
$$+ \int_{z_1}^{H_{mbl}} \frac{\partial \bar{n}}{\partial t} dz + \int_{z_1}^{H_{mbl}} \frac{\partial \bar{j}_x}{\partial x} dz$$
$$+ \int_{z_1}^{H_{mbl}} \frac{\partial \bar{j}_y}{\partial y} dz - \int_{z_1}^{H_{mbl}} (\bar{g}_V - \bar{l}_V) dz. \qquad (2.8\text{-}13)$$

The quantity $f(z_1)$ is the mean gross upward flux of SSA particles of a given size through a horizontal plane at height z_1; that is, the mean gross rate at which SSA particles of this size move upward through the bottom of a column of unit cross-sectional area bounded below by the horizontal plane $z = z_1$. For $z_1 = 0$, (2.8-13) is an explicit representation of the interfacial SSA production flux f_{int}; however, as noted above, the inability to determine the dry deposition flux at the surface from concentration measurements at a greater height precludes use of (2.8-13) in determining f_{int} (§2.7.2). For z_1 chosen as ~10 m, typical of the height at which measurements are taken, (2.8-13) describes f_{eff}, the effective SSA production flux (§2.1.6). The first term on the right hand side of (2.8-13) represents the (downward) SSA dry deposition flux through the height z_1, determined from concentration measurements at height z_{ref} (generally taken to be equal to z_1)—this is the mean gross flux of SSA particles out through the bottom of this column. The second term represents the net vertical flux of SSA particles through the top of the column resulting from mixing and entrainment of the marine boundary layer with the free troposphere. The next term is the time rate of change of the mean SSA number concentration, integrated over the column. The next two terms represent the change in along-wind flux in the along-wind direction and the change in the cross-wind flux in the cross-wind direction, respectively, and account for the effects of horizontal inhomogeneity; gusts and large-scale fluctuating winds will contribute to these terms, as will changes in the marine boundary layer height with location. The final term in (2.8-13) represents the contributions from volumetric gain and loss terms (i.e., wet deposition; §2.7.1); as noted in §2.8.1, this is often neglected because measurements are made during dry conditions in which removal by precipitation is not locally important.

Eq. 2.8-13 provides the basis for several methods of determining the effective SSA production flux from measurements outlined in §1.2 and discussed in detail in §3, and explicitly shows the physical basis for the terms that must be taken into account in doing so. The steady state dry deposition method (§3.1) assumes steady state conditions, horizontal homogeneity in both the along-wind and cross-wind directions, negligible gain and loss terms, and negligible entrainment of the free troposphere. If these conditions are satisfied, all terms on the right hand side of (2.8-13) except the first can be omitted and the equation becomes

$$f_{eff} = \bar{n} \times v_d,$$

which describes the balance between production and removal through dry deposition.

The concentration buildup method (§3.3) assumes stationary conditions (that is, at any fixed location the concentration does not change with time), horizontal homogeneity in the cross-wind direction, negligible gain and loss terms, and negligible entrainment of the free troposphere. For these conditions being satisfied, all terms on the right hand side of (2.8-13) except the first and the fourth can be omitted, resulting in

$$f_{eff} = \int \left(\frac{\partial \bar{j}_x}{\partial x} \right) dz + \bar{n} \times v_d.$$

For this situation, SSA production is not balanced by dry deposition but is given by the sum of the vertical integral of the gradient of the along-wind flux in the along-wind direction and the SSA dry deposition flux. Whatever is produced goes into increasing the SSA concentration (and thus providing an along-wind gradient of the flux in the along-wind direction) or is removed by dry deposition.

Micrometeorological methods (§3.5) are based on measurements of the turbulent (vertical) SSA flux $\overline{n'w'}$ and the mean SSA number concentration. Under conditions for which there is no mean vertical flow, Brownian motion and slip flows other than gravitational sedimentation are negligible (as is typically the situation at measurement heights near 10 m), and the fluctuating component of the gravitational sedimentation velocity can be neglected, this turbulent vertical flux can be obtained from (2.8-9) as the sum of the effective net SSA flux and the SSA flux due to gravitational sedimentation. As the effective net SSA flux is given by the difference between the effective SSA production flux and the effective SSA dry deposition flux (Eq. 2.8-11), it follows that

$$f_{eff} = \overline{n'w'} + \bar{n} \times v_d - \bar{n} \times v_{term}; \qquad (2.8\text{-}14)$$

that is, the effective SSA production flux is equal to the turbulent flux plus the difference of the dry deposition flux and the flux due to gravitational sedimentation. This equation serves as a basis for micrometeorological methods of determining the effective SSA production flux (§3.5).

The statistical wet deposition method (§3.9) is based on the assumption that all SSA particles in the atmosphere at the time of measurement have been produced since the last rainfall event, which occurred, on average, a time τ_{wet} previous. In this method dry deposition and vertical entrainment are neglected, and it is assumed that horizontal homogeneity exists in both the along-wind and cross-wind directions and that no volumetric gain and loss processes have occurred during the intervening time subsequent to the last precipitation event. Under these conditions, the only term on the right hand side that contributes to the effective SSA production flux is the vertical integral of the time rate of change of the number concentration. In this method the time derivative $\partial \bar{n}/\partial t$ is replaced by a difference $\Delta\bar{n}/\Delta t$ on the assumption that production is constant with time, and the concentration is assumed independent of height, so that it can be taken out of the integral. In this situation, replacing Δt with τ_{wet} and substituting multiplication by H_{mbl} for the integral over height yields

$$f_{\text{eff}} = \frac{\bar{n} H_{\text{mbl}}}{\tau_{\text{wet}}}.$$

Despite the considerable differences in the approaches that form the bases for the several methods of determining the effective SSA production flux $f_{\text{eff}}(r_{80})$ discussed in detail in §3, ultimately these approaches rest on the (2.8-13), which serves as a unifying concept relating SSA concentrations and fluxes with the effective SSA production flux, which is the bottom boundary condition for the equation.

2.9. VERTICAL DISTRIBUTION OF SSA PARTICLES

The vertical distribution of SSA particle concentration in the marine boundary layer is of key importance to understanding SSA and its behavior. This distribution determines the extent to which concentration measurements at any one height are representative of those at other heights, and to which they can be used to infer vertical integrals of SSA properties such as the SSA optical thickness τ_{sp} (§2.1.5.3). Additionally, knowledge of the height dependence of the concentration of SSA particles of a given r_{80} provides insight into the lifetimes of these particles and the extent to which they are mixed in the atmosphere, permitting evaluation of their importance for various atmospheric processes and allowing a quantitative estimate to be made for $\Psi_{\text{f}}(r_{80})$, the ratio of the effective SSA production flux to the interfacial SSA production flux (§2.1.6), and the dependence of this quantity on particle size.

This section examines processes controlling the vertical distribution of SSA particles in the marine atmosphere and inferences that can be drawn from knowledge of this distribution. The extent to which an SSA particle of a given r_{80} has attained its equilibrium radius at the height of measurement is examined (§2.9.1). Whether or not this equilibration has occurred has important consequences for the determination of the value of r_{80} for such a particle from measurement of its geometric radius and affects determination of $n(r_{80})$; it also has implications for the atmospheric residence times of large SSA particles and the vertical distribution of the concentration of such particles. The upward entrainment of SSA particles from near the sea surface by turbulent eddies is examined, and an upper limit is obtained for the sizes of SSA particles that have an appreciable chance of being entrained upward (§2.9.2). An expression is obtained for the height dependence of the steady state SSA concentration in the surface layer (§2.9.3). This expression is used to examine the height dependence of SSA concentration as a function of particle size in the lowest ~20 m of the marine boundary layer and the extent to which SSA particles are mixed throughout this height, and thus the extent to which measurements of SSA concentration and properties depends upon the height of measurement (§2.9.4). This expression is also used to examine the extent to which SSA particles are mixed throughout the marine boundary layer, allowing an independent estimate of the upper limit on the size of SSA particles that are vertically mixed to any appreciable extent (§2.9.5). Additionally, the temporal development of the steady state SSA concentration is investigated, and times scales characterizing the approach to steady state with respect to dry deposition are estimated (§2.9.6). These time scales can be used to evaluate the validity of applying the steady state dry deposition method of determining the effective SSA production flux (§3.1) and are also pertinent to the concentration buildup method (§3.3).

Although the multitude of conditions that might be encountered over the oceans precludes the possibility of general results that are accurate in all situations, estimates that are obtained for typical values of wind speed, RH, and other factors are relatively insensitive to the values of these controlling variables, and thus the conclusions reached here are expected to hold for a wide range of conditions. Field measurements of the vertical distribution of SSA concentration and its dependence on particle size are examined to determine the extent to which they support these conclusions. Few systematic investigations of this nature have been reported, and often only one or at most a few measurements (sometimes characterized as "typical") have been presented from individual investigations (often without estimates of uncertainties), limiting knowledge of variability under nominally the same conditions which may be due to differences in sea state, differences in air mass history, or the like. Additionally, there are concerns with the interpretation of these measurements, such as the simplistic representation employed for the structure of the marine boundary layer (§2.4) and the extent to which steady state conditions apply during the measurement period. Nonetheless, a rather consistent picture is obtained of the vertical distribution of size-dependent SSA concentration and the factors controlling this distribution.

2.9.1. Equilibration With Respect to Relative Humidity

Whether the geometric radius of an SSA particle is near its equilibrium radius, or conversely, near its formation radius, depends on how long the particle has been in the atmosphere relative to the time characterizing the approach of the radius of the particle to its equilibrium value at the ambient RH. The radius of an SSA particle that has been in the atmosphere much longer than this characteristic time will be approximately equal to its equilibrium value, whereas the radius of an SSA particle that has been in the atmosphere much less than this characteristic time will be near its formation value.

The time characterizing the approach of the radius of a recently formed SSA particle to its equilibrium value at 80% RH, $\tau_{98,80}$ (Eq. 2.5-6), increases rapidly with increasing particle size, from ~8 s for $r_{80} = 10$ μm to nearly a minute for $r_{80} = 25$ μm (at 20 °C). Sea salt aerosol particles detected within a meter or so of the sea surface that have been recently produced will have had little time to equilibrate with the ambient RH, and thus measurements within a meter or so of the sea surface that yield drop radii greater than a few micrometers often reflect values near formation radii. Jet drops, for instance, are ejected with such great velocities that they attain their maximum ejection heights ($\lesssim 20$ cm) before they experience any appreciable change in radius, and thus retain radii very near their formation values during their ascent (§4.3.2). However, equilibration is expected to have occurred for most small and medium SSA particles at heights more than several meters above the sea surface.

It has been assumed by many investigators, at least implicitly, that all SSA particles will have attained their equilibrium radii with respect to the local RH by the time of measurement near 10 m, and thus that all measured radii at this height represent equilibrium values (§4.1.4.2). Although this assumption is valid for sufficiently small SSA particles, it becomes increasingly questionable as particle size increases, and for sufficiently large SSA particles it certainly fails to hold. An estimate of the range of particle sizes for which this assumption is valid can be obtained by considering $z_{98,80}$, the distance an SSA particle, initially at its formation radius, falls under the influence of gravity in still air with ambient RH of 80% during its characteristic time of equilibration $\tau_{98,80}$ given by (2.5-6). For SSA particles with geometric radii less than ~40 μm, both $\tau_{98,80}$ and v_{term} (Eq. 2.6-8) are nearly directly proportional to the square of the radius; thus $z_{98,80}$ is nearly directly proportional to the

fourth power of the radius [*Eriksson*, 1959]. If the change in radius (and thus in v_{term}) with time is taken into account, then at 20 °C the value of $z_{98,80}$ is given approximately by

$$\frac{z_{98,80}}{\text{m}} \approx \left(\frac{r_{80}}{14\,\mu\text{m}} \right)^4 . \qquad (2.9\text{-}1)$$

This equation can be inverted to yield $r_{80,eq}$, the maximum value of r_{80} for which equilibration can be expected to have occurred. Sea salt aerosol particles with $r_{80} \lesssim r_{80,eq}$ will typically have attained radii near their equilibrium values by the time of measurement whereas SSA particles with $r_{80} \gtrsim r_{80,eq}$ typically will not have done so (Table 2).

The value of $r_{80,eq}$ for $z = 10$ m, a typical height of measurement, is approximately 25 μm and varies little over a wide range of conditions. Although at 0 °C the value of $\tau_{98,80}$ is nearly twice that at 20 °C, the value of $r_{80,eq}$ (again for $z = 10$ m) is reduced by only about 20% to $r_{80,eq} \approx 20$ μm (v_{term} is only ~6% greater at 0 °C than at 20 °C and this difference can be neglected). Likewise, for ambient RH values of 70% and 90% (at 20 °C), the times required for equilibration (that is, for a drop initially at RH 98% to attain a radius within 10% of its equilibrium value at the ambient RH; §2.5.4) differ from that given by (2.5-6) for 80% RH by factors of approximately 0.7 and 2, respectively, but the corresponding critical radii (for $z = 10$ m) change little, and corresponding values of $r_{80,eq}$ are near 28 μm and 21 μm, respectively. Additionally, $r_{80,eq}$ is rather insensitive to the chosen height; for $z = 20$ m it is only ~20% greater (i.e., $r_{80,eq} \approx 30$ μm) than at $z = 10$ m. The rapid increase in both v_{term} (and thus the vertical velocity required to entrain such a particle upward) and $\tau_{98,80}$ (and thus the required duration of this velocity) with increasing r_{80}, and the resultant r_{80}^4 dependence of the height in (2.9-1), ensure that over a wide range of conditions there is a strong decrease near $r_{80} \approx 25$ μm in the probability that the radius of an SSA particle measured at a height near 10 m will have attained a value near its equilibrium radius. This natural, rather sharp demarcation between SSA particles that do and do not equilibrate with respect to the ambient RH while in the atmosphere supports the choice of $r_{80} \approx 25$ μm as the division between medium and large SSA particles (§2.1.2). Because Stokes' Law (Eq. 2.6-8) overestimates v_{term} (§2.6.3) and because ventilation (§2.5.4) increases the rate of moisture transfer, thus reducing $\tau_{98,80}$, (2.9-1) becomes increasingly inaccurate as particle size increases; however, these considerations do not appreciably alter the value of $r_{80,eq}$ obtained.

For r_{meas}, the radius of an SSA particle measured at a height near 10 m, to be near its equilibrium radius, the particle must not only have remained in the atmosphere for a time sufficiently long to have equilibrated with respect to this ambient RH, but also during this time it must have experienced a sustained upward velocity sufficiently large not just to keep it from falling back into the sea, but additionally to have transported it to the height of measurement [*cf.*, *Van Eijk et al.*, 2001; *Hoppel et al.*, 2002]. For SSA particles with $r_{80} \lesssim 25$ µm, these requirements are easily met under nearly all realistic conditions. For instance, for an SSA particle with $r_{80} = 20$ µm to attain a radius within 10% of its equilibrium value (assuming an ambient RH of 80% at 20 °C) it must remain in the atmosphere only ~30 s, during which it must be transported only an additional ~4 m above the height of measurement to account for the distance it would have fallen due to gravity during this time. Moreover, it is unlikely that a particle with $r_{80} = 20$ µm will not have attained a radius near its equilibrium value at the time of measurement near 10 m, as this would require that the measurement occurred within a very short time ($\lesssim 30$ s) after the particle was formed, and additionally would require an extremely rapid upward transport of the drop from the sea surface to 10 m. As the mean atmospheric residence time of an SSA particle with $r_{80} = 20$ µm is expected to be much greater than this time (~10^3 s based on estimates obtained in §2.9.6), the probability that an SSA particle of this size measured at 10 m has been in the atmosphere less than 30 s is extremely small, and for smaller SSA particles the corresponding probability is even smaller.

With increasing particle size, however, these requirements become increasingly difficult to satisfy, and the probability that an SSA particle measured at 10 m has been in the atmosphere for a time equal to or greater than its equilibration time with respect to RH becomes increasingly small. An SSA particle with r_{80} of (30, 40, 50) µm, having initial terminal velocity (that is, the terminal velocity of the particle at its formation radius) of (0.33, 0.50, 0.65) m s^{-1} (Eq. 2.6-11), has an equilibration time $\tau_{98,80}$ (assuming an ambient RH of 80% at 20 °C) of (60, 90, 130) s, during which it will fall from gravity a distance of roughly (15, 40, 80) m (these values take into account the increased rate of response the particle with respect to RH due to ventilation and the deviation of the terminal velocity from that given by Stokes' Law). Therefore, for such a particle to have attained a radius near its equilibrium value, it must have experienced a sustained upward velocity of greater than (0.4, 0.55, 0.7) m s^{-1} throughout its equilibration time, or one at least initially greater than its initial terminal velocity but for a longer time. From these considerations it is evident that the probability that SSA particles of these sizes are entrained to 10 m is very low because of the prolonged sustained upward velocities near the sea surface that are required, and that this probability decreases rapidly with increasing particle size. Additionally, it is much more likely that the small fraction of SSA particles of these sizes produced at the sea surface that are entrained to a height near 10 m would have done so in relatively short times, as large

upward velocities of the required magnitudes near the sea surface are uncommon and become increasingly so as their duration increases. Thus, with increasing particle size greater than $r_{80} \approx 25$ µm it becomes increasingly improbable that an SSA particle at 10 m has been in the atmosphere for a time comparable to or greater than its equilibration time, and it becomes increasingly likely that its geometric radius is more representative of its formation radius than of its equilibrium radius (Table 2).

The correspondence between the radius of an SSA particle measured at a height of 10 m above the sea surface r_{meas} and its equilibrium radius r_{80} is illustrated schematically in Fig. 11 for ambient RH 80%. Here it is assumed that the SSA particles are produced from seawater of salinity 35 for which the formation radius $r_{form} \approx 2r_{80}$ (§2.5.3). The upper panel shows the value of the measured radius at 10 m and 80% RH of an SSA particle with a given r_{80}, with distributions representing the probability of occurrence of r_{meas} for a given r_{80}. Under these conditions the measured radius must satisfy the inequality $r_{80} < r_{meas} < 2r_{80}$. The value of r_{meas} will be toward the lower part of this range (i.e., near

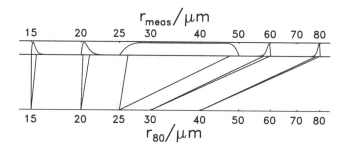

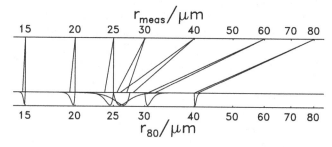

Figure 11. Correspondence of SSA particle radius measured at 10 m above the sea surface at RH 80% r_{meas} and equilibrium radius of an SSA particle at 80% RH r_{80}, for SSA particles that are produced from seawater of salinity 35 for which the formation radius $r_{form} \approx 2r_{80}$ (§2.5.3). Upper panel shows measured radius of an SSA particle at 10 m and RH 80% (upper axis) as a function of its equilibrium radius r_{80} (lower axis). Lower panel shows radius of an SSA particle in equilibrium at 80% RH r_{80} (lower axis) as a function of its measured radius at 10 m and 80% RH r_{meas}. Distributions schematically indicate range of probable values.

r_{80}) for nearly complete equilibration, and near the upper part of this range (i.e., near $2r_{80}$) if little equilibration has occurred. For SSA particles with $r_{80} \lesssim 25$ μm (25 μm being equal to the value of $r_{80,eq}$ defined above), nearly complete equilibration will typically have occurred and $r_{meas} \approx r_{80}$, with the extent of equilibration increasing with decreasing r_{80}. Conversely, for SSA particles with $r_{80} \gtrsim 25$ μm, little equilibration will typically have occurred and $r_{meas} \approx 2r_{80}$. This situation is illustrated by the probability distributions on the r_{meas} axis at values slightly less than $2r_{80}$ (as all SSA particles will have decreased their radii from their formation values to some extent), which narrow with increasing r_{80}, as it is increasingly unlikely that larger particles have remained in the atmosphere for any appreciable time. Sea salt aerosol particles with $r_{80} \approx 25$ μm will have equilibrated to various extents, depending on the time they have spent in the atmosphere, and thus would be expected to correspond to a range of r_{meas} of nearly a factor of two between ~25 μm and ~50 μm depending on whether they are sampled late or early in their sojourn in the atmosphere.

The lower panel of Fig. 11 shows the value of r_{80} corresponding to a given measured radius at 10 m above the sea surface at ambient RH 80%, with distributions schematically representing the probability that this value of r_{meas} corresponds to a specific r_{80}. Under these conditions the value of r_{80} must satisfy the inequality $r_{meas}/2 < r_{80} < r_{meas}$. The value of r_{80} will be toward the upper part of this range (i.e., near r_{meas}) if nearly complete equilibration has occurred and near the lower part of this range (i.e., near $r_{meas}/2$) if little equilibration has occurred. Values of r_{meas} that are less than 25 μm most likely correspond to equilibrium values and $r_{meas} \approx r_{80}$, with this situation becoming more probable with decreasing r_{meas}. Values of r_{meas} that are greater than 50 μm most likely are close to formation values and $r_{meas}/2 \approx r_{80}$, with this situation becoming more probable with increasing r_{meas}. Measured radii between ~25 μm and ~50 μm all correspond to r_{80} near 25 μm.

Although the correspondence of r_{meas} to r_{80} is one-to-many (that is, a single value of r_{meas} can correspond to a range of r_{80} between $r_{meas}/2$ and r_{meas} depending on the extent of equilibration), in fact it is essentially one-to-one over most of the range of r_{80} or r_{meas}, except for the range of r_{meas} from ~25 μm to ~50 μm where the correspondence is many-to-one, as a single value of r_{80} can correspond to a range of r_{meas} between r_{80} and $2r_{80}$. Likewise although the correspondence of r_{80} to r_{meas} is one-to-many, as a single value of r_{80} can correspond to a range of r_{meas} between r_{80} and $2r_{80}$, it is essentially one-to-one except for r_{80} near 25 μm. Based on these considerations SSA particles with $r_{80} < 25$ μm—small and medium SSA particles—that are more than a meter or so above the sea surface are treated throughout this review as if they instantaneously attain their equilibrium radii, and SSA particles with $r_{80} > 25$ μm—large

SSA particles—are treated as if they retain their formation radii during the entire time they spend in the atmosphere, implying that measured radii of these particles both near the surface and at heights near 10 m will be close to their formation values. Sea salt aerosol particles within the range of measured radii of 25-50 μm correspond to a narrow range of r_{80} near 25 μm. These assumptions allow the size of an SSA particle to be determined solely by its r_{80} value and the ambient RH (§2.1.1). Any errors associated with this approach are of little consequence with regard to the effective SSA production flux, as large SSA particles do not contribute substantially to this flux; likewise, no serious concerns arise from this approach with regard to the interfacial SSA production flux, as particles in the affected size range $r_{80} = 25$-50 μm are of thought to contribute little to properties of interest such as heat and momentum transfer between the oceans and the atmosphere, which are thought to be dominated by even larger particles.

There are several consequences of failure to take into account the lack of equilibration of large SSA particles. The assumption that the few large SSA particles that are measured at heights near 10 m are in equilibrium with the ambient RH (and thus that their measured radii are equilibrium values) would result in an overestimation of r_{80} ascribed to these particles. Additionally, as $n(r_{80})$ is typically a decreasing function of r_{80} for particles of these sizes (§4.1.4.3), at a given r_{80} the SSA number concentration would also be overestimated, and concentrations of radius, surface area, and volume (or mass), which are typically calculated from number concentrations, would be overestimated by an even greater amount, in addition to referring to incorrect sizes. A further consequence is that whereas measured radii less than ~25 μm (at heights more than a few meters above the sea surface) almost certainly correspond to equilibrium values and measured radii greater than ~50 μm almost certainly correspond to formation values, any measured radius between these values corresponds to a narrow range of r_{80} near 25 μm. Thus the assumption that all SSA particles have equilibrated with respect to the ambient RH by the time of measurement would result in size distributions of SSA concentration for SSA particles in the range of measured radii 25 μm to 50 μm that would be spread over a larger range of r_{80} than actual, and the values of $n(r_{80})$ would be underestimated, in addition to referring to a range of (incorrect) sizes rather than a narrow range of r_{80} near 25 μm.

2.9.2. Upward Entrainment of SSA Particles From Near the Sea Surface

Consideration of the behavior of SSA particles immediately subsequent to their production at the sea surface yields insight into the sizes of particles that have an appreciable chance of being entrained upward as opposed to directly returning to the sea. If the upper value of this size range is denoted by $r_{80,ent}$, then only a small fraction of SSA particles with $r_{80} \gtrsim r_{80,ent}$ produced at the sea surface will be initially entrained upward. Additionally, an even smaller fraction of these particles will be further entrained upward and attain a height of 10 m, contributing to the effective SSA production flux. This latter fraction, equal to the quantity $\Psi_f(r_{80})$, the ratio of the effective SSA production flux to the interfacial SSA production flux (§2.1.6), decreases to a value much less than unity as r_{80} increases toward $r_{80,ent}$.

For an SSA particle to be entrained upward immediately subsequent to its production at the sea surface, it must experience sustained upward motion sufficiently great not only to overcome the gravitational force acting upon it, characterized by its terminal velocity v_{term}, but also to transport it further upward. As there can be no mean component of the vertical velocity of the air at the surface (other than Stefan flow, §2.6.1, which is negligible in the present context), the required upward motion can be provided only by turbulent eddies. The magnitude of this upward motion is characterized by the root-mean-squared value of the vertical component of the wind speed σ_w (§2.4), it being assumed that measurements of σ_w typically made at heights of several meters above the sea surface are representative of the values at lower heights. As the geometric radius of an SSA particle immediately subsequent to production is nearly equal to its formation radius (§2.9.1), the gravitational terminal velocity of an SSA particle is evaluated at this formation radius for purposes of estimating $r_{80,ent}$. Whether an SSA particle will be entrained upward can be assessed by considering the size of an SSA particle, denoted $r_{80,ent}$, for which this gravitational terminal velocity is equal to σ_w. The approximation $\sigma_w \approx 0.05 U_{10}$ (§2.4) yields $r_{80,ent} = 40$ µm at $U_{10} = 10$ m s^{-1} (using 2.6-11 to evaluate the terminal velocity of SSA particles of this size). An SSA particle with $r_{80} \ll r_{80,ent}$ has high likelihood of being entrained upward, whereas the likelihood of entrainment decreases greatly as r_{80} approaches $r_{80,ent}$. The value of $r_{80,ent}$ depends somewhat on wind speed, being near 25 µm and 55 µm at $U_{10} = 5$ m s^{-1} and 15 m s^{-1}, respectively. However, entrainment upward from near the sea surface requires that σ_w be substantially greater than the initial terminal velocity; otherwise the particle will not gain height with respect to the sea surface. Additionally, the short time drops of this size remain in the atmosphere before falling back to the sea surface, and during which they must encounter eddies of sufficient strength to entrain them upward, further reduces the probability that they will be further entrained upward (this argument is examined further in §4.3.1.4). Therefore the upper limit of r_{80} for which upward entrainment would be expected must be substantially less than $r_{80,ent}$, i.e., substantially less than 40 µm at $U_{10} = 10$ m s^{-1}.

Similar conclusions have been reached by other investigators. *Toba* [1961b] and *Koga and Toba* [1981] concluded that $r_{80,ent} \approx 35$ µm for U_{10} near 12 m s^{-1}, but they based their arguments on the assumption that both the eddy diffusivity and mean horizontal velocity were independent of height. Both *Byutner* [1978 (cited in *Bortkovskii*, 1987, p. 104)] and *Wu* [1982a], the latter by analogy with the sedimentation of particles in a channel [*Sumer*, 1977], argued that the size of the particles that are expected to be entrained vertically is determined by the criterion $\kappa u_* > v_{term}$ (this result is also obtained below; §2.9.3). As $\kappa \approx 0.40$ (§2.4), this criterion is more stringent than the one obtained above and yields $r_{80,ent} \approx 16$ µm at $U_{10} = 10$ m s^{-1} (under the assumption that SSA particles retain their formation radii near the sea surface), consistent with the value of $r_{80,ent} \approx 12$ µm determined from field measurements of *Wu et al.* [1984] for U_{10} of 6 m s^{-1} and 7.5 m s^{-1}. This result is also consistent with the value $r_{80,ent} = 15$-20 µm for $U_{10} = 10$ m s^{-1} obtained by *Bortkovskii* [1987, pp. 104-108], who took into account the time for drops to accommodate to fluctuating wind speeds resulting from turbulent eddies.

Therefore, based on the arguments presented in this section, and supported by results of experimental and other theoretical investigations discussed above, it can be concluded that under most oceanic conditions the upper limit for the value of r_{80} for which SSA particles have an appreciable chance of being entrained upward from near the sea is substantially less than 40 µm. Additionally, the upper size limit of SSA particles that have any appreciable probability of being further entrained upward and attaining a height of 10 m, and thus of contributing to the effective SSA production flux (this upper limit is also the demarcation between medium SSA particles and large SSA particles; §2.1.2), is also well less than 40 µm (Table 2).

2.9.3. Steady State Vertical SSA Concentration Profile in the Surface Layer

Consideration of the steady state vertical profile of the size-dependent mean concentration of SSA particles $n(r_{80})$ (the overbar denoting temporal average is omitted hereinafter, but should be understood), yields information on the efficiency with which SSA particles of a given size are entrained upward into the marine atmosphere and on the heights to which these particles are mixed. In this section the steady state vertical profile of the concentration of SSA particles of a given r_{80} is obtained, and in the next several sections conclusions that can be drawn from this profile are investigated. However, no assumption is made that such a steady state profile typically exists. Whether or not steady state has been attained at the time of measurement depends on the time required for the establishment of this profile, which depends strongly on particle size and on meteorological conditions. An estimate for this time, and its dependence on r_{80}, is presented in §2.9.6.

The mechanisms that control the motion of SSA particles in the surface layer in the absence of wet removal, and thus govern the extent of vertical mixing of these particles and the heights they attain, are turbulent diffusion, parameterized by the eddy diffusion coefficient D_{eddy}, which is assumed to be independent of particle size (§2.6.1), and gravitational sedimentation, parameterized by v_{term} (§2.6.2, §2.6.3). The value of D_{eddy} at a given height depends on u_* and the atmospheric stability (§2.4), or equivalently on U_{10} and the atmospheric stability (§2.3.2), whereas v_{term} depends on the geometric radius of the particle and its density, both of which are determined uniquely by r_{80} and RH (§2.9.1). Thus, the height dependence of the concentration of SSA particles of a given r_{80} in the surface layer, under conditions of steady state, is determined by U_{10}, RH, and the atmospheric stability [*Toba*, 1965b]. For a given atmospheric stability, the efficiency of entrainment, the extent of mixing, and the heights to which SSA particles of a given r_{80} are mixed, all increase with increasing wind speed (which leads to increasing D_{eddy}) and decrease with increasing r_{80} and increasing RH (both of which lead to increasing v_{term}). Additionally, for a given r_{80} and wind speed, unstable atmospheric conditions result in greater mixing and more efficient entrainment, and stable atmospheric conditions result in less mixing and less efficient entrainment, than conditions of neutral stability.

The equation for the steady state concentration profile for SSA particles of a given r_{80} can be obtained from (2.8-8) and (2.8-9) under the assumption that the mean net vertical flux is zero:

$$D_{eddy} \frac{\partial n}{\partial z} = -v_{term} n, \qquad (2.9\text{-}2)$$

where v_{term} is defined to be positive downward. This equation, also known as the Rouse equation (after *Rouse* [1937], although it was discussed earlier by *Schmidt* [1925] and *O'Brien* [1933], as discussed in *Bennett et al.* [1998], for example), can be solved to yield

$$\frac{n(z)}{n(z_1)} = \exp\left\{ -\int_{z_1}^{z} \frac{v_{term}}{D_{eddy}} \, dz \right\} \qquad (2.9\text{-}3)$$

for a given height z_1. If RH depends on height then so does v_{term}, but if RH remains roughly constant for heights between z_1 and z then v_{term} can be taken out of the integral with little sacrifice in accuracy. An exponential dependence of $n(r_{80})$ on height (sometimes explicitly assumed) results from the assumption that both RH (and thus v_{term} for a given r_{80}) and D_{eddy} are independent of height [*Toba*, 1965a].

However, for neutrally stable conditions, D_{eddy} (Eq. 2.4-2) is directly proportional to z and the resulting size-dependent concentration profile is described by a power law in height:

$$\frac{n(z)}{n(z_1)} = \left(\frac{z}{z_1} \right)^{\left(\frac{-v_{term}}{\kappa u_*} \right)}. \qquad (2.9\text{-}4)$$

This expression is used below to describe the dependence of the concentration of SSA particles of a given size on height. Eq. 2.9-3 can also be integrated for other expressions for D_{eddy} (for instance, those that approach a constant value with increasing height or that become zero at the top of the marine boundary layer), but the results obtained using these expressions do not differ greatly from those given by (2.9-4), and the conclusions reached below remain qualitatively the same.

If a characteristic mixing velocity v_{mix} is defined by $v_{mix} = \kappa u_*$, then the ratio of the concentrations of SSA particles of a given r_{80} at any two heights z and z_1 given by (2.9-4) depends only on the ratio v_{term}/v_{mix}. For $U_{10} = 10$ m s^{-1}, the characteristic mixing velocity $v_{mix} \approx 14$ cm s^{-1}. Similarly, if a characteristic mixing radius r_{mix} is defined as the radius of an SSA particle for which $v_{term} = v_{mix}$, then for the size range for which this terminal velocity is given by Stokes' Law (Eq. 2.6-8), $v_{term}/v_{mix} = (r/r_{mix})^2$, and the ratio of the number concentrations at any two heights depends only on the ratio r/r_{mix}. The quantity r_{mix} defined in this manner refers to the geometric radius of an SSA particle, which is not necessarily its value of r_{80}; situations in which it refers to a value of the radius other than r_{80} (as it will for sufficiently large SSA particles) are examined below. For a given U_{10}, r_{mix} is given approximately by

$$\frac{r_{mix}}{\mu m} \approx 10 \left(\frac{U_{10}}{\text{m s}^{-1}} \right)^{\frac{1}{2}}; \qquad (2.9\text{-}5)$$

for $U_{10} = 10$ m s^{-1}, $r_{mix} \approx 32$ μm. The ratio r/r_{mix}, or equivalently, v_{term}/v_{mix}, describes the relative importance of the effects of gravity and turbulent diffusion on the motion of an SSA particle. Sea salt aerosol particles with $r \prec r_{mix}$ ($v_{term} \prec v_{mix}$), for which the effect of gravity is much less than that of turbulent diffusion (the symbol $\prec$ was introduced in §2.1.5.2 to denote considerably less than), are expected to become well mixed over appreciable heights, whereas SSA particles with $r \gtrsim r_{mix}$ ($v_{term} \gtrsim v_{mix}$), for which the effect of gravity is comparable to or greater than that of turbulent diffusion, are expected to be distributed over much lesser heights and to have concentrations that decreasing rapidly with increasing height.

The value of r_{mix} depends somewhat on meteorological conditions (i.e., wind speed, atmospheric stability, and RH), but it varies little over a wide range of conditions typical of those in the marine atmosphere. At 80% RH r_{mix} increases from near 23 μm to near 40 μm as U_{10} increases from 5 m s^{-1} to 15 m s^{-1}. Similarly, r_{mix} depends on RH through the dependencies of the gravitational terminal velocity of an SSA particle of a given r_{80} on the geometric radius and the density of the particle (Eq. 2.6-7), but at $U_{10} = 10$ m s^{-1}, r_{mix} decreases only from ~34 μm to ~28 μm as the ambient RH increases from 70% to 90%. Thus the value $r_{mix} \approx 32$ μm can be taken as generally valid and is used below.

From these considerations it can be concluded that with increasing particle size the extent to which SSA particles are mixed vertically decreases abruptly near $r_{80} \approx 32$ μm as a result of the power law concentration profile described by (2.9-4). Additionally, because of the failure of large SSA particles to equilibrate to the ambient RH (§2.9.1), their terminal velocities are greater than those calculated under the assumption that had equilibrated, resulting in an even greater decrease in vertical mixing of SSA particles near this size. It is important to note that although the value $r_{mix} \approx 32$ μm obtained here is near the value $r_{80,eq} \approx 25$ μm obtained for an upper limit of drop equilibration with respect to RH in §2.9.1, and consistent with the upper limit of less than 40 μm for $r_{80,ent}$ obtained in §2.9.2, these values are based on very different physical considerations.

2.9.4. Height Dependence of SSA Concentration in the Lowest Portion of the Marine Boundary Layer

The steady state SSA concentration profile from heights of several meters to 20 m or so above the sea surface is of key importance. Consideration of this profile yields insight into the extent to which measurements of size distributions and integral properties of SSA concentration (§4.1) at one height are representative of those at other heights, the efficiency with which SSA particles are entrained upward into the marine atmosphere, and the size range of SSA particles that are expected to contribute appreciably to the effective SSA production flux. Consideration of this profile can also be used to obtain a quantitative estimate of the size dependence of Ψ_f, the ratio of the effective SSA production flux to the interfacial SSA production flux (§2.1.2).

The steady state size-dependent SSA concentration profile calculated from (2.9-4) under the assumption of neutral atmospheric stability is presented in Fig. 12 (upper panel) for SSA particles of several r_{80} as the ratio $n(z)/n(5$ m$)$ for U_{10} values of 5, 10, and 15 m s^{-1} at RH 80%. As measurements of size distributions and properties of SSA concentration in the marine atmosphere are typically made at heights between 5 m and 20 m, consideration of $n(z)/n(5$ m$)$ allows

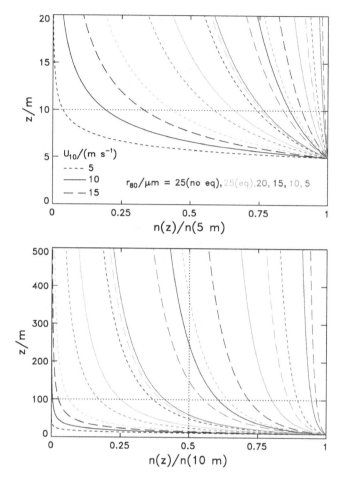

Figure 12. Steady state dependence of the concentration of SSA particles with height above the sea surface shown as ratio of concentration at height z, $n(z)$ to that at 5 m, $n(5$ m$)$ (upper panel) and to that at 10 m, $n(10$ m$)$ (lower panel) for SSA particles having indicated values of r_{80} in equilibrium at 80% RH, evaluated using (2.9-4). Particles are assumed to be acted upon only by turbulent diffusion and gravity under neutrally stable atmospheric conditions for indicated values of wind speed at 10 m U_{10}. Also shown are the concentration profiles of SSA particles with $r_{80} = 25$ μm under assumption that these particles retain their radius at formation corresponding to 98% RH, (i.e., physical radius equal to 50 μm). Horizontal dotted line at 10 m shown in upper panel. Intersection of vertical dotted line at 0.5 in lower panel indicates value of z_{50}, the height at which the concentration is 50% that at 10 m.

evaluation of the extent to which differences in reported measurements of these quantities can be attributed to differences in height. It is assumed that particles with $r_{80} \lesssim 25$ μm have completely equilibrated with respect to the ambient RH of 80%, and thus that their geometric radii are equal to their r_{80} values (§2.9.1). Concentration profiles for particles with $r_{80} = 25$ μm are presented both for the situation for which complete equilibrium has occurred (i.e., geometric radius equal to 25 µm) and for the situation for which no equilibration has occurred (i.e., geometric radius equal to 50 µm). These results are similar to those reported by *Burk* [1984] and by *Stramska* [1987] for SSA particles with r_{80} from 2 µm to 20 µm based on numerical models of vertical transport in the marine boundary layer.

For SSA particles with $r_{80} \lesssim 10$ µm, the concentration ratio $n(20 \text{ m})/n(5 \text{ m}) > 0.75$ for $U_{10} \gtrsim 5$ m s^{-1} (Fig. 12); hence little variation of concentration with height from 5 m to 20 m is expected under most conditions. With increasing particle size this concentration ratio decreases, but even for SSA particles as large as $r_{80} = 20$ µm the ratio $n(20 \text{ m})/n(5 \text{ m}) \gtrsim 0.5$ for $U_{10} \gtrsim 7.5$ m s^{-1}. However, as r_{80} approaches 25 µm there is a sharp decrease in this ratio, both because the effect of gravity becomes comparable to that of turbulent diffusion (that is, r_{80} approaches the value $r_{mix} \approx 32$ µm; §2.9.3) and because it becomes increasingly unlikely that such particles will have attained their equilibrium radii during their residence time in the atmosphere (that is, r_{80} approaches the value $r_{80,eq} \approx 25$ µm; §2.9.1), resulting in their gravitational velocities being even greater than those calculated under the assumption of equilibration. At $U_{10} = 10$ m s^{-1}, the ratio $n(20 \text{ m})/n(5 \text{ m})$ is near 0.4 for particles with $r_{80} = 25$ µm that have attained their equilibrium radii at ambient RH 80%, but this ratio is near 0.03 for particles with $r_{80} = 25$ µm that retain their formation radii.

The vertical gradient of the steady state SSA number concentration can be obtained from (2.9-2) as $d\ln n/dz = -v_{term}/(\kappa u_* z)$. For a height of 10 m at $U_{10} = 10$ m s^{-1}, this equation yields a decrease in the concentration of approximately $[r_{80}/(10 \text{ µm})]^2$ percent for each 1 m increase in height. This corresponds to a decrease in concentration of ~6% per meter for SSA particles with $r_{80} \approx 25$ µm that are assumed to have attained their equilibrium radii at 80% RH (i.e., complete equilibration has occurred), and less for smaller particles. Concentrations of large SSA particles, which are assumed to have retained their formation radii (i.e., no equilibration has occurred), exhibit a much larger gradient, given approximately by $[r_{80}/(5 \text{ µm})]^2$ percent for each 1 m increase in height. This corresponds to a decrease in concentration of ~25% per meter for SSA particles with $r_{80} = 25$ µm at their formation radii, and greater for larger particles. Thus measured number concentrations of small and medium SSA particles are expected to be relatively insensitive to the actual height of measurement so long as it is near 10 m, and differences in measurements of $n(r_{80})$ for SSA particles with $r_{80} \lesssim 25$ µm (and hence of N, R, and A, for which the dominant contributions are from particles in this size range) are unlikely to be due to differences in the height at which they were made. To some extent this conclusion applies also to measurements of SSA mass concentration M, depending on the size range of SSA particles that

are included in the measurements, although as larger SSA particles provide a relatively greater contribution to M than smaller particles, this quantity would be expected to exhibit a greater dependence on height than quantities such as N that are dominated by smaller SSA particles.

Relatively few measurements of the dependence of the concentration of SSA particles of a given r_{80} on height from 5 m to 20 m above the sea surface have been reported in open ocean conditions (measurements at coasts are not considered here, as the pertinence of such measurements to open ocean conditions are subject to question because of the effects of surf spray, different wind conditions at the shore, and the like). Shipboard measurements of number concentrations of SSA particles with r_{80} from ~3 µm to ~7 µm at wind speeds up to 8.4 m s^{-1} reported by *Chaen* [1973] exhibited little dependence with height from 1 or 2 m to 9 m above the sea surface, perhaps at most a 20% decrease over these heights in some instances. Number concentrations of SSA particles with various r_{80} in the approximate range 2.5 µm to 20 µm at wind speeds up to nearly 13 m s^{-1} for heights of 2.5 or 3 m to 13 m were also reported by *Chaen* [1973]; although they exhibited more variability (and in some instances an increase of up to ~50% between 6 m and 13 m), they are also broadly consistent with the above conclusions. Ratios of number concentrations of marine aerosol particles measured at 30 m above the sea surface to those at 10 m reported by *Exton et al.* [1986] for both a "shore" region (defined to be the first 10 km out to sea, and thus including the surf zone) and a "sea" region (from 10 km to 20 km from shore) were roughly 75%, 80%, 90%, and 90-95% for ambient radii in overlapping size ranges 1-24 µm, 1-6 µm, 0.5-8 µm, and 0.25-4 µm, respectively. Wind speeds and relative humidities were not presented, nor were estimates of the uncertainties in these values, and some of the measured particles probably consisted of substances other than sea salt, but nonetheless these results are consistent with the conclusions reached above. Measurements of the size distribution of the number concentration of SSA particles with ambient radii from ~5 µm to ~45 µm and of light-scattering coefficient (which is nearly proportional to the concentration of total surface area; §2.1.5.3; §4.1.4.5) reported by *de Leeuw* [1986a, 1986b, 1987, 1989a, 1990a] for wind speeds to near 14 m s^{-1} were remarkably constant with height from a few meters above the sea surface to up to ~10 m and in some instances ~20 m (the measurements within a few meters of the sea surface are examined below; §4.2). As discussed in §2.9.1, because of the time required for equilibration with respect to RH, a given measured radius can correspond to a range of possible r_{80} values, depending on the measured radius and the height; most SSA particles with measured radii less than ~25 µm at more than a few meters above the sea surface would probably have attained their equilibrium radii by the time of measurement, whereas the largest particles would probably not have. Thus it can be concluded that most of the SSA particles measured by *de Leeuw* had $r_{80} \lesssim 25 \ \mu m$. These results also support the conclusion that under typical conditions concentrations of small and medium SSA particles are roughly independent of height from a few meters to ~20 m above the sea surface.

Consideration of the ratio of the steady state number concentration of SSA particles of a given r_{80} at 10 m to that at a height very near the surface also yields an estimate for the ratio of the effective SSA production flux to the interfacial SSA production flux Ψ_f (§2.1.6) and its dependence on r_{80} and U_{10}. The choice for the lower height is arbitrary, but the values 0.1 m and 0.01 m are considered here. These heights are typical of those to which jet drops are ejected (§4.2.1), and it can be argued that drops (both bubble-produced and spume) that do not attain one or the other of these heights contribute little to SSA production important on large scales or to geophysical processes such as moisture, momentum, and heat exchange between the ocean and the atmosphere, and thus for most applications would not be included in what constitutes the interfacial SSA production flux. It is thus assumed that the steady state SSA concentration ratios $n(10 \ m)/n(0.1 \ m)$ and $n(10 \ m)/n(0.01 \ m)$ exhibit the same qualitative dependence on r_{80} and U_{10} as does Ψ_f. These ratios, calculated using (2.9-4) for several wind speeds, are shown in Fig. 13 as a function of r_{80} on the assumption that SSA particles have attained their equilibrium radii at 80% RH. These concentration ratios are qualitatively similar to each other, and the dependence on the choice of the lower height is not a strong one over most of the range of r_{80} up to 25 μm. For SSA particles with $r_{80} \lesssim 3 \ \mu m$ the ratios are nearly unity, as particles in this size range would be well mixed over the lowest 10 m (at least) above the ocean. With increasing r_{80} the concentration ratios exhibit a large decrease near ~5 μm to ~10 μm, depending on wind speed, and at $r_{80} = 25 \ \mu m$ they have become extremely small (~0.06 and ~0.015 for 0.1 m and 0.01 m, respectively, at $U_{10} = 10 \ m \ s^{-1}$), as only a very small fraction of the particles of this size that are present at the lower heights are transported up to 10 m. For SSA particles with $r_{80} \gtrsim 25 \ \mu m$ the concentration ratios are even less than those shown in Fig. 13 because such particles will not have equilibrated to the ambient RH and their radii will remain near formation values for most of the time these particles spend in the atmosphere. It is expected that Ψ_f exhibits a similar trend with r_{80}, being near unity for small SSA particles and decreasing from this value to near zero as r_{80} increases from 1 μm to 25 μm, implying that only small and medium SSA particles contribute appreciably to the effective production flux (Table 2). This estimate is discussed further in §5.1.2.

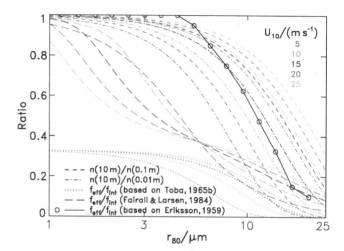

Figure 13. Ratio of steady state SSA concentration and of production flux at 10 m to values near the surface as a function of particle radius in equilibrium at 80% RH, r_{80}, for indicated values of wind speed at 10 m, U_{10}. Ratios of concentration at 10 m to that at 0.1 m, $n(10 \ m)/n(0.1 \ m)$, and to that at 0.01 m, $n(10 \ m)/n(0.01 \ m)$, are presented for conditions of neutral atmospheric stability under steady state conditions which particles are assumed to be acted upon only by turbulent diffusion and gravity, evaluated using (2.9-4). Production flux ratio is shown as $\Psi_f(r_{80})$, the ratio of the effective SSA production flux f_{eff} to the interfacial production flux f_{int}, evaluated from (4.6-9) following *Toba* [1965b] and from (4.6-10) following *Fairall and Larsen* [1984], as discussed in §4.6.5. Also shown is estimate for ratio of f_{eff} to f_{int} based on *Eriksson* [1959], discussed in §5.5.1.

2.9.5. Vertical Mixing of SSA Particles

The extent to which SSA particles are mixed in the marine atmosphere can further be examined by considering the steady state size-dependent SSA concentration profiles for heights greater than 10 m. These profiles, calculated from (2.9-4) under the assumption of neutrally stable atmospheric conditions, are presented in Fig. 12 (lower panel) for several r_{80} in terms of $n(z)/n(10 \ m)$ for U_{10} values of 5, 10, and 15 m s^{-1} at 80% RH. As before, it is assumed that particles with $r_{80} < 25 \ \mu m$ have completely equilibrated with respect to this ambient RH, and concentration profiles for particles with $r_{80} = 25 \ \mu m$ are presented both under the assumption that complete equilibration has occurred and under the assumption that no equilibration has occurred. As discussed in §2.9.3, the expression for the eddy diffusion coefficient D_{eddy} (Eq. 2.4-2) that was used in the derivation of (2.9-4) is valid only for the surface layer, which extends only to several tens of meters above the sea surface, but as calculations involving more realistic values of D_{eddy} yield qualitatively similar results, the conclusions in this section remain valid.

Under most conditions, the concentrations of SSA particles with $r_{80} \lesssim 5$ µm exhibit little variation with height over the entire marine boundary layer, with $n(500\text{ m})/n(10\text{ m}) \gtrsim 0.9$ for SSA particles in this size range at $U_{10} = 10$ m s^{-1} (Fig. 12). Thus, these particles can be considered well mixed over the marine boundary layer, and the column burden (i.e., vertical integral) of the concentration of these particles can be approximated by the product of $n(10\text{ m})$ and H_{mbl} (§2.1.2, §2.1.5). With increasing particle size, the decrease in concentration with increasing height becomes more pronounced (Fig. 12). For example, at $U_{10} = 10$ m s^{-1} the concentration ratio $n(100\text{m})/n(10\text{m})$ for SSA particles with $r_{80} = (5, 10, 15, 20, 25)$ µm that have completely equilibrated with respect to an ambient RH of 80% is equal to (0.95, 0.8, 0.6, 0.4, 0.25), whereas for SSA particles with $r_{80} = 25$ µm that retain their formation radii this ratio is less than 0.004. The heights to which SSA particles are mixed can be characterized by z_{50}, the height at which the steady state number concentration is 50% of its value at 10 m; thus z_{50} is defined by the equation $n(z_{50})/n(10\text{ m}) = 0.5$. For SSA particles with $r_{80} = (10, 15, 20, 25)$ µm that have completely equilibrated to an ambient RH of 80% at $U_{10} = 10$ m s^{-1}, $z_{50} \approx (1.2 \cdot 10^4, 230, 60, 30)$ m, whereas for SSA particles with $r_{80} = 25$ µm that retain their formation radii, $z_{50} \approx 13$ m. Although the value of 12 km for z_{50} for $r_{80} = 10$ µm is clearly well above the region for which this analysis applies, these results indicate that under typical conditions in a neutrally stable atmosphere SSA particles with r_{80} up to ~10 µm are expected to be nearly uniformly mixed throughout the marine boundary layer. Sea salt aerosol particles with r_{80} up to ~25 µm are mixed throughout the surface layer, with mixing heights well greater than the height at which measurements are commonly made (~10 m). However, with increasing particle size the extent of mixing, and the mixing heights, decrease rapidly. For SSA particles with $r_{80} \gtrsim 25$ µm, which are not expected to achieve RH equilibration (§2.9.1), the extent of mixing is quite restricted, and the concentrations of these particles exhibit large vertical gradients. These results lend further confirmation to the utility of separating medium SSA particles and large SSA particles at $r_{80} \approx 25$ µm (Table 2).

These conclusions hold for a wide range of conditions. For a given r_{80}, v_{term} increases with increasing RH, but differs from its value at 80% RH by less than 30% for RH between 50% and 90% (Fig. 7). Higher wind speeds will entrain larger particles upward more efficiently, but even for gale force winds of $U_{10} = 20$ m s^{-1}, which corresponds to $u_* \approx 90$ cm s^{-1} and $v_{mix} \approx 36$ cm s^{-1} (where C_D is taken to be 0.002 to account for the increase in the wind stress coefficient with wind speed; §2.3.2), the concentration ratio $n(100\text{ m})/n(10\text{ m})$ remains quite low for large SSA particles—for an SSA particle with $r_{80} = 40$ µm assumed to have retained its formation radius, $n(100\text{ m})/n(10\text{ m}) \approx 0.04$,

and z_{50} is only ~17 m (where Eq. 2.6-11 was used for the terminal velocity). Additionally, although the times required for small SSA particles ($r_{80} \lesssim 1$ µm) to attain their steady state concentrations with regard to turbulent diffusion and gravitational sedimentation are so long that it is unlikely that such steady state conditions ever exist (§2.9.6), the conclusion that these particles are well mixed throughout the marine boundary layer is expected to remain valid because of the minimal role played by gravity in their motion and because of the rapid mixing of the marine boundary layer characterized by τ_{conv}, which is of order 10^3 s (§2.4).

These conclusions are consistent with calculations of *Burk* [1984], who investigated the mixing of SSA particles in the marine boundary layer numerically using a model that took into account turbulent diffusion (using a second-moment closure method) and dry deposition (which included the effects of gravitational sedimentation and of inertial impaction across the viscous sublayer), and also effects of the geostrophic flow and the change in direction of the wind with height. He concluded that particles with $r_{80} \lesssim 10$ µm exhibit little change in concentration throughout the lowest several hundred meters above the sea surface.

The size-dependent steady state concentration profile was also calculated for various wind speeds by *Junge* [1957b, 1963, pp. 126-129], but his results cannot be directly applied. The expression he used for D_{eddy}, equivalent to $\kappa U_5 z$ for the lowest several tens of meters, yields values that are too large by roughly a factor of 30 (*cf.* Eq. 2.4-2); likewise, the value of v_{mix} defined in §2.9.3, which would be equivalent to κU_5 for this value of D_{eddy}, would be overestimated by the same amount. However, Fig. 6 of *Junge* [1957b] implies that the values he used for D_{eddy} and thus for v_{mix} were too high by only a factor of about 7; as r_{mix} is proportional to $v_{mix}^{1/2}$ (§2.9.3), the values of r_{mix} resulting from *Junge's* analysis would be too great by only a factor near 2.5. Thus, *Junge's* conclusions that SSA particles with (geometric) radii 5-10 µm are not appreciably affected by gravity but that the upward diffusion of particles with radii greater than ~20 µm is limited by gravity in a "rather sharp manner," would remain valid using radii that are a factor of ~2.5 smaller; that is, that SSA particles with radii 2-4 µm would not be appreciably affected by gravity, whereas upward diffusion of SSA particles with radii 8 µm would be limited by gravity—a result qualitatively similar to the result obtained above. Additionally, *Junge* found r_{mix} to be fairly insensitive to wind speed, as also concluded here.

The discussion so far in this section has assumed near-neutral atmospheric stability; although neutral stability is typical over the ocean, stable or unstable conditions can result in diminished mixing or in enhanced mixing, respectively, the latter due to convection, which acts in addition to mechanical mixing to transport particles upward. For stability conditions different from neutral, D_{eddy} exhibits a dependence on z that differs from that given by (2.4-2), thus modifying (2.9-4) and affecting the steady state vertical distribution of the SSA particle number concentration. These effects, and their dependence on particle size, were investigated numerically by *Goroch et al.* [1980], who demonstrated that mixing in stable conditions is generally decreased more than it is enhanced in unstable conditions. *Goroch et al.* concluded that at high wind speeds the turbulent mixing is governed primarily by wind shear, and thus that stability plays little role in the vertical concentration profile, although at low wind speeds the effects of stability become more important. However, as SSA production is thought to be minimal or nonexistent at low wind speeds, this result is of limited relevance to considerations of SSA production. *Davidson and Schütz* [1983], based on more than 400 shipboard measurements in the North Atlantic, reported that number concentrations of SSA particles with ambient radii near 1 µm were less for stable conditions, and greater for unstable conditions, than for neutral conditions, for the same value of u_* (and thus presumably the same production flux), consistent with the model of *Goroch et al.* Further, they reported that the departures in concentrations from those expected for neutral conditions decreased with increasing u_*, as would be expected because of the increase in turbulent diffusive mixing. However, the amounts by which the concentrations differed from those under neutral conditions exceeded the differences expected from the analysis of *Goroch et al.* by factors of 10 or more, and there are several reasons these results are not applicable to the present discussion. Because of the low terminal velocity of an SSA particle with radius 1 µm ($\sim$0.014 cm s^{-1}; §2.6.2), even for the lowest value of u_* for which results were reported, $\sim$3 cm s^{-1} (corresponding roughly to $U_{10} \approx 1$ m s^{-1} for neutral stability), the decrease in concentration expected for the maximum stability encountered would be only $\sim$2.5% relative to neutral stability, well less than the spread of their measurements. Furthermore, it is unlikely that concentrations of these particles had attained their steady state values under the actions of turbulent diffusion and gravity, as the time required for this would be several weeks (based on arguments presented in §2.9.6, especially Table 8). Hence it is likely that other factors must have come into play besides the effects of stability.

Several investigations based on aircraft measurements have reported the vertical distribution of SSA particles in and above the marine boundary layer (typically demarcated by the presence of a large temperature inversion), in both coastal and open ocean locations. Size distributions of SSA concentration determined at various heights were presented for particles with r_{80} up to $\sim$10 µm by *Woodcock and Gifford*

[1949] and *Woodcock* [1962], and for particles with r_{80} up to $\sim$20 µm by *Woodcock* [1952, 1953], for several coastal and open ocean locations. These concentrations exhibited either little change or else a slight decrease with height up to cloud base (which varied from between 150 and 300 m during stable conditions to nearly 2000 m in unstable conditions), above which there was a large decrease in concentration for SSA particles of all sizes. Other studies [*Lodge*, 1955; *Durbin and White*, 1961; *Dinger et al.*, 1970] reported various results for the dependence of SSA concentration on height, but these investigations involved heights to several thousand meters, with poor vertical resolution in the lowest hundred meters. Both *Durbin and White* [1961] and *Dinger et al.* [1970] noted that for the most part SSA particles were confined to the lowest few kilometers of the marine atmosphere, with number concentrations above this height much less than those below. Based on measurements near the Pacific coast of North America, *Junge et al.* [1969] concluded that SSA was virtually absent above $\sim$2 km. *Woodcock et al.* [1971] reported that the mean ratio of the "quantity of salt" (presumably SSA dry mass) at 50 m above sea level to that at 550 m under trade wind conditions in well mixed air near Hawaii, based on 15 separate analyses, was near unity, with a standard deviation of 0.3. Vertical distributions of SSA particles with r_{80} from approximately 1 µm to 8 µm at a coastal region in Australia reported by *King and Maher* [1976, Fig. 4] exhibited little decrease with height (less than a factor of two) from 200 m to $\sim$1200 m, but decreased rapidly above the height of the marine boundary layer at $\sim$1500 m (the wind speed at which these measurements were made was not reported). *King and Maher* fitted their results to an exponential dependence on height with a characteristic height (at which the concentration should be less than that near the surface by a factor of e) of roughly 1800 m, independent of particle size, although uncertainties in the data were not presented, and the scatter in the data do not preclude other fits. Concentrations of aerosol particles with r_{amb} from 0.1 µm to $\sim$3 µm reported by *Patterson et al.* [1980] were more than an order of magnitude greater in the marine boundary layer than in the free troposphere above. Measurements from a coastal region in the North Sea presented by *Park et al.* [1990, Fig. 1] demonstrated that size-dependent number concentrations of particles with r_{amb} up to $\sim$8 µm were roughly uniform with height over the marine boundary layer (up to $\sim$1400 m), above which they decreased rapidly. A large number of measurements of the total concentration of marine aerosol particles in the size range $r_{amb} \approx 0.25$-1.6 µm in the Southwest Pacific over heights of up to several kilometers were reported by *Kristament et al.* [1993, Fig. 3], who fitted the data, which exhibited much scatter, to an exponential dependence in height with a characteristic height of 1900 m. The investigators noted a marked difference between the concentrations in the marine boundary layer and those in the free troposphere above, with boundary layer values being two orders of magnitude greater than free troposphere values. They also commented on "striking instances where a marked temperature inversion coincided with a very sharp drop in aerosol number concentration." Vertical profiles of number concentrations of SSA particles in three range of ambient radii, $r_{amb} \approx$ 1-12 μm, 3.5-12 μm, and 7.5-12 μm, measured at least 50 km off the east coast of North Carolina presented by *Reid et al.* [2001, Fig. 3] for wind speeds of 8 m s^{-1} and 11 m s^{-1} were roughly independent of height up to ~800 m, the top of the marine boundary layer, above which they were much less. Number concentrations of SSA particles of these sizes for 14 m s^{-1} were roughly independent of height up to ~600 m and attained much greater values (by up to an order of magnitude) at heights from ~600 m to ~1000 m (near the top of the marine boundary layer), above which they decreased rapidly. As these sizes referred to ambient radii, much of the increase in concentrations at higher elevations can be attributed to inclusion of smaller SSA particles which would have grown into this size range at the higher RH values at these elevations. *Maring et al.* [2003] reported that no SSA particles were detected above the marine boundary layer in measurements near Puerto Rico.

Taken as a whole, these measurements confirm the conclusions reached above that small SSA particles and medium SSA particles up to $r_{80} \approx$ 5-10 μm are nearly uniformly distributed throughout the marine boundary layer and exhibit much lower concentrations in the free troposphere above.

2.9.6. Temporal Development of SSA Concentration

The temporal behavior of the number concentration of SSA particles of a given r_{80} in the marine atmosphere determines the rate at which this concentration approaches its steady state value with respect to dry deposition and the rate at which it responds to changing conditions. These considerations are important for the steady state dry deposition method (§3.1) and the concentration buildup method (§3.3). They provide insight into the correlation that might be expected between measured concentrations of SSA mass (§4.1.2) or number (§4.1.3) and local wind speed, and into the extent to which measured concentrations of SSA particles reflect local production at the time of measurement. Additionally, they yield estimates for the mean atmospheric lifetime of SSA particles as a function of size.

It is important to distinguish between the time required for SSA particles of a given size to become mixed throughout their characteristic mixing heights and the time required for concentrations of these particles to attain their steady state values with respect to removal by dry deposition. This is especially important for small SSA particles (i.e., $r_{80} \lesssim$ 1 μm), in whose behavior gravity plays little role. Such particles become nearly uniformly mixed throughout the marine boundary layer within a time of order 10^3 s (§2.4), whereas the time required for the establishment of steady state concentrations against removal through dry deposition for SSA particles with r_{80} = 1 μm is estimated below as ~10^6 s. The consequences of failure to take into account the time required for vertical mixing on estimated times required for SSA concentrations to attain their steady state values with respect to dry deposition depend on the assumptions used in examination of the temporal behavior of these concentrations and on the criterion used to determine when steady state concentrations have been attained, and thus different models may yield different results. Failure to take into account the time of mixing will likely underestimate the time required for attainment of steady state values if this assessment is based on concentrations in the upper portion of the mixing height, but it may overestimate the time if it is based on concentrations in the lower portion of the marine boundary layer. However, these effects are not expected to alter the basic conclusions reached below.

In this section the temporal behavior of $n(r_{80})$ is examined using a simple model that assumes that SSA particles, once formed, are instantaneously (that is, on a shorter time scale than that over which the column burden of concentration changes appreciably) and uniformly mixed over a height H_{mix}, which depends only on particle size and wind speed (which is taken as independent of height and equal to its value at 10 m, U_{10}, throughout the marine boundary layer). Smaller particles ($r_{80} \lesssim$ 5-10 μm) are expected to be well mixed over the entire marine boundary layer and thus for these particles H_{mix} is taken to be H_{mbl} = 0.5 km, whereas steady state mixing heights of larger particles are lower, and for these particles H_{mix} is taken to be z_{50}, the height at which the concentration is 50% of its 10 m value (§2.9.5). Horizontally homogeneous conditions are assumed in both the along-wind and cross-wind directions, and entrainment of the free troposphere and gain and loss terms are neglected, i.e., no wet deposition; thus estimates for the characteristic times obtained below are those for removal through dry deposition alone and do not necessarily reflect mean atmospheric residence times. In this situation the only mechanisms that act to change the number concentration of SSA particles with a given r_{80} are production, characterized by the size-dependent effective production flux f_{eff}, which is assumed to be uniform over the ocean (but which may vary with time), and removal by dry deposition, parameterized by the size-dependent dry deposition velocity v_d, which is assumed to act sufficiently slowly that it does not appreciably alter the vertical concentration profile (§2.7.2). Under these assumptions, (2.8-13) becomes

$$H_{mix} \frac{dn}{dt} = f_{eff} - v_d n \qquad (2.9\text{-}6)$$

(a similar equation can also be obtained by considering the column burden of the number concentration). If f_{eff} is independent of time t for $t > 0$, then the solution to (2.9-6) can be written as $n(t) = n_\infty - (n_\infty - n_0)\exp(-t/\tau_{dry})$, where the steady state value for $n(t)$ is given by $n_\infty = f_{eff}/v_d$, and the characteristic time τ_{dry} is defined by

$$\tau_{dry} = \frac{H_{mix}}{v_d}; \qquad (2.9\text{-}7)$$

the values of v_d and τ_{dry} correspond to the conditions (e.g., wind speed) characterizing f_{eff}. This characteristic time τ_{dry} for the establishment of the steady state concentration under the actions of turbulent diffusion and dry deposition is the same as the mean atmospheric residence time for SSA particles of a given r_{80} against removal through dry deposition defined in §2.7.2.3, for which values are presented in Table 8 for several r_{80} values at $U_{10} = 10$ m s^{-1}. For an extreme situation in which production commences at $t = 0$ into an atmosphere initially free of SSA particles (i.e., $n_0 = 0$), the temporal behavior of the concentration of SSA particles of a given r_{80} is described by $n(t) = n_\infty [1 - \exp(-t/\tau_{dry})]$, where τ_{dry} corresponds to the conditions (e.g., wind speed) characterizing the production flux. The concentration $n(t)$ initially increases nearly linearly with time at a rate $dn/dt = n_\infty/\tau_{dry}$, and attains nearly two-thirds of its steady state value by time $t = \tau_{dry}$ and 90% of its steady state value by $t \approx 2.3\tau_{dry}$. Equivalently, for the opposite extreme situation in which production instantaneously ceases at $t = 0$ (i.e., $f_{eff} = 0$ for $t > 0$), the temporal behavior of the SSA concentration is described by $n(t) = n_0 \exp(-t/\tau_{dry})$, where the value of τ_{dry} corresponds to the final conditions (i.e., the wind speed accompanying the lack of production). The concentration $n(t)$ initially decreases with time at a rate $dn/dt = -n_0/\tau_{dry}$, and decreases to just over one-third of its initial value by $t = \tau_{dry}$ and to 10% of its initial value at $t \approx 2.3\tau_{dry}$.

There are several implications of these results. The rate of increase in SSA concentration following an abrupt increase in wind speed (and hence production) is greater than the rate of decrease following an abrupt decrease in wind speed. Because v_d increases, and thus τ_{dry} decreases, with increasing wind speed (§4.6.2), the value of τ_{dry} characterizing the former situation (i.e., increasing wind speed) is less than that characterizing the latter, and the rate of increase is thus

greater. Additionally, for small SSA particles for which τ_{dry} is much greater than τ_{wet}, it is unlikely that steady state concentrations will be attained; in this situation $n_0 < n_\infty$ and again the initial rate of decrease in SSA concentration following a rapid decrease in production would be less than the situation for steady state conditions had been attained. Thus for small SSA particles for which values of τ_{dry} are days to weeks—so long that it can confidently be concluded that concentrations of such particles will rarely if ever have attained their steady state values with respect to dry deposition, it is expected that following a rapid decrease in SSA production the rate of decrease of concentrations measured near 10 m above the sea surface will be low, and that these concentrations will remain near their previous values for a considerable time. Likewise, the assumption that SSA concentrations will approach zero will rarely if ever be satisfied for many particle sizes of interest. One implication of this result is that the value of the SSA number concentration N, which is dominated by small SSA particles ($r_{80} < 1$ μm), would persist for an extended period (weeks) after the wind speed had dropped to zero. Additionally, little correlation can be expected between current local wind speed and quantities such as N to which small SSA particles make the dominant contribution. Likewise, times required for concentrations of SSA particles with $r_{80} \gtrsim 5\text{-}10$ μm to attain their steady state values with respect to dry deposition are order hours, and thus a much greater correlation would be expected between local wind speed and quantities such as M to which these particles make the dominant contribution.

Another estimate of the time characterizing the approach of the concentration of SSA particles of a given size to its steady state value for the situation in which vertical mixing is assumed to occur much more rapidly than times over which the column burden of concentration changes appreciably was presented by *Hoppel et al.* [2002]. For small SSA particles these characteristic times approach the value H_{mbl}/v_{term} (instead of the value H_{mbl}/v_d presented above), whereas for larger particles the characteristic times depend on an arbitrary height above the sea surface δ at which SSA particles are assumed to be produced. With the choice $\delta = 1$ m considered by *Hoppel et al.*, the characteristic times for $r_{80} = 5$, 10, 15, 20, and 25 μm at $U_{10} = 10$ m s^{-1} are approximately 35 h, 6 h, 1.5 h, 20 min, and 5 min, respectively, whereas with the choice $\delta = 10$ m, the respective times are approximately 35 h, 7.5 h, 2.3 h, 50 min, and 20 min. Although the functional form of this estimate differs from that resulting from the model proposed above (for which values are presented in Table 8), for $r_{80} \gtrsim 10$ μm the estimates for the two choices of δ and the estimates presented above are comparable.

This analysis can also be applied to the concentration buildup method (§3.3) under the assumption that the wind is blowing from the land to the sea with constant speed U that is independent of height. In this situation, the concentration of SSA particles of a given size at a fixed location is independent of time but increases toward its steady state value with along-wind distance from the coast x. When viewed in a frame of reference moving with the wind (i.e., in a Lagrangian sense), this situation is equivalent to the one above in which production commences at $t = 0$ (corresponding to $x = 0$) into an atmosphere initially free of SSA particles and the number concentration increases with increasing time (corresponding to distance from the coast). The substitution $t = x/U$ into (2.9-6) and its solution describes this situation in terms of x and defines a characteristic distance

$$X - \frac{UH_{mix}}{v_d} - U\tau_{dry};   \qquad (2.9\text{-}8)$$

at $x = X$, the concentration has attained nearly two-thirds of its steady state value.

The characteristic time for removal through dry deposition τ_{dry} and the characteristic distance X are strongly dependent on r_{80} (§2.7.2.3), both though their dependence on v_d (§4.6.2) and through the dependence of X on H_{mix} (Eq. 2.9-7). Additionally, these quantities depend on U_{10} (taken to be the same as U for this analysis) not simply through the relation $X = U\tau_{dry}$, but also because U_{10} affects both v_d and H_{mix} for a given particle size. The dependencies of τ_{dry} and X on particle size are illustrated by the values presented in Table 8 for $U_{10} = 10$ m s^{-1}, based on values of v_d obtained below (§4.6.2), and using for H_{mix} the lesser of $H_{mbl} = 0.5$ km and z_{50} (§2.9.5). Although the estimates presented should be considered as accurate only to a factor of 3 or so because of their dependence on the model used to estimate v_d, the values used for z_{50} (which are based on the expression for D_{eddy} given by Eq. 2.4-2), wind speed, RH, the height of the marine boundary layer, and the like, they nevertheless allow important conclusions to be drawn concerning the realization of steady state for the concentrations of SSA particles, and by extension the applicability of the steady state dry deposition method to determine the SSA production flux (§3.1).

Times for concentrations of SSA particles to attain their steady state values with respect to dry deposition decrease rapidly with increasing particle size, ranging from days to weeks for small SSA particles to hours to minutes for medium SSA particles (Table 8). These same times characterize the times required for concentrations of SSA particles to respond to changing production (i.e., wind speed), although with changing wind speed the SSA production flux also is expected to change, and there are additional factors such as times required for ocean state to respond to the new conditions.

Sea salt aerosol particles with r_{80} less than a few micrometers require quite long times and distances (weeks and thousands of kilometers) to attain their steady state concentrations with respect to removal through dry deposition, and hence these steady state concentrations are rarely if ever realized. As these times are much larger than the time required for mixing throughout the marine boundary layer ($\sim 10^3$ s; §2.4), particles of these sizes will become nearly uniformly mixed throughout the marine boundary layer on a time scale much shorter than that required for reaching steady state (Table 2). Wet deposition provides the dominant contribution to removal for particles of these sizes in nearly all circumstances of appreciable precipitation (§2.7.3). After a wet removal event, concentrations of these particles will increase at a rate proportional to the magnitude of their production flux until they are removed by a subsequent wet removal event. As the characteristic time τ_{dry} for these particles decreases with increasing wind speed (because of the dependence of τ_{dry} on v_d; H_{mix} being equal to H_{mbl} and thus assumed independent of wind speed), the response of concentrations of these small particles to decreasing wind is expected to be slower than to increasing wind, as noted above. With increasing r_{80} a marked decrease in H_{mix} occurs for r_{80} near 10 µm, as particles of this size and larger are not mixed over the entire marine boundary layer (§2.9.5). Consequently, there is a decrease in τ_{dry} and X because of both greater v_d and smaller H_{mix}. With further increase in r_{80} the times required for particles to attain their steady state concentrations with respect to removal through dry deposition are sufficiently small (hours to minutes) and the distances sufficiently short (several hundred kilometers to several kilometers) that steady state conditions are increasingly likely to be realized in many oceanic situations. As SSA particle size becomes sufficiently large and removal becomes increasingly rapid, the time required for vertical mixing can no longer be considered negligible compared to τ_{dry}, resulting in the time required for steady state to be attained being somewhat greater than H_{mix}/v_d. However, this would occur only for particles so large that their characteristic removal times are a few hours or less, and concentrations of such particles will rapidly attain their steady state values in many situations; thus this consideration does not appreciably alter any of the conclusions reached using this model.

The temporal behavior of the vertical profile of SSA number concentration has previously been considered by several investigators. This behavior was investigated numerically by *Junge* [1957a] for particles sufficiently small that the effect of gravity could be neglected (thus this investigation essentially determined the rate of turbulent mixing), using an expression for D_{eddy} that approaches an asymptotic value of ~ 30 m^2 s^{-1} with increasing height (which *Junge* argued was "determined by the intensity of convection"), approximately equal to the value given by (2.4-2) for D_{eddy} at 400 m for $U_{10} = 5$ m s^{-1} (although not explicitly stated, this was apparently the wind speed used). Although steady state solutions do not exist in this model (as there is no removal mechanism), *Junge* concluded that the concentrations approached a value nearly independent of height over a mixing height of 2 km after a day or so. Substitution of $z = 2$ km and $D_{eddy} = 30$ m^2 s^{-1} into the estimate presented for τ_{mix} in §2.4 (Eq. 2.4.4), $H_{mbl}^2/(2D_{eddy})$, yields the value 18.5 h, consistent with *Junge's* conclusions. However, based on the values presented, *Junge's* expression for the eddy diffusion coefficient in the lowest several tens of meters[1] is equivalent to $D_{eddy} \approx 0.05U_5z$, a factor of 3-4 higher than given by (2.4-2). If the rate at which the concentration at a given height responds to surface production is limited by diffusion in the lowest portion of the marine boundary layer, then the times required for steady state to be attained would be underestimated by the same factor of 3-4, and the time required for the profile to become nearly uniform with height would be several days or more instead of the value of one day given by *Junge*. However, as noted in §2.4, large-scale mixing of the marine boundary layer occurs by convection in addition to turbulent diffusion, and times characterizing this mixing would be of order 10^4 s for this marine boundary layer height.

The response of SSA concentration to changing wind speed was examined numerically by *Burk* [1984] using a model described above (§2.9.5) that took into account turbulent diffusion, dry deposition, and effects of the geostrophic flow and the change in direction of the wind with height. *Burk* found that when wind speed increased linearly from 4.5 m s^{-1} to 9.2 m s^{-1} over a 3 hour period from steady state conditions, concentrations of SSA particles with r_{80} from 1 μm to 25 μm were near (within ~40% of) their new steady state values after ~6 h to 9 h, whereas when wind speed decreased linearly from 9.2 m s^{-1} to 4.5 m s^{-1} over 3 hours, concentrations of SSA particles with $r_{80} \lesssim$ 10 μm remained much closer to their original values than to their new steady state values after 12 hours, and the size-dependence of the number concentration of larger SSA particles did not resemble either its new steady state value or its old one. These results are consistent with the conclusion reached above that the response of SSA concentration to decreasing wind is slower than its response to increasing wind. *Burk* further concluded that SSA particles with $r_{80} \gtrsim$ 14 μm have sufficiently large terminal velocities that their concentrations respond rapidly to variations in surface

production (with a time lag of hours), but that the residence times of smaller particles are too large to respond quickly to such variations. This conclusion is also consistent with that reached above that the characteristic response time for concentrations of SSA particles with $r_{80} \gtrsim$ 10-15 μm is of order hours.

The temporal development of the concentration profile was examined also by *Stramska* [1987] using a numerical model in which a constant eddy diffusion coefficient D_{eddy} was used for heights from 30 m to the top of the marine boundary layer at 1500 m. For wind speeds from 6 m s^{-1} to 34.2 m s^{-1} and RH of 70% and 90%, values were presented for the times required for the concentrations of SSA particles with r_{80} of 2, 6, 10, 14, and 20 μm to approach what she referred to as a "quasi-equilibrium state," "quasi-stationary state," or "steady state," the onset of which she took as the SSA concentration (at what height was not stated) changing by less than 2% per hour. Concentration profiles over the lowest 20 m over the ocean were also presented. The times required to attain this quasi-equilibrium state decreased with increasing particle size from 20-25 h for $r_{80} = 2$ μm to 3-9 h for $r_{80} = 20$ μm. For a given SSA particle size and wind speed, the times were equal or slightly less at RH 90% than at RH 70%. For a given RH the time required for concentrations of SSA particles with $r_{80} \geq 10$ μm to reach quasi-equilibrium increased with increasing wind speed (by up to a factor of 2 over the above wind speed range), whereas for particles with r_{80} of 2 μm and 6 μm these times were roughly independent of wind speed. It is important to note that the quasi-equilibrium times presented by *Stramska* [1987] are not the times required for concentrations to attain their steady state values with respect to removal by dry deposition. The arbitrary criterion for attainment of steady state, that the concentration change by less than 2% per hour, is especially inappropriate for small SSA particles given the long times (weeks) for their concentrations to reach their steady state values under removal by dry deposition. For such particles, under the assumption that vertical mixing over the marine boundary layer occurs much more rapidly than other pertinent time scales, concentrations resulting from a constant production flux would initially increase linearly (as noted above), and the 2% per hour criterion would be satisfied after ~50 h, despite the concentration being more than an order of magnitude less than its steady state value with respect to dry deposition given by the solution to (2.9-6). The discrepancy between 50 h and the values of ~25 h reported by *Stramska* for SSA particles with

---

[1] The value given for the constant prefactor of the "Austausch coefficient"—the product of the eddy diffusion coefficient and the density of the air—was incorrectly stated in the expression presented by *Junge*; it should have been $6 \cdot 10^{-5}$ and not $6 \cdot 10^5$; however, it appears that he used the correct value in the calculations.

$r_{80} = 2$ μm may result from the time required for particles of this size to become mixed over the marine boundary layer. For $U_{10} = 10$ m s^{-1} under neutrally stable atmospheric conditions, the value of the eddy diffusion coefficient used for $z > 30$ m was $D_{eddy} \approx 4.3$ m^2 s^{-1}, according to which the mixing time for $H_{mbl} = 1500$ m (approximated by $H_{mbl}^2/(2D_{eddy})$, as in §2.4, Eq. 2.4.4) would be approximately $2.6 \cdot 10^5$ s (i.e., 3 days)—much longer than the estimate of $\sim 10^3$-10^4 s based on the convective mixing velocity using this value of H_{mbl}. For SSA particles with r_{80} of 10, 15, and 20 μm at $U_{10} = 10$ m s^{-1}, the 2% per hour criterion yields times of approximately 12, 8, and 5 h, respectively, according to the model of *Stramska*; these values can be compared to the values of approximately 17, 5, and 0.6 h, respectively, for this criterion obtained with (2.9-6) and values of τ_{dry} from Table 8. However, for SSA particles with $r_{80} \gtrsim 10$ μm, the requirement that the concentration change by less than 2% per hour is nearly equivalent to steady state conditions. The discrepancy for $r_{80} = 20$ μm is perhaps due to *Stramska's* use of a constant eddy diffusivity for $z > 30$ m and the role played by the finite mixing time.

The temporal development of the SSA concentration profile was examined by *Hoppel et al.* [2002] using a numerical model in which SSA particles were assumed to be produced at height 1 m and the eddy diffusivity D_{eddy} was assumed to be nearly directly proportional to $\kappa u_* z$ at low heights above the sea surface (similar to that given by Eq. 2.4-2), reach at maximum at 500 m, and decrease to zero at the height of the marine boundary layer at 1 km. Assessment of steady state conditions for a given r_{80} was based on the concentration of SSA particles averaged over the lowest 2 m above the ocean. At $U_{10} = 10$ m s^{-1} and for 80% RH throughout the marine boundary layer, SSA particles with $r_{80} \gtrsim 12$ μm had essentially attained their steady state values after 12 h, whereas SSA particles with $r_{80} \approx 6$ μm required 4 d to attain 90% of their steady state concentrations, and SSA particles with $r_{80} = 1$ μm attained only 6% of their steady state values after 4 d. These times are greater than the estimates of *Hoppel et al.* discussed above partly because of the greater value assumed for H_{mbl} (1 km vs. 0.5 km; this would be expected to contribute a factor of two), partly because of the lower values of D_{eddy} over much of the marine boundary layer (which would be expected to contribute roughly another factor of two), and perhaps because of the earlier assumption that vertical mixing occurred instantaneously.

There are surprisingly few measurements of the temporal behavior of concentrations of SSA particles in the marine boundary layer with which to compare the results of the above analyses. Concentrations of SSA particles with ambient radii 1-3.5 μm and 3.5-8 μm at wind speeds of 4, 8, and 12 m s^{-1} were measured from aircraft at heights from 30 m to over 1000 m, the height of the marine boundary layer, at distances up to 50 km from the coast during conditions of offshore winds by *Reid et al.* [2001]. For wind speeds of 4 m s^{-1} there was little if any increase in SSA concentration in the along-wind direction and hardly any vertical structure, probably because there was little if any SSA production at this wind speed. For the other two wind speeds, SSA concentrations at the lowest height at which they were measured, and the height at which a given value of the SSA concentration was attained, generally increased with increasing distance from the coast. Concentrations of SSA particles with $r_{amb} = 1$-3.5 μm appear to have approached a value that varied little with distance at 50 km from the coast, although these particles were not uniformly mixed over the marine boundary layer at this location and exhibited a noticeable vertical profile. The times required to cover a distance of 50 km for wind speeds of 8 m s^{-1} and 12 m s^{-1} are approximately 1.7 h and 1.2 h, respectively, comparable to the estimates of the time required for mixing (of order an hour); hence it is likely that over this distance these particles (and those with $r_{amb} = 3.5$-8 μm) would not have had time to become well mixed over the marine boundary layer. Differences between the behavior of the marine boundary layer near coastal regions and that over the open ocean may have also played a role in these results. Additionally, based on estimates of the characteristic times and distances required for concentrations of these particles to approach their steady state values (Table 8), it can confidently be concluded that concentrations of SSA particles of these sizes were far from their steady state values with respect to dry deposition over the measurement region.

Although the characteristic times obtained from the several numerical models depend on the particular model and its assumptions and conditions, the results are for the most part consistent with those obtained from the simple arguments presented above. Times required for concentrations of small SSA particles to attain their steady state values against removal by dry deposition are of order weeks—far too long to expect that steady state conditions would ever occur, although concentrations of these particles would rapidly attain nearly uniform vertical profiles whose magnitude changes little with increasing time (< 2% per hour after 50 h). With increasing particle size the times required for SSA concentrations to approach their steady state values with respect to dry deposition rapidly decrease, both because of decreasing mixing height and increasing dry deposition velocity, from order days for SSA particles with $r_{80} = 1$ μm, to order hours for SSA particles with r_{80} in the approximate range 5-20 μm, to order minutes for larger SSA particles. Thus over a wide range of SSA particle sizes the assumption that SSA concentrations are near their steady state values is likely to be satisfied.

2.9.7. Summary

Several different lines of reasoning all converge to the same general conclusions concerning the behavior of SSA particles in the marine environment, and support the classification of SSA particles into three size ranges (§2.1.2) with critical r_{80} values of around 1 μm and 25 μm for the differentiation of small, medium, and large SSA particles based on this behavior. The demarcation of small and medium SSA particles at $r_{80} \approx 1$ μm is based on the importance of removal mechanisms of these particles in the atmosphere, with wet deposition being dominant for small SSA particles. The demarcation of medium and large SSA particles near $r_{80} \approx 25$ μm is rather sharp and results from several physical considerations: equilibration with respect to the ambient RH, vertical entrainment of SSA particles immediately subsequent to their production at the sea surface, and the extent of vertical mixing of SSA particles into the marine atmosphere.

Quantitative estimates have been obtained for several quantities pertinent to SSA and its production: $r_{80,eq}$, the maximum size of SSA particles that are expected to have attained their equilibrium radii before measurement at a height near 10 m (§2.9.1); $r_{80,ent}$, the maximum size of SSA particles that have an appreciable probability of being entrained upward from near the sea surface immediately subsequent to their production (§2.9.2); the dependence of Ψ_f on r_{80} (§2.9.4); the extent to which SSA particles of a given r_{80} are mixed in the marine boundary layer (§2.9.5); the mean atmospheric residence times of SSA particles of a given size; the times required for concentrations of SSA particles of a given size to attain their steady state profiles with respect to dry deposition (§2.9.6), and the dependencies of these quantities on wind speed U_{10}. Although the exact values of these quantities depend somewhat on wind speed, ambient RH, and atmospheric stability, the r_{80}^4 dependence of the distance fallen by an SSA particle before its radius equilibrates with respect to the ambient RH (2.9-1), and the power law dependence of the concentration on height in (2.9-4), ensure that the critical sizes do not vary much from the values presented above for 80% RH and $U_{10} = 10$ m s^{-1}. These conclusions are thus expected to hold over a wide range of conditions.

Small SSA particles ($r_{80} \lesssim 1$ μm) will have attained their equilibrium radii with respect to the ambient RH by the time of measurement at nearly any height. The role of gravity in the fate of these particles is negligible; a particle with radius 1 μm requires over a week to fall 100 m in still air (§2.6.2). Sea salt aerosol particles in this size range are rapidly mixed over the entire marine boundary layer, and their atmospheric lifetimes against removal by dry deposition are long—of order weeks, with removal therefore occurring almost

entirely by wet deposition; hence concentrations of these particles rarely if ever attain their steady state values against dry deposition. Sea salt aerosol particles in this size range spend sufficient time in the atmosphere to be important with regard to atmospheric chemistry, cloud formation, light scattering, and other processes of interest (Table 2). The effective and interfacial SSA production fluxes of these particles are very nearly equal.

Medium SSA particles (1 μm $\lesssim r_{80} \lesssim 25$ μm), depending on their size, may not have attained their equilibrium radii with respect to RH by the time of measurement within a few meters of the surface, although most particles in this size range will have equilibrated by the time of measurement at 10 m. With increasing particle size, the role of gravity in the fate of these particles becomes increasingly important; an SSA particle with geometric radius 10 μm requires ~2 h to fall 100 m in still air, whereas one with geometric radius 25 μm requires 20 min. Depending on their size, medium SSA particles will become mixed over heights of tens to hundreds of meters—much or most of the marine boundary layer. Times required for concentrations of SSA particles of these sizes to attain their steady state values with respect to dry deposition are minutes to days, again depending on their size; consequently it is probable that under many circumstances a balance between surface production and removal by dry deposition is attained. Particles in this size range spend sufficient time in the atmosphere to play important roles in many processes of interest (Table 2). Steady state concentrations of medium SSA particles exhibit rather weak dependencies on height from ~5 m to ~20 m; at $U_{10} = 10$ m s^{-1} concentrations of particles with $r_{80} \lesssim 10$ μm are expected to vary by less than ~15% over this range of heights, and those with $r_{80} \lesssim 20$ μm by less than ~40% (§2.9.4). Thus the exact height at which concentrations of SSA particles in this size range are measured is typically not of great importance as long as it is in the range 5-20 m. For SSA particles in the lower portion of this size range, the effective SSA production flux is nearly equal to the interfacial SSA production flux, but with increasing size the effective SSA production flux becomes increasingly less than the interfacial SSA production flux, and the ratio of these fluxes becomes much less than unity at $r_{80} = 25$ μm.

Large SSA particles ($r_{80} \gtrsim 25$ μm) will rarely have attained their equilibrium radii with respect to RH during their brief sojourn in the atmosphere. Consequently, the assumption that the radii of these particles have equilibrated by the time of measurement would result in substantial overestimation of both the sizes and concentrations of such particles (§4.1.4). As particles in this size range retain radii near their formation values for some or most of the time they spend in the atmosphere, their ability to be entrained upward is further reduced. The motion of SSA particles in this size range is controlled largely by gravity; an SSA particle with $r_{80} = 25$ μm that retains its formation radius (i.e., its geometric radius is 50 μm) requires only 5 min to fall 100 m in still air. Although large SSA particles are occasionally detected at heights of 10 m and above, they represent a small fraction of the number of particles of these sizes produced at the surface, with this fraction decreasing rapidly with increasing particle size and with increasing height. These particles are thus confined mainly to the lowest several meters above the sea surface, and they do not contribute to the effective SSA production flux. Their atmospheric residence times are of order minutes to seconds, and thus they do not spend sufficient time in the atmosphere to be important for most atmospheric processes of interest (Table 2).

3. Methods of Determining Size-Dependent Sea Salt Aerosol Production Fluxes

The nine methods enumerated in §1.2 that have been used or might in principle be used to determine either the interfacial SSA production flux or the effective SSA production flux (Table 9) are examined in this section: the steady state dry deposition method, the whitecap method, the concentration buildup method, the bubble method, micrometeorological methods, the along-wind flux method, the direct observation method, the vertical impaction method, and the statistical wet deposition method. The information necessary to calculate the pertinent SSA production flux according to each method is specified, the assumptions and conditions required for each to yield accurate results are identified, and the concerns with each are discussed. Application of these methods to estimate SSA production fluxes is examined in §5. Methods that yield the effective SSA production flux (steady state dry deposition method, concentration buildup method, micrometeorological methods, and the statistical wet deposition method), which depend upon measurements at ~10 m or higher and are restricted to small and medium SSA particles, are discussed based on considerations of the behavior of the size-dependent number concentration presented in §2.8.3.

3.1. STEADY STATE DRY DEPOSITION METHOD

The steady state dry deposition method determines the size-dependent effective SSA production flux from field measurements of the mean size-dependent concentrations of SSA particles at a given reference height z_{ref}, $\bar{n}(r_{80}, z_{ref})$ (§4.1.4), and model-derived values of the size-dependent dry deposition velocity $v_d(r_{80}, z_1, z_{ref})$ (§2.7.2.1; §4.6), under the assumption of local balance—that at the time and location of measurement the effective production flux of SSA particles of a given r_{80} is equal to the removal flux of these particles through dry deposition:

$$f_{eff}(r_{80}, z_1) = \bar{n}(r_{80}, z_{ref}) \times v_d(r_{80}, z_1, z_{ref}). \quad (3.1\text{-}1)$$

Sea Salt Aerosol Production: Mechanisms, Methods, Measurements, and Models
Geophysical Monograph Series 152
Copyright 2004 by the American Geophysical Union.
10.1029/152GM03

This approach was used as early as 1954 [*Moore and Mason*, 1954] and later by *Eriksson* [1959], *Blanchard* [1963, pp. 128-132], and *Toba* [1965a,b, 1966]. More recently it has been used by *Chaen* [1973], *Fairall* and coworkers [*Fairall et al.*, 1982, 1983; *Fairall and Larsen*, 1984; *Fairall et al.*, 1994; *Miller*, 1987], *Iida et al.* [1992], *Smith et al.* [1993], and *Smith and Harrison* [1998]. Estimates of the effective SSA production flux obtained using this method are examined in §5.2.

This assumption of local balance is essential to this method. Implicit in this assumption is that dry deposition provides the dominant removal mechanism of SSA particles within a specified size range, and that steady state with respect to dry deposition exists during the atmospheric residence time of SSA particles of this size (§2.9.5). This assumption further implies that the local meteorological conditions at the time of measurement, under which particles are removed by dry deposition, are similar to those responsible for their production, and furthermore that SSA particles at the time and location of measurement have experienced these same conditions throughout their residence times in the atmosphere. Otherwise, the spatial and temporal separation between production and removal of SSA particles, and the response times of production of SSA particles and of their concentrations to changing meteorological conditions relative to their atmospheric residence times prior to measurement, result in a situation in which production and removal are not in balance, and the method fails. This assumed balance of production and removal through dry deposition results in a time-independent vertical concentration profile and zero mean net flux through any height z above the surface, i.e., $\bar{j}_z(z) = 0$ for any z. The production flux at a height z_1 can then be determined from (2.8-13), which relates the SSA production flux to processes that control SSA concentration. The height z_{ref} to which (3.1-1) refers is typically taken to be 10 m, and z_1 is the height to which the production flux pertains; for $z_1 \approx 10$ m this method yields the effective SSA production flux (§2.1.6). For $z_1 = 0$ this flux would correspond to the interfacial SSA production flux, but this approach requires that further assumptions be made to account for SSA particles that do not attain the height of measurement but nonetheless contribute to the dry deposition at the surface (§2.1.6; §2.7.2). Some attempts of

Table 9.   Methods for Determining Size-Dependent Sea Salt Aerosol Production Fluxes

Method	Flux[a]	Particle Size[b]	Drop Type[c]	Key Required Quantities
Steady state dry deposition	Eff	M	J	Size-dependent number concentration Size-dependent dry deposition velocity
Whitecap	Int	S, M	F, J	Size-dependent laboratory whitecap production flux Oceanic whitecap ratio
Concentration buildup	Eff	S, M	F, J	Size-dependent number concentration as a function of along-wind distance and height
Bubble	Int	S, M	F, J	Sizes and number of drops produced per bubble Size distributions of bubble concentration Bubble rise velocities
Micrometeorological	Eff	S	F	Size-dependent number concentration Size-dependent vertical fluxes at ~10 m
Along-wind flux	Int	L (S[d])	J, S (F)	Size-dependent along-wind flux near sea surface Mean horizontal distance traveled
Direct observation	Int	L	S	Count from video
Vertical impaction	Int	M, L	J, S	Size-dependent vertical flux near sea surface
Statistical wet deposition	Eff	S	F, J	Size-dependent number concentration Mean time between precipitation events

[a] Effective (Eff) or Interfacial (Int) production flux.
[b] Particle size refers to that for which the method is most applicable; S, M, and L denote small, medium, and large SSA particles, respectively (§2.1.2).
[c] Drop type refers to that for which the method is most applicable; F, J, and S denote film, jet, and spume drops, respectively.
[d] For small SSA particles, this method is essentially the same as the statistical wet deposition method.

this nature that have used SSA concentration measurements at heights of several meters above the sea surface to infer the interfacial SSA production flux are discussed in §4.6.5, but in practice the steady state dry deposition method is restricted to determining the effective SSA production flux.

3.1.1. Conditions

Successful implementation of this method requires the satisfaction of several conditions to ensure that the effects of processes other than production on the number concentration of SSA particles of a given size over the ocean are small compared to those of dry deposition. These conditions, which can be determined by examination of the terms in (2.8-13) that describe the effects of these other processes and that have been omitted in deriving (3.1-1), are:

1. The entrainment condition—negligible entrainment and mixing of the free troposphere,
2. The stationarity condition—negligible time-dependence of the mean size-dependent SSA number concentration,
3. The horizontal homogeneity condition—no appreciable gradients in either the along-wind or the cross-wind direction, and
4. The volumetric condition—negligible volumetric gain and loss terms (principally wet deposition).

These conditions can be expressed as inequalities stating that the size-dependent dry deposition velocity $v_d(r_{80})$ (§2.7.2) be considerably greater than velocities characterizing

these other processes; equivalently, they can be expressed as inequalities stating that the mean size-dependent atmospheric residence time of SSA particles with respect to dry deposition $\tau_{dry}(r_{80})$ (§2.7.2.3; §2.9.6) be considerably less than times characterizing these other processes. As v_d and τ_{dry} depend strongly on r_{80} (this dependence should be understood where not explicitly stated in the discussion that follows), as may these other characteristic velocities and times, the satisfaction of these conditions likewise depends strongly on r_{80}. As v_d is an increasing function of r_{80} (§4.6.2), and likewise τ_{dry} is a decreasing function of r_{80}, for $r_{80} \gtrsim 0.1$ μm (comprising nearly the entire range of SSA particle sizes; §2.1.1), these inequalities result in a lower bound on the value of r_{80} (which depends to some extent on environmental and meteorological conditions) for which this method can be accurately applied. Additionally, because of the dependencies of v_d (and τ_{dry}) and these other characteristic velocities on meteorological factors, satisfaction of these conditions for a given r_{80} can vary greatly with time and location. These conditions are examined below, and rough estimates for the characteristic velocities and times are presented. Similar considerations apply to other methods of determining the effective SSA production flux based on (2.8-13).

3.1.1.1. Entrainment condition. The entrainment condition requires that the decrease in the number concentration of SSA particles of a given r_{80} caused by entrainment of free tropospheric air into the marine boundary layer (§2.4) be negligible compared to the decrease caused by removal through dry deposition. Entrainment of free tropospheric air results in an increase in the height of the marine boundary layer. As the concentration of SSA particles in the free troposphere is generally negligible compared to that in the marine boundary layer below (§2.9.5), this entrainment results in a dilution of SSA particles of sizes that are mixed throughout this layer (although it will not affect concentrations of particles that are not well mixed to the height of this layer), providing a mechanism other than dry deposition that acts to decrease the concentration of SSA particles. If this loss rate were not taken into account, then the effective SSA production flux determined from (3.1-1) would be erroneously low. The quantity characterizing the rate of entrainment of the free troposphere is the entrainment velocity v_{ent}, with associated characteristic time τ_{ent} (§2.4). The effects of this entrainment and the resultant change of the marine boundary layer height with time on the concentration of SSA particles of a given r_{80} that are mixed throughout the marine boundary layer are manifested in (2.8-13) in the term $\bar{n}(r_{80})v_{ent}$, whereas the corresponding term for dry deposition is $\bar{n}(r_{80})v_d(r_{80})$. Thus the entrainment condition can be expressed as $v_{ent} \prec v_d$, or equivalently, $\tau_{dry} \prec \tau_{ent}$ (where the symbol $\prec$ was defined in §2.1.5.2 to denote "considerably less than"). Typical values of v_{ent} are in the range 0.3-0.8 cm s^{-1}, resulting in values of τ_{ent} in the range ~0.5 to ~2 d for a marine boundary layer height of 0.5 km (§2.4). Thus the condition that the dry deposition velocity be considerably greater than v_{ent} imposes a lower bound on the value of r_{80} of between 2 μm and 5 μm (Table 8; §4.6.2). Some formulations of the dry deposition method (§5.2.1.3) explicitly include the effect of entrainment and employ the sum of the dry deposition velocity and the entrainment velocity instead of the dry deposition velocity in (3.1-1). Such an approach is valid only for SSA particles which are mixed to the height of the marine boundary layer and in theory will overestimate the effective production flux of larger SSA particles which are not mixed to this height (and thus will not be affected by entrainment of the free troposphere), but for these larger SSA particles the dry deposition velocity would be substantially greater than the entrainment velocity, so the effect of this overestimation is negligible.

3.1.1.2. Stationarity condition. The stationarity condition requires that the vertical integral of the time rate of change of number concentration of SSA particles of a given r_{80} be negligible compared to the dry deposition flux of these particles. The equation describing this time rate of change for the situation in which only the effective SSA production flux and dry deposition operate was presented in §2.9.6 (Eq. 2.9-6), where it was demonstrated that the time characterizing the approach of the number concentration of SSA particles of a given r_{80}

to its steady state value is $\tau_{dry}(r_{80})$, the mean atmospheric residence time for these particles with respect to dry deposition. Therefore, satisfaction of the stationarity condition for SSA particles of a given r_{80} requires that the system must have been in steady state for a time well greater than $\tau_{dry}(r_{80})$ (§2.9.6), and additionally that factors that might appreciably affect the balance between SSA production and dry deposition must have remained constant (or nearly so) for at least this time; this argument was put forth by *Blanchard* [1963, p. 132] as the criterion for steady state. For SSA particles that are nearly uniformly mixed throughout the marine boundary layer, the vertical integral of the rate of change of the SSA number concentration can be written as $H_{mbl}(dn/dt)$, from which a characteristic velocity v_c describing the rate of change with time of this concentration can be defined as $v_c(r_{80}) = H_{mbl}(dn/dt)/n = H_{mbl}(d\ln n/dt)$, with associated characteristic time $\tau_c(r_{80}) = 1/(d\ln n/dt)$. The stationarity condition thus requires that $v_c \prec v_d$, or $\tau_{dry} \prec \tau_c$.

This velocity v_c can also be thought of as a piston velocity. For example, the value $v_c = +0.01$ (−0.01) cm s^{-1}, approximately equal to +10 (−10) m d^{-1}, corresponds to a daily increase (decrease) in the number of SSA particles in the atmosphere in a vertical column above a unit area of the sea surface equal to the number of such particles in a vertical region nearly 10 m thick. For constant H_{mbl} this increase (decrease) in SSA particle number would result in a corresponding increase (decrease) in SSA particle number concentration distributed over the height of the marine boundary layer. As a numerical example, a rate of increase of the number concentration of SSA particles of a given r_{80} of only 10% per day, which might seem indicative of steady state conditions (and would certainly be difficult to detect), would result in $v_c \approx 0.05$ cm s^{-1} (for $H_{mbl} = 0.5$ km), or $\tau_c \approx 10^6$ s. Thus for SSA particles whose dry deposition velocity is equal to this value of v_c (0.05 cm s^{-1}), the assumption that this rate of change is sufficiently small that the system can be considered to be in steady state (that is, that the term describing the vertical integral of the rate of change of number concentration can be neglected) would result in an error in the estimated effective SSA production flux of 100%. Such a dry deposition velocity is typical for SSA particles with r_{80} near 1 μm (Table 8; §4.6.2); for smaller SSA particles, which have even lower values of v_d, the error would be even greater.

The requirement that local wind speed and other pertinent quantities have remained constant for at least a time greater than the mean atmospheric residence time for SSA particles of the size under consideration is essential for local balance; otherwise the meteorological conditions under which the particles were produced cannot be assumed to be the same as those under which they are removed. This requirement places stringent limitations on the range of r_{80} for which this method can be applied. For a typical wind speed $U_{10} = 10$ m s^{-1} SSA particles are transported nearly 1000 km in a day; thus for SSA particles with $\tau_{dry} \approx 3$ d, for example, this condition requires that wind speed (and direction) and other factors have been constant over nearly 3000 km of ocean. This requirement also places a lower limit on the fetch for which this method can be satisfactorily applied for a given r_{80} (*cf.* arguments in §2.9.6). For instance, for SSA particles with $\tau_{dry} \gtrsim 1$ d, corresponding roughly to $r_{80} \lesssim 3$ μm for typical conditions (Table 8), this condition will not be satisfied by any measurement with fetch less than nearly 1000 km ($\approx U_{10}\tau_{dry}$). However, even this requirement is not fully sufficient, as it does not ensure that the factors affecting the particles during their transit to the measurement location have remained constant over this time. This requirement is certainly not satisfied when a strong front or a different air mass passes a given location, but it is in general difficult to determine the time over which the wind speed and other factors that affect SSA production and concentration have remained constant, and it is also difficult to specify the inaccuracy resulting from a given change of meteorological conditions.

Perhaps even more important for the satisfaction of steady state for SSA particles with given r_{80} is the requirement that the aerosol has not experienced wet deposition within a time $\tau_{dry}(r_{80})$ previous to measurement. Wet deposition occurs intermittently and sporadically, and may have a large impact on the concentration of SSA particles of all sizes because of its efficiency as a removal mechanism (§2.7.1). As noted in §2.8.3, even though wet deposition might not occur during the time of measurement, if it has occurred within a time less than the mean atmospheric residence time of SSA particles of a given r_{80} with respect to dry deposition, $\tau_{dry}(r_{80})$, then the concentration of these particles will not have attained its steady state value and will still be increasing with time (*cf.* arguments in §2.9.6). Consequently, under these circumstances, steady state conditions cannot be assumed, nor will SSA production be balanced by dry deposition; the SSA concentration would be less than it would be under steady state conditions, hence the SSA production flux would be underestimated.

The stationarity condition for wet deposition can be described by the inequality $\tau_{dry} \prec \tau_{wet}$, where τ_{wet} is the time since the air parcel under consideration last experienced appreciable wet removal (§2.7.1). In particular situations this time can be determined, at least in principle, by analysis of back trajectories from satellite observations, for example, but if this time is not known, then the value presented in §2.7.1 for the mean residence time of an SSA particle in the marine boundary layer with respect to removal by precipitation, $\tau_{wet} \approx 3$ d, provides a reasonable first estimate for this quantity. The requirement of stationarity with regard to wet deposition, although perhaps less stringent than other requirements, is perhaps of greater consequence; whereas it might be argued that other processes merely alter the relative importance of dry deposition

on SSA concentrations, precipitation events result in buildup from very small concentrations and cannot be neglected.

3.1.1.3. Horizontal homogeneity condition. The horizontal homogeneity condition requires that inhomogeneities in either the along-wind or cross-wind direction not appreciably interfere with the balance between SSA production and removal by dry deposition; that is, the vertical integral of the gradient of the along-wind flux of SSA in the along-wind direction and the vertical integral of the cross-wind flux of SSA in the cross-wind direction must be negligible compared to the dry deposition flux. Inhomogeneities can result in SSA particles being advected into or out of a given region, increasing or reducing SSA concentration and violating the requirement that production be balanced by removal through dry deposition. Such inhomogeneities can result from spatial gradients in concentrations or wind speeds, such as might result from gusts, unsteady flow, or changing meteorological conditions, and as noted in §2.8.3, any resulting change in SSA concentration is manifested as an apparent flux.

It is difficult to quantify the effects of horizontal inhomogeneities, but inequalities are presented here for two situations, both involving inhomogeneity in the along-wind direction. One is a situation in which the height of the marine boundary layer remains constant with time but increases with distance x in the along-wind direction. This situation, when viewed in a reference frame moving with the mean wind speed at the top of the marine boundary layer U (which for purposes of the present argument can be taken as equal to U_{10} because of the relatively weak dependence of wind speed on height; §2.4), is analogous to one in which there is an entrainment velocity equal to $U_{10}(dH_{mbl}/dx)$, with corresponding characteristic time $H_{mbl}/[U_{10}(dH_{mbl}/dx)]$. If this velocity and characteristic time are denoted as v_{inh} and τ_{inh}, respectively, then the horizontal homogeneity condition for this situation becomes $v_{int} \prec v_{d}$, or $\tau_{dry} \prec \tau_{inh}$. As a numerical example, at $U_{10} = 10$ m s^{-1} a gradient in dH_{mbl}/dx of only 1 m per km yields $v_{inh} = 1$ cm s^{-1}, or $\tau_{inh} \approx 0.5$ d for $H_{mbl} = 0.5$ km (a similar argument was presented in *Slinn et al.* [1983]). Another situation, similar to the one described above with regard to the stationarity condition, is one in which there is an along-wind gradient of the number concentration of SSA particles of a given r_{80} that are well mixed throughout the marine boundary layer. According to (2.8-13), this gradient provides a contribution to the effective SSA production flux of $H_{mbl}U_{10}(dn/dx)$, for which a characteristic velocity can be defined as $v_{inh} = H_{mbl}U_{10}[(dn/dx)/n]$, and a characteristic time as $\tau_{inh} = [U_{10}(d\ln n/dx)]^{-1}$. The horizontal homogeneity condition is again given by $v_{int} \prec v_{d}$, or $\tau_{dry} \prec \tau_{inh}$. An SSA concentration gradient of only 1% per 100 km at $U_{10} = 10$ m s^{-1}, corresponding to a rate of change of the concentration at a fixed location of only about 10% per day, yields

$v_{inh} \approx 0.1$ cm s^{-1} and $\tau_{inh} \approx 10^6$ s, again imposing a lower limit of r_{80} between 1 μm and 2 μm on the size of SSA particles for which this condition can generally be satisfied.

3.1.1.4. Volumetric condition.

The volumetric condition is that wet deposition not occur during the measurement period; in theory coagulation and fracture might also act as volumetric gain and loss terms for SSA particles of a given r_{80}, but these mechanisms are unimportant for SSA in the marine atmosphere (§2.8.1; §4.1.4.5; §4.7.4). Although this condition is typically met, the importance of wet deposition is potentially much greater, as discussed above in regard to the stationarity condition.

3.1.1.5. Summary.

These four conditions are not necessarily independent, and it is often difficult to unambiguously assign a given effect or process to a specific condition. For instance, entrainment of the free troposphere and the decrease in the mean SSA number concentration with time may be correlated [Fairall et al., 1982, 1983], with the net result being that the overall effect of these two processes is less than that of either one individually. In other words, entrainment of the free troposphere and the resulting decrease with time of the SSA number concentration could occur even if SSA production and dry deposition were equal to each other. However, such a fortuitous cancellation cannot be assumed, and if it is not demonstrated that the terms that are omitted in obtaining the production flux from (2.8-13) are not appreciable, then the resulting production flux might be in substantial error.

3.1.2. Discussion

There are several concerns with this method, both conceptual and practical. The satisfaction of the conditions required for the successful application of this method is sometimes difficult to ascertain, as this requires accurate knowledge of the back trajectories and previous meteorological conditions experienced by SSA particles subsequent to their production. Additionally, it is difficult to determine the accuracy of modeled SSA dry deposition velocities over the oceans, and experimental results that would aid in this determination are lacking (§4.6.4).

Based on the inequalities presented for the required conditions, it can be confidently concluded that this method cannot be accurately applied to small SSA particles (i.e., $r_{80} \lesssim$ 1 μm) or to medium SSA particles with r_{80} up to at least ~3 μm because of their large atmospheric residence times with respect to dry deposition (days to weeks), over which factors that may influence SSA concentration cannot be considered constant, and during which wet deposition will almost certainly have occurred. Similar conclusions were reached by Blanchard [1963, p. 132] and Toba [1965a], who suggested

that concentrations of SSA particles with $r_{80} \lesssim 5$ μm would be unlikely to have attained steady state. As this method is based on measurements near 10 m and thus determines the effective SSA production flux, it also does not apply to large SSA particles. However, for a moderately wide range of SSA particle sizes from $r_{80} \approx 5$ μm up to ~25 μm this method can be satisfactorily applied under the appropriate circumstances, albeit with recognition that as measured size distributions of SSA concentrations vary considerably (§4.1.4), even for the same wind speed, the effective SSA production flux (assumed proportional to the measured number concentration; Eq. 3.1-1) will vary by at least a similar amount.

This method has the advantage of simplicity, as it requires knowledge only of the size-dependent dry deposition velocity to determine the effective SSA production flux from measured size distributions of concentration (§4.1.4). Additionally, as the measured concentrations include SSA particles produced by all mechanisms, the flux obtained using this method includes contributions both from drops from bursting bubbles and from spume drops.

3.2. WHITECAP METHOD

The whitecap method, first suggested by Blanchard [1963, p. 128], determines the size-dependent oceanic interfacial SSA production flux from measurements of SSA production from laboratory simulated whitecaps (§4.3.4) or from the surf zone (§4.3.5) together with field measurements of the oceanic whitecap ratio W (§4.5.1), the fraction of the sea surface covered by whitecaps. The size-dependent SSA production flux over whitecap area only, $f_{wc}(r_{80})$, is determined, from which the size-dependent interfacial SSA production flux over the ocean, $f_{int}(r_{80})$ (including both whitecap area and non-whitecap area), is calculated as the product of $f_{wc}(r_{80})$ and W:

$$f_{int}(r_{80}) = f_{wc}(r_{80}) \times W(U_{10}). \qquad (3.2\text{-}1)$$

This approach assumes that the size-dependent interfacial SSA production flux per unit white area is independent of the nature and extent of the white area and of its method of production, and hence is the same for laboratory whitecaps and active surf zone as it is for oceanic whitecaps. In other words, it is assumed that the same number of SSA particles in a given size range is produced per unit time from a given amount of any white area of a water surface. Despite the fact that W undoubtedly depends on many meteorological and environmental factors, in practice it is usually presented as a function solely of U_{10}; this implies that the wind speed determines only the magnitude but not the shape of the size distribution of interfacial SSA production flux. Estimates of the interfacial SSA production flux obtained using this method are examined in §5.3.

Two methods of generating laboratory whitecaps have been employed: (1) the continuous laboratory whitecap method, such as that produced by a continuous waterfall from a weir—a "steady state breaking wave" [*Cipriano and Blanchard*, 1981; *Cipriano et al.*, 1983]—or that created by a swarm of bubbles produced by forcing air through a frit [*Mårtensson et al.*, 2003], and (2) the discrete laboratory whitecap method, in which individual breaking waves are formed by the collisions of two moving parcels of water in a wave tank [*Monahan et al.*, 1980, 1982, 1983b]. These two methods are discussed and compared by *Woolf* [1985] and by *Monahan* [1986], and experiments employing these methods are examined in §4.3.4.

The continuous laboratory whitecap method relies on measurement of the size-dependent production rate of SSA particles, $p(r_{80})$, the number of particles in a unit logarithmic interval of r_{80} produced per unit time from the laboratory whitecap of area A_{wc} (§4.3.4.1). The flux of SSA particles over whitecap area is calculated as

$$f_{wc}(r_{80}) = \frac{p(r_{80})}{A_{wc}}. \tag{3.2-2}$$

The discrete laboratory whitecap method relies on measurement of the increase in $\Delta n(r_{80})$, the number concentration of SSA particles per unit logarithmic interval of r_{80}, resulting from a single laboratory breaking wave with initial white area $A_{wc,0}$ enclosed in a tank of air volume V. Under the assumption that the SSA particles are uniformly distributed throughout the volume of the tank before measurement, the number of SSA particles produced by a single laboratory breaking wave, per unit logarithmic interval of r_{80}, is given by $\Delta n(r_{80})V$, and hence the total number of SSA particles produced per unit logarithmic interval of r_{80} per unit initial whitecap area is obtained as $\Delta n(r_{80})V/A_{wc,0}$. In addition to the assumption stated above that all white areas are equally efficient producers of SSA particles (that is, that the total number of SSA particles of any given size produced by a whitecap during its lifetime is directly proportional to the initial whitecap area), this variant of the method further assumes that the area of any whitecap decreases exponentially with a characteristic decay time τ_{wc}, which is the same for both laboratory and oceanic whitecaps, independent of their size or their method of production [*Monahan and Zietlow*, 1969; *Monahan*, 1971; *Monahan et al.*, 1982, 1983b, 1986]. This latter assumption allows τ_{wc} to be determined from laboratory measurements [*Monahan and Zietlow*, 1969; *Monahan et al.*, 1980, 1982; *Stramska et al.*, 1990]. The time dependence of the area of an individual whitecap is thus described by $A_{wc}(t) = A_{wc,0}\exp(-t/\tau_{wc})$, from which it follows that $\Delta n(r_{80})V = f_{wc}(r_{80})\int A_{wc}(t)dt$, which in turn is

equal to $f_{wc}(r_{80})A_{wc,0}\tau_{wc}$. Hence the flux of particles over whitecap area is given by

$$f_{wc}(r_{80}) = \frac{\Delta n(r_{80})V}{A_{wc,0}\tau_{wc}}. \tag{3.2-3}$$

Again this quantity, together with W, is used in calculating the size-dependent interfacial SSA production flux over the oceans by (3.2-1).

The surf zone variant of this method relies on measurements of the change in size-dependent SSA number concentration $\Delta n(r_{80})$ produced by a surf zone of along-wind extent L during conditions of onshore wind U (that is, wind blowing from the sea to the land) under the assumption that the SSA particles produced by the surf zone are uniformly distributed throughout an air parcel of height H and do not fall out before measurement. This allows f_{wc} to be calculated as

$$f_{wc}(r_{80}) = \frac{\Delta n(r_{80})HU}{L}. \tag{3.2-4}$$

Among other meteorological and environmental factors, the interfacial SSA production flux is expected to depend on the water temperature (§2.3.5), as temperature affects bubble rise speeds, gas exchange properties, and possibly production, and thus the size distribution of the SSA particles produced. Therefore, in principle, the application of laboratory results to oceanic conditions requires knowledge of the sea surface temperature and of the temperature dependence of $f_{wc}(r_{80})$ determined from laboratory tank studies [*Monahan*, 1986], as well as the dependence of W on U_{10} and other controlling factors. In practice, however, these other effects are typically not included.

A strength of this method as it has been applied using laboratory whitecaps is that it employs the size distribution of drops produced from bubble swarms (as distinct from individual bubbles) under controlled conditions, permitting in principle systematic examination of the dependence of this size distribution on controlling factors. Additionally this method (in theory) allows determination of the production flux of small SSA particles ($r_{80} \lesssim 1$ μm) without interference from pre-existing particles or from non-SSA particles. The fact that several different methods of producing laboratory whitecaps have been employed allows comparison of the results from the several approaches—the shapes and magnitudes of the resulting SSA particle size distributions, permitting assessment of possible errors resulting from the laboratory whitecaps not being representative of whitecaps in the open ocean.

There are, however, several concerns with this method. A fundamental concern is the extent to which laboratory or surf zone whitecaps simulate the wave breaking process and bubble spectra in the ocean, and hence the size distribution of the resulting flux of SSA particles (§4.3.4.4). Specifically, the assumption that all white areas are equally productive for particles of all sizes (and hence that oceanic breaking waves can be accurately simulated by laboratory whitecaps) is open to question and has not been verified. This concern arises in large part from the limited surface area and depth of the laboratory waves compared to those of actual oceanic whitecaps. As employed thus far these surface areas (and depths) range from $\sim 3 \cdot 10^{-4}$ m^2 (~ 0.04 m) in the frit method to ~ 0.02 m^2 (~ 0.3 m) in the continuous waterfall method to ~ 0.5 m^2 (~ 0.7 m) in the colliding waves method (4.3.4). The assumption inherent in (3.2-1) that the size dependence of the interfacial production flux is independent of wind speed (which further implies that the shape of the size distribution of bubble concentration is the same for all wind speeds) is likewise open to question. The dependence of whitecap ratio on controlling factors has not been established (§4.5), and measurements of W for nominally the same conditions, i.e., wind speed, exhibit variability of an order of magnitude or more (§4.5.2).

A further concern with the continuous whitecap method involving bubbles produced by forcing air through a frit is that although this method yields the interfacial production flux for small SSA particles, it is difficult to account for the sizes of the drops produced. These drops appear to be too small to be jet drops (§4.3.2), yet there appear to be too few bubbles sufficiently large to produce film drops (§4.3.3). A possible explanation for this discrepancy is offered in §4.3.4.1.

A further concern with the discrete breaking wave method is the assumption that the decay of whitecap area is accurately described by an exponential function of time, with characteristic decay independent of the size of the whitecap. This assumption is not supported by the data (§4.5.4), and it would be expected that larger whitecaps, which entrain bubbles deeper, would have longer decay times than smaller whitecaps.

Another concern with laboratory determinations of the whitecap method is that they do not actually determine the interfacial SSA production flux, but rather the flux of those drops that are detected by the measurement device employed (that is, the assumption of uniform mixing throughout the volume is not always satisfied). As this method has been applied to determine the production flux only of drops with $r_{80} \lesssim 10$ µm, it appears to be more pertinent to determination of the effective SSA production flux than to the interfacial SSA production flux. However, the methods by which the laboratory whitecaps have been produced and the lack of any wind over the water preclude the use of these investigations to simulate the upward entrainment of drops subsequent to their formation and hence to determine the effective SSA production flux. Furthermore, this method detects only drops produced by the bursting of bubbles and not spume drops produced by tearing of wave crests. Spume drops, which

comprise an increasingly greater fraction of SSA particles produced with increasing particle size, may be more important than bubble-produced drops for many applications of the interfacial SSA production flux, thereby limiting the utility of this approach for determination of the interfacial production flux of medium and large SSA particles, for which the effective SSA production flux and the interfacial SSA production flux differ substantially.

A concern with the surf zone variant of this method is that the mechanisms of wave breaking and SSA production in the surf zone are different from those in the open ocean, and the chemical composition of SSA produced in the surf zone has been reported to differ from that produced over the open ocean. Additionally, results are expected to depend on sea bottom topography and other factors that are specific to location, leading to concerns regarding the extent to which any such measurement can be considered to be applicable to breaking waves over the open ocean. These concerns are examined in §4.3.5.

3.3. CONCENTRATION BUILDUP METHOD

The concentration buildup method [*Reid et al.*, 2001] determines the size-dependent effective SSA production flux from the increase in size-dependent column burden (i.e., vertical integral) of SSA number concentration with increasing along-wind distance x under conditions for which SSA production is not locally balanced by removal by dry deposition. To date this method has been applied only under conditions of off-shore winds (winds blowing from the land to the sea) near a coast (Fig. 14). Under these conditions the column burden of the size-dependent SSA concentration increases with increasing distance over the sea from the coast until local balance between SSA production and removal through dry deposition is approached at a distance that depends strongly on particle size (§2.9.6). It is in the region before this balance occurs that the increase in the column burden of SSA number concentration can be used to determine the effective SSA production flux by this method.

This method assumes that SSA production is independent of x, but it can of course depend on the wind speed. The effective SSA production flux at a height z_1 for stationary conditions, under the assumption that the situation is homogeneous in the cross-wind direction (i.e., no y-dependence) and that there are no volumetric gain and loss terms, is obtained from (2.8-13) as

$$f_{\text{eff}}(r_{80}, z_1) = \int_{z_1}^{z_2} \frac{\partial \overline{j}_x}{\partial x} dz + \overline{n}(r_{80}, z_{\text{ref}}) v_{\text{d}}(r_{80}, z_1, z_{\text{ref}}) \quad (3.3\text{-}1)$$

where z_2 is greater than the height through which particles have been mixed (this is a slight variant of Eq. 1 in *Reid et al.*

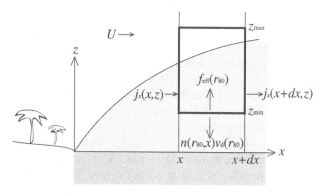

Figure 14. Schematic of the concentration buildup method. It is assumed that the wind blows from the land (at left) to the ocean, that SSA production is uniform in the cross-wind (i.e., y-) direction, and that SSA production is uniform in the along-wind (i.e., x-) direction. The height through which SSA particles of a given size are mixed to any appreciable extent is denoted by the light shaded region. Box denotes a sample volume at a fixed location used in deriving expression for $f_{eff}(r_{80})$ with sides at x and $x + dx$, bottom at z_{min} (the lowest height at which SSA concentrations are measured), and top at z_{max} (greater than the height to which SSA particles have been mixed to any appreciable extent). The rates at which SSA particles of a given size enter and exit the sides of the box, per unit distance in the y-direction, are $\int j_x(x, z)dz$ and $\int j_x(x + dx, z)dz$, respectively, where the integrals are from z_{min} to z_{max}. The rates at which SSA particles of the given size enter and exit the bottom of the box, per unit distance in the y-direction, are $f_{eff}(r_{80})dx$ and $n(r_{80}, x)v_d(r_{80})dx$, respectively. There is no exchange of SSA particles of the given size through the top of the box.

[2001], in which a Lagrangian approach was employed—that is, the time derivative in that investigation was meant to be taken in a frame of reference moving with the mean wind speed). It is further assumed that the concentration increases sufficiently rapidly in the along-wind direction that the only term of importance in $\overline{j}_x$ is $\overline{n}\,\overline{u}$ (§2.8.2). Because stationary conditions are assumed, the mean wind speed can be moved in front of the derivative (that is, $\partial \overline{j}_x / \partial x$ can be replaced by $\overline{u}\,\partial \overline{n} / \partial x$). If it is further assumed that the wind speed exhibits little dependence on z over heights from z_1 to z_2, then $\overline{u}$ can be taken out of the integral, resulting in

$$f_{eff}(r_{80}, z_1) = \overline{u}\frac{\partial}{\partial x}\left(\int_{z_1}^{z_2} \overline{n}(r_{80})\,dz\right) + \overline{n}(r_{80}, z_{ref})v_d(r_{80}, z_1, z_{ref}). \quad (3.3\text{-}2)$$

Physically, this means that SSA particles that attain heights sufficiently great that they contribute to the effective SSA production flux either increase the vertical integral of their number concentration as a function of along-wind distance or are removed through dry deposition.

This method requires sufficiently dense measurement coverage (requiring aircraft) in the vertical and along-wind directions to determine both the size-dependent average number concentration $\overline{n}$ and its column burden $\int \overline{n}\,dz$ as a function of x with sufficient accuracy that the derivative of this column burden with respect to x can be calculated. If $\overline{u}$ is also accurately measured as a function of z then (3.3-1) can be used instead of (3.3-2). Additionally, the dry deposition velocity v_d must be known, or it must be established that it is sufficiently small that dry deposition can be neglected for SSA particles in the size range of interest. This method includes contributions from both bubble-produced drops and spume drops (although under most circumstances few spume drops attain sufficient heights to be sampled at the lowest height at which aircraft measurements are taken).

An extreme application of this method is a so-called "one-point" method in which a single measurement of the size-dependent SSA number concentration at a fixed location in the ocean at a distance X from the coast in the along-wind direction is made during conditions of off-shore winds with wind speed U. This approach could in principle be applied to determine the production flux of small SSA particles, which are nearly uniformly mixed over the marine boundary layer of height H_{mbl} (§2.9.5). Particles in this size range are minimally affected by dry deposition (§2.7.2) and the column burden of their number concentration is limited by the fetch (§2.9.6). If the derivative in (3.3-2) is replaced by a difference, and if it is further assumed that dry deposition and the contribution from the surf zone to the concentration of these particles can be neglected, then the SSA production flux (either effective or interfacial, as these are nearly the same for SSA particles in this size range) can be obtained as

$$f_{eff}(r_{80}) = \frac{Un(r_{80})H_{mbl}}{X} \quad (3.3\text{-}3)$$

In principle this method could also be applied at a fixed location in the ocean after a precipitation event if it could be established in some way that the air mass being advected to this location was initially free of SSA particles, and if the distance over the ocean over which particles had traveled since production could be determined (for instance, using a relation between the time since precipitation and the wind speed); similar discussions are presented in §3.6 and §3.9.

A fundamental concern with this method as it has been applied is the extent to which the wave breaking process, and thus the production of SSA particles, in the near-shore region is representative of that in the open ocean, as the near-shore region may not contain a fully developed wind-wave spectrum, especially under conditions of off-shore winds. An additional concern is the extent to which the marine atmospheric boundary layer in the near shore region is representative of that over the open ocean. Other concerns pertinent to its implementation near the coast include contributions from continental aerosol particles and from SSA particles produced in the surf zone, for which there is not a counterpart in the open ocean; such contributions can lead to errors if they are not taken into account.

3.4. BUBBLE METHOD

The bubble method, proposed by *Eriksson* [1959] and *Blanchard* [1963, p. 159], determines the size-dependent interfacial SSA production flux of bubble-produced (i.e., jet and film) drops from knowledge of the size-dependent number flux of bubbles bursting at the sea surface and of the number and sizes of drops produced by a bursting bubble of a given size. Key assumptions are that each bubble rises and bursts at the surface independently of other bubbles and that the numbers and sizes of drops produced by a bursting bubble at the ocean surface are the same as those produced from an individual laboratory bubble of the same size. If upon bursting an individual bubble of radius R_{bub} produces (probabilistically) a spectrum of drops described by $s(r_{80}, R_{bub})$, the number of drops per unit logarithmic interval of r_{80}, then the size-dependent interfacial production flux of SSA particles produced by a swarm of bubbles bursting (independently) at the surface, whose flux is described by $f_{bub}(R_{bub})$ (the number of bubbles bursting per unit logarithmic interval of R_{bub} per unit area per unit time) is given by

$$f_{int}(r_{80}) = \int f_{bub}(R_{bub})s(r_{80}, R_{bub})d\log R_{bub}. \quad (3.4\text{-}1)$$

This method has been employed by *Wu* [1989a; 1992a] and *Zhuang et al.* [1993] to calculate the oceanic interfacial SSA production flux (§5.5) and by *Cipriano and Blanchard* [1981] to compare the measured size distribution of drops from a continuous laboratory whitecap with that calculated from the measured size distribution of bubble concentration (§4.3.4.1). This procedure has also been inverted to calculate the size distribution of bubble concentration very near the sea surface from knowledge of the size-dependent SSA production flux inferred from laboratory measurements using the whitecap method [*Monahan*, 1988a,b; *Monahan and Woolf*, 1988; *Monahan and Lu*, 1990; *Monahan*, 1993].

No measurements have been reported for the size-dependent flux of the bubbles bursting at the sea surface, $f_{bub}(R_{bub}, 0)$, but this quantity is typically inferred from knowledge of $n_{bub}(R_{bub}, 0) = dN_{bub}(R_{bub}, 0)/d\log R_{bub}$, the size-dependent number concentration of bubbles very near the surface ($z = 0$), and $v_{bub}(R_{bub})$, the terminal velocity of a bubble of radius R_{bub} (it being assumed that all bubbles are moving at their terminal velocities when reaching the surface),

these quantities being related by $f_{bub}(R_{bub}) = n_{bub}(R_{bub}, 0) \times v_{bub}(R_{bub})$. With this relation, (3.4-1) becomes

$$f_{int}(r_{80}) = \int n_{bub}(R_{bub}, 0)v_{bub}(R_{bub})s(r_{80}, R_{bub})d\log R_{bub}. \quad (3.4\text{-}2)$$

Application of the method using (3.4-2) thus requires knowledge of the size-dependent number concentration of bubbles at the surface (§4.4.1.2), the terminal velocity of a bubble of radius R_{bub} (§4.4.2.1); and the number of drops per unit logarithmic interval of r_{80} produced per bubble as a function of R_{bub} (§4.3.2, §4.3.3).

A simplifying assumption is sometimes made that for both jet drops and film drops there is a one-to-one relation between r_{80} and R_{bub}; that is, a bubble with given radius R_{bub} produces only drops having a specific r_{80}, and all drops of a given r_{80} are produced only from bubbles with a specific R_{bub}. Under this assumption, for each drop type (i.e., film or jet) the quantity $s(r_{80}, R_{bub})$ can be replaced by $s(r_{80})$, the number of drops of r_{80} produced by a bubble with the corresponding R_{bub}, and the integral in (3.4-2) can be eliminated, allowing the production flux for each drop type to be determined as the product of the near-surface bubble concentration, the bubble rise speed, and the number of drops of the given size produced per bubble burst, all expressed in terms of r_{80}:

$$f_{int}(r_{80}) = n_{bub}(r_{80}, 0)v_{bub}(r_{80})s(r_{80}). \quad (3.4\text{-}3)$$

The interfacial SSA production flux is then given by the sum of the interfacial production flux for film drops and that for jet drops.

There are several concerns with this method, stemming both from the assumptions necessary for its application and from the incomplete knowledge of some of the required quantities. Additionally, as with the whitecap method, this approach determines the interfacial production flux of only bubble-produced drops and not spume drops, which because of their larger size are the most important for many applications of the interfacial SSA production flux (§2.1.7).

A fundamental concern is the assumption that each bubble rises independently of other bubbles and, upon reaching the surface, also bursts independently, allowing laboratory determinations of rise velocities and of drop production from individual bubbles to be applied. Swarm effects may result in the rise velocity of a bubble in a bubble plume differing from that of an individual bubble of the same size, and effects of clustering and screening, which would be difficult to quantify, raise questions regarding the independence of bubble bursting at the sea surface, and hence the extension of these results to oceanic conditions just after a breaking wave when the majority of drops are produced (similar considerations are discussed in §4.3.4.4).

The assumption is also made that the flux of bubbles of a given R_{bub} that burst at the surface is equal to the product of their concentration at some depth (typically the shallowest depth at which bubble concentrations are measured) and their terminal velocity in still water, and thus that the size distribution of bubble concentration does not change between the depth of the measurement and the surface. However, bubble concentrations at depths do not relate directly to those at the surface, rather the size distribution evolves with time because of turbulence, size-dependent rise velocities, and gas exchange between bubbles and the surrounding water (§4.4.2), the latter effect leading to changes in the sizes of individual bubbles with time and possibly to dissolution of sufficiently small bubbles. Estimates for the size at which this occurs differ (§4.4.2.2), with resultant uncertainty in the interfacial SSA production flux over a range of SSA particle sizes. Bubble rise velocities and gas-exchange properties depend not only on the size of the bubble, but also on temperature, which is typically not taken into account in formulations using this method, and on other factors, such as the near-surface turbulence of the ocean water and the presence and nature of surface-active organic substances which can adhere to the bubble; these factors can vary spatially and temporally and would be difficult to know on scales necessary for modeling (§4.4.2).

Accurate estimation of $f_{int}(r_{80})$ using this method is hindered also by incomplete knowledge of some of the required quantities. Measurements of size distributions of bubble concentrations in the ocean show vastly disparate results (§4.4.1.2), and there are few measurements in the size range important for film drop production (§4.3.3). Additionally, the dependence of size distributions of bubble concentration on controlling factors is poorly known. Sizes and numbers of film and jet drops per bubble have been measured only under laboratory conditions and only using single bubbles bursting at still, horizontal surfaces (§4.4.1,§4.4.2). Results for film drops from different investigators differ widely (§4.4.2), and the size distribution of film drops over a wide range of r_{80} has not been reported. Additionally, some of the results for jet drops (§4.4.1) exhibit a surprising temperature dependence [*Spiel*, 1994a,b, 1997a], calling into question the applicability of measurements made over only limited temperature range. The choice of how many jet drops are produced per bubble (§4.3.2.1) is typically not addressed; the validity of counting all drops, even those that rise only 1 mm or less from the surface, is questionable, as these drops are not likely to be entrained upward and thus not likely to contribute to the effective SSA production flux (§4.3.4.2); likewise they contribute little to air-sea fluxes of moisture, heat, or momentum, or to other applications for which the interfacial SSA production flux is desired.

3.5. MICROMETEOROLOGICAL METHODS

Micrometeorological methods, such as eddy correlation, eddy accumulation, and gradient (or profile) methods, are well established techniques for measurement of the vertical fluxes (in either direction) of momentum, heat, water vapor, gases, and, to a limited extent, particles, caused by turbulent transport. Although most measurements (especially for particles) have been made over land, in principle these methods might also be used to measure particle fluxes over the oceans, and for this reason these methods are discussed here. Only a few micrometeorological measurements of particle fluxes over the ocean have been reported (§5.6), all involving the eddy correlation method [*Wesely et al.*, 1982; *Sievering et al.*, 1982; *Schmidt et al.*, 1983; *Katsaros et al.*, 1987; *Nilsson and Rannik*, 2001; *Nilsson et al.*, 2001], though with little information on the dependence of the flux on particle size, wind speed, or other controlling factors.

The size-dependent effective SSA production flux at a height z_1, under the assumptions that the mean vertical wind is zero, Brownian motion can be neglected, and the only contribution to the vertical component of the slip velocity is that due to gravitational sedimentation, is given by (2.8-14) as

$$
\begin{aligned}
f_{eff}(r_{80}, z_1) &= \overline{n'(r_{80}, z_1) \times w'(z_1)} \\
&+ \overline{n}(r_{80}, z_1) \times [v_d(r_{80}, z_1) - \overline{v}_{term}(r_{80})]
\end{aligned} \quad (3.5\text{-}1)
$$

where $w'(z_1)$ is the fluctuating component of the vertical wind speed (as the mean concentration is typically measured at the same height at which the effective SSA production flux is desired, the dependence of v_d on z_{ref} is omitted).

Under conditions of local balance between SSA production and removal through dry deposition, the effective net flux $\overline{j}_z(z_1)$ is zero and under such circumstances the turbulent flux of SSA particles of a given size is equal (by Eq. 2.8-9) to the gravitational flux of those particles (again under the assumptions of no mean vertical wind, negligible Brownian diffusion, and negligible contributions to the slip flow from mechanisms other than gravity). Under these circumstances f_{eff} would equal the effective SSA dry deposition flux $f_d(z_1)$ and the method would reduce to the steady state dry deposition method (§3.1). Thus, micrometeorological methods are useful when production is not balanced by removal through dry deposition. There are two extreme situations of this: (1) SSA production is negligible and dry deposition provides the dominant contribution to the vertical flux, and (2) SSA production greatly exceeds the dry deposition flux, which can typically be neglected. In the first situation the effective net SSA flux would be downward and equal in magnitude to the SSA dry deposition flux, and micrometeorological measurements would determine the dry deposition velocity; however, over the ocean such instances, involving no production of SSA particles, are necessarily restricted to low wind speeds, and hence these measurements would be of little use in determining the SSA production flux. In the second situation, if it could be ascertained in some manner that the difference between the effective SSA dry deposition flux and the gravitational sedimentation flux (i.e., the second term on the right hand side of Eq. 3.5-1) is negligible, or else has such minor effect that it can be treated using modeled dry deposition velocities, then the effective SSA production flux would be nearly equal to, and in the same direction as, the turbulent flux, which is the measured quantity. This situation is expected to occur only for small SSA particles, for which removal occurs almost entirely through wet deposition; thus the applicability of micro-meteorological methods to determine the effective SSA production flux is generally restricted to this size range.

The several micrometeorological methods and measurement requirements are presented here, followed by a discussion of concerns common to these approaches.

3.5.1. Eddy Correlation

Eddy correlation (also known as eddy covariance) techniques consist of simultaneous measurement of both the fluctuating component of the vertical wind speed and the fluctuating component of a quantity of interest q, from which the vertical flux associated with this quantity can be inferred. For example, q may be the size-dependent SSA number concentration or the some other property of this concentration from which the size-dependent flux of SSA particle number can be determined, such as the size-dependent light-scattering coefficient. The mean vertical flux associated with the quantity q can be decomposed (§2.8.2) into $\overline{qw} = \overline{q}\,\overline{w} + \overline{q'w'}$. Typically the fluxes in other directions are also measured, and a coordinate rotation is used to remove any mean flow in the vertical direction, permitting correction of any misalignment of the sensor with the vertical. Such a procedure allows determination of $\overline{q'w'}$, the vertical flux due to turbulent diffusion. However, the mean vertical wind speed $\overline{w}$ is often not zero (although it is generally too small to measure) because of vertical gradients of RH (leading to a water vapor flux and thus Stefan flow; §2.6.1) or temperature (leading to a heat flux), and corrections for these effects must be made [*Bakan*, 1978; *Jones and Smith*, 1978; *Smith and Jones*, 1979; *Webb et al.*, 1980; *Wesely et al.*, 1981]. The reason that Stefan flow is important at all heights in this situation and only near the surface in the steady state dry deposition method is that the eddy correlation method involves the net vertical flux, whereas the steady state dry deposition method involves the gross vertical flux (§2.1.6), which is typically much larger

than the contribution of the flux due to Stefan flow, except possibly near the surface (§2.6.1).

The time over which the measurements are averaged should be long enough that individual gusts or breaking waves do not influence the results and so that sufficient data can be taken to achieve meaningful averages, but not so long that environmental conditions change appreciably; for eddy correlation measurements of gases this time is typically around 30 min. Such averaging also serves to remove sensor noise. Because the vertical component of the wind speed can be measured with a high signal-to-noise ratio and because random instrumental noise would be uncorrelated with the fluctuating vertical component of the wind, the signal-to-noise ratio of the apparatus that detects the quantity of interest need not be extremely high [*Jones et al.*, 1978]. Sensor stability is not required for the entire sampling period, but only for time scales longer then the period corresponding to the lowest frequency [*Jones et al.*, 1978]; in practice low-frequency trends, due either to sensor stability or meteorological conditions, can be removed later [*Rannik and Vesala*, 1999]. The sensor response should be sufficiently rapid to include the smallest eddies that can contribute appreciably to the flux; typically the frequencies measured range from 0.02-10 Hz. Physical separation of the sensor for wind speed and that for particle concentration should be minimized, otherwise the flux will be underestimated [*Moore*, 1986; *Lee and Black*, 1994; *Kristensen et al.*, 1997]. Depending on the sampling method, attenuation of the concentration or damping of fluctuations in sampling tubes [*Lenschow and Raupach*, 1991; *Massman*, 1991] and other possible changes in the concentration, such as diffusive losses of particles to tubing walls by Brownian motion, can also affect the results [*Wesely et al.*, 1985; *Buzorius et al.*, 1998].

3.5.2. Eddy Accumulation

The eddy accumulation method [*Desjardins*, 1972 (cited in *Pattey et al.*, 1993); *Desjardins*, 1977b; *Hicks and McMillen*, 1984; *Speer et al.*, 1985] consists of conditional sampling in which the upwardly moving eddies and the downwardly moving eddies are sampled separately with a volume flow rate proportional to the magnitude of the vertical component of their velocity. The net flux associated with a quantity q is therefore the difference of the upward and downward fluxes: $\overline{qw} = \overline{qw}_{up} - \overline{qw}_{down}$. In practice this is determined using

$$\overline{qw} = \frac{\int q\,dV_{up} - \int q\,dV_{down}}{A\Delta t}, \qquad (3.5\text{-}2)$$

where A is the cross-sectional area of the sampling inlet, Δt is the sampling time, and V_{up} and V_{down} are the volumes of the upwardly moving eddies (i.e., updrafts) and the downwardly moving eddies (i.e., downdrafts), respectively. Fast-response sensors for particles are not required, and the samples can be collected and analyzed later, though fast-response proportional sample valves are required. In determining the size-dependent production flux it is important to ensure that the size distribution of the particles sampled does not change between sampling and analysis (because of coagulation, for instance). A correction can be made for sample volume imbalance, such as might occur if the sensor were misaligned or the detector measuring the updrafts had a slightly different area than that measuring the downdrafts. This method is difficult to implement, however, not only because of the difficulty in controlling the flow rate with the required speed, accuracy, and dynamic range [*Dabberdt et al.*, 1993], but also because of the requirement of sampling linearity with vertical wind speed. In addition, accurate results are difficult to achieve as they involve taking the difference of two numbers of roughly equal value [*Businger*, 1986].

3.5.3. Relaxed Eddy Accumulation

The relaxed eddy accumulation method [*Businger and Oncley*, 1990; *Oncley et al.*, 1993; *Pattey et al.*, 1993; *Nie et al.*, 1995] overcomes some of the sampling difficulties of the eddy accumulation method by collecting samples of upwardly moving eddies and downwardly moving eddies at a constant rate, rather than at a rate proportional to the velocity. The vertical turbulent flux associated with a quantity q is determined under the assumption that it is proportional to the product of the root-mean-squared value of the vertical component of the wind speed σ_w (§2.4) and the difference in the mean values of q between the upward and downward samples:

$$\overline{qw} = b\sigma_w(\overline{q}_{up} - \overline{q}_{down}), \qquad (3.5\text{-}3)$$

where b is a coefficient (usually determined experimentally and typically around 0.6) that depends on the probability distribution of w. It is important that the mean value of w that is measured be zero (requiring careful alignment), as biased measurements cannot be corrected later, though in practice the effect of a nonzero mean value is slight. Again, high accuracy is required in the measurements of $\overline{q}_{up}$ and $\overline{q}_{down}$, as these quantities are often roughly equal, and it is necessary to ensure that the size distribution does not change between when the particles are sampled and when they are analyzed. In addition to these concerns there are possible errors due to the time lag between the change of sign of w and the sampling of the air. To overcome this difficulty, a threshold velocity is sometimes used, resulting in a deadband (a time

interval about the occurrence of $w' = 0$) during which no sampling occurs, and only those updrafts and downdrafts with speeds greater this threshold are sampled. This procedure necessitates a correction, the value of which depends on the ratio of the threshold velocity and σ_w [*Businger and Oncley*, 1990; *Oncley et al.*, 1993; *Pattey et al.*, 1993].

3.5.4. Gradient Methods

The gradient, or profile, method, determines the size-dependent effective SSA production flux $f_{eff}(r_{80})$ from the difference in the measured SSA number concentrations at two heights $\Delta n(r_{80})$, based on the assumption that f_{eff} and Δn are directly proportional to each other [*Eriksson*, 1959; *Gillette et al.*, 1972; *Porch and Gillette*, 1977; *Sievering*, 1981, 1982; *Garland and Cox*, 1982]. It is further assumed that the constant of proportionality between f_{eff} and Δn can be determined from the flux of momentum, heat, or water vapor, which is taken to be proportional to the difference in horizontal velocity, temperature, or water vapor concentration at the same two heights, the underlying assumption being that the turbulent processes responsible for the transport of particles are the same as those responsible for the transport of momentum, heat, or water vapor. Thus if the turbulent diffusion coefficient K of a quantity with concentration q is defined in terms of the vertical flux of that quantity F_q by the equation $F_q = K(dq/dz)$, then under the assumption that this same diffusivity K applies also to SSA particles, the effective SSA production flux would be given by

$$f_{eff}(r_{80}) = F_q \frac{\Delta n(r_{80})}{\Delta q}. \qquad (3.5\text{-}4)$$

Meaningful results from this method are difficult to obtain, as expected fluxes of SSA particles, and thus Δn, are generally quite small, and hence again large fractional error can arise from subtraction of two numbers of roughly equal value. These measurements are expected to work better in other situations, for instance, for gases that have a large flux toward the surface (and thus a large concentration gradient), or water vapor, which can have a large flux away from the surface. Also, the basic assumption of this method, that upward turbulent transport of SSA particles is the same as that of other quantities, might not be valid; for example, differences have been reported between the covariance spectra of vertical velocities with particle concentrations and with temperature or water vapor [*Katen and Hubbe*; 1985; *Duan et al.*, 1988].

3.5.5. Concerns with Micrometeorological Measurements

Concerns with eddy correlation measurements, many of which are common to all micrometeorological measurements, are examined in *Fairall* [1984], *Businger* [1986], *Moore* [1986], *Wesely* [1986], *Sievering* [1987], *Nicholson* [1988], and *Gallagher et al.* [1997]. Oceanic measurements encounter additional difficulties [*S. Smith et al.*, 1991]; sea spray is hostile to electronic and optical devices and can interfere with measurements [*Friehe and Schmitt*, 1976; *Davidson et al.*, 1978; *Schmitt et al.*, 1978; *Fairall et al.*, 1979; *DeCosmo et al.*, 1996], and providing a stationary, stable, horizontal platform removed from coastal influences is obviously difficult.

An intrinsic concern of all micrometeorological methods to determine the flux of SSA particles is that such particles may represent only a small fraction of the total number present in the marine atmosphere (§4.1.1.1), thus detection methods that do not discriminate between particles of different composition can sometimes lead to spurious results. For instance, a net downward flux can result from the deposition of particles from continental sources, for which there is no source at the ocean surface [*Sievering et al.*, 1982]. However, in principle this would not be a problem if dry deposition of these other particles could accurately be accounted for. Such a correction typically involves estimating the dry deposition flux from measured concentrations and modeled dry deposition velocities; adding this estimate to the measured net flux would account for the contribution to the net flux from dry deposition of both SSA and non-SSA particles yielding the correct result for the effective SSA production flux. In practice, however, the uncertainty associated with such an estimate including uncertainty arising from incomplete knowledge of the relative abundance and size-dependent composition of non-SSA particles and the composition- and size-dependent dry deposition velocities of these particles might lead to large uncertainty in the inferred effective SSA production flux.

There are several concerns regarding the character of the turbulent wind velocity. Correct orientation of the sample devices is necessary, and ensuring that these devices or platforms (including ships) do not affect either the turbulent or mean components of the flow is also crucial (as the turbulent properties of the flow can be affected by a sampling device even when the mean flow pattern is not seriously affected), placing restrictions on the size and placement of sampling devices and platforms [*Jones et al.*, 1978; *Mollo-Christensen*, 1979; *Wyngaard*, 1981; *Dyer*, 1981; *Högström*, 1982; *Businger*, 1986; *Wyngaard*, 1988a,b; *Bradley et al.*, 1991; *Oost*, 1993; *Yelland et al.*, 1998; *Edson et al.*, 1998; *Yelland et al.*, 2002]. Gusts or any horizontal non-uniformity in upwind conditions can result in apparent vertical fluxes (§2.8.3); additionally, horizontal gusts can affect measurements if the sampling device is non-isokinetic or if the anemometer does not respond linearly to the wind speed [*Desjardins*, 1977a; *Fairall*, 1984; *Katen and Hubbe*,

1985; *Sievering*, 1987]. The existence of waves will result in the lower surface not being horizontal and in vertical flow components with a time scale corresponding to that of the waves, of order seconds.

Gradients and fluctuations of relative humidity, which are ubiquitous over the ocean, have several effects in addition to the Stefan flow and associated mean vertical velocity component (§2.6.1). If RH is correlated with the fluctuating component of the vertical velocity, w', as would be expected if RH decreases with increasing height from the surface, then particle growth with RH can manifest itself as a vertical flux, yielding incorrect results because of particle growth into or out of size bins. These effects might differ with different instruments (this is examined further in §5.6), and also because the response time of larger particles to reach their equilibrium radii may be such that these particles do not fully follow the turbulent fluctuations in velocity [*Wesely et al.*, 1982; *Sievering et al.*, 1982; *Schmidt et al.*, 1983; *Fairall*, 1984; *Lee and Wesely*, 1989; *Buzorius et al.*, 2000; *Kowalski*, 2001].

Small net fluxes are typically expected, making it difficult to establish that the conditions necessary for the application of the methods are satisfied; that is, that the turbulent flux is the dominant contribution to vertical transport and that the flux due to dry deposition is sufficiently small that it can be accurately accounted for. Error in dry deposition flux would result in incorrect estimation of the effective SSA production flux; dry deposition velocities are poorly known, especially for smaller particles for which these methods are mostly used. As SSA number concentrations are typically much less than aerosol particle number concentrations over land (where micrometeorological methods are generally applied), sampling statistics due to low number counts impose restrictions on the sampling time required, or, equivalently, the total number of particles that must be sampled, to achieve a certain confidence level, depending on mean wind speed, the height, and the dry deposition velocity of the particle size of interest [*Fairall*, 1984]. In practice this restricts these methods to small particles ($r_{80} < 1$ μm), as concentrations of larger SSA particles are typically far too low to allow reasonable sampling times. Additionally, low concentration even of small SSA particles has resulted in poor size resolution, with fluxes determined comprising a fairly wide range of particle sizes.

3.6. ALONG-WIND FLUX METHOD

The along-wind flux method, initially presented in a simplified form by *Wu* [1993], determines the interfacial production flux of SSA particles from measurements of their along-wind fluxes and/or concentrations very near the sea surface (typically within ~1 m). Laboratory measurements have also been used. Although in principle this method could be applied directly, in practice it has most often been used in conjunction with other approaches (§5.7).

The size-dependent along-wind flux (i.e., in the x-direction) of SSA particles at height z, $j_x(r_{80}, z)$, is the number of particles per unit logarithmic interval of r_{80} at height z passing through a unit area of a reference plane perpendicular to both the mean wind and the plane of the mean ocean surface per unit time (§2.1.6). The total rate at which particles in this size range pass through a unit width of this reference plane (i.e., in the y-direction, perpendicular to the mean flow) is given by the vertical integral of the along-wind flux $\int j_x(r_{80}, z)dz$, where the integral is over the range of z from the surface ($z = 0$) to a height $z_{\max}$ sufficiently large to include the dominant contribution of the along-wind flux. For conditions under which the production flux of SSA particles at the surface and meteorological conditions do not change appreciably over the mean time the particles remain in the atmosphere (i.e., steady state), this integral is equal to the total rate at which SSA particles that pass through the reference plane are produced at the sea surface, per unit width in the cross-wind (y-) direction. Thus if $p(r_{80}, x)$ is the probability that an SSA particle of a given r_{80} produced at the sea surface at a distance x upwind of the measurement plane travels a horizontal distance in the along-wind direction greater than or equal to x (and thus passes through the reference plane), then

$$\int_0^{z_{\max}} j_x(r_{80}, z)\, dz = \int_0^{\infty} f_{\text{int}}(r_{80}, x) p(r_{80}, x)\, dx. \quad (3.6\text{-}1)$$

For $f_{\text{int}}(r_{80}, x)$ independent of x (which is expected to hold in steady state over the ocean, at least in a statistical sense, although it may not be valid in laboratory studies), $f_{\text{int}}(r_{80})$ can be taken out of the integral. The integral $\int p(r_{80}, x)dx$ (where the limits are from zero to as large as necessary to include the vast majority of SSA particles passing through the reference plane) is equal to the mean horizontal distance traveled by an SSA particle of given r_{80} produced at the surface $\overline{L}_p(r_{80})$. Thus,

$$f_{\text{int}}(r_{80}) = \frac{\displaystyle\int_0^{z_{\max}} j_x(r_{80}, z)\, dz}{\overline{L}_p(r_{80})}. \quad (3.6\text{-}2)$$

This method is illustrated in Fig. 15 for the situation in which all SSA particles of a given r_{80} travel the same distance L.

In wind tunnel experiments for which the fetch is much smaller than $\overline{L}_p(r_{80})$, in which situation nearly all particles

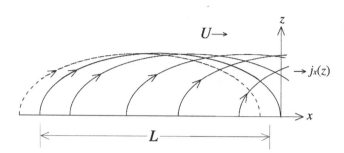

Figure 15. Simplified schematic of the along-wind flux method in which it is assumed that all SSA particles of a given size travel the same distance L, that SSA production is uniform in the cross-wind (i.e., y-) direction, and that SSA production is uniform in the along-wind (i.e., x-) direction. The rate at which SSA particles of the given size pass through a plane perpendicular to the wind, per unit distance in the y-direction, is $\int j_x(z)dz$, where the integral is over heights sufficient to include the bulk of the contribution of the along-wind flux. The rate at which these SSA particles are produced at the sea surface is $f_{\text{int}}(r_{80}) \times L$.

produced between the tunnel entrance and the plane of measurement pass through this plane and contribute to the along-wind flux, the integral in (3.6-2) should be divided by the fetch instead of by $\overline{L}_p(r_{80})$; this further requires the assumption that the production flux is independent of the distance from the tunnel entrance. For situations in which this assumption of spatially uniform production is not satisfied and for intermediate fetches the situation is more complex.

The quantity $\overline{L}_p(r_{80})$ is also equal to the harmonic mean—the reciprocal of the mean of the reciprocals—of the horizontal distances traveled by all particles with r_{80} that pass through the measurement plane; i.e.,

$$\overline{L}_p = \left(\overline{L_m^{-1}}\right)^{-1},$$

where the subscripts "p" and "m" refer to those particles produced and measured, respectively. This identity, apparently not previously recognized, may have applications in population theory or in reservoir theory [*Eriksson*, 1971; *Bolin and Rodhe*, 1973; *Rodhe*, 1992], where the analogy would be that the harmonic mean of the ages of particles in a reservoir at any given time, under steady state conditions, is equal to the mean transit time of the particles entering the reservoir.

The quantity $j_x(r_{80}, z)$ is also equal (Eq. 2.1-28) to the product of the number concentration of SSA particles with given r_{80} at height z, $n(r_{80}, z)$, and the mean horizontal velocity component of all SSA particles with given r_{80} at this height that are measured $\bar{u}_m(r_{80}, z)$, i.e., $j_z(r_{80}, z) = n(r_{80}, z)\,\bar{u}_m(r_{80}, z)$, Equivalently, $\bar{u}_m(r_{80}, z)$ is the harmonic mean of the horizontal velocity components of all particles of this size at this height that are produced at the sea surface,

$$\bar{u}_m = \left(\overline{u_p^{-1}}\right)^{-1}.$$

Thus $f_{int}(r_{80})$ can also be determined from the vertical integral of the product of the number concentration and the mean horizontal velocity as

$$f_{int}(r_{80}) = \frac{\displaystyle\int_0^{z_{max}} n(r_{80}, z)\bar{u}_m(r_{80}, z)\,dz}{\bar{L}_p(r_{80})}. \qquad (3.6\text{-}3)$$

An important consequence of (3.6-2) and (3.6-3) is that $f_{int}(r_{80})$ will not in general exhibit the same dependence on r_{80} as either $j_x(r_{80}, z)$ or $n(r_{80}, z)$ at a given value of z (or their integrals over z), because of the dependencies of $\bar{L}_p(r_{80})$ and $\bar{u}_m(r_{80}, z)$ on r_{80}. Nevertheless, it has typically (though incorrectly) been assumed that $f_{int}(r_{80})$ is directly proportional to $n(r_{80}, z)$ at a given value of z near the ocean surface, with the constant of proportionality (or equivalently, the magnitude of f_{int}) being determined from the value of $f_{int}(r_{80})$ at a given r_{80} obtained by another method. As the along-wind flux method has been the basis for several determinations of $f_{int}(r_{80})$, it is necessary to examine this approach in greater detail.

The applicability of this method depends strongly on particle size. It has been applied mainly to large SSA particles, whose radii at the time of measurement are nearly the same as at formation because of the long times required for equilibration with respect to RH (§2.5.4; §2.9.1). As the mean atmospheric residence times of large SSA particles are of order seconds and as such particles generally attain heights of only a few meters above the sea surface (§2.9.2, §2.9.4), performing measurements over sufficient heights to include the dominant contribution to the quantity $\int j_x(r_{80}, z)dz$ might be possible. One approach would be to determine the rate at which SSA particles impact upon a strip of small width (so as to minimally affect the flow) that is held vertical and perpendicular to the flow. The quantity $\bar{u}_m(r_{80}, z)$ for large SSA particles is in general not equal to the mean wind speed at this height U_z because the time required for these particles to approach a horizontal velocity equal to that of the mean flow (§2.6.4) may be greater than the time they have been in the atmosphere, and may even be greater than the time the particles remain in the atmosphere. For instance, an SSA particle with $r_{80} = 250$ μm (which will have a radius near 500 μm

during its entire residence in the atmosphere) that is instantaneously injected into a horizontal flow of 10 m s^{-1} requires more than 1 s to attain 90% of this horizontal velocity, during which time it travels more than 3 m; during this same time it would fall (in still air) approximately 3 m under the influence of gravity (§2.6.4). However, the quantity $\bar{u}_m(r_{80}, z)$ could conceivably be measured (using streak photography, for instance) independently of $n(r_{80}, z)$.

A fundamental concern with this method for large SSA particles is the determination of $\bar{L}_p(r_{80})$. This quantity depends strongly on particle size and would be difficult to determine from measurements. This quantity also depends upon meteorological conditions, especially wind speed, and for large SSA particles especially, on conditions very near the sea surface (i.e., within ~1 m). For these particles modeling would be difficult because of the difficulty in determining flow fields at such heights under conditions for which these particles would be produced—just after breaking waves or near the tops of wave crests from which spume drops are being torn. Furthermore, it is implicit in this method that production occurs at the surface of the ocean, which is assumed to be a horizontal plane, but spume drops are formed at wave crests during conditions under which the sea is not horizontal, and under such conditions the height above the sea surface is not well defined.

Medium SSA particles are mixed over much or most of the marine boundary layer (§2.9.5), depending on particle size. These particles rapidly attain a horizontal velocity equal to that of the mean flow (§2.6.3); thus $\bar{u}_m(r_{80}, z)$ can be set equal to the mean wind speed at height z, U_z, permitting evaluation of the integral in (3.7-4) if the number concentration is known as a function of height. However, estimation of $\bar{L}_p(r_{80})$ is still required. Sea salt aerosol particles with $r_{80} \lesssim 10$ μm are distributed over most of the marine boundary layer (§2.9.4), and estimates for values of $\bar{L}_p(r_{80})$ (§2.9.6, Table 8) are hundreds to thousands of kilometers, over which conditions may have changed, violating the assumption of steady state necessary for this method (that is, the conditions under which the particles cross the reference plane and those under which the particles were produced might differ substantially). Using the arguments of §2.9.6, $\bar{L}_p(r_{80})$ could be estimated as the characteristic distance with respect to removal by dry deposition $X(r_{80})$ (Eq. 2.9-8), but with this assumption the method essentially reduces to the steady state dry deposition method (§3.1).

Application of the along-wind flux method to small SSA particles is also problematic because of the difficulty in estimating $\bar{L}_p(r_{80})$. Small SSA particles are generally well mixed over the height of the marine boundary layer (§2.9.5), and they attain the velocity of the surrounding air nearly instantaneously (§2.6.2); thus to a first approximation the dependence of $n(r_{80}, z)$ upon z and the dependence of $\bar{u}_m(r_{80}, z)$ upon r_{80} can be omitted, with the result that the along-wind flux of small SSA particles of a given size at any height is given by $n(r_{80})U_z$. Additionally, because the mean horizontal velocity is typically a slowly varying function of height at heights more than a few meters above the ocean surface, U_z can be replaced by U_{10} with little loss in accuracy, making the integrand in (3.6-3) independent of z (this simplified analysis neglects any change in direction of the mean wind speed with height, §2.4, although this does not affect the arguments presented here). With these assumptions, the interfacial SSA production flux is given approximately by

$$f_{\text{int}}(r_{80}) \approx \frac{n(r_{80})U_{10}H_{\text{mbl}}}{\bar{L}_p(r_{80})}; \qquad (3.6\text{-}4)$$

this equation is essentially the same as (3.3-3) derived above for one-point variant of the concentration buildup method (§3.3). As discussed in §2.7.1, small SSA particles are removed primarily through wet deposition. If the time since the last appreciable precipitation event were known then the mean distanced traveled by the particles $\bar{L}_p(r_{80})$, which would be independent of r_{80}, could be determined. However, these quantities are generally not known, precluding use of this approach for small SSA particles. Alternatively, statistical estimates for these quantities might be used, in which situation the method becomes identical to the statistical wet deposition method presented below (§3.9).

In summary, the along-wind flux method can in principle be applied to estimate the interfacial production flux of large SSA particles, as the range of heights required is small, but errors may result if measurements do not cover this range of heights sufficiently densely to accurately determine the total along-wind flux, and accurate estimation of $\bar{L}_p(r_{80})$ is problematic. There have been few oceanic measurements of the along-wind flux of large SSA particles near the sea surface (§4.2), especially under conditions for which spume drop production is expected to be prevalent, and most applications of this method rely heavily on results from laboratory measurements of near-surface along-wind fluxes, of which there have also been relatively few (§4.3.6.1). However, the extent to which drop production in small wave tanks can accurately reflect drop production over the ocean, especially for large particles which are mostly formed by tearing of wave crests, is highly questionable (§4.3.6.2). To date this method has been used to determine only the shape of the size dependence of the interfacial SSA production flux, with the absolute magnitude determined by other arguments (§5.7), but based on the considerations presented above (i.e., neglect of the size dependence of $\bar{L}_p(r_{80})$), these determinations are in error.

3.7. DIRECT OBSERVATION METHOD

In principle the size-dependent interfacial SSA production flux might be determined by direct photographic or video observation of the flux of SSA particles as they leave the sea surface or of upwardly moving SSA particles in close proximity to the sea surface. No such field measurements have been reported. Drop production was observed by *Koga* [1981; see also *Koga and Toba*, 1981] in the laboratory using still photographs (§4.3.6.1) but these investigators did not present quantitative measurements of the interfacial SSA production flux. The production of freshwater drops with r_{80} (calculated as $0.5r_{\text{form}}$; §2.1.1) greater than ~200 μm was recorded in the laboratory using video by *Anguelova et al.* [1999] (§4.3.6.1), who presented a model by which these laboratory production fluxes could be transformed to oceanic interfacial SSA production fluxes; these results are discussed in §5.8.

Although this method would provide direct determination of the interfacial SSA production flux, current limitations on photographic resolution restrict its application to determination of large SSA particles. Implementation of this method would be a challenge under field conditions for which an appreciable production flux of large SSA particles would be expected, including keeping the instrument afloat, free of spray, and focused at the sea surface during breaking waves.

3.8. VERTICAL IMPACTION METHOD

The vertical impaction method determines the size-dependent interfacial SSA production flux by measuring the rate of impaction of SSA particles of a given size upon the underside of a sampling substrate kept horizontal near the sea surface. Calculations presumably based on a variant of this method were presented by *Monahan* [1968] for one particle size at two wind speeds, and this method was briefly discussed by *Bortkovskii* [1987, pp. 41-45], who reported a few measurements (§5.9). As employed in the latter study, a device was mounted typically 15 cm above the sea surface on a float tethered to a research vessel and maintained upwind of the boat to avoid influences of the boat on wind and waves. According to *Bortkovskii*, the device was usable in a variety of sea conditions including breaking waves.

This method has the advantages of being very direct and of relying solely on field measurements, and it has the ability to provide information on the size-dependent interfacial SSA production flux for a variety of meteorological conditions. Because of the low impaction efficiency of small SSA particles, this method is applicable only to medium and large SSA particles. Concerns with this method are its ability to detect drops that are not ejected to the height of the device (15 cm is approximately the maximum height to which jet drops are ejected in still air; §4.3.2.3), its ability to detect spume drops, which are torn off wave crests at rather high wind speed and are transported nearly horizontally, its ability to be used in the vicinity of breaking waves (which are responsible for the majority of SSA particles), and possible interference of the flow pattern and thus the flux of SSA particles at the location of the device. This method has seen very little use, and few measurements are available (§5.9).

3.9. STATISTICAL WET DEPOSITION METHOD

The statistical wet deposition method yields a rough estimate of the effective SSA production flux for those particles for which wet deposition provides the dominant contribution to removal. Although under some circumstances wet deposition is the dominant removal mechanism of SSA particles of all sizes, under most circumstances it is the primary removal mechanism for small SSA particles, i.e., those with $r_{80} \lesssim 1$ μm, making this method most applicable to this size range.

As noted in §2.7.1, if the time that had elapsed since the air parcel containing the aerosol last experienced an appreciable precipitation event were known, and if the air that was present had not been compromised by admixture of air that had not experienced such wet deposition, then for SSA particles of sizes for which the removal mechanism is predominantly wet deposition the effective production flux $f_{eff}(r_{80})$ could be determined from measurements of their concentration under the assumption that virtually all of the SSA particles of a given r_{80} present in the atmosphere at the time of measurement had been produced subsequent to the last precipitation event. This method also assumes that SSA particles retain their identities and are not so transformed by cloud processing, uptake and exchange of gases, and the like that they can still be considered as identifiable SSA particles. As small SSA particles are expected to be well mixed over the marine boundary layer of height H_{mbl} (§2.9.5), the column burden of the number concentration of SSA particles of a given r_{80} (that is, the number of these particles in the atmosphere above a unit area of the sea surface) can be approximated by $n(r_{80})H_{mbl}$ (§2.1.5). This column burden is equal to the number of SSA particles of this size that have been produced per unit area of the sea surface since the last precipitation event, given by the product of the time since this event and the effective production flux $f_{eff}(r_{80})$. The mean time since the last precipitation event depends on the statistics of the frequency distribution of the times since the most recent precipitation event; if these times followed a Poisson distribution then the mean time would be equal to τ_{wet}, whereas if precipitation occurred regularly at intervals of τ_{wet} then the mean time would be equal to $\tau_{wet}/2$. As typically this time is not known, it can be taken (to within a factor of two or so) as equal to τ_{wet}, the mean

time between precipitation events, for which the value $\tau_{wet} \approx 3$ d was given in §2.7.1. This approach is the statistical wet deposition method.

Equating the column burden of the number concentration with the product of the production flux and τ_{wet} yields

$$f_{eff}(r_{80}) \approx \frac{n(r_{80})H_{mbl}}{\tau_{wet}} \qquad (3.9\text{-}1)$$

(this equation can also be obtained from Eq. 3.6-4 upon substitution of $\bar{L}_p(r_{80}) = U_{10}\tau_{wet}$). Similar arguments relating total SSA number concentration and total number production flux based on mean particle residence time (though not necessarily based on wet deposition) have been advanced earlier [e.g., Mason, 1957a; Junge, 1957a; Blanchard, 1969; Kritz and Rancher, 1980; Cipriano et al., 1983, 1987]. As this method provides only a rough estimate of the SSA production flux (to within a factor of 3 or so), H_{mbl} can be taken as 0.5 km to within the accuracy of this approach, allowing estimates of the production flux to be obtained from measurements of the SSA number concentration only. This results in the relation

$$\frac{f_{eff}(r_{80})}{\mathrm{m^{-2}s^{-1}}} = 2 \cdot 10^3 \frac{n(r_{80})}{\mathrm{cm^{-3}}} \qquad (3.9\text{-}2)$$

which is evaluated in §4.1.4.5. As $\Psi_f(r_{80})$, the ratio of the effective SSA production flux $f_{eff}(r_{80})$ to the interfacial SSA production flux $f_{int}(r_{80})$, is nearly unity for small SSA particles (§2.1.2; §2.9.4), the same relation also provides an estimate for $f_{int}(r_{80})$. In addition to providing an estimate for these SSA production fluxes, a further important consequence of this relationship, as noted in §2.7.3, is that within the assumptions that wet deposition is the dominant removal mechanism and that τ_{wet} is independent of r_{80}, the size dependence of the production flux of small SSA particles is the same as that of their number concentration. This method cannot provide any information on the wind-speed dependence of the SSA production flux, as those SSA particles that are measured were likely produced far away under conditions different from local conditions at the time of measurement.

3.10. SUMMARY OF METHODS FOR FLUX DETERMINATION

Nine independent methods by which the size-dependent SSA production flux can be determined have been presented, as summarized in Table 9. Methods based on measurements pertaining to processes occurring at the surface such as whitecaps, drop production from individual bubbles, or bubble populations at the sea surface (the whitecap method and the bubble method), and those involving near-surface measurements of SSA concentration or flux (the along-wind flux method, the direct observation method, and the vertical impaction method), yield the interfacial SSA production flux, whereas methods based on measurements of size-dependent number concentrations of SSA particles at heights near 10 m or greater (the steady state dry deposition method, the concentration buildup method, micrometeorological methods, and the statistical wet deposition method) necessarily yield the effective SSA production flux. Each of the methods has its strengths and weaknesses, and no single method is free of concerns. Additionally, each method is restricted to a limited range of SSA particle sizes (Table 2, Table 9).

For small SSA particles ($r_{80} \lesssim 1$ μm) the interfacial and effective production fluxes are nearly the same, i.e., Ψ_f is near unity (§2.1.2). The methods best suited to determine fluxes of SSA particles in this size range are the concentration buildup method, micrometeorological methods, and the statistical wet deposition method. The concentration buildup method has been used for small SSA particles but only to a limited extent and for a limited range of sizes; it could conceivably be extended to smaller sizes. Micrometeorological methods have seen some application to SSA particles, but have provided little information on the size dependence of the SSA production flux. The statistical wet deposition method can provide a long-term (e.g., monthly) average size distribution of the SSA production flux but provides no information on the dependence on meteorological factors. Other methods are not applicable to this size range. The steady state dry deposition method cannot be applied, as steady state conditions will not be realized for concentrations of SSA particles in this size range because of their long residence times (§2.9.6, §4.6) and because the probable source of their removal is wet deposition (§2.7.3). The whitecap method has been applied to SSA particles in this size range using a continuous laboratory whitecap formed by forcing air through a frit, but the applicability of this approach is doubtful, as noted in §3.2 (also §4.3.4.1). The along-wind flux method when applied to small SSA particles is essentially the same as the statistical wet deposition method (§3.7). The bubble method cannot currently be used in this size range in view of the present lack of knowledge concerning the production of film drops, which are thought to comprise the majority of small SSA particles (§4.3.3). The direct observation method and vertical impaction method are not practical for this size range.

For medium SSA particles (1 μm $\lesssim r_{80} \lesssim 25$ μm) the ratio of the effective SSA production flux to the interfacial SSA production flux Ψ_f decreases from near unity to much less than unity as r_{80} increases throughout this size range. Several methods are suited to determine one or the other of these fluxes. The steady state dry deposition method and the concentration buildup method can be applied to determine the effective SSA production flux over much of this size range. The whitecap method and the bubble method can be used to estimate the interfacial SSA production flux over some or all of this size range, but neither method includes the contributions of spume drops, which may be important in the upper part of this size range. The vertical impaction method could also be used to determine the interfacial SSA production flux in this range, and has been to a very limited extent. Micrometeorological methods are not well suited because of the relatively low concentrations of SSA particles in this size range. The direct observation method is not practical for SSA particles in this size range because of technical limitations. The along-wind flux method is not practical because of the large range of heights over which measurements must be made. The statistical wet deposition method cannot be applied, as wet deposition does not provide the dominant removal for most particles in this size range.

For large SSA particles ($r_{80} \gtrsim 25$ μm) only the interfacial SSA production flux is important, as these particles do not remain long in the atmosphere and few attain heights more than a few meters above the sea surface and thus do not contribute to the effective SSA production flux. Of the five methods that determine the interfacial production flux, the along-wind flux method, the direct observation method, and the vertical impaction method can be used to determine the interfacial production flux of large SSA particles; the whitecap method and bubble method are based on measurements from bursting bubbles and do not include contributions from spume drops, which are thought to provide the dominant contribution to the SSA production flux over much of this size range. However, the along-wind flux method has typically required other methods to determine the magnitude of the interfacial SSA production flux, and the size-distribution of this flux has been based on incorrect assumptions. The direct observation method and the vertical impaction method have seen only limited application.

Estimates of the effective and interfacial SSA production fluxes based on the methods described in this section are examined in §5.

4. Measurements and Models of Quantities Required to Evaluate Sea Salt Aerosol Production Fluxes

This section reviews field and laboratory measurements and model-based estimates of quantities required to calculate interfacial and effective SSA production fluxes according to the methods outlined in §3.

Field measurements of SSA concentration, and inferences of the dependencies of SSA concentration on controlling factors, are examined in §4.1. Measurements of total concentrations of SSA mass and number and of size distributions of SSA concentration are examined and compared, and conclusions are drawn based on these measurements regarding the importance of SSA to various atmospheric processes. In addition to being of fundamental importance in describing the amount of SSA in the atmosphere, these quantities are required to evaluate the effective SSA production flux according to the steady state dry deposition method (§3.1) and the statistical wet deposition method (§3.9), they are employed as a check on estimates of SSA production fluxes, and they are used to determine the dependencies of the effective SSA production flux on meteorological quantities.

Measurements of near-surface SSA concentrations and fluxes over the ocean, mostly of large SSA particles, are examined in §4.2. These measurements have been used in formulations of the interfacial SSA production flux based on the along-wind flux method (§3.6).

Laboratory measurements pertinent to SSA production are examined in §4.3. Measurements of the number, sizes, and ejection heights of jet and film drops from individual bursting bubbles and the dependencies of these quantities on various factors (§4.3.1, §4.3.2, §4.3.3) are used to determine the interfacial SSA production flux according to the bubble method (§3.4). Measurements of SSA production from laboratory simulated whitecaps (§4.3.4) and from surf zones (§4.3.5) are used to calculate the interfacial SSA production flux according to the whitecap method (§3.3). Measurements of near-surface SSA fluxes from wind/wave tanks (§4.3.6) are used to determine the interfacial SSA production flux using the along-wind flux method (§3.6).

Bubbles and their relation to SSA production are examined in §4.4; specifically field measurements of the size distribution of bubble concentration in the oceans and its dependencies on controlling factors, and theoretical and laboratory investigations of the dynamics of individual bubbles. This information, in conjunction with results from experiments involving drop formation from individual bubbles (§4.3), is used to evaluate the interfacial SSA production flux according to the bubble method (§3.4).

Research on oceanic whitecaps is examined in §4.5, specifically investigations on the whitecap coverage of the ocean and its dependence on controlling factors and on the temporal behavior of individual whitecaps. This information is required to extrapolate laboratory measurements of drop production from simulated whitecaps (§4.3.4), thereby allowing estimates to be made of the interfacial SSA production flux over the oceans by the whitecap method (§3.2).

Dry deposition models and measurements are examined in §4.6. Dry deposition velocities derived from these models are used in conjunction with size distributions of SSA number concentration to determine the effective SSA production flux according the steady state dry deposition method (§3.1) and in determination of the effective SSA production flux by the concentration buildup method (§3.3).

Minor mechanisms of SSA production, such as wave breaking at coasts, production by rain and snow, and production from bursting of bubbles formed other than by breaking waves, are examined in §4.7. Fracture of SSA particles upon drying, which would result in a change in the size distribution of SSA particles in the atmosphere, is also examined.

Throughout this section it is the explicit intent to present measurements for each quantity of interest from a large number of sources. This approach yields the range of values that have been reported and that can typically be expected for a given quantity, and permits estimation of the variability expected for nominally the same conditions. This variability provides a sense of the extent to which any one data set is representative and can be generalized to other locations and conditions. Additionally, this approach provides a method of ascertaining the controlling meteorological and environmental factors and the dependence of quantities of interest on these factors. Examination of a large body of data also permits identification of outliers, which may be accurate

Sea Salt Aerosol Production: Mechanisms, Methods, Measurements, and Models
Geophysical Monograph Series 152
Copyright 2004 by the American Geophysical Union.
10.1029/152GM04

and indicative of an unusual situation worthy of further exploration or alternatively may be indicative of some measurement error or artifact. Either way flagging such outliers identifies potential concerns that must be resolved.

4.1. SEA SALT AEROSOL CONCENTRATIONS

Concentrations of SSA particles (§2.1.4) and properties of these concentrations (§2.1.5) are of fundamental importance in characterizing SSA in the marine atmosphere, and knowledge of these concentrations derives ultimately from measurements. Measurements yield information, both in an absolute sense and relative to other components of the marine atmosphere, on the concentrations of the number, surface area, mass, and other integral properties of sea salt aerosol at the time and location of measurement, and on the size distribution of SSA concentration, knowledge of which identifies the size ranges of SSA particles that contribute most to properties of interest. Measurements also provide information on the relations of these quantities to factors thought to be controlling, key among which is wind speed. By extension, these measurements and relations are used to infer these quantities at other locations and times. The results of these measurements are also used to constrain models of SSA production and transport; to quantify the role of SSA in atmospheric processes of interest (§2.1.5), such as the ability of SSA particles to act as cloud drop nuclei (§2.1.5.1), provide a sink for condensable gases (§2.1.5.2), and scatter light, thus affecting the radiation budget of Earth (§2.1.5.3); and to calculate the effective SSA production flux using the dry deposition method (§3.1) and the statistical wet deposition method (§3.9).

Pertinent concentrations are N, the total number concentration of SSA particles (§4.1.3); R, the total concentration of SSA radius; A, the total concentration of SSA surface area (which under the approximation $Q_s = 2$ is equal to twice the light-scattering coefficient σ_{sp}; §2.1.5.3); M, the total dry sea salt mass concentration (§4.1.2); and $n(r_{80})$, the size distribution of SSA number concentration (§4.1.4). The results of measurements of these quantities, whose duration may be as short as minutes or as long as days, refer to mean values in the Reynolds-averaged sense of §2.8.2 rather than to fluctuating values, and the overbar is omitted in describing them. The quantities N and M, although they may in principle be influenced by RH, do not require a value of the RH for their specification, in contrast to the size-dependent and total concentrations of SSA radius, surface area, and volume (§2.1.4.1); for specification of these latter quantities, the value of 80% RH, as is used throughout this review, is chosen here as well.

Measurement techniques that have been employed to determine SSA concentrations and their differential and integral properties are described, the ranges of values typically reported for these concentrations are presented, and the relative contribution of SSA particles of different sizes to these concentrations is examined. This section also examines attempts to determine the dependencies of these concentrations on meteorological and environmental factors based upon these measurements, and various parameterizations that have been proposed to describe these dependencies.

4.1.1. General Considerations

In principle any integral property of sea salt aerosol can be determined from knowledge of the size distribution of the SSA number concentration at a given RH by integration of the pertinent size-dependent weighting factor $W(r)$ over this distribution (§2.1.5). In practice integral properties of the marine aerosol concentration are commonly measured directly, using an instrument or method pertinent to the property of interest, such as optical particle counters to determine the total aerosol number concentration or nephelometers to determine the aerosol light-scattering coefficient. However, such measurement techniques are for the most part not specific to sea salt, and although SSA is a major component of the marine aerosol, it is by no means the only component, and under some circumstances it may not be the dominant component in some size ranges (and therefore may not provide the main contribution to the quantity under consideration). Even far from land continental influences may be important (especially in the Northern Hemisphere) because of industrial emissions, terrestrial sources such as biomass burning and soil dust, and gas-to-particle conversion of volatile gases; additionally, gas-to-particle conversion can occur for volatile gases of oceanic origin. Failure to account for these other components when analyzing results of measurements that are not specific to sea salt and which thus include the contribution of other particles can result in incorrect inference of the concentration of SSA particles and properties of these concentrations. Thus an important consideration for SSA particles with regard to their role in atmospheric phenomena and to measurements of marine aerosol particles is their relative abundance in the marine atmosphere. This is examined in §4.1.1.1.

Measurements of SSA concentration and associated differential and integral properties (§2.1.5) require sampling, identifying, counting, and/or sizing SSA particles and their properties. Measurements of different properties require techniques of widely different complexity. For example, measurement of the total SSA dry mass concentration in principle requires only collection of all marine aerosol particles in a given volume of air and some chemically or physically specific method of determining the SSA mass sampled. In contrast, measurement of the total SSA number concentration requires identifying particles as SSA particles as well as counting individual particles. Measurement of size distributions of SSA particle concentration requires identification and sizing individual particles, and counting the number of SSA particles in several size ranges. Although these procedures may appear straightforward, measurements of SSA particle sizes, of concentrations (and fluxes) of SSA particles of a given size, and of integral properties of size distributions of SSA concentration (and flux), such as the total concentrations of SSA number and dry mass, are subject to several concerns, some of which are common to various sampling techniques, that limit the ability of such measurements to accurately determine the quantity of interest. Several of the concerns presented in §3.5.5 with regard to micrometeorological techniques apply here as well. Examination of some of these concerns provides insight into the confidence that can be placed in reported measurements. These topics are examined in §4.1.1.2.

Other concerns that must be addressed with regard to implementation and interpretation of measurements of SSA concentrations are the variability of measurements of a given quantity and the representativeness of any one measurement or set of measurements, and the dependencies of SSA concentration on height, wind speed, and other influencing factors. Consideration of these topics allows assessment of the confidence that can be placed in any conclusions regarding the dependence of the concentration of SSA particles of a given size (or of integral properties of the SSA concentration) on factors thought to be controlling. These topics are examined in §4.1.1.3.

Many topics in this section are pertinent also to measurements of SSA fluxes and of bubble concentrations.

4.1.1.1. Relative abundance of SSA particles compared to other marine aerosol particles.   The relative abundance of SSA particles is defined to be the fraction of the number concentration ϕ_N, volume concentration ϕ_V, or dry mass concentration ϕ_M, of marine aerosol particles in a given size range that derives from sea salt. Particle size is classified here in terms of r_{80}; to allow comparison with SSA particles it is assumed here that all particles have the same RH dependence, thus for particles other than SSA particles r_{80} is taken to be twice the dry radius, as discussed above (§2.1.1; §2.5.3). Several investigations that have determined the relative abundance of SSA particles in various size ranges are summarized in Table 10, and the results are presented in Fig. 16; similar trends are illustrated in Plate 1 of *Guazzotti et al.* [2001], for example, and are depicted schematically in Fig. 5 of *Chuang et al.* [2000]. In selecting data obvious instances of continental and surf influence have been avoided, although in some instances samples were obtained

at coastal locations. It should not be assumed that SSA particles were detected over the entire range of particle sizes indicated by the horizontal range bars in Fig. 16, as in many instances the size range sampled was quite wide. Additionally, over such wide ranges it is possible that SSA particles can provide the dominant contribution to dry mass concentration (i.e., ϕ_M near unity) while at the same time providing only a minor contribution to number concentration (i.e., ϕ_N much less than unity).

Sea salt aerosol particles with $r_{80} < 0.1$ μm have been reported by a sufficiently large number of investigators that these reports cannot be dismissed as measurement artifacts. Sodium-containing particles, presumably sea salt, with r_{80} as small as 0.06 μm were detected using single-particle sodium flame photometry by *Hobbs* [1971a] near the coast of Washington state and by *Radke et al.* [1976] in Barrow, Alaska. Individual SSA particles as small as $r_{80} = 0.06$ μm were detected by *Mészáros & Vissy* [1974] in the Southern Atlantic and Indian Oceans, as small as $r_{amb} = 0.03$ μm by *Gras and Ayers* [1983] at Cape Grim, Tasmania, and as small as $r_{80} = 0.08$ μm by *Bigg et al.* [1995] in the Southern Indian Ocean, all using electron microscopy. Sea salt aerosol particles with r_{80} of 0.035, 0.05, and 0.075 μm in the Pacific and Southern Oceans were reported by *Berg et al.* [1998] based on hygroscopic growth factors determined using a tandem differential mobility analyzer. Sodium-containing particles (again presumably SSA particles) as small as $r_{80} = 0.03$ μm were reported by *Sellegri et al.* [2001] from chemical analysis of filter samples collected in the Mediterranean using an impactor. Sea salt aerosol particles as small as $r_{80} = 0.01$ μm were reported by *Clarke et al.* [2003] based on thermal volatility measurements at a coastal location in Hawaii. Additionally, some laboratory investigations have concluded that SSA particles as small as $r_{80} = 0.01$ μm are produced by bursting bubbles (§4.3.3.1).

Despite their demonstrated presence in the marine atmosphere, SSA particles with $r_{80} < 0.1$ μm are typically not abundant, and many investigations reported absence of SSA particles in this size range. Additionally, such particles, when they are present, generally comprise only a small fraction of marine aerosol particles in this size range. Sea salt aerosol particles with $r_{amb} = 0.03$ at Cape Grim were reported by *Gras and Ayers* [1983] to comprise less than 1% of the number concentration of marine aerosol particles at this size, and few SSA particles with $r_{80} < 0.1$ μm were detected by *Bigg et al.* [1995] in the Southern Indian Ocean, both locations where contributions from continental sources are expected to be quite low. Fewer than 5% of the particles with $r_{80} \lesssim 0.1$ μm in the Pacific sampled by *Hoppel and Frick* [1990] contained sea salt. Sea salt aerosol particles with $r_{80} = 0.035$ μm were present in only 8 of the 745 samples collected in the Pacific and Southern Oceans by

Table 10. Measurements of Relative Abundance of Sea Salt Aerosol Particles in the Marine Boundary Layer

Reference	Location	Technique	$r_{80}/\mu m$ Range[a]	Sea Salt Fraction by Number (ϕ_N), Volume (ϕ_V), or Mass (ϕ_M)[b]
Mészáros & Vissy, 1974	Southern Atlantic and Indian Oceans	electron and optical microscopy of individual particles	0.06-0.2 0.2-2 > 2	$\phi_N \approx 0.06$ (0.004-0.14)[c] $\phi_N \approx 0.18$ (0.023-0.51)[c] $\phi_N \approx 0.76$ (0.38-1.0)[c]
Bigg, 1980	Cape Grim, Tasmania	electron microscopy of individual particles	> 1	$\phi_N > 0.9$
Sievering et al., 1982	Coastal site near Miami	electron microscopy of individual particles, chemical analysis of filter samples	< 0.3 0.15-0.5 0.75-1.5 > 1.5 > 3	$\phi_M = 0.06$[d] $\phi_N \approx 0.5$[d] $\phi_N \approx 0.5$[d] $\phi_N \approx 0.6$[d] $\phi_M = 0.61$[d]
Gras & Ayers, 1983	Cape Grim, Tasmania	electron microscopy of individual particles	0.03 0.025-5 0.3-5	$\phi_N < 0.01$ $\phi_N < 0.05; \phi_M > 0.95$[e] $\phi_N = 1.0$
Clarke et al., 1987	Equatorial Pacific	thermal volatility and optical particle counter	0.1-0.4 > 1	$\phi_M < 0.01$[f] $\phi_M \approx 1.0$
Jennings & O'Dowd, 1990[g]	Coastal site at Mace Head, Ireland	thermal volatility and optical particle counter	0.09-0.2 0.15-0.3 0.24-0.84 0.6-3	$\phi_N = 0; \phi_V = 0$[h] $\phi_N = 0; \phi_V \approx 0$[h] $\phi_N = 0.18$[h]; $\phi_V = 0.49$[i] $\phi_N = 0.77$[i]; $\phi_V = 0.82$[i]
O'Dowd & Smith, 1993[g,j]	North Atlantic	thermal volatility and optical particle counter	0.1-0.24 0.25-0.5 0.54-2.9	$\phi_N = 0.41$[k] $\phi_N = 0.61$[k] $\phi_N = 0.7$[l]
Quinn et al., 1996	South Pacific	chemical analysis of filter samples	0.1-0.5 0.5-6	$\phi_M \approx 0.55$ (0.1-0.98)[m] $\phi_M \gtrsim 0.95$[m]
McInness et al., 1997	Southern Pacific	electron microscopy of individual particles	0.1-1 1-6	$\phi_N = 0.04$-0.13[n] $\phi_N = 0.86$-1.0[o]
Murphy et al., 1997; *Murphy et al.*, 1998b[g]	Cape Grim, Tasmania	single particle mass spectrometry	0.16-3	$\phi_N > 0.9$
Kreidenweis et al., 1998[g]	Macquarie Island (South Pacific)	thermal volatility and electron microscopy of individual particles	0.08-0.2 0.2-1	$\phi_N = 0.2$-0.65[p] $\phi_N = 0.7$-0.9[q]
Berg et al., 1998	Pacific and Southern Oceans	hygroscopic growth factor in tandem differential mobility analyzer	0.035 0.05 0.075 0.15 0.165	$\phi_N = 0.15 \pm 0.11$ when present (8 of 745 samples)[r] $\phi_N \approx 0.15 \pm 0.11$ (in one case > 0.5) when present (39 of 868 samples)[r] $\phi_N \approx 0.12 \pm 0.07$ when present (35 of 269 samples)[r] $\phi_N \approx 0.35 \pm 0.25$ (and in one case 0.82) when present (225 of 579 samples)[r] $\phi_N \approx 0.09 \pm 0.05$ when present (39 of 297 samples)[r]
Quinn et al., 1998	Southern Ocean	chemical analysis of filter samples	0.18-0.25 0.25-0.38 0.38-10	$\phi_M \approx 0.5$ (0.1-0.7)[s] $\phi_M \approx 0.7$ (0.3-0.8)[s] $\phi_M \approx 0.99$ (0.98-1.0)

Table 10. (Continued)

Reference	Location	Technique	$r_{80}/\mu m$ Range[a]	Sea Salt Fraction by Number (ϕ_N), Volume (ϕ_V), or Mass (ϕ_M)[b]
Murphy et al., 1998a[g,t]	Cape Grim, Tasmania, Macquarie Island (South Pacific)	electron microscopy and mass spectrometry of individual particles	> 0.13 0.05-0.15	$\phi_N > 0.9$ $\phi_N = 0.05$-0.25 (Cape Grim); $\phi_N = 0.05$-0.47 (Macquarie Island)[u]
Ebert et al., 2000	Island of Helgoland (Germany), North Sea	electron microscopy of individual particles	0.04-0.2 0.2-0.8 0.8-3 3-11	$\phi_N \approx 0.4$ (0.13-0.66)[v] $\phi_N \approx 0.44$ (0.15-0.7)[v] $\phi_N \approx 0.5$ (0.26-0.7)[v] $\phi_N \approx 0.6$ (0.42-0.78)[v]
Putaud et al., 2000[f]	Tennerife (North Atlantic)	chemical analysis of filter samples	< 1	$\phi_M = 0.21$ (0.19-0.31)[w]
Chabas & Lefèvre, 2000	Coastal site in Greece	electron microscopy of individual particles	> 1	$\phi_N \approx 0.55$ (0.33-0.8)[x]
Sellegri et al., 2001	Mediterranean	chemical analysis of filter samples	< 0.15 > 1	$\phi_M \leq 0.3$[y] $\phi_M > 0.8$

Relative abundances of sea salt number ϕ_N, volume ϕ_V, and mass ϕ_M are presented as a function of r_{80} in Fig. 16.

[a] Values of r_{80} are approximate and have been converted from sizes presented (which included aerodynamic, ambient, and dry radii) on the assumption particles were sea salt (that is, in some instances dry sizes were presented, in which case the relation $r_{80} = 2r_{dry}$ was used). When RH was not presented it was assumed to be 80%; as the equilibrium radii of SSA particles vary less than 20% from r_{80} over the RH range 60 to 90% (§2.5.3), this assumption should result in little difference in most instances.

[b] Number fraction denotes the fraction of particles containing sea salt; as particles may be internally mixed, these particles may also contain other substances. Mass fraction refers to dry aerosol mass and does not include the contribution of water.

[c] Remainder was mainly ammonium sulfate/bisulfate and insoluble particles probably from Africa.

[d] Remainder was soil and anthropogenic substances.

[e] Most particles were ammonium sulfate.

[f] Most mass was sulfuric acid or ammonium sulfate/bisulfate.

[g] For maritime air masses.

[h] Most of remainder was ammonium sulfate/bisulfate.

[i] Most of remainder was crustal material.

[j] For conditions of high winds (14-17 m s^{-1}).

[k] Most of remainder was sulfuric acid and dust.

[l] Most of remainder was unidentified.

[m] Most remaining mass was non-sea salt sulfate and associated ammonium.

[n] Most others were ammonium sulfate and acidic sulfate.

[o] Most others were mineral particles.

[p] Most others were sulfate, carbonaceous, or mineral particles.

[q] Most of remainder was carbonaceous and mineral particles.

[r] Non-sea salt sulfate occurred in 100% of the samples.

[s] Most of remainder was non-sea salt sulfate.

[t] This contains results presented in *McInness et al.* [1997], *Murphy et al.* [1997, 1998b], *Quinn et al.* [1998], *Kreidenweis et al.* [1998], *Berg et al.* [1998].

[u] Most other particles were sulfate.

[v] Most of remainder was anthropogenic, crustal, or biological.

[w] Most of remainder was non-sea salt sulfate and organic carbon.

[x] Most others were crustal, anthropogenic, or biogenic.

[y] Non-sea salt sulfate comprised 20 to 70% of the mass of analyzed ionic species of particles in this size range, the other main components being ammonium and nitrate.

Berg et al. [1998], who reported that SSA particles with $r_{80} = 0.05$ μm occurred in only 39 of 868 samples, and SSA particles with $r_{80} = 0.075$ μm occurred in only 35 of 269 samples; for each of these sizes SSA particles comprised only ~15% of the number concentration when they were present (Table 10). An impediment to determination of the relative contribution of SSA particles in the size range $r_{80} \lesssim 0.1$ μm is that SSA particles can rapidly become internally mixed by exchange and uptake of gases (§2.1.5.2; §2.5.2), in the process possibly changing size and losing their identity as SSA particles, and complicating attribution of such particles as SSA particles.

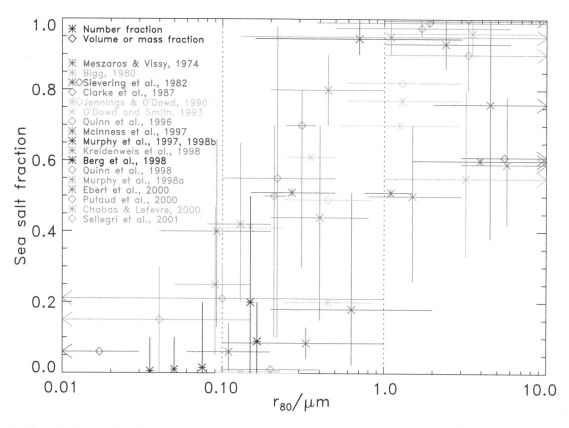

Figure 16. Sea salt fraction of marine aerosol by number (ϕ_N), volume (ϕ_V), or mass (ϕ_M) at locations in the marine boundary layer mainly free from obvious surf effects and continental influence as a function of r_{80}, summarized in Table 10. Horizontal ranges indicate r_{80} range for technique employed, converted from reported sizes under the assumption that the RH dependence of all marine aerosol particles is the same as that of SSA particles; symbols are placed at the geometric mean of the radius range. For r_{80} range extending to 0.01 μm or 10 μm, sample may have contained particles less than or greater than that value (designated by arrow). Vertical ranges indicate spread of reported values over indicated range of r_{80}, with symbols placed at the arithmetic mean of the values. In some instances for which upper or lower ranges were not specified the mean value was plotted using the upper or lower limit of the graph. In some instances lines are offset slightly to avoid overlap. Dotted vertical lines are shown at r_{80} values of 0.1 μm and 1 μm.

In the size range 0.1 μm $\lesssim r_{80} \lesssim$ 1 μm, characterizing small SSA particles, the relative abundance of SSA particles exhibits much variability, both spatial and temporal, caused both by the variation in abundance of SSA particles and by the large amount of variation in sources and sinks, and hence abundance, of non-SSA particles. Many of the non-SSA particles in this size range result from secondary production mechanisms that have controlling factors different from those controlling production of SSA particles; hence the relative abundance of SSA particles in this size range is likely to depend on many factors, complicating attempts to infer the controlling factors for concentrations of SSA. In marine conditions not appreciably influenced by anthropogenic or continental sources, sea salt may be an important component of the marine aerosol in this size range, but non-sea salt sulfates often provide an appreciable or even

dominant contribution. In continentally-influenced situations, anthropogenic substances frequently dominate the number concentration and in some instances also the mass concentration in this size range. Consequently, it is difficult to draw general conclusions regarding the relative abundance of SSA particles in this size range, and there is a potential for misattribution of particles as SSA in measurements that do not discriminate particles by composition (§4.1.1.2; §4.1.4.2); this is especially applicable to optical particle counters. Additionally, depending on the size range of interest, SSA particles may provide the dominant contribution to the mass fraction and a minor contribution to the number fraction of the marine aerosol at the same time and location, as noted above.

In the size range 1 μm $\lesssim r_{80} \lesssim$ 25 μm, characterizing medium SSA particles (§2.1.2), sea salt is typically the dominant contributor to the marine aerosol, in terms of both number and mass, except during sporadic dust events [*Prospero*, 1979; *Li et al.*, 1996; *Prospero*, 1999; *Reid et al.*, 2003b]. Hence, measurement of aerosol dry mass concentration by gravimetric analysis of a sample collected on a filter, for example, in principle yields a good determination of M provided the measurement is taken during times of low dust loading.

In the size range $r_{80} \gtrsim 25$ μm, characterizing large SSA particles, sea salt provides the overwhelming majority of both the number and the mass concentrations in nearly all situations. However, the concentration of SSA particles in this size range at heights more than a few meters above the sea surface is typically quite low, and these particles generally contribute little to quantities measured at heights near 10 m.

The above considerations permit some general conclusions to be reached regarding the relative abundance of SSA as a component of the marine aerosol as a function of particle size. These conclusions allow the relative importance of SSA to a given process to be ascertained in some instances, and they determine for a given size range whether or not measured marine aerosol particles can be accurately assumed to be SSA particles. For the most part, the marine aerosol in the size range $r_{80} < 0.1$ μm consists largely of material other than sea salt, and SSA particles in this size range generally contribute little to SSA properties of interest (§4.1.4.3). Thus, a practical lower limit for the sizes of SSA particles can by taken as $r_{80} \approx 0.1$ μm, albeit with recognition that SSA particles this size and smaller have been detected in marine aerosol samples. For 0.1 μm $< r_{80} < 1$ μm the sea salt fraction can be quite variable, whereas for $r_{80} > 1$ μm the marine aerosol is typically dominated by sea salt. Therefore, the number concentration of marine aerosol particles is generally dominated by substances other than sea salt, even in clean marine conditions. In clean marine conditions, the contribution of SSA particles to the radius concentration of the marine aerosol may be important, and SSA particles may provide the dominant contribution to the concentration of surface area, and thus to light scattering by aerosol particles. As noted above, the mass concentration of the marine aerosol is generally dominated by the SSA mass concentration.

4.1.1.2. Measurement concerns. Associated with any measurement of SSA concentration is a number of issues regarding how accurately the measurement represents the quantity desired. Some of these concerns are common to most techniques, such as perturbation of the flow by the sampling device, inlet and internal losses, particle bounce and fracture upon impaction with a sampling substrate or walls of a sampling device (especially of concern for dry particles), and non-isokinetic sampling, all of which can result in counting efficiencies greater than or less than unity, depending on the particular instrument, the particle size, and the wind speed and direction. Consequently, these factors can have a large effect on measured concentrations. Additionally, site-specific characteristics such as bow-generated sea spray can affect samples, and measurements at coastal sites can be affected by local surf conditions, for example. In this section some of the techniques that are typically employed to measure SSA particle concentration and concerns specific to these techniques are examined (a comprehensive review of particle sampling techniques is presented in *McMurry* [2000]). Concerns specific to measurement of SSA particle size and size distributions of SSA concentration are addressed in §4.1.4.1.

A key issue in characterizing SSA is distinguishing it from other components of the marine aerosol so that physical measurements, which for the most part do not discriminate particles by composition, can confidently be interpreted as representing SSA or alternatively be questioned. Considerable effort has been directed over several decades to identifying the constituents of the marine aerosol, in part directed at determining anthropogenic influences. Principal techniques are bulk chemical analysis, generally size-segregated, and physical methods, such as thermal volatility, microscopic analysis, and sodium flame photometry. Bulk chemical analysis requires sample collection, which often results in a limited range of particle sizes because of exclusion of large particles due to inlet designs. One limitation of bulk chemical analysis is that it gives no indication of the extent to which the aerosol particles comprising the sample were internally or externally mixed. Some physical methods take advantage of greater volatility of particles formed by gas-to-particle conversion, such as sulfate and organic substances, to ascribe residual material after heating to sea salt (and soot and crustal materials); sea salt can be distinguished from other nonvolatile substances such as mineral dust by further heating which volatilizes the sea salt [e.g., *Dinger et al.*, 1970; *Pueschel et al.*, 1973; *Clarke et al.*, 1987; *Jennings and O'Dowd*, 1990; *Clarke*, 1991; *O'Dowd and Smith*, 1993; *Clarke and Porter*, 1993; *Lowe et al.*, 1996; *O'Dowd et al.*, 1997a; *Kreidenweis et al.*, 1998; *O'Dowd et al.*, 1999b]. Scanning electron microscopy (SEM) and transmission electron microscopy (TEM) provide identification of composition and size [e.g., *Mészáros and Vissy*, 1974; *Bigg*, 1980; *Sievering et al.*, 1982; *Gras and Ayers*, 1983; *Andreae et al.*, 1986; *Parungo et al.*, 1986a,b, 1987; *Pósfai et al.*, 1994, 1995; *McInnes et al.*, 1994, 1997; *Kreidenweis et al.*, 1998], although there may be changes in morphology and composition during sample collection and preparation, and quantitative information is often difficult to obtain by this method. Sodium flame-photometry [*Soudain*, 1951 (cited in *Vonnegut and Neubauer*, 1953); *Vonnegut and Neubauer*, 1953; *Woodcock and Spencer*, 1957; *Radke and Hobbs*, 1969; *Hobbs*, 1971; *Bodhaine and Pueschel*, 1972; *Slinn et al.*, 1983; *Hegg et al.*, 1995; *Clark et al.*, 2001; *Campuzano-Jost et al.*, 2003] determines the sodium content in SSA particles, from which the values of r_{80} can be inferred, thus allowing determination of the size distribution of SSA concentration. However, this method has been little used. Substantial advances have been made in recent years through single-particle mass spectrometry [e.g., *Murphy et al.*, 1997, 1998b; *Guazzotti et al.*, 2001], which can identify the composition of particles (albeit with substantial limitations on quantitative determination), determine their size, obtain the size distribution of specific substances, and distinguish internal vs. external mixtures. This approach is still in its infancy, so there is not a large data base from which to draw generalizations.

In contrast to these techniques, many of the measurements of sizes, size distributions, and concentrations of SSA particles have employed techniques such as optical particle counters that are based on physical properties and do not distinguish particles by composition (although some of these techniques have been used in conjunction with chemical or other analyses that can determine composition). Consequently, their use may lead to incorrect results if care is not taken to ensure that only sea salt particles are detected. As small SSA particles (i.e., $r_{80} \lesssim 1$ μm) often comprise only a small fraction of the particles present in the marine aerosol in this size range (§4.1.1.1), failure to distinguish particles by composition will especially affect measurements of concentrations of small SSA particles and of the SSA number concentration N. Additionally, inclusion of particles other than sea salt will complicate attempts to quantify the dependence of SSA concentration on wind speed and other meteorological factors, as these dependencies may be very different for these other particles, which have different production and removal mechanisms and different behaviors from SSA particles under similar conditions.

Further issues regarding the ability to accurately determine SSA particle concentrations are the particle size ranges, sampling rates, and efficiencies of measurement techniques. For example, limitations of sampled particle size will affect measurements in different ways, depending on the concentration property under consideration; different lower sampling limits can result in differences in measured number concentration, and different upper sampling limits can result in differences in measured light-scattering and mass concentrations and in quantities inferred from these measurements.

The behavior of an SSA particle with respect to RH also affects measurements of sizes and concentrations of such particles. Collection efficiencies of small particles are highly sensitive to the physical (i.e., geometric) size of the particle and thus for a given r_{80} this efficiency is strongly dependent on RH. Additionally, the physical size of an SSA particle can be affected by RH gradients within sampling instruments, such as those caused by heating. Measurements in which SSA particles are sampled at RH values between the efflorescence RH and the deliquescence RH, or near one of these values, can lead to ambiguities in the reported size (§2.5.5). Dry SSA particles are especially susceptible to bounce or fracture, and determination of their size using optical techniques can lead to errors, as such techniques are typically calibrated using homogeneous spheres (§2.5.5). Failure to take into account the time required for an SSA particle to respond to changes in RH can also lead to incorrect inferences of the particle size; this is especially important for large SSA particles, which may have not achieved their equilibrium radius at the time of measurement (§2.9.1, §4.1.4.1), but may also apply to smaller particles, depending on sampling conditions (such as height) and instrument conditions.

4.1.1.3. Considerations regarding interpretation of SSA concentration measurements. An important consideration in interpreting measurements of SSA concentration and its dependencies on controlling factors is the extent to which SSA concentration at a particular location or time, usually encompassing a narrow range of quantities such as water temperature, wind history, and the like, is representative of SSA concentration at other locations, seasons, and environmental conditions. Interpretation of measurements of SSA concentration is thus greatly enhanced when the measurements are accompanied by a description of sampling conditions (height, location, particle size range sampled, and the like) and analysis procedures; specification of RH, wind speed, atmospheric stability, height of the marine boundary layer, and other pertinent meteorological conditions; estimates of uncertainty due to experimental technique; and presentation of the variability in measured values under nominally identical conditions. This auxiliary information, especially the characterization of the spatial and temporal variability, provides some indication of the reliability of the data and of the extent to which the measurements can be considered representative. Additionally, such information allows inference of the dependencies of the concentration of SSA particles of a given size on factors thought to be controlling.

In some instances this auxiliary information is not available although the data themselves are pertinent to SSA concentrations. This situation has often occurred with studies designed to investigate the aerosol present in a region and not necessarily to examine oceanic SSA production, possibly limiting attention paid to characterizing SSA and controlling factors. For example, sea salt sodium has often been used to determine sea salt sulfate for purposes of calculating non-sea salt sulfate.

A key consideration is measurement location relative to coasts and shorelines. Measurements of SSA concentration in coastal regions are prevalent, as it is more convenient and less costly to perform measurements from or near shore than from shipboard or aircraft over the open ocean, and permanent sampling stations allow measurements to be taken continuously for long periods of time and under many different meteorological conditions. However, measurements at coastal locations may encounter several confounding influences not experienced by those over the open ocean: surf spray, effects of the coast on the wind profile and on the entrainment of particles upward, differences in wave breaking in shallow coastal waters, and the like. Additionally, continental aerosol can strongly influence results of measurements that do not distinguish particles by composition, such as determination of number concentration using optical particle counters. Consequently, measurements taken from shore may yield results different from those taken over the open ocean under otherwise similar conditions.

Factors thought to be important in influencing SSA production and concentration (§2.3) are wind speed (characterized by U_{10}), RH, atmospheric stability, mixing layer height (typically assumed to be the height of the marine boundary layer), fetch, and the time since the last rainfall, plus the time history of these factors experienced by the air mass containing the aerosol particles. Of these, wind speed is expected to be a major factor, if not the dominant one, and wind speed at the time and location of measurement has been used by virtually all investigators to classify measurements of SSA number and mass concentrations and of size distributions of SSA concentration, despite recognition of the strong influence of other meteorological and environmental factors and prior history. The variance in SSA concentration that is not accounted for by U_{10} has thus often been treated as random, rather than being assigned to these other controlling factors [Borisenkov and Kuznetsov, 1978], with estimates of uncertainties provided by the range of measured values for the same nominal conditions (i.e., wind speed). However, even if wind speed were the primary factor controlling SSA production, it does not follow that the same dependence would be manifested in SSA concentration because of the confounding influences associated with dilution, advection, and removal of SSA particles. Key to developing understanding of the relation between SSA concentration and factors controlling this concentration is the recognition that the SSA that is present at a given time and location results from a set of prior processes: production, entrainment, transport, and removal. Consequently, factors that affect any of these processes also affect SSA concentration, and interpretation of measurements that fails to account for these processes and attempts simply to relate the aerosol present to current and local conditions may result in large variability.

Because SSA concentration is an integral over the history of the particles once formed, interpretation of the dependence of SSA concentration on meteorological conditions would likely be improved by taking prior history into account, although the pertinent prior conditions are difficult to determine and to represent by variables with which the concentration can be correlated.

The mean atmospheric residence times of SSA particles depends strongly on their size—from minutes to hours for SSA particles with $r_{80} \gtrsim 10$ µm, up to several days for SSA particles with $r_{80} \lesssim 1$ µm (§2.9.6; Table 8). For this reason, the period of time prior to measurement that affects concentrations of SSA particles also depends strongly on particle size. A consequence of this dependence on SSA particle size is that concentrations of larger SSA particles ($r_{80} \gtrsim 10$ µm), and M, to which these larger particles contribute appreciably, might be expected to be better correlated with meteorological conditions at the time of measurement, and concentrations of small SSA particles ($r_{80} \lesssim 1$ µm), and N, which is dominated by these particles, more poorly correlated. A further consequence of the dependence of the mean atmospheric residence time on particle size is that the size distribution of SSA particles that are present at a given time and location of measurement is expected to differ from the size distribution of the SSA particles that are being produced at that time and location, and also from the size distribution of the SSA particles that were produced at earlier times.

A related consideration is the height dependence of the concentration of SSA particles of a given size. Knowledge of the vertical distribution of the size-dependent SSA concentration provides information on the size range of particles that are expected to remain in the atmosphere long enough to participate in various atmospheric processes. Sea salt aerosol particles with $r_{80} \lesssim 1$ µm, which may remain in the atmosphere up to several days, generally achieve nearly uniform concentrations over the entire marine boundary layer (§2.9.5), over which mixing occurs on a time scale of order an hour (§2.4). Consequently the SSA number concentration N, being dominated by these small particles, might also be expected to be independent of the height of the measurement within this layer (although both N and the concentrations of small SSA particles of a given r_{80} may depend on the marine boundary layer height). Additionally, measurements at heights of several meters to several tens of meters of quantities that are dominated by small SSA particles (such as N) might be expected to accurately represent those quantities over the entire marine boundary layer (§2.9.5), so that column burdens (i.e., vertical integrals) of these quantities can accurately be approximated by the product of their concentrations at heights near 10 m and the height of the marine boundary layer (§2.1.5). This conclusion applies also to medium SSA particles with $r_{80} \lesssim 10$ µm and to properties that are controlled by these particles, such as the concentration of SSA radius R. As particle size increases, gravity plays an increasingly greater role in removal. As the ability of larger SSA particles to be entrained upward diminishes as particle size increases, concentrations of SSA particles with $r_{80} \gtrsim 10$ μm, or of quantities such as M, for which these particles contribute appreciably, are expected to decrease with height over the marine boundary layer (§2.9.5; Fig. 12). However, as concentrations of SSA particles with $r_{80} \lesssim 25$ μm are not expected to differ substantially between 5 m and 20 m above the sea surface (except for the largest of these particles; §2.9.4; Fig. 10), differences in measurements of concentration of such particles are expected to be small over this range of heights. Large SSA particles ($r_{80} \gtrsim 25$ μm) are not mixed vertically to any appreciable extent, being present mainly at heights below several meters, and concentrations of these particles are expected to exhibit large vertical gradients, with the decrease in concentration being much more pronounced with increasing particle size (§2.9.5). Consequently, measurements of large SSA particles at heights near 10 m are expected to exhibit much variation, depending on the exact height of measurement. A further consequence of all this is that size distributions of SSA concentration are expected to demonstrate a somewhat complex behavior, with the shape of the distribution varying with height.

With these considerations in mind, the available data on SSA number and mass concentrations and the size distributions of SSA concentration, classified in terms of the wind speed U_{10}, are examined below.

4.1.2. Sea Salt Aerosol Mass Concentration

The dry mass of sea salt per unit volume of air M is perhaps the most widely used measure of SSA abundance in the marine atmosphere. Knowledge of M allows determination of the volume concentration of SSA at a given RH and provides a consistency check on measured size distributions of SSA concentration presented in §4.1.4. Additionally, together with the mass scattering efficiency (§2.1.5.3) and the dependence of r/r_{80} on RH (§2.5.3), M is used to estimate the SSA contribution to aerosol optical properties and to aerosol radiative forcing, that is, the aerosol influence on the amount of solar radiation absorbed by the Earth-atmosphere system [e.g., *Haywood et al.*, 1999]. The quantity M is in principle easily measured and has been frequently reported, and numerous investigators have attempted to quantify its dependence on wind speed, often with the presumption that this relation will reflect the wind-speed dependence of SSA production as well. In this section reported measurements of M and proposed parameterizations relating M to U_{10} are examined.

4.1.2.1. Measurements. Although a variety of methods have been used to determine M, the typical approach consists of collecting aerosol particles from a known volume of air on filters that are subsequently analyzed for sodium, from which the value of M is calculated as 3.26 times the sodium mass concentration (Table 6). Sodium is well suited to characterize SSA as it is not abundant in continental aerosols and, unlike chloride, is not volatilized by acidification. Alternatively, if the size distribution of the SSA number concentration $n(r_{80})$ is known at a given RH, M can be calculated (§2.1.4.1) as

$$M = \frac{4\pi}{3}\rho_{ss}\left(\frac{r_{dry}}{r}\right)^3 \int r^3 n(r_{80})d\log r_{80}, \qquad (4.1\text{-}1)$$

where r is the radius at the ambient relative humidity, ρ_{ss} is the density of dry sea salt, and the ratio r_{dry}/r is a known function of RH (§2.5.3).

The main contribution to M at heights at which measurements are typically made (~10 m) is expected to be from medium SSA particles, although large SSA particles may contribute somewhat, depending on the situation and measurement height and technique. For the canonical SSA size distribution of concentration introduced in §2.1.5 (which was based on measurements of size distributions of SSA concentration; §4.1.4.4), more than half of the contribution to M is from SSA particles with r_{80} within about a factor of two of 7 μm; i.e., approximately 3.5 μm to 15 μm (Fig. 2). As SSA particles of these sizes have relatively short residence times against dry deposition, ranging from roughly a day to an hour depending on size and meteorological conditions (§2.9.6; Table 8), it is reasonable to expect some correlation of M with local conditions responsible for production, e.g., wind speed (§2.9.6). Additionally, little decrease in the number concentration of SSA particles with $r_{80} \lesssim 15$ μm is expected over heights of 5-15 m (§2.9.5; Fig. 12), typical of heights at which most measurements have been made. At greater heights a slight decrease in M with increasing height is expected, as SSA particles toward the upper end of this size range exhibit decreasing concentrations with increasing height (§2.9.5; Fig. 12).

Measurements of M that have been reported together with wind speed are shown in Fig. 17, as are fits given by the investigators to their data. These measurements and fits are summarized in Table 11. Although different measurements may not refer to the same quantity because of differences in measurement techniques (such as heights and cutoff sizes), most investigators have discussed and fitted M as if it referred to an unambiguous and well defined physical property, and it is treated as such here. For many of the measurements, sampling periods were a day or more, over which time the wind speed may have varied substantially; for these measurements the mean wind speed over the sampling interval was used. Some other data sets [*Gautier*, 1899; *Jacobs*, 1937; *Fournier d'Albe*, 1951; *Moore*, 1952; *Junge*, 1954; *Woodcock and Spencer*, 1957; *Metnieks*, 1958; *Monahan*, 1968, 1973; *Frank et al.*, 1972; *Patterson et al.*, 1980; *Raemdonck et al.*, 1986; *Sellegri et al.*, 2001; *Maring et al.*, 2003] have been omitted because of few data or indication of probable contamination by continental aerosol, or because the measurements were made either very close to the surface (i.e., within a meter or so) or at high altitudes (~1 km or higher). Measurements of SSA mass concentration in the surf zone have not been included; such concentrations can exceed those characteristic of the open ocean by one to two orders of magnitude [*e.g., Duce and Woodcock*, 1971; *McKay et al.*, 1994; *de Leeuw et al.*, 2000]. The values of column burdens (i.e., vertical integrals) of M reported by *Krishna Moorthy et al.* [1997] (also not shown) are too high by roughly a factor of 6, as these investigators appear to have mistakenly calculated mass concentrations using the radius at 70% RH instead of the dry radius. Expressions relating M and U_{10} that were presented without data [*Gravenhorst*, 1978; *Várhelyi and Gravenhorst*, 1983; *Quinn et al.*, 1998; *Quinn and Coffman*, 1999] and fits to data taken by others [*Blanchard and Woodcock*, 1980; *Erickson et al.*, 1986; *Jaenicke*, 1988] are also not shown.

There is an immense amount of variation in reported values of M—more than two orders of magnitude in the data sets as a whole (and yet another order of magnitude if outliers are included), and more than an order of magnitude for any narrow range of wind speed. The majority of measurements of M in Fig. 17 fall within the range 5-50 µg m^{-3}, consistent with measurements from other investigations for which wind speed was not reported[1]. This range of M corresponds to a total SSA volume concentration at 80% RH of approximately 15-150 µm^3 cm^{-3} by (2.5-3), resulting in a volume mixing ratio of SSA particles in air of (0.15-1.5) · 10^{-10}. At 98% RH the volume mixing ratio would be about an order of magnitude greater, of order 10^{-9} or less. This range of M also corresponds to a total concentration of dry SSA mass and associated water mass at 80% RH of approximately 20-200 µg m^{-3} (§2.5.3), resulting in a mass mixing ratio of SSA particles in air of approximately (0.2-2) · 10^{-7}. The low values typically observed for volume and mass mixing ratios of SSA particles support the conclusions (§2, §2.8) that these particles are present in sufficiently

low concentrations that their effect on the surrounding atmosphere is negligible.

To provide a sense of the magnitude of M, the values in Fig. 17 may be compared with the maximum concentrations of particulate matter specified in air quality standards. For example, current U.S. Environmental Protection Agency standards require that the concentrations of PM-10—particulate matter (PM) with aerodynamic diameter less than 10 µm at ambient RH (corresponding roughly to $r_{amb} \lesssim$ 5 µm for SSA particles)—at RH 20-45% (which would result in dry SSA particles; §2.5.4) not exceed 150 µg m^{-3} for a 24-hour average and 50 µg m^{-3} for an annual average [*EPA*, 2002].

Most of the tabulated measurements of M were made at wind speeds in the range 3-15 m s^{-1}, and some of the data sets [*Barger and Garrett*, 1970; *Blanchard and Syzdek*, 1972b; *Kritz and Rancher*, 1980; *Kulkarni et al.*, 1982; *McDonald et al.*, 1982] covered an even smaller range of wind speeds (Table 11)—too narrow upon which to base inferences concerning the correlation of M with wind speed. For wind speeds from 3 m s^{-1} to 15 m s^{-1} the data are contained mostly within the shaded region shown in Fig. 17, which encompasses an order of magnitude in M. There is generally a slight overall increase in M with increasing U_{10}, but over this range of wind speeds the mean increase is much less than the scatter in reported values of M at any one wind speed. For wind speeds greater than ~15 m s^{-1} the data do not appear to congregate within a narrow range of values but exhibit more scatter, with an increasingly large fraction of the reported values exceeding 50 µg m^{-3}. These results are in contrast to those expected (§2.9.6) under the assumption that wind speed is the dominant controlling factor for SSA production, which implies that the correlation between SSA concentration and wind speed increases with increasing wind speed because of the decrease in atmospheric residence time with increasing particle size.

Some of the spread in measured values of M can be attributed to differences among different data sets, specifically the location, sampling technique, and sampling height. For example, the data of *Exton et al.* [1985, 1986] and those of *Smith et al.* [1989], which were obtained at the same coastal location, are for the most part higher than other reported data (Fig. 11). *Exton et al.* [1986] attributed their higher values to differences in experimental technique, but did not elaborate. *Exton et al.* [1985] suggested that they obtained higher values because whitecapping is likely to be more extensive

---

[1] e.g., *Junge et al.* [1969]; *Duce and Woodcock* [1971]; *Hoffman and Duce* [1972]; *Hoffman et al.* [1977]; *Prospero* [1979]; *Andreae* [1982]; *Tomasi and Prodi* [1982]; *Chamberlain* [1983]; *Parungo et al.* [1986b]; *Okita et al.* [1986]; *Parungo et al.* [1987]; *Sievering et al.* [1987]; *Arimoto et al.* [1992]; *François et al.* [1995]; *Yoshizumi and Asakuno* [1996]; *Li et al.* [1996]; *Carmichael et al.* [1997]; *Li-Jones et al.* [1998]; *Quinn et al.* [1998]; *Bates et al.* [1998]; *Quinn and Coffman* [1999]; *Chiapello et al.* [1999]; *Ebert et al.* [2000]; *Andreae et al.* [2003].

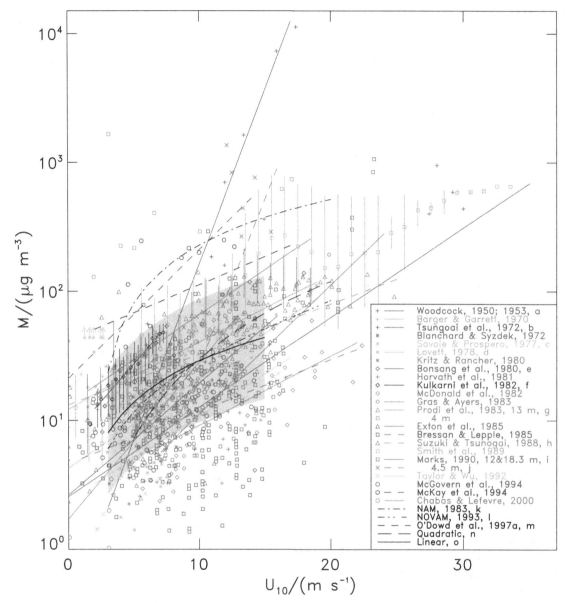

Figure 17. Measurements of total SSA dry mass concentration M, mostly in the lower marine boundary layer, as a function of wind speed at 10 m above sea surface, U_{10}, averaged over the sample period. Individual data sets are described in Table 11. Symbols denote measurements; symbols with vertical range bars denote means and standard deviations. Symbols with upward pointing arrows denote values above the maximum on the figure on which they were presented [Fig. 7 of *Savoie and Prospero*, 1977]. Data of *Exton et al.* [1985] and *Smith et al.* [1989] are offset slightly from each other for greater clarity. Lines denote fits by investigators to their data except where noted in Table 11. Black lines and curves denote results of parameterizations of size distributions of SSA concentration discussed in §4.1.4.4: NAM, the original Navy Aerosol Model [*Gathman*, 1983a,b], NOVAM, Navy Ocean Vertical Aerosol Model [*Gathman and Davidson*, 1993], *O'Dowd et al.* [1997a], and quadratic (based on canonical SSA size distribution of concentration; §2.1.5; §4.1.4.4). Linear relation denotes an empirical linear dependence on wind speed; shaded gray area is a factor of $\overset{\times}{\div}$ 3 about this relation. Notes to Fig. 17:

[a] Fit to data of *Woodcock* [1953] is that presented by *Tsunogai et al.* [1972].

[b] Data shown for *Tsunogai et al.* [1972] consist of 55 values from Fig. 5 of that reference; line is fit to these data by those investigators.

[c] Data shown for *Savoie and Prospero* [1977] are those from Barbados and Sal Island (Cape Verde Islands); data from Miami, not presented in that reference, were stated to be lower by roughly a factor of three at any given wind speed.

[d] Data shown for *Lovett* [1978] consist of 65 values from Fig. 2 of that reference representing a single cruise; fit is that given for all data (Table 7 of that reference).

in the shallow waters over the continental shelf (where their measurements were conducted) than in the open ocean; such a suggestion would thus call into question the applicability of their results to other locations (specifically the open ocean). It appears that the values of M from data sets shown in Fig. 17 from coastal sites may be slightly greater overall than those from open ocean locations, perhaps because of surf zone contributions, although the scatter in the data makes this difficult to determine (this is discussed further below with regard to measurements of *Daniels* [1989]). *Exton et al.* [1985] also suggested that because the optical detectors they used detected all particles, and not just SSA particles, their values of M could have been overestimated, although it is doubtful that this was an important factor, as it is highly unlikely that particles other than sea salt were present sufficient concentrations in the size ranges that provided the majority of the contribution to measured mass concentration (particles with r_{amb} in the ranges 0.3-8 μm and 8-16 μm provided roughly the same contribution) to make an appreciable difference. As the data sets of *Exton et al.* [1985, 1986] and *Smith et al.* [1989] were the only ones out of those summarized in Table 11 for which values of M were calculated from measurements of size distributions of SSA concentration employing optical techniques (using Eq. 4.1-1) as opposed to SSA mass concentrations determined from samples collected on filters, it is possible that overestimation of SSA particle radii by the optical detectors could have resulted in erroneously high values of M (*Reid et al.* [2003a], concluded that optical particle counters, even though properly calibrated, may overestimate the size of dust particles by a factor of more than two). An overestimation of only 25% in measured radii of SSA particles would have resulted in an overestimation of M calculated by (4.1-1) by a factor of two. Alternatively, many of the measurements made using

filters may result in underestimation of M due to omission of large SSA particles because of sampling limitations.

Some of the differences among data sets can be attributed to differences in the upper limit of measured particle size, as M is strongly influenced by larger particles. For instance, for the canonical SSA size distribution of concentration presented in §2.1.5, measurements at 80% RH for which the upper limit of radius is (5, 10, 15) μm, which corresponds to an aerodynamic radius of approximately (5.5, 11, 16.5) μm, would sample nearly (30, 70, 80)% of the particle mass (the size distribution of the mass concentration for this canonical distribution is shown in Fig. 2). The upper limit of measured particle size is often determined by the sampling technique and the devices controlling the flow to the collecting device (filter or impactor), and differences in this quantity can result in large differences. Investigations in which different impactors were compared results in differences in measured values of M by typically a factor of 2 [*François et al.*, 1995; *Howell et al.*, 1998], much of which was attributed to differences in the largest sizes of SSA particles sampled. However, estimates of this cutoff limit (and accompanying inlet losses) are rarely reported. One exception is the data set of *Gras and Ayers* [1983], which consisted only of particles for which r_{amb} was less than ~5 μm; values of M reported in their study appear to be consistently lower than most others (Fig. 17), probably because only smaller particles were sampled. These measurements were made at a tower 94 m above sea level, but it is unlikely that measurement height was responsible for their low value, as little variation in the concentrations of particles of the sizes they measured would be expected between 10 m (typical of the height of other measurements) and the height of the tower (§2.9.5; Fig. 12).

Qualitatively different dependencies of M upon sampling height have been reported in different investigations.

[e] Data for *Bonsang et al.* [1980] taken from Tables 1 and 2 of that reference; Fig. 5a of that reference presented a different number of data, some of which had greatly different values.

[f] Data of *Kulkarni et al.* [1982] shown are 231 values averaged over wind speed from their site M.

[g] Fit is to *Prodi et al.* [1983] 30 min filter data at 13 m; fit to 24 h filter data is similar.

[h] Wind speed used for plotting data and for fit of *Suzuki and Tsunogai* [1988] is the component of the wind speed measured at Mt. Kamui (585 m above mean sea level) along the direction 285° (only non-negative values are presented).

[i] Fit to *Marks* [1990] for 12 m and 18.3 m data is for dry conditions at 12 m; fits for wet conditions at 12 m and wet and dry conditions at 18.3 m are similar.

[j] Fit to *Marks* [1990] for 4.5 m is for dry conditions at 4.5 m; fit for wet conditions at 4.5 m is similar.

[k] NAM (the original Navy Aerosol Model [*Gathman*, 1983a,b]) expression (Tables 14, 15):
$M/(μg\ m^{-3}) = 2.2[U_{10,avg}/(m\ s^{-1}) - 2.2] + 27[U_{10}/(m\ s^{-1}) - 2.2]$, is evaluated for $U_{10,avg} = U_{10}$.

[l] NOVAM (Navy Ocean Vertical Aerosol Model [*Gathman and Davidson*, 1983]) expression (Tables 14, 15):
$M/(μg\ m^{-3}) = 2.2[U_{10,avg}/(m\ s^{-1}) - 2.2] + 2.8exp[0.14U_{10}/(m\ s^{-1})]$, is evaluated for $U_{10,avg} = U_{10}$.

[m] *O'Dowd et al.* [1997a] expression is $M/(μg\ m^{-3}) = 0.11exp[0.22U_{10}/(m\ s^{-1})] + 41exp[0.097U_{10}/(m\ s^{-1})] + 0.65exp[0.16U_{10}/(m\ s^{-1})]$; only the middle term on the right hand side provides any appreciable contribution.

[n] Quadratic relation, $M/(μg\ m^{-3}) = 0.3[U_{10}/(m\ s^{-1})]^2$, results from canonical SSA size distribution of concentration (§2.1.5; §4.1.4.4).

[o] Linear relation is $M/(μg\ m^{-3}) = 3U_{10}/(m\ s^{-1}) - 1$. Shaded band denotes a factor of ≩ 3 above and below this relation.

Table 11. Measurements of Total Dry Sea Salt Aerosol Mass Concentration M, and Fit Parameters to $\ln M = a_M U_{10} + \ln M_0$

Reference	Location	Technique	Sample Duration	$U_{10}/$ (m s^{-1}) Range	$a_M/$ (s m^{-1})[a]	$M_0/$ (μg m^{-3})[a]
Woodcock, 1953	cloud base (640 m) from aircraft flights mostly in Hawaii; high wind data from ~40 m tower at Florida coast	38 samples collected by impaction upon glass slides sized with microscope ($r_{80} > 1.5$ μm)	up to several hours	≤ 35[b]	0.16[c] 0.152[d]	2.6[c] 2.82[d]
Barger & Garrett, 1970	20 m tower on coast in Hawaii	12 filters weighed	1 h to 1 d	1-8.5	0.135	12.2
Tsunogai et al., 1972[f,g]	12 m above sea surface on cruise in Pacific Ocean	189 air samples drawn through distilled water or filter paper analyzed for Na	20-50 h	3-17	0.62[g]	0.33[g]
Blanchard & Syzdek, 1972b	18 m tower on coast in Hawaii	46 samples collected on rotating and stationary wires analyzed for Na	1-2 h	4.5-8		
Savoie & Prospero, 1977[h]	25 m above sea level on Barbados, Sal Island (Cape Verde Islands), and Miami (the latter two 500 m inland)	215 filters analyzed for Na	24 h	2-11		
Lovett, 1978[i]	5, 10, 15 m above sea level on eleven cruises in North Atlantic	1821 filters analyzed for Na	up to several hours	≤ 19[j]	0.16 ± 0.06[i] 0.20[k]	4.3 ⚹ 2.2[i,l] 3.0[k]
Kritz & Rancher, 1980[m]	8 m above sea level on cruise in Gulf of Guinea	24 filters analyzed for Na	12 h	3-9		
Bonsang et al., 1980[n]	Cruises in Indian, Atlantic Oceans	27 filters analyzed for Na	12 h	≤ 18	0.2	2.6 ⚹ 1.6[l]
Horváth et al., 1981[m]	14 m above sea level on a cruise in Pacific, Indian Oceans	22 filters analyzed for Na	~60 h	4.5-16.6		
Kulkarni et al., 1982[o,p]	3 m high 0.25 km inland and 1.2 m high 1.8 km inland from Indian coast	hundreds of filter papers analyzed for conductivity	8-16 h	2-7[o]	0.27[p]	7.2 ⚹ 1.5[l,p]
McDonald et al., 1982	10 m tower at Florida coast, 20 m tower at Enewetak coast (South Pacific)	7 impactor samples analyzed for Na	2-12 d	3.4-9.9		
Gras & Ayers, 1983	94 m tower at Cape Grim, Tasmania	57 size distributions from spectrometer for $r_{amb} < 5$ μm		6-22	0.124	2.6

Table 11. (Continued)

Reference	Location	Technique	Sample Duration	$U_{10}/$ (m s⁻¹) Range	$a_M/$ (s m⁻¹)[a]	$M_0/$ (μg m⁻³)[a]
Prodi et al., 1983[q]	4 m, 13 m above sea level on cruise in Mediterranean Sea, Red Sea, Indian Ocean	24 h filters (13) at 13 m analyzed for Na; 30 min filters (20) at 13 m and 8 impactor samples at 4 m sized under microscope using Cl reaction	30 min and 24 h for filters, 160 s for impactor	2-19	0.13[r] 0.12[s]	5.9[r] 12.2[s]
Exton et al., 1985[t]	10 m tower (14 m above high water mark) and 2 m tower (data not used) on coast of Outer Hebrides Island	calculated from volume concentrations determined from sizes measured using optical detectors, $r_{amb} < 16$ μm		≤ 19	0.16	13.3
Bressan & Lepple, 1985; *Lepple et al.,* 1983	16 m above sea level on cruises in South Atlantic	253 filters analyzed for Na	20 min	1-21	0.093	4.6
Suzuki & Tsunogai, 1988[m,u]	8 m above sea level on coast at Okushiri Island in the Japan Sea	186 filters analyzed for Na	24 h	≤ 25[u]	0.09[u]	13.2[u]
Smith et al., 1989[v]	10 m tower (14 m above high water mark) on coast of Outer Hebrides Island	calculated from volume concentrations determined from sizes measured using optical detectors, $r_{amb} < 23.5$ μm		≤ 35	0.33[w] 0.38[x]	1.6[w] 0.47[x]
Marks, 1990	4.5, 12, 18.3 m above sea level on platform 9 km offshore in North Sea	750 filters from impactor analyzed for Na	0.5-3 h	7-24	0.23[y] 0.59[z]	1.1[y] 0.073[z]
Taylor & Wu, 1992	8.2 m above sea level on pier on South Carolina coast	27 filters analyzed for Na	0.1-0.7 h	2-17	0.15	2.4
McGovern et al., 1994	30 m above sea level (4 m above ground) on Irish coast	weighing of 85 impactor samples for $r_{amb} = 0.5$-5 μm	3-7 d	2-14		
McKay et al., 1994	1.3-1.5 m above ground near high tide line on English coast	8 filters analyzed for Na	2-4 h	1.6-13	0.23	21.1
Chabas & Lefèvre, 2000	10 m above sea level on coast at Delos, Greece	156 filters analyzed for Na	12 h	0-13	0.28	1.7

Table 11. (Continued)

Reported values of mass concentration M are presented as a function of U_{10} in Fig. 17.

Size ranges of particles sampled and inlet cutoff values were typically not stated.

[a] Values of a_M and M_0 are from fits of $\ln M = a_M U_{10} + \ln M_0$. Unless otherwise noted, fits are those presented by investigators to their data.

[b] *Woodcock* [1953] presented data for 6 Beaufort wind forces; see discussion in §4.1.2.1.

[c] Values from *Tsunogai et al.* [1972]; this fit shown in Fig. 17.

[d] Values from *Jaenicke* [1988].

[e] Values determined from fit in Fig. 2 of *Barger and Garrett* [1970].

[f] *Tsunogai et al.* [1972, p. 5286] incorrectly stated that the ordinate of their graph had units µg m^{-3} instead of mg m^{-3} (as correctly stated in the caption of Fig. 5 of that reference). The expression in the text, however, was in µg m^{-3}.

[g] Data of *Tsunogai et al.* [1972] shown in Fig. 17 and upon which the fit is based are 55 values presented in their Fig. 5.

[h] Data of *Savoie and Prospero* [1977] shown in Fig. 17 are from Barbados and Sal Island (Cape Verde Islands); data from Miami were not presented but were stated to be roughly a factor of 3 lower.

[i] Data of *Lovett* [1978] shown in Fig. 17 consist of 65 values from his Fig. 2 representing a single cruise. Fit shown in Fig. 17 is that to data from all cruises.

[j] Highest wind speed not stated; 19 m s^{-1} is the highest value shown in Fig. 2 of *Lovett* [1978].

[k] Values for 65 data shown in Fig. 2 of *Lovett* [1978].

[l] The symbol ⚄ denotes a multiplicative uncertainty in M_0, corresponding to an additive uncertainty (i.e., ±) of one standard deviation in $\ln M_0$, the value of $\ln M$ corresponding to $U_{10} = 0$ when fitted to the form $\ln M = a U_{10} + \ln M_0$.

[m] Na mass concentrations were converted to sea salt mass concentrations using a factor of 3.26 (Table 6); *Kritz and Rancher* [1980] used a factor of 3.07.

[n] Data of *Bonsang et al.* [1980] shown in Fig. 17 are from their Tables 1 and 2; their Fig. 5a presented a different number of data, some of which had very different values. Na mass concentrations were converted to sea salt mass concentrations using a factor of 3.26 (Table 6).

[o] *Kulkarni et al.* [1982] reported the mass concentration of the conductivity-equivalent NaCl; as this is within 2% of the sea salt mass concentration, no correction was made. Their wind speeds refer to 6 m elevation at 1.8 km inland; no corrections were made.

[p] Data and fit of *Kulkarni et al.* [1982] shown in Fig. 17 are for 213 values from their site M.

[q] *Prodi et al.* [1983, p.10963] incorrectly stated their concentration to be in g m^{-3}; the values presented were in µg m^{-3}. Also, they incorrectly stated the fit for data of *Tsunogai et al.* [1972]: 8.01 should be 1.11.

[r] Values stated by *Prodi et al.* [1983] for 24 h filter data.

[s] Values stated by *Prodi et al.* [1983] for 30 min filter data; this is the fit shown in Fig. 17.

[t] Only maritime air mass data used. Data from 2 m tower not used. *Exton et al.* [1985] incorrectly stated units on their Fig. 9 as µm^3 m^{-3}; they should be µm^3 cm^{-3}. *Exton et al.* [1986] gave a slightly different fit but used only some of the data used in *Exton et al.* [1985].

[u] The wind speed used for the data and fit of *Suzuki and Tsunogai* [1988] shown in Fig. 17 is the component of the wind speed measured at Mt. Kamui (585 m above mean sea level) along the direction 285°, this being typical of the angle of the winter monsoon, and the center of the sector open to the sea at the sampling location. Only non-negative wind speed values are shown in Fig. 17.

[v] Only maritime air mass data used. The value of the coefficient a (which is used differently there than it is here) for Size band 2 in Table 2 of *Smith et al.* [1989] was incorrectly stated as 0.404; it should be 0.0404; see *Bigg et al.* [1995, p. 57].

[w] Fit to data with 16 µm < r_{amb} < 23.5 µm.

[x] Fit to data with 8 µm < r_{amb} < 16 µm.

[y] Values presented for *Marks* [1990] for dry conditions at 12 m; values for wet conditions at 12 m and dry and wet conditions at 18.3 m are similar.

[z] Values presented by *Marks* [1990] for dry conditions at 4.5 m; values for wet conditions at 4.5 m are similar.

However, as the effect of height on measured values of M depends strongly on the size range of SSA particles that contribute appreciably to measurements of this quantity, generalizations regarding the height dependence of M cannot be made without knowledge of the upper limit of particle sizes sampled.

Some investigators reported the dependence of M on height from measurements to hundreds of meters or more. Values of M resulting from measurements from ~65 m to over 2000 m over the ocean in the Caribbean reported by *Woodcock and Spencer* [1957] for wind speeds of 5-10 m s^{-1} generally exhibited a decrease with increasing height but sometimes showed a slight increase between ~65 m and ~500 m. Measurements of M reported by *Woodcock* [1962] from widely different geographical locations for wind speeds up to 15 m s^{-1} exhibited a roughly linear decrease with increasing height between ~65 m and ~1500 m, with both the value of M at the surface and the rate of decrease with height increasing with increasing wind speed. Virtually no sea salt mass was detected above heights of ~2500 m at any wind speed. As noted in §2.9.5, *Woodcock et al.* [1971] reported that the ratio of values of M at 50 m and at 550 m, when the cumulus base was near 620 m, was near unity. Vertical distributions of M over the oceans in the Caribbean reported by *Gordon et al.* [1977] exhibited differences from day to day. On some days a nearly uniform profile existed from 76 m to ~400 m, whereas on others there was a decrease of nearly a factor of two in M between 76 m and 229 m; no wind speeds, temperature profiles, or other meteorological data that could have accounted for these differences were reported. Values of M at ~330 m and ~640 m over the ocean in the Caribbean reported by *Maring et al.* [2003] were roughly 25% and 15%, respectively, of those of values measured at 10 m.

Other studies have investigated the variation of M with height within a few tens of meters over the ocean. No significant variation in M between heights of 5 m and 15 m was reported by *Lovett* [1978] from shipboard measurements, whereas shipboard measurements of M by *Prodi et al.* [1983] at 4 m obtained using an impactor were an order of magnitude higher than those at 13 m obtained using filters (although the 4 m measurements exhibited a much lesser dependence on wind speed); the difference was attributed to the difference in heights. Values of M measured by *Blanchard et al.* [1984] from a tower at heights near 14 m and 19 m at a coastal site in Hawaii at wind speeds 5-8 m s^{-1} were a factor of from 2 to 4 higher than those measured at heights of 30 m during aircraft flights from 5 km to 10 km from the coast. In several instances very little decrease in M was observed from 30 m up to nearly 1 km, whereas in others there was a large decrease between 200 m and 400 m. These investigators concluded that most of the decrease probably resulted from SSA particles with $r_{80} \gtrsim 11$ μm that did not attain heights more than about 20 m above the sea surface, although they noted that atmospheric stability played a large role in determining the vertical profile of M. An abrupt decrease in M between 10 m and 30 m at coastal sites in Hawaii was reported also by *Daniels* [1989], but the locations that exhibited the greatest decrease with height also yielded values of M that were clearly affected by surf-produced SSA, as these values ranged from 80 μg m^{-3} to over 200 μg m^{-3} at U_{10} near 10 m s^{-1}—nearly a factor of six greater than those at a beach tens of kilometers away, and much greater than most of those in Fig. 17. Values of M reported by *Daniels* remained fairly constant with height above 30 m up to 130 m at both locations. It is thus likely that the abrupt decrease in M at heights below 30 m that was reported by *Blanchard et al.* and *Daniels* resulted from artifacts of the measurement locations and may have been due in part to contribution from surf spray. Values of M at a platform in the North Sea at 4.5 m reported by *Marks* [1990] were in some instances up to an order of magnitude higher than those at 12 m and 18 m, and they exhibited a stronger dependence on wind speed, although in other instances there was little or no difference in M over these heights. From these studies it may be concluded that although some variation in M may be due to differences in height, this variation cannot account for the vast amount of scatter in Fig. 17. Factors other than wind speed that might affect M have therefore also been examined by several investigators.

Correlation of M with the height of the mixed layer was determined by *Lovett* [1978] not to be significant; because M is controlled by larger particles which are not expected to be mixed over the marine boundary layer height (§2.9.4), this result is not surprising. Both *Lovett* [1978] and *Exton et al.* [1985] reported that they could find no clear dependence

of M on RH. A possible effect of time variation of the wind speed on the relationship between M and U_{10} was suggested both by *Lovett* [1978] based on fits to his data and by *Warner and Hoover* [1985] based on the argument that the mass concentration does not respond instantaneously to changes in wind speed. For a given wind speed at the time of measurement, *Bressan and Lepple* [1985] reported higher values of M (by nearly 50%) during times of decreasing wind speeds than during times of increasing wind speeds. These investigators also noted that although M can increase quite rapidly with increasing wind speed, it decays more slowly, 10-16% per hour, following decrease in wind speed; they attributed this behavior to the relatively long residence time of SSA particles compared to the time characterizing substantial changes in the wind speed. Such results are consistent with conclusions obtained in §2.9.6, and the characteristic (*e*-folding) time of 6-10 h corresponds roughly to values for SSA particles with r_{80} in the range 5-10 μm at $U_{10} = 10$ m s^{-1} (Table 8). An increased dependence on wind speed with higher sea surface temperatures was reported by *Bressan and Lepple* [1985], but as different sea temperatures were associated with different geographical regions (as opposed to seasonal differences at a given location), this increased dependence may result from location-dependent factors. These investigators also reported that for the coldest sea surface temperatures (< 5 °C) there was virtually no correlation between measured values of M and wind speed. There appears to be no significant correlation between sea temperature (based on location of measurement) and values of M in Fig. 17. None of these other factors, whose effects span a factor of only about 1.5, can account for the large variations seen in Fig. 17.

As the scatter in the data from any one set of measurements is generally at least as great as the difference between data sets, the scatter cannot be ascribed entirely, or even largely, to differences in measurement techniques or locations. Much of this variability reflects the variability characteristic of the mass concentration of SSA in the marine atmosphere (although differences in measurement techniques, such as size range sampled or analytic technique, also contribute somewhat to the inter-investigation spread in the data), and must therefore be viewed as a consequence of variation in the production of SSA and/or its removal, as modulated by transport. For example, at a given wind speed a much lower SSA mass concentration would be expected shortly after a rain event than after a prolonged period free of precipitation.

Several data sets in Fig. 17 stand out from the others. In particular, the data set of *Tsunogai et al.* [1972], from the Pacific Ocean, differs markedly from most of the other data sets in Fig. 17, with M exhibiting a much stronger dependence on wind speed and having values up to several orders of magnitude greater at a given wind speed. *Lovett* [1978] suggested that the differences might have been caused by sea spray produced by the motion of the ship, implying that the results of *Tsunogai et al.* were due to measurement artifact. *Gong et al.* [1997a] suggested that the long sampling times (20-50 h) of *Tsunogai et al.* resulted in an overestimation of M and presented a model supporting this contention, but this model yields an overestimation of less than a factor of two, not the several orders of magnitude required to explain the discrepancy. The use of such extremely long sample times (Table 11), over which the wind speed and other meteorological quantities may vary considerably, calls into question the relevance of the reported wind speeds, but still the high values of M and its large wind-speed dependence remain unexplained.

In contrast to the data of *Tsunogai et al.* [1972], the large data set of *Savoie and Prospero* [1977] from Sal Island (Cape Verde Islands) and Barbados exhibits essentially no dependence on wind speed for U_{10} between 2 m s^{-1} and 11 m s^{-1}. The linear correlation coefficient of daily-average values of M and daily-average wind speeds for 148 measurements is $r_{corr} = -0.065$ (i.e., a slight decrease in M with increasing U_{10}), not significant at the 95% confidence level; when six anomalous data were excluded the correlation coefficient increased to $r_{corr} = 0.4$, but such a correlation still accounts for only about 16% ($= r_{corr}^2$) of the variance in the data. *Savoie and Prospero* noted that the most obvious manifestation of this correlation is an increase in the minimum expected value of M with increasing wind speed.

Values of M at the coast in Hawaii at 10 m reported by *Daniels* [1989] did not correlate with U_{10}, perhaps because of the contribution of surf spray, which is not expected to exhibit the same dependence on wind speed as that of spray produced by breaking waves in the open ocean.

The paucity of data at high wind speeds (i.e., $U_{10} \gtrsim$ 20 m s^{-1}) reflects the fact that wind speeds of this magnitude occur infrequently—less than a few percent of the time for most areas of the ocean (§2.3.1). Furthermore, these high wind conditions are often isolated transient occurrences that would likely not represent steady state (§2.3.1); the data of *Woodcock* [1950b, 1953] in Fig. 17 at the highest wind speeds were taken on a single day, as were most of the high wind speed data of *Smith et al.* [1989]. These two data sets deserve comment.

The data presented by *Woodcock* [1953] consist of measurements taken at cloud base (altitude around 640 m) from an aircraft in Hawaii at reported Beaufort wind forces of 1, 3, 4, 5, and 7 (corresponding roughly to 10 m wind speeds of 1, 4, 7, 9, and 16 m s^{-1}, respectively) and four measurements taken at a height of 38 m from a lighthouse at the Florida coast during a hurricane. In the original paper describing the latter set of measurements [*Woodcock*, 1950b], the Beaufort

force was given as 10-11, and the wind speeds corresponding to the four measurements were listed as 27.4, 28, 29.2, and 30 m s^{-1}; subsequently *Woodcock* [1953] presented these same data with the Beaufort force given as 12, which corresponds to $U_{10} = 35$ m s^{-1}. It would be expected that the concentrations reported by *Woodcock* for $U_{10} \leq 16$ m s^{-1}, being at much greater heights, would be less than those from most other data sets, and indeed this appears to be the case. However, any fit to the data of *Woodcock* would be strongly affected by the values at the highest wind speeds, as the other data are at wind speeds of less than half of this value (the fit shown in Fig. 17 to the data of *Woodcock* [1953] was given by *Tsunogai et al.* [1972] based on the data presented in *Woodcock* [1953], which had a stated Beaufort force of 12, or a wind speed of around 35 m s^{-1}, for the high wind speed data; the expression presented by *Erickson et al.* [1986] was based on data presented in *Woodcock* [1950b], which listed wind speeds in the range 27-30 m s^{-1}; and the expression presented by *Jaenicke* [1988], was based on data presented in *Blanchard and Woodcock* [1980], which also showed the high wind speed data as having wind speed 35 m s^{-1}). Thus, it would seem questionable to infer the wind-speed dependence of M (or any other quantity, for that matter) from a combination of data taken near the surface (38 m) during a hurricane and data taken at much greater heights (~ 640 m) under very different conditions.

The data of *Smith et al.* [1989] were obtained at a coastal location in Scotland. Most of the data at the highest wind speeds ($U_{10} \gtrsim 20$ m s^{-1}) were obtained during the passage of a single tropical cyclone. Values of M shown in Fig. 17 for *Smith et al.* were obtained by combining the SSA volume concentrations presented in that paper for $r_{amb} = 0.3$-8 μm and $r_{amb} = 8$-16 μm, and converting to dry mass concentrations using the relative humidities shown in Fig. 6 of that paper. The decrease in the mass concentration with increasing wind speed near $U_{10} = 15$ m s^{-1}, which is not evident in the graphs of volumetric concentration presented in the original paper, arises from the conversion of volume concentration to mass concentration that takes into account the rapid increase in RH from around 70% to around 90% that occurred over a period of several minutes. The diminished size of the range bars (denoting standard deviation) for the very highest wind speeds results from the few data at these wind speeds having been obtained under essentially the same conditions during a single event. *Smith et al.* [1989] suggested that their reported values of M may have been seriously underestimated because of undersampling of larger particles (only particles with $r_{amb} < 23.5$ μm were sampled). The wind-speed dependence of M for $U_{10} >$ 15 m s^{-1} is much weaker than that for lower wind speeds. Although the apparent agreement between this data set and that of *Woodcock* [1953] at wind speeds greater than 30 m s^{-1}

might be considered encouraging, it would seem unwise to place a great deal of confidence in this agreement for the reasons noted. The data of *Marks* [1990] at wind speeds to just over 20 m s^{-1}, which are similar in magnitude to those of *Woodcock* and of *Smith et al.*, exhibit a much greater dependence on wind speed.

4.1.2.2. Parameterizations. Several parameterizations have been proposed relating M and U_{10}. A power law relationship between M and U_{10} in which both the coefficient and the exponent are functions of the height z above the sea surface was proposed by *Blanchard and Woodcock* [1980] for heights from 1 m to 300 m and wind speeds from 3.5 m s^{-1} to 14 m s^{-1}, based on the average value of measurements of *Chaen* [1973] at 1 m, averages of measurements reported by several researchers at heights near 10 m, and measurements of *Woodcock* [1962 and previously unpublished] at heights near 60 m, 175 m, and 300 m. For $z = 10$ m this relation reduces to $M/(\mu g\ m^{-3}) \approx 0.66[U_{10}/(m\ s^{-1})]^{1.64}$ and for $U_{10} = 10$ m s^{-1} it reduces to $M/(\mu g\ m^{-3}) \approx 43(z/m)^{-0.18}$. The latter expression yields $M \approx 28\ \mu g\ m^{-3}$ at $z = 10$ m, and values of M that vary from this by less than a factor of two over heights from 1 m to 300 m. The marked decrease in M between 20 m and 30 m reported by *Blanchard et al.* [1984] and *Daniels* [1989] discussed above (§4.1.2.1) is not captured by this relation. The few data upon which this relation is based, the slight dependence on height it yields, and the extreme scatter in the data reported near 10 m (Fig. 17), raise questions concerning the confidence that can be placed in this parameterization.

Most investigators have fitted their data to a linear relation between the logarithm of the SSA mass concentration and the wind speed:

$$\ln M = a_M U_{10} + \ln M_0 \qquad (4.1\text{-}2)$$

which is equivalent to an exponential dependence of M on wind speed, $M = M_0 \exp(a_M U_{10})$; M_0 is the SSA mass concentration that would be expected to occur at $U_{10} = 0$ according to this expression. This relation was introduced by *Tsunogai et al.* [1972] to compare their results with those of *Toba* [1961a,b], who had fitted his laboratory values of the size-dependent interfacial SSA production flux to an exponential function of wind speed for laboratory wind speeds corresponding to U_{10} from 15 m s^{-1} to 23 m s^{-1} and later [*Toba*, 1965b] concluded that a similar expression described the size-dependent interfacial SSA production flux over the ocean. Eq. 4.1-2 has subsequently become the standard means of describing the dependence of M on wind speed, and the expressions used by various investigators to fit their data shown in Table 11 are all of this form. It should be noted, however, that the coefficients a_M and M_0 resulting

from the (linear) least squares fit of data to (4.1-2), the typical method by which these coefficients are obtained, will in general differ from those resulting from a (nonlinear) least squares fit of the same data to $M = M_0 \exp(a_M U_{10})$, which would tend to result in a greater dependence on wind speed because of greater weighting of data at higher wind speeds. Additionally, employing a least squares fit of $\ln M$ to U_{10} implicitly assumes that for a given wind speed the values (and uncertainties) of M are lognormally distributed, and that the appropriate mean value over many individual measurements is the geometric mean of the different M values [*Limpert et al.*, 2001], an assumption for which there is no physical basis.

Several investigators have reported statistical uncertainties associated with the coefficients a_M and M_0 characterizing the spread of the data about the regression line when fitted to (4.1-2). Analogous to the symbol $\pm$ ("plus or minus") denoting an additive uncertainty, the symbol $⨰$ ("times or divided by") denotes a multiplicative uncertainty. For an additive uncertainty (such as a standard deviation) in $\ln M_0$ of $\pm\varepsilon$, the corresponding multiplicative factor describing the uncertainty in M_0 is $⨰ \exp(\varepsilon)$. The reported multiplicative uncertainties in M_0 in Table 11 range from 1.5 to 2.2, indicative of a spread of the data on both sides of the regression line by such a factor. Such a spread is a measure of the variability of the data after the variation associated wind speed is taken into account. The variation in M at any given wind speed is much greater than this in several of the data sets and, as noted above, greater yet in the data as a whole.

The quantity $\ln M$ was fitted to a linear function of U_{10} by *Bressan and Lepple* [1985], who reported a linear correlation coefficient of $r_{corr} = 0.33$ for 253 data points, implying that around 10% of the variance in $\ln M$ can be attributed to wind speed. They also presented linear correlation coefficients between $\ln M$ and U_{10} and between M and powers of U_{10} when the values of each of the pairs of their data (U_{10}, M) were independently ranked by order, obtaining the highest correlation coefficient for $\ln M$ and U_{10}, from which they claimed that the distribution of $\ln M$ could be determined from distributions of wind speed. However, such a procedure removes any connection between concurrent measurements of M and U_{10} and thus it is difficult to attribute physical significance to these correlations.

A widely used relation proposed by *Erickson et al.* [1986] for the dependence of the SSA mass concentration at 15 m on "local surface wind speed," presumably U_{10}, consists of an expression of the form (4.1-2) with different values of a_M and M_0 in two wind speed ranges. For $U_{10} \leq 15$ m s^{-1}, the expression ($a_M = 0.16$ s m^{-1}, $M_0 = 4.3\ \mu g\ m^{-3}$) is the same as that presented by *Lovett* [1978] for his data (Fig. 17), whereas for $U_{10} > 15$ m s^{-1} the expression ($a_M = 0.13$ s m^{-1},

$M_0 = 6.6$ µg m^{-3}) was based on measurements of *Lovett* [1978], *Lepple et al.* [1983; the same data as given in *Bressan and Lepple*, 1985], and *Woodcock* [1950b]. Little justification can be given for the use of different relationships between M and U_{10} in different wind speed ranges, for several reasons. First, there are few data for $U_{10} > 15$ m s^{-1} upon which the fit for the upper wind speed range was based, most of which comprise only a narrow range of wind speeds; only for the measurements of *Woodcock* discussed above did U_{10} exceed 21 m s^{-1}. Second, the difference in the expressions for the two wind speed ranges is very small; extrapolation of the expression presented for the lower wind speed range to 21 m s^{-1} yields a value that differs from that given by the expression for the upper wind speed range by only 20%, and at 35 m s^{-1} by less than a factor of two. Finally, the scatter in the data at any given wind speed as shown in Fig. 17 is much greater than the factor of 5 increase in M given by these expressions as U_{10} increases from 5 m s^{-1} to 15 m s^{-1}.

Values of M at different locations obtained using a chemical transport model were fitted to the form (4.1-2) by *Gong et al.* [1997a], resulting in a narrow range of values of a_M, 0.20 s m^{-1} to 0.26 s m^{-1}, but a rather wide range of M_0, from 4.1 µg m^{-3} to 22.2 µg m^{-3}. Comparing their model results (their Fig. 7) with fits of several of the other data sets discussed above, *Gong et al.* concluded that although the wind-speed dependencies of M were similar, the substantial variability in the constant M_0 as a function of location suggested possible dependence on salinity or on meteorological conditions other than wind speed, or differences in measurement techniques or sampling times. However, salinity differences are too slight to account for more than a minuscule portion of the large differences (§2.5.1). Moreover, a dependence of M_0 on meteorological conditions other than wind speed would imply that M cannot be accurately described or parameterized by wind speed alone. *Vignati et al.* [2001] also fitted their modeled SSA mass concentrations to (4.1-2), but the wind speed range was not stated, nor were the model results shown to allow determination of how well they were represented by the fit.

Several formulations of the form (4.1-2) proposed for M were reviewed by *Gong et al.* [1997a, 1998] and *Fitzgerald* [1991], but these investigators presented only fits to the data, which fail to convey the extremely large variability in the measured values (Fig. 17), and in some instances the fits were extrapolated far beyond the range of measurements.

Although most investigators have plotted M vs. U_{10} on a linear-logarithmic graph and have fitted the observations to a straight line according to (4.1-2), presenting (and fitting) the data in this manner can hardly be construed as evidence that the data are accurately described by an exponential relation and certainly does not provide evidence of a physical basis for this relationship (as has been often implied). Even under the assumption of a fairly tight relation between M and U_{10} (and Fig. 17 shows this assumption clearly not to be satisfied), it has not been demonstrated that an exponential dependence of M on U_{10} represents the data better than do a variety of alternative expressions; some data sets are in fact better fitted (as characterized, for instance, by the root-mean-squared error or by the linear correlation coefficient) by alternative functional forms such as linear or quadratic relationships between M and U_{10}. For an expression of the form (4.1-2) described by $a_M = 0.2$ s m^{-1}, typical of several of those that have been proposed (Table 11), a linear expression of the form $M = \alpha_M U_{10} + \beta_M$ or a quadratic expression of the form $M = \gamma_M (U_{10})^2$ can be determined that yields values of M that differ from those given by the exponential expression by no more than 25% over the wind speed range 5-15 m s^{-1}. Given the spread in the data, any two relations that agree to within this amount can be considered virtually identical. The relation between M and U_{10} in Fig. 17 (solid black curve) about which the shaded region is drawn, which arguably represents the data as well as any of the other proposed expressions, is in fact a linear one.

Despite the fact that the wind speed dependence of M can be equally well fitted by a variety of functional forms, the supposition that M varies exponentially with U_{10} is pervasive in the literature. In many papers (and some reviews on SSA) this relationship has been treated as if it had been conclusively demonstrated, and it is so commonly accepted that it has sometimes been used even when it clearly does not represent the data at all well (e.g., Fig. 4 of *McKay et al.* [1994]). Furthermore, although an individual data set might perhaps be well represented by an exponential relation (which is rarely demonstrated), the values of a_M for different data sets listed in Table 11 vary greatly, from less than 0.1 s m^{-1} for the data of *Bressan and Lepple* [1985] to 0.59 s m^{-1} for the data of *Marks* [1990] for dry conditions at 4.5 m, and 0.62 s m^{-1} for the data of *Tsunogai et al.* [1972]. The range of values of a_M is even more pronounced when other data are fitted to (4.1-2): the data sets of *Savoie and Prospero* [1977], *Prodi et al.* [1983; impactor data], *McGovern et al.* [1994], and *Kritz and Rancher* [1980] yield values of a_M of 0.02, 0.05, 0.06, and 0.08 s m^{-1}, respectively. As U_{10} increases from 5 m s^{-1} to 15 m s^{-1} these values of a_M yield increases in M by factors of 1.2, 1.6, 1.8, and 2.2, respectively; however, such changes are minuscule compared to the spread of the data. *Lepple et al.* [1983; see also *Bressan and Lepple*, 1985] reported that when they used only data from a 5 day subset of their 16 day data set, the slope a_M of their fit (0.15 s m^{-1}) was nearly twice the value obtained (0.093 s m^{-1}) when using data for the entire data set. Values of a_M reported by *Quinn and Coffman* [1999] for $r_{80} \lesssim 0.6$ µm and for 0.6 µm $\lesssim r_{80} \lesssim 6$ µm for different cruises, mostly in the

Southern Pacific, vary from 0.1 s m^{-1} to 0.4 s m^{-1}, with linear correlation coefficients varying from 0.2 to 0.78. The large variation in the values of a_M that have been obtained further calls into question the validity and the utility of an exponential relationship between M and U_{10}.

There is, of course, no physical basis for any of the relations proposed. Although the linear correlation between ln$M$ and U_{10} may be positive and significant at a given confidence level, such a correlation does not provide any evidence of an exponential dependence of M on wind speed (nor should the existence of a nonzero correlation between any two functions of M and U_{10} be construed as a basis for assuming any functional dependence or physical mechanism relating them), but demonstrates only that M in general increases with increasing U_{10} (examples of vastly different data sets exhibiting very different function forms but possessing the same linear regression lines and linear correlation coefficients are presented in *Anscombe* [1973]). Most reported correlation coefficients are fairly low and thus account for only a small fraction of the variation of M with wind speed. Additionally, there is no *a priori* reason to expect that the same functional form for the wind-speed dependence applies over such a wide range of U_{10} (if indeed there is any reason at all to expect any wind-speed dependence); as whitecapping is minimal for U_{10} below around 5 m s^{-1} (§4.5.1), different wind-speed dependencies of SSA production would be expected for wind speeds lower and higher than this value, and different production mechanisms—bubble bursting (jet and film drops) and wave tearing (spume drops)—may exhibit different wind-speed dependencies. Because transport and removal mechanisms are undoubtedly complicated functions of wind speed and RH in addition to other factors, the possibility that a simple wind-speed dependence (a single exponential, for example) will exist for the mass (or number) concentration is highly unlikely.

Attempts to represent M as a function solely of U_{10} implicitly assume that U_{10} is the primary factor determining M, and that M rapidly responds to changes in U_{10}. However, the low correlation coefficients and the large amount of scatter that exists in the data in Fig. 17, coupled with the fact that the overall increase in the data over the wind speed range 3 m s^{-1} to 15 m s^{-1} is less than the scatter in the data at any one wind speed, make it evident that U_{10} is not the only factor controlling M, and possibly not the dominant one. Indeed, even the fact that $M_0 \neq 0$ implies that M does not instantaneously respond to changes in U_{10}; if it did then M should be zero for $U_{10} = 0$, as no SSA production would occur. Wind speed history and removal history determine what is generated and transported to a location, and much of the correlation of M with wind speed may be due to the correlation between the wind speed at the time of measurement and the wind speed experienced by that air parcel at a previous time when the particles were produced, given the residence times of several hours to a day or so for the size range of SSA particles providing the dominant contribution to M.

4.1.2.3. Global maps.   Global distributions of dry SSA mass concentrations over the ocean have been presented in several studies. Global maps of seasonal (3 month) averages of M were presented by *Erickson et al.* [1986] based on their expressions relating M and U_{10} (discussed above) and on monthly mean wind speeds and standard deviations, with values of M ranging from 10 µg m^{-3} to 45 µg m^{-3}, roughly twice those that would have been obtained using mean wind speeds alone. This range is consistent with the majority of values in Fig. 17. Maps were also presented by *Exton et al.* [1986] of "volumetric ... loadings (µg m^{-3}) ... calculated using $V_L = 0.17 U_{10} + 4.1$" (stated in the caption of Fig. 18 of that reference) for the North Atlantic in January and July, based on monthly wind speed averages[1]. The values presented on the maps, which range from 150 to 600, almost certainly refer instead to volume concentration in units µm^3 cm^{-3}, corresponding to values of M ranging from 35 µg m^{-3} to 135 µg m^{-3}. These values expressed as M are much greater than the great majority of the measured SSA mass concentrations shown in Fig. 17 but are consistent with the measurements reported by *Exton et al.* [1986] on which they were based. A global map of two-year average values of M was presented by *Genthon* [1992b] using wind speeds obtained from a general circulation model and the expression relating M and U_{10} given by *Erickson et al.* [1986]; values are comparable to those presented by *Erickson et al.* [1986]. Global maps of the contribution to the SSA mass concentration from particles with r_{80} in the ranges 0.26-0.50 µm and 8-16 µm were presented by *Gong et al.* [1998] for June based on the results of their global climate model and on the production flux expression of *Monahan et al.* [1986] discussed in §5.3.1. A global map of the average annual SSA mass concentration was presented by *Takemura et al.* [2000] from a general circulation model

---

[1] The caption to Fig. 18 of *Exton et al.* [1986] incorrectly stated the relationship, which was given in the text as ln$V_L = 0.17U_{10} + 4.1$. As the data were taken at an average RH of near 85%, the relationship corresponds roughly to ln[M/(µg m^{-3})] = 0.17U_{10}/(m s^{-1}) + ln(14.3). One of these maps was presented also by *Latham and Smith* [1990], where the caption states that the quantities presented are volumetric loadings in µg m^{-3}, but the values are the same as those in *Exton et al.* [1986], i.e., volume concentrations in units µm^3 cm^{-3}.

based on the expression relating M and U_{10} given by *Erickson et al.* [1986]; values are comparable to those presented by *Erickson et al.* [1986]. Annual average values of M as a function of latitude presented by *Heintzenberg et al.* [2000], based on several data sources, varied from near 7 μg m^{-3} to 15 μg m^{-3} (toward the low end of the range of values comprising the majority of the data in Fig. 17), with the highest values occurring from 45° to 60° S and from 45° to 75° N. Global maps of M for January and July and annual average values were presented by *Grini et al.* [2002] based on a chemical transport model; values over most of the oceans ranged from 10 μg m^{-3} to near 45 μg m^{-3}. A global map of the annual average sea salt mass column burden (i.e., the vertical integral of M) was also presented; values ranged from approximately 0.005-0.045 g m^{-2} over most of the oceans. Global maps of sea salt mass column burden presented by *Chin et al.* [2002] as an annual average for 1990 based on their global climate model, and by *Dobbie et al.* [2003] as three-month averages based on tracer model simulations of others, ranged from approximately 0.01-0.06 g m^{-2} over most of the ocean. Under the assumption that the SSA mass is uniformly mixed over a height of 0.5 km, the ranges for the sea salt mass column burdens of *Grini et al.* and of *Chin et al.* and *Dobbie et al.* would correspond to values of M ranging from 10-90 μg m^{-3} and 20-120 μg m^{-3}, respectively. However, M is expected to decrease with increasing height, resulting in greater values of M at low elevations (i.e., near 10 m); the upper portions of these ranges are greater than the majority of measured values in Fig. 17.

As these maps exhibit similar ranges in M to those in Fig. 17 and are for the most part consistent with the data reported in that figure, they provide little new information. Additionally, they do not capture the large spread of values reported in M for a given wind speed; hence little confidence can be placed in them.

4.1.2.4. Summary. In summary, although in general M increases with increasing U_{10}, there is a vast amount of spread in values of M that have been reported at any one wind speed—over an order of magnitude, even when the data of *Tsunogai et al.* [1972] are not considered—greater than the general increase in M over the wind speed range covered by most of the data. Much of this scatter appears to be intrinsic in the data and cannot be attributed solely to differences in technique. Attempts to quantify the dependence of M on U_{10} and other factors have not been successful. The large amount of scatter calls into question the confidence that can be placed in the use of any of the relations that have been proposed, or for that matter any relation between M and U_{10} for geophysical purposes; for the data sets as a whole, no single parameterization adequately represents the range of values of M that have been reported. Additionally,

the use of any such relationship in models and for descriptive and predictive purposes implies a precision that is difficult to justify. This, coupled with the narrow wind speed range of several of the data sets, the lack of physical basis for any of the proposed relations, and the inevitable influence of prior conditions of M at a given time and location, suggest that further efforts to refine a relation between M and U_{10} would be fruitless.

4.1.3. Sea Salt Aerosol Number Concentration

The total SSA number concentration N is an important property of SSA, influencing the dynamics of the marine aerosol and the microphysics of maritime clouds and precipitation (§2.1.5.1). In view of recent interest in influences of anthropogenic aerosols on marine clouds and precipitation (with concomitant influence on climate change), the number concentration of SSA particles assumes an important role as a property of the background aerosol to which influences of anthropogenic aerosols are to be compared. Number concentration of SSA particles is especially important in consideration of the Twomey effect (§1.1), which depends on the ratio of the number concentration of cloud drops in clouds perturbed by anthropogenic aerosols to that in unperturbed clouds [*Twomey*, 1977b, 1991; *Charlson et al.*, 1992; *Schwarz and Slingo*, 1996].

The SSA number concentration is dominated by small SSA particles, i.e., those with $r_{80} < 1$ μm (Table 2; §4.1.4.3). The abundance of SSA particles relative to all marine aerosol particles in this size range is quite variable, and particles other than sea salt are frequently dominant in this size range (§4.1.1.1), influencing the dynamics of the marine aerosol and complicating attempts to measure N and to correlate it with wind speed and other meteorological factors. The use of physically-based devices such as optical particle counters, impactors, or mobility analyzers, which do not discriminate different types of particles, will often yield values of N that are erroneously high unless they are used in conjunction with techniques such as chemical analysis, electron-microscopy, or thermal volatility that distinguish particles by composition. Moreover, even these techniques will not yield accurate estimates for N if they do not count sufficiently small sizes to include the great majority of SSA particles. Results obtained from different sampling techniques may differ, in part because of differences in minimum cutoff radius. As with M, the value of N at a given time and location is a consequence not only of local SSA production but also of prior history of transport and removal. Because of the long atmospheric residence times of SSA particles in the size range that contributes most greatly to N (days to weeks; §2.7.1; §2.9.6), the correlation between N and local conditions (i.e., wind speed) at the time of measurement is expected to be even less than between M and local conditions.

The partial number concentration δN (i.e., the number concentration within a given size range; §2.1.3) has also often been reported. This quantity is determined from measurements using instruments that classify particles by size. For small intervals of $\delta\log r_{80}$, $\delta N \approx n(r_{80})\delta\log r_{80}$ (§2.1.3), where a single value of r_{80} characterizes the size of all particles in the interval. For larger intervals, however, this situation does not hold. An additional complication is that in many instances the size interval is specified by values not of r_{80} but rather of r_{amb}, which depends on RH as well as r_{80} (§2.1.4.1); conversion to an interval of r_{80} (or $\log r_{80}$) is not possible unless this RH is specified.

4.1.3.1. Measurements. Values reported for the SSA particle number concentration N in the marine atmosphere vary greatly, from well less than 1 cm^{-3} to greater than 200 cm^{-3}. Number concentrations of SSA particles with $r_{80} > 0.15$ μm ranging from as low as 0.02 cm^{-3} to 4.5 cm^{-3} were measured by *Metnieks* [1958] on the west coast of Ireland using a cascade impactor and a spot test for chloride. Number concentrations of SSA particles with $r_{80} \gtrsim 0.15$ μm of 15-20 cm^{-3} were measured by *Woodcock* [1972] by collection on glass slides for ocean waters near Hawaii and Alaska. Number concentrations of SSA particles with $r_{80} \gtrsim 0.06$ μm, determined by electron microscopy of samples collected on filters by *Mészáros and Vissy* [1974], ranged from less than 1 cm^{-3} in the Atlantic Ocean near the equator to nearly 10 cm^{-3} in the Indian Ocean; the differences in concentrations were attributed to differences in weather conditions. Number concentrations of chloride-containing particles with $r_{80} \gtrsim 0.1$ μm determined by chemical analysis of filters, reported by *Prodi at al.* [1983] from shipboard measurements in the Mediterranean, Red Sea, and Indian Ocean, were in the range 1-10 cm^{-3}; the investigators suggested that these values might have been underestimated because of the loss of chlorine from SSA particles (§2.5.2). Values for N measured by *Gras and Ayers* [1983] at Cape Grim, Tasmania, based on electron microscopy of filter samples for particles with $r_{amb} \gtrsim 0.025$ μm, were also in the range 1-10 cm^{-3}. Number concentrations of SSA particles with r_{80} in the range 0.01-3 μm from roughly 2 cm^{-3} to 80 cm^{-3} were reported by *O'Dowd and Smith* [1993], based on thermal volatility techniques, during a cruise in the North Atlantic. Values of N reported by *Kreidenweis et al.* [1998] for Macquarie Island in the Southern Ocean, determined by thermal volatility and electron microscopy, generally ranged from about 50 cm^{-3} to as great as 150 cm^{-3}, and up to 171 cm^{-3} in one instance for wind speed 17 m s^{-1}. Number concentrations of marine aerosol particles comprised predominantly of sea salt reported by *Bates et al.* [1998] from

shipboard measurements in the Southern Ocean ranged from roughly 1 cm^{-3} to 55 cm^{-3}. The number concentration of particles containing sea salt reported by *Murphy et al.* [1998a,b] from measurements at Macquarie Island and Cape Grim, and from shipboard measurements in the Southern Ocean, ranged from ~30 cm^{-3} to greater than 100 cm^{-3}. Values of N determined by *O'Dowd et al.* [1999b] using thermal volatility were near 7 cm^{-3} in the Eastern Pacific and 72 cm^{-3} in the Northeast Atlantic. Values of N determined by *Guazzotti et al.* [2001] in the Indian Ocean using single-particle mass spectrometry ranged from less than 1 cm^{-3} to 5 cm^{-3}. Number concentrations of SSA particles with $r_{80} \gtrsim 0.01$ μm of nearly 250 cm^{-3} were reported by *Clarke et al.* [2003] in Hawaii, also based on a thermal volatility technique.

In contrast to the above values of the total SSA number concentration N, total marine aerosol number concentrations are typically at least several hundred per cm^3 in clean marine conditions [e.g., *Gras and Ayers*, 1983; *Gras*, 1995; *Heintzenberg et al.*, 2000; *O'Dowd et al.*, 2001], and values in continentally-influenced situations can be much greater [e.g., *Hoppel et al.*, 1985, 1990; *van Dingenen et al.*, 1995; *Cover et al.*, 1998]. Hence, as noted above (§4.1.1.1), the relative abundance of SSA particles is typically low compared to all marine particles in terms of number. Values of N rarely exceed 100 cm^{-3}; consequently, the mean distance between SSA particles in the marine atmosphere is typically less than 0.2 cm. As most SSA particles are in the size range $r_{80} \lesssim 1$ μm (§4.1.1.3), the distance characterizing their mean separation is more than a thousand times the distance characterizing their mean size. Such a result supports the assumptions made throughout this review (§2; §2.8) that coagulation of SSA particles is expected to be negligible and that SSA particles do not interact with one another to any appreciable extent.

4.1.3.2. Dependence on wind speed. Numerous investigators have attempted to quantify the dependence of N or δN on local wind speed at the time of measurement, and it has been suggested by *O'Dowd et al.* [1999b] that the correlation between N and U_{10} implies that wind speed can be used as a predictor for SSA number concentration. In this section, investigations on the relation between N or δN and local wind speed are examined, and the extent to which total or partial number concentrations of SSA particles can be determined from wind speed is evaluated.

Various functional forms have been proposed to relate N or δN and U_{10}. A power law in wind speed was used to parameterize δN in each of three size ranges to wind speed by *Exton et al.* [1985], who noted that power laws are also commonly used to express the dependence of whitecap coverage on wind speed (§4.5.2); however they suggested that an exponential relation would be better because it would

Table 12. Measurements of Total Sea Salt Aerosol Number Concentration N or Partial Sea Salt Aerosol Number Concentration δN, and Fit Parameters to $\ln N = a_N U_{10} + \ln N_0$ (and Similarly for δN)

Reference	SSA Only?	Location, Height, Technique	a_N/ (s m^{-1})^a	N_0 or δN_0/ (cm^{-3})^a	Size Range/μm
Woodcock, 1953[b]	Y	Florida coast (~40 m tower), Hawaii (aircraft at ~640 m); impaction	0.085[c]	0.20[c]	$0.9 < r_{80}$
Toba & Chaen, 1973	Y	East China Sea (6 m, shipboard); impaction	0.54[d]	$4.2 \cdot 10^{-5}$ [d]	$8 < r_{80}$
Parungo et al., 1986b	Y	Pacific Ocean (15 m, shipboard); impaction	0.10[c,e]	1.23[c,e]	$0.5 < r_{80}$
Exton et al., 1985[f,g]	N	Scotland coast (tower 14 m above sea level); optical	0.09[c] 0.13 0.22[c]		$r_{amb} = 0.1\text{-}0.3$ $r_{amb} = 0.3\text{-}8$ $r_{amb} = 8\text{-}16$
Exton et al., 1986[f,h]	N	Scotland coast (tower 14 m above sea level); optical	0.12 0.2	33 3.8	$r_{amb} = 0.1\text{-}0.25$ $r_{amb} = 0.25\text{-}16$
Smith et al., 1989[f,i]	N	Scotland coast (tower 14 m above sea level); optical	0.066 0.093[j] 0.23 0.1 0.026 0.061 0.1 0.3 0.39 **0.081[c]**	56 16 0.45 0.54 0.43 0.24 0.021 $4.3 \cdot 10^{-4}$ $1.5 \cdot 10^{-5}$ **69[c]**	$r_{amb} = 0.09\text{-}0.125$ $r_{amb} = 0.125\text{-}0.25$ $r_{amb} = 0.25\text{-}0.5$ $r_{amb} = 0.5\text{-}1$ $r_{amb} = 1\text{-}2$ $r_{amb} = 2\text{-}4$ $r_{amb} = 4\text{-}8$ $r_{amb} = 8\text{-}16$ $r_{amb} = 16\text{-}23.5$ **$r_{amb} = 0.09\text{-}23.5$**
Marks, 1990[g,k]	Y	North Sea (12 m platform); impaction	0.12[c] 0.23 0.24 0.23 0.23 0.23[c]		$r_{amb} = 0.245\text{-}0.475$ $r_{amb} = 0.475\text{-}0.75$ $r_{amb} = 0.75\text{-}1.5$ $r_{amb} = 1.5\text{-}3.6$ $r_{amb} = 3.6\text{-}15$ $r_{amb} = 0.475\text{-}15$
O'Dowd & Smith, 1993[f,l]	Y	Northeast Atlantic (18 m, shipboard); thermal volatility	0.21 0.23 0.23 0.21 **0.22[c]**	0.83 0.42 0.47 0.24[m] **2.0[c]**	$r_{80} = 0.1\text{-}0.2$ $r_{80} = 0.2\text{-}0.3$ $r_{80} = 0.38\text{-}0.84$ $r_{80} = 0.8\text{-}3$ **$0.1 < r_{80}$**
Bigg et al., 1995	N	South Indian Ocean (15 m, shipboard); CN, CCN counters	0.085[c]	6.0[c]	$r_{80} = 0.003\text{-}3$
Gras, 1995	N	Cape Grim, Tasmania coast (104 m above sea level); optical	~0.18[n]	1[o]	mode centered at $r_{amb} = 0.26\text{-}0.3$
Despiau et al., 1996	N	Mediterranean coast, 10 m; optical	0.04-0.1[p]	55	$r_{amb} = 0.25\text{-}5$
Bates et al., 1998	N	South Pacific (18 m, shipboard); aerodynamic	0.26 0.10[d]	1[o] 5.4[d]	$r_{80} = 0.3\text{-}5$
Nilsson et al., 2001	N	Arctic Ocean (12.5 m, shipboard); CN counter	0.23	47	$r_{80} = 0.01\text{-}5$

Table 12. (Continued)

Number concentration N is presented as a function of U_{10} in Fig. 18. Note sums for *Smith et al.* [1989] and for *O'Dowd and Smith* [1993] shown in bold type.

[a] Values of a_N and N_0 are from fits of $\ln N = a_N U_{10} + \ln N_0$ (similarly for δN). Unless otherwise noted, fits are those presented by investigators to their data.

[b] *Woodcock's* lower sampling limit of $r_{99} = 2.4$ μm, corresponding to $r_{80} = 0.9$ μm, was incorrectly stated by *Bigg et al.* [1995] as $r_{dry} = 0.9$ μm.

[c] Fit parameters from *Bigg et al.* [1995].

[d] Fitted by present authors.

[e] Fit based only on data taken south of latitude 19° S.

[f] Only maritime air mass data used.

[g] Based on wind-speed dependence of volume or mass concentration (that is, *Bigg et al.*, 1995, took a_N equal to a_M).

[h] *Exton et al.* [1986] contained only some of the data presented in *Exton et al.* [1985].

[i] Measurements of *Smith et al.* [1989] of partial marine aerosol number concentration δN in several ranges of r_{amb} and their fits are presented as a function of U_{10} in Fig. 19. Values of a_N from fits of $\ln(\delta N) = a_N U_{10} + \ln(\delta N_0)$ reported by *Smith et al.* [1989] are presented as a function of r_{amb} in Fig. 20.

[j] This value was incorrectly stated in *Smith et al.* [1989, Table 2]; see *Bigg et al.* [1995, p. 57].

[k] Values for dry conditions.

[l] Measurements of *O'Dowd and Smith* [1993] of partial SSA number concentration δN in several ranges of r_{80} and their fits are presented as a function of U_{10} in Fig. 21.

[m] Inlet losses reported to be from 15% to 50% over this size range by *O'Dowd and Smith* [1993].

[n] Values were presented by *Gras* [1995] for each of the four seasons.

[o] The value $N_0 = 1$ cm^{-3} was implicitly imposed because the data were fitted to $\ln(N/\text{cm}^{-3}) = a_N U_{10}/(\text{m s}^{-1})$.

[p] Values were presented for low and high winds of continental and marine origin (however, for winds of marine origin particles were of continental origin and for winds of continental origin particles were of marine origin).

allow a nonzero value at zero wind speed. Of course the requirement of a nonzero δN at zero wind speed (if indeed the wind speed is ever truly zero) is hardly a justification for a particular functional form, as this requirement could also be met by a variety of other expressions, such as addition of a constant to a power law, for example. Number concentrations of SSA particles in various size ranges were fitted to both power laws and exponential functions of wind speed by *de Leeuw* [1986b], who concluded, based on similar correlation coefficients for the two types of relations, that the choice between the two should be made on physical arguments (he did not decide on one over the other).

As with M, the wind speed dependence of the total SSA number concentration N has typically been expressed as a linear relation between the logarithm of N and the wind speed:

$$\ln N = a_N U_{10} + \ln N_0, \qquad (4.1\text{-}3)$$

where according to this relationship N_0 is the value of the total SSA number concentration that would occur at $U_{10} = 0$ (likewise for the partial number concentration δN). Eq. 4.1-3, equivalent to $N = N_0 \exp(a_N U_{10})$, has become the standard method of describing the wind-speed dependence on N (and δN), and parameterizatons have typically been of this form (Table 12). As with M, this functional form has virtually assumed the status of a geophysical law, as opposed to being recognized as, at best, an expression of correlation, and the same lack of physical basis for fitting the data in this form applies. The large variation in the values of both a_N and N_0 for these data sets (Table 12) suggests that it is useful to examine the individual data sets from which these values were obtained, the conclusions that have

been reached using these data sets, and the extent to which such conclusions can be supported when the data as a whole are considered.

As both SSA number and mass concentration, N and M respectively (and δN and δM), are often fitted to an exponential dependence on wind speed, examination of the implications of the wind-speed dependencies (characterized by a_N and a_M, respectively) being equal or different is warranted. Both δN and δM can be expressed as integrals over the size distribution (§2.1.4):

$$\delta N(U_{10}) = \int n(r_{80}, U_{10}) \, d\log r_{80}$$

and

$$\delta M(U_{10}) = \frac{4\pi\rho_{ss}}{3\cdot 8} \int r_{80}{}^3 n(r_{80}, U_{10}) \, d\log r_{80},$$

where an assumed dependence of n on U_{10} is explicitly noted (the factor of 8 in the denominator is the cube of r_{80}/r_{dry}). If the limits of integration cover the entire range of r_{80} then these quantities refer to the total quantities $N(U_{10})$ and $M(U_{10})$. By the intermediate value theorem, the ratio $\delta M/\delta N$ is given by

$$\frac{\delta M}{\delta N} = \frac{4\pi\rho_{dry}}{3\cdot 8}(r_{80}^*)^3,$$

where r_{80}^* is some value of r_{80} in the range of integration. If this range is sufficiently small that $r_{80}{}^3$ and $n(r_{80}, U_{10})$ vary little over the range, then r_{80}^* characterizes the size of any SSA particle with r_{80} in the interval, and the ratio $\delta M/\delta N$ is nearly constant, independent of wind speed (or any other variable); this result holds for any wind-speed dependence of $n(r_{80})$. However, for large intervals this conclusion does not in general hold, as concentrations of SSA particles in different size ranges could possess different wind-speed dependencies. Likewise it can be concluded that the quantities N and M should not in general be expected to exhibit the same dependence on wind speed. If it is assumed that the dependencies of N and M on wind speed are given by (4.1-2) and (4.1-3), respectively, then

$$(a_{\mathrm{M}} - a_{\mathrm{N}})U_{10} = \ln\left(\frac{\int r_{80}^3 n(r_{80}, U_{10})\, d\log r_{80}}{\int n(r_{80}, U_{10})\, d\log r_{80}}\right) - \ln\left(\frac{M_0}{N_0}\frac{3\cdot 8}{4\pi\rho_{\mathrm{dry}}}\right).$$

$$(4.1\text{-}4)$$

If $a_{\mathrm{M}} = a_{\mathrm{N}}$ then the shape of the size distribution of concentration $n(r_{80}, U_{10})$ must be independent of wind speed (although the magnitude may still depend on wind speed), and conversely, if the shape of $n(r_{80})$ is independent of wind speed then $a_{\mathrm{M}} = a_{\mathrm{N}}$. Alternatively, if $a_{\mathrm{M}} \neq a_{\mathrm{N}}$, then the shape of the size distribution depends on wind speed; i.e., $n = n(r_{80}, U_{10})$. Specifically, for $a_{\mathrm{M}} > a_{\mathrm{N}}$, the number concentration of larger SSA particles increases with increasing wind speed faster than that of smaller SSA particles.

As an extension of the survey by *Bigg et al.* [1995] assessing the wind-speed dependence of N, measurements of N as a function of wind speed at the time of measurement U_{10} are displayed in Fig. 18 (and summarized in Table 12). *Bigg et al.* considered several data sets [*Woodcock*, 1950b, 1952, 1953; *Exton et al.*, 1985; *Parungo et al.*, 1986b; *Smith et al.*, 1989; *Marks*, 1990; and *O'Dowd and Smith*, 1993] and fits to these data sets of the form (4.1-3). *Bigg et al.* also presented fits to their own shipboard measurements of concentrations of cloud condensation nuclei (CCN) at supersaturations of 0.1% and 0.6% (corresponding to $r_{80} \gtrsim 0.12$ μm and $r_{80} \gtrsim 0.035$ μm, respectively, for SSA particles, according to Eq. 2.1-12) and of condensation nuclei (CN; corresponding to $r_{80} \gtrsim 0.003$ μm for SSA particles) in the Indian Ocean. Several of these fits and others listed in Table 12 are also shown in Fig. 18. *Bigg et al.* argued that for considerations of climate change the dependence of particle number concentrations on wind speed is more important than their absolute magnitudes, and that although particles other than sea salt may have contributed substantially to measured values of N in some instances, the similarity of the values of a_{N} (~ 0.08 s m^{-1}) for most of these data sets was indicative of a wind-speed dependence of SSA concentration. Implicit in

the use of different size ranges in the various studies and in the application of wind-speed dependencies of M to this analysis is the assumption that the shape of the size-dependent SSA number concentration is independent of wind speed, and thus (Eq. 4.1-4) that $a_{\mathrm{N}} = a_{\mathrm{M}}$.

Several concerns may be noted with the arguments of *Bigg et al.* regarding both the extent to which the actual data are represented by expressions of the form (4.1-3) and the data used in their analysis. Regarding the choice of fitting function, for many data sets functional forms other than a linear relation between the logarithm of N and U_{10} represent the data equally well (or poorly). For example, a linear relation $N = \alpha_{\mathrm{N}} U_{10} + \beta_{\mathrm{N}}$ can be found that approximates the fit given by *Bigg et al.* [1995] to their data within 25% for U_{10} from 5 m s^{-1} to 27 m s^{-1}; this fit is shown in Fig. 18.

With respect to the data themselves, numerous concerns can also be raised. Combining *Woodcock's* data at the highest wind speeds with those at lower wind speeds is potentially misleading, as discussed in §4.1.2.1, because the measurements were made under quite different conditions and at very different heights above the sea surface. Additionally, *Woodcock's* data at the lower wind speeds are not well represented by the fit shown in Fig. 18. The value of a_{N} from the fit determined by *Bigg et al.* for the number concentrations of *Woodcock* (0.085 s m^{-1}) is roughly half of that of a_{M} (0.16 s m^{-1}) describing the wind-speed dependence of the mass concentration of these data (Table 11); this would imply that concentrations of larger SSA particles increase more rapidly with increasing U_{10} than do concentrations of smaller SSA particles (which indeed they might). The data of *Woodcock* [1950b, 1952, 1953], who sampled only particles with $r_{80} > 0.9$ μm (and in some instances only particles with $r_{80} > 1.5$ μm), were used by *Toba* [1966] to generate global seasonal maps of SSA particle number for these sizes, both for the sea surface and for a height of 1 km. However, such maps are of little utility because of the limited range of SSA particle sizes, which is expected to include only a small fraction of the number of SSA particles (concentrations of SSA particles of these sizes over most of the oceans were less than 1 cm^{-3}), and because of the large variability both in reported SSA number concentrations from various investigations and in the inferred dependence of N on wind speed.

The data set of *Exton et al.* [1985] consisted of volume concentrations for marine aerosol particles with r_{amb} in three size ranges: 0.1-0.3 μm, 0.3-8 μm, and 8-16 μm. *Exton et al.* [1986] reported most of these same data and presented fits of partial number concentrations of marine aerosol particles in two size ranges of the form (4.1-3), whose sum is shown in Fig. 18. As discussed below with regard to the data of

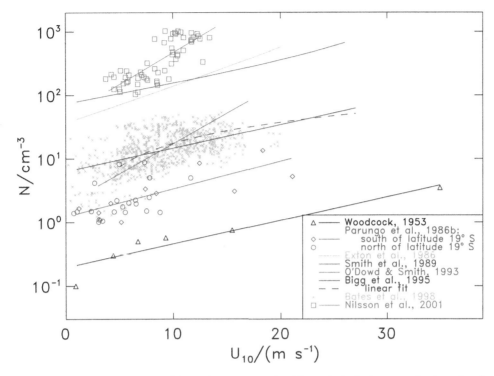

Figure 18. Measurements of marine aerosol or SSA number concentration N, mostly in the lower marine boundary layer, as function of wind speed at 10 m above the sea surface, U_{10}. Symbols denote measurements; solid lines denote fits of $\ln N = a_N U_{10} + \ln N_0$, or sums of fits of $\ln(\delta N) = a_N U_{10} + \ln(\delta N_0)$ over several size ranges, to indicated data sets. Details of the measurements are presented in Table 12. Data of *Woodcock* [1953] consisted only of SSA particles with $r_{80} > 0.9$ µm. Fits to data of *Woodcock* [1953] and *Parungo et al.* [1986b] are from *Bigg et al.* [1995]. Fits to *Exton et al.* [1986], *Smith et al.* [1989], and *O'Dowd and Smith* [1993] are sums of fits by these investigators to partial marine aerosol concentrations for maritime conditions only, covering the size range $r_{amb} = 0.25$-16 µm for *Exton et al.*, $r_{amb} = 0.09$-23 µm for *Smith et al.*, and $r_{80} = 0.1$-3 µm for *O'Dowd and Smith* (reported inlet losses of 15% to 50% for the size range $r_{80} = 0.8$-3 µm were not taken into account, but would have had minimal effect on the result here). Dashed curve denotes a linear fit ($N = \alpha_N U_{10} + \beta_N$) by present authors that closely approximates the logarithmic fit of *Bigg et al.* [1995]. Fit to *Nilsson et al.* [2001] is by those investigators.

Smith et al. [1989] taken at the same location, most of the contribution to the partial number concentration in the smallest size range was probably from non-SSA particles. Considerably different values of a_N were obtained for each size range, indicative of a dependence of the shape of the size distribution of the marine aerosol concentration on wind speed, as discussed above. The data of *Marks* [1990] considered by *Bigg et al.* consisted of measurements of δM for five size ranges of r_{amb}, for which *Bigg et al.* assumed that $a_N = a_M$. However, as discussed above (§4.1.2.2), expressions of the form (4.1-2) do not accurately represent the wind-speed dependence of M (or V), and thus by extension expressions of the form (4.1-3) are not expected to accurately represent the wind-speed dependence of δN. Number concentrations of marine aerosol particles with $r_{80} > 0.5$ µm were reported by *Parungo et al.* [1986b], who stated that

particles in this size range were comprised mostly of sea salt. Only the data for latitudes south of 19° S were fitted by *Bigg et al.* [1995], who argued that these data would provide a more accurate dependence of SSA production on wind speed than the data from the rest of the cruise because of the high winds encountered in the more southerly latitudes. In fact, reported wind speed U_{10} exceeded 8.5 m s^{-1} for only four of these measurements. As well, the data from all latitudes exhibited much greater scatter than those south of 19° S (Fig. 18), raising questions regarding the justification for excluding a subset of the data.

An extensive data set consisting of number concentrations of marine aerosol particles at a location on the western coast of Scotland was reported by *Smith et al.* [1989]. Partial number concentrations of marine aerosol particles were measured using optical detectors in nine size ranges of r_{amb} from

0.09-23.5 μm, for wind speeds up to 26 m s^{-1} for the smallest three size ranges ($r_{amb} < 0.5$ μm) and up to 34 m s^{-1} for the other size ranges, and fits of the form (4.1-3) for each size range were presented[1]. Some of these measurements, averaged over 1 m s^{-1} intervals (which were presented for only four of the nine size ranges), as well as the fits, are shown in Fig. 19. Several features are evident. The partial number concentrations for the different size ranges vary over many orders of magnitude, with the partial concentrations of smaller particles being much more numerous than larger ones; the sum of the fits to the partial number concentrations of the three smallest size ranges ($r_{amb} = 0.09$-0.5 μm) comprises over 98% of the total ($r_{amb} < 23.5$ μm) over the entire range of wind speeds (Fig. 19). However, as the optical detectors employed did not discriminate by composition, particles other than sea salt would also have been detected and counted, especially at the coastal site of the measurements. In fact, it is highly likely that particles other than SSA particles provided the dominant contribution to the measured number concentrations in these size ranges, as the values reported for these size ranges (based on the fits provided) were much greater than those generally reported for partial number concentrations of SSA particles (§4.1.3.1). The presence of very small ($r_{amb} < 0.01$ μm) marine aerosol particles was noted by *Smith et al.* [1989], who suggested they resulted from "malodorous sulphurous emissions from the rotting seaweed" being converted to sulfate particles, which subsequently coagulated and accreted onto existing particles, increasing the number concentrations of the particles with $r_{amb} < 0.5$ μm (the smallest three size ranges). They also noted an increase in the number concentration of the smallest particles during daylight hours, suggestive of the influence of photochemistry. In this context, the observation of *Jennings and O'Dowd* [1990] is pertinent, namely that 85-95% of the volume of marine aerosol particles with $r_{80} < 0.4$ μm at a coastal site roughly 500 km distant from the measurement site of *Smith et al.* consisted of ammonium sulfate or bisulfate. These considerations raise questions as to the pertinence of the lower three size ranges of *Smith et al.* to SSA number concentrations in open ocean conditions and the reported dependence of these concentrations on wind speed.

The partial number concentrations of marine aerosol particles reported by *Smith et al.* [1989] also exhibited very different wind-speed dependencies in different size ranges and wind speed ranges (Fig. 19; here the lines denote fits by *Smith et al.* to the form Eq. 4.1-3). Despite expectation of increasing partial number concentrations with increasing

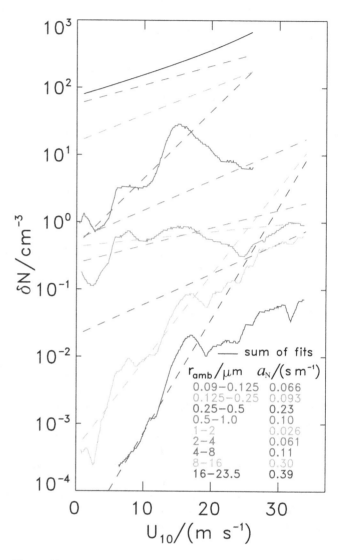

Figure 19. Partial number concentration of marine aerosol δN measured at a location on the Scotland coast 14 m above sea level by *Smith et al.* [1989, Fig. 8, Table 2] for indicated range of ambient radius r_{amb}, as a function of wind speed at 10 m above the sea surface, U_{10}. Individual measurements were binned into 1 m s^{-1} wind speed intervals and averaged. Also shown (dashed lines) are fits by *Smith et al.* of $\ln(\delta N) = a_N U_{10} + \ln(\delta N_0)$ to the several binned data sets; for several radius ranges only the fits were presented, not the measurements. Solid curve at top of figure is sum of fitted expressions.

wind speed, the number concentration of particles in the size range $r_{amb} = 0.25$-0.5 μm in fact decreased with increasing wind speed above 15 m s^{-1}, and the number concentration in the size range $r_{amb} = 2$-4 μm was nearly independent of wind speed from 5 m s^{-1} to 34 m s^{-1}. The number concentrations

---

[1] As noted by *Bigg et al.* [1995, p. 57], the value of $a = 0.404$ given in Table 2 of *Smith et al.* [1989] as the coefficient of U_{10} in the expression for $\log_{10}N$ for the size range 0.125 μm $< r_{amb} < 0.25$ μm should be 0.0404, corresponding to $a_N = 0.093$ s m^{-1}.

for the two largest size ranges (r_{amb} = 8-16 µm and r_{amb} = 16-23.5 µm) exhibited markedly different wind-speed dependencies above and below 15 m s⁻¹; for these size ranges the quantity $d\ln(\delta N)/dU_{10}$ was nearly twice as great for U_{10} less than 15 m s⁻¹ as for U_{10} greater than 15 m s⁻¹. Results such as these not only indicate that expressions of the form (4.1-3) yield poor representation of the data over the entire wind speed range, but also demonstrate that the assumption that the partial number concentration increases at the same exponential rate for all wind speeds can result in great overestimation. For example, the fit of *Smith et al.* to the number concentration of marine aerosol particles in the size range r_{amb} = 0.25-0.5 µm yields a value more than an order of magnitude greater than the measured value at U_{10} = 26 m s⁻¹ (the highest wind speed for which data were obtained), and the fits for the particle number concentrations of marine aerosol particles in the size ranges r_{amb} = 8-16 µm and r_{amb} = 16-23.5 µm yield values more than one and two orders of magnitude, respectively, greater than the measured values at 34 m s⁻¹. Although the fits might be the "best" (in some sense), the relative frequency of occurrence of lower wind speeds is so great that any such fit will be dominated by the data at these low wind speeds. In addition, the values of a_N reported by *Smith et al.*, which are meant to quantify the dependence of δN upon U_{10}, exhibit a non-monotonic dependence on particle size (Table 12; Fig. 20), increasing from a_N = 0.066 s m⁻¹ for r_{amb} = 0.09-0.125 µm to a_N = 0.23 s m⁻¹ for r_{amb} = 0.25-0.5 µm, then decreasing by nearly an order of magnitude to a_N = 0.026 s m⁻¹ for r_{amb} = 1-2 µm (the latter value yields an increase in δN of only 70% for an increase in U_{10} from 10 m s⁻¹ to 30 m s⁻¹), and again increasing more than an order of magnitude to a_N = 0.39 s m⁻¹ for r_{amb} = 16-23.5 µm. Such behavior provides further strong indication that the partial number concentrations for the smaller sizes ($r_{amb} \lesssim$ 1-2 µm) were dominated by particles other than sea salt and calls into questions conclusions based on the assumption that the observed wind-speed dependencies are those of SSA concentrations.

High correlation between partial number concentrations of SSA particles and local wind speed at the time of measurement was exhibited in the data set of *O'Dowd and Smith* [1993], who during a cruise in the Northeast Atlantic employed a thermal analysis technique to separate particles by the temperature at which they volatilize, allowing concentrations of SSA particles to be determined with confidence that the measurements were not influenced by constituents other than sea salt (Fig. 21). *O'Dowd and Smith* fitted their partial SSA number concentration in four different size ranges to the expression (4.1-3), resulting in values of a_N near 0.2 s m⁻¹ and high linear correlation coefficients (r_{corr} near 0.85) between $\ln(\delta N)$ and U_{10} in each size range (Table 12; Fig. 21). Despite the apparent high correlation,

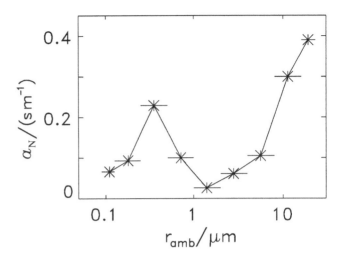

Figure 20. Values of coefficient a_N from fits of $\ln(\delta N)$ = $a_N U_{10} + \ln(\delta N_0)$ to measured values of marine aerosol partial number concentration δN at a location on the Scotland coast 14 m above sea level by *Smith et al.* [1989, Fig. 8, Table 2], as a function of geometric midpoint of the range of ambient radius r_{amb}.

the root-mean-squared uncertainty in $\ln N$ is approximately 0.5 in each size range, corresponding to a multiplicative uncertainty in N of approximately ⨯̸ 1.7.

Size distributions of marine aerosol particle concentration were measured by *Bates et al.* [1998] during a cruise in the South Pacific using a differential mobility analyzer. They reported that based on chemical sampling of the aerosol [*Quinn and Coffman*, 1998] the so-called "coarse mode," containing marine aerosol particles with r_{80} > 0.3 µm, consisted primarily of SSA particles. They reported low correlation between $\ln N$ and wind speed ($r_{corr}^2 \approx 0.32$) and argued that local winds do not necessarily reflect the wind history experienced by the particles, concluding that local instantaneous wind speed is not a good indicator of SSA number concentration in regions of rapidly varying wind speed.

The total number concentration of marine aerosol particles with 0.01 µm < r_{80} < 5 µm over wind speeds of 3.5 m s⁻¹ to 13.5 m s⁻¹ in the Arctic was measured by *Nilsson et al.* [2001], who fitted their measurements to an expression of the form (4.1-3), yielding a_N = 0.23 s m⁻¹ (Fig. 18). The linear correlation coefficient of $\ln N$ and U_{10} for their data was approximately 0.8, and the root-mean-squared error of their fit was approximately ± 0.5 in $\ln N$, corresponding to a multiplicative uncertainty of ⨯̸ 1.7 in N; both the correlation and the uncertainty in $\ln N$ are nearly the same as those found by *O'Dowd and Smith* [1993] to their data, as noted above. As the number concentrations reported by *Nilsson et al.* ranged from roughly 100 cm⁻³ to 1000 cm⁻³, much greater (typically by an order of magnitude or more) than

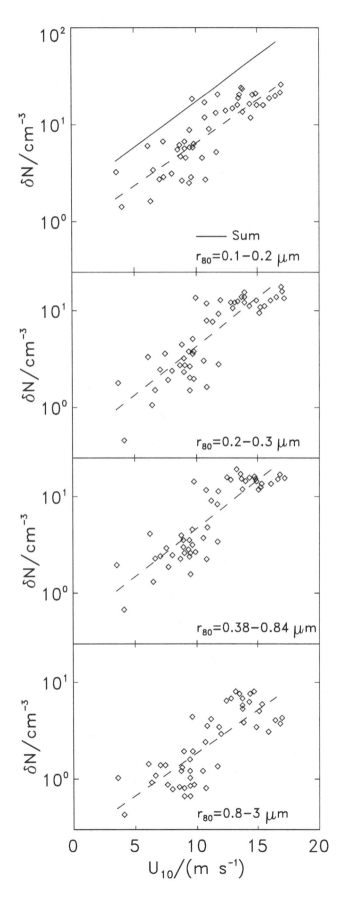

most other reported values of N for SSA particles (§4.1.3.1), it seems likely that particles other than sea salt provided the majority of the contribution to these measurements.

Several other investigators have reported measurements of δN and its wind-speed dependence. Number concentrations of SSA particles with $r_{80} > 5$ μm in various size ranges were fitted to the form (4.1-3) by *de Leeuw* [1986b], resulting in values of a_N ranging from less than 0.1 s m^{-1} to more than 0.4 s m^{-1} and showing no systematic trends with r_{80} or with height from 0.2 m to 11 m. Partial number concentrations of marine aerosol particles with $r_{amb} = 0.25$-5 μm in the Mediterranean for different wind speed ranges and wind directions were fitted to the form (4.1-3) by *Despiau et al.* [1996], with values of a_N ranging from 0.04 s m^{-1} to 0.1 s m^{-1}. However, as there was no speciation of the particles, it is uncertain what fraction of them were sea salt. Additionally, these investigators reported the puzzling conclusion that winds of marine origin resulted in particles mainly of continental origin and vice versa. Number concentration of aerosol particles with $r_{amb} > 0.5$ μm at a Danish coastal site was empirically fitted by *Vignati et al.* [1999b] to an expression of the form (4.1-3) for various fetches, yielding values of a_N of −0.42 (i.e., N decreasing with increasing U_{10}), 0.23, and 0.09 s m^{-1} for short (1.5-2 km), intermediate (2-15 km), and long (>15-20 km) fetches, respectively. The mass concentrations for these samples were low (< 7.5 μg m^{-3}), and the results may have been affected by the proximity to the coast and by the inclusion of particles other than sea salt.

Alternative expressions for the dependence of N upon U_{10} have been proposed. Number concentrations of marine aerosol particles in the "coarse" mode, dominated by sea salt, were fitted to the form $\ln N = a_N U_{10}$ by *Gras* [1995] and by *Bates et al.* [1998]. However, this expression requires that N_0, the SSA number concentration at zero wind speed, be equal to 1 cm^{-3}, an arbitrary requirement for which there is no basis either in theory or in the data themselves. If the data of *Bates et al.* are fitted to (4.1-3) to allow for $\ln(N_0/\text{cm}^{-3})$ to have a value different from zero, then the value of a_N is reduced (Table 12) by more than a factor of 2.5, and the fit (as characterized by the root-mean-squared error) is improved considerably (the data of *Gras* were not presented). *Flamant et al.* [1998], modeling the optical properties of SSA based on measurements reported

Figure 21. Sea salt aerosol partial number concentration δN in the Northeast Atlantic measured shipboard at 18 m above the sea surface for indicated range of radius at 80% RH r_{80} by *O'Dowd and Smith* [1993, Fig. 7], as a function of wind speed at 10 m above the sea surface, U_{10}. Dashed lines are fits of $\ln(\delta N) = a_N U_{10} + \ln(\delta N_0)$ for the several size ranges. Solid curve (top panel) is sum of the several fitted expressions.

in *Blanchard and Woodcock* [1980], *Exton et al.* [1985], and *Smith et al.* [1989], assumed that N exhibited different dependencies on U_{10} in different wind speed ranges: constant for $U_{10} < 3$ m s^{-1}, increasing linearly with U_{10} between 3 m s^{-1} and 10 m s^{-1}, and increasing as $(U_{10})^3$ at higher wind speeds, but there is little justification for this parameterization.

Several investigators have reported little or no correlation of number concentrations of small SSA particles (which contribute most to N) with wind speed. The number concentration of SSA particles with $r_{80} > 0.2$ μm at Cape Grim was reported by *Covert et al.* [1998] to be nearly independent of local wind speed. Likewise, the number concentration of SSA particles with $r_{80} = 0.15$ μm in the South Pacific was found by *Berg et al.* [1998] to be poorly correlated with local wind speed ($r_{corr} = 0.08$) for wind speeds up to at least 20 m s^{-1}. The low correlation between number concentrations of SSA particles with $r_{80} > 0.3$ μm in the South Pacific and wind speed reported by *Bates et al.* [1998] has already been noted. Little dependence of the mass concentration of SSA particles with $r_{amb} < 0.5$ μm with wind speed (averaged over the sampling time of 6 to 15 hours) in the Mediterranean was reported by *Sellegri et al.* [2001]. Findings such as these raise questions regarding generalization of the wind-speed dependence of SSA number concentration from one data set to other locations and conditions. These findings also make it difficult to accept the conclusion of *Bigg et al.* [1995] that the number concentration of SSA particles in the marine boundary layer can be characterized by a single wind-speed dependence.

4.1.3.3. Dependence on other factors. The possible dependence of N on factors other than wind speed has been considered by several investigators, but no strong controlling factors have been identified. Relative humidity was found to have virtually no effect on measured values of marine aerosol number concentration in most circumstances by *Exton et al.* [1985, 1986], although they noted that it can affect the partial number concentration of particles in a given ambient size range as smaller particles which are more numerous "grow" into a given size range with increasing RH. The effect of mixing layer height might also affect N; if SSA particles are uniformly mixed over a certain height, as is assumed for smaller SSA particles, then their concentration should depend not only on their production flux at the surface but also on the volume of air through which they are dispersed, described in terms of a mixing layer height [*Exton et al.*, 1985, 1986; *Park and Exton*, 1986; *Park et al.*, 1990; *Bigg et al.*, 1995]. Two mixing heights were considered by *Park et al.* [1990]: a convective mixing layer height, defined as the height to which a parcel of air at sea surface temperature and local RH would rise under its own buoyancy, and a turbulent (or "mechanical") mixing layer height, calculated

from local wind speed and the local roughness length (§2.4); the larger of the two mixing heights being the relevant one. Measurements of number concentrations of marine aerosol particles presented by these researchers generally exhibited a decrease with increased mixing layer height, but it was noted that the relation between these concentrations and mixing height may have been obscured because factors such as wind speed or atmospheric stability can affect both the dilution of the aerosol once produced (as represented by the mixing layer height) and the production and subsequent entrainment of the aerosol particles upward. Other factors certainly affect N; specifically, the time since the air parcel containing the aerosol last experienced a precipitation event of any appreciable magnitude and duration is expected to be a dominant factor, as N will continue to increase for small SSA particles (which provide the dominant contribution to N) until these particles are removed by wet deposition (§2.7.1). Much of the spread in reported values of N may be attributable to differences in this time, but evidence for this possibility is lacking, as this time is difficult to determine and has rarely been reported.

4.1.3.4. Implications for climate change. Based on correlations between partial number concentration and wind speed reported by *Smith et al.* [1989], *Latham and Smith* [1990] proposed a negative feedback mechanism on climate change resulting from an increase in N that might be expected due to an increase in wind speed associated with greenhouse warming. They argued that a doubling of the mean wind speed U_{10} from 10 m s^{-1} to 20 m s^{-1} would result in an increase in N by an order of magnitude, and that an increase in U_{10} to 30 m s^{-1} would result in an increase in N by two orders of magnitude, with resultant cooling influence on climate due to enhancement of cloud reflectivity by SSA. Such a doubling or tripling of the climatological mean wind speed would in itself have a major climatological impact, with resultant changes on the global budgets of energy, angular momentum, and wind stress, increasing the kinetic energy of the wind over the ocean by a factor of four or eight and increasing the wind stress by one to two orders of magnitude. Such perturbations can hardly be considered minor and extend far beyond the range for which extrapolation of previously determined trends can be justified. Aside from these issues, there are also major difficulties associated with *Latham and Smith's* hypothesis as it relates to SSA. They argued that the most important range for SSA particles that act as cloud condensation nuclei is $r_{amb} = 0.25$-0.5 μm, for which the wind-speed dependence is described by $a_N = 0.23$ s m^{-1} (Table 12). However, the value of a_N for this size range given by *Smith et al.* [1989] was much greater than that for any other size range in their study except the largest two ($r_{amb} > 8$ μm) and certainly not typical for small particles

(Fig. 20; Table 12); the values of a_N given for the next smaller and next larger size ranges were 0.093 s m^{-1} and 0.10 s m^{-1}, respectively—less than half of the value used by *Latham and Smith*. In addition, and more importantly, the assumptions that the particles measured by *Smith et al.* in the size range $r_{amb} = 0.25$-0.5 μm were predominantly SSA particles and that the wind-speed dependence of the concentration of these particles is typical of the wind-speed dependence of concentrations of SSA particles over the open ocean must be questioned for the reasons noted above. This concern holds all the more in view of the possible influence of the emissions from rotting seaweed noted by *Smith et al.* and the results of *Jennings and O'Dowd* [1990] indicating that the predominant aerosol component may have been ammonium sulfate or bisulfate. Finally, the assumption that the exponential dependence of partial number concentration on wind speed for the size range $r_{amb} = 0.25$-0.5 μm extends to wind speeds up to 30 m s^{-1} is not supported by the data of *Smith et al.* [1989], which actually exhibited a decrease in partial number concentration with increasing wind speed for $U_{10} > 15$ m s^{-1} (Fig. 19). If the trend of the data for $U_{10} > 15$ m s^{-1} continued to $U_{10} = 30$ m s^{-1} and applied to N, then the value of N at this wind speed would hardly be greater than that at $U_{10} = 10$ m s^{-1}, and certainly not two orders of magnitude greater, as assumed by *Latham and Smith*. When all of these concerns are taken into consideration, it is difficult to place much credence in this hypothesis.

The effect of a global increase in wind speed on the concentration of cloud condensation nuclei (CCN) was examined also by *Bigg et al.* [1995], who concluded that an increase of 10% in the mean oceanic wind speed (which they took to be 5 m s^{-1}) would result in an increase of 4.4% in the CCN concentration, under the assumption that the relationship between CCN concentration and wind speed is unaffected by other factors. The effect of increased wind speed on SSA mass loading was examined also by *Dobbie et al.* [2003], who concluded, based on the expression of *Erickson et al.* [1986] (§4.1.2.2), that an increase of 20% in the mean surface wind (which they took to be 6 m s^{-1}) would result in an approximately 20% increase in M, and a corresponding 20% increase in the SSA optical thickness and influence on climate by light scattering (§2.1.5.3). While the premises of these hypotheses are much less extreme than those of *Latham and Smith* [1990], still, given the wide range of values that have been reported for N (and M) and for the wind-speed dependencies of these quantities, it is difficult to place much confidence in such projections.

4.1.3.5. Summary. Accurate determination of SSA number concentration N in the marine boundary layer is much more difficult than of SSA mass concentration M because N is dominated by small SSA particles (with $r_{80} \lesssim 1$ μm), with the resultant likely contribution of abundant long-lived non-SSA particles of continental and marine origin to measured total particle number concentrations. Even when care is taken to ensure that measured particles are predominantly SSA particles, determination of correlation of N with wind speed U_{10} at the time and location of measurement is much more problematic than for M because of the long residence times of the small particles dominating N in contrast to the shorter residence times of larger particles dominating M, resulting in lower expected correlation of N with conditions of production than for M.

Correlations between ln$N$ and U_{10}, while often accounting for an appreciable fraction of the variance in measured values of N at various wind speeds, result in expressions for N that typically exhibit departures from the observations by factors of 2-3 or more in a given data set, and by substantially more when compared to data of other investigators. The expectation that the total SSA number concentration N or the partial SSA number concentration δN is well represented by an exponential dependence on wind speed with constant parameters over the entire range of wind speeds has been shown for several data sets not to be supported by the data, with departures from observed number concentrations by an order of magnitude or more. Moreover, it is difficult to confidently conclude that any given parameterization will accurately describe the wind-speed dependence of N and can be used as a predictor of N from wind speed, and all the more as an indicator of SSA production and its dependence on wind speed.

4.1.4. Sea Salt Aerosol Concentration Size Distributions

The size distribution of SSA number concentration (§2.1.4), the number of SSA particles per unit volume within a narrow interval of some well defined variable characterizing particle size (such as r_{80} or the dry particle mass m_{dry}), constitutes a complete description of this aerosol. This is a consequence of the assumption, made throughout this review, that the solute composition of all SSA particles, independent of their size, is that of seawater (§2.5.2) which is constant over the oceans (§2.5.1). Knowledge of this size distribution permits examination of the relative and absolute contributions of SSA particles of different sizes to SSA properties and processes of interest. The size distribution of SSA concentration at a time and location of observation, like total SSA mass and number concentrations, is a consequence of prior production, removal, and transport processes, all of which are size-dependent; to the extent that these transport and removal processes can be accurately modeled, the size distribution of SSA concentration yields information on size-dependent SSA production. This is the basis for determination of the size-dependent effective SSA production flux using the steady state dry deposition method (§3.1) and the statistical wet deposition method (§3.9). Comparison of measured and modeled size distributions of SSA concentration is used to evaluate the ability of chemical transport models to represent production, transport, and removal of this nearly conservative atmospheric tracer (§6). In this section measurements of size-dependent SSA concentration and interpretations of such measurements are examined.

4.1.4.1. Determination of size distributions of SSA concentration.   The determination of SSA concentration size distributions involves identifying particles as being SSA particles, classifying these particles into size ranges, and measuring the amount of some property of the aerosol (such as number concentration or mass concentration) in each size range. Correlation of size distributions of SSA concentration with meteorological parameters further requires determining these size distributions under a variety of conditions. In the limit of a large number of classes of small size range the distribution approaches a continuous function of the radius variable, here r_{80}. For size distributions determined as $n(r_{80}) \equiv dN/d\log r_{80}$, SSA particles are classified into intervals of $\log r_{80}$ and the number in each size range is counted. Most reported size distributions of SSA concentration, regardless of the representation in which they were reported, were originally measured by counting the number of SSA particles in different size intervals (and thus were determined in a representation such as dN/dr, $dN/d\ln r$, $dN/d\log r$, dN/dD, dN/dm_{dry}, or $dN/d\log m_{dry}$, where D is the particle diameter and m_{dry} is the dry particle mass) and then converted by calculation to the representation in which they were presented (such as dV/dr, $dV/d\ln r$, or a cumulative distribution).

Although in principle determination of size distributions of SSA concentration is straightforward, there are numerous nontrivial concerns associated with the conduct and interpretation of such measurements, involving sampling of marine aerosol particles, identification of particles as SSA particles, the determination of SSA particle size, and the enumeration of SSA particles in a given size range, that limit the accuracy of any such determination. Additionally, although conversion of a size distribution of SSA concentration from one representation to another is ostensibly straightforward (§2.1.4.1), even such conversion can introduce ambiguities and errors because of failure to explicitly report the representation used to describe the size distribution of the concentration (§2.1.4) and the quantity used to describe particle size. Not infrequently the representation has not been clearly specified, and in some instances the

representation has been ambiguously stated as "the number concentration," which could refer to dN/dr, $dN/d\log r$, $dN/d\ln r$, or $dN/d\log m_{dry}$, to name but a few possibilities (similar considerations regarding size distributions of SSA production fluxes were noted by *Andreas et al.* [2001]). This situation has sometimes led to misinterpretation by later investigators; examples are noted in several places below (§4.1.4.2; §4.2). In many instances it is difficult to determine even what is meant by the "size" of a particle. Statements such as "the size of the particle was 1 μm" or "the particle was larger than 1 μm" are prevalent, as are references to "submicron" particles, for instance, and it is often not clear if these sizes refers to radius or diameter. Additionally, it is often not clearly specified whether this radius or diameter refers to the value at formation, at a reference RH such as 80%, in equilibrium with the ambient RH or with the RH of the sampling device, or of the dry particle (as would be the situation for a particle under an electron microscope). Incorrect assumptions regarding to which of these a reported value of r refers can lead to large errors not only in the inference of r_{80} (or other measures of size) by subsequent investigators, resulting in incorrect location of size distributions of concentration in r_{80} space (i.e., their horizontal location in a graph of $n(r_{80})$), but also in the magnitude of the concentration when converted from one representation to another (i.e., the vertical location of individual measurements and the shape of the size distribution).

Even if the size characterizing SSA particles and the representation used for the size distribution of SSA concentration are clearly specified, conversion of a size distribution to the representation $n(r_{80})$ can still introduce uncertainty for several reasons. The finite width of the size intervals limits the accuracy with which a size distribution of SSA concentration can be specified because of the variation of SSA particle properties over the interval; for instance, the volumes and masses of SSA particles vary by nearly an order of magnitude over a size bin that encompasses a range of radii of a factor of 2 ($\Delta \log r_{80} = 0.3$), and there can be a large difference between the results obtained by interpolating dV/dr between two radii before converting to the representation $n(r_{80})$ and converting dV/dr to the representation $n(r_{80})$ and then interpolating between two radii. Additionally, although the endpoints of radius intervals are readily transformed between r_{80} and $\log r_{80}$, the midpoint of an interval specified in terms of r_{80} (i.e., the arithmetic mean of the endpoints) does not transform to the midpoint of the corresponding interval specified in terms of $\log r_{80}$ (the geometric mean of the endpoints), the difference between these values increasing with increasing interval width Δr_{80} and leading to some ambiguity as to the best value to use to represent the location of the interval. Further concerns arise when size intervals are specified by a single inequality; for instance, the interval defined by $r_{80} < 1$ μm presents no problems for a size distribution in the representation $dN(r_{80})/dr_{80}$, but transforming this interval to the representation $n(r_{80}) \equiv dN(r_{80})/d\log r_{80}$ necessitates an estimate of the lower bound on r_{80} that in turn affects not only the value of r_{80} used to represent this size range but also the value of $n(r_{80})$ if the number concentration of SSA particles in the interval is to be conserved.

Several methods are commonly used to determine sizes of SSA particles and the number of SSA particles in a given size range [determination of particle size is reviewed in *Reid et al.*, 2003a]. Particles may be collected onto a surface by impaction and later sized and counted, using microscopy or chemical methods, for example. In some such techniques the size of a particle is determined by the size of the crater it leaves, yielding the geometric radius of the particle at the time of impaction; these methods require knowledge of the ambient RH to convert geometric radii to r_{80} values. Chemical methods typically determine the amount of chloride in the particle by the size of the spot left by the reaction of the particle residue with a reagent, requiring calibration with particles containing a known amount of chloride. Sizes of SSA particles that may have lost some of their chloride may be underestimated by such methods; this is of concern especially for small SSA particles (§2.5.2).

Aerodynamic particle sizers and impactors classify particles by their aerodynamic radius (which for a spherical particle in the Stokes regime is greater than the geometric radius by the square root of particle density in units g cm^{-3}; §2.6.2). Aerodynamic particle sizes typically classify particles into numerous size ranges of small width, whereas impactors typically consist of several stages of fairly large size intervals, from which size distributions of concentration can be determined by measurement of the mass on each stage (or of the mass of Na, making the measurement specific to SSA particles). However, inversion of impactor data, necessary to obtain size distributions of concentration, requires knowledge of the efficiency and cutoff radius of each stage of the impactor, which in turn requires careful calibration and knowledge of particle density (for nonspherical particles, such as would be expected for dry particles, a shape factor, requiring knowledge of particle shape, is also necessary). Such a procedure additionally requires some assumption of the shape of the size distribution or of radii to assign to each of the stages. The several stages are typically characterized by 50% cutoff radii, implying that half of the number of particles with radii equal to one of these values are collected by that stage and half pass to the next stage. Depending on the sharpness of the cutoff, much of the mass on any given stage may be contributed by particles larger than the cutoff radius. As

these cutoff radii typically increase by a factor of two between stages, accurate determination of size distributions of concentration is difficult. A comparison of three types of impactors [*Howell et al.*, 1998] revealed differences of a factor of two or greater for mass concentrations of SSA particles in a given size range.

Differential mobility analyzers (DMAs) utilize the ability of charged particles to move across the mean flow in an applied electrical field to classify particles by their mobility radius, which for a spherical particle is equal to the geometric radius in the Stokes range (§2.6.2). Size distributions of concentration are calculated under assumptions of charging efficiency. However, such methods are not specific to SSA unless used in conjunction with other techniques such as thermal volatility, and typically the operating range of DMAs does not extend to sufficiently large particles to provide entire size distributions of SSA concentration.

Optical methods of sizing particles utilize the wavelength-dependent light-scattering properties of SSA particles (§2.1.5.2), and most such determinations of particle size rest on the assumptions that the particles are spherical and homogeneous (neither of which is an accurate assumption for a dry SSA particle; §2.5.5), and thus that the light scattering by an SSA particle for a given wavelength depends only on its geometric radius and its index of refraction (§2.1.5.3), both of which are determined by its value of r_{80} and RH. However, calibration is usually performed using particles with an index of refraction that differs from that of SSA particles at the RH of the measurement, and sensitivity to index of refraction can result in errors in radius of up to 50% [*Liu and Daum*, 2000]. Even optical detectors that are calibrated can yield very different results when compared against each other, as discussed below. Additionally, many optical techniques suffer from so-called Mie resonances [*Pinnick and Auvermann*, 1979], leading to ambiguities in the size of a particle determined by light scattering, resulting from the fact that the light-scattering cross section is not a monotonic function of the geometric radius of the particle (§2.1.5.2). The resulting multi-valuedness of the light-scattering cross section with respect to geometric radius often occurs in the size range of interest (radii from several tenths of a micrometer to several micrometers) and thus results in uncertainties in the inversion of light-scattering data to size distributions of concentration. Approaches to deal with the latter problem range from ignoring it, to using larger size intervals (which reduces instrument resolution), to rather complex procedures. The method proposed by *Schacher et al.* [1981a], for example, which has been widely used by other investigators, involves rejecting data in bins affected by such ambiguities and smoothing the resulting data by fitting them to a 7th order polynomial in log (dN/dr) vs. log$r$, with two fictitious points added at log(r/μm) equal to

±1.5 (i.e., at 0.032 μm and 32 μm). Such a procedure not only results in problems at the extreme ranges of the data but also raises questions regarding the accuracy of the smoothed fit. The lack of specificity of optical techniques to sea salt can result in inclusion of particles other than SSA particles; this is especially problematic for smaller sizes ($r_{80} \lesssim 1$ μm) because of the prevalence of particles of other types in this size range (§4.1.1.1). Additionally, difficulties may occur in measurements of concentrations of larger particles (and thus values of M determined from these) when using optical detectors at coastal locations because of pollen, insects, saltation of sand particles, and other non-SSA particles [e.g., *Park et al.*, 1990].

Once SSA particles are classified by size, the size distribution of number concentration is obtained typically by counting the number of SSA particles in a known volume of air in each of several given size ranges. The accuracy of such a measurement can be limited by several factors (§4.1.1.2). Sampling and retention efficiencies different from unity resulting from inlet and internal losses, non-isokinetic sampling, and bounce, can greatly affect measured size distributions. Impaction efficiencies generally depend strongly on particle size and decrease with decreasing particle size, often restricting the range of SSA particle radii that can be measured. Fracture of dry particles during impaction can of course lead to incorrect determination of the number of SSA particles in a given size range.

Determination of size distributions of SSA concentration is affected by relative humidity in several ways because of the effects of RH on the sizes and behavior of SSA particles (§2.5.3, §2.5.4, §2.5.5). Knowledge of ambient RH at the time of sampling is essential to relate the geometric size of an SSA particle to its value of r_{80} and to convert among different representations of size distributions (e.g., from dV/dr to $n(r_{80})$ or vice versa), as discussed below (§4.1.4.2). However, errors can arise even if the ambient RH is accurately known. Some measurements involve heating the aerosol before sampling, causing the RH to differ from its ambient value; as the time required for an SSA particle to attain its new equilibrium radius in these situations depends strongly on its value of r_{80} and on sampling conditions (§2.5.4), the extent to which equilibration has been attained may also depend on particle size, complicating inference of r_{80} from measured radii. Exposing SSA to RH values below the efflorescence value of sea salt (~45% RH) can result in dry particles that are not spherical and not necessarily homogeneous or isotropic and which thus exhibit light scattering and aerodynamic properties different from those of liquid drops (§2.5.5). Such particles can also suffer rebound or fracture upon impaction with sampling devices or walls of sampling tubes. Exposing the aerosol to RH between the efflorescence and deliquescence values for SSA particles

(~45% and ~75% RH, respectively), or near one of these values, can result in ambiguity (§2.5.5), as both wet and dry particles of different r_{80} can possess the same geometric size or can behave similarly (i.e., have similar responses to a measurement technique), leading to them being treated as if they had the same value of r_{80}.

Accurate determination of size distributions of SSA concentration is especially problematic for large SSA particles. Because of the long times required for equilibration with respect to ambient RH, large SSA particles present anywhere in the marine atmosphere, and sufficiently large medium SSA particles within a few meters of the sea surface which have been recently produced, will probably not have equilibrated to the ambient RH. Consequently, measured radii for these particles will not reflect their equilibrium radii but instead may be closer to formation values (§2.9.1), resulting in overestimation of r_{80} by a factor of two or more, depending on RH, and, in turn, in incorrect size distributions of SSA concentration in the representation $n(r_{80})$. Measured radii of SSA particles that are less than ~25 μm at heights greater than a few meters above the sea surface are probably near equilibrium radii, whereas measured radii at these heights that are greater than ~50 μm are probably closer to formation values; measured radii between ~25 μm and ~50 μm probably correspond to a narrow range of r_{80} near 25 μm (§2.9.1; Fig. 11). Thus, for a size distribution of SSA concentration for which all measured radii are less than ~25 μm, the conversion of particle sizes to r_{80} is unambiguous, as the measured radii are very near equilibrium values, and determination of the size distribution of concentration for these particles is also unambiguous. Similarly, for a size distribution for which all measured radii are greater than ~50 μm conversion of particle sizes to r_{80} values is also unambiguous, as these radii are very near formation values; determination of the size distribution of concentration is likewise unambiguous in this situation. In contrast to these two situations, determination of size distributions of SSA concentration that include measured radii totally within the range ~25 μm to ~50 μm is problematic. As all measured radii within this range correspond to a narrow range of r_{80} near 25 μm, the value of $d\log r_{80}$ is difficult to determine, leading to corresponding uncertainty in evaluation of $n(r_{80}) \equiv dN/d\log r_{80}$. Additionally, as $\Delta\log r_{80}$ will be much smaller than $\Delta\log r_{\text{meas}}$, actual values of $n(r_{80})$ will be considerably greater than reported values. Likewise, determination of size distributions of SSA concentration that include measured radii both within the range 25-50 μm and outside this range is also problematic, as no single transformation can be applied to convert these measured radii to r_{80} values.

The assumption that all measured radii correspond to equilibrium values with respect to the ambient RH leads to overestimation of r_{80} for SSA particles with measured radii greater than ~50 μm (§2.9.1). This assumption will typically result in size distributions of concentration that are skewed and that decrease with increasing radius less rapidly than the actual size distributions. Size distributions of the concentration of SSA particles with measured radii greater than ~50 μm originally presented in the representation $n(r_{80})$ under the assumption of equilibration will have the correct shape (i.e., size dependence), but will be shifted horizontally to erroneously high values of r_{80}. As size distributions of SSA number concentration typically decrease with increasing radius for large SSA particles, this situation would result in overestimation of number concentration, and much greater overestimation of surface area and mass concentrations, of such particles. These issues do not seem to be widely appreciated despite numerous investigators having reported concentration size distributions of SSA particles with apparent r_{80} ranging up to several tens of micrometers or more. Conversely, the assumption that all measured radii correspond to their formation values results in underestimation of r_{80} for SSA particles with measured radii smaller than this range (§2.9.1), and resultant errors in the conversion of the concentration size distribution to other representations.

Even if these problems are resolved, uncertainty can still arise from counting statistics if few particles are sampled in a given size range for a given set of conditions such as wind speed. This concern is especially pertinent to larger SSA particles because of their low concentrations, and it therefore has important consequences in consideration of concentrations of SSA area or volume (or mass), to which the dominant contribution often comes from these larger particles. Possible fixes include increasing the sampling volume or flow rate to allow more particles to be counted, but this may not be possible because of the measurement device or techniques; some optical detectors have a fixed sampling area through which particles must pass to be sized or detected, and impactors generally have a fixed sampling rate. Similarly, increasing the sampling time is often not possible for a given set of conditions (such as a given wind speed), as these conditions may occur only for limited times. Increasing the size intervals into which the particles are sorted results in the loss of fine structure and hence lower resolution of the size dependence of the concentration. Furthermore, this may not help solve the problem at hand; for example, extending the range of interest from $r_{80} > 10$ μm to $r_{80} > 5$ μm provides little additional information on the concentration of SSA particles in the original size range. Likewise, classifying the parameters describing meteorological conditions into less restrictive ranges (such as wider wind speed intervals) yields more particles in each size range, but obscures the dependencies of

concentrations on these parameters and hinders attempts to quantify them.

Estimates of both measurement uncertainty and of variability under ostensibly the same meteorological conditions allow inference of the accuracy and representativeness of a given size distribution and thus of the confidence that can be placed in resulting parameterizations of size distributions of SSA concentration in terms of meteorological variables. Additionally, these estimates provide guidance as to whether or not a reported size distribution of SSA concentration can be considered anomalous, or whether or not two size distributions can be said to agree (or to differ). Unfortunately, such estimates are often not presented. The number of significant digits to which data and parameterizations of these data are reported can rarely be taken as a measure of this uncertainty and variability, as data are commonly presented with much greater implied precision that can be justified by their scatter. Consequently, alternative approaches are necessary to estimate uncertainties and variability from published data.

Estimates of the variability in reported size distributions of SSA concentration can be obtained by comparing determinations obtained by different investigators under the same nominal conditions; this is done below (§4.1.4.2). Estimates of the uncertainty in reported size distributions of SSA concentration can be obtained from calibration data, but such data are typically not reported. Another approach to estimating this uncertainty is by comparing simultaneous determinations of size distributions of SSA concentration using different devices. Such comparisons have been reported by several investigators.

Chaen [1973] reported impaction efficiencies of SSA particles as a function of r_{80} and of wind speed on films of different widths, with typical variability for a given set of conditions being a factor of three or more. For marine aerosol particles with r_{80} between 0.3 μm and 0.5 μm, *Mészáros and Vissy* [1974] reported that the concentration determined by electron microscopy was 20% higher than that determined by optical microscopy, a difference they considered satisfactory. Number concentrations of SSA particles with radii (at ambient RH) between ~5 μm and ~15 μm measured simultaneously with rotating impactor rods and using optical methods by *de Leeuw* [1986a] were reported to be within a factor of three of each other more than 90% of the time, a situation the investigator characterized as being "good agreement" and "acceptable" when taking into account the variations observed between different optical particle counters. Differences of up to an order of magnitude over the size range $r_{amb} \approx 0.5\text{-}10$ μm were observed in a recent field comparison of four optical detectors that had been previously calibrated in the laboratory [*de Leeuw et al.*, 2000], and at any given size the difference between any two of the instruments was typically greater than a factor of two. In another recent study *Reid et al.* [2001] reported simultaneous measurements of size distributions of number concentration in which particle diameters less than 2 μm referred to diameters of dried particles (equivalent to r_{80} values) whereas those for diameters greater than 2 μm referred to diameters of particles at the ambient RH of 70-80% (approximately $2r_{80}$), as obtained with two different optical detectors (Passive Cavity Aerosol Spectrometer Probe, PCASP, and Forward Scattering Spectrometer Probe, FSSP, respectively). They stated that over the range of overlap (diameters between 2 and 2.5 μm) the concentrations obtained from the two devices agreed to ~25%, but from their Fig. 2 it appears that for the three size classes for which the concentration size distributions overlap (r_{80} approximately 1.2 μm, 1.4 μm, and 1.7 μm) the agreement was considerably poorer, with the concentrations from the PCASP generally a factor of 1.4 to 2.5 times greater than those obtained using the FSSP, and in some instances more.

Measurement uncertainties are rarely documented and discussed and are thus rarely taken into account in presentations of size distributions of SSA concentration measured using such instruments or in calculations of quantities derived from such measurements. However, based on the above observations it must be concluded that uncertainties in size distributions of SSA concentration that arise both from uncertainties in particle size and uncertainties in amplitude are generally at least several tens of percent, and frequently considerably greater.

4.1.4.2. Measurements. Measured size distributions of SSA concentration over the oceans, described in Table 13, are presented in Figs. 22a-f for six different wind speed ranges as logarithmic-logarithmic graphs of number concentration in the representation $n(r_{80})$ ($\equiv dN/d\log r_{80}$) vs. r_{80} and of surface area concentration at 80% RH in the representation $a_{80}(r_{80})$ ($\equiv dA_{80}/d\log r_{80}$) vs. r_{80}, for r_{80} from 0.05 μm to 100 μm. The two representations are not independent determinations as they are derived from the same data and hence contain the same information. In each of these figures the two graphs are presented using the same logarithmic abscissa scale. All six pairs of graphs are plotted using the same axis scales to permit comparison among different wind speed ranges. Presentation of the quantities $n(r_{80})$ and $a_{80}(r_{80})$ on a linear scale with r_{80} on a logarithmic scale would result in so-called "equal-area" plots (§2.1.3), in which the contribution to either of these quantities in any interval of r_{80} would be proportional to the area under the curve over this interval; however, $n(r_{80})$ ranges over many orders of magnitude, and the quantities must be plotted using a logarithmic ordinate scale to adequately present the data over such a range. Presenting size distributions of SSA concentration in the representations $n(r_{80})$ and $a_{80}(r_{80})$ illustrates the relative importance of SSA particles of a given size to cloud drop production (§2.1.5.1) and light scattering, respectively (§2.1.5.3); the relative importance of SSA particles of a given size to other properties of SSA concentration can readily be inferred from these.

As wind speed is the quantity most often used to describe and classify size distributions of SSA concentration, it is chosen here as well, although it is expected that substantial variability will remain even after classifying size distributions in this manner because of the dependencies of size distributions of SSA concentration on other factors, as discussed in §4.1.1.3. For purposes of this presentation, the SSA concentration size distributions have been segregated into six ranges of the 10 m wind speed U_{10}: a) 0-4 m s^{-1}, b) 5-7 m s^{-1}, c) 8-11 m s^{-1}, d) 12-14 m s^{-1}, e) 15-20 m s^{-1}, f) 25-30 m s^{-1}. Size distributions of SSA concentration that could not be classified in this manner [*Junge et al.*, 1969; *Blifford*, 1970; *Woodcock and Duce*, 1972; *DeLuisi et al.*, 1972; *Schacher et al.*, 1981a; *Tomasi and Prodi*, 1982; *Gras and Ayers*, 1983; *Parungo et al.*, 1987; *Kim et al.*, 1990; *McGovern et al.*, 1994; *Quinn et al.*, 1996; *Bates et al.*, 1998; *Quinn and Coffman*, 1999; *Campusano-Jost et al.*, 2003] are broadly consistent with the values shown here; these include measurements for which no wind speed was given, or data encompassing a wide range of wind speeds. The data presented in Fig. 22 comprise a large fraction of the size distributions of SSA concentration that have been reported. As with *M* and *N*, only by considering data sets from different investigators, taken at different times and locations, is it possible to estimate the variability in SSA size distributions of concentration and thus ascertain the extent to which generalizations can be made and to which confidence can be placed in any one set of measurements or in any parameterization of the data.

In preparing the graphs, every effort was made to use raw data. However, in some instances the data sets had been smoothed by the original investigators, and it is questionable to what extent the true size distribution is reflected in the smoothed fit. Additionally, every effort was made to include only data representing mostly or entirely SSA particles; data that may contain substantial contributions from particles other than sea salt are noted and discussed. For this reason, concentrations are presented only for $r_{80} > 0.4$ μm for measurements that were not obtained by methods specific to sea salt because of the probable influence of non-SSA particles for lower values of r_{80} (§4.1.1.1; Fig. 16). No effort appears to have been made by the original investigators to establish that the measurements were made under conditions in which SSA concentrations represented their steady state values with respect to dry deposition, and certainly no such effort

Table 13. Measurements of Size Distributions of Sea Salt Aerosol Concentration

Reference	Location	Technique	r_{80}/μm Range[a]	U_{10}/(m s^{-1})
Fournier d'Albe, 1951	Mediterranean	impaction, microscopy	1-6[b]	2.5, 3.5, 9
Woodcock, 1952	Massachusetts and Florida coasts	impaction, microscopy	1-25	5, 6, 9
Woodcock, 1953[c]	Florida coast	impaction, microscopy	2-25	29
Metnieks, 1958	Ireland coast	impaction, chemical analysis	0.15-11	1-2
Chaen, 1973[d]	Indian Ocean, East China Sea	impaction, chemical analysis	2-48	0.8-16.6
Preobrazhenskii, 1973	North Atlantic	impaction, microscopy	1-45[e]	6-10[f], 15-25
Mészáros & Vissy, 1974[g]	Southern Atlantic and Indian Oceans	impaction, electron microscopy	0.1-50	6, 7, 9, 12
Hughes & Richter, 1980	California coast	optical	0.225-14.7[h]	6, 8.5
Prodi et al., 1983	Mediterranean	impaction, chemical analysis	0.15-4	5
Monahan et al., 1983a	North Sea	optical	0.8-10	3[i]
Fairall et al., 1983	North Sea, California coast	optical	0.8-15	6, 9, 11, 13, 15, 18
Exton et al., 1985[j]	Scotland coast	optical	0.4-20[h]	0-5, 5-10, 10-15, >15
de Leeuw, 1986a,b	North Atlantic	impaction, microscopy	6-25[k]	5.5, 7, 8.5, 13, 17
de Leeuw, 1987	North Sea	impaction, microscopy	6-25[k]	9, 12
Smith et al., 1989[l]	Scotland coast	optical	0.4-20[h]	12, 20, 25, 30
de Leeuw, 1990a	North Sea	impaction, microscopy	3-60[m]	25
M. Smith et al., 1991	Scotland coast	optical	0.4-25[h]	3.4, 6
Taylor & Wu, 1992	North Carolina coast	optical	~25-~32[n]	13.6, 14.9, 16.0
O'Dowd & Smith, 1993	Northeast Atlantic	thermal volatility	0.1-2.5	17
Smith et al., 1993	Scotland coast	optical	1-22[o]	0, 5, 10, 15, 20, 25, 30
Lowe et al., 1996	Scotland coast	optical, thermal volatility	1-30[p]	< 5, 14
O'Dowd et al., 1997a	North Atlantic	optical, thermal volatility	0.1-75[q]	6, 8, 9, 17
O'Dowd et al., 1999b	North Atlantic	optical, thermal volatility	0.1-1.2[r]	1-2, 17
Reid et al., 2001	Atlantic	optical	1-20	5, 8, 11, 14
O'Dowd et al., 2001	Irish coast	optical	0.4-75[h,s]	6, 10

Size distributions of SSA concentration, most of which were measured at 5-20 m above the sea surface, are presented in Fig. 22a-f as a function of r_{80} for several ranges of U_{10}. Size distributions of SSA concentrations measured at 6 m above the sea surface by *Chaen* [1973] at Beaufort wind force 5 (corresponding approximately to $U_{10} = 8.0$-10.7 m s^{-1}) are presented in Fig. 23.

[a] Approximate range of r_{80} for which data were reported; this may vary somewhat for different data sets by the same investigators. In many cases ambient radii were presented and conversion to r_{80} values is based on mean or typical RH values, or on RH values determined from graphs presented along with the data.

[b] Data for $r_{80} < 1$ μm are not used as particles in this size range probably consisted mostly of substances other than sea salt.

[c] Size distribution is based on fit of curve for Beaufort wind force 12 in Fig. 1 of *Woodcock* [1953]. These data were measured from a tower ~40 m above sea level (§4.1.2.1).

[d] Averages of size distributions of SSA concentration presented by *Chaen* [1973] for two cruises (KH-69-3 and KH-70-3) at Beaufort wind force ranges 2 to 6 and individual size distributions for $U_{10} = 14.4$ m s^{-1} and 16.6 m s^{-1} are presented in Fig. 22 in the corresponding wind speed range. Individual size distributions of SSA concentration at 6 m for each cruise at Beaufort wind force 5, corresponding to $U_{10} = 8.0$-10.7 m s^{-1}, are presented in Fig. 23.

[e] Radius range presented by *Preobrazhenskii* [1973] for $U_{10} = 15$-25 m s^{-1} extended up to 45 μm for measurements at 7 m, although RH was not specified. Values are presented in Fig.22d under the assumptions that drops did and did not equilibrate with respect to RH at 80%, and thus that the radii presented were r_{80} or formation values, respectively.

[f] Wind speed range was stated by *Preobrazhenskii* [1973] both as 6-10 m s^{-1} and as 7-12 m s^{-1}.

Table 13. (Continued)

[g] Data have not been modified as in *E. Mészáros* [1982]; concentrations for $r_{80} > 10$ µm are probably overestimated because of non-isokinetic sampling conditions (§4.1.4.2 §4.1.4.3).

[h] Only data for $r_{80} > 0.4$ µm are used in Fig. 22; data for smaller sizes are unlikely to represent sea salt.

[i] Only the 3 m s^{-1} JASIN data reported in *Monahan et al.* [1983a] are presented here; other JASIN data at different wind speeds are contained in data of *Fairall et al.* [1983] presented here.

[j] RH values were not presented by *Exton et al.* [1985], but 80% was stated as typical and this value is assumed and used here to convert data.

[k] *De Leeuw* [1986a,b, 1987] reported radii up to ~45 µm under the assumption that they had equilibrated to the ambient RH; as discussed in §4.1.4.2 they probably had not equilibrated and thus corresponded to r_{80} near 25 µm.

[l] RH values are inferred from Fig. 6 of *Smith et al.* [1989]. Data for wind speeds 25 m s^{-1} and 30 m s^{-1} were also included in *Smith et al.* [1993].

[m] Radii are converted to r_{80} values on the assumption that diameters presented in *de Leeuw* [1990a] were formation values (and not values in equilibrium at 65% RH, as he assumed); this certainly underestimates r_{80} values of smaller particles (which will have equilibrated by the time of measurement).

[n] Data used for *Taylor and Wu* [1992] are from Fig. 4 of that reference; in some instances these differed somewhat from those in Fig. 2 of that reference, which supposedly contain the same data (see discussion in §4.1.4.2). Radius range presented by them extended from 50 µm to ~160 µm, but *Taylor and Wu* stated that as their data "were generally collected at humidities below the equilibrium relative humidity for the seawater concentration, the size of each droplet was thereby adjusted to that at the sea surface according to a procedure suggested by *Fitzgerald* [1975]," implying that they assumed that the drops had attained their equilibrium radii with respect to the ambient RH of 65%. However, as the range 50-160 µm for formation radii corresponds to r_{80} values of 24-75 µm (taking into account the fact that the mean salinity in the ocean near their sampling location had salinity 28.5, for which $r_{\mathrm{form}}/r_{80} \approx 2.1$; §2.5.3), and most of SSA particles in this size range are large SSA particles which would not have equilibrated, these radii may be overestimated by a factor of 2.4, resulting in the upper limit of the range of r_{80} being ~32 µm. Values are presented in Figs. 22d and 22e under the assumptions that drops did and did not equilibrate with respect to RH at 80%, and thus that their radii were formation values or were too large by a factor of 2.4, respectively. See further discussion in §4.1.4.2.

[o] Data for wind speeds 25 m s^{-1} and 30 m s^{-1} included those from *Smith et al.* [1989].

[p] RH values were not presented by *Lowe et al.* [1996], but it was stated that "results refer to wet particles" and it is assumed here in converting the data that the radii they presented were those in equilibrium with 80% RH. This almost certainly overestimates the values of r_{80} for the largest SSA particles, which would have been near their formation values, and the upper range of r_{80} was probably near 30 µm.

[q] *O'Dowd et al.* [1997a] presented size distributions of SSA concentration "normalized to 80% relative humidity," presumably implying that the particles had attained their equilibrium values at the ambient RH. Values are presented in Fig. 22 under this assumption. This almost certainly overestimates the values of r_{80} for the largest SSA particles, which would have been near their formation values, and the upper range of r_{80} was probably near 75 µm.

[r] Measurements were taken near cloud base (~400 m), but the investigators stated that the SSA was well mixed throughout the marine boundary layer; thus concentrations are expected to be representative of those near 10 m.

[s] Radii are converted to r_{80} values under the assumption that ambient RH was 98% (values were reported to be 97-99%) and that radii presented, which ranged up to 150 µm, were formation radii.

was made here. As noted in §2.9.6, whether or not steady state with respect to dry deposition has been achieved depends strongly on particle size and on prior meteorological conditions experienced by the air mass, most importantly time since prior precipitation. In compiling data from coastal regions, attempts were made to select data that reflected only maritime air masses (i.e., continental air masses were excluded), and some measurements that were thought not to represent concentrations typical of the marine atmosphere are not included, although in some instances other measurements from the same investigation are included [e.g., *Fournier d'Albe*, 1951; *Metnieks*, 1958; *Prodi et al.*, 1983]. Reported size distributions of SSA concentration that were certainly or probably strongly influenced by surf effects [*Woodcock et al.*, 1963; *Podzimek*, 1984; *Stramska*, 1987; *McKay et al.*, 1994; *Zieliński and Zieliński*, 1996; *Hooper and Martin*, 1999; *de Leeuw et al.*, 2000; *Reid et al.*, 2001] are also not included, as these would provide little information on the oceanic SSA production flux and are probably strongly dependent on geographical location and sampling location with respect to the coast. For example, the size distribution of SSA concentration reported by *Woodcock*

et al. [1963] from coastal surf at a height of 1 m above sea level for $U_{10} = 4$ m s^{-1} was 1½ to 2 orders of magnitude greater than the largest size distributions shown for $U_{10} = 5$-7 m s^{-1} in Fig. 22b (§4.3.5).

An extraordinarily rich data set was presented by *Chaen* [1973], consisting of more than one hundred size distributions of SSA concentration determined by chemical analysis of samples collected by impaction, and covering wind speeds up to 17 m s^{-1} and r_{80} up to 48 µm. *Chaen* also presented mean values of size distributions of SSA concentration at Beaufort wind forces 2 to 7 (§2.3.1) for each of two cruises (KH-69-3 and KH-70-3). For Beaufort wind forces 2 to 6 these mean distributions are presented in Fig. 22 in the corresponding wind speed range; the two size distributions at Beaufort wind force 7 are presented in Figs. 22d and 22e in the corresponding wind speed range. The size distributions of SSA concentration presented by *Chaen* [1973] at 6 m for each cruise at Beaufort wind force 5, corresponding to $U_{10} = 8.0$-10.7 m s^{-1}, are shown in Fig. 23. The spread among these data for a narrow range of wind speeds, roughly an order of magnitude, illustrates the variability that can be expected for size distributions of SSA

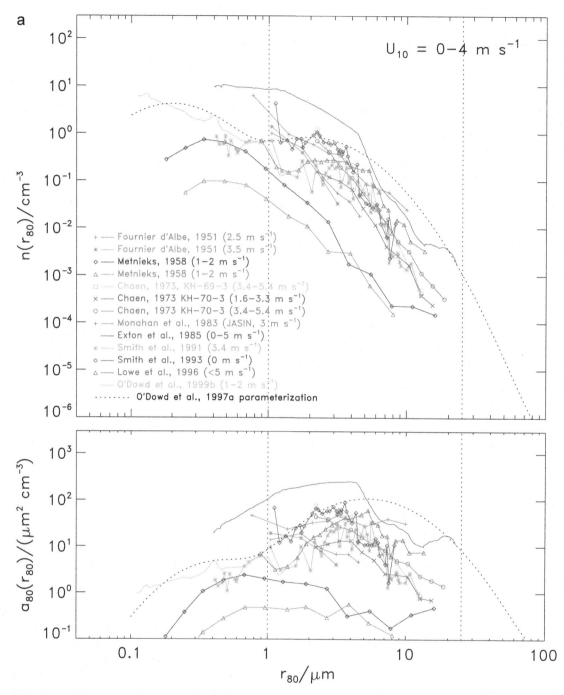

a

$U_{10} = 0-4$ m s^{-1}

+ —— Fournier d'Albe, 1951 (2.5 m s^{-1})
* —— Fournier d'Albe, 1951 (3.5 m s^{-1})
◇ —— Metnieks, 1958 (1–2 m s^{-1})
△ —— Metnieks, 1958 (1–2 m s^{-1})
□ —— Chaen, 1973, KH-69-3 (3.4–5.4 m s^{-1})
× —— Chaen, 1973 KH-70-3 (1.6–3.3 m s^{-1})
○ —— Chaen, 1973 KH-70-3 (3.4–5.4 m s^{-1})
+ —— Monahan et al., 1983 (JASIN, 3 m s^{-1})
—— Exton et al., 1985 (0–5 m s^{-1})
* —— Smith et al., 1991 (3.4 m s^{-1})
◇ —— Smith et al., 1993 (0 m s^{-1})
△ —— Lowe et al., 1996 (<5 m s^{-1})
—— O'Dowd et al., 1999b (1–2 m s^{-1})
······ O'Dowd et al., 1997a parameterization

Figure 22. Measured size distributions of SSA concentration, represented as number concentration $n(r_{80}) \equiv dN(r_{80})/d\log r_{80}$ (upper panels), and as area concentration at 80% RH, $a_{80}(r_{80}) \equiv dA_{80}(r_{80})/d\log r_{80}$ (lower panels), as a function of r_{80} in several ranges of the wind speed at 10 m above the sea surface, U_{10}: *a*) 0-4 m s^{-1}, *b*) 5-7 m s^{-1}, *c*) 8-11 m s^{-1}, *d*) 12-14 m s^{-1}, *e*) 15-20 m s^{-1}, *f*) 25-30 m s^{-1}. All graphs are presented on the same axes to facilitate comparison among different wind speed ranges. Data and assumptions made for conversion of reported size distributions of SSA concentration to $n(r_{80})$ are summarized in Table 13 and discussed in §4.1.4.2. Measurements of *Preobrazhenskii* [1973] at 7 m for wind speed range 15-25 m s^{-1} are presented in Fig. 22e both under the assumption of equilibration to RH 80% (and thus that the reported radii were r_{80} values) and under the assumption of no equilibration (and thus that the reported radii were formation radii). Measurements of *Taylor and Wu* [1992] are presented in Figs. 22d and 22e both under the assumption of equilibration to RH 65% (and thus that the reported radii were formation values) and under the assumption of no equilibration (and thus that the reported radii were too large by a factor of up to 2.4). Measurements for which equilibration has

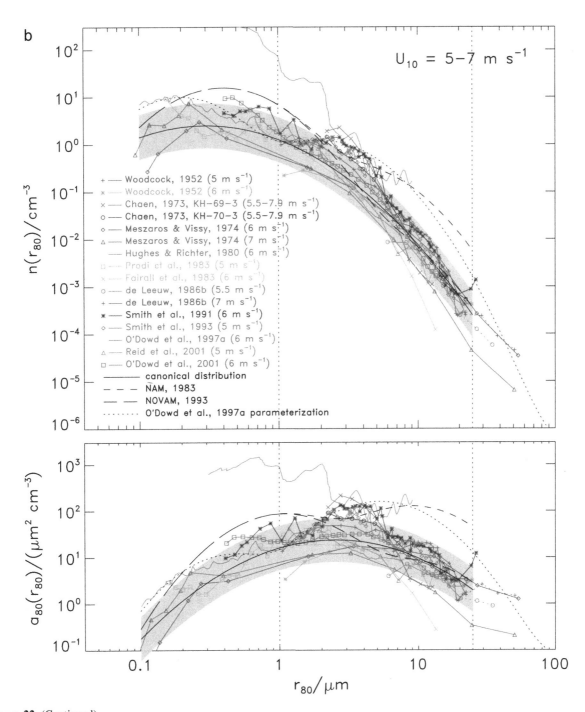

Figure 22. (Continued).
or has not been assumed that are known to be incorrect in certain size ranges are connected by dotted lines. Parameterizations are evaluated at the midpoint of the wind speed range. Canonical distribution is an empirical lognormal distribution of the form (2.1-9) with $n_0/\text{cm}^{-3} = 0.07[U_{10}/(\text{m s}^{-1})]^2$, $r'_{80} = 0.3$ μm and $\sigma = 2.8$; shaded bands denote a factor of $\lesssim 3$ about this distribution. NAM, the original Navy Aerosol Model [*Gathman*, 1983a,b], and NOVAM, Navy Ocean Vertical Aerosol Model [*Gathman and Davidson*, 1993], are based on a large number of data sets. *O'Dowd et al.* [1997a] parameterization is a sum of three lognormal distributions. Two parameterization by *O'Dowd et al.* [1997a] are shown in Fig. 22e; parameterization 1 uses the value of n_0 for the spume drop given by the proposed expression; parameterization 2 uses the value given in Fig. 2a of that reference (§4.1.4.4). Vertical dotted lines at r_{80} values of 1 μm and 25 μm demarcate small, medium, and large SSA particles (§2.1.2).

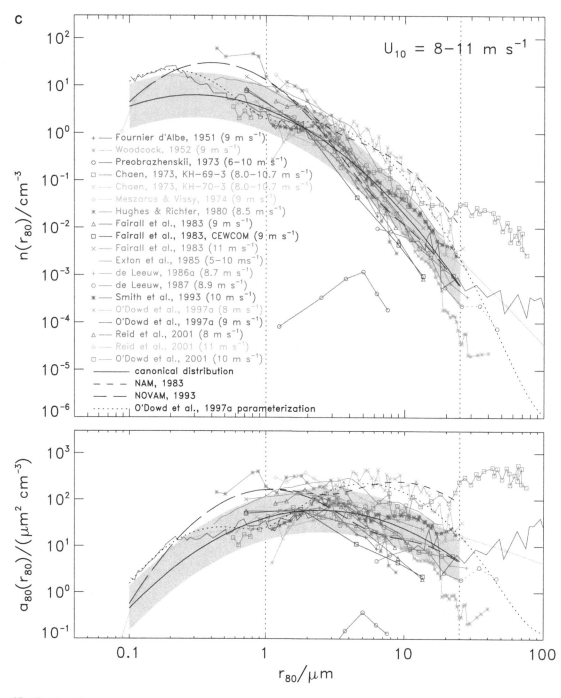

C

$U_{10} = 8-11 \ m \ s^{-1}$

+ —— Fournier d'Albe, 1951 (9 m s⁻¹)
* ---- Woodcock, 1952 (9 m s⁻¹)
○ —— Preobrazhenskii, 1973 (6–10 m s⁻¹)
□ —— Chaen, 1973, KH-69-3 (8.0–10.7 m s⁻¹)
□ —— Chaen, 1973, KH-70-3 (8.0–10.7 m s⁻¹)
◇ —— Meszaros & Vissy, 1974 (9 m s⁻¹)
* —— Hughes & Richter, 1980 (8.5 m s⁻¹)
△ —— Fairall et al., 1983 (9 m s⁻¹)
□ —— Fairall et al., 1983, CEWCOM (9 m s⁻¹)
× —— Fairall et al., 1983 (11 m s⁻¹)
—— Exton et al., 1985 (5–10 ms⁻¹)
+ —— de Leeuw, 1986a (8.7 m s⁻¹)
○ —— de Leeuw, 1987 (8.9 m s⁻¹)
* —— Smith et al., 1993 (10 m s⁻¹)
× —— O'Dowd et al., 1997a (8 m s⁻¹)
—— O'Dowd et al., 1997a (9 m s⁻¹)
△ —— Reid et al., 2001 (8 m s⁻¹)
○ —— Reid et al., 2001 (11 m s⁻¹)
□ —— O'Dowd et al., 2001 (10 m s⁻¹)
—— canonical distribution
– – – NAM, 1983
— — NOVAM, 1993
········· O'Dowd et al., 1997a parameterization

Figure 22. (Continued).

concentration that were measured and analyzed by the same investigator using the same equipment.

Seven size distributions of SSA concentration obtained by Na determination of samples collected by a multistage impactor for average wind speeds ranging from 3.4 m s⁻¹ to 9.9 m s⁻¹ were reported by *McDonald et al.* [1982]. These

data have been used by *Erickson and Duce* [1988] to investigate possible increase in characteristic SSA particle sizes with increasing wind speed (§4.1.4.4). For the most part these size distributions appear to be consistent with those shown in Fig. 22, but concentrations of SSA particles with $r_{80} \lesssim 0.5 \ \mu m$ were in some instances greater by up to an

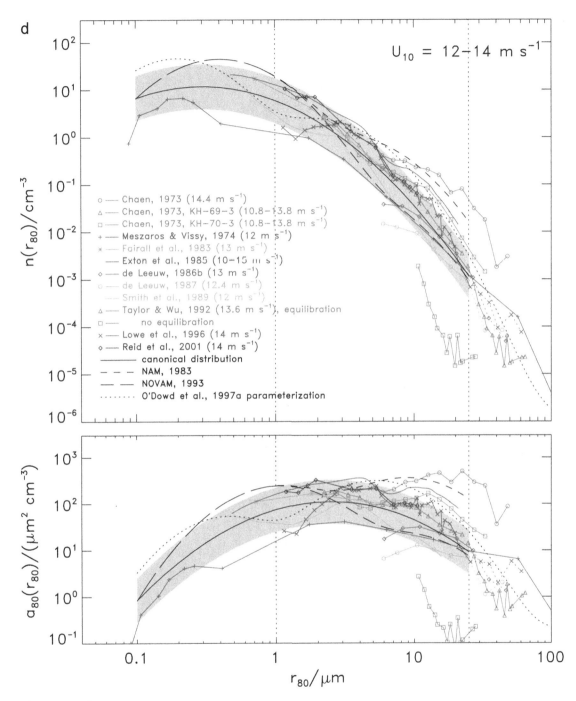

d

$U_{10} = 12-14 \ m \ s^{-1}$

o —— Chaen, 1973 (14.4 m s^{-1})
△ —— Chaen, 1973, KH-69-3 (10.8-13.8 m s^{-1})
□ —— Chaen, 1973, KH-70-3 (10.8-13.8 m s^{-1})
+ —— Meszaros & Vissy, 1974 (12 m s^{-1})
✳ —— Fairall et al., 1983 (13 m s^{-1})
  —— Exton et al., 1985 (10-15 m s^{-1})
◇ —— de Leeuw, 1986b (13 m s^{-1})
○ —— de Leeuw, 1987 (12.4 m s^{-1})
  —— Smith et al., 1989 (12 m s^{-1})
△ —— Taylor & Wu, 1992 (13.6 m s^{-1}), equilibration
□ ——   no equilibration
✕ —— Lowe et al., 1996 (14 m s^{-1})
◇ —— Reid et al., 2001 (14 m s^{-1})
  —— canonical distribution
  ---- NAM, 1983
  —— NOVAM, 1993
  ······ O'Dowd et al., 1997a parameterization

Figure 22. (Continued).

order of magnitude, possibly because of incorporation of mass from SSA particles larger than the specified cutoff radii, as discussed above (§4.1.4.2). Because size distributions resulting from the inversion of these data were not reported, and because of the difficulty in assigning values of radii that accurately represent each size bin, these size

distributions of SSA concentration are not presented in Fig. 22.

Most size distributions of SSA concentration presented in Fig. 22 were obtained from measurements at heights from 5 m to 20 m above the sea surface and thus represent SSA particles that contribute to the effective production

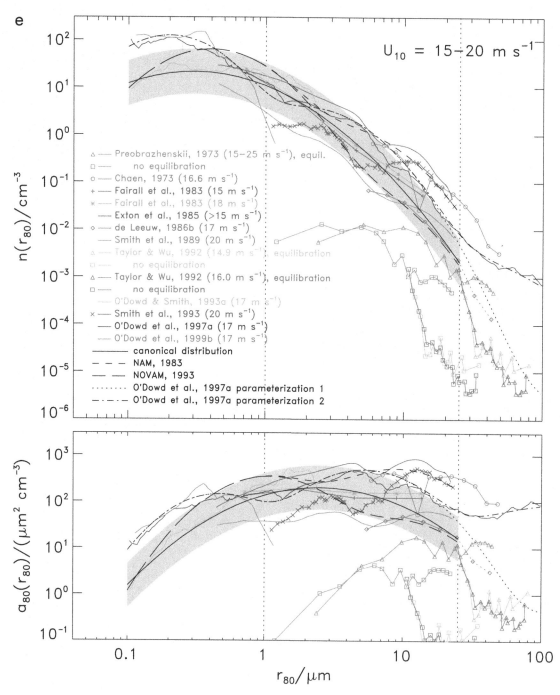

Figure 22. (Continued).

flux. Based on the conclusions reached in §2.9.4 and §4.1.1.3, little variation is expected for concentrations of SSA particles with r_{80} up to ~25 μm over this range of heights. Size distributions of SSA concentration from *Chaen* [1973] presented in Fig. 23 were measured at 6 m above the sea surface; the differences between the values of these

concentrations and those at 10 m are also expected to be small. The measurements of *Woodcock* [1953] at 29 m s⁻¹ (Fig. 22*f*) were made on a tower near 40 m above sea level at a coastal location (§4.1.2.1). Size distributions of concentrations of SSA particles very near the sea surface (i.e., within a meter or so), examined in §4.2, are not included,

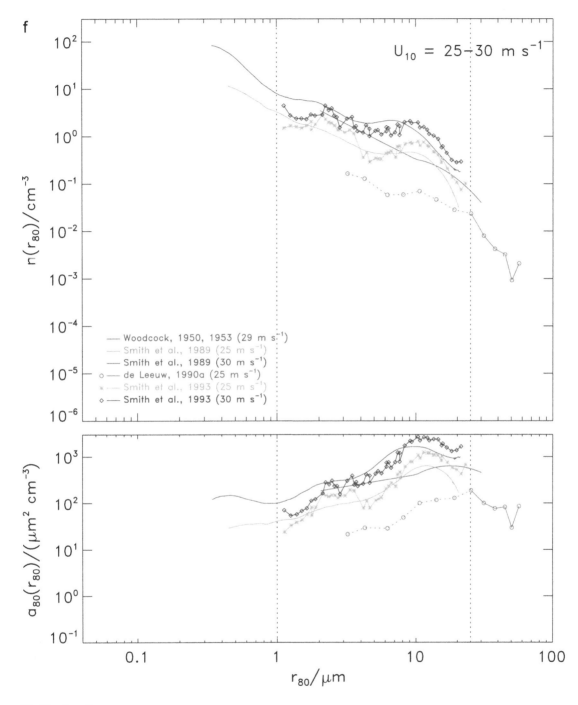

f

$U_{10} = 25-30 \text{ m s}^{-1}$

—— Woodcock, 1950, 1953 (29 m s^{-1})
—— Smith et al., 1989 (25 m s^{-1})
—— Smith et al., 1989 (30 m s^{-1})
○—— de Leeuw, 1990a (25 m s^{-1})
✳······ Smith et al., 1993 (25 m s^{-1})
◇—— Smith et al., 1993 (30 m s^{-1})

Figure 22. (Continued).

nor are laboratory measurements (§4.3). Additionally, in most instances only data taken in the lower part of the marine boundary layer are presented. Consequently, size distributions of SSA concentration reported by *Woodcock and Gifford* [1949], *Woodcock* [1952], and *Woodcock* [1953], for example, measured at heights of many tens to

several hundreds of meters are not included (these size distributions were discussed in §2.9.5, and for the most part are consistent with those shown here), nor are the size distribution of concentration reported by *Hughes* [1980] for a height of over 100 m nor those of *Patterson et al.* [1980] at heights near 1 km, for example. The size distributions of SSA

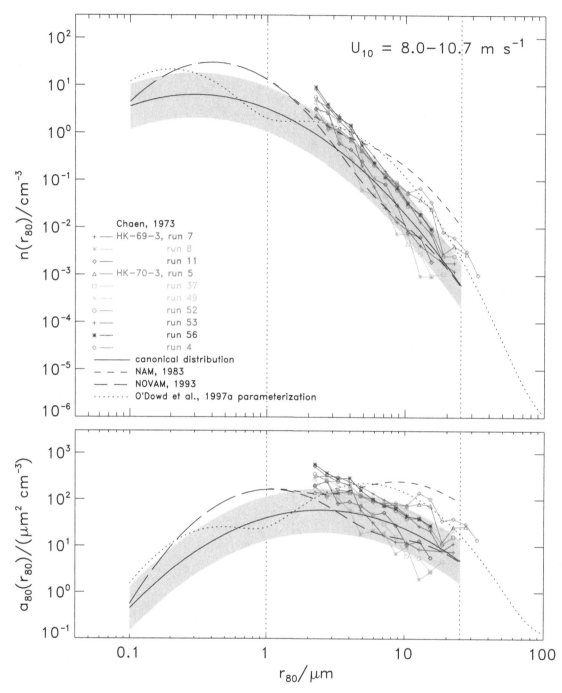

Figure 23. Measured size distributions of SSA concentration at 6 m above the sea surface presented by *Chaen* [1973] (cruises KH-69-3 and KH-70-3) for Beaufort wind speed range 5 (8.0-10.7 m s^{-1}), represented as number concentration $n(r_{80}) \equiv dN(r_{80})/d\log r_{80}$ (upper panel), and as area concentration at 80% RH, $a_{80}(r_{80}) \equiv dA_{80}(r_{80})/d\log r_{80}$ (lower panel), as a function of r_{80} on the same axes as in Fig. 22. Parameterizations are evaluated at $U_{10} = 9.5$ m s^{-1} and are the same as in Fig. 22c.

concentration reported by *O'Dowd et al.* [1999b] measured at several hundred meters are included in Fig. 22 as they provide useful information on SSA particles in the size range 0.1-1 μm where there are few data; the investigators indicated that the marine boundary layer was well mixed up to the level at which measurements were made, and it is thus expected that these measurements are representative of those at lower heights.

The measurements presented in Fig. 22 were taken in a variety of locations, some coastal and some open ocean. Some of the reported size distributions of SSA concentration are averages from different locations or different times, whereas others are individual determinations at a single location over a relatively short period of time. Several data sets are from the same group or same location, and it is possible that some of the data in Fig. 22 are not independent (for example, individual size distributions presented by an investigator may be included in average size distributions presented later by the same investigator or group). Several data sets have rather coarse resolution in radius, precluding detection of any fine structure in the distribution. Some of the SSA concentration size distributions were described as "typical," whereas others represent averages over long periods of time and over various conditions such as RH, with no indication of how much a given distribution varied with time or over what time scale. As in many instances particle sizes were determined at ambient RH, which was not reported with the data, mean values of the RH often had to be used to convert the radii to r_{80} values; any variability in RH would lead to error in this conversion. Many of the size determinations were obtained using optical detectors, though usually with little or no discussion of how Mie resonances were treated or of what assumptions were made concerning the index of refraction of the SSA particles.

Despite the appearance of an abundance of data, there is actually a paucity of reported size distributions of SSA concentration, especially considering the large number of studies and field programs that have measured aerosols in the marine atmosphere. As noted above, the size distributions of SSA concentration presented in Fig. 22 comprise a large fraction of those reported. Additionally, many of these size distributions were from only a few research groups, often from the same geographical location. Relatively few size-dependent concentrations of small SSA particles ($r_{80} \lesssim 1$ µm) have been reported, and most of these are at lower wind speeds, i.e., $U_{10} \lesssim 7$ m s^{-1}. Relatively few SSA concentration size distributions are available at very low wind speeds (Fig. 22a); such distributions are important, for example, to considerations of aerosol indirect forcing. Likewise few data are available at the highest wind speeds, i.e., $U_{10} \gtrsim 20$ m s^{-1}, and some of the size distributions of SSA concentration presented in Fig. 22f are not independent. A further concerns with respect to understanding the spread of the data is the absence in general of information other than wind speed characterizing the measurement situations. Key among such information would be the time following a precipitation event, which although generally not reported along with SSA concentration size distributions, would be important to considerations of

precipitation scavenging and to the concentration buildup method (§3.3) and the statistical wet deposition method (§3.9).

All of the size distributions of SSA concentration that are presented in Fig. 22 were determined by counting particles in a given size range and then converted by the original investigators to the representation in which they were reported. These SSA concentration size distributions were subsequently converted for presentation in Fig. 22 by the present authors to the representations $n(r_{80})$ and $a_{80}(r_{80})$ using relations such as those in §2.1.4.1; this procedure also includes converting reported sizes of SSA particles to r_{80} values. For size distributions of SSA concentration originally determined and reported as number concentrations as a function of the mass of sea salt in the particle m_{dry}, such as those of *Woodcock* [1952, 1953] and *Chaen* [1973], or as a function of the dry radius r_{dry}, such as those of *Metnieks* [1958], *Mészáros and Vissy* [1974], *O'Dowd and Smith* [1993a], and *O'Dowd et al.* [1999b], conversion of particle sizes to r_{80} is unambiguous (Eq. 2.5-2) and conversion of size distributions of SSA concentration to the representation $n(r_{80})$ is also ostensibly unambiguous, although it is not always straightforward. For example, a size distribution of SSA concentration at $U_{10} = 29$ m s^{-1} was presented by *Woodcock* as a function of "weight of sea salt particles (µµgrams)" in three different representations: the "distribution of number and weight" in terms of the number per liter of air [1950b], the "distribution-function curves" in terms of "N(m^{-3} µµgram^{-1})" [1952], and the "size distribution" in terms of "sea salt particles (no. m^{-3}) greater than indicated weight" [1953]. The second representation is equivalent to dN/dm_{dry} and the third to $N - N(m_{dry})$, where $N(m_{dry})$ is the cumulative distribution (§2.1.4.1) in terms of the dry mass of an SSA particle, but his first representation is unclear; as values are not monotonically decreasing it cannot be related to the cumulative distribution as in *Woodcock* [1953], but it does not appear to match the values presented in *Woodcock* [1952]. Another size distribution of SSA concentration was presented in the same figure in *Woodcock* [1950b] which also is not monotonically increasing but does not appear to match comparable size distributions of SSA concentration in the representation dN/dm_{dry}. The size distribution of SSA concentration presented in Fig. 22f for $U_{10} = 29$ m s^{-1} is based on that presented in *Woodcock* [1953, Fig. 1].

The size distributions presented by *Chaen* [1973] were also not clearly specified, as the quantity θ (cm^{-3}), which he used to represent "the number concentration of sea salt particles," was meant to represent the number concentration of SSA particles for interval of m_{dry} given by $\Delta \log m_{dry} = 0.25$ (corresponding to $\theta = 0.25 \times dN/d\log m_{dry}$, equivalent to $n(r_{80})/12$ in some instances and for interval of m_{dry} given by

$\Delta\log m_{dry} = 0.5$ (corresponding to $\theta = 0.5 \times dN/d\log m_{dry}$, equivalent to $n(r_{80})/6$) in other instances.

The concentration units reported in *Preobrazhenskii* [1973] were also not clearly specified, being stated as "1/m³," but based on values for the total SSA number concentration reported elsewhere in that reference, the measurements refer to the number of SSA particles per cubic meter of air per 5 μm diameter interval, i.e., $5 \times dN(D)/dD$ (in units $m^{-3} \mu m^{-1}$) and not $dN(D)/dD$ as interpreted by some investigators [*Wang and Street*, 1978a; *Wu*, 1981a; *Mestayer and Lefauconnier*, 1988; *Wu*, 1990d][1].

Several investigators reported size distributions of SSA concentration as a function of ambient radius and also provided the ambient RH (or the mean RH over the sample period); this information allows conversion to r_{80} values under the assumption of RH equilibration. In the several instances [*Exton et al.*, 1985; *Lowe et al.*, 1996] in which ambient RH was not specified but in which it seemed certain that reported radii did not refer to those of dry particles, a value of 80% RH is assumed here. However, as discussed above, conversion between size distributions of SSA concentration is not entirely free of concerns; in several instances uncertainty results either in the description of SSA particle size, the size distribution of SSA concentration, or both. These uncertainties arise primarily because of the lack of equilibration of large SSA particles with respect to RH (§2.9.1).

It has been argued throughout this review that large SSA particles (i.e., $r_{80} \gtrsim 25$ μm), because of their short atmospheric residence times and limited vertical distribution, do not contribute to the effective SSA production flux. However, several of the size distributions of SSA concentration presented in Fig. 22 extend into the range of large SSA particles, and it is important to present available data for this size range to quantify the abundance of these particles and to determine the extent to which the observations support the argument that their contribution to the effective SSA production flux can be neglected. As discussed in §2.9.1, most large SSA particles will not have equilibrated at the time of measurement at any height above the sea surface, and most measured radii greater than ~50 μm are thus probably closer to formation values than to equilibrium values at the ambient RH. For a known RH, conversion to $n(r_{80})$ of size distributions of SSA concentration determined as a function of measured radius is unambiguous for measured radii less than ~25 μm (§4.1.4.1), but for larger SSA particles ambiguity can arise because of lack of complete

equilibration with respect to RH, and consequently errors can arise in conversion of size distributions of SSA concentration as a function of measured radius to those as a function of r_{80}. Conversion of size distributions of SSA concentration that contained measured radii both less than and greater than ~25 μm requires an additional assumption as to whether or not the larger particles had equilibrated with respect to the ambient RH, and this assumption, and choice of the method of converting the data, can lead to errors in the size distribution of SSA concentration in one or another of these size ranges (§4.1.4.2). In these situations a choice must be made as to the range of r_{80} for which the information would be most useful, and the appropriate transformation applied, with the possibility that the sizes and size distributions of concentration of the SSA particles in other ranges of r_{80} will be incorrect.

In general, the choice made in presenting the size distributions of SSA concentration was to use the conversion that appeared to be correct for the majority of the data (although in some instances other choices were made, depending on the number of data in a given size range and their importance). Thus except as noted, size distributions of SSA concentration for which measured radii extended only slightly greater than 25 μm are presented in Fig. 22 under the assumption that equilibration was complete for all SSA particles, and thus that measured radii corresponded to those in equilibrium with the ambient RH; for these data sets both r_{80} values and concentrations of large SSA particles are overestimated in Fig. 22. Likewise, size distributions of SSA concentration for which most of the measured radii are larger than 25 μm are presented in Fig. 22 under the assumption that no equilibration had occurred and that measured radii correspond to those at formation; this underestimates r_{80} values for SSA particles with $r_{80} \lesssim 25$ μm. The consequences of these choices are illustrated by the following discussion of the SSA concentration size distribution reported by *Preobrazhenskii* [1973] at 7 m for strong winds ($U_{10} = 15\text{-}25$ m s^{-1}) and of the SSA concentration size distributions reported by *Taylor and Wu* [1992].

A size distribution of the concentration of SSA particles with reported radii from 5 μm to greater than 45 μm at 7 m above the sea surface for wind speeds $U_{10} = 15\text{-}25$ m s^{-1} was measured by *Preobrazhenskii* [1973], who collected particles by impaction upon glass slides which were later photographed under a microscope. Radii of sufficiently small SSA particles ($r_{80} \lesssim 25$ μm) had probably equilibrated to the ambient RH (which was not reported) by the time of

---

[1] The size distributions of SSA concentrations reported by *Preobrazhenskii* [1973] also appear to have been incorrectly interpreted in some other references [*de Leeuw*, 1986a; *Iida et al.*, 1992; *de Leeuw*, 1993], although they appear to have been correctly interpreted by *Koga and Toba* [1981], *Wu* [1982a], *Wu et al.* [1984], and *Stramska* [1987].

collection, whereas larger SSA particles probably had not equilibrated to any appreciable extent, and their radii were probably closer to formation values (§2.9.1; Fig. 11). However, *Preobrazhenskii* did not specify whether the reported radii referred to those at the time the drops were sampled, corresponded to radii at a fixed RH when the drops were later photographed, or had been converted to formation radii. This size distribution of SSA concentration is presented in Fig. 22e both under the assumption that the reported radii correspond to formation values (i.e., that no equilibration had occurred) and under the assumption that the reported radii correspond to r_{80} values (i.e., that total equilibration to RH 80% had occurred). The large difference between the resulting values of $n(r_{80})$ demonstrates that the assumption chosen can have a large effect on both the magnitude and the shape of the resulting size distribution $n(r_{80})$, and even more so in the representation $a_{80}(r_{80})$. The units *Preobrazhenskii* used to represent his size distributions were discussed above, and other issues with these measurements are discussed below (§4.1.4.3). In fact, the reported size distribution of SSA concentration probably corresponded to SSA particles with r_{80} ranging from only 2.5 μm to near 25 μm, with those SSA particles with reported radii greater than ~25 μm corresponding to values of r_{80} near 25 μm. Thus, the actual SSA concentration size distribution corresponding to his measurements is probably similar in shape to the one presented in Fig. 22e for r_{80} up to ~25 μm, with the number concentrations of SSA particles in the last several size intervals he presented being contained in a narrow range of r_{80} near 25 μm, thus resulting in increased values of these concentrations because of smaller intervals of $d\log r_{80}$.

Size distributions of SSA concentration reported by *Taylor and Wu* [1992] were obtained by optically sizing and counting particles at ambient RH near 65%. These investigators stated that the range of geometric radii that could be measured was 25 μm to 250 μm. Size distributions of SSA concentration were presented in Fig. 2 and Fig. 4 of that reference; although these two figures ostensibly present the same data, there are a few values present in Fig. 2 that are absent from Fig. 4, and the values of the radii presented in Fig. 2a for $U_{10} = 13.6$ m s^{-1} were approximately 50% lower than those in Fig. 4a for this wind speed. As the range of radii in their Fig. 4a was closest to that for the other wind speeds, the data from that figure were used to generate the size distribution of SSA concentration presented in Fig. 22. *Taylor and Wu* stated that "the size of each droplet was thereby adjusted to that of the sea surface," presumably implying that the measured radii were assumed to be in equilibrium at 65% RH and were converted to (and reported as) formation radii. Because the salinity of the ocean near their measurement site was 28.5, for which $r_{form}/r_{80} \approx 2.1$ (§2.5.3), and because $r_{80}/r_{65} \approx 1.15$ (§2.5.3), the value of the

formation radii would be a factor of 2.4 greater than the measured radii, which *Taylor and Wu* apparently assumed were equilibrium values. As these investigators reported radii ranging from approximately 60 μm to 150 μm, the measured radii would have ranged from approximately 25 μm to 60 μm. Particles with measured radii near the upper part of this range almost certainly had not equilibrated by the time of measurement but instead probably retained radii near their formation values (§2.9.1; Fig. 11), implying that their r_{80} values were near 30 μm. Thus the conversion by *Taylor and Wu* would have resulted in reported formation radii that were greater than actual formation radii by a factor of 2.4. On the other hand, particles with measured radii near the lower part of this range almost certainly had equilibrated to the ambient RH by the time of measurement, and these particles would have values of r_{80} of near 25 μm. Thus most of the particles would have had r_{80} values in the narrow range 25-30 μm. Radii ranging from approximately 50 μm to 160 μm, if they were formation radii, would correspond to a range of r_{80} from approximately 24 μm to 75 μm. On the other hand, if these particles had not equilibrated to 65% RH at the time of measurement, it would be incorrect to convert measured radii to formation values, and hence the radii at formation reported by *Taylor and Wu* would be too high by a factor of up to ~2.4. If the reported values are corrected by the maximum amount (i.e., a factor of 2.4), then the corresponding range of r_{80} values would be approximately 10 μm to 32 μm. Consequently, these size distributions are presented in Figs. 22d and 22e both under the assumption that the particles had completely equilibrated to 65% RH, as assumed by the original investigators (and thus that the radii they presented were formation radii), and under the assumption that no equilibration had occurred (and thus that the measured radii were formation values, and that conversion to formation radii resulted in r_{80} values that were too large by a factor of 2.4). As with the size distribution of SSA concentration of *Preobrazhenskii* [1973] presented in Fig. 22d, the resulting size distributions of SSA concentration of *Taylor and Wu* [1992] presented in Figs. 22d and 22e differ greatly.

There is an apparent paradox that results from this analysis. If it is assumed that complete equilibration had occurred, then for most of the range of r_{80} calculated under this assumption (24 μm to 75 μm) it is unlikely that equilibration to the ambient RH of 65% would have occurred within the atmospheric residence time of the particles. Alternatively, for the size distributions of SSA concentration presented in Figs. 22d and 22e under the assumption of equilibration, all the values of r_{80} are greater than ~25 μm—too large for this assumption to be valid. On the other hand, if it is assumed that no equilibration had occurred, then the range of r_{80} obtained under this assumption (10 μm to 32 μm) is such that equilibration is likely to have occurred. Alternatively, for the size distributions of SSA concentration presented in Figs. 22d and 22e under the assumption of no equilibration nearly all the values of r_{80} are less than ~32 μm, and for most of this size range equilibration would have occurred. The resolution of this apparent paradox is that the size distributions of SSA concentration corresponding to these measurements probably start at the low end of those presented in Fig. 22 under the assumption of equilibration, i.e., near $r_{80} = 25$ μm, and extend only to the largest values of r_{80} on the size distributions presented in Fig. 22 under the assumption of no equilibration, i.e., near $r_{80} = 32$ μm. Thus the range of r_{80} measured by *Taylor and Wu* probably covered only a factor of near 1.3 (25 μm to 32 μm) and not an order of magnitude, as they stated (25 μm to 250 μm). Additionally, over most of this range the values of $n(r_{80})$ are underestimated, as they correspond to values of r_{80} within a size interval of $\Delta \log r_{80}$ much smaller than those resulting from the assumption of complete equilibration or of no equilibration.

Size distributions of SSA concentration reported by some other investigators also contained measured radii spanning the critical value of ~25 μm. For these size distributions a choice was made in converting and presenting them here; this procedure necessarily leads to incorrect values in one or another range of r_{80}. Those data for which the choice leads to incorrect values are denoted by dotted lines in Fig. 22.

The size distributions of *de Leeuw* [1986a,b, 1987], which included measured radii up to ~45 μm, were reported as size distributions of SSA number concentration under the assumption that all SSA particles had attained their equilibrium radii at the ambient RH. As the majority of measured radii were less than ~25 μm this assumption is likely valid, and these size distributions of SSA concentration are presented in Fig. 22 under this assumption. However, the largest particles probably had not fully equilibrated, with resultant overestimation of sizes and concentrations for these particles, and the largest value of r_{80} was probably near 25 μm. The size distribution of SSA concentration of *de Leeuw* [1990a] for $U_{10} = 25$ m s^{-1} was reported as a size distribution of volume concentration under the assumption that all particles had equilibrated to the ambient RH of 65%. Reported ambient (i.e., measured) radii ranged from 6.5 μm to over 100 μm, which under the assumption of equilibration at 65% RH correspond to r_{80} from 7.5 μm to 115 μm. However, the largest of these particles almost certainly had not equilibrated with respect to RH, and their radii were probably close to those at formation, a factor of 2.3 greater than the reported values. Consequently, the reported volume concentration almost certainly referred to the incorrect particle sizes. In the representation $dN/d\log r_{80}$ such an incorrect classification of particle radii would result in the size distribution of concentration being shifted horizontally

(a constant factor on a log scale), retaining its shape and vertical location; in other representations both the shape and vertical location would also be incorrect. As this concentration size distribution was determined by counting particles in a given size range, it can be converted to the representation $n(r_{80})$ under the assumption that the radii correspond to formation values, and it is presented in Fig. 22f under this assumption. Thus the largest value of r_{80} was near 60 μm. Although this treatment underestimates the value of r_{80} of smaller SSA particles (which would have equilibrated) by up to a factor of 2.3, this choice was made because of the paucity of data for large SSA particles at high wind speeds, these data being the only ones available in this wind speed range ($U_{10} = 25$-30 μm).

The size distribution of SSA concentration of *Lowe et al.* [1996] for 14 m s^{-1}, which those investigators described as "typical" for high wind speed conditions, contained radii of up to 60 μm that referred to "wet particles." This size distribution is presented in Fig. 22d under the assumption that all particles had equilibrated to the ambient RH, assumed here to be 80%; this presentation overestimates the values of r_{80} and of $n(r_{80})$ for the largest particles, and the upper limit of r_{80} for these measurements was probably near $r_{80} = 30$ μm. The reported median and maximum values of A and V for coarse mode particles in maritime air (which were most likely dominated by SSA particles) are almost certainly in error, as they are much greater than the values of these quantities evaluated for their typical size distributions, and far greater than other reported values for these quantities (§4.1.4.3). For instance, the median volume concentrations for winter and summer campaigns correspond to values of M of 2650 μg m^{-3} and 1400 μg m^{-3}, respectively, and the maximum values correspond to a value of M of more than $8.5 \cdot 10^4$ μg m^{-3}—nearly three orders of magnitude greater than the upper range of values for this quantity in Fig. 17.

The size distributions of SSA concentration of *O'Dowd et al.* [1997a] were obtained from measurements of particles with ambient radii up to 150 μm that were "normalized to 80% relative humidity," presumably on the assumption that the particles had attained their equilibrium values at the ambient RH, and they are presented in Fig. 22 under this assumption. However, as the largest of these particles almost certainly had not equilibrated with respect to the ambient RH by the time of measurement, the values of r_{80} and $n(r_{80})$ are thus overestimated (and even more so for $a_{80}(r_{80})$), and the upper limit of r_{80} was probably near 75 μm.

The size distribution of SSA concentration presented by *O'Dowd et al.* [2001] for $U_{10} = 10$ m s^{-1}, which they described as "typical" (only one size distribution at this wind speed was presented), resulted from measurements of particles with radii up to 150 μm. As the ambient RH was reported to be 97-99%, this size distribution is presented in

Fig. 22c under the assumption that these radii referred to values at formation (i.e., 98% RH); over the range 97-99% RH the equilibrium radii vary by only ~30% from those at 98% RH.

4.1.4.3. Features of size distributions of SSA concentration.

Despite the concerns and caveats that have been noted, the data in Fig. 22 present a rather coherent picture of the size distribution of SSA concentration. For the most part, the overall shapes of the size distributions of SSA concentration for 0.1 μm $\lesssim r_{80} \lesssim$ 25 μm within any given wind speed range are similar over a range of wind speeds up to $U_{10} = 20$ m s^{-1}, as indicated by the shaded gray regions in Fig. 22b-e, and the great majority of the data fall within a swath of only about an order of magnitude, or a factor of $\lesssim 3$ about the central curve (this is perhaps more readily seen on the graphs of $a_{80}(r_{80})$, which are presented on the same vertical scale).

In addition to considerations of data within any given wind speed range, a rather coherent picture emerges of the dependence of the size distributions of SSA concentration on wind speed, a sense of which may be gained by comparing the panels for the several wind speed ranges. As the value of U_{10}/(m s^{-1}) at the midpoint of the wind speed ranges in Fig. 22b-e increases from 6 to 9.5 to 13 to 17.5, the central value of the gray shaded region increases by successive factors of 2.5, 1.9, and 1.8, representing an overall increase of nearly an order of magnitude in SSA concentration over this range of U_{10}. While on the one hand this order-of-magnitude increase is quite substantial and appears to be indicative of the strong role played by wind speed in SSA production, on the other hand, this increase is roughly the same as the spread of the data within any individual wind speed range.

The order-of-magnitude spread in the data in each of the several rather narrow ranges of U_{10} at any value of r_{80} seems to be indicative of the variability in SSA concentrations such as $n(r_{80})$ or $a_{80}(r_{80})$, and likewise for other extensive properties of SSA, that is inherent over much of the world's oceans and that is attributable to controlling factors other than wind speed at the time of measurement. It would not seem possible to decrease the width of the shaded gray region and still encompass the majority of the data. This is seen also in Fig. 23, which displays size distributions of SSA concentration measured by a single investigator [Chaen, 1973] over a rather narrow range of wind speeds ($U_{10} = 8.0$-10.7 m s^{-1}). The implication of this variation is that local wind speed at the time of measurement is not the sole controlling factor in determining concentration values, and possibly not even the dominant factor; similar conclusions were drawn above (§4.1.2.2; §4.1.3.2) from the wind-speed dependencies of the mass concentration M and the

number concentration N. Such a large amount of scatter at any one wind speed, and the large amount of overlap among different wind speeds, precludes justification of any parameterization of size distributions of SSA concentration or of other extensive properties of SSA as a function solely of U_{10} (§4.1.4.4) that could be used at any given time and location with confidence better than a factor of $\lesssim 3$.

The expectation prevalent throughout the literature on sea salt aerosol that local wind speed is the dominant controlling factor of SSA production would imply that the correlation of SSA concentration and local wind speed at the time of measurement should increase with increasing particle size because of the decrease in mean atmospheric residence time with increasing particle size. Hence, concentrations of larger SSA particles should provide a good indication of local SSA production. However, this expectation is not borne out by the size distributions presented in Fig. 22, which do not exhibit noticeably less scatter with increasing particle size.

In the lowest wind speed range ($U_{10} = 0$-4 m s^{-1}) no effort was made to present a shaded region that would encompass most of the data, given the large spread of the data. However, other than the two size distributions of SSA concentration reported by Metnieks [1958] that are substantially lower than the others and the one by Exton et al. [1985] that is substantially higher than the others, the vast majority of the data are encompassed by the gray shaded region presented for the wind speed range $U_{10} = 5$-7 m s^{-1}. The fact that SSA concentrations at wind speeds 0-4 m s^{-1}, under which conditions breaking waves are rare if not entirely absent (and thus for which essentially no SSA production would be expected), are comparable to those at higher wind speeds is testimony to the importance of prior production on the aerosol that is present at a given time and location of measurement and to the fact that local wind speed at the time of measurement is not a good predictor of SSA concentration. In the highest wind speed range ($U_{10} = 25$-30 m s^{-1}) the data were so sparse, and from so few groups, that a shaded region encompassing the majority of the data could not be confidently drawn; in fact most of the data sets for the highest wind speeds are from a single group [e.g., Smith et al., 1989, and Smith et al., 1993] and are not independent.

As the shapes of the size distributions of SSA concentration presented in Fig. 22 are for the most part similar in different wind speed ranges, this size dependence can be examined independently of wind speed.

Relatively few concentration size distributions of SSA particles have been reported that extend much below $r_{80} = 1$ μm at $U_{10} > 7$ m s^{-1}, making it difficult to draw general conclusions regarding the dependence of concentrations of small SSA particles (i.e., $r_{80} \lesssim 1$ μm) on size and on wind speed. Several reported size distributions of marine aerosol concentrations [e.g., *Exton et al.*, 1985; *M. Smith et al.*, 1989, 1991] exhibit increasing values of $n(r_{80})$ with decreasing r_{80} near 0.1 μm (resulting also in substantial contributions to the total concentration of surface area from particles of these sizes), but as these size distributions were determined using optical detectors which are not specific to sea salt, this contribution probably consisted in large part of particles other than sea salt. For this reason, size distributions of SSA concentrations based on these measurements presented in Fig. 22 were not extended to r_{80} below 0.4 μm. Importantly, measurements by *O'Dowd et al.* [1997a] using optical detectors together with a thermal (volatility) technique to distinguish sea salt particles from other types yielded much lower concentrations of SSA particles of these sizes than measurements that were not specific to SSA particles.

Very few size distributions of SSA concentration have been reported for $r_{80} \lesssim 0.1$ μm, perhaps in part because of the overwhelming abundance of other types of particles in this size range (§4.1.1.1; Fig. 16). The average of seven size distribution of SSA concentration presented by *Gras and Ayers* [1983], which they stated to be similar to each other, exhibited a maximum value (in terms of $n(r_{80})$) near 10 cm^{-3} at $r_{amb} \approx 0.1$ μm, and decreased rapidly with decreasing r_{amb} to ~0.3 cm^{-3} for $r_{amb} \approx 0.03$ μm (these size distributions were not presented in Fig. 22 as no wind speeds were stated). Measurements of size distributions of marine aerosol concentration reported by *Clarke et al.* [2003] at a coastal location in Hawaii for which surf zone influences were removed revealed the presence of large numbers of particles (~100 cm^{-3}) remaining after heating to 360 °C, suggesting that these particles probably contained sodium (and hence were SSA particles). These particles were strongly internally mixed, as indicated by the marked difference in the size distributions of concentration before and after heating, and it appears that sea salt contributed only a small fraction of the mass of most particles. One size distribution of SSA concentration was presented, extending down to $r_{80} \approx 0.01$ μm, with $n(r_{80})$ exhibiting a broad maximum of ~100 cm^{-3} from $r_{80} = 0.02$-0.06 μm and decreasing rapidly outside of this range. In contrast to these sets of measurements, many other measurements of marine aerosol concentrations reported that few if any SSA particles with $r_{80} \lesssim 0.1$ μm were present, as noted in §4.1.1.1. Typically measurements of size distributions of number concentration of marine aerosol particles exhibit an increase with decreasing particle size near radius 0.1 μm, with values of the total number concentration being several hundred per cm^{-3} even for clean marine conditions, as noted above (§4.1.3.1). Based on these considerations it must be concluded that in most situations SSA particles with $r_{80} < 0.1$ μm contribute little to N, and even less to quantities such as R_{80}, A_{80}, and V_{80} (or M).

The size distribution of SSA concentration in the representation $n(r_{80})$ generally exhibits a peak between 0.2 μm and 0.5 μm, with maximum values of $n(r_{80})$ ranging from near 1 cm^{-3} to 10 cm^{-3} at moderate wind speeds and from 10 cm^{-3} to 100 cm^{-3} for higher wind speeds (Fig. 22). The vast majority of the contribution to N is from small SSA particles (i.e., $r_{80} \lesssim 1$ μm). As this size range consists mostly of particles with r_{80} ranging from 0.1 to 1 μm, the value of $\Delta\log r_{80}$ is approximately unity; thus values of N calculated from size distributions of SSA concentration have roughly these same numerical values (i.e., 1 cm^{-3} to 100 cm^{-3}), consistent with reported values discussed in §4.1.3.1. With increasing r_{80} greater than 0.2-0.5 μm, $n(r_{80})$ decreases, with values at moderate wind speeds ranging from 1 cm^{-3} to 10 cm^{-3} at $r_{80} = 1$ μm to near 10^{-3} cm^{-3} at $r_{80} = 25$ μm (although there is much variability at this size). Medium SSA particles (i.e., 1 μm $\lesssim r_{80} \lesssim 25$ μm) contribute little to N, but provide substantial contributions to R_{80} and to A_{80}, and they provide the dominant contribution to V_{80} and M.

The representation $a_{80}(r_{80})$ exhibits a broad maximum at r_{80} of several micrometers, with typical maximum values ranging from 10 μm^2 cm^{-3} to 100 μm^2 cm^{-3} at moderate wind speeds, and up to several hundred μm^2 cm^{-3} at higher wind speeds. The majority of the contribution to $a_{80}(r_{80})$ is from SSA particles with r_{80} from somewhat less than 1 μm to somewhat greater than 10 μm, again comprising an interval characterized by $\Delta\log r_{80}$ of approximately unity; thus numerical values of A_{80} will also be in the range 10 μm^2 cm^{-3} to several hundred μm^2 cm^{-3}. Based on the shapes of the representations $n(r_{80})$ and $a_{80}(r_{80})$, $dR_{80}/d\log r_{80}$ is expected to exhibit a maximum in the vicinity of $r_{80} \approx 1$ μm, having maximum values from about 1 μm cm^{-3} to 10 μm cm^{-3}, with appreciable contributions to R_{80} from SSA particles with r_{80} ranging from several tenths of a micrometer to several micrometers. Likewise, the representation $v_{80}(r_{80})$ (equivalently $m(r_{80})$) is expected to exhibit a maximum at r_{80} near 10 μm with $v_{80}(r_{80})$ ranging from roughly 20 μm^3 cm^{-3} to 200 μm^3 cm^{-3} (and values of $m(r_{80})$ ranging from ~7 μg m^{-3} to 75 μg m^{-3}), with appreciable contributions over much of the range comprising medium SSA particles (i.e., $r_{80} \approx$ 1-25 μm). The corresponding values for V_{80} and M range from 20 μm^3 cm^{-3} to 200 μm^3 cm^{-3} and ~7 μg m^{-3} to 75 μg m^{-3}, respectively—consistent with range of values of M presented in Fig. 12 and discussed in §4.1.2.1 (~5-50 μg m^{-3}).

There are few measurements of size distributions of SSA concentration that extend much above $r_{80} \approx 25$ μm, and values that have been reported differ greatly—often by several orders of magnitude. Some of this difference can probably be attributed to sampling techniques, heights, and the like, but little confidence can be placed in most of these measurements because of issues with RH equilibration (§4.1.4.2);

furthermore, concentrations of SSA particles in this size range generally suffer from large uncertainties resulting from small numbers of particles counted. However, based on the size distributions of SSA concentration presented in Fig. 22 it can be concluded that the number concentration of large SSA particles (i.e., $r_{80} \gtrsim 25$ µm) at typical heights at which measurements are made is so low that under most circumstances these particles would contribute little to properties of interest with regard to atmospheric chemistry, light scattering, and the like. This is not surprising, as few large particles are expected to be entrained to sufficient heights and in sufficient numbers to be detected (§2.9.2). The presence of large SSA particles at heights of several meters or more was attributed by *Bortkovskii* [1987, p. 107] to disturbing influences of ship hulls and coasts on the wind and wave fields, implying that the actual concentration of these particles in the marine atmosphere at heights of ~10 m would be even less than those reported.

A further concern is that as particle size increases, the consequences of counting statistics become increasingly important, as fewer particles are available for sampling. In particular, the low concentrations of large SSA particles frustrate attempts to characterize them in a quantitatively meaningful way. Consequently, there are large uncertainties in number concentrations for the largest SSA particles, and correspondingly large uncertainties of SSA surface area and volume (or mass) concentrations calculated from number concentrations, that result in low accuracy of these concentrations, often at sizes that ostensibly provide the dominant contribution. Some of the size distributions of SSA concentration in Fig. 22 exhibit substantial noise, likely resulting from small numbers of particles in size intervals. In several instances (such as the size distributions of SSA concentration of *Woodcock* [1952]) the concentrations of both number and surface area appear to increase at the largest sizes; such increases arise in part from conversion from representations such as *dN/dr*, but they are also indicative of large uncertainty in concentrations at these sizes.

Knowledge of the uncertainty resulting from counting statistics permits evaluation of the confidence that can be placed in reported size distributions of SSA concentration and in conclusions that can be drawn from these data. Although not often reported, in some instances estimates for this uncertainty can be obtained. *Chaen* [1973] noted when fewer than 10 particles were present in the sample counted for a given size interval; this situation usually occurred for $r_{80} \gtrsim 5$ µm at U_{10} near 5 m s^{-1}, increasing to $r_{80} \gtrsim 15$ µm at U_{10} near 12 m s^{-1}. *Exton et al.* [1985] reported that for concentrations of large particles (no range was specified but the size range containing the largest particles was $r_{amb} = 16$-23.5 µm) the variations from different 15 min samples

were as great as 100% of the mean values. The statistical sampling errors reported by *de Leeuw* [1990a] for the size distribution at $U_{10} = 25$ m s^{-1} varied from 14% to 70% for concentrations of particles with reported radii from 50 µm to 100 µm. Descriptions of measurement techniques permit sampled particle rates for other investigations to be calculated from reported size distributions of SSA concentration. For instance, only 1 to 2 particles per minute with measured radii greater than ~25 µm would have been sampled by *de Leeuw* [1986a,b, 1987], and reported sampling times of 4-7 min would have resulted in ~10 or fewer particles of these sizes. Roughly one particle per minute in the size interval of reported radius near 50 µm would have been sampled by *Taylor and Wu* [1992]. Approximately one particle every three minutes for the size interval of reported radius near 20 µm, and approximately one particle per hour for the size interval of reported radius near 50 µm, would have been sampled by *Lowe et al.* [1996]. These values reflect the low concentrations of large SSA particles and call into question any parameterizations of their concentration and its dependence on wind speed.

Of interest is whether there is any evidence of multiple modes in size distributions of SSA concentration that might be indicative of different production mechanisms (film, jet, spume), as has been suggested by some investigators [*Mészáros and Vissy*, 1974, and *Patterson et al.*, 1980; *O'Dowd et al.*, 1997a]. In several instances indication of systematic trends with wind speed may be seen in the data of a single group. For example, a relative increase in the concentration of larger SSA particles with increasing wind speed was reported by *Woodcock* [1953] (though with no indication modal behavior), but as this result was based primarily on measurements taken at heights of hundreds of meters it might be explained by the increase in vertical mixing resulting from higher wind speed (§2.9.5). Size distributions of SSA concentration reported by *Chaen* [1973] exhibited proportionally greater increases in concentrations of SSA particles with $r_{80} \gtrsim 5$-10 µm than in concentrations of smaller SSA particles at wind speeds $U_{10} \gtrsim 11$ m s^{-1}. *Exton et al.* [1985] suggested that spume drop production could account for the increase in the volume concentration of SSA particles with $r_{80} \gtrsim 8$ µm with increasing wind speed at $U_{10} \approx 11$ m s^{-1}. Concentrations of particles with $r_{amb} > 4$ µm reported by *Smith et al.* [1993] exhibited a much greater dependence on wind speed than those of smaller particles, with concentrations of particles with $r_{amb} \gtrsim 7$ µm exhibiting a marked increase as wind speeds increased above 15 m s^{-1}. Additionally, for the two largest wind speed ranges ($U_{10} = 15$-20 m s^{-1} and $U_{10} = 25$-30m s^{-1}) there is some suggestion of a disproportionate increase in the concentrations of SSA particles with $r_{80} \gtrsim 10$ µm relative to those at lower wind speeds (Figs. 22*e* and 22*f*). Such an increase might be indicative of spume drop production at these higher wind speeds, but alternatively it might simply represent enhanced vertical entrainment of particles in these sizes to the height of measurement. While there is evidence of modal behavior in some of the size distributions of SSA concentration at these high wind speeds, there are few data and the conditions under which these data were taken often represented transient occurrences, which do not appear to be representative of other situations at the same wind speed (this was discussed in §4.1.2.2 with regard to measurements of M by *Woodcock* [1953], and by *Smith et al.* [1989] at high wind speeds). Moreover, a relative increase in the concentrations of larger SSA particles with increasing wind speed (which might imply spume drop production) is not apparent in the data for $U_{10} < 20$ m s^{-1} as a whole. Thus, based on the data considered, it would be difficult to argue for evidence of spume drop production for SSA particles with $r_{80} \lesssim 25$ μm (small and medium SSA particles) at most wind speeds encountered over the ocean.

Several size distributions of SSA concentration differ markedly from the majority of the others. Those of *Preobrazhenskii* [1973] at 6-10 m s^{-1} (Fig. 22c) and at 15-25 m s^{-1} (Fig. 22e) are anomalous, exhibiting SSA concentrations that are several orders of magnitude lower than nearly all other values (this does not appear to have been discussed previously despite numerous citations to that study). Some of the size distributions of SSA concentration reported by *A. Mészáros and Vissy* [1974], especially the those at 12 m s^{-1} (Fig. 22d) and that at 7 m s^{-1} (Fig. 22b) for $r_{80} > 10$ μm, are somewhat lower than those of other investigations. The latter situation is at even greater variance than it might appear, as the concentrations of these particles were probably overestimated [*E. Mészáros*, 1982], possibly by up to an order of magnitude, because of non-isokinetic sampling (although no correction for this is made in Fig. 22). The size distribution of SSA concentration reported by these investigators at 6 m s^{-1}, while qualitatively similar to that at 7 m s^{-1} (Fig. 22b), is greater by as much as a factor of 5 for $r_{80} \gtrsim 3$ μm. Some of this difference may be the result of these size distributions being averages of data taken at different locations with very different sea temperatures, but there are too few systematic data to draw conclusions about the dependence of SSA particle concentrations on such factors. The concentrations for $r_{80} < 2$ μm reported by *Hughes and Richter* [1980] at 6 m s^{-1} (Fig. 22b) are markedly greater than most other reported measurements; these investigators commented on possible uncertainties in the measured relative humidity and wind speed, and it is also possible that continental aerosol particles contributed to their measurements.

The SSA concentration size distribution presented by *O'Dowd et al.* [1997a] for $U_{10} = 8$ m s^{-1} (Fig. 22c), which they characterized as "typical," is much greater than most of the others in this wind speed range, and in fact is greater than

the SSA concentration size distribution they presented for $U_{10} = 9$ m s^{-1} by nearly an order of magnitude over most of the range of r_{80} from 3 μm to 25 μm. The size distribution of SSA concentration presented by *O'Dowd et al.* [2001] for $U_{10} = 10$ m s^{-1}, which these investigators also characterized as "typical" (only one size distribution at this wind speed was presented) yields concentrations of SSA particles with $r_{80} \gtrsim 25$ μm that considerably greater than the few other determinations in that wind speed range (Fig. 22c). As the ambient RH at which these data were obtained was 97-99%, the radii of these particles would have remained near their formation values (i.e., in equilibrium at 98% RH) for the entire time they spent in the atmosphere (§4.1.4.2), and these data were presented in Fig. 22 under this assumption. However, under this assumption, the SSA volume concentration they reported would correspond to a dry SSA mass concentration of nearly 800 μg m^{-3}, far greater than those reported in Fig. 17 at this wind speed (had the particles equilibrated to RH 99% instead of RH 98% the value of M would have been nearly 400 μg m^{-3}, still much greater than most of the values in Fig. 17). Additionally, based on the reported concentration of SSA surface area, the light-scattering coefficient would have been $\sigma_{sp} \approx 700$ Mm^{-1} (under the approximation $Q_s = 2$, which is valid for the range of SSA particles that provided most of this contribution; §2.1.5.3), implying that a beam of light would experience attenuation of 50% over a horizontal distance of 1 km. Moreover, it is unlikely that many such particles could attain the height of measurement (10 m) in the concentrations reported, as the eddy diffusivity at this wind speed would be insufficient to overcome the large gravitational terminal velocities of such particles (§2.9.4); SSA particles with r_{80} values of (25, 50, 75) μm, having geometric radii of (50, 100, 150) μm, would fall 10 m because of gravity in (40, 15, 10) s, and thus would not remain long in the atmosphere. Therefore it is most unlikely that this distribution can be considered typical, as characterized by these investigators.

4.1.4.4. Parameterizations. Several expressions have been proposed to parameterize size distributions of SSA concentration as a function of wind speed. However, the large spread in $n(r_{80})$ within any one wind speed range—roughly an order of magnitude for most sizes—and the large amount of overlap among different wind speed ranges, together with the knowledge that SSA concentration is influenced by factors other than local wind speed at the time of measurement, suggest that a precise parameterization based on only wind speed should not be expected. Nonetheless, parameterizations of size-dependent SSA concentration as a function solely of wind speed may be useful for scoping purposes (e.g., as a basis for comparison with concentrations of particles other than sea salt or as a basis for estimating properties of SSA concentration such as light-scattering coefficients), although variability of size distributions of SSA concentration at any given wind speed must be recognized and taken into account.

Several different functional forms have been used to parameterize the size dependence of the concentration of marine aerosol and of SSA, the most commonly used ones being power law distributions, gamma or modified gamma distributions, and lognormal distributions (§2.1.4.2). Power laws were formerly common [e.g., *Moore and Mason*, 1954], and have more recently been used [*Toba and Chaen*, 1973; *Lai and Shemdin*, 1974; *Wu*, 1979a; *Wu et al.*, 1984; *Wu*, 1990d, 1993] to represent the size distribution of concentrations of large SSA particles very near the surface (§4.3.6.1). One such expression, $n(r_{80}) = cr_{80}^{-3}$, was proposed by *Chaen* [1973] to describe his data; over the range $r_{80} = 2\text{-}25$ μm the data presented in Fig. 23 for $U_{10} = 8.0\text{-}10.7$ m s^{-1} can described by this relation within an uncertainty of $\lessgtr 3$. Another expression, equivalent to $n(r_{80}) = cr_{80}^{-6}$, was proposed by *Taylor and Wu* [1992] to describe their data, but based on the discussion in §4.1.4.2 concerning the lack of equilibrium with respect to RH for most of the particles they measured, the range of r_{80} for their data (i.e., 24-32 μm) appears too narrow to justify such a formulation. More generally, it is clear from Fig. 22 that power law expressions, which would be straight lines on such a log-log plot, do not accurately represent the majority of the data over any appreciable range of r_{80}.

Gamma distributions and modified gamma distributions have been widely used to parameterize marine aerosol and SSA size distributions of concentration, often with regard to considerations of light transmission through the atmosphere [*Eriksson*, 1959; *Deirmendjian*, 1964; *Barnhardt and Streete*, 1970; *DeLuisi et al.*, 1972; *Wells et al.*, 1977; *Borisenkov and Kuznetsov*, 1978; *Hughes*, 1980; *Hughes and Richter*, 1980; *Davidson et al.*, 1982; *Goroch et al.*, 1982; *Bortkovskii*, 1987; *Dobbie et al.*, 2003]. Several such expressions include a wind-speed dependence of the parameters that govern the shape of the size distribution, allowing a relative increase in the concentration of larger particles with increasing wind speed. However, most of these formulations were based on rather limited data sets. An expression of this form was proposed by *Taylor and Wu* [1992], who suggested that it would apply to r_{80} less than 1 μm despite the fact that their data did not extend below $r_{80} \approx 25$ μm (§4.1.4.2) and exhibited no indication of the maximum in dV/dr near $r_{80} \approx 2$ μm shown in Fig. 4 of that reference. Although sums of gamma or modified gamma distributions can provide accurate approximations to measured size distributions of SSA concentration, they are currently not often used for this purpose.

The functional form most commonly used to parameterize size distributions of SSA concentration is a lognormal distribution or a sum of lognormal distributions (§2.1.4.2). In most parameterizations of this form the dependence on wind speed is contained only in the coefficient(s) n_0 representing the amplitude(s) of a given mode, with values of r'_{80} (which represents the location of the peak of the mode) and σ (which represents its width) independent of wind speed. With multiple lognormal modes even though the location r'_{80} and width σ of each mode may be independent of wind speed, the shape of the overall size distribution may vary with wind speed because of different wind-speed dependencies in values of n_0 for the different modes. Three formulations of size distributions of SSA concentration that consist of several lognormal modes are the original Navy Aerosol Model [*Gathman*, 1983a,b], a different version of this model [*Gathman*, 1989; *Gerber*, 1991] that was later incorporated into the Navy Ocean Vertical Aerosol Model [*Gathman and Davidson*, 1993], and an expression proposed by *O'Dowd et al.* [1997a]. These formulations are shown in Fig. *22b-e*. Values of n_0, r'_{80} and σ for each of the modes of these formulations for the representation $n(r_{80})$ are presented in Table 14, as are the locations of the maximum values of $dR_{80}(r_{80})/d\log r_{80}$, $a_{80}(r_{80})$, and $v_{80}(r_{80})$ (or $m(r_{80})$) for each of the modes; other representations can be obtained using the relations in Table 3. Values of total integral properties of SSA concentration N, R_{80}, A_{80}, V_{80}, and M for each mode of these formulations at $U_{10} = 10$ m s^{-1} are presented in Table 15; values at other wind speeds can readily be calculated from the wind-speed dependencies given in Table 14.

The original Navy Aerosol Model [*Gathman*, 1983a,b] was developed to investigate infrared and optical transmission in the marine boundary layer. This formulation was based upon measurements of marine aerosol concentrations (mostly in unpublished or internal reports) obtained from shipboard, mainly in the North Atlantic, and from towers off the California coast and in the North Sea. These data included concentrations of particles with r_{amb} to greater than 20 μm and wind speeds to over 20 m s^{-1} (though not necessarily all from the same study); the lack of adequate data for large particles was noted. The size distribution of marine aerosol concentration was presented as the sum of three lognormal modes with parameters that could be evaluated in terms of readily available meteorological or environmental quantities: wind speed, RH, and visibility or radon concentration (to provide an estimate of the continental influence of the air mass). The first mode represents a background continental aerosol; as it is not relevant to this discussion, it is not considered further. The second mode represents a stationary marine component: aerosol with a long residence time assumed to have been produced by prior wind conditions;

Table 14. Lognormal Expressions Proposed to Parameterize Size Distributions of Sea Salt Aerosol Concentration

Parameterization	Mode	n_0/cm^{-3}	$r'_{80}/\mu\text{m}$	σ	Location of Maximum $/\mu\text{m}$		
					$\dfrac{dR_{80}}{d\log r_{80}}$	a_{80}	v_{80}
Navy Aerosol Model	2nd	$4.15[U_{10,\text{avg}}/(\text{m s}^{-1}) - 2.2]$	0.4	2.0	0.66	1.1	1.8
[*Gathman*, 1983a,b]	3rd	$0.090[U_{10}/(\text{m s}^{-1}) - 2.2]$	3.3	2.0	5.4	9	15
Navy Ocean Vertical	2nd	$4.15[U_{10,\text{avg}}/(\text{m s}^{-1}) - 2.2]$	0.4	2.0	0.66	1.1	1.8
Aerosol Model	3rd	$0.0093 \exp[0.14 \cdot U_{10}/(\text{m s}^{-1})]$	3.3	2.0	5.4	9	15
[*Gathman and Davidson*, 1993]							
O'Dowd et al., 1997	film	$2.7 \exp[0.22 \cdot U_{10}/(\text{m s}^{-1})]$	0.2	1.9	0.3	0.5	0.7
	jet	$0.68 \exp[0.097 \cdot U_{10}/(\text{m s}^{-1})]$	2.0	2.0	3	5	8.5
	spume	$1.2 \cdot 10^{-6} \exp[0.16 \cdot U_{10}/(\text{m s}^{-1})]$	12	3.0	40	130	450
Canonical distribution		$0.07[U_{10}/(\text{m s}^{-1})]^2$	0.3	2.8	0.87	2.5	7.2

Values of n_0, r'_{80} and σ for each mode of lognormal expressions (of the form Eq. 2.1-9) proposed to parameterize SSA size distributions of concentration described in §4.1.4.4 in representation $n(r_{80}) \equiv dN/d\log r_{80}$, and location of maximum values of SSA size distributions of concentration in alternative representations $dR_{80}(r_{80})/d\log r_{80}$, $a_{80}(r_{80})$, and $v_{80}(r_{80})$ (or $m(r_{80})$).

the amplitude of this mode was parameterized in terms of the average of the local wind speed (at 10 m) over the previous 24 hours, $U_{10,\text{avg}}$. The third mode represents SSA particles produced locally at the sea surface; the amplitude of this mode was parameterized by the wind speed at the time and location of measurement, U_{10}. Thus the latter two modes were intended to represent SSA particles. The effect of RH was taken into account by assuming all particles behaved as if they were sea salt. Size-dependent SSA concentrations

obtained from this formulation are shown in Fig. 22b-e (evaluated using $U_{10,\text{avg}} = U_{10}$) over the radius range $r_{80} = 0.1\text{-}25\ \mu\text{m}$ (the range to which this formulation applies was not specified; *Gathman* [1983b] presented size distributions extending to much larger sizes). For $r_{80} \lesssim 1\ \mu\text{m}$ and for $r_{80} \gtrsim 5\ \mu\text{m}$ this formulation yields size distributions of SSA concentration greater than most measured values by more than an order of magnitude, especially so at low wind speeds. The SSA mass concentration M based on this

Table 15. Values of Total Integral Properties of Sea Salt Aerosol Concentration According to Proposed Parameterizations of Size Distributions at $U_{10} = 10\ \text{m s}^{-1}$

Parameterization	Mode	N/cm^{-3}	$R_{80}/(\mu\text{m cm}^{-3})$	$A_{80}/(\mu\text{m}^2\text{ cm}^{-3})$	$V_{80}/(\mu\text{m}^3\text{ cm}^{-3})$	$M/(\mu\text{g m}^{-3})$
Navy Aerosol Model	2nd	25	13	140	65	17
[*Gathman*, 1983a,b][a]	3rd	0.54	2.3	200	780	210
	sum	**25**	**15**	**340**	**850**	**230**
Navy Ocean Vertical	2nd	25	13	140	64	17
Aerosol Model	3rd	0.029	0.12	11	41	11
[*Gathman and Davidson*, 1993][a]	**sum**	**25**	**13**	**150**	**110**	**28**
O'Dowd et al., 1997a	film	17	4.2	20	3.7	1
	jet	1.4	3.4	180	400	110
	spume	$7.4 \cdot 10^{-6}$	$1.6 \cdot 10^{-4}$	0.15	12	3.2
	sum	**18**	**7.6**	**200**	**420**	**110**
Canonical distribution		7.6^b	4.0^b	76^c	110^c	30^c

Values are rounded to two significant digits. Values at other wind speeds can be calculated using the wind speed dependencies presented in Table 14.
[a] $U_{10,\text{avg}}$ is taken equal to U_{10} to evaluate properties for these parameterizations.
[b] For lower limit of integration $r_{80} = 0.1\ \mu\text{m}$, the values of N and R_{80} are reduced by approximately 15% and 5%, respectively.
[c] For upper limit of integration $r_{80} = 25\ \mu\text{m}$, the values of A_{80} and V_{80} (and M) are reduced by approximately 1% and 10%, respectively.

formulation, shown in Fig. 17 over the wind speed range 3 m s^{-1} to 20 m s^{-1} (taking $U_{10,avg} = U_{10}$), is likewise much greater than most reported values. As more than 95% of the SSA mass concentration obtained with this formulation is provided by SSA particles with $r_{80} < 25$ µm, the overestimation of M cannot be due to lack of RH equilibration. These comparisons indicate that this formulation does not accurately represent SSA size distributions of concentration.

A subsequent formulation [*Gathman*, 1989; *Gerber*, 1991], based mainly on measurements of concentrations of larger SSA particles reported by *de Leeuw* [1986b], used a different wind speed dependence for the amplitude of the third mode (representing freshly produced SSA particles) compared to the original Navy Aerosol Model discussed above. This formulation was later incorporated into the Navy Ocean Vertical Aerosol Model [*Gathman and Davidson*, 1993], and is used in the current Navy Aerosol Model, NAM-6 [*Zeisse*, 1999; *A. Van Eijk*, personal communication, 2004]. For wind speeds 5-20 m s^{-1} the amplitude of this mode (Table 14) is less than that of the corresponding mode in the original Navy Aerosol Model by a roughly a factor of 15. *Hughes* [1987] also proposed modifying the third mode by retaining the same functional dependence on U_{10} but decreasing the amplitude by a factor of 14, based also on measurements of *de Leeuw* [1986b]. The location and shape of both modes and the amplitude of the second mode remained the same as in the original Navy Aerosol Model, and for $r_{80} \lesssim 2$ µm the two formulations are essentially the same. Size distributions of SSA concentration according to the Navy Ocean Vertical Aerosol Model, shown in Fig. 22b-e over the range $r_{80} = 0.1$-25 µm (again evaluated using $U_{10,avg} = U_{10}$), are much more consistent with measured size distributions of SSA concentration for $r_{80} \gtrsim 5$ µm, but still appear to overestimate most measured values for $r_{80} \lesssim 1$ µm (and thus will result in higher values of N and R_{80}; Table 15). The SSA mass concentration M based on this formulation (again evaluated using $U_{10,avg} = U_{10}$), which consists of roughly equal contributions from each mode, is much more consistent with measured values than that based on the original Navy Aerosol Model (Fig. 17).

The use of average wind speed, or wind speed history, to parameterize SSA size distributions of concentration was criticized by *Reid et al.* [2001] as being unphysical, especially in midlatitudes, as such a parameterization is in essence forecasting on persistence. However, as SSA concentrations are the consequence of prior production, transport, and removal processes, parameterization of SSA concentration in terms of the average wind speed at the location of measurement $U_{10,avg}$ represents an attempt to account for these influences, and thus is perhaps somewhat better suited as a basis for such parameterization than is U_{10} at the time and location of measurement. Still, it would

seem that parameterization in terms of this average wind speed at the location of measurement might not adequately account for these influences, as this average wind speed is not necessarily the same as the wind speed responsible for production and subsequent transport and removal of the aerosol, especially if there has been a substantial change in U_{10} over the averaging time at the location of measurement.

The formulation of *O'Dowd et al.* [1997a] parameterizes the size distribution of SSA concentration over the range $r_{80} = 0.1$-150 µm as a sum of three lognormal modes, denoted the "film drop", "jet drop", and "spume drop" modes; thus,

$$n(r_{80}) = n_{film}(r_{80}) + n_{jet}(r_{80}) + n_{spume}(r_{80}).$$

Such an assignment of different production mechanisms to different modes had previously been advanced by *Mészáros and Vissy* [1974] and *Patterson et al.* [1980]. The form used by *O'Dowd et al.* to represent the size distributions of SSA concentration was not explicitly stated, but as graphs of $dN/d\log r$ were presented and as the quantities they presented characterizing their parameterization are consistent with this form, it is assumed that the values given for the location of the peak of the mode referred to the representation $n(r_{80})$. This formulation was based on measurements at wind speeds up to 17 m s^{-1} during a cruise in the Northeast Atlantic; it was stated by *O'Dowd et al.* [1999a] as applicable to the wind speed range $U_{10} = 2$-17.5 m s^{-1}. A thermal volatility technique was used for small ($r_{80} \lesssim 3$ µm) particles to ensure that only sea salt particles were included; the same data were used by *O'Dowd and Smith* [1993] as the basis for the dependence of δN on wind speed (§4.1.3.2; Fig. 21). Few details were presented for the sampling and analysis of the larger particles, which were sized and counted using optical particle counters. Apparently particles with geometric radii up to 150 µm were measured, and evidently the investigators assumed that these particles were in equilibrium with the ambient relative humidity, as implied by their presentation in their Fig. 2a of "the observed size distributions normalized to 80% relative humidity"; however, such equilibrium was almost certainly not established, and the largest values of r_{80} probably did not exceed 75 µm.

The size distributions of SSA concentration obtained from the formulation of *O'Dowd et al.* [1997a] are shown in Figs. 22a-e (the two parameterizations in Fig. 22e are discussed below). Size distributions of SSA concentration resulting from this formulation are systematically greater (by up to an order of magnitude) than most measured values reported in other investigations, both for $r_{80} \lesssim 0.5$ µm (the range of SSA particle size that contributes the most to N) and for

$r_{80} \gtrsim 2$ µm (the range that contributes most to M). The concentration of SSA mass M according to this formulation is likewise greater than most measured values (Fig. 17). As according to this formulation approximately 95% of the contribution to M is from SSA particles with $r_{80} < 25$ µm, this overestimation of M cannot be attributed to contributions of large SSA particles that may or may not have equilibrated. For these reasons it must be concluded that this formulation does not accurately represent the data as a whole.

The formulation of *O'Dowd et al.* [1997a] for the film drop and jet drop modes was extended to lower (< 2 m s^{-1}) and higher (> 17.5 m s^{-1}) wind speeds by *Jones et al.* [2001] for use in a climate model, based in part on fitting results of *O'Dowd et al.* [1999a], so that as wind speed approached zero the values of n_0 for these modes approached zero, and so that at high wind speed these values of n_0 asymptotically approached values that are respectively 70% and 20% greater than their values at 17.5 m s^{-1}. There is a concern with the assumption that the concentration approaches zero as the wind speed approaches zero, as this would imply that the wind speed has been zero for sufficiently long that the existing concentration has decreased to near zero; as noted in §2.9.6, the characteristic time describing the rate of decrease of SSA concentration when the wind speed has rapidly decreased to a much lower value is of order days for SSA particles with r_{80} less than a few micrometers. Additionally, there is concern with the extrapolation to higher wind speeds, as there are few data in this wind speed range. There is also no apparent physical justification for the assumption that with increasing wind speed the concentrations of SSA particles of a given size in the film and jet drop modes approach a constant value.

There is a major discrepancy in the formulation proposed by *O'Dowd et al.* [1997a] concerning the value of the amplitude of the spume drop mode: at $U_{10} = 17$ m s^{-1} the formula presented on p. 76 of that reference yields $n_0 = 1.8 \cdot 10^{-5}$ cm^{-3}, whereas the value given for N for the spume drop mode in Fig. 2a of that reference (which is consistent with the fit shown on the graph presented there for this wind speed) yields $n_0 = 4.1 \cdot 10^{-3}$ cm^{-3}—more than a factor of 200 greater. The formula came from the wind speed relationship for the full data set [*C. O'Dowd*, personal communication, 1999] whereas the value for N at $U_{10} = 17$ m s^{-1} presented in Fig. 2a of that reference resulted from a lognormal fit of a size distribution of SSA concentration at 17.5 m s^{-1}. The two parameterizations are shown in Fig. 22e, denoted parameterization 1 and parameterization 2, respectively. Such a large discrepancy raises questions regarding the wind speed-dependence of the amplitude of this mode.

A key concern with the formulation of *O'Dowd et al.* [1997a] is the apportionment of the measured size-dependent SSA concentrations into several modes and attribution of these modes to different production mechanisms (i.e., film drops, jet drops, and spume drops). Although such an attribution (or that proposed previously by *Mészáros and Vissy* [1974] or by *Patterson et al.* [1980]) would be attractive, there would not seem to be any way to ascribe the SSA particles in a particular size range to a particular production mechanism solely from measurements of SSA concentrations. Arguably if these size distributions clearly exhibited modal behavior and if the relative amplitudes of different modes changed with increasing wind speed, then an explanation that the different modes arise from different production mechanisms might be offered, but this would seem hard to demonstrate from SSA concentration measurements, which, as noted previously, are the result of prior size-dependent production, transport, and removal processes. In any event there is little evidence in Fig. 22 of systematic change in the shape of the size distribution of SSA concentration with increasing wind speed that would justify multiple modes. Moreover there are specific concerns with the expressions for the concentrations of SSA particles in the several modes proposed by *O'Dowd et al.* based on consideration of the relative amplitudes of these modes. For instance, the contribution of the film drop mode to total number concentration N is virtually the same as the sum of the concentrations δN in the four size ranges presented by *O'Dowd and Smith* [1993] (Fig. 21) despite the fact that these size ranges also included roughly half the size range of the jet drop mode and more than half the number concentration from this mode.

There are several further concerns with this formulation with regard to the spume drop term given by the wind speed formula (parameterization 1). An overriding concern is that SSA particles with $r_{80} \gtrsim 25$ µm would not have equilibrated with respect to the ambient RH by the time of measurement, as presumably assumed by *O'Dowd et al.* [1997a], and thus that concentrations of these larger SSA particles would be attributed to erroneously high values of r_{80}. As a result $n(r_{80})$ would be greatly overestimated at these higher values of r_{80} because of the rapid decrease in $n(r_{80})$ with increasing r_{80} given by this formulation. Despite the almost certain overestimation of the concentrations resulting from this mode, even if these erroneously high values are used the contribution of the spume drop mode to concentration properties such as N, R, A, and V and M is negligible over the entire range of wind speeds for which this formulation is valid, as is shown in Table 15 at $U_{10} = 10$ m s^{-1}. For example, at $r_{80} = 12$ µm, the location of the maximum value of $n(r_{80})$ for the spume drop mode, the contribution to the number concentration of SSA particles from this mode is exceeded by that from the jet drop mode by a factor of more than 10^4 at $U_{10} = 10$ m s^{-1}, and by nearly $7 \cdot 10^3$ at $U_{10} = 17$ m s^{-1} (the highest wind speed of the measurements). Even using the higher value of n_0 from parameterization 2 at $U_{10} = 17$ m s^{-1}, the contribution to the SSA number concentration from the jet drop mode at $r_{80} = 12$ µm would exceed that from the spume drop mode by a factor of 30, and the contribution to SSA surface area concentration from the jet drop mode would exceed that from the spume drop mode by nearly a factor of 4. It would therefore seem difficult not only to accurately determine the parameters characterizing this mode, but even to infer the existence of such a mode. At $U_{10} = 17$ m s^{-1} using the value of n_0 from parameterization 2, the spume drop mode would provide the dominant contribution to the dry SSA mass concentration M: just over 2200 µg m^{-3} (compared to just over 200 µg m^{-3} for the jet drop mode); this value is much greater than nearly all measured values presented in Fig. 17. However, the maximum values of $a_{80}(r_{80})$ and $m(r_{80})$ (or $v_{80}(r_{80})$) for this mode—130 µm and 450 µm, respectively (Table 14), are not only far beyond the range of SSA particles that can be assumed to have equilibrated, but also near or beyond the largest sizes of SSA particles that were measured by *O'Dowd et al.*, and far beyond the range for which confidence can be placed in a formulation based on data for which the peak in $n(r_{80})$ occurred at $r_{80} = 12$ µm (§2.1.4.1). These arguments thus raise further serious questions regarding parameterization 2 of *O'Dowd et al.* [1997a].

A further consequence of use of the formulation proposed by *O'Dowd et al.* [1997a] is that the estimated concentrations of the large SSA particles are so small that few of these particles would have been counted. According to parameterization 2, at $U_{10} = 17$ m s^{-1} the concentration of particles per unit logarithmic interval at $r_{80} = 50$ µm is $7.3 \cdot 10^{-5}$ cm^{-3} in the jet drop mode and $1.7 \cdot 10^{-3}$ cm^{-3} in the spume drop mode, and according to parameterization 1 at this same wind speed these concentrations are $7.3 \cdot 10^{-5}$ cm^{-3} and $7.9 \cdot 10^{-6}$ cm^{-3}, respectively. Sea salt aerosol particles in this size range would have radii nearly equal to their formation radii, ~100 µm, at the time of measurement (§2.9.1). A Particle Measuring Systems instrument OAP, which was used to measure particles of this size, detects particles with radii from 97.5 µm to 102.5 µm ($\Delta \log r = 0.022$) in a channel for which the sample area is 6.71 mm^2 [*Particle Measuring Systems*, personal communication, 1999]. A wind speed of 17 m s^{-1} would result in a volume flow rate of 114 cm^3 s^{-1}, which would yield a sampling rate in this channel of less than one jet drop per hour and roughly one spume drop every four minutes. At 10 m s^{-1} the sampling rate in this channel would be roughly one jet drop every five hours and (using parameterization 1) one spume drop every three days. For particles with $r_{80} > 50$ µm, the concentrations would be even lower, and the sampling rates even less. Such low concentration rates raise questions both over the accuracy of the reported concentrations and over the uncertainty due to counting statistics.

The lack of discussion of the methods by which sizes of large SSA particles were converted to their r_{80} values, the negligible contributions of these particles to the estimated concentrations of SSA number, surface area, and mass for most wind speeds, the large discrepancy between the reported value of n_0 for the spume drop mode at $U_{10} = 17$ m s^{-1} and that calculated according to the formulation, and the few large SSA particles that would have been sampled under most circumstances, certainly raise questions over the confidence that can be placed in the parameters for the modes proposed by *O'Dowd et al.* Likewise, consideration of the wind-speed dependence of total SSA number concentration N (§4.1.3.2) raises similar questions regarding the functional form of the size dependencies of the concentrations of small and medium SSA particles and their dependencies on wind speed.

Other parameterizations of size distributions of SSA concentration, such as those presented graphically by *Woodcock* [1952] and by *Blanchard and Woodcock* [1980] to data of *Woodcock* [1953] at heights of 600-800 m, and that presented by *Chaen* [1973], are roughly consistent with the data presented in Fig. 22 but do not readily lend themselves to determination of local integral properties of the size distributions, such as N, R_{80}, A_{80}, V_{80}, and M. *Chaen* [1973] fitted his measured size distributions of SSA number concentrations to different wind-speed dependencies for different ranges of r_{80} using the dimensionless group $g T_s^2 u_* / v_a$ (§2.3.5), where g is the acceleration due to gravity, T_s is the period of the significant wave, u_* is the wind friction velocity, and v_a is the kinematic viscosity of air, but in view of the strong correlation of T_s with U_{10} there appears to be little advantage to use of such a parameterization (§2.3.5). The parameterization of *Davidson et al.* [1984], which consisted of a single lognormal mode with r'_{80} increasing with increasing wind speed, was developed in an attempt to describe the size distributions of SSA concentration presented in *Fairall et al.* [1983], but no comparison with measurements was presented. The expression presented is almost certainly incorrect; the value of r_{m0} in Table 2 of *Davidson et al.* [1984] should probably be $0.088 + 0.026 U_{10}$ (not $0.088 + 0.26 U_{10}$ as stated) and the value of N_0 should probably be $0.0012 U_{10}^{3.5} / r_m^3$ (not $0.0012 U_{10}^{3.5} / r_{m0}^3$), but even with these corrections this parameterization does not yield an accurate representation of the data of *Fairall et al.* or of other measurements in Fig. 22.

A parameterization was proposed by *Erickson and Duce* [1988] for the increase in what they termed the mass-mean radius (§2.1.5) at 80% RH, $r_{mmr,80}$, with increasing wind speed at 15 m above the sea surface U_{15}:

$$\frac{r_{mmr,80}}{\mu m} = 0.422 \frac{U_{15}}{m\ s^{-1}} + 2.12.$$

According to this relation the value of $r_{mmr,80}$ increases from ~4 μm to ~8 μm as U_{15} increases from 5 s^{-1} to 15 m s^{-1}. This relation was based on size distributions of SSA concentration reported by *Woodcock* [1953] and those of *McDonald et al.* [1982] discussed in §4.1.4.2 which had been "corrected" by adding sufficient mass on the first stage of their impactor to result in a lognormal distribution. However, there are several concerns with this relationship. The data of *Woodcock* were mostly taken near cloud base; the height dependence of the concentration of larger SSA particles would be expected to exhibit a strong dependence on wind speed because of increased vertical mixing of such particles with increasing wind speed (§2.9.5). Additionally, *McDonald et al.* stated that their size distributions of SSA concentration, which were obtained from a multistage impactor with the 50% cutoff for the first stage at $r_{80} \approx 3.3$ μm, did not exhibit an increase of mass-median radius at 80% RH with increasing wind speed; only when the SSA concentration size distributions were "corrected" was such an increase present. As no such increase in characteristic size of SSA particles with increasing wind speed is apparent in the data as a whole presented in Fig. 22 for wind speeds less than 25 m s^{-1}, it would seem that little confidence could be placed in such a parameterization based on a few data sets.[1]

For the reasons noted, none of the formulations examined above have been found to yield adequate parameterizations of size distributions of SSA concentration at a given wind speed. Moreover, the nature of the wind-speed dependencies of most formulations typically has little physical basis, and often these dependencies can be represented equally well (or equally poorly) by exponential or power law functions of wind speed, as discussed in relation to the wind-speed dependencies of M (§4.1.2.2) and N (§4.1.3.2). In view of these concerns, and in view also of the inherent spread in reported size distributions of SSA concentration that must necessarily accompany any parameterization of these data, the utility of such a parameterization, together with the fact that the great majority of the available data have been assembled and can readily be compared, suggests that yet another attempt may be warranted. Here the guiding principles are that the expression be as simple as possible and that it more or less faithfully represent the vast majority of the

measurements that can be considered credible (rathers than the data of a single investigation or group). These criteria impose a range of r_{80} and of U_{10} for which any such formulation is applicable and permit its uncertainty to be inferred.

An expression that meets these criteria reasonably well over the ranges $r_{80} = 0.1$-25 μm and $U_{10} = 5$-20 m s^{-1} consists of a single lognormal mode of the form (2.1-9), with amplitude $n_0/cm^{-3} = 0.07[U_{10}/(m\ s^{-1})]^2$, and with $r'_{80} = 0.3$ μm and $\sigma = 2.8$, both independent of wind speed. This formulation is shown in Fig. 22*b-e* for r_{80} in the range 0.1-25 μm evaluated at the central value of U_{10} for each wind speed range. The shaded regions, which encompass the majority of the measurements, extend a factor of $\lesssim 3$ about the central values of the parameterization in each wind speed range, suggesting the uncertainty associated with this parameterization. This factor is somewhat greater than the increase in SSA concentration with increasing wind speed between successive wind speed intervals in Fig. 22, for which this parameterization yields an increase in concentration of roughly a factor of 2; equivalently, over the range of wind speeds within a given wind speed interval the concentration of SSA particles of a given size obtained from this parameterization is about a factor of 1.4 above or below central value of the parameterization. Still, the overall increase in n_0 over the wind speed range $U_{10} = 5$-20 m s^{-1}, a factor of 16, is appreciably greater than the spread in the data at any given wind speed, as noted in §4.1.4.3.

This parameterization, introduced in §2.1.5 and denoted there as the canonical SSA size distribution of concentration, is used throughout this review for evaluating properties of SSA concentration (e.g., §2.1.5; Fig. 2; Fig. 3; Fig. 5). The use of this canonical distribution for such evaluations gains support from the comparisons of this parameterization with observed size distributions of SSA concentration in Fig. 22. These comparisons further suggest that an uncertainty $\lesssim 3$ be associated with extensive properties of SSA (§2.1.5) derived from this parameterization such as N, R_{80}, A_{80}, V_{80}, M, the uptake rate constant for reactive gases or smaller marine aerosol particles τ_u^{-1} (§2.1.5.2), or the light-scattering coefficient σ_{sp} (§2.1.5.3). Coincidentally, this canonical size distribution evaluated at $U_{10} = 8$ m s^{-1} is remarkably similar to the graph presented by *Junge* [1963,

---

[1] The relation presented by *Erickson and Duce* [1988] for the mass-mean radius yields values very similar to those of mass-median radii reported by *McDonald et al.* [1982] for the "corrected" data, and in fact it is likely that *Erickson and Duce* meant this quantity to refer to mass-median radius as it did in *Slinn and Slinn* [1980] to which these investigators referred. According to the parameters of the lognormal size distributions to which the data were fitted by *Erickson and Duce* ($\sigma = 3$) and the relations presented in §2.1.5, the mass-mean radius would be nearly a factor of two larger than the mass-median radius. This situation has led to confusion by subsequent investigators. For example, *Genthon* [1992a,b] referred to the quantity r_{mmr} in the above equation as the mean mass radius, which would be more than an order of magnitude less than the mass-mean radius, and more than a factor of 6 less than the mass-median radius, whereas *Schulz et al.* [2004] referred to this same quantity as the mass-median radius.

Fig. 26] for the size distribution of SSA concentration, both in shape and in magnitude [see also *Junge et al.*, 1969].

Values of N, R_{80}, A_{80}, V_{80}, and M that result from the canonical SSA size distribution of concentration at $U_{10} = 10$ m s^{-1} are presented in Table 15, and the quadratic wind-speed dependence can be used to determine values at other wind speeds. Over the range of wind speeds $U_{10} = 5$-15 m s^{-1}, the approximate ranges of N, R_{80}, A_{80}, V_{80}, and M that result from the canonical SSA size distribution of concentration are 2-20 cm^{-3}, 1-10 μm cm^{-3}, 20-200 μm^2 cm^{-3}, 25-250 μm^3 cm^{-3}, and 7-75 μg m^{-3}, respectively. This range of values of N is consistent with reported values for this quantity (Fig. 18, Fig. 19; Fig. 21; §4.1.3.1), although it does not extend to the upper limit of such values (more than 200 cm^{-3}; §4.1.3.1). The range of values of M resulting from this expression, shown by the quadratic expression in Fig. 17, is also consistent with measured values (§4.1.2.1).

It is stressed that this parameterization, like any parameterization of size distributions of SSA concentration, must be viewed as purely empirical, and no physical basis is implied for the lognormal form of the size dependence of the concentration at a given wind speed, the parameters characterizing the size dependence, or the wind-speed dependence of the concentration. Still use of this parameterization to represent size distributions of SSA concentration and their dependence on wind speed can be justified because it represents the data as a whole reasonably well, yields values of total integral SSA concentration properties (Table 15) that are consistent with reported values for these quantities, readily allows evaluation of these and other SSA concentration properties of interest, and yields an estimate of the range of variability that might be expected for such properties. As this formulation consists of a single mode for which only the amplitude (characterized by n_0), but not the location (characterized by r'_{80}) or the width (characterized by σ), depends on wind speed, neither the location of the maximum of the mode nor its shape in any representation (Table 14) depends on wind speed (in contrast to formulations consisting of two or more modes whose amplitudes have different wind-speed dependencies); likewise, any intensive property of SSA concentration (§2.1.5) is independent of wind speed. According to this formulation the approximate values at 80% RH for the effective radius $r_{eff,80}$, the mass-mean radius, and the mean mass radius (§2.1.5), are 4.2 μm, 12 μm, and 1.5 μm, respectively, all of these quantities independent of wind speed. The full width at half maximum (§2.1.4.2) for this parameterization is given by $\Delta_{1/2}\log r_{80} \approx 1.05$, also independent of wind speed, for each of the representations listed in Table 14. Thus to good approximation the total integral quantities N, R, A, and V (or M) are numerically equal to the maximum values of the corresponding distributions (§2.1.4.2).

This parameterization is shown in Fig. 22 only down to $r_{80} = 0.1$ μm, as there are insufficient data to extend it to lower sizes and as the shape of the size distribution of SSA concentration below this size is not well known. Based on considerations presented in §4.1.1.1 and on the few data presented in Fig. 22 near $r_{80} = 0.1$ μm, number concentrations of smaller SSA particles appear to be relatively low compared to the number concentration of SSA particles greater than this size and certainly compared to the number concentration of non-SSA particles in this size range (§4.1.1.1; Fig. 16). Consequently, $r_{80} = 0.1$ μm has been taken as a practical lower limit of SSA particles that must be considered under most circumstances (§2.1.1; §4.1.1.1). Similarly, the parameterization is shown in Fig. 22 only up to $r_{80} = 25$ μm, as this was taken as the upper limit of SSA particles that need be considered with regard to the effective SSA production flux. According to this parameterization if extrapolated, the region $r_{80} < 0.1$ μm would contribute approximately 15% of the SSA number concentration and 5% of the SSA radius concentration; similarly the region $r_{80} > 25$ μm would contribute approximately 1% of the concentration of SSA surface area and 10% of the concentration of SSA volume or mass (Table 15). Compared with the uncertainty of ⨝3 associated with the magnitude of $n(r_{80})$ according to this formulation these contributions to these several integral quantities are negligible.

The canonical size distribution of SSA concentration is used in the next section to evaluate SSA concentration properties of interest and to draw general conclusions about these properties.

4.1.4.5. Implications. The size distributions of SSA concentration presented in Figs. 22 and 23 have important implications on geophysical processes of interest. These implications are examined using the canonical SSA size distribution of concentration presented above at $U_{10} = 10$ m s^{-1} and RH 80%, values representative of those encountered in the marine environment. As noted above, extensive properties of SSA concentration evaluated in this way must be considered uncertain to a factor of 3 in either direction. Additionally, over the range of wind speeds comprising the majority of situations over the ocean, values of the SSA concentration, and thus values of extensive properties of this concentration, range over an interval of a factor of nearly 3 above and below those evaluated at $U_{10} = 10$ m s^{-1}. Nonetheless, useful conclusions can be obtained concerning the relative importance of SSA particles in different size ranges and their relative importance to atmospheric properties and processes of interest.

There are generally few SSA particles with $r_{80} < 0.1$ μm relative to larger SSA particles and to non-SSA particles in this size range, and SSA particles in this size range contribute little to most SSA concentration properties of interest (§4.1.1.1; §4.1.4.3). Realization of this low relative abundance led to the choice of $r_{80} = 0.1$ μm as the practical lower limit of SSA particle sizes that need be considered in most situations (§2.1.1; §4.1.1.1). Such a choice justifies several important assumptions and simplifying approximations made throughout this review. For example, to good approximation the effect of surface tension (Kelvin effect) on the equilibrium radii of SSA particles in the marine atmosphere can be neglected, resulting in overestimation of the radii of SSA particles with $r_{80} > 0.1$ μm by less than 10% at values of RH less than 98% (§2.5.3). Consequently, the geometric size of an SSA particle depends only on RH and is independent of r_{80}, and can thus be expressed as $r/r_{80} = \zeta(rh)$; §2.5.3. Likewise, non-continuum corrections pertaining to the rate of condensation of water (§2.5.4) and the rate of uptake of trace gases with accommodation coefficient near unity (§2.1.5.2; Fig. 3) on SSA particles can be neglected with little loss in accuracy. Additionally, the rate of uptake by SSA particles of condensable gases with high accommodation coefficient, which characterizes the ability of SSA to act as a sink for these gases and to possibly suppress nucleation, may be taken as directly proportional to the concentration of SSA radius R rather than the concentration of SSA surface area A (§2.1.5.2; Fig. 3); the same considerations apply to uptake of smaller marine particles by SSA. Additionally, as SSA particles with $r_{80} \gtrsim 0.1$ μm activate to form cloud drops at a supersaturations of 0.1% or less, the vast majority of SSA particles by number, containing the overwhelming majority of the SSA mass, will function as cloud condensation nuclei under most conditions for which clouds are formed in the marine boundary layer (§2.1.5.1; Fig. 2).

It is expected that the dominant removal mechanism for small SSA particles is wet deposition (§2.7.3), and that for precipitation of appreciable intensity and duration the efficiency of wet removal is near unity for particles in this size range, and more importantly, nearly independent of particle size (§2.7.1). Consequently, the size dependence of the number concentration of these particles should be nearly the same as the size dependence of their production flux (§3.9). Further, the size distributions of SSA concentration for small SSA particles can be used to infer their production flux (either interfacial or effective, as these are expected to be nearly the same for SSA particles in this size range; §2.1.6) by the statistical wet deposition method (§3.9). As this method cannot provide information on the dependence of the production flux on factors such as wind speed, it is evaluated here using the canonical SSA size distribution of concentration at a typical wind speed $U_{10} = 10$ m s^{-1}, with $H_{mbl} = 0.5$ km and $\tau_{wet} = 3$ d. From these values, the effective SSA production flux $f_{eff}(r_{80}) \equiv dF_{eff}/d\log r_{80}$ can be determined from (3.9-1) or (3.9-2) as a lognormal having the same size dependence as this canonical distribution, i.e., a lognormal with maximum value at $r'_{80} = 0.3$ μm and $\sigma = 2.8$, having amplitude $1.5 \cdot 10^4$ m^{-2} s^{-1}. Over the range of small SSA particles, i.e., $r_{80} = 0.1$-1 μm, which spans roughly an order of magnitude centered about the location of the maximum of $n(r_{80})$, the SSA number concentration $n(r_{80})$ varies between its maximum value and half of this maximum value. If the mean of these values (i.e., 0.75 times the maximum value of $n(r_{80})$) is taken, then within the uncertainties inherent in the statistical wet deposition method (§3.9) and in the canonical SSA size distribution of concentration (§4.1.4.4), the value of $f_{eff}(r_{80})$ is a constant over this size range, roughly 10^4 m^{-2} s^{-1}. As small particles dominate N, this also provides a first estimate of the total SSA number production flux. As this estimate is based on a value of $n(r_{80})$ that yields $N \approx 7$ cm^{-3} and a mean atmospheric residence time $\tau_{wet} = 3$ d, such a production flux yields an increase in the SSA number concentration of about 2 cm^{-3} per day.

The dominant contribution to R, the total concentration of SSA radius, is provided from SSA particles with $r_{80} \approx$ 0.3-3 μm. Consequently, the role of SSA particles in uptake of trace gases and smaller SSA particles is often directly proportional to R. As discussed in §2.1.5.2, SSA particles can provide a sink for condensable gases and for smaller aerosol particles, and in clean marine environments SSA particles may provide the majority of this sink. As a consequence, SSA could play an important role in controlling new particle formation in the marine atmosphere by suppression of nucleation, with implications regarding the composition and size distribution of concentration of the marine aerosol, as illustrated by the following example. According to an expression proposed by *Napari et al.* [2002] for ternary nucleation of sulfuric acid (H_2SO_4), water, and ammonia (NH_3), for RH 50% (values for RH 80% were not presented) and NH_3 mixing ratio of 100 nmol mol^{-1}, the rate J of new particle formation becomes appreciable (1 cm^{-3} s^{-1}) for sulfuric acid concentration $[H_2SO_4] = 3 \cdot 10^6$ cm^{-3}, increasing sharply with increasing H_2SO_4 concentration according to $d\log J/d\log[H_2SO_4] \approx 7$. The concentration of H_2SO_4 is governed by its rate of production by reaction of OH with SO_2 and its rate of loss by diffusion to and attachment to the surface of existing aerosol particles (assumed here to be exclusively SSA particles). The rate of production of H_2SO_4 concentration is given by $k_2[OH][SO_2]$, where k_2 is the second-order reaction rate constant, approximately equal to $9 \cdot 10^{-13}$ cm^3 s^{-1} [*Sander et al.*, 2003]. For atmospheric mixing ratio of SO_2 taken to be 0.02 nmol mol^{-1} [e.g., *Andreae et al.*, 1988; *Nguyen et al.*, 1992; *De Bruyn et al.*, 1998], roughly equivalent to a concentration of $5 \cdot 10^8$ cm^{-3}, and characteristic midday OH concentration $3 \cdot 10^6$ cm^{-3} [e.g., *Mari et al.*, 1998; *Crawford et al.*, 2003], the production rate of

H_2SO_4 is 1300 cm^{-3} s^{-1}. The effective first-order rate constant for irreversible uptake of H_2SO_4 by SSA is $\tau_u^{-1} = 4\pi D_g R$ (§2.1.5.2; Table 5). For D_g, the diffusion coefficient of sulfuric acid monomer in air, taken as 0.1 cm^2 s^{-1} (§2.1.5.2) and $R_{80} = 4$ μm cm^{-3} (Table 15), the characteristic time for this uptake is given by (2.1-19) as $\tau_u \approx 0.5$ h. The resulting steady state H_2SO_4 concentration (which is inversely proportional to R) is approximately $3 \cdot 10^6$ cm^{-3}, coincidentally equal to the value noted above for which the rate of nucleation becomes appreciable. Thus for H_2SO_4 production much greater or R much lower, there would be appreciable nucleation by this mechanism, whereas for H_2SO_4 production much lower or R much greater, nucleation would be suppressed by the diffusional sink to SSA particles. While a quantitative assessment of the SSA influence would depend on details of the situation, this example demonstrates the potentially important role of SSA in suppression of new particle formation in the marine atmosphere.

Similar considerations apply to scavenging of smaller marine aerosol particles by SSA. The rate constant for uptake of these particles by SSA is given as $\tau_u^{-1} = 4\pi D_s R$, where D_s, the diffusivity of such a smaller particle of radius r_s, is proportional to r_s^{-2} (§2.1.5.2). Evaluating the rate of uptake using (2.1-20), again taking a typical value of $R = 4$ μm cm^{-3}, yields a characteristic time $\tau_u/\text{h} \approx 3.5(r_s/\text{nm})^2$. Thus the characteristic time for scavenging of smaller particles with $r_s = (1, 5, 10)$ nm by SSA is approximately (4 h, 4 d, 2 wks), implying that SSA might be important in the evolution of the size distribution of freshly nucleated marine aerosol particles, but that once the particle radius has grown to several nanometers diffusional loss to SSA is unimportant.

This same analysis can be applied to investigate the possibility that appreciable numbers of very small SSA particles (r_{80} from 0.01 μm to 0.1 μm) that might be produced escape detection because they rapidly coagulate with each other or onto larger SSA particles, and thus do not establish appreciable concentrations in the marine atmosphere. Under the assumption that all of the small SSA particles have radius $r_s = 0.01$ μm (this assumption yields an upper bound on the diffusivity and hence a lower bound on the characteristic uptake time) and again taking $R = 4$ μm cm^{-3}, the characteristic time for uptake of these small particles on larger SSA particles is obtained from (2.1-20) as ~2 weeks—far too long for such uptake to play a role in limiting their concentration. Likewise, the time required for the concentration of small SSA particles, each assumed (as above) to have $r_{80} = 0.01$ μm, to be reduced to 10% of its initial value n_0 through self-coagulation can be estimated (following *Friedlander* [2000, pp. 192-195] and *Fuchs* [1964, p. 294]) as $2 \cdot 10^{10}$ s/(n_0/cm^{-3}). Even for an initial concentration of 10^4 cm^{-3} (much greater than measured SSA concentrations and than typical CCN concentrations in clean marine air;

§4.1.1.1) this characteristic time is nearly one month—again far too great for such a mechanism to be important. This result justifies the neglect of coagulation as a loss mechanism of SSA particles (§2.8.1), and demonstrates that the reason so few SSA particles with $r_{80} \lesssim 0.1$ μm have been detected in the marine atmosphere cannot be attributed to their being rapidly scavenged by larger SSA particles.

The dominant contribution to the total concentration of SSA surface area A is provided by SSA particles in the size range $r_{80} \approx 1$-10 μm (Figs. 22b-e). Hence $Q_s = 2$ is an accurate approximation for the light-scattering efficiency over the range of visible wavelengths (Fig. 5), with the result that the total SSA light-scattering coefficient σ_{sp} is nearly directly proportional to A, being related by $\sigma_{sp} \approx A/2$ (§2.1.5.3). At $U_{10} = 10$ m s^{-1} the canonical SSA size distribution of concentration yields $A_{80} \sim 80$ μm^2 cm^{-3} (Table 15), and thus the light-scattering coefficient at 80% RH $\sigma_{sp,80} \approx 40$ μm^2 cm^{-3}, or equivalently, ~40 Mm^{-1}. As $\sigma_{sp,80}$ is an extensive property, meaning that it is directly proportional to the magnitude of the size distribution of the concentration (§2.1.5), the factor of three uncertainty in this magnitude yields values of $\sigma_{sp,80}$ that corresponds roughly to the range 12-120 Mm^{-1}. This value of light-scattering coefficient compares with the (strongly wavelength-dependent) Rayleigh scattering coefficient of air σ_R [*Rayleigh*, 1871, 1899], which is approximately 12 Mm^{-1} for $\lambda = 0.55$ μm (at the peak of the solar spectrum and the peak of visual response of the human eye) and increases from approximately 4.5 Mm^{-1} to 45 Mm^{-1} as λ increases from 0.7 μm to 0.4 μm [*Penndorf*, 1957]. For low $\sigma_{sp,80}$, scattering by SSA particles and by Rayleigh scattering from air molecules are comparable, whereas for high $\sigma_{sp,80}$ scattering by SSA particles dominates. In the absence of other aerosol substances (and also of absorbing gases such as NO_2), extinction (which results almost exclusively from scattering in this situation, as there is little absorption by air or by SSA particles over these wavelengths) is determined by these two contributions, $\sigma_{ext} = \sigma_R + \sigma_{sp,80}$. The influence of $\sigma_{sp,80}$ on extinction in turn influences visual range x_{VR}, which is given approximately by $x_{VR} \approx 4/\sigma_{ext}$ [*Koschmieder*, 1924a,b; *Middleton*, 1952, p. 105]. At $\lambda = 0.55$ μm, the value $\sigma_{sp,80} = 12$ Mm^{-1} yields $\sigma_{ext} = 24$ Mm^{-1} and $x_{VR} \approx 170$ km, and the value $\sigma_{sp,80} = 120$ Mm^{-1} yields $\sigma_{ext} = 130$ Mm^{-1} and $x_{VR} \approx 30$ km (these values would vary according to the actual σ_{sp} characterizing a given situation, and of course would be influenced by other aerosol species). The visual range in the lowest few meters above the sea surface might be considerably less than the foregoing values because of higher RH and because of the presence of larger SSA particles that do not contribute to the above estimates.

As noted in §2.1.5.3, a further consequence of the approximation $Q_s = 2$ is that the Ångström exponent α is equal to zero, independent of wavelength. The value of α calculated in that section for the canonical SSA size distribution of concentration was approximately -0.09 to within a few percent over the range of visible wavelengths (0.4-0.7 μm), very close to zero and consistent with field measurements noted there.

Sea salt aerosol particles in the size range providing the dominant contribution to both A and σ_{sp} (i.e., $r_{80} \approx 1$-10 μm) are expected to be well mixed over the marine boundary layer (§2.9.5). For this situation, with a marine boundary layer of height $H_{mbl} \approx 0.5$ km (§2.4) and RH taken to be 80% throughout, the above value of $A_{80} = 80$ μm^2 cm^{-3} results in a total SSA surface area over the oceans of roughly 4% of the sea surface; that is, a total SSA surface area of ~0.04 m^2 above each square meter of the sea surface. Additionally, the corresponding SSA optical thickness $\tau_{sp,80}$ (i.e., the vertical integral of the light-scattering coefficient; §2.1.5.3) can be approximated as the product of $\sigma_{sp,80}$ and H_{mbl}; i.e., $\tau_{sp,80} \approx \sigma_{sp,80}H_{mbl}$. The above estimate $\sigma_{sp,80} \approx 40 \cdot 10^{-6}$ m^{-1}, again with the choice $H_{mbl} = 0.5$ km, results in an estimate for the optical thickness at 80% RH due to SSA of $\tau_{sp,80} \approx 0.02$, consistent with reported values of the minimum total marine aerosol optical thickness in clean marine environments [*Forgan*, 1987; *Durkee et al.*, 1991; *Deuzé et al.*, 1999; *Chiapello et al.*, 1999; *Kaufmann et al.*, 2001; *Kuśmierczyk-Michulec et al.*, 2001; *Wilson and Forgan*, 2002; *Smirnov et al.*, 2002, 2003a,b].

Values of σ_{sp} and τ_{sp} are affected by RH mainly through its influence on the size of an SSA particle with given r_{80}, as the index of refraction changes little with RH for liquid solution drops (Fig. 4). Thus under the approximation $Q_s = 2$, σ_{sp} is expected to increase with increasing RH (for the RH range above the efflorescence value; §2.5.5) by the square of the factor by which the radius increases, i.e., $[\zeta(rh)]^2$; §2.5.3. The effect of RH on τ_{sp} may be somewhat more complicated; with increasing RH the gravitational terminal velocity of an SSA particle with a given r_{80} increases, reducing the extent of vertical mixing; thus the dependence of τ_{sp} on RH is expected to be somewhat less than that of σ_{sp}. The windspeed dependencies of σ_{sp} and τ_{sp} have been described by several investigators as exponential functions of wind speed [*Hoppel et al.*, 1990; *Krishna Moorthy et al.*, 1997; *Satheesh et al.*, 1999; *Wilson and Forgan*, 2002; *Kleefeld et al.*, 2002; *Satheesh*, 2002; *Dobbie et al.*, 2003; *Satheesh and Lubin*, 2003]; however, as these quantities are extensive properties of the concentration the same concerns apply as noted with regard to other extensive concentration properties such as M (§4.1.2) and N or δN (§4.1.3), and the quadratic wind-speed dependence of the canonical SSA size distribution of concentration can be presumed to satisfactorily describe the wind-speed dependence of these quantities.

For a size distribution of SSA concentration that has a shape that is independent of wind speed, and in particular the canonical SSA size distribution of concentration, intensive SSA properties such as the mass scattering efficiency σ_{sp}/M (§2.1.5.3) and the effective radius $r_{eff} = 3V/A$ (§2.1.3) would be expected to exhibit values that are constant and independent of wind speed. Any variation in these quantities would thus result from variations in the shape of the size distribution of SSA concentration and not from the atmospheric concentration of SSA. Moreover, such quantities would be expected not to exhibit anywhere near the factor of $\lessgtr 3$ variability that characterizes size distributions of SSA concentration and extensive properties calculated from these distributions. For the canonical SSA size distribution of concentration at 80% RH, within the approximation $Q_s = 2$, the mass scattering efficiency at 80% RH $\sigma_{sp,80}/M$ is equal to 1.3 m^2 g^{-1} (§2.1.5.3) and the effective radius at 80% RH $r_{eff,80}$ is equal to 4.2 μm (§2.1.5), both independent of wind speed. A caveat to these conclusions, however, is the sensitivity of determination of these quantities to instrumentally imposed upper limits of particle size in measurement of σ_{sp}, A, V, or M. In fact, a further but important consequence of the fact that the size range of SSA particles that contribute substantially to σ_{sp} extends to $r_{80} \approx 10$ μm is that measurements of light scattering by SSA that do not include the contribution of particles over the entire range of r_{80} up to and greater than this size will underestimate σ_{sp} with resultant underestimation of τ_{sp}; this situation applies to many of the measurements that have been reported for these quantities (for example, *Quinn and coworkers* [*Quinn et al.*, 1995, 1996; *Quinn and Coffman*, 1998] measured marine aerosol particles only up to $r_{80} \approx 5$ μm, and *Hegg et al.* [1996] only up to $r_{80} \approx 3$ μm). Moreover, measurements that do not adequately sample the SSA particle size range that provides the dominant contribution to M may overestimate the mass scattering efficiency even though they might also underestimate $\sigma_{sp,80}$.

Recognition of the SSA mass scattering efficiency σ_{sp}/M as an inherent property of SSA has motivated efforts to determine the value of this quantity, thereby allowing the SSA light-scattering coefficient σ_{sp} to be estimated in situations of interest from the measured SSA dry mass concentration M. Often measurements of light-scattering coefficient of SSA have been made at sufficiently low RH that the SSA particles are dry and are converted to obtain light-scattering coefficients at 80% RH from the ratio $\sigma_{sp,80}/\sigma_{sp,dry}$ (and similarly for other RH of interest). Values of this ratio ranging from 1.8 to 3.2 have been reported [*Pueschel et al.*, 1969; *Covert et al.*, 1972; *Charlson et al.*, 1984; *McInnes et al.*, 1998; *Carrico et al.*, 1998; *Li-Jones et al.*, 1998; *Chin et al.*, 2002]. With the choice 2.5, reported values of mass scattering efficiencies converted to 80% RH, $\sigma_{sp,80}/M$, range from 1 m^2 g^{-1} to 2.3 m^2 g^{-1} [*Li et al.*, 1996; *Quinn et al.*, 1996, 1998; *Chiapello et al.*, 1999; *Quinn and Coffman*, 1999; *Kuśmierczyk-Michulec et al.*, 2001], and values greater than

10 m^2 g^{-1} have been reported for limited size ranges. These differences would appear to be due in large part to insufficient radius range in sampling for σ_{sp} and M.

Values of SSA mass scattering efficiency used in models estimating the effect of SSA on light scattering in the atmosphere vary by an even greater amount—more than an order of magnitude—from 0.2 m^2 g^{-1} to 2.5 m^2 g^{-1} [*Andreae*, 1995; *Tegen et al.*, 1997; *Haywood et al.*, 1999; *Tegen et al.*, 2000; *Chin et al.*, 2002; *Penner et al.*, 2002]. Consequently model results and their conclusions on the annual average global direct radiative forcing due to SSA also vary greatly: -0.08 W m^{-2} [*Andreae*, 1995], -2 W m^{-2} (with instantaneous values as great as -10 W m^{-2}) [*Winter and Chýlek*, 1997], -1.6 W m^{-2} to -5.2 W m^{-2} [*Haywood et al.*, 1999], -0.54 W m^{-2} [*Jacobson*, 2001], -6.2 W m^{-2} [*Satheesh*, 2002], -1.1 W m^{-2} (-2.2 W m^{-2} when the effects of clouds are not taken into account) [*Grini et al.*, 2002], and -0.15 W m^{-2} [*Dobbie et al.*, 2003], the negative sign signifying that there is a net cooling effect. For comparison, the greenhouse forcing due to enhanced CO_2 concentrations over the industrial period is estimated as $+1.6$ W m^{-2} [*Ramaswamy et al.*, 2001]; hence uncertainty in modeled values of climate forcing due to SSA could result in large uncertainty in modeling climate forcing of anthropogenic influences.

As noted above, much of this variation appears to be due to inadequate consideration of the role of the upper radius cutoff in determination of σ_{sp}, M, or both. This situation is illustrated in Fig. 24, in which the mass scattering efficiency $\sigma_{sp,80}/M$ is plotted against the value of r_{80} corresponding to the upper limit of SSA particles contributing to the quantities $\sigma_{sp,80}$ and M. Values of this quantity resulting from the approximation $Q_s = 2$ are also shown. According to this calculation the value of the mass scattering efficiency for the canonical SSA size distribution of concentration when all particles up to $r_{80} = 25$ μm are used to calculate σ_{sp} and M is approximately 1.6 m^2 g^{-1} (1.4 m^2 g^{-1} for $Q_s = 2$), whereas the value of this quantity when SSA particles only up to $r_{80} \approx 1$ μm are included is more than 7 times greater, and that for $r_{80} \approx 5$ μm is nearly a factor of 2 greater. Thus measurements that include only part of the size range of SSA particles present in the atmosphere may greatly overestimate the mass scattering efficiency. Additionally, the sharp increase in this quantity with decreasing upper limit of r_{80} near 1 μm implies that there will be a strong dependence in determinations of this quantity on RH, sampling efficiencies, and other factors that would alter the upper sampling limit.

A further implication of the shape of the size distribution of SSA concentration is that the dominant contribution to the total SSA volume concentration at 80% RH, V_{80}, or equivalently the total SSA mass concentration M, at heights near 10 m above the sea surface is predominantly from SSA particles with r_{80} from several micrometers up to almost 25 μm

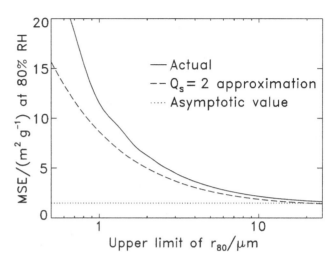

Figure 24. Dependence of SSA mass scattering efficiency (MSE), the ratio of light-scattering coefficient σ_{sp} (§2.1.5.3) to dry mass concentration M, on upper size limit r_{80}, for both quantities evaluated for the canonical SSA size distribution of concentration at 80% RH. Calculation is for wavelength $\lambda = 0.532$ μm. Dashed line indicates value resulting from the approximation $Q_s = 2$. Dotted line indicates asymptotic value, that is, the value of the mass scattering efficiency when the full size range of SSA particles is considered.

(Fig. 2), with nearly half of the contribution being provided from SSA particles with $r_{80} \gtrsim 7$ μm. One consequence of this result is that the upper sampling size limit in sampling SSA particles, if appreciably below $r_{80} \approx 25$ μm (the value of r_{80} that encompasses ~90% of the SSA dry mass concentration), would be expected to appreciably decrease the measured value of M below the actual value (§4.1.2).

The contribution of large SSA particles (i.e., $r_{80} \gtrsim 25$ μm) to concentration properties at typical measurements heights near 10 m is small (Fig. 22). Additionally, because of their large masses and correspondingly large gravitational sedimentation velocities (§2.6.3), such particles remain in the atmosphere for only short times, and their vertical profiles are such that their concentrations decrease rapidly with increasing height; thus the contributions of these particles to the column burdens of concentration properties will be small. This is consistent with the conclusions reached earlier (§2.1.6) that large SSA particles contribute little to the effective SSA production flux and thus play little role in atmospheric phenomena such as light scattering and geochemical cycling.

4.2. NEAR-SURFACE FIELD MEASUREMENTS PERTINENT TO SSA PRODUCTION

The fact that SSA particles are produced at the sea surface suggests that measurements made within the lowest few meters above the ocean might be pertinent to SSA production and to determination of the interfacial SSA production flux. This section examines studies reporting such measurements (with the exception of the direct vertical flux measurements of *Bortkovskii* [1987], which are examined in §5.9) for the insight they provide on the importance of different production mechanisms of SSA particles (§2.2) and on dependencies of SSA production on wind speed and other meteorological factors. Additionally, some of these measurements have been used as a basis for the along-wind flux method (§3.6). In some instances SSA concentrations have been measured directly (such as by flash photography), whereas in others SSA concentrations were inferred using (2.1-29) from measurements of the along-wind flux $j_x(r_{80})$, determined either optically or by impaction, under the assumption that the horizontal velocities of SSA particles are equal to the mean wind speed at the height of measurement. As SSA particles are expected for the most part to retain their formation radii when measured near the sea surface (§2.9.1; §4.1.4.2), values of SSA particle sizes are presented here in terms of r_{80} converted on the assumption that measured radii are equal to formation radii.

The number concentration of seawater drops with $r_{80} > 22.5$ μm at 13 cm above the instantaneous sea surface was measured by *Monahan* [1968] using a photographic system on a raft, for wind speeds at 47 cm above the sea surface $U_{0.47}$ up to 10.5 m s^{-1}; this wind speed corresponds to U_{10} near 14.5 m s^{-1} according to (2.4-1) with $C_D = 0.0013$ (§2.3.1; §2.4). Much greater concentrations of drops with $r_{80} > 22.5$ μm and of drops with $r_{80} > 45.5$ μm were reported when $U_{0.47}$ exceeded a critical value of ~8.5 m s^{-1} (corresponding to $U_{10} \approx 12$ m s^{-1} for $C_D = 0.0013$, although *Wu* [1971a] concluded that it corresponds to $U_{10} = 15$ m s^{-1}), a wind speed appreciably greater than commonly suggested thresholds for whitecapping (§4.5) and other air-sea interactions. It was reported that drops sometimes occurred temporally in clusters, implying that in these instances most of the detected drops derived from individual events (such as breaking waves or whitecaps) and might have included spume drops; however, considerable variation was seen in the extent of clustering. For U_{10} near 11 m s^{-1} and 16 m s^{-1}, size distributions of SSA number concentration were presented for r_{80} from ~30 μm to ~300 μm. These distributions are described roughly by $n(r_{80}) \propto r_{80}^{-0.5}$, indicating that throughout this size range the size-dependent concentrations of SSA surface area and volume at 80% RH, $a_{80}(r_{80})$ and $v_{80}(r_{80})$, respectively, increase with increasing r_{80}. Based on these measurements, *Monahan* [1968] calculated the interfacial production flux for SSA particles with $r_{80} = 75$ μm; these calculations are discussed below in §5.9.

While sorting out quantities and units for reporting size distributions of SSA concentration has been a problem throughout the preparation of this review (e.g., §4.1.4.2; §4.1.4.3), an example that seems particularly egregious is the treatment by later investigators of the size distributions of SSA number concentration reported by *Monahan* [1968, Figs. 2 and 3] as "α_i in cm^{-4}" vs. radius, where α referred to size distributions of the form $dN(r)/dr$. Of eleven later references in which these data were graphed and compared to other concentration measurements, five [*Wang and Street*, 1978a; *Wu*, 1979a; *Koga and Toba*, 1981; *Wu*, 1982a; *Wu et al.*, 1984] plotted the values *Monahan* used for radius as diameter, and these references plus five others [*Lai and Shemdin*, 1974; *de Leeuw*, 1986a; *Wu*, 1990d; *Iida et al.*, 1992; *de Leeuw*, 1993] incorrectly assumed that the quantity α referred to size distributions of the form dN/dD. One particularly confused instance is found in *Koga and Toba* [1981], who stated "the droplet size distribution α (number per unit volume per unit increment in radius) is given by θ divided by its size range width Δd: $\alpha \equiv \theta/\Delta d$," leaving it ambiguous whether their size distribution referred to $dN(r)/dr$ or $dN(D)/dD$. Of the eleven references examined, only *Bortkovskii* [1987] correctly interpreted both the x- and y-axes as given in *Monahan* [1968].

Size-dependent along-wind fluxes of drops with radii from 2.5 μm to greater than 200 μm were measured by *Preobrazhenskii* [1973] by impaction on oiled glass plates for moderate winds (stated both as $U_{10} = 6$-10 m s^{-1} and $U_{10} = 7$-12 m s^{-1}) and strong winds ($U_{10} = 15$-25 m s^{-1}) at wave-crest level (1.5 m for moderate winds, 2 m for strong winds) and at 4 m and 7 m above the mean sea surface. Reported size-dependent SSA number concentrations were calculated from these measurements using (2.1-29) with U taken as the wind speed at the measurement height. The maximum radii of drops detected at heights of 7 m, 4 m, and at the wave-crest level (1.5 m for moderate winds, 2 m for strong winds) were 15 μm, 40-45 μm, and 65-75 μm, respectively, for moderate winds, and 45 μm, 100 μm, and 250-1000 μm, respectively, for strong winds. Ambiguities in the representation of these data were discussed in §4.1.4.2. As also noted in §4.1.4.2, whether or not these radii referred to geometric radii at the time the drops were sampled, equilibrium radii at the ambient RH in the laboratory when the sizes were measured, or formation radii was not specified; the consequences of this uncertainty were examined in §4.1.4.2 and §4.1.4.3 and illustrated in Fig. 22 for the measurements at 7 m. Sufficiently small drops measured at 7 m had probably been in the atmosphere sufficiently long that their radii had equilibrated to the ambient RH (which was not reported), whereas larger drops, especially those measured near wave crests, probably had not, and radii of these drops were probably close to formation values (§2.9.1); the situation for intermediate sizes was examined in §4.1.4.2. As with the concentration data of *Monahan* [1968], the measurements at the lowest heights indicated an increase in $a_{80}(r_{80})$ and $v_{80}(r_{80})$ with increasing r_{80} for all drop sizes. *Preobrazhenskii* reported that the concentrations decreased with increasing height, albeit only slightly between heights of 2 m and 4 m. He also fitted the total SSA dry mass concentration, M, for the two wind speed ranges to an exponential decrease with height, but there is far too much scatter in the data to support any particular functional form. Additionally, as the concentrations at 7 m were roughly two orders of magnitude lower than the 10 m concentrations reported by other investigators at comparable wind speeds (Fig. 22), they must be considered suspect (§4.1.4.3), leaving the data at other heights suspect as well.

Using an optical detection system consisting of a laser and a phototransistor mounted on a raft, *Wu et al.* [1984] measured the size-dependent along-wind flux of seawater drops with $r_{80} = 10\text{-}130$ μm at heights from 13 cm to 130 cm above the instantaneous sea surface at U_{10} of 6 m s^{-1} to 8 m s^{-1} in the mouth of Delaware Bay. Discrete temporal clusters of drops were observed at the same period as the peak of the frequency spectrum of the ocean waves. The total along-wind flux of SSA volume exhibited a slight decrease with increasing height. Size distributions of the frequency of occurrence of drops with r_{80} from 12.5-130 μm, presumably proportional to $dJ_x(r_{80})/dr_{80}$, were reported for heights less than 20 cm and for heights greater than 20 cm; these size distributions were fitted to different power laws in different size ranges. For heights less than 20 cm, the frequencies were fitted to expressions equivalent to $j_x(r_{80})$ ($\equiv dJ_x(r_{80})/d\log r_{80}$) $\propto r_{80}^{-1.8}$ for $r_{80} = 37.5\text{-}100$ μm and $j_x(r_{80}) \propto r_{80}^{-7}$ for $r_{80} > 100$ μm; for heights greater than 20 cm the exponents were the same, but the ranges were $r_{80} = 12.5\text{-}50$ μm and $r_{80} > 50$ μm, respectively. The investigators argued that as 20 cm is approximately the maximum ejection height of jet drops (§4.3.2.3), the size distribution of this frequency of occurrence for lower heights corresponds to the size distribution of the interfacial SSA production flux, and further that the vertical distribution of drops with $r_{80} < 12.5$ μm is more or less uniform (for these wind speeds), whereas the vertical distribution of larger drops is determined by their ejection heights (this is very similar to the criterion $v_{Stk}/(\kappa u_*) \lesssim 1$ for the upward entrainment of drops discussed in §2.9.3). *Wu et al.* argued that the same power law (with exponent 1.8) was an appropriate fit to earlier data, and fitted the size dependence of SSA concentrations determined from along-wind flux measurements reported by *Monahan* [1968] at a height of 13 cm, converted using (2.1-29) with U taken as the wind velocity at that height (these data were stated to be for $U_{10} = 11$ m s^{-1}, although presumably this wind speed should have referred to $U_{0.47} = 11$ m s^{-1}) and the size dependence of concentrations reported by *Preobrazhenskii* [1973] for moderate wind

speeds at heights of 1.5-2 m and 4 m to $n(r_{80}) \propto r_{80}^{-1.8}$ for $r_{80} > 37.5$ μm for $z < 20$ cm (and for $r_{80} > 12.5$ μm for $z > 20$ cm). *Wu et al.* also concluded that the data of *Monahan* and of *Preobrazhenskii* for smaller sizes could be fitted to $n(r_{80})$ equal to a constant, independent of r_{80}. However, because of the few data presented by *Monahan*—only four data for $z < 20$ cm for each of two wind speeds (for which *Wu et al.* had taken the radii as diameters, as noted above), which do not appear to be well described by either of the forms for $n(r_{80})$—and because of the concerns noted in §4.1.4.3 with the data of *Preobrazhenskii*, it is difficult to give much credence to the power laws proposed by *Wu et al.*

Values of the size dependence of the along-wind flux represented as $j_x(r_{80})$ determined from measurements reported in Fig. 4 of *Wu et al.* [1984] for $U_{10} = 6$ m s^{-1} at heights $z = 50$ cm and $z = 90$ cm are shown in Fig. 25, as are the power law expressions discussed above. The data are plotted in arbitrary units as only frequencies of occurrence were presented by *Wu et al.* (hence these graphs illustrate only the shape, but not the magnitude, of the size distribution of the along-wind SSA flux). Also shown in Fig. 25 are values of $j_x(r_{80})$, also in arbitrary units, calculated from Fig. 8a of that reference; these data were stated to be an example of a spectrum of the along-wind volumetric flux at $U_{10} = 6$ m s^{-1} for which the absolute magnitude was reported (but not the height to which this flux pertained). The data for $r_{80} > 50$ μm appear to be well fitted by a power law of the form $j_x(r_{80}) \propto r_{80}^{-13}$ (also shown in Fig. 25), but any data set that exhibits a decrease of nearly four orders of magnitude over a factor of two increase in radius raises concerns about the effects of uncertainties in measured sizes, counting statistics, and the like. As the wind speed did not exceed 8 m s^{-1}, the extent to which these measurements can be extrapolated to higher wind speeds is questionable. These measurements are examined further along with laboratory measurements of near-surface along-wind SSA fluxes in §4.3.6.

Measurements of size-dependent number concentrations of SSA particles with r_{80} up to ~25 μm and of total light-scattering coefficient (nearly proportional to the total SSA surface area concentration; §2.1.5.3) as a function of height above the sea surface were reported at wind speeds up to 14 m s^{-1} by *de Leeuw* [1986a,b, 1987, 1989a, 1990a]. These measurements confirm the expectation (§2.9.4) that for heights from a few meters above the sea surface up to ~20 m the concentration of SSA particles in this size range is nearly independent of height. For $U_{10} \gtrsim 7.5$ m s^{-1} the concentrations sometimes exhibited one or two maxima at lower heights, typically 1-2 m above the instantaneous sea surface. *de Leeuw* [1990a] suggested that the maximum at the lowest height might be an artifact of the sampling method (because the sampler spends more time in wave troughs than at wave

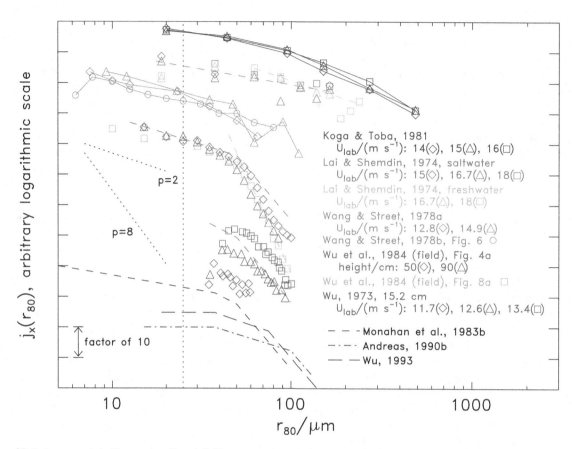

Figure 25. Laboratory (wind/wave tunnel) and field measurements and parameterizations of the size dependence of the along-wind flux in the representation $j_x(r_{80}) \equiv dJ_x(r_{80})/d\log r_{80}$, with arbitrary amplitude, as a function of r_{80}. Here the y-value of any given point is meaningful only relative to the other points of the same distribution; data have been grouped by investigation and spread out along in the y-direction for ease of visualization and comparison. Each tick on the y-axis denotes a factor of 10. All data are from laboratory experiments except the field measurements of *Wu et al.* [1984]. All laboratory experiments employed freshwater except for saltwater measurements of *Lai and Shemdin* [1974]. Power law functions of the form $j_x(r_{80}) \propto r_{80}^{-p}$ as given in the original investigation are shown for several of the data sets (colored dashed lines); solid colored lines connect data points to guide the eye. Dotted lines indicating power law functions with $p = 2$ and $p = 8$ are also shown. Data of *Koga and Toba* [1981] are from Fig. 8 of that reference; those at $U_{lab} = 14$ m s^{-1} were measured at height 13.5 cm above the water surface, those for $U_{lab} = 15$ m s^{-1} at 15 cm, and those at $U_{lab} = 16$ m s^{-1} at 20 cm. Data of *Lai and Shemdin* [1974] at U_{lab} of 15.0, 16.7, and 18.0 m s^{-1} are from Figs. 7 and 11 of that reference for height 13 cm above the water surface; $p = 1$ for freshwater and $p = 2$ for saltwater. Data of *Wang and Street* [1978a] at U_{lab} of 12.8 and 14.9 m s^{-1} are from Fig. 8 of that reference for height 10.2 cm above the water surface; data of *Wang and Street* [1978b] from Fig. 6 of that reference are for height 10.2 cm above the water surface and fetch 12.3 m at laboratory wind speed 14.5 m s^{-1}. Field measurements of *Wu et al.* [1984] from Fig. 4a of that reference are for heights of 50 cm and 90 cm above the sea surface at $U_{10} = 6$ m s^{-1}, and measurements from Fig. 8a of that reference are at $U_{10} = 6$ m s^{-1} (height not specified); $p = 1.8$ for r_{80} from 12.5 μm to 50 μm and $p = 7$ for r_{80} from 50 μm to 100 μm, are also shown. A power law function with $p = 13$ by the present authors is also shown. Data of *Wu* [1973] for height 15.2 cm above the water surface at U_{lab} of 11.72, 12.62, and 13.43 m s^{-1} are from Fig. 8 of that reference; $p = 3$ for r_{80} from 35 μm to 50 μm and $p = 7$ for r_{80} from 50 μm to 100 μm. Also shown are piecewise continuous power law functions with different values of p in different ranges of r_{80}, as proposed by *Monahan et al.* [1983b, 1986], *Andreas* [1990b], and *Wu* [1993]. Vertical dotted line at $r_{80} = 25$ μm demarcates medium and large SSA particles.

crests), but that the maximum near 2 m was not an artifact, and he accounted for this maximum by a wave-rotor model in which particles are temporarily trapped in eddies formed by the separation of flow at the stagnation points on breaking waves [*de Leeuw*, 1986b, 1987, 1989a; this was discussed

further in *Wu*, 1990c and *de Leeuw*, 1990b]. Total SSA surface area concentration increased with decreasing height below ~2 m for $U_{10} = 7.7$ m s^{-1}, exhibited maxima at ~1.5 m for $U_{10} = 12$ m s^{-1}, and appeared nearly independent of height for $U_{10} = 14.2$ m s^{-1} (in one place in the paper it was stated that profiles similar to that observed at $U_{10} = 7.7$ m s^{-1} were uncommon, although later it was stated that these three types of profiles occurred with comparable frequency). Time series of the total SSA surface area concentration at heights of 2, 4, 8, and 12 m above the mean sea surface exhibited fluctuations that decreased in amplitude with increasing height and were correlated with the motion of the underlying water surface. Unlike measurements of SSA concentrations at heights near 10 m, for which it can be assumed that nearly all SSA particles with $r_{80} \lesssim 25$ μm will have attained their equilibrium radii, measurements at heights within a meter or two of the sea surface will include both particles that have attained their equilibrium radii and those more recently formed that have radii near their formation values, confounding attempts to determine size distributions of SSA concentrations as a function of r_{80}. No indication of spume drop formation was indicated by these measurements for SSA particles as large as $r_{80} \approx 15$ μm at wind speeds up to $U_{10} = 14$ m s^{-1}; concentrations of larger particles were too low to draw conclusions.

4.3. LABORATORY STUDIES PERTINENT TO SSA PRODUCTION

This section examines laboratory investigations of SSA production. Drops formed by bursts of individual bubbles are examined in §4.3.1, §4.3.2, and §4.3.3; specifically the numbers, sizes, and heights to which these drops are ejected, and the factors that control these quantities. These quantities are required for evaluation of the interfacial SSA production flux according to the bubble method (§3.4). Sea salt aerosol production from laboratory whitecaps (that is, aggregates of large numbers of closely packed bursting bubbles) is examined in §4.3.4, and SSA production inferred from surf zone measurements is examined in §4.3.5. These measurements have resulted in formulations for the size-dependent interfacial SSA production flux over whitecap area $f_{wc}(r_{80})$, which is required to evaluate the interfacial SSA production flux according to the whitecap method (§3.2); these formulations are also examined in §4.3.5. Laboratory measurements of along-wind fluxes within ~1 m of the surface, which have been used to evaluate the interfacial SSA production flux according to the along-wind flux method (§3.6), are examined in §4.3.6. Implications of results from the above sections to SSA production over the ocean are examined in §4.3.7.

4.3.1. Drops from Bursting Bubbles

As drops from bursting bubbles are believed to be the major source of SSA particles (§2.2), understanding the process by which these drops are formed is important to quantifying their production and describing their role as SSA particles. Additionally, determining the interfacial SSA production flux from the size-dependent flux of bubbles bursting at the surface of the ocean using the bubble method (§3.4) requires knowledge of the number of drops produced per bubble as a function of bubble size, together with the size distribution and (maximum) ejection heights of these drops. Knowledge of these quantities is also important to consideration of entrainment by eddies and contributions of drops to the effective SSA production flux. The following sections review the available data on bubble-produced drops pertinent to SSA production, as a function of bubble size, temperature, and other potentially controlling factors. Most investigations of bubble-produced drops and their properties have employed either individual bubbles or a stream of bubbles arriving at the surface and bursting at a constant rate; in the case of a stream of bubbles, the effect of any change in surface conditions experienced by a bubble caused by the previous bubbles is thought to be minor. A potential concern with this approach is that drop production under actual conditions, when large numbers of closely packed bubbles arrive and burst at the surface of the ocean following a breaking wave, is likely to be quite different from drop production by single bursting bubbles under laboratory conditions, but no measurements of drop production as a function of bubble size have been reported for actual breaking waves.

Two types of drops from bursting bubbles have long been known (§2.2): jet drops, produced from the column of water formed by the collapse of the bubble cavity, and film drops, fragments of the collapsed bubble cap (Fig. 6). Drops formed by the breakup of liquid jets that become unstable have long been studied [*Plateau*, 1873 (cited in *Stuhlman*, 1932); *Rayleigh*, 1879; *Tyler and Richardson*, 1925; *Castleman*, 1931; *Tyler and Watkin*, 1932; *Tyler*, 1933; *Schweitzer*, 1937; *Merrington and Richardson*, 1947; *Rumscheidt and Mason*, 1962; *Goedde and Yuen*, 1970; a recent review is presented in *Lin and Reitz*, 1998]. Remarkable photographs of jet drops formed by splashes were made over 100 years ago [*Worthington*, 1963], and more recent photographs [*Edgerton and Killian*, 1979, pp. 22-25] clearly illustrate the process of jet drop formation. As early as 1927 *Foulk* [cited in *Foulk*, 1932] noted the "well-known fact that a bubble bursting on the surface of a liquid frequently throws a droplet of the liquid upward, sometimes to a height of 4 or 5 in.," and stated further "the droplet thrown upward is not part of the ruptured bubble film as is commonly supposed," continuing on to describe what is now known as jet drop formation. Drops formed by the bursting of bubble films were discussed by *Chapman* [1938b] and *Facy* [1951b], and following the introduction of high-speed photography to study drops formed by bursting bubbles, the occurrence of both jet drops and film drops became appreciated by the oceanographic and atmospheric research communities [*Woodcock et al.*, 1953; *Knelman et al.*, 1954; *Kientzler et al.*, 1954; *Blanchard*, 1954a; *Newitt*, 1954; *Newitt et al.*, 1954; *Mason*, 1954; *Moore and Mason*, 1954].

The process of drop formation by a bursting bubble occurs as follows (Fig. 6). When a bubble arrives at the surface of a liquid it initially overshoots its equilibrium position. If the surface contains a surface-active substance—which will nearly always be the situation for water or seawater, the bubble relaxes within several milliseconds to its equilibrium position, a portion of which (the bubble cavity) is below the equilibrium height of the surface on account of internal pressure in the bubble due to surface tension of the film, where it remains for a time varying from milliseconds to several seconds, depending on the nature and amount of material on the surface of the liquid (and on the surface of the bubble, which may have scavenged material during its ascent). The bubble film (also called cap or dome) becomes thinner over time and ultimately bursts; depending on the size of the bubble, this bursting may yield fragments, denoted film drops, which are projected at various angles with respect to the surface, from vertical to horizontal. This bursting occurs in tens of microseconds, and the velocity of the ruptured film, and of the film drops produced, may be of order 10 m s^{-1} [*Spiel*, 1998]. Over a time of order milliseconds, the cavity formed by the bubble fills with water on account of surface tension, with accelerations of the water surface being 10^3 to 10^6 times the acceleration due to gravity. This process results in the formation of a small vertical column, or jet, in the center of the cavity left by the bubble. This jet is composed of the water from a thin layer on the surface of the cavity, having thickness of order 5-10% of the bubble diameter [*Worthington*, 1963; *Blanchard*, 1963, pp. 195-196; *MacIntyre*, 1968, 1972], and it is generally believed that jet drops are generated from a thinner layer of the ocean surface than are film drops. The jet rises roughly one bubble diameter above the surface and becomes unstable if its length exceeds its circumference (which occurs for sufficiently small bubbles), in which situation one to several jet drops are ejected vertically over several tens of microseconds. These jet drops have ejection (i.e., initial) velocities up to tens of meters per second and rapidly attain their maximum heights, which depend on the drop radius and ejection velocity, and on the wind near the surface. Drops that are ejected above the viscous boundary layer (of order 0.1-1 mm; §2.4) may be entrained upward into the atmosphere by turbulent eddies (and thus contribute to the sea salt aerosol), depending on their size and on the wind speed and atmospheric stability; however drops with $r_{80} \gtrsim 25$ μm are unlikely to remain airborne longer than a second under typical wind speeds unless they encounter upward-moving

eddies and are entrained upward (§2.9.2; §4.3.2.4). Bubbles with radius $R_{bub} \lesssim 0.5$ mm, are believed to produce only jet drops, and bubbles with $R_{bub} \gtrsim 3$ mm only film drops, with bubbles of intermediate size producing both jet and film drops. Examples illustrating the transition from jet drop production to film drop production (and the increase in jet drop size) with increasing bubble size can be found in *Garner et al.* [1954], *Newitt et al.* [1954], and *Tomaides and Whitby* [1976]. For bubbles that produce both types of drops, jet drops are generally considerably larger than film drops but film drops are more numerous.

Whereas the formation mechanism of jet drops appears to be established, the means by which film drops are produced remains unresolved. Several different patterns of ejection of film drops have been reported, including a cloud of drops, a single vertical line of drops, a segment of a spherical cap, a completely capped vortex ring, and a family of drops projected at a low angle to the horizontal [*Knelman et al.*, 1954; *Garner et al.*, 1954; *Newitt et al.*, 1954; *Mason*, 1954, 1957a; *Blanchard*, 1963; *Day*, 1964, 1967; *Paterson and Spillane*, 1969; *MacIntyre*, 1972; *Resch et al.*, 1986; *Resch*, 1986; *Iacono and Blanchard*, 1987; *Spiel*, 1997b, 1998; *Mason*, 2001]. Several mechanisms of bubble bursting and drop formation have been proposed to explain these patterns (which appear to be correlated to some extent with bubble size), but there is too much photographic evidence showing multiple mechanisms to suppose that a single explanation will account for all of the observations [*Facy*, 1951b; *Newitt*, 1954; *Newitt et al.*, 1954; *Mason*, 1954, 1957a; *Ranz*, 1959; *Blanchard*, 1963; *Day*, 1964, 1967; *MacIntyre*, 1972; *Tomaides and Whitby*, 1976; *Azbel and Lipias*, 1983; *Resch et al.*, 1986; *Iacono and Blanchard*, 1987; *Chalmers and Bavarian*, 1991; *Spiel*, 1997b, 1998; *Mason*, 2001]. The occurrence of different ejection patterns appears to be variable and not subject to experimental control; for example, *Blanchard* [personal communication, 2001] reported that whereas one day he might observe a vortex ring, a week or so later, with no apparent change in the conditions of the experiment, he would observe a vertical line of drops. Evidence for the several mechanisms, much of which is contradictory, is not examined in detail here, but two mechanisms for which photographic evidence has been presented are the rupture of the entire bubble film, and the bursting of the film at a single location, the latter resulting in a torus (formed by the rolling up of the film) which then either becomes unstable and breaks into film drops or collides with the surface of the liquid, likewise forming film drops. The mechanism suggested by *Tomaides and Whitby* [1976], that the bubble film bursts in several places, resulting in concentric tori of different sizes, lacks supporting evidence. Another proposed mechanism [*Azbel and Liapis*, 1983; *Koch et al.*, 2000; *Reinke et al.*, 2001], that bursting is caused by a resonance between oscillations of the bubble film and of the gas in the bubble, can be discounted, as critical film thicknesses predicted by this theory are at most only several molecules thick, far too small to be realistic or to account for the mass of the film drops reported by several investigators, and several orders of magnitude smaller than estimated and reported film thicknesses, which are of order 1 μm for bubbles in seawater [*Toba*, 1959; *MacIntyre*, 1972, 1974b; *Spiel*, 1998; *Chalmers and Bavarian*, 1991].

The primary factor controlling film and jet drop production is bubble size; additional controlling factors are the surface tension, viscosity, and density of the liquid (and thus its temperature). Of these factors, only the viscosity varies appreciably with temperature for water or seawater (§2.5.1). However, the surface tension is affected by surface-active materials, and consequently the nature and amount of surface-active materials on the surface of the bubble and on the surface of the liquid may affect drop formation. Because bubbles resting at an air-water interface depart somewhat from sphericity with larger bubbles exhibiting greater departure, bubble size is described here (and presumably in the references cited, although this is not always explicitly stated) by R_{bub}, the volume-equivalent radius; that is, the radius of a sphere having the same volume as that of the bubble [*Toba*, 1959]. As a bubble with an equivalent radius of 4 mm has a maximum horizontal diameter just over 15% greater, and a maximum vertical extent around 25% less, than the volume-equivalent diameter, and as bubbles of this size and greater typically have low concentrations in the ocean, most bubbles are well approximated as spherical. Bursting behavior is believed to be similar in fresh water and seawater; reported differences are attributed to the position of the bubble at the surface when it bursts, which is strongly influenced by the presence or absence of surface-active materials on the surface of the liquid or collected by the bubble during its ascent, as discussed below.

In addition to their importance as SSA particles, drops from bursting bubbles are important in industry[1] and in nuclear engineering and safety[2]. Drops from bursting bubbles can transport material across water-air interfaces, often in concentrations greatly exceeding those in the bulk liquid, the extent to which depends on the sizes and type of drops. Such transported material includes organic substances[3], radioactivity[4], pollen [*Valencia*, 1967] and other particulate matter [*Quinn et al.*, 1975; *Cloke et al.*, 1991], and bacteria and viruses[5]. This transport thus can play a role in health and hygiene[6]. Drops from bursting bubbles have been used to generate aerosol particles of a given size [*Blanchard*, 1954b; *Tomaides and Whitby*, 1976] and as a means of investigating the composition of the sea surface microlayer[7]. The electrical charge on bubble-produced drops, which differs between film drops and jet drops and has important implications for the electric flux charge over the oceans, has been the subject of several investigations[8], and has been used for the collection of drops [*Blanchard and Syzdek*, 1975, 1982, 1990b] and determination of whether they are film drops or jet drops [*Woolf et al.*, 1987].

4.3.2. Jet Drops

Despite theoretical and numerical studies on jet drop production from bursting bubbles [*Stuhlman*, 1932; *Davis*, 1940; *Newitt et al.*, 1954; *Toba*, 1959; *Blanchard*, 1963; *MacIntyre*, 1968, 1972; *Spiel*, 1992; *Boulton-Stone and*

---

[1] *Foulk* [1932]; *Davis* [1940]; *Garner et al.* [1954]; *Newitt et al.*, [1954]; *Azbel and Liapis* [1983]; *Papachristodoulou et al.* [1985]; *Kuo and Wang* [1999, 2002].

[2] *Yamomoto et al.* [1968]; *Ginsberg* [1985]; *Starflinger et al.* [1995]; *Koch et al.* [2000] and references therein.

[3] *Woodcock* [1955]; *Wilson* [1959a]; *Baylor et al.* [1962]; *Sutcliffe et al.* [1963]; *Blanchard* [1963, 1964]; *Garrett* [1968]; *Batoosingh et al.* [1969]; *Baier* [1972]; *Pueschel and van Valin* [1974]; *Bezdek and Carlucci* [1974]; *Blanchard and Syzdek* [1974a]; *Hoffman and Duce* [1976]; *Blanchard* [1978]; *Tusseau et al.* [1980]; *Gershey* [1983a,b]; *Cloke et al.* [1991]; *Tseng et al.* [1992]; *Cini et al.* [1994]; *Saint-Louis et al.* [2004].

[4] *Belot et al.* [1982]; *Cambray and Eakins* [1982]; *Walker et al.* [1986]; *McKay and Pettenden* [1990]; *Nelis et al.* [1994]; *McKay et al.* [1994].

[5] *Horrocks* [1907]; *Woodcock* [1948]; *Anderson et al.* [1952]; *Woodcock* [1955]; *Blanchard and Syzdek* [1970, 1972a]; *Bezdek and Carlucci* [1972]; *Aubert* [1974]; *Blanchard and Syzdek* [1974a,b, 1975]; *Baylor et al.* [1977a,b]; *Blanchard* [1978]; *Blanchard and Syzdek* [1978a]; *Blanchard et al.* [1981]; *Blanchard and Syzdek* [1982]; *Blanchard* [1983]; *Burger and Bennett* [1985]; *Blanchard* [1989b]; *Blanchard and Syzdek* [1990b]; *Cincinelli et al.* [2001]; *Pósfai et al.* [2003]; *Saint-Louis et al.* [2004].

[6] *Horrocks* [1907]; *Jacobs* [1939a]; *Woodcock* [1948]; *Anderson et al.* [1952]; *Woodcock* [1955]; *Blanchard and Syzdek* [1974a]; *Blanchard* [1989b]; *Wangwongwatana et al.* [1990 and references therein].

[7] *MacIntyre* [1968, 1972]; *Bezdek and Carlucci* [1972, 1974]; *Fasching et al.* [1974]; *MacIntyre* [1974b]; *Pattenden et al.* [1981]; *Gershey* [1983a,b]; *Sakai* [1989]; *Cloke et al.* [1991].

[8] *Chapman* [1937, 1938a,b]; *Blanchard* [1955]; *Loeb* [1958, pp. 119-122]; *Blanchard* [1958, 1963, 1966]; *Blanchard* [1967, p. 117, pp. 126-127]; *Iribarne and Mason* [1967]; *Jonas and Mason* [1968]; *Medrow and Chao*, 1971; *Blanchard* [1985b]; *Woolf et al.* [1987].

Blake, 1993; *Dekker and de Leeuw*, 1993; *Koch et al.*, 2000; *Reinke et al.*, 2001], most current understanding derives from laboratory experiments involving drops from a stream of monodisperse bubbles bursting individually at a calm water surface, summarized in Table 16. In this section the results of these experiments are examined, especially those concerning the number (§4.3.2.1), sizes (§4.3.2.2), and maximum heights and corresponding ejection (i.e., initial) velocities (§4.3.2.3) of the drops produced, and the factors upon which these quantities depend. Additionally, the extent to which results of these laboratory experiments can be applied to the bursting process under oceanic conditions is examined.

4.3.2.1. Number of jet drops produced by a bursting bubble. The mean number of jet drops per bubble N_{jet} reported from laboratory investigations of single bubbles bursting at a calm water surface (distilled water, tap water, pond water, seawater, or NaCl or KCl solutions) is shown in Fig. 26 as a function of R_{bub}. For studies involving seawater, drop sizes are reported as r_{80}; for studies involving liquids other than seawater, drop sizes are reported as an effective r_{80} evaluated as $r_{80} = 0.5r_{form}$ (§2.1.1). It is possible that some of the reported data underestimate the number of drops produced. For instance, *Garner et al.* [1954] collected drops at a height of 1 cm and *Newitt et al.* [1954] collected drops at various heights above the water surface

Table 16. Laboratory Investigations of Jet Drop Production from Individual Bubbles

Reference	R_{bub}/mm	Liquid	Quantities Reported[a]	T/ °C
Stuhlman, 1932	0.1-1.8	water, benzene	h	21
Kientzler et al., 1954	0.1-0.9	water, seawater	r_{80}	21-25
Garner et al., 1954	0.38-2.3	water, benzene, ethyl alcohol	r_{80}	~0
Newitt et al., 1954	1.6-2.7	water	r_{80}, h	25, 35, 45
Boyce, 1954	0.1-2.4	seawater	r_{80}, h	20
Moore & Mason, 1954	0.15-2.15	seawater	r_{80}	not stated
Blanchard & Woodcock, 1957 (see also *Blanchard*, 1989a)	0.05-1.45	seawater	r_{80}, h	22-26
Hayami & Toba, 1958 (see also *Toba*, 1961a,b)	0.55-3	water, seawater	N_{jet}, r_{80}, h	4, 16, 30
Blanchard, 1963 (see also *Blanchard*, 1989a)	0.11-0.45	water, seawater	r_{80}, h	~4
Bezdek, 1971	0.15-0.85[b]	seawater	h vs. r_{80}	room
Medrow & Chao, 1971	0.1-0.3	NaCl, KCl solutions	r_{80}	23, 50
Blanchard & Syzdek, 1972a	0.35	water with nutrient broth and bacteria	effects of bubble age on h	22-26
Blanchard & Syzdek, 1975	0.185	water with nutrient broth and bacteria	r_{80} for all drops for R_{bub} = 0.185 mm	room
Tomaides & Whitby, 1976	0.7, 2.75	NaCl solution	r_{80}	room
Detwiler & Blanchard, 1978	0.4-0.63	water, salt solutions	effects of bubble age on h	room
Blanchard & Hoffman, 1978	0.45	seawater	effects of bubble age on h for R_{bub} = 0.45 mm	20-22
Tedesco & Blanchard, 1979	0.05-0.3	water, pond water	r_{80}, h	22-24
Sakai, 1989	0.56-1.8	water, MgCl$_2$ solution	r_{80}	30
Spiel, 1992	0.24-2.9	water	v_0	room
Spiel, 1994a	0.35-1.5	water	N_{jet}, r_{80}	17-23
Spiel, 1994b	0.35-1.5	water, seawater	N_{jet}, r_{80}	18-25
Spiel, 1995	0.35-1.5	water, seawater	v_0	20-22
Spiel, 1997a	0.35-1.5	seawater	N_{jet}, r_{80}, v_0	27-29
Koch et al., 2000 (and references therein)[c]	0.05-2	water	N_{jet}, r_{80}	room

Dependence on R_{bub} of mean number of jet drops per bubble N_{jet}, r_{80}, and ejection height h are presented in Figs. 26, 27, and 28, respectively.
[a] N_{jet} is the mean number of jet drops per bubble, h the height to which they are ejected, and v_0 their ejection (i.e., initial) velocity. For drops from bubbles in media other than seawater, values of r_{80} are reported as $r_{80} = 0.5r_{form}$.
[b] Range of R_{bub} calculated from range of r_{80} using (4.3-3).
[c] *Koch et al.* [2000] reported results of models and of laboratory investigations of others.

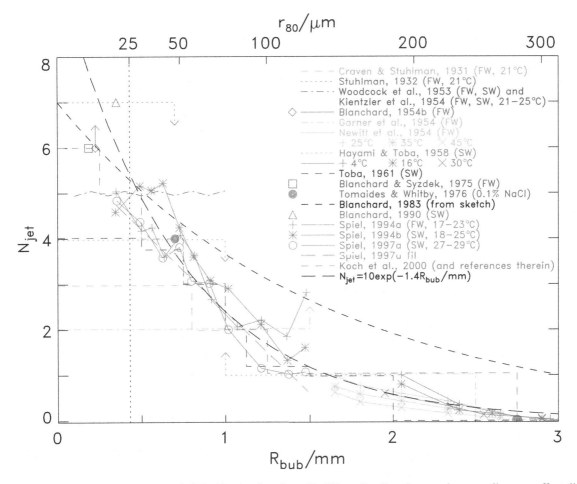

Figure 26. Mean number of jet drops per bubble N_{jet} as a function of bubble radius R_{bub}. In some instances lines are offset slightly for visibility. Arrows denote values that are greater than or equal to (or less than or equal to) value shown. Also shown are the relations presented graphically in *Blanchard* [1983] and in *Spiel* [1997a] and expression (4.3-1). Upper axis shows values of r_{80} for jet drops based on (4.3-3). Data of *Garner et al.* [1954] are for drops collected at a height of 1 cm above the water surface and those of *Newitt et al.* [1954] for drops collected at a height of 6.4 mm (the latter investigators collected drops at various heights). There is a discrepancy in *Spiel* [1997a] between the value of N_{jet} for $R_{bub} = 0.353$ mm obtained from summing the values of $p(i)$ from Table 1 of that reference (4.84) and the value shown in Fig. 6 of that reference (4.35); the value should probably be the one from the table [*D. Spiel*, personal communication, 2000] and is the one used here. Vertical dotted line at $r_{80} = 25$ μm demarcates medium and large SSA particles.

(the data of *Newitt et al.* in Fig. 26 are for a height of 0.64 cm); drops ejected to lower heights would not have been counted. Some of the data represented in Fig. 26 were taken from statements such as "about 5 drops," or "more than 3 drops," and only recently have there been any systematic studies for bubble sizes for which $N_{jet} > 1$. It has generally been assumed that $N_{jet} = 7$ for small bubbles ($R_{bub} \lesssim 0.5$ mm), decreasing as R_{bub} increases until $N_{jet} = 0$ for R_{bub} near 3 mm; a value of $N_{jet} = 5$ is often chosen as typical. However, several investigations have demonstrated that more than seven drops can be produced by a single bubble. *Blanchard* [1983] reported that up to 10 drops could be produced by bursting of a single

bubble, and *Spiel* [1994a,b, 1997a] reported that up to 15 drops could be produced by individual bubbles in distilled water, and more than 7 drops by individual bubbles in seawater (although this occurred very infrequently). Additionally, *Spiel* [1992, 1994b] reported that sometimes a train of numerous closely spaced drops results from a bubble burst and noted that *Blanchard* had also observed such behavior. It is expected that extremely small bubbles do not eject drops because the jet does not extend to sufficient height to become unstable or to eject a drop, although no estimate has been suggested for a limiting bubble radius below which drop production does not occur.

Several relations between N_{jet} and R_{bub} have been proposed. A rather heuristic argument that $N_{jet} \propto R_{bub}^{-1.9}$ was presented by *Tomaides and Whitby* [1976]. A relation represented graphically by *Blanchard* [1983, Fig. 6], corresponding to

$$N_{jet} = 7\exp\left(-0.65\frac{R_{bub}}{mm}\right),$$

was probably meant only as a rough approximation; the expression

$$N_{jet} = 10\exp\left(-1.4\frac{R_{bub}}{mm}\right) \qquad (4.3\text{-}1)$$

appears to fit the majority of the data better. These latter two relations are shown in Fig. 26. Probabilities that a bubble of a given size would eject a certain number of jet drops were presented by *Spiel* [1994a,b, 1997a], who determined that N_{jet} decreased with increasing R_{bub} over the range 0.35 mm to 1.5 mm. These probability tables reveal some peculiar patterns, specifically for the 6th drop in the sequence of jet drops ejected, implying much complexity in the jet drop production process. A relation based on these tables corresponding to

$$N_{jet} = 6.0 - 3.6\left(\frac{R_{bub}}{mm}\right),$$

which was presented graphically [*Spiel*, 1997a, Fig. 6], is also shown in Fig. 26. Based on this same data a more complicated expression (not shown) was proposed by *Wu* [2002b] which approaches the value $N_{jet} = 7$ at $R_{bub} = 0$, but this expression differs little from that of *Spiel* over the range $R_{bub} = 0.35\text{-}1.5$ mm.

The few data available on the temperature dependence of N_{jet}, all for bubbles so large ($R_{bub} > 1.55$ mm) that $N_{jet} \leq 1$, indicate that N_{jet} decreases with increasing water temperature [*Newitt et al.*, 1954; *Hayami and Toba*, 1958]. It is expected that the temperature dependence of N_{jet} is manifested through the viscosity, surface tension, and density, which control the behavior and shape of the jet (and in turn determine the number of jet drops formed); thus studies involving different liquids could shed light on the importance of these properties. However, few liquids other than water and seawater have been investigated; bubbles with $R_{bub} < 0.75$ mm in benzene at 22 °C may produce up to 4 jet drops as opposed to 7 for water at 21 °C [*Stuhlman*, 1932], and bubbles in water, ethyl alcohol, and benzene formed more than one jet drop for $R_{bub} > 1.5$ mm and no drops for $R_{bub} > 2.5$ mm [*Garner et al.*, 1954], although no difference

in N_{jet} among the substances was reported. *Newitt et al.* [1954] argued that a longer jet is formed at lower temperatures because of higher resistance to break-up due to higher viscosity and aided by higher surface tension. These investigators concluded from tests on other liquids that the number of drops is determined primarily by the surface tension, being proportional to its square root, but no supporting evidence was presented. *Papachristodoulou et al.* [1985] argued that the critical bubble radius above which jet drops are not produced depended only on the surface tension and the density of the fluid, with viscosity playing a secondary role, but also provided no evidence for this assertion. As the temperature dependence of surface tension of water (or seawater) is quite small, around 0.2% per °C (§2.5.1), it seems unlikely that differences in surface tension due to temperature play any appreciable role in jet drop formation in the ocean, although differences in surface tension due to the amount and nature of surface-active materials may be important, as discussed below.

Although it has generally been assumed that as R_{bub} decreases toward zero the number of jet drops produced per bubble approaches some value between 5 and 10 (as do the parameterizations for N_{jet} as a function of R_{bub} discussed above and shown in Fig. 26), sufficiently small bubbles probably lack sufficient energy to eject jet drops. Additionally, although the number of jet drops from a bursting bubble is an important quantity pertinent to SSA production, other factors merit consideration, such as drop sizes and the heights to which drops are ejected, and thus the number of drops that remain airborne for any substantial amount of time as opposed to rapidly falling back to the sea surface without interacting with the atmosphere to any appreciable extent (such as by exchanging heat or mass). These considerations are examined below in §4.3.2.4.

4.3.2.2. Sizes of jet drops. Sizes of jet drops vary over a wide range. At one extreme, several investigations [*Garner et al.*, 1954; *Newitt et al.*, 1954; *Hayami and Toba*, 1958] have reported jet drops with $r_{80} > 200$ μm; at the other extreme, jet drops as small as $r_{80} = 0.5$ μm have been observed [*Blanchard*, 1954b], and evidence for jet drops as small as $r_{80} = 0.25$ μm has been reported [*Woolf et al.*, 1987]. *Blanchard* [1989a] reported observing bubbles so small that the top jet drop (i.e., the first of the drops when more than one are produced) was ejected only 0.04 cm, from which he concluded (based on extrapolation from larger bubbles) that such drops were produced from bubbles with $R_{bub} \approx 0.005$ mm and had $r_{80} \approx 0.25$ μm (the 1.3 power relationship between R_{bub} and r_{80} discussed below yields $r_{80} \approx 0.08$ μm for bubbles of this size). However, few jet drops of this size are expected to be produced at seawater surfaces, as bubbles of this size would dissolve in a matter of seconds (§4.4.2.2) and thus would not survive to the surface and burst. Additionally, only a small fraction of drops ejected to such a low height would be expected to be entrained upward into the atmosphere (§4.3.2.4).

The radii, or average radii, of jet drops produced by single bubbles bursting at a calm liquid surface, determined by laboratory investigations for which data from more than one bubble size were presented, are shown in Fig. 27 as a function of R_{bub} (as noted above, for studies involving freshwater, values of r_{80} are reported as half the radius at formation). Values presented by *Koch et al.* [2000] from computer models are also shown. The ranges of jet drop sizes given in *Toba* [1961a,b] from experiments in tap water are consistent with the values in Fig. 27. In many of these investigations the uncertainties and range of r_{80} for a given R_{bub} and

temperature were not provided, but typically were up to 25%. Relatively few measurements have been conducted for $R_{bub} < 0.3$ mm, the only systematic studies being that of *Medrow and Chao* [1971] for NaCl and KCl solutions and that of *Tedesco and Blanchard* [1979] for fresh water and pond water (each at only a single temperature), and the few radii reported for seawater drops exhibit no discernible trend with R_{bub} in this range. A horizontal dotted line at $r_{80} = 25$ μm, demarcating the size between medium and large SSA particles (and thus the upper limit of the sizes of drops expected to contribute to the effective SSA production flux; §2.1.6), demonstrates that most jet drops smaller than this size are produced by bubbles with $R_{bub} < 0.5$ mm, excluding those drops in the small mode found by *Spiel* (discussed below); if these drops are included, then bubbles with

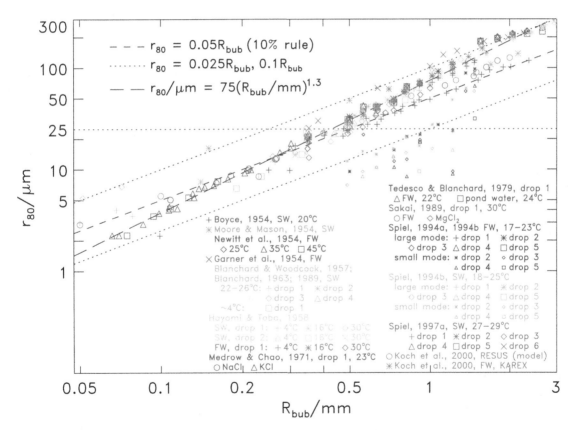

Figure 27. Radii of jet drops at 80% RH r_{80} as a function of bubble radius R_{bub} from investigations summarized in Table 16. Also shown are the 10% rule (Eq. 4.3-2) and lines denoting half this value and twice this value, and the relation given by (4.3-3). Data of *Garner et al.* [1954] are for drops collected at a height of 1 cm above the water surface, and those for *Newitt et al.* [1954] are Sauter mean radii (ratio of the third moment with respect to radius to the second moment) of drops collected at a height of 0.64 cm (these investigators collected drops at various heights). The freshwater data from Fig. 9 of *Spiel* [1994b] were used instead of those from Fig. 7 of *Spiel* [1994a]; there were only minor differences between them regarding whether certain of the later drops (predominantly from the largest two bubble sizes) should be placed in the large drop mode or the small drop mode; *Spiel* [1994b] stated that based on their ejection velocities these drops should be placed in the small drop mode. Results from computer models reported by *Koch et al.* [2000] are also shown. Horizontal dotted line at $r_{80} = 25$ μm demarcates medium and large SSA particles.

R_{bub} as large as 1.5 mm and possibly larger (1.5 mm was the largest bubble radius considered by *Spiel*) must be considered in relation to jet drop production with regard to the effective SSA production flux.

Reported sizes of jet drops other than the top drop exhibit considerable variability, though little systematic investigation has been reported until recently. The sizes of the top five jet drops in freshwater and seawater for bubbles with R_{bub} from 0.35 mm to 1.5 mm were measured by *Spiel* [1994a,b, 1995, 1997a], who noted an unusual dependence on temperature—a bimodal distribution for the sizes of later jet drops in the sequence for both seawater and distilled water at temperatures near 20 °C. In one mode, termed the larger mode, the radii of the later drops are roughly the same as that of the top drop (except those of the third drops which were around 30% less than those of the other drops for $R_{bub} < 650$ μm); these radii are consistent with most others presented in Fig. 27. In a previously unreported mode, termed the small mode, drops other than the top drop have smaller radii (typically by a factor of 2 to 3, although again similar to each other) and greater ejection velocities (by roughly a factor of 2) than the top drop. Not all of the later drops from a given bubble are necessarily in the same mode. Average values of r_{80} for drops in each mode are shown separately in Fig. 27 as a function of R_{bub}. Occasional production of drops smaller than those normally encountered was also reported by *Kientzler et al.* [1954], and *Newitt et al.* [1954] suggested that during the breakup of the jet so-called "satellites," or smaller drops between two larger jet drops, are also formed [*Worthington*, 1963; *Iribarne and Mason*, 1967; *Spiel*, 1997a; *Koch et al.*, 2000]; this is a well known phenomenon in the breakup of jets not formed by bubbles [*Rayleigh*, 1891; *Castleman*, 1931; *Rumscheidt and Mason*, 1962; *Donnely and Glaberson*, 1966; *Blanchard*, 1967, Plate IV; *Goedde and Yuen*, 1970; *Edgerton and Killian*, 1979, p. 27; *Qian et al.*, 1986]. Such findings prompt concerns that earlier investigations [*Garner et al.*, 1954; *Newitt et al*, 1954; *Tomaides and Whitby*, 1976], in which drops from bursting bubbles were classified as "small" and "large," assumed to be film drops and jet drops, respectively, may have included some jet drops of this smaller mode in the film drop category.

Several relations have been proposed between jet drop size and R_{bub}. Graphs relating the radii of the top two jet drops from bubbles in seawater and R_{bub} for $R_{bub} = 0.04$ mm to 2 mm were presented by *Blanchard and Woodcock* [1957], and graphs relating the radius of the top jet drop from bubbles in seawater and R_{bub}, extending to R_{bub} as low as nearly 0.005 mm for 4 °C and down to $R_{bub} = 0.002$ mm for 22-26 °C were presented by *Blanchard* [1963, Fig. 9]. However, such representations involved substantial extrapolation, as there were no data for the top drop from bubbles with $R_{bub} < 0.05$ mm, only 2 data for $R_{bub} < 0.25$ mm at 4 °C,

and only 4 data for $R_{bub} < 0.25$ mm at 22-26 °C [*Blanchard*, 1989a]. There is therefore little basis for expressions in this size range such as those presented by *MacIntyre* [1970] and *Monahan* [1988a] (presumably applicable to 22-26 °C) based on *Blanchard's* figure. A rule of thumb that has long been used—to the point where it is often accepted as a hard and fast rule—is that the radius of a jet drop at formation is ten percent of the radius of the parent bubble; as r_{80} is half the radius at formation, this relation is equivalent to

$$r_{80} = 0.05 R_{bub}. \qquad (4.3\text{-}2)$$

This "10% rule," first stated by *Kientzler et al.* [1954], is shown in Fig. 27 by a dashed line; also shown are lines representing radii that are twice this size and half this size. Although this relation provides a good representation of the data of *Boyce* [1954] and a fair representation of the other data for bubbles with R_{bub} in the range 0.15-0.5 mm, it generally overestimates the radii of jet drops by over 50% at $R_{bub} \approx 0.05$ mm (the smallest bubble sizes in Fig. 27; although the majority of the data in this size range are from a single investigation, that of *Tedesco and Blanchard* [1979]), and it underestimates the radii of jet drops at $R_{bub} \approx 3$ mm (the largest bubble sizes in Fig. 27) by roughly a factor of two. Additionally, this relation clearly does not apply to the small mode reported by *Spiel* [1994a,b, 1995, 1997a]. *Blanchard and Syzdek* [1975] concluded that the 10% rule applies only to the top two jet drops, and *Blanchard* [1989a] concluded that although this relation holds reasonably well for the top drop for small bubbles ($R_{bub} \lesssim 0.25$ mm) in seawater at 22 °C (based on data in *Blanchard and Woodcock* [1957]), it overestimates the radii of the top drops from bubbles in seawater at 4 °C, and it doesn't apply for the later jet drops in the sequence. *Blanchard* [1989a] observed that no graph existed justifying this widely-accepted rule and stressed the paucity of pertinent data.

Several investigators have proposed power law relations of the form $r_{80} \propto R_{bub}{}^p$, with p typically in the range 1.2-1.5 [*Blanchard*, 1963, p. 111; *Tomaides and Whitby*, 1976; *Tedesco and Blanchard*, 1979; *Spiel*, 1994b, 1997a; *Koch et al.*, 2000], and the relationships presented graphically by *Blanchard and Woodcock* [1957, Fig. 2] for the top and second drops are well approximated by a relation of this form for $R_{bub} \gtrsim 0.2$ mm. A power law corresponding to

$$\frac{r_{80}}{\mu m} = 75 \left(\frac{R_{bub}}{mm} \right)^{1.3} \qquad (4.3\text{-}3)$$

is also shown in Fig. 27. For R_{bub} in the range 0.125-0.43 mm, the values of r_{80} determined from (4.3-2), which range from ~6 μm to ~22 μm, are within 20% of those determined from (4.3-3), which range from ~5 μm to ~25 μm. As the latter expression provides a better overall representation of the data (excluding the small drop mode of *Spiel*), it is used below. A more complicated expression (not shown) proposed by *Wu* [2002b], based on data of *Spiel* [1997a] over the range $R_{bub} = 0.35$-1.5 mm, differs from (4.3-3) by no more than 15% over this range.

In addition to R_{bub}, the surface tension, viscosity, and density of the liquid are expected to be important factors in governing the sizes of the jet drops produced. *Garner et al.* [1954] reported that benzene and ethyl alcohol produced jet drops that are roughly 15% and 30%, respectively, smaller than those formed from water for the same R_{bub} because of reduced surface tension, and these investigators further stated that larger jet drops would be formed in substances with higher viscosity. However, extensive investigations involving jet drops formed in other media, which might elucidate the roles of surface tension and viscosity in determining jet drop size, have not been reported, and only a few different temperatures have been studied, with contrasting results. *Newitt et al.* [1954] reported that the average radius of jet drops from large bubbles ($R_{bub} = 1.55$-2.65 mm) in water decreased with increasing temperature by a few tenths of a percent per °C for temperatures of 25, 35, and 45 °C. By contrast, the data of *Hayami and Toba* [1958] for $R_{bub} >$ 0.8 mm at 4, 16, and 30 °C in seawater, and those of *Blanchard* [1963, Fig. 9] at ~4 °C and 20-22 °C in seawater, displayed the opposite trend, with top jet drop radii increasing with increasing temperature by around 0.5% per °C, from which *Blanchard* [1963, p. 111; 1983, 1989a] concluded that drop size increased with increasing temperature for all bubble sizes. *Medrow and Chao* [1971] reported that at 50 °C the drop sizes displayed less uniformity than at 23 °C, but implied that the mean sizes were the same. A more unusual dependence on temperature is the bimodal size distribution for later jet drops in the sequence reported by *Spiel* [1994a,b, 1995] at temperatures near 20 °C. At higher temperatures (27-29 °C), the larger mode was essentially unchanged, but the smaller mode was mostly absent, although it could be repeatedly turned on or off by adjusting the temperature [*Spiel*, 1997a]. This behavior was attributed by *Spiel* [1997a] to the effects of difference in surface tension and/or viscosity between the two temperatures (about 1.5% and 15%, respectively) on the dynamics of the jet breaking up into drops. The average radius of the top three jet drops reported by *Spiel* [1994a,b, 1997] was about 3% lower at these higher temperatures than that of the larger mode at 20-22 °C over the range of bubble sizes considered (0.35-1.5 mm), yielding a temperature dependence of comparable magnitude to that reported by *Hayami and Toba* [1958] and *Blanchard* [1963, Fig. 9], but of opposite sign. Other than four data

at ~4 °C reported by *Blanchard* [1963, 1989a], all other data on sizes of jet drops with $r_{80} \lesssim 25$ μm are for temperatures near 20 °C or higher. In view of the narrow temperature range for most data on jet drop radii, the vast difference in jet drop size distributions between temperatures of ~20 °C and ~28 °C reported by *Spiel*, and the weak dependence of mean jet drop size on temperature reported by previous investigators, it is difficult to draw confident conclusions on the dependence of jet drop radii for a given R_{bub} on temperature.

4.3.2.3. Ejection heights of jet drops.   Jet drops produced by bubbles bursting at a calm water or seawater surface attain maximum heights of up to 20 cm in still air. These ejection heights determined by laboratory investigations are shown in Fig. 28 as a function of R_{bub}; the upper axis shows corresponding values of r_{80} based on (4.3-3). The vertical dotted line at $r_{80} = 25$ μm identifies the upper size range of medium SSA particles, and thus the upper size limit of drops that are expected to remain in the atmosphere sufficiently long to contribute to the effective SSA production flux. The data of *Bezdek* [1971], who reported (maximum) ejection height as a function of jet drop size for $r_{80} = 6.5$-60 μm in seawater at 22 °C, are also shown (at values of R_{bub} calculated from Eq. 4.3-3). Ejection velocities for both large and small mode drops produced by bubbles bursting in distilled water and seawater were reported by *Spiel* [1995, 1997a] as a function of R_{bub}; ejection heights shown in Fig. 28 were calculated from averages values of these velocities and mean drop sizes reported in *Spiel* [1994a,b], taking into account deceleration due to drag and gravity. The ranges of ejection velocities for a given bubble size were not reported, but ejection velocities as a function of drop size [Figs. 14 and 15 of *Spiel*, 1995] exhibited a large amount of scatter for drops other than the top drop. Additionally, sizes of later drops in the sequence, and of drops in the small mode, exhibited much scatter, and ejection heights calculated from average values of drop radii and ejection velocity only roughly approximate average ejection heights because ejection height is not linearly related to drop size and ejection velocity. There are few data for $R_{bub} < 0.2$ mm ($r_{80} \lesssim 10$ μm) other than those of *Tedesco and Blanchard* [1979], who reported results for one temperature each in distilled water and in pond water.

The maximum height attained by the top jet drop (Fig. 28) increases with increasing R_{bub} (or r_{80}) to a maximum of nearly 20 cm for $R_{bub} \approx 0.7$-1 mm ($r_{80} \approx 50$-80 μm). As R_{bub} (or r_{80}) further increases, the ejection height decreases, approaching zero for bubbles so large that they no longer produce jet drops (Fig. 26), i.e., at $R_{bub} \approx 3$ mm ($r_{80} \approx 300$ μm). In contrast to these results, *Mestayer and Lefauconnier* [1988], using bubbles formed by spray bubblers at the bottom of a wind/wave tunnel containing freshwater, reported that a

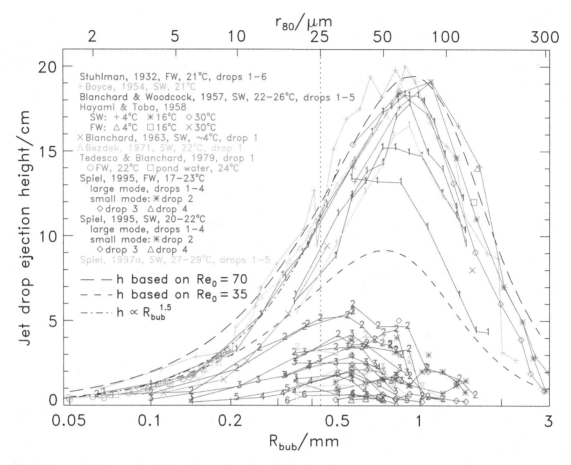

Figure 28. Maximum heights attained by jet drops ejected in still air from bubbles with radius R_{bub} from investigations summarized in Table 16. In most instances data from the same investigation are connected to aid viewing. Also shown are heights resulting from the relation $Re_0 = 70$, $Re_0 = 35$, and a relation in which the height is directly proportional to the 1.5 power of R_{bub} (with an empirically chosen constant of proportionality). Upper axis shows values of r_{80} based on (4.3-3); data of *Bezdek* [1971] were converted using this relation. Heights for *Spiel* [1995] were calculated using ejection velocities presented in that reference and sizes of drops from a given R_{bub} from *Spiel* [1994a]. Vertical dotted line at $r_{80} = 25$ μm demarcates medium and large SSA particles.

non-negligible fraction of drops with r_{80} (evaluated as $0.5r_{form}$; §2.1.1) from 1.5 μm to 6.25 μm were ejected higher than 20 cm, and that a large fraction of drops with $r_{80} > 0.5$ μm were ejected higher than 10 cm; the investigators stated that probably no drops were ejected to 30 cm. They believed that most drops with $r_{80} > 2.5$ μm were jet drops and that most smaller drops were film drops, and suggested that these results contradict results such as those shown in Fig. 28, although no explanation was offered for the contradiction. However, this discrepancy is easily resolved. Based on the dimensions of the bubblers used in their experiments (30 cm × 36 cm), the air flow employed in these studies (1 m³ h⁻¹) would have resulted in a vertical air velocity of 0.28 cm s⁻¹ above the surface of the water, which from (2.6-8) corresponds to the terminal velocity of a drop with

$r \approx 5$ μm. Hence, it is likely that the drops observed at heights greater than those in Fig. 28 resulted from upward transport by the air from the bubblers and that their results are not inconsistent with those shown in Fig. 28. It is uncertain to what extent this vertical air flow would have affected other of their results.

In the sequence of jet drops produced by the burst of a single bubble the maximum heights attained by later drops are appreciably less than that of the top drop. *Stuhlman* [1932] concluded that the ejection heights of drops from bubbles bursting in freshwater decrease exponentially according to the number of the drop in the sequence, with the ejection height of each drop in the sequence being roughly 65% that of the previous drop. Average ejection heights for the entire sequence of jet drops from bubbles with $R_{bub} = 0.185$ mm were reported by *Blanchard and Syzdek* [1975] as 3, 2, 1, 0.8, 0.5, and 0.1 cm (in order of ejection). *Blanchard* [1989a] suggested that the ejection heights for the second and third drops are 2/3 and 1/3 that of the top drop, respectively. *Spiel* [1992, 1995, 1997a] reported that for what he termed the large mode, the ejection velocities of the 2nd and 3rd drops were several times lower, and those of the 4th and 5th drops an order of magnitude or more lower, than those of the top drops from bubbles with R_{bub} from 0.35 mm to 1.5 mm. *Spiel* [1995] also reported that for a given R_{bub} the ejection velocities of the drops in the small mode were nearly twice as great as those of the large mode, although this still resulted in lower maximum heights attained for these drops because of their smaller sizes.

The maximum height h attained by a jet drop is determined by its radius, ejection velocity v_0, and the forces acting upon it—gravity and drag (§2.6.1), and can be calculated using (2.6-14). This drag force is initially greater than the gravitational force by up to several orders of magnitude, and the initial Reynolds number, $Re_0 = 2v_0r_{98}/v_a$ (§2.6.3), is typically much greater than unity, in which situation Stokes' Law (which assumes the drag force is proportional to the velocity) is not applicable and a different relation is required to describe the drop's motion (§2.6.3). Various expressions for the drag coefficient as a function of Reynolds number $C_d(Re)$ that have been used [*Newitt et al.*, 1954; *Blanchard*, 1963, Fig. 11; *Wu*, 1979a; *Rouault et al.*, 1991; *Spiel*, 1992, 1995] yield very similar results for the maximum heights attained in still air for a given drop size and ejection velocity. The maximum heights attained by drops bursting at the surface of the ocean are expected to be somewhat less than this maximum height in still air because of the additional drag due to the horizontal component of the wind velocity [*Wu*, 1979a; this conclusion holds despite the use of incorrect equations in that study, as noted in §2.6.3]; for example, the maximum height attained by a jet drop with $r_{80} = 75$ μm is reduced from its value of ~19 cm in still air (Fig. 28) to ~15 cm for $U_{10} = 10$ m s^{-1}. The possibility arises also that the force on the drop caused by such large initial deceleration is so large as to cause its shape to deviate from spherical, and thus have a drag force different from that assumed. However, for drops with the sizes and ejection heights reported in Figs. 27 and 28, i.e., $r_{80} \gtrsim 1$ μm, the deviation from spherical should be small. In an attempt to explain why the aerodynamic radii of fluid drops reported by *Baron* [1986] and *Griffiths et al.* [1986] measured using an aerodynamic particle sizer (which sizes particles based on their response to changing flow conditions) were smaller than expected based on the known radii, *Bartley et al.* [2000] modeled the distortion of drops under velocities similar to those observed for jet drops, but with much greater accelerations, and found the deformation for water drops to be less than one percent.

To achieve the maximum heights observed for jet drops requires ejection velocities up to tens of meters per second; such velocities are supported by photographs and other measurements. For instance, photographs presented in *Kientzler et al.* [1954] indicate an ejection velocity of at least 3 m s^{-1} for a drop from a bubble with $R_{bub} = 0.75$ mm (for which $r_{80} \approx 50$ μm, according to Eq. 4.3-3); it is possible that consecutive pictures show different drops, in which case the actual ejection velocity would be much greater. Photographs presented in *Blanchard* [1967, Plate VII and pp. 73-74] for bubbles with $R_{bub} = 0.85$ mm (for which $r_{80} \approx 60$ μm) yield similar values. According to *Blanchard* [1963, pp. 113-114], the ejection velocity of the top drop from a bubble with $R_{bub} = 0.035$ mm in seawater at 4 °C would be 80 m s^{-1}; this velocity is probably not accurate (it was apparently based on a value of $r_{80} = 0.85$ μm from Fig. 9 of that reference and a maximum height of 0.2 cm from Fig. 3 of that same reference; these values are based on extrapolations beyond the smallest bubble sizes for which *Blanchard* reported ejection heights and sizes of drops at this temperature—$R_{bub} = 0.11$ mm and $R_{bub} = 0.17$ mm, respectively), but nonetheless it is probably of the correct magnitude. Graphs of ejection velocity as a function of R_{bub} for temperatures of 4 °C and 22-26 °C presented by *Blanchard* [1963, Fig. 12], based on calculations using measured maximum heights, exhibited generally decreasing ejection velocities with increasing R_{bub}. There is a flattening of the curve, and possibly a local minimum in the ejection velocity, in the vicinity of $R_{bub} = 0.1$ mm (corresponding to $r_{80} \approx 4$ μm) for temperatures of 22-26 °C; however, as few data were reported for bubbles of this size or smaller (only three drop radii and one ejection height), it is difficult to justify this conclusion (the symbols in Fig. 12 of *Blanchard* [1963] do not correspond to measured values but presumably represent calculated values based on fits presented graphically in Figs. 3 and 9 of that reference). Likewise, it is difficult to place much confidence in the discontinuity in ejection velocity at drop radii near 30 μm reported by *Rouault et al.* [1991] based on jet drop radii and ejection heights from *Blanchard* [1963]. The ejection velocity for the top drop from a bubble with $R_{bub} = 0.35$ mm in seawater at 27-29 °C measured by *Spiel* [1995, 1997a] was near 7 m s^{-1}, decreasing with increasing bubble size and with the number of the jet drop in the sequence. The ejection velocities of the top jet drops reported by *Spiel* are roughly equal to those reported by *Blanchard* [1963, Fig. 12] at $R_{bub} = 1.5$ mm (i.e., nearly 2 m s^{-1}), becoming increasingly less with decreasing R_{bub}, until they are roughly half as large as those reported by *Blanchard* at $R_{bub} = 0.35$ mm (i.e., ~7 m s^{-1} vs. ~15 m s^{-1}). *Spiel* extrapolated his fit of the ejection velocity of the top jet drop to ~11 m s^{-1} as R_{bub} approaches zero; if valid, this result would yield lower ejection heights than those reported by *Blanchard*

[1963, 1989a] for drops from bubbles with $R_{bub} < 0.35$ mm. However, there are few data for ejection velocities of small jet drops (*Spiel* [1995, 1997a] did not measure ejection velocities for bubbles with $R_{bub} < 0.35$ mm), and little confidence can be placed in extrapolations of ejection velocities calculated from maximum heights because of the few data on the radii of jet drops from small bubbles and because of the extreme sensitivity of ejection velocity to drop size. For example, to eject a drop with $r_{80} = 2$ μm to a height of 0.5 cm requires an ejection velocity of $v_0 = 46$ m s^{-1}, whereas to eject a drop with $r_{80} = 3$ μm to this height requires an ejection velocity of only $v_0 = 18$ m s^{-1}.

Because of their large ejection velocities, jet drops will initially experience enhanced evaporation (at roughly three times the rate of stationary drops at the same temperature and relative humidity). However, the times required for drops to attain their maximum heights are small: less than 0.07 s for drops with $r_{80} < 25$ μm, increasing to a maximum time of around 0.16 s for drops with $r_{80} = 75\text{-}125$ μm (based on an assumed initial Reynolds number $Re_0 = 70$ discussed below). These times are for the top jet drops in the sequence; later drops, which are not ejected as high, will attain their maximum heights in still less time. As these times are more than two orders of magnitude less than the characteristic times for change of size with respect to RH (§2.5.4; Eq. 2.5-6) over the entire range of sizes of jet drops considered, it can be concluded that jet drops experience little change in radius during their ascent and can accurately be treated as if they retain their formation radii, equal to r_{98}. Additionally, because the RH in the lowest several centimeters above the ocean surface is expected to differ little from 98%, there is even less change in size for small drops which are ejected to low heights.

Several relationships between the maximum height h attained by a jet drop and R_{bub} have been proposed. Such a relationship, together with knowledge of the dependence of r_{80} on R_{bub}, would allow determination of the smallest sizes of jet drops that can be ejected to a given height (§4.3.2.4). However, as noted above, the relationship between r_{80} and R_{bub} is not well established for small drops; thus, any relation between h and r_{80} necessarily contains much uncertainty in this size range. A relationship of the form $h \propto R_{bub}^{3/2}$ was suggested by *Stuhlman* [1932] for the maximum heights of the top jet drop from bubbles with $R_{bub} < 0.5$ mm; this relation, shown in Fig. 28 for an empirically chosen constant, provides a good representation of *Stuhlman's* data in this size range. The maximum heights of the top jet drops with $r_{80} < 15$ μm (four values) in seawater reported by *Bezdek* [1971] are well fitted by $h \propto r_{80}^2$, although larger drops do not follow this relationship. *Tedesco and Blanchard* [1979] graphically fitted their data for the top jet drop in distilled water to $h \propto R_{bub}^{1.6}$ and $r_{80} \propto R_{bub}^{1.4}$, implying that

$h \propto r_{80}^{1.14}$, which is consistent with the fit they presented, $h/\text{cm} = 0.32(r_{80}/\mu\text{m}) - 0.185$, to within 10% over the size range of their drops. An empirical formulation that yields a remarkably good representation of the maximum heights attained by jet drops over the entire range of R_{bub} for which these heights have been measured (Fig. 28) is obtained (using Eq. 4.3-3 to relate R_{bub} and r_{80}) by assuming that the initial Reynolds number of the jet drop Re_0 is equal to 70 (which for a given r_{80} yields a value of v_0, from which the maximum height attained can be determined from Eq. 2.6-14). The square of Re_0 is proportional to the ratio of the initial kinetic energy to r_{80}. The value $Re_0 = 70$ overestimates the maximum heights of smaller drops ($r_{80} \lesssim 10$ μm) in freshwater and pond water reported by *Tedesco and Blanchard* [1979], which are better fitted (again using Eq. 4.3-3 to relate R_{bub} and r_{80}) with the assumption $Re_0 = 35$ (Fig. 28), which for a given r_{80} corresponds to half the ejection velocities, or one-fourth the initial kinetic energies, as does the assumption $Re_0 = 70$. Ejection velocities calculated using $Re_0 = 35$ are given by

$$\frac{v_0}{\text{m s}^{-1}} \approx \frac{130 \, \mu\text{m}}{r_{80}},$$

resulting in values of tens of meters per second for drops with $r_{80} \lesssim 10$ μm. For drops with $r_{80} \lesssim 25$ μm, an empirical relation yielding a good approximation for the maximum heights obtained by jet drops is given by

$$\frac{h}{\text{cm}} \approx \left(\frac{r_{80}}{20 \, \mu\text{m}} \right) (Re_0)^{0.5}$$

which yields $h/\text{cm} = 0.4(r_{80}/\mu\text{m})$ for $Re_0 = 70$ and $h/\text{cm} = 0.3(r_{80}/\mu\text{m})$ for $Re_0 = 35$; this latter relation results in ejection heights similar to those given by the fit of *Tedesco and Blanchard* [1979] discussed above.

Sources of the kinetic energy of the jet drops were investigated by *Toba* [1959], *Blanchard* [1963, pp. 115-119], and *Starflinger et al.* [1995]. *Blanchard* concluded that the main source of energy used to eject jet drops is the surface energy in the interface between the bubble and the liquid, which varies roughly as R_{bub}^2 for bubbles sufficiently small to produce jet drops in the size range of interest [*Blanchard*, 1963, pp.115-119]. He concluded that although the energy stored in the compressed air (roughly a factor of two less than the surface energy) was of sufficient magnitude, it could not be a source of energy for the drops, as assumed by several previous investigators [*Stuhlman*, 1932; *Davis*, 1940; *Garner et al.*, 1954; *Newitt et al.*, 1954; *Tomaides and Whitby*, 1976], because the drops were ejected before this energy could be released. The gravitational potential energy from the cavity resulting from the bubble and electrical energy stored in the bubble film were both far too small to provide any substantial contribution to the jet drop energy. *Blanchard* concluded that 10-20% of the surface energy is utilized by the jet drops, mostly by the top drop. Somewhat similar conclusions had been reached by *Hayami and Toba* [1958] and *Toba* [1959], but they inferred the initial kinetic energies of jet drops from their gravitational potential energies at their maximum heights, and as up to 90% of the initial energy for drops of the sizes they considered ($r_{80} >$ 50 μm) is lost due to viscous drag, their estimates greatly underestimated the initial energies of the drops. For a drop with $r_{80} = 20$ μm, all but about 1% of the initial kinetic energy is lost to drag. *MacIntyre* [1968] concluded that roughly the same amount of energy was lost to viscosity in the formation of the jet as was carried off by the jet drops (because of the strong dependence of the viscosity of water on temperature, this effect could have important consequences for the temperature dependence of drop production). Kinetic energies of each of the drops, and the average total kinetic energy of all drops, for a given R_{bub} from 0.35 mm to 1.5 mm were presented by *Spiel* [1995], who concluded that more than 80% of the total kinetic energy of all the drops was carried off by the top drop, roughly an order of magnitude more than is carried off by the second drop.

The heights to which the top jet drops are ejected were reported to be quite reproducible for periods of hours in several laboratory investigations [*Blanchard*, 1954b, 1958; *Blanchard and Syzdek*, 1975; *MacIntyre*, 1978]. *Blanchard* [1954b, 1958] described these heights in water and seawater to be "remarkably" and "amazingly" reproducible. Ejection heights for all six jet drops from bubbles with $R_{bub} = 0.185$ μm in water containing nutrient broth were demonstrated to be remarkably uniform over a two hour interval by *Blanchard and Syzdek* [1975], who additionally reported that the ejection heights of the top three jet drops from bubbles of this size in water containing bacteria remained constant (roughly 2.8, 2.3, and 1.1 cm) to within 0.02 cm over a five hour period. *MacIntyre* [1978] reported that jet drops from seawater were produced with such repeatability from a stream of bubbles that he was able to see them at the same location with a strobe light, and that this behavior was found for a wide range of surface (cleanliness) conditions.

In contrast to the above results, other investigators have indicated substantial differences between the ejection heights of top jet drops from the same bubble in water and those in seawater. Erratic jet drop heights in pond water were reported by *Blanchard and Syzdek* [1978a] and in reasonably pure (distilled-deionized-distilled) water by *Detwiler and Blanchard* [1978], although tap water and

saltwater solutions gave consistent, repeatable results. The ejection heights of distilled water drops reported by *Stuhlman* [1932] from bubbles with $R_{bub} > 0.5$ mm (Fig. 28) were lower than those reported for seawater drops by *Blanchard and Woodcock* [1957] and *Blanchard* [1963]. The ejection heights of distilled water drops are often considerably less than those of seawater drops for R_{bub} from 0.25 mm to 0.5 mm, according to *Blanchard and Syzdek* [1978a] (although *Blanchard* [1989a] stated that ejection heights of drops with $R_{bub} < 0.175$ mm reported by *Tedesco and Blanchard* [1979] for distilled water were less than those for seawater, this conclusion is based on only two seawater data, of which only one had a lower ejection height than the height of a drop ejected in distilled water).

The differences between jet drop ejection heights in freshwater and those in seawater for the same value of R_{bub} have been attributed not to any intrinsic differences in the mechanism of bursting in these two media, but to the position of bubbles at the surface when they burst [*Blanchard and Syzdek*, 1978a]. Bubbles in pure water, upon reaching the surface, typically rise above it and burst immediately, whereas bubbles in seawater generally come to equilibrium and remain at the surface for a second or more before breaking [*Kientzler et al.*, 1954; *Blanchard*, 1963, p. 111; *Blanchard and Syzdek*, 1978a]; thus the position of the bubble when it bursts is related to its residence time on the surface. According to *Blanchard* [1983], the ejection heights of the drops from bubbles in distilled water that burst after remaining on the surface for several seconds are the same as those from bubbles in seawater. Bubbles that are higher in the water, and thus have a smaller cavity, contain less energy available for jet drop ejection [*Blanchard and Syzdek*, 1978a; *Blanchard*, 1983; *Blanchard*, 1990]. *Blanchard* [personal communication, 2001] stated that he had observed bubbles that produced no jet drops when bursting at a position at the surface much higher than equilibrium. The position of the bubble on the surface may also account for why some investigators have reported that jet drops from bubbles that burst immediately upon arriving at clean water surfaces may be projected in directions other than vertical [*Day*, 1964, Fig. 3; *MacIntyre*, 1978], although *Blanchard* [personal communication, 2001], who has observed tens of thousands of bubble bursts, stated that he had not observed jet drops ejected at angles other than vertical.

The residence time of a bubble on the surface depends on several factors, including the amount and nature of surface-active material in the water [*Newitt et al.*, 1954; *Kientzler et al.*, 1954; *Blanchard*, 1963; *Day*, 1964; *Garrett*, 1967a; *Blanchard*, 1990], the RH and the speed of the air over the surface [*Burger and Blanchard*, 1983], and the bubble size [*Bikerman*, 1968; *Zheng et al.*, 1983b; *Struthwolf and Blanchard*, 1984]. For distilled water the residence time has been found to decrease with increasing bubble size, whereas for seawater the opposite trend has been found [*Struthwolf and Blanchard*, 1984]. In addition to these considerations, at sea the residence time of a bubble at the surface would be expected to be strongly affected by the presence of other bubbles, especially just after a wave breaks. The residence time is also strongly affected by the bubble age—the time the bubble spends in the liquid before reaching the surface [*Hardy*, 1925; *Garrett*, 1967a; *Blanchard*, 1963; *Blanchard and Syzdek*, 1972a; *Detwiler and Blanchard*, 1978; *Blanchard and Hoffman*, 1978; *Blanchard*, 1990], which depends on the size of the bubble, its starting depth, and the temperature of the water (through its effect on viscosity; §4.4.2.1). Increased bubble age allows more time for the bubble to accumulate surface-active material [*Blanchard and Syzdek*, 1972a], which usually (though not always [*Blanchard*, 1963, p. 139, Fig. 22]) results in decreased ejection heights.

It would be expected that surface-active materials would alter the surface tension and thus the amount of energy available for drop ejection, as well as the partitioning of this energy among the various drops [*Blanchard*, 1963; *Blanchard and Syzdek*, 1972a; *Morelli et al.*, 1974; *Blanchard and Hoffman*, 1978; *Detwiler and Blanchard*, 1978]. As the column of water that forms the jet drops consists of water from a very thin layer at the surface of the bubble cavity, even a small amount of surface-active material can affect its instability and breakup. Not only the amount of organic material, but also the type can play a role [*Blanchard*, 1963; *Blanchard and Hoffman*, 1978], and the effect of organic substances on jet drop ejections heights depends also on the size of the bubble [*Blanchard and Hoffman*, 1978]. No consistent pattern of the effects of surface-active materials on jet drop ejection has been identified. The ejection height of the top jet drop reported by *Blanchard* [1963, p. 139, Fig. 22] generally decreased with increasing bubble age, but this behavior depended on the time the artificial sea water had been in the experimental apparatus. An abrupt decrease in the height to which the top drop was ejected when the bubble age was changed was reported by *Blanchard and Syzdek* [1975], although the ejection height of the third drop in the sequence increased, implying a difference in the mechanism of bursting and the partitioning of energy rather than merely a reduction in the amount of energy available. The decrease in ejection height is generally a function of bubble age, size, and the amount of surface-active organic material present [*Blanchard*, 1983]; but *Blanchard and Syzdek* [1972a] reported that when a clean bubble bursts at an interface with an organic coating, the effect of the coating is to increase the ejection height of the top drop, though they stated that the reason for this was unknown to them. They also reported that the radius of the

top jet drop decreases with increasing bubble age, though not uniformly, and that although the ejection height of the top drop generally decreased with increasing bubble age, that of the second drop increased. *Blanchard and Syzdek* [1972a] argued that in natural waters the concentration of organic matter is often sufficient to produce substantial decreases in the ejection height of the top drop and concluded that the relation of ejection heights to bubble size, such as shown in Fig. 28, probably does not hold for most bubbles in natural waters. From experiments involving large numbers of bubbles produced by bubbling air through seawater, *Morelli et al.* [1974] reported that addition of oleic acid (a long-chain fatty acid, a prototypical surfactant and the main component of olive oil) to the water surface resulted in a reduction of the SSA mass production rate (and hence the rate of production of jet drops) by nearly an order of magnitude, which they attributed to the reduction of surface tension which resulted in less energy available for bursting. *Blanchard* [1990] reported that the presence of organic surface-active monolayers caused some of the later jet drops in the sequence to coalesce, resulting in fewer but larger jet drops; *Rayleigh* [1896, p. 369] had previously reported that drops from a vertical jet can collide with one another. However, the prevalence of this effect in the ocean is uncertain. As with many other effects of surface-active materials on SSA production, their effects on jet drop production appear highly variable and difficult to quantify.

Few data have been reported that would aid in determining the temperature dependence of jet drop ejection heights. Ejection heights for jet drops from bubbles with $R_{bub} > 0.6$ mm at 4, 16, and 30 °C were reported by *Hayami and Toba* [1958], who reported a shift in the curve describing ejection heights to smaller bubble sizes with increasing temperatures, which they attributed to lower surface tension at higher temperatures, allowing bubbles to rise higher in warmer water and thus changing the mechanics of the jet drop formation [*Toba*, 1959]. The only other data on jet drop ejection heights or ejection velocities that have been reported for temperatures below ~20 °C are five ejection heights at 4 °C [*Blanchard*, 1963, pp. 109-111; 1989a], which for the same value of R_{bub} were lower than those for drops at higher temperatures. These lower heights were attributed by *Blanchard* [1989a] to the increase in viscosity with decreasing temperature, the change in surface tension being too small to play a role. Ejection velocities for jet drops in seawater at 27-29 °C averaged 6% larger than those at 20-22 °C [*Spiel*, 1997a], but as the drop sizes were around 3% smaller, the maximum heights attained would be about the same for these two temperature ranges.

Ejection heights of jet drops bursting in liquids other than water or seawater have been reported in a few studies. The maximum heights of jet drops from bubbles bursting in benzene reported by *Stuhlman* [1932] were roughly half those in water for the same R_{bub} (although the heights attained by the second drops were roughly similar), from which he concluded that surface energy is the dominant factor (other than R_{bub}) controlling ejection heights and bursting. Ejection heights of jet drops from bubbles bursting in benzene and ethyl alcohol were reported by *Garner et al.* [1954] to be lower than those for water, but no quantitative results were presented.

4.3.2.4. Implications for jet drop production and entrainment over the oceans.   Knowledge of the number and sizes of jet drops produced by a bubble bursting at the sea surface is required to determine the interfacial SSA production flux according to the bubble method (§3.4). This value has typically been taken to be 5 in models and formulations, and based on Fig. 26, this would appear to be a reasonable assumption for bubbles that produce jet drops as large as $r_{80} = 50$ μm. However, the number of jet drops produced overestimates the number of jet drops that are ejected to sufficient heights to remain airborne for any appreciable time and thus contribute to the interfacial SSA production flux in any meaningful way or to moisture and momentum transfer between the ocean and the atmosphere. Likewise, it overestimates the number of jet drops that attain heights sufficiently great and remain airborne sufficiently long to be further entrained upward into the atmosphere and thus contribute to the effective SSA production flux. For example, a drop with $r_{80} = 10$ μm, which is ejected to a height no greater than 3 cm (Fig. 28) and would return to the surface in roughly 0.5 s with little change in radius in still air, would contribute little to air-sea interaction; drops of this size that are ejected to much lower heights (as would smaller drops or later drops in the sequence) would contribute even less. Based on the results presented in Figs. 26, 27, and 28, some conclusions can be reached concerning the numbers and sizes of jet drops that have an appreciable probability of remaining airborne for appreciable times.

There are three factors that might provide a lower limit on the size of jet drops that are formed and that have any appreciable probability of being entrained upward into the atmosphere. Possibly the limiting factor is the size of bubbles that burst at the surface [*Cipriano et al.*, 1983]; small bubbles rise extremely slowly and frequently dissolve before they arrive at the surface (§4.4.2.2), even in water supersaturated with air, because of the additional pressure of surface tension and the hydrostatic pressure of water above the bubble [*Blanchard and Woodcock*, 1957]. According to a model of *Blanchard and Woodcock* [1957], bubbles with $R_{bub} = 0.01$ mm, which according to (4.3-3) would produce jet drops with $r_{80} \approx 0.2$ μm, dissolve completely in 10-20 s at a depth of 10 cm in seawater which is saturated with gas (such

bubbles would require minutes to rise to the surface from this depth); thus few such bubbles would typically survive to the surface to produce jet drops. This is examined further in §4.4.2.2.

Another factor limiting the probability that a jet drop of a given size is entrained upward into the atmosphere is the maximum height attained by the drop [*Cipriano et al.*, 1983]; jet drops that are not ejected above the height of the viscous sublayer (typically 0.1-1 mm; §2.4) are unlikely to be further entrained upward by turbulent eddies, as noted by several investigators [*Eriksson*, 1959; *Cipriano and Blanchard*, 1981; *Cipriano et al.*, 1983]. For the sizes of bubbles that produce jet drops with $r_{80} \lesssim 25$ μm ($R_{bub} \lesssim 0.4$ mm), the ejection height decreases with decreasing R_{bub} (Fig. 28), as does jet drop size (Fig. 27), and although the paucity of data for jet drops from small bubbles makes it difficult to determine accurate relations between these quantities and thus to estimate the lower limit of the size of top jet drops that attain a given height, an estimate of $r_{80} = (0.03, 0.3)$ μm for $h = (0.1, 1)$ mm can be obtained from the relation $h/cm = 0.3(r_{80}/μm)$ based on $Re_0 = 35$ (§4.3.2.3). If bubbles that produce drops of this size produce more than one drop, heights of later drops are considerably less than those of the top drop, and these later drops are thus even less likely to be entrained upward.

A final factor that might limit the sizes of jet drops produced is the energy available for their production. In their equilibrium position, small bubbles reside mostly below the surface, and the surface energy available to the drop is nearly proportional to their surface area. For $Re_0 = 35$ (and using Eq. 4.3-3), the fraction of the surface energy of the bubble given to the top drop increases with decreasing drop size, from 25% for $r_{80} = 1$ μm to 50% for $r_{80} = 0.25$ μm. Loss of energy to friction during the jet drop production process (estimated by *MacIntyre* [1968] to be 50%), though uncertain, implies that there may be a lower limit of r_{80} for jet drops of several tenths of a micrometer, below which there would be insufficient energy left to produce a drop. Thus (4.3-1) is probably not valid down to $R_{bub} = 0$; instead the mean number of jet drops produced per bubble N_{jet} first increases with decreasing R_{bub} according to this relation, then for sufficiently small R_{bub} decreases because there is insufficient energy available for jet drop production. Based on these arguments, it can be concluded that not only are few jet drops with r_{80} values less than several tenths of a micrometer produced, but that few of those that are produced are ejected to sufficient heights to contribute to the interfacial (and likewise the effective) SSA production flux. As discussed below, film drops are commonly produced in this size range, so the above arguments do not provide a lower limit on the sizes of drops from bursting bubbles that need be considered, only on the sizes of jet drops.

The time required for a jet drop of a given r_{80} to attain its maximum height and to fall back to the sea surface provides an estimate of the time during which it must encounter an upward moving eddy if it is to be entrained upward and contribute to the interfacial SSA production flux (a similar argument was presented in §2.9.2). As discussed in §4.3.2.3, jet drops over nearly the entire size range for which they have been measured attain their maximum heights before they experience any appreciable change in size due to evaporation, and thus they retain their formation radii during their ascent. Additionally, jet drops with $r_{80} \gtrsim 5$ μm falling in still air retain their formation radii during their descent, which generally requires much more time than does their ascent. The time required for a top jet drop with $r_{80} = (10, 15, 20, 25)$ μm to attain its maximum height of $(5, 8, 10, 12)$ cm and to fall back to the sea surface in still air, under the assumption $Re_0 = 70$, is approximately $(1, 0.7, 0.6, 0.5)$ s. However, under realistic conditions the horizontal component of the wind results in additional drag (§4.3.2.3) and thus in heights that are somewhat lower than those shown in Fig. 28 for still air. Additionally, the assumption $Re_0 = 70$ overestimates the maximum height for top jet drops toward the lower part of this size range; the assumption $Re_0 = 35$ yields values that are closer to those shown in Fig. 28 for $r_{80} \lesssim 20$ μm, resulting in even shorter times required to fall back to the sea surface during which the drops can encounter upward moving eddies. Later jet drops in the sequence, which have radii comparable to the top drop but are ejected to much lower heights (Fig. 28), have even less time available to encounter upward moving eddies, and thus have even lower probability of being entrained upward. Therefore the mean number of jet drops per bubble that are entrained upward and contribute to the interfacial SSA production flux is much less than that given by the expressions for N_{jet} presented in §4.3.2.1, and quite possibly less than unity.

4.3.3. Film Drops

From the perspective of SSA production the pertinent quantities of interest for film drop production are, as with jet drop production, the number, sizes, and ejections heights of the drops per bubble burst as a function of bubble size. However, compared to jet drop production, for which different investigations obtained rather similar numbers, sizes, and ejection heights (Figs. 26, 27, 28), film drop production exhibits much greater variability. For example, *Blanchard* [1963, p. 124] reported that on occasion the number of film drops produced per bubble burst would suddenly change by a factor of ten for no obvious reason and that similar fluctuations were often noted from one bubble to the next. The wide range of values reported from different investigations for the mean number of film drops produced for a given bubble

size, and the large standard deviations reported for these values from any one investigation, strongly suggest that film drop production may occur through several different mechanisms and, additionally, that experimental conditions and techniques play a role in determining which mechanism operates, with resultant concern about the relevance of such experiments to film drop production in the ocean. This section examines the results of laboratory experiments on film drop production from bubbles bursting individually at a calm water surface, summarized in Table 17, specifically the available data on the number, sizes, and ejection heights of film drops as a function of bubble size.

4.3.3.1. Sizes and ejection heights of film drops. Sizes of film drops vary greatly. There is evidence that film drops can be produced with r_{80} as small as nearly 0.01 μm, and possibly smaller [*Twomey*, 1960; *Blanchard*, 1963; *Day*, 1964, 1967; *Cipriano and Blanchard*, 1981; *Cipriano et al.*, 1983; *Resch and Afeti*, 1992]. At the other extreme, film drops as large as $r_{80} = 125$ μm have been reported from bubbles in seawater by *Afeti and Resch* [1990] and by *Spiel* [1998], and film drops with sizes corresponding to $r_{80} > 170$ μm were reported from bubbles in distilled water by *Afeti and Resch* [1990]. Thus the possible range of radii of film drops spans over four orders of magnitude. Film drops covering a considerable portion of this size range can be produced from a single bubble. *Resch and Afeti* [1991] reported that film drops with r_{80} from 0.2-10 μm can be produced in comparable numbers to those with $r_{80} > 10$ μm for seawater bubbles with R_{bub} from 0.5-3 mm, and *Resch and Afeti* [1992] concluded that seawater bubbles with $R_{bub} \approx 1$ mm produce film drops with $r_{80} = 0.014$-0.2 μm in numbers several times those of larger drops, with the majority of film drops having $r_{80} < 0.043$ μm. Other investigators have concluded that a large fraction of the number of film drops from bursting bubbles have r_{80} less than 1 μm, and the majority of film drops detected by *Cipriano et al.* [1983] from a laboratory whitecap involving large numbers of bubbles bursting in seawater had $r_{80} < 0.02$ μm (§4.3.4.1).

Although some investigations have reported film drop size distributions in various media [*Blanchard and Syzdek*, 1975; *Tomaides and Whitby*, 1976; *Blanchard and Syzdek*, 1982; *Resch et al.*, 1986; *Afeti and Resch*, 1990; *Resch and Afeti*, 1991, 1992; *Spiel*, 1998], all studies covered limited ranges of r_{80} (typically one to one and a half orders of magnitude), and several examined at most a few bubble sizes; most investigations involved only water and saltwater or seawater. Several film drop size distributions represented as $dN_{film}/d\log r_{80}$, comprising a substantial fraction of those reported in the literature, are presented in Fig. 29 to indicate the ranges in size and number of film drops observed (as above, r_{80} values of film drops produced with freshwater

Table 17. Laboratory Investigations of Film Drop Production from Individual Bubbles

Reference	R_{bub}/mm	Liquid	Measurement Technique	r_{80}/μm Range Detected
Mason, 1954 (see also *Moore & Mason*, 1954)	0.25-1.5	water, seawater	multiple bubbles bursting individually into expansion chamber	> 0.1
Garner et al., 1954	0.53, 1.15, 3.52	water, benzene, ethyl alcohol	collection on glass slides 1 cm above burst	$\gtrsim$ 2
Newitt et al., 1954	1.55-2.65	water	collection on MgO-coated slides ≥ 0.64 cm above burst	$\gtrsim$ 2
Mason, 1957a	0.125-1.075	water, seawater	multiple bubbles bursting individually into expansion chamber	> 0.1
Blanchard, 1963	0.25-3	water, NaCl solution, seawater	individual bubbles bursting in diffusion cloud chamber or expansion chamber	$\gtrsim$ 0.01 (est.)
Day, 1964 (see also *Day*, 1967)	0.05-1.95	water, NaCl solution, seawater	individual bubbles bursting in diffusion cloud chamber	$\gtrsim$ 0.01 (est.)
Paterson & Spillane, 1969	0.9	seawater with varying amounts of surface-active substances	individual bubbles bursting in diffusion cloud chamber	$\gtrsim$ 0.01 (est.)
Blanchard & Syzdek, 1975	0.37	water with nutrient broth	collection on MgO-coated slides	> 0.5
Tomaides & Whitby, 1976	0.7, 2.75	NaCl solution	collection on membrane, sized by electron microscopy	> 0.5
Blanchard & Syzdek, 1982	0.85	water with bacteria	collection on gelatin-coated slides	> 0.25
Blanchard & Syzdek, 1988	0.5-3.15	pond water, NaCl solution, seawater	multiple bubbles bursting individually in condensation nucleus counter	~0.005-2
Afeti & Resch, 1990 (see also *Resch*, 1986 and *Resch et al.*, 1986)	2, 3, 4, 5	water, seawater	laser holography of individual bubble bursting	> 2.5
Resch & Afeti, 1991[a] (see also *Afeti & Resch*, 1990)	0.5-3.1	seawater	two optical counters with individual bubble bursting	0.2-10
Resch & Afeti, 1992	0.8-2.85	seawater	multiple bubbles bursting individually in condensation nucleus counter (with particle size selector) or optical counter	0.014-0.2 for condensation nucleus counter with particle size selector, 0.2-5 in six size ranges for optical counter
Spiel, 1998[b] (see also *Spiel*, 1997b)	1.5-6.3	seawater	individual bubbles bursting collected on MgO-coated sheets observed under microscope and with optical counter	> 1.25 for number; > 2.5 for sizes

Several measurements of the size distribution of film drops $dN_{film}/d\log r_{80}$ per bubble burst from individual bubbles as a function of r_{80} are presented in Fig. 29 for several values of R_{bub}. Values of mean number of film drops per bubble N_{film} for seawater are presented in Fig. 30 as a function of R_{bub}.
[a] Lower limit of range for Laser Holography in Table 1 of *Resch and Afeti* [1991] should be 40 μm.
[b] Film drop ejection velocities (and speeds of films rolling up) were determined photographically.

and other media different from seawater have been evaluated as $r_{80} = 0.5 r_{form}$ to allow comparison with drops produced with seawater; §2.1.1). Fig. 29 is an equal-area graph, in which the relative contribution to the number of

film drops between any two values of r_{80} is proportional to the area under the curve between these two values (§2.1.3). A parameterization for film drop size distributions proposed by *Wu* [2001], based on data of *Resch and Afeti* [1991] and

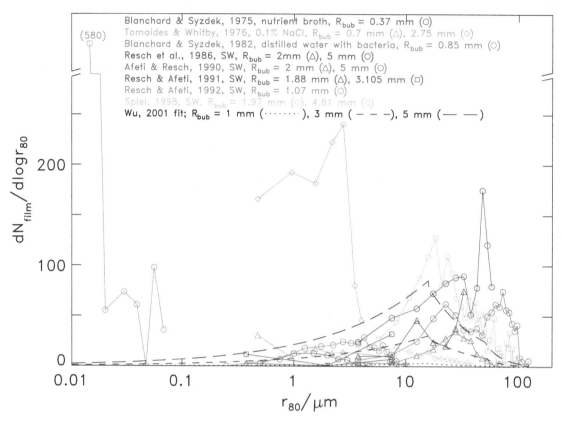

Figure 29. Size distributions $dN_{film}(r_{80})/d\log r_{80}$ of film drops ejected from individual bubbles as a function of r_{80}. Fits of *Wu* [2001] are also presented for three values of R_{bub}. The value corresponding to the lowest radius presented by *Resch and Afeti* [1992] is highly uncertain, being determined from a diffusion battery in the range $r_{80} \approx 0.014$-0.015 μm.

of *Spiel* [1998] and consisting of different expressions in each of three different ranges of r_{80}, is also shown in Fig. 29 for several values of R_{bub}.

Overall, there are too few data which cover too limited a size range for a "typical" film drop size distribution to be determined for a given R_{bub}. No size distribution covering the full size range of film drops even from a single bubble has been reported; as noted above, several investigations have concluded that most film drops are extremely small, i.e., $r_{80} \ll 1$ μm, although frequently more specific information on the sizes of these drops is limited. The only distribution covering any appreciable portion of the size range $r_{80} < 1$ μm is that of *Resch and Afeti* [1992] shown in Fig. 29, and there are no size-resolved measurements for film drops with r_{80} from ~0.1 to 0.5 μm—the size range containing the majority of SSA particles by number in the marine atmosphere (§4.1.4). Additionally, as noted, there is often considerable variability from one bubble to the next; this is illustrated by the differences in the size distributions and maximum sizes reported by *Resch et al.* [1986] for a single bubble burst and those reported by *Afeti and Resch*

[1990] for mean values of the number of film drops produced by 14 and 8 bubble bursts for $R_{bub} = 2$ mm and 5 mm, respectively; for some bubble sizes the number of film drops from a single bubble burst was up to 9 times greater than the mean number for the same bubble size. Although some investigations have concluded that the mean film drop size for bubbles in seawater (and distilled water) increases with increasing bubble size [*Resch et al.*, 1986; *Afeti and Resch*, 1990; *Spiel*, 1998], these investigations considered only film drops with $r_{80} > 2.5$ μm. Although film drops are generally smaller than jet drops for a given bubble size, the wide range of sizes reported and the variety of patterns observed for film drops, and the intrinsic variability in the film drop production process, create difficulties in specifying a typical film drop size, quantifying the dependence of film drop numbers and sizes on bubble size or other factors, and comparing the results of different investigations that detect film drops in different size ranges.

The maximum heights attained by film drops have generally been reported to range from several millimeters to a few centimeters. The number of film drops with radii at formation greater than 10 μm (corresponding to $r_{80} > 5$ μm for seawater drops) reported by *Newitt et al.* [1954] from bubbles with R_{bub} from 1.555 mm to 2.65 mm bursting in pure water, for instance, decreased greatly between heights of 0.64 and 1.91 cm (typically by well over an order of magnitude), and no drops were reported for heights greater than 4.44 cm. *Blanchard* [1963, p. 98] reported that film drops produced in a diffusion chamber were never ejected to heights greater than 1 cm. However, the variety of film drop patterns that have been observed makes it difficult to characterize the height to which film drops are ejected, and too few data exist to allow quantitative determination of this height as a function of drop size or bubble size. Additionally, the experimental techniques employed to detect film drops often preclude determination of the heights under conditions that are not strongly affected by the experiment itself. For example, small film drops bursting in a diffusion chamber with strong RH and temperature gradients rapidly increase in size (permitting their detection), but this growth affects the dynamics of these drops and the heights they attain. In the experiments of *Spiel* [1998], the majority of the film drops were ejected at angles below the horizontal; not only does this make the determination of the heights they obtain moot, but it implies that these drops would not be important for SSA production, unless they form secondary drops upon collision with the water surface (§4.7.3) as suggested by *Spiel* [1998] and *Leifer et al.* [2000a].

4.3.3.2. Number of film drops produced by a bursting bubble. Values of N_{film}, the mean number of film drops produced per bubble, determined from laboratory experiments covering a range of bubble sizes in seawater are shown in Fig. 30 as a function of R_{bub}. Values of N_{film} reported by *Mason* [1957a] were stated to be valid for the indicated range of R_{bub}. The data of *Blanchard* [1963] on the lower axis correspond to values of N_{film} that are less than 0.1 (and may be zero). Shown for comparison is the value of N_{jet} given by (4.3-1). Results of any one investigation often exhibit considerable scatter; in the data of *Blanchard* [1963], for instance, more than an order of magnitude at the same R_{bub}. The standard deviations (not shown), when reported, were often substantial; *Day* [1964], for instance, found standard deviations to be roughly one third of the mean values, and standard deviations reported by *Blanchard and Syzdek* [1988], *Afeti and Resch* [1990], and *Spiel* [1998] were typically two-thirds or more of the mean values.

Most of the data suggest that N_{film} generally increases with increasing R_{bub}, but for the same R_{bub} values from the several investigations range some two orders of magnitude. Some of the differences (and the large standard deviations of the data) are undoubtedly due to intrinsic variability in film drop production itself, but some can be attributed to the experimental techniques employed. Different techniques

detected only limited ranges of sizes of film drops, and additionally they may have affected the types and numbers of film drops produced by changing the bursting process. Hence discussion of some of the experiments is necessary.

The results of *Mason* [1954, 1957a] obtained from measurements using a condensation chamber are anomalous in that values of N_{film} are much greater than reported by other investigators and, in contrast to later investigations, N_{film} was reported to be in dependent of R_{bub} over the indicated range of bubble sizes. *Mason* [1971, p. 78] argued that these results held [as did *Mason*, 2001] and had been confirmed by *Twomey* [1960], who reported 120-160 very small (presumably film) drops with r_{80} down to 0.01 μm in a diffusion chamber from bubbles with R_{bub} between 0.25 mm and 1 mm bursting in seawater. However, as the experiments of *Twomey* involved bubbles formed by forcing air through a frit and not individual bubbles, and as the size distribution of the bubbles was not reported, these experiments cannot be considered as confirmation of *Mason's* results (especially the lack of dependence of N_{film} on R_{bub}). *Twomey* [1971] later reported that when filtered air was bubbled through NaCl solutions or seawater about 100 cloud condensation nuclei with $r_{80} \gtrsim 0.015$ μm (presumably film drops) were produced by each bubble, but experimental details were not reported. It was suggested by *Day* [1964] that *Mason's* results could be explained if it was assumed that many small bubbles coalesced at the surface before bursting, resulting in drops from larger bubbles, but *Mason* did not describe his experiments in enough detail to allow this to be determined, nor did he present data to support his results. Subsequently *Iribarne and Mason* [1967] stated that the number of film drops increases rapidly with R_{bub}. Thus the results of *Mason* [1954, 1957a] should probably be discounted.

The investigations of *Blanchard* [1963, Fig. 15] and *Day* [1964, 1967] utilized diffusion chambers that could detect drops with r_{80} at least as small as 0.01 μm. The results of these two investigations are similar in that the maximum values reported by *Blanchard* are roughly equal to the mean values reported by *Day* (Fig. 30). However, *Day* [1964] considered film drops only from bubbles that burst with no perceptible waiting time on the surface and employed a liquid temperature near 0 °C, whereas *Blanchard* included film drops from all bubbles [*Blanchard*, 1983] and employed a liquid temperature of around 15 °C [*Blanchard*, 1963, p. 98]. As noted earlier with reference to jet drop production (§4.3.2.3), the residence time of a bubble on the surface is expected to affect the bursting process and thus film drop production too (this is discussed further below); this surface residence time can be affected by the strong gradients of temperature and RH present in diffusion chambers. The values of N_{film} from these investigations are larger than later reported values, and *Day* [1967], based on improved

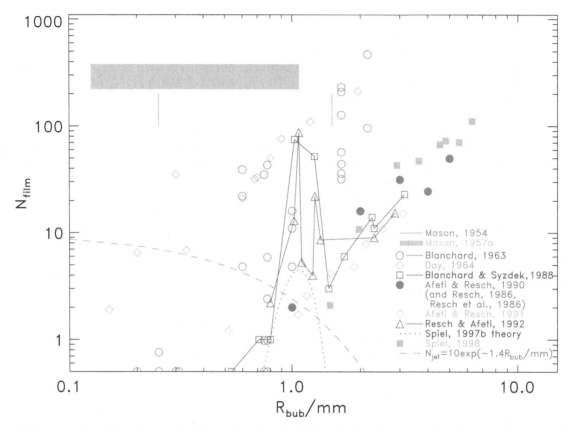

Figure 30. Mean number of film drops per bubble as a function of bubble radius R_{bub} in seawater from investigations summarized in Table 17. Symbols on lower axis from *Blanchard* [1963] denote values for which only an upper bound of 0.1 was reported. Data of *Day* [1964] do not include family of drops projected horizontally, and only data for bubbles which burst with no perceptible waiting time on the surface were included. Results of *Afeti and Resch* [1990] shown include datum at $R_{bub} = 1$ mm from *Resch* [1986], *Resch et al.* [1986]. Data of *Blanchard and Syzdek* [1988] and *Resch and Afeti* [1992] are connected to aid seeing the peaks exhibited by these data. Filled symbols denote results of experiments that detected only film drops with $r_{80} > 2.5$ μm [*Resch*, 1986; *Resch et al.*, 1986; *Afeti and Resch*, 1990] or those with $r_{80} > 1.25$ μm [*Spiel*, 1998]. Most of the film drops detected by *Spiel* [1998] were ejected at angles below the horizontal. Dotted line is N_{film} according to theory of *Spiel* [1997b]. Also shown for comparison (dashed line) is N_{jet} according to (4.3-1).

techniques and using artificial seawater, probably 3% (by mass) NaCl solution, concluded that the actual values of N_{film} are even greater than those reported earlier [*Day*, 1964]. Some of the discrepancies between results of these and later investigations can be explained by the size ranges of film drops detected. Several of the later investigations could detect only larger film drops; those of *Resch* [1986], *Resch et al.* [1986], and *Afeti and Resch* [1990] could detect only film drops with $r_{80} > 2.5$ μm, and that of *Spiel* [1998] could detect only film drops with $r_{80} > 1.25$ μm (data from these investigations are denoted by filled symbols in Fig. 30). Most of the film drops detected by *Spiel* [1998] had been ejected at angles below the horizontal; such drops, if they occurred in other experiments, probably would not have been detected or included (*Day* [1964], for instance, specifically stated that his results did not include film drops ejected

at low angles to the horizontal). However, the investigations of *Blanchard and Syzdek* [1988], *Resch and Afeti* [1991], and *Resch and Afeti* [1992], which covered a wide range of bubble sizes and included very small drops, resulted in values of N_{film} that are substantially lower than those of *Blanchard* [1963] and *Day* [1964]; thus factors other than sizes of film drops detected are required to explain these differences.

A remarkable result reported by *Blanchard and Syzdek* [1988] is a large peak in N_{film} for seawater bubbles near $R_{bub} = 1$-1.5 mm, averaging near 75 (although this value varied considerably) compared to less than 5 for bubbles that were only slightly larger or smaller (the peak value of N_{film} was much greater—near 280—for bubbles in a NaCl solution). Although this report prompted much discussion about the existence of this peak and its importance

[*Wu*, 1990a,b; *Blanchard and Syzdek*, 1990a; *Resch and Afeti*, 1991; *Wu*, 1992a, 1994a], it was confirmed by *Resch and Afeti* [1992] that such a peak occurs at $R_{bub} = 1.07$ mm with a value for N_{film} similar to that reported by *Blanchard and Syzdek* [1988]. *Resch and Afeti* [1992] concluded that 75% of the film drops from bubbles of this size have $r_{80} < 0.075$ µm. The data of *Resch and Afeti* indicate another, smaller peak near $R_{bub} = 1.255$ mm (Fig. 30) which they did not discuss. Such peaks even further support the suggestion that multiple mechanisms operate to form film drops. A hypothesis has been proposed to explain the peak at $R_{bub} = 1.07$ mm in which the bubble film upon bursting rolls up as a toroid that collides with the surface [*Spiel*, 1997b], but this theory results in a value of N_{film} near 5 at this location (Fig. 30), many times less than the values measured by *Blanchard and Syzdek* [1988] and by *Resch and Afeti* [1992]. *Spiel* [1998] later argued that this peak bubble size was too small to even produce film drops, and that the drops that were observed were not really film drops at all, but were formed when the toroid, formed by the rolling up of the film after it burst, collided with the film edge, creating splash drops or secondary bubbles which then produced drops.

Given a narrow peak (or peaks) in the number of film drops produced as a function of R_{bub}, the question naturally arises of how thoroughly the bubble size range must be explored in order to ascertain whether peaks of this sort occur at other bubble radii. If the peak reflects a resonance of some sort, then modes might occur at other bubble radii, but if the peaks are sufficiently narrow then even a systematic search may have difficulty finding them.

Several expressions have been proposed relating N_{film} to R_{bub} or to other quantities such as the film area, A_{film}, which might be the relevant quantity if film drops are formed by the shattering of the entire film, or the film radius, R_{film}, which might be the relevant quantity if film drops are formed by the collision of the toroid formed by the bubble film as it rolls up with the surface of the liquid; in either situation film thickness would also be important. However, not only is the utility of these formulations questionable because of the extremely large range of reported values of N_{film} for a given R_{bub} (Fig. 30) and the fact that bubbles may burst without being in equilibrium at the surface (making determination of A_{film} difficult), but many of the proposed relations have been based on erroneous conclusions regarding the relations between R_{film} or A_{film} and R_{bub}. Hence it is necessary to briefly review these relations.

The shape of a bubble with given (volume-equivalent) radius R_{bub} resting in equilibrium at the surface of a liquid (Fig. 31), and thus the values of both A_{film} and R_{film}, depend only on the ratio R_{bub}/a, where the capillary length $a = [2\sigma/(\rho g)]^{1/2}$ depends only on properties of the liquid: σ, the surface tension of the air-liquid interface, and ρ the

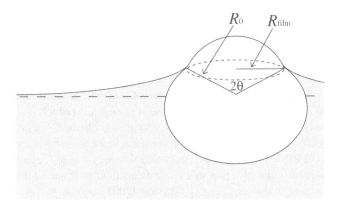

Figure 31. Bubble resting on surface of water, after *Toba* [1959]. The bubble film, or cap, subtends an angle 2θ and has a radius of curvature R_0. The intersection of this bubble film with the water surface forms a circle with radius R_{film}. R_{bub}, not shown, is the volume-equivalent radius, defined such that the volume of the bubble is equal to $(4\pi/3)R_{bub}^3$.

liquid density (g is the acceleration due to gravity). For seawater of salinity 35 at 20 °C, $a \approx 3.8$ mm, and it varies by less than 2% over the temperature range 0-20 °C (the value of a for pure water is ~1% greater at a given temperature). Based on geometrical considerations (Fig. 31) $R_{film} = R_0\sin\theta$, and $A_{film} = 2\pi R_0^2(1 - \cos\theta)$, where R_0 is the radius of curvature of the bubble cap and θ is the semi-angle subtended by the cap from its center of curvature (also dependent on R_{bub}/a); R_0 is greater than R_{bub} by a factor that is nearly 2 for small bubbles, and around 1.5 for larger bubbles. Values of quantities permitting calculation of R_{film} or A_{film} for a given R_{bub} were presented by *Toba* [1959] for 8 bubble radii from ~0.4 mm to ~4 mm, and later by *Princen* [1963] for bubble radii from ~0.9 mm up to ~7 mm. In the range of overlap, these two sets of calculations are nearly identical; approximate relations describing the film radius and area are given by $R_{film}/mm \approx 0.4(R_{bub}/mm)^{1.9}$ and $A_{film}/mm^2 \approx 0.5(R_{bub}/mm)^{3.8}$ for R_{bub} near 1 mm, and $R_{film}/mm \approx 0.8(R_{bub}/mm)^{1.2}$ and $A_{film}/mm^2 \approx 1.8(R_{bub}/mm)^{2.6}$ for R_{bub} near 6 mm. These results were experimentally verified by *Princen and Mason* [1965]. However, the fit presented graphically by *Day* [1967] to *Toba's* results for A_{film} is markedly inconsistent with these results, and the expression proposed by *MacIntyre* [1972] to describe *Toba's* results for A_{film} yields values nearly a factor of three too low. The expressions of *Voßnacke et al.* [1996 (cited in *Koch et al.*, 2000)] yield values of R_{film} that are 8-25% less, and values of A_{film} that are roughly 20% less, than those given by *Toba* [1959]; as these quantities results were presented in a nondimensional form (i.e., in terms of R_{bub}/a) by both investigators, the differences are not due to temperature.

Facy [1951b] modeled a bubble film as a hemispherical cap (and thus assumed that $A_{film} \propto R_{film}^2$ and $R_{film} \propto R_{bub}$), concluding that $N_{film} \propto R_{bub}^{1/2}$ for a given film thickness. *Toba* [1961a,b] concluded from experiments of *Newitt et al.* [1954] that $N_{film} \propto A_{film}$. *Blanchard* [1963, pp. 126-127] concluded that the maximum value of N_{film} was proportional to $A_{film}^{1/2}$ and argued that because $A_{film} \propto R_{film}^2$ (which it is not) it follows that $N_{film} \propto R_{film}$, which he claimed (incorrectly) was the same conclusion as reached earlier by *Facy* [1951b]. *Day* [1964] also concluded that $N_{film} \propto A_{film}^{1/2}$, and argued (correctly) that this is not equivalent to *Blanchard's* statement that $N_{film} \propto R_{film}$. *Mason* [2001] argued that for *Day's* data $N_{film} \propto R_{bub}^2$, and stated (incorrectly) that this result was equivalent to $N_{film} \propto A_{film}$. The fits presented graphically in *Day* [1967], however, indicate that $N_{film} \propto A_{film}^{1/3}$ (these fits also indicate that $N_{film} \propto R_{bub}^{3/2}$, but, as noted above, the relation between A_{film} and R_{bub} indicated by this figure is very different from the results of *Toba* [1959], which presumably were used to relate these quantities, as they were in *Day* [1964]). As *Day* [1964] considered only bubbles that burst immediately upon reaching the surface and which were thus higher than their equilibrium positions (discussed further below), his attempts to relate his measurements of N_{film} to equilibrium bubble properties given by *Toba* [1959], such as A_{film}, seem misguided. *Tomaides and Whitby* [1976] argued that $N_{film} \propto R_{bub}$, but the values they stated are not consistent with this relationship; they also stated that $N_{film} \propto R_{film}$. *Wu* [1989a] parameterized the few seawater data of *Resch et al.* [1986], which consisted of a total of seven bubble bursts for four different bubble radii, by an expression of the form $N_{film} \propto R_{bub}^{2.15}$, and *Wu* [1990b] represented selected data of *Blanchard and Syzdek* [1988] (which did not include the peak in N_{film} discussed above) by an expression with the same power law. *Wu* [1994a, 2001] later presented several other expressions, concluding that the data are consistent with $N_{film} \propto R_{bub}^2$, which he claimed has a physical basis because $A_{film} \propto R_{bub}^2$ (which is not correct, as discussed above); he also concluded that the total surface area of the film drops produced was a constant fraction of the surface area of the parent bubble. *Resch and Afeti* [1991] fitted their data for film drops in the size range $0.2 \ \mu m < r_{80} < 125 \ \mu m$ to $N_{film} \propto R_{bub}^{5/3}$, and *Spiel* [1998] reported that N_{film} (for drops with $r_{80} > 1.25 \ \mu m$) increases linearly with R_{bub}, from zero at $R_{bub} < 1.25$ mm (near where the peak occurs) to around 100 for $R_{bub} = 6$ mm, but no physical basis was attributed to either expression by these investigators. Because of the variety of patterns of ejection of film drops that have been reported, the wide range of sizes of film drops, the vast discrepancies in N_{film} among various investigations and the large scatter in the results of any one investigation, the existence of the peaks reported by *Blanchard and Syzdek* [1988] and the lack of

knowledge concerning them, together with the possibility (discussed above) of other peaks, any proposed functional relation between N_{film} and R_{bub} based on the data currently available (Fig. 30) would seem to be at best an order-of-magnitude approximation of limited utility.

A related issue is the smallest bubble size for which film drops are produced. Film drops from bubbles as small as $R_{bub} = 0.53$ mm in distilled water were reported by *Garner et al.* [1954], and according to *Mason* [1957a] bubbles as small as $R_{bub} = 0.125$ mm in seawater produced hundreds of film drops (however, as discussed above, this result is questionable). Film drops from bubbles as small as $R_{bub} = 0.15$ mm in seawater were reported by *Day* [1964], but in much smaller numbers than from larger bubbles. Up to 100 film drops per bubble was reported by *Paterson and Spillane* [1969] for bubbles in seawater with $R_{bub} = 0.9$ mm (the only size they investigated), and an average of 20 film drops per bubble from bubbles with $R_{bub} = 0.7$ mm in a 0.1% (by mass) NaCl solution was reported by *Tomaides and Whitby* [1976]. Other results, however, indicate that few if any film drops are produced by bubbles with radii less than around 1 mm. *Spiel* [1998] concluded that no film drops could be produced by bubbles with $R_{bub} < 1.2$ mm that are in equilibrium with the surface, and his observations supported this conclusion, although they were based on film drops with $r_{80} > 1.25 \ \mu m$.

4.3.3.3. Factors affecting film drop production from individual bubbles.   As with jet drop production, it is expected that factors such as temperature and surface-active materials affect film drop production, although few systematic studies to investigate these effects have been conducted. Additionally, the liquid medium in which bubbles burst may affect film drop production, but no systematic investigations of this nature have been reported. *Garner et al.* [1954] stated that increased liquid viscosity resulted in both decreased sizes and decreased numbers of film drops produced, but no evidence was provided. *Newitt et al.* [1954] reported that an increase in temperature from 25 °C to 45 °C resulted in a large decrease (by a factor of ~2.5) in the number and small decrease (~10%) in the mean sizes of film drops with $r > 10 \ \mu m$ detected at a height of 0.64 cm above a pure water surface.

Several investigations have compared film drop production between pure water and NaCl solutions or seawater, but the differences in techniques employed make comparisons between different experiments difficult. Evaporation of small drops can strongly affect the results; pure water drops with radii at formation less than 1 μm would evaporate completely in much less than a second when exposed to 99% RH at 20 °C, and in even less time for lower RH (§2.5.3). Complete evaporation of small distilled water drops can explain the results of *Mason* [1954, 1957a], who reported observing small film drops from bubbles bursting in seawater but not in distilled water, and may explain the results of *Day* [1964], who observed fewer film drops from bubbles bursting in distilled water than in seawater or in a 3% (presumably by mass) NaCl solution. For larger drops, for which evaporation is not an important consideration, results are still quite inconsistent. Values of N_{film} for bubbles with $R_{bub} = 1.55$-2.65 mm bursting in pure water reported by *Newitt et al.* [1954] are only slightly less than the lowest ones shown in Fig. 30 for bubbles of similar sizes in seawater, but that investigation allowed detection of film drops only with formation radii greater than 10 μm, corresponding to $r_{80} > 5$ μm for seawater (drops in seawater or NaCl solutions were not investigated); evaporation would have had a minimal effect on the sizes of these drops before detection. Film drops from bubbles with R_{bub} equal to 0.25 mm and 0.65 mm in distilled water were observed by *Blanchard* [1963, p. 122], but no drops were observed from bubbles of these sizes in seawater or in a 3.5% (by mass) NaCl solution; additionally, more film drops were produced in distilled water than in seawater or in NaCl solutions for bubbles with $R_{bub} = 2.2$ mm. Despite these differences, *Blanchard* [1963, p. 124] concluded that for $R_{bub} \gtrsim 0.75$ mm there are no significant differences in film drop production among distilled water, seawater, and a 3.5% (by mass) NaCl solution. *Day* [1964] reported that small bubbles ($R_{bub} \lesssim 0.75$ mm) in NaCl solution produced more film drops, and that larger bubbles produced roughly comparable numbers of film drops, than did bubbles of the same R_{bub} in seawater, although he did not quantify these results. *Resch* [1986] and *Resch et al.* [1986] reported that roughly twice as many film drops, which were on average slightly smaller, were produced in distilled water than in seawater under similar conditions, although these results were based on not more than two bursting events in freshwater and in saltwater for each of 5 bubble radii. *Afeti and Resch* [1990], based on 8-16 bubble bursts for each of 4 different radii in distilled water and in seawater, reported the opposite result, that fewer film drops, with larger average radii, were produced by bubbles bursting in distilled water than in seawater. For a given R_{bub}, a strong dependence of N_{film} on the bubble rise distance (the vertical distance traversed by the bubble in the fluid) in a 3% (presumably by mass) NaCl solution was reported by *Blanchard and Syzdek* [1988], but they observed practically no dependence in seawater; intermediate results were reported for pond water. Additionally, these investigators reported that the number of film drops produced for $R_{bub} = 1.05$ mm (the location of the peak discussed above) was around 75 for seawater compared to 100 to 280 (depending on bubble rise distance) for the NaCl solution.

A factor that would appear to account for much of the variability in film drop production, the patterns of film drops, and differences in N_{film} among different investigations, is the residence time of the bubble on the liquid surface before it bursts. Bubbles with very short residence times are higher above the surface of the liquid, and thus have greater film areas, when they burst than bubbles that have had time to attain their equilibrium positions. Additionally, there is less time for the films of bubbles with shorter surface residence times to drain, resulting in a greater volume of liquid available for film drops. Both of these factors would be expected to affect film drop production and hence the number and sizes of the film drops produced. *Day* [1964] concluded that the residence time of a bubble on the surface is an important factor in determining the number of film drops it produces, and that this residence time may depend on the cleanliness of the surface. As noted above in reference to jet drops, other factors affecting the residence time of a bubble on the surface of a liquid are its size, its age (the time it spends rising to the surface), the RH and the speed of the air over the surface, and the presence of other bubbles. Additionally, the amount and nature of surface-active material in the liquid play a major role in all aspects of film drop production, including changing the surface tension of the bubble film and thus the energy available for film drop production.

The experimental techniques themselves affect the bursting process and surely account for some of the variability and different patterns that have been observed by influencing the position of the bubble on the surface, its residence time, or the thickness of the bubble film. For instance, the conditions experienced by bubbles in diffusion chambers, with strong temperature and RH gradients (atypical of those experienced by bubbles bursting at the ocean surface), would be expected to affect thinning and bursting of the bubble film. *Blanchard and Syzdek* [1988], using a diffusion chamber similar to that used by *Blanchard* [1963], demonstrated that N_{film} depends on the residence time of the bubble on the surface, which in turn is influenced by temperature and water vapor pressure gradients in the chamber. Difference in the number and sizes of film drops produced by bubbles in distilled water and those in seawater (or NaCl solutions) have been attributed to differences in their residence times, but with conflicting results. The greater numbers and lower sizes (on average) of film drops detected from bubbles in seawater than from those in distilled water by *Afeti and Resch* [1990] were attributed by them to the longer surface residence times of bubbles in seawater (or NaCl solutions) than those in distilled water, allowing more time for the films of these bubbles to drain and become thinner. However, seawater bubbles, bursting at equilibrium positions, have films with less surface area than bubbles of the same R_{bub} in distilled water that burst above their equilibrium position. This was the reason given by *Resch et al.* [1986]

(based on a single bubble burst for each of several bubble sizes) for the lower numbers and larger sizes of film drops in seawater than in distilled water, and the reason given by *Blanchard* [1963, p. 122] for the greater numbers of drops produced in distilled water than in seawater. *Blanchard* [1963, p. 122] stated that this conclusion was demonstrated by the fact that bubbles in distilled water produced substantially more film drops when they burst immediately than when they had a residence time on the surface of several seconds. However, both explanations would apply to such bubbles, and it can be concluded only that the bursting process itself differs between bubbles with thicker films which are above their equilibrium positions, and bubbles with thinner films which are at their equilibrium positions.

A major contributor to the variability in film drop production is the presence of surface-active materials, which can have large effects even in very low concentrations (§2.3.4). *Blanchard* [1963, p. 122] concluded that the presence or absence of an organic surface-active film is the most important factor controlling the production of film drops, and reported that merely touching a liquid surface with a finger can eliminate film drop production completely. *Paterson and Spillane* [1969] reported that the mean number of film drops produced by an individual seawater bubble with $R_{bub} = 0.9$ mm is greatly decreased by small amounts of a surface film. However, from experiments involving large numbers of bubbles produced by bubbling air through seawater, *Morelli et al.* [1974] reported that the production rate of drops with $r_{80} \lesssim 1$ μm, presumably film drops, was not appreciably affected by the presence of a surface film of oleic acid. The strong dependence of N_{film} on the bubble rise distance reported by *Blanchard and Syzdek* [1988] in a NaCl solution, and lack of such a dependence in seawater, can be explained by the amount of surface-active materials collected on the surface of the bubble during its ascent; bubbles in NaCl solution that is nearly free of surface-active materials being more affected, and thus exhibiting a stronger dependence on bubble rise distance, than bubbles in seawater that contains greater amounts of surface-active materials.

In summary, too little is known concerning film drop production from individual bubbles bursting at a calm liquid interface to estimate the magnitudes or trends in the influences of possible governing factors such as temperature or surface-active materials on the film drop production process or on the number and sizes of film drops produced.

4.3.3.4. Implications for film drop production and entrainment over the oceans. Compared to jet drop production, data on film drop production by individual bursting bubbles present a much less coherent picture, and it is difficult to place much confidence in parameterizations of quantities such as N_{film} (Fig. 30), or in estimates of typical values of

r_{80} or of the distribution of sizes of film drops produced by a bubble with given R_{bub}. The situation for aggregates of large numbers of closely packed bubbles bursting at the ocean surface would appear to be less certain yet. Data on the sizes and maximum ejection heights of film drops are scant, but these heights are generally considerably less than those attained by jet drops; thus, the conclusions obtained in §4.3.2.4 on the maximum size of jet drops that are likely to be entrained upward apply also to film drops. At the other end of the size range, it is believed that most SSA drops with $r_{80} \lesssim 1$ μm are probably film drops as they are unlikely to be jet drops for the reasons stated in §4.3.2.4 (this is also the conclusion of several studies involving laboratory whitecaps; §4.3.4.4), and it appears that under some circumstances a large majority of film drops produced might be very small, i.e., $r_{80} < 0.1$ μm. Additionally, as argued by several investigators [e.g., *Cipriano and Blanchard*, 1981; *Cipriano et al.*, 1983], it is likely that the majority of the small (i.e., $r_{80} < 1$ μm) SSA particles detected under oceanic conditions are film drops, as jet drops of this size are unlikely to be produced and entrained upward for reasons discussed in §4.3.2.4.

The peak (or peaks) of N_{film} at R_{bub} in the range 1-1.5 mm (and possibly at other values of R_{bub}) may be important, depending on its width and height, and also on concentrations of bubbles with those radii (bubble concentrations in the ocean are examined in §4.4.1.2). However, the lack of current knowledge concerning the height and width of such peaks, not to mention the number and location of possible other peaks, precludes determination of their importance to oceanic SSA production.

In conclusion, despite the above studies of film drop production, much remains to be learned about the number, size distribution, and maximum ejection heights of film drops produced per bubble as a function of R_{bub}, and of the factors that control these quantities. Certainly the large variability in N_{film}, as manifested in the range of values in Fig. 30, for example, raises concern over the quantitative use of such results to estimate SSA production by the bubble method (§5.5.1).

4.3.4. Investigations with Laboratory Whitecaps

Laboratory experiments involving simulated whitecaps or breaking waves have been performed to investigate SSA production and phenomena such as whitecap lifetime and decay. It might be expected that such experiments more accurately simulate processes that occur during SSA production at the ocean surface than do experiments involving bubbles bursting individual at a flat, calm water surface. Additionally, such experiments allow control of temperature and other factors, permitting insight to be gained into the possible dependencies of SSA production, whitecap lifetime, and other pertinent quantities on these factors. The results of these investigations have been used to determined $f_{wc}(r_{80})$—the size-dependent SSA production flux over whitecap area, which together with estimates of the oceanic whitecap ratio W have been used to calculate the oceanic interfacial SSA production flux according to the whitecap method (§3.2). Two types of laboratory whitecaps have been employed: (1) a continuous, or steady state, whitecap formed by a continuous waterfall or by bubbling air through a frit (i.e., a glass filter), and (2) a discrete whitecap formed by the collision of two parcels of water. These studies, and the extent to which they can be used to determine oceanic SSA production fluxes, are examined in this section, and production fluxes over whitecap area based on these results are presented and compared in Fig. 32.

4.3.4.1. Sea salt aerosol production from continuous laboratory whitecaps. The continuous laboratory whitecap method (§3.2) relies on measurement of the size-dependent production rate of SSA particles $p(r_{80})$—the number of particles in a unit logarithmic interval of r_{80} produced per unit time from a laboratory whitecap of area A_{wc}. The flux of SSA particles over laboratory whitecap area is then calculated from (3.2.2) as $f_{wc}(r_{80}) = p(r_{80})/A_{wc}$. This method has been applied by *Cipriano and Blanchard* [1981] and *Cipriano et al.* [1983] using a continuous waterfall and by *Mårtensson et al.* [2003] using air forced through a frit. Additionally, a continuous laboratory whitecap has been used by *Cipriano and Blanchard* [1981] to determine the interfacial production fluxes of jet drops and film drops using

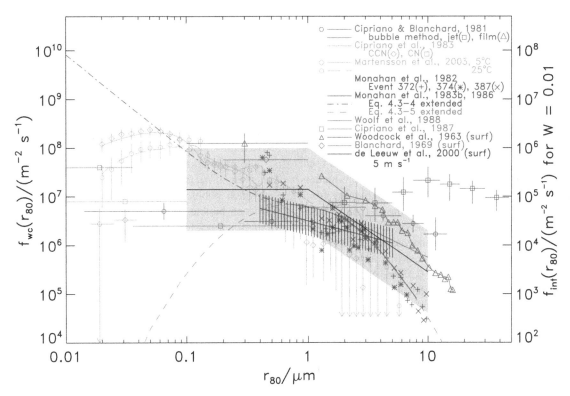

Figure 32. Measurements of the size-dependent SSA production flux over white area $f_{wc}(r_{80})$, left axis, from laboratory simulated whitecaps and from surf areas. Vertical bars denote uncertainties stated by investigators; arrows denote uncertainties that extend further. Horizontal bars denote r_{80} intervals, which in several instances had to be chosen somewhat arbitrarily; as discussed in §4.3.4.1, $f_{wc}(r_{80})$ is relatively insensitive to this choice. The data of *Cipriano and Blanchard* [1981] are measurements of SSA production and estimates from measurements of bubble concentrations. Lines for *Mårtensson et al.* [2003] denote fits by original investigators. The formulation of *Monahan et al.* [1983b, 1986] (Eq. 4.3-4) and the lognormal fit to this formulation (Eq. 4.3-5) have been extended to values of r_{80} well beyond their range of validity. The fit presented by *Woolf et al.* [1988] is given by (4.3-6). The data of *Woodcock et al.* [1963] and *Blanchard* [1969] are from measurements of SSA concentrations over an active surf zone. The formulation of *de Leeuw et al.* [2000] is evaluated at $U_{10} = 5$ m s^{-1}, after correction by multiplication by a factor of 10^6 (as discussed in §4.3.5), shown with uncertainties of a factor of $\lesssim 2$. Shaded band drawn to encompass the bulk of the data denotes a factor of $\lesssim 7$ in $f_{wc}(r_{80})$; central line is constant for $r_{80} = 0.1$-1 μm and proportional to $r_{80}^{-1.7}$ for $r_{80} = 1$-10 μm. Right axis yields the values of the interfacial SSA production flux $f_{int}(r_{80})$ for whitecap ratio $W = 0.01$; at a given value of W the uncertainty in f_{int} will be substantially greater than that in f_{wc} because of the uncertainty associated with W (§5.3).

the bubble method (§3.4) from measurements of the bubble flux below a laboratory simulated whitecap together with assumptions regarding bubble rise speeds and the number and sizes of drops produced per bubble burst (§5.5.1).

Production of bubbles and SSA particles from a continuous whitecap of area 0.02 m^2 was investigated by *Cipriano and Blanchard* [1981]. A so-called steady state breaking wave was formed by a continuous waterfall in which ~400 cm^3 s^{-1} of seawater at 26 °C overflowed from a weir 0.3 m above a tank of radius 0.25 m and depth ~0.3 m to form a white area of radius ~0.08 m. Splash drops (§4.7.2) would also have been formed in addition to bubble-produced drops, but these probably were larger and contributed little to the total number of SSA particles produced. The investigators assumed that as the volume of air entrained by the waterfall increased, the bubble spectrum would approach a characteristic shape reflecting that of the bubble spectrum produced by a whitecap at sea (implying that the resulting size distribution of SSA production would also reflect that from an oceanic whitecap). However, results were reported only for a single set of experimental conditions (i.e., volume flow rate of water, height of the waterfall, and the like), and no indication was given that other conditions had been employed to determine whether the size distributions of the bubble concentration and the SSA production flux were independent of these conditions. Filtered air was used in these experiments, and the size-dependent SSA production rate was measured for SSA particles with r_{80} ranging from 0.3 μm to 14 μm in five size intervals using an optical particle counter. Additionally, the total rate of SSA particle production was determined using a condensation nuclei (CN) counter, from which it was determined that approximately half of the drops had r_{80} less than 0.3 μm. Measurements with a diffusion battery indicated that particles were produced with r_{80} as small as ~0.014 μm. The SSA production flux over whitecap area $f_{wc}(r_{80})$ is shown in Fig. 32, with the single datum for the production flux of drops with $r_{80} < 0.3$ μm, covering the range 0.014-0.3 μm, shown at the geometric mean of the endpoints of this region, $r_{80} = 0.065$ μm. The increase in the error bars for larger drops was attributed to the difficulty in evaluating sedimentation loss.

For this data set, as for several others treated below, somewhat arbitrary estimates are required for the range of values of r_{80} corresponding to a given measurement, such as that of the CN concentration. However, as $f_{wc}(r_{80})$ is defined to be the size-dependent SSA production flux per unit whitecap area per interval of $\log r_{80}$, the consequences of this choice on the value of f_{wc} are relatively minor compared to the much greater differences in estimates of f_{wc} from different investigations. For example, if the r_{80} range in the previous example were chosen as 0.014-1 μm instead of 0.014-0.3 μm, the location of the data point would be $r_{80} = 0.12$ μm instead

of 0.065 μm, and as the value of $\Delta \log r_{80}$ would be 1.85 instead of 1.33, the value of f_{wc} would be decreased by only ~30%. As the ordinate in Fig. 32 is a logarithmic scale covering 6 orders of magnitude, such a difference, corresponding to ~0.14 orders of magnitude, is of minor importance.

Size distributions of bubble concentration for R_{bub} from 0.05 mm to 4 mm at distances 0, 0.02, 0.07, and 0.12 m from the center of the upwelling bubble plume were also measured by *Cipriano and Blanchard* [1981] using photographic methods. A strong dependence of the shape of the size distribution of the bubble concentration on the distance from the center of the upwelling plume was observed, specifically a rapid depletion of bubbles with $R_{bub} > 1$ mm as the distance from the center of the bubble plume increased to greater than 0.07 m, although the concentration of bubbles with R_{bub} from 0.05 mm to 0.15 mm was roughly constant with distance from the center of the bubble plume up to 0.12 m (near the tank edge). The rate at which bubbles reached the surface of the tank was calculated from the mean size-dependent bubble concentration averaged over the entire bubble plume [*Cipriano and Blanchard*, 1982] using known bubble rise speeds. The total rate of air entrainment was determined to be around 125 cm^3 s^{-1} (nearly one-third the volume flow rate of water), with more than 95% being converted into bubbles with $R_{bub} > 0.5$ mm.

The production rate of jet drops was estimated for seven intervals of r_{80} from nearly 1 μm to 50 μm based on the rate at which bubbles reached the surface and the relationship between bubble and jet drop sizes presented by *Blanchard and Woodcock* [1957, Fig. 2], which is similar to (4.3-3) for $R_{bub} \gtrsim 0.2$ mm, on the assumption that each bubble produced 5 jet drops; the jet drop production flux over whitecap area calculated from these results is presented in Fig. 32. The investigators concluded that the jet drop formation mechanism was sufficient to account for the number of observed SSA particles with $r_{80} > 5$ μm, but argued that most of the smaller drops would be film drops, as bubbles of the size required to produce jet drops of these sizes would probably not be present in sufficient numbers and would rapidly dissolve before reaching the surface (§4.3.2.4; §4.4.2.2). However, at $r_{80} \approx 10$ μm the estimate of the SSA production flux over whitecap area f_{wc} calculated from the bubble flux is nearly an order of magnitude greater than the measured value of f_{wc}. Part of this discrepancy could be due to larger drops not having been entrained upward and hence not having been detected. Additionally, as discussed in §4.3.2.4, the estimate of 5 jet drops per bubble almost certainly overestimates the number of jet drops that attain sufficient heights that they contribute in any appreciable way to SSA and its role in atmospheric phenomena, and is greater than the number of jet drops produced for $R_{bub} > 1$ mm (Fig. 26). Furthermore, the estimated value of f_{wc} was calculated based on the assumption that all bubbles burst independently at the surface, and hence that there would have been no clustering or screening effects; such an assumption is questionable for this experiment, and it is likely that these effects would have influenced the results and may account for some of the difference.

The rate of film drop production likewise was estimated from the rate at which bubbles reach the surface using the maximum values of N_{film} (Fig. 30) reported by *Blanchard* [1963, Fig. 15]. The production flux of film drops over white area is also shown in Fig. 32 at $r_{80} = 0.3$ µm, near the geometric mean of the somewhat arbitrarily chosen r_{80} range of 0.1-1 µm. The investigators concluded that the film drop production rate calculated in this manner was sufficient to account for the production rate of smaller drops ($r_{80} <$ 2.5 µm), and furthermore that film drops must be the source of the drops with $r_{80} \lesssim 0.3$ µm detected by the condensation nucleus counter. From streak photography it was determined that large bubbles moved as fluid spheres rather than rigid spheres (§4.4.2.1); specific sizes were not mentioned, but presumably this statement referred to bubbles sufficiently large to produce film drops. It was thus concluded that these bubbles had not adsorbed enough organic surface films to affect their motion, and hence that they would have burst immediately upon reaching the surface and thus would have produced the maximum number of film drops (§4.3.3.3). The estimate of the film drop production flux over whitecap area calculated from the bubble flux was roughly an order of magnitude greater than the observed production flux of small SSA particles; this difference was attributed to changes in drop production due to bubble interference, differences in film drop production between bubbles bursting at a calm surface and those bursting at an agitated surface, coalescence of drops once formed, and screening and coalescence of bubbles. *Cipriano et al.* [1983] noted that if the average of the number of film drops per bubble were used instead of the maximum number then the estimated and observed production fluxes of small drops would have agreed much more closely (to within a factor of 2). *Cipriano and Blanchard* [1981] concluded that both types of drops contribute substantially to the SSA production flux of drops with r_{80} from 2.5 µm to 5 µm.

The same experimental apparatus was used by *Cipriano et al.* [1983] to investigate the production of condensation nuclei (CN) and cloud condensation nuclei (CCN) from a steady state breaking wave in real and artificial seawater using a variety of instrumentation, including an electrical aerosol analyzer, several CN counters, a diffusion battery, a thermal gradient diffusion chamber, and three optical particle counters. As measured total CN concentrations were several-fold greater than concentrations of particles with $r_{80} \gtrsim 0.1$ µm determined optically, it was concluded that the vast majority of the drops produced had $r_{80} \lesssim 0.1$ µm.

An estimate of f_{wc} obtained from measurements of the CN concentration, on the assumption that the radii of these particles are in the interval $r_{80} = 0.01$-0.1 µm, is plotted in Fig. 32 at the geometric mean of this interval. Measurements using a diffusion battery indicated production of SSA particles with r_{80} less than 0.01 µm. Information on the size distribution of the production flux, shown in Fig. 32, is inferred from measurements involving a thermal gradient diffusion chamber, for which CCN concentrations were presented for supersaturations of 0.25, 0.5, 1.0, and 2.0%; according to (2.1-11) these supersaturations correspond to values of $r_{80,s}$, (that is, the minimum values of r_{80} of SSA particles that activate) equal to 0.62, 0.39, 0.25, and 0.16 µm, respectively ($r_{80,s}$ values stated by *Cipriano et al.* were approximately 25% lower). The diffusion chamber measurements yielded the unphysical result that CCN production at 0.25% supersaturation was ~15% lower than that at 0.5% supersaturation, although uncertainties in CCN production at each of these supersaturations exceeded the difference between them. The investigators reported that the total CCN production at 2.0% supersaturation was approximately one-third the total CN count; this would imply that two-thirds of the number of SSA particles produced would have had $r_{80} \lesssim 0.016$ µm. However, it was suggested that the thermal gradient diffusion chamber may have yielded erroneously low values, implying a smaller fraction of the particles in this size range (and values of f_{wc} greater than those shown in Fig. 32 for the CCN measurements of *Cipriano et al.* [1983]); the investigators argued that this would not completely account for the difference and therefore that these measurements also indicated the presence of SSA particles with $r_{80} \lesssim 0.016$ µm.

In an experiment with a laboratory whitecap of area $3 \cdot 10^{-4}$ m² produced by bubbling filtered air through a frit located 0.04 m below the water surface, *Mårtensson et al.* [2003] reported size-dependent drop production in synthetic seawater for salinities 0, 9.2, and 33 and temperatures −2, 5, 15, and 23-25 °C. Salinity 9.2 was chosen as typical of the Baltic Sea, whereas 33 is near that of most of the ocean surface waters (§2.5.1). The size distribution of bubble concentration resulting from the frit, measured over the range $R_{bub} = 0.015$-0.62 mm and presented as $dN_{bub}/d\log R_{bub}$ for each of the four temperatures at salinity 33, exhibited a peak near $R_{bub} = 0.1$-0.15 mm at each temperature, with the number concentration of larger bubbles decreasing rapidly with increasing R_{bub}. There was also a large change in the size distributions of bubble concentration between 5 °C and 15 °C, with the size distribution at the lower temperature exhibiting a minimum near $R_{bub} = 0.05$ mm and having much lower (by a factor of up to 5) concentrations of bubbles with $R_{bub} \lesssim 0.1$ mm. Size distributions of SSA production flux were calculated from measurements of SSA particles in

11 size intervals in the range $r_{80} = 0.02\text{-}0.135$ μm using a differential mobility analyzer and in 30 intervals in the range $r_{80} = 0.135\text{-}20$ μm using an optical particle counter. Treatment of Mie ambiguities (§4.1.4.1), which would have occurred in the size range of interest, was not discussed, nor was any comparison of sizes or concentrations of SSA particles obtained with the two instruments near $r_{80} = 0.135$ μm. Values of the size-dependent SSA production flux over whitecap area f_{wc} determined from individual 30 min experiments are shown in Fig. 32 for salinity 33 and temperatures 5 °C and 25 °C. With increasing temperature over the range −2 °C to 25 °C, $f_{wc}(r_{80})$ decreased by up to a factor of ~5 in the size range $r_{80} \lesssim 0.1$ μm and increased by a comparable factor in the size range 0.3 μm $\lesssim r_{80} \lesssim 2$ μm (data for larger sizes exhibited large uncertainties). However, it was not established that the temperature dependence of f_{wc} was not an artifact of the bubble production mechanism at the frit, given the strong temperature dependence of the bubble concentration size distribution reported in this experiment and the observed temperature dependence reported in several other experiments involving formation, coalescence, and shattering of bubbles at frits or orifices [*Lessard and Zieminski*, 1971; *Craig et al.*, 1993a; *Slauenwhite and Johnson*, 1999], calling into question the relevance of such a dependence to oceanic SSA production.

Measurements for r_{80} greater than a few micrometers exhibited large uncertainty, and consequently the investigators presented a formulation for f_{wc} for r_{80} only up to 2.8 μm. This parameterization consisted of a linear function of temperature with coefficients being different quartic functions of r_{80} in each of three size ranges: 0.02-0.145 μm, 0.145-0.419 μm, and 0.419-2.8 μm. There appears to be no physical basis, nor one from their data, for the choice of these size ranges, each of which contains one-third of the number of intervals for which particle sizes were measured. Additionally, the coefficients a_j of the temperature in the expression for the production flux for a given size interval (shown in Fig. 8 of that reference) exhibited an abrupt change with increasing r_{80} near 0.1 μm, from $(-5 \pm 1) \cdot 10^6$ m^{-2} s^{-1} K^{-1} for smaller sizes to $(0.5 \pm 1) \cdot 10^6$ m^{-2} s^{-1} K^{-1} for larger ones; this would appear to provide a more natural basis for the choice of size range. Moreover, according to the parameterization presented the values of f_{wc} change discontinuously from each size range to the next. Values of $f_{wc}(r_{80})$ from this parameterization are also shown in Fig. 32 for salinity 33 and temperatures 5 °C and 25 °C.

It is difficult to reconcile the results of *Mårtensson et al.* [2003] with what is known concerning jet drop and film drop production. These investigators noted the presence of two modes in the size-dependent SSA production flux over whitecap area in the representation $f_{wc}(r_{80}) \equiv dF_{wc}/d\log r_{80}$: one centered at $r_{80} \approx 0.1$ μm and one at $r_{80} \approx 2.5$ μm

(although given the uncertainty in the data for the larger drop sizes it would seem difficult to conclude the existence of this latter mode), which they suggested indicated different formation processes. It would be expected that the vast majority of the drops with $r_{80} \lesssim 1$ μm would be film drops, as few jet drops of these sizes (down to $r_{80} = 0.02$ μm) would be produced according to the arguments presented in §4.3.2.4. However, there were few bubbles with $R_{bub} \gtrsim 0.4$ mm, the range of bubble sizes thought to be required for film drop production (§4.3.3.2; Fig. 30). One explanation of these results is that they are an artifact of the experimental technique. The volume flow rate of air employed (10 mL min^{-1}) would have resulted in an upward air velocity of ~0.55 cm s^{-1} over the white area, equal to the terminal velocity of an SSA particle with $r \approx 2$ μm, and more than sufficient to have carried nearly all the SSA particles produced upward, many of which would not otherwise have been ejected to sufficient heights to be entrained (§4.3.2.4). Additionally, many of the small bubbles, which would have dissolved before reaching the surface under oceanic conditions, would have been forced upward toward the surface with the other bubbles at velocities greater than their terminal velocities in still water, allowing them to have survived to the surface where they could burst and produce jet drops. These issues raise concern over the applicability of the results of this investigation to oceanic SSA production.

4.3.4.2. Sea salt aerosol production from discrete laboratory whitecaps. Laboratory whitecaps formed by the collision of two parcels of water which were caused to move toward each other have been used in a series of studies by *Monahan* and coworkers to investigate the temporal behavior of freshwater and seawater laboratory whitecaps [*Monahan and Zietlow*, 1969; *Monahan et al.*, 1980, 1982; *Stramska et al.*, 1990], SSA production from laboratory whitecaps [*Monahan et al.*, 1980, 1982, 1983b, 1986; *Cipriano et al.*, 1987; *Woolf et al.*, 1988; *Bowyer et al.*, 1990], differences between film and jet drops [*Woolf et al.*, 1987], charge production [*Bowyer et al.*, 1990], and the dependencies of SSA production on temperature and gas saturation [*Bowyer*, 1984; *Monahan et al.*, 1984a; *Bowyer et al.*, 1990; *Stramska et al.*, 1990]. Results of these investigations have been used to calculate the size-dependent SSA production flux over laboratory whitecap area $f_{wc}(r_{80})$ using (3.2.3) from the number of SSA particles per unit logarithmic interval of r_{80} produced by a single breaking wave (typically determined as the product of the change in size-dependent concentration of SSA particles $\Delta n(r_{80})$ and the volume of air enclosed in the tank V), the initial white area $A_{wc,0}$, and the exponential decay time of the whitecap τ_{wc} as $f_{wc} = \Delta n(r_{80})V/(A_{wc,0}\tau_{wc})$.

The discrete laboratory whitecap method (§3.2) of determining the interfacial SSA production flux rests on the explicit assumption that area of any whitecap, independent of its size and means of production, decreases exponentially with the same characteristic decay time τ_{wc}. This decay time has been determined in several investigations, typically from photographs or video recordings of a whitecap by fitting the (natural) logarithm of the amount of white area to a linear function of time, the negative of the slope being $1/\tau_{wc}$. In a study using a tank 1.5 m × 0.3 m with mean water depth 0.16 m values of τ_{wc} were determined by *Monahan and Zietlow* [1969] as 3.85 s and 2.54 s, respectively, for whitecaps in NaCl solution (the concentration was stated as 35%, but as the experiment was meant to simulate seawater, the concentration was presumably 3.5%) at 28 °C and in freshwater at 22 °C. The initial whitecap area $A_{wc,0}$ was determined to be ~0.03 m^2 from extrapolation of the fit of the logarithm of the amount of white area vs. time. For measurements with freshwater for which only rafts of bubbles with area greater than $1.6 \cdot 10^{-6}$ m^2 were taken into account, τ_{wc} was determined to be 2.10 s; an area $1.6 \cdot 10^{-6}$ m^2, corresponding to a circular area with radius 0.7 mm, is much smaller that the sizes typically associated with whitecaps and is more characteristic of sizes of individual bubbles. The influence of temperature on the decay times was not reported. The difference between τ_{wc} for freshwater and that for saltwater was used to explain the greater whitecap ratios in seawater than in freshwater at the same wind speeds [*Monahan and Zietlow*, 1969; *Monahan*, 1970, 1971]; this is discussed further in §4.5.3.

In later work with a tank 3 m × 0.6 m with mean water depth 0.66 m *Monahan et al.* [1980] reported τ_{wc} = 4.7 s for the characteristic exponential decay time of whitecaps in seawater (events 123, 125, 126, 127, 129, and 130), although the data did not appear to be well represented by an exponential decrease with time. The water temperature was not reported but presumably was around 17 °C. Based on extrapolation $A_{wc,0}$ was determined to be 0.056 m^2; this corresponds to a circular area of radius 0.13 m.

Later *Monahan et al.* [1982], with seawater in the same tank, measured the decay of foamy area determined from cine-film recordings of four breaking wave events (events 139-142); again water temperature was not specified. Although the investigators noted that exponential decay was "not readily apparent" for the one example presented (breaking wave event 140; Fig. 2 of *Monahan et al.* [1982]), they nonetheless fitted the amount of white area for the first 5 s and that for the next 9 s to exponentially decreasing functions of time, with characteristic times reported as 1.98 s and 3.53 s, respectively. Because according to these characteristic times the whitecap area would have decreased by 92% during the first 5 s, this period of rapid decay would appear to be the most important for SSA production. Despite this, the characteristic decay time of the second period, 3.53 s,

has been used to infer the oceanic production flux in *Monahan et al.* [1982] and in many subsequent papers [e.g., *Monahan et al.*, 1983b, 1986; *Monahan*, 1986, 1988a; *Monahan and Lu*, 1990]. Based on extrapolation of the amount of white area using the value τ_{wc} = 3.53 s (obtained for times greater than 5 s), the initial white area was reported as $A_{wc,0}$ = 0.349 m^2 (corresponding to a circular area of radius 0.33 m), and these values were assumed to hold for all breaking wave events. The investigators argued that the choice of τ_{wc} was not critical, as the pertinent quantity for application of the whitecap method—the product of $A_{wc,0}$ and τ_{wc}—differed by only ~20% for the two time periods. The value of $A_{wc,0}$ has generally been assumed to be the same for all breaking wave events from this tank or similar tanks, but no evidence was presented to allow determination of the range of values that can be expected for this quantity, and *Bowyer et al.* [1990] reported that the value of $A_{wc,0}$ was strongly dependent on the heights of the two parcels of water that collide to form the laboratory whitecap.

From measurements in a tank of the same dimensions as that used by *Monahan et al.* [1980], *Stramska et al.* [1990] determined that the exponential decay time τ_{wc} for seawater increased linearly with water temperature from 2.30 s at 12 °C to 4.45 s at 20 °C.

Measurements of size-dependent SSA production from a laboratory breaking wave were made by *Monahan et al.* [1982]. The change in the size-dependent concentration in the enclosure due to the decay of the whitecap formed by a breaking wave, $\Delta n(r_{80})$, was determined for three breaking wave events (numbered 372, 374, and 387) in seawater in which the air in the enclosure above the tank (V = 1.85 m^3) was passed through an optical particle counter at 0.63 m above the mean water level for 180 s after the water parcel collision to size and count the particles produced by the simulated breaking wave. Over the approximate radius range 1 μm to 10 μm (which, as RH during these experiments was between 56% and 67%, was stated as corresponding to an r_{80} range of approximately 0.8 μm to 8 μm), the ratio of the signal to the background particle count was generally greater than 50%; presumably the high background relative to the signal was due largely to the use of unfiltered air to transport the aerosol in the hood through the detectors and to residual SSA production from previous experiments. Water temperature was not reported. The investigators concluded that the true values of $f_{wc}(r_{80})$ were probably between 0.8 and 2 times the values resulting from these measurements.

Monahan et al. [1983b, 1986] reported that subsequent experiments (which were not described) in the same tank but with volume V = 1.55 m^3 revealed that particle concentrations at 0.015 m above the mean water surface were much greater than those at 0.07 m, from which the investigators concluded that not all of the particles produced in the previous experiments had been detected. Based on the assumption that the concentration of SSA particles decreased exponentially with height, a correction factor by which the previously derived flux was to be multiplied was determined to be $1 + 0.057(r_{80}/\mu m)^{1.05}$. This factor varies from 1.05 to 1.5 over the range $r_{80} = 0.8$-8 μm; a similar correction factor proposed by *Spillane et al.* [1986] based on the same data, $1 + 0.033(r_{80}/\mu m)^{1.179}$, is less than the above correction factor by from 2% to 8% over the range $r_{80} = 0.8$-8 μm. The values of f_{wc} for the three breaking wave events described above, multiplied by the first correction factor, are shown in Fig. 32. In view of the scatter in the original flux data [Fig. 6 of *Monahan et al.*, 1982] and in the data that were fitted to determine the correction factor [Fig. 5 of *Monahan et al.*, 1983b], the implied precision (based on the number of significant figures) in this factor and perhaps even the need for such a correction factor at all hardly seem justified.

Based on these results, *Monahan et al.* [1983b, 1986] presented a formulation for the size-dependent SSA production flux over the range of r_{80} from 0.8-8 μm (later it was reported by *Stramska et al.* [1990] that this production flux represented that of jet drops only, based on experiments of *Woolf et al.* [1987] described in §4.3.4.3). This formulation, equivalent to

$$\frac{f_{wc}(r_{80})}{m^{-2}s^{-1}} = 8.2 \cdot 10^5 \left(\frac{r_{80}}{\mu m}\right)^{-2} \left[1 + 0.057\left(\frac{r_{80}}{\mu m}\right)^{1.05}\right]$$
$$\times 10^{1.19 \exp\left\{-2.4\left[0.38 - \log_{10}\left(\frac{r_{80}}{\mu m}\right)\right]^2\right\}}, \quad (4.3\text{-}4)$$

is shown in Fig. 32. The r_{80} dependence in (4.3-4) results from the logarithm of the size-dependent volume flux over whitecap area in the representation $\log(dF^V(r_{80})/dr_{80})$ being fitted to a Gaussian function of $\log r_{80}$; however, no point of inflection (which would distinguish a Gaussian from a quadratic) was apparent in the data. A quadratic relation between the logarithm of the size-dependent production flux of SSA volume over whitecap area and $\log r_{80}$ would have resulted in a lognormal relationship between the size-dependent volume flux over whitecap area and r_{80}, and consequently a lognormal dependence of f_{wc} on r_{80} (§2.1.4.2); such a relation would also more readily display the dependence of f_{wc} on r_{80}. A lognormal expression that differs from (4.3-4) by less than 8% over the entire range of measurements is

$$\frac{f_{wc}(r_{80})}{m^{-2}s^{-1}} = 7.0 \cdot 10^6 \exp\left\{-\frac{1}{2}\left[\frac{\ln(r_{80}/0.75\mu m)}{\ln 2.2}\right]^2\right\} \quad (4.3\text{-}5)$$

(a similar expression was proposed by *Capaldo et al.* [1999]). Despite the fact that extrapolation of (4.3-4) and (4.3-4) below $r_{80} = 0.8$ μm, the range of validity of (4.3-4) stated by *Monahan et al.* [1983b, 1986], cannot be justified (especially in the absence of theoretical basis for the fitting function), and despite the fact that extrapolations of the two expressions differ by many orders of magnitude (Fig. 32), one or another of these expressions has nonetheless been used by subsequent investigators as a basis for comparison of other laboratory measurements [*Mårtensson et al.*, 2003] to r_{80} as low as 0.02 μm and as the basis for calculation of SSA production fluxes to r_{80} below 0.01 μm in some instances (§6.2). Consequences of such extrapolations are examined in §6.2.

Based on subsequent experiments in which filtered air was used, *Monahan* and coworkers [*Woolf et al.*, 1988] determined the size-dependent SSA production flux over white area using a different method of accounting for the vertical gradient of SSA particle concentration (a factor of τ_{wc} was omitted from the denominator in the expression for $dF_0(r)/dr$ on p. 182 of that reference). Using the same values of $A_{wc,0}$ and τ_{wc} as those of *Monahan et al.* [1982], *Woolf et al.* presented an expression for the size-dependent SSA production flux over the range of r_{80} from 0.8-10 μm equivalent to

$$\frac{f_{wc}(r_{80})}{m^{-2}s^{-1}} = 10^{\{6.79 - 0.49L - 1.08L^2 + 0.527L^3\}}, \quad (4.3\text{-}6)$$

where $L = \log_{10}(r_{80}/\mu m)$; this is also shown in Fig. 32. A similar formulation but with a value $\tau_{wc} = 4.27$ s instead of 3.53 s was proposed by *Monahan and Woolf* [1988]. Although the expression proposed by *Woolf et al.* (Eq. 4.3-6) yields values of f_{wc} that differ by less than 10% from that given by *Monahan et al.* [1986] (Eq. 4.3-4) for r_{80} from 0.8 μm to 2 μm, it departs substantially from that expression for $r_{80} \gtrsim 2$ μm, yielding values of f_{wc} nearly an order of magnitude greater for $r_{80} = 8$ μm. As (4.3-6) does not readily illustrate the dependence of f_{wc} on r_{80}, it is useful to observe that this expression can be closely approximated by a lognormal expression

$$\frac{f_{wc}(r_{80})}{m^{-2}s^{-1}} = 1.3 \cdot 10^7 \exp\left\{-\frac{1}{2}\left[\frac{\ln(r_{80}/0.12\mu m)}{\ln 5.8}\right]^2\right\}$$

(maximum departure less than 6% over the range 0.8-10 μm) or a power law expression

$$\frac{f_{wc}(r_{80})}{m^{-2}s^{-1}} = 8.6 \cdot 10^6 \left(\frac{r_{80}}{\mu m}\right)^{-1.2}$$

(maximum departure less than 3% over the range 2-10 μm). The choice of formulation between that of *Monahan et al.* [1983b, 1986] and that of *Woolf et al.* [1988] has a substantial influence on the total fluxes of surface area and volume (or mass), and calls into question estimates of global production of SSA mass (Table 1) based on these formulations. According to (4.3-4) the fluxes $dF^A(r_{80})/d\log r_{80}$ and $dF^V(r_{80})/d\log r_{80}$ exhibit maxima at r_{80} of 2.6 μm and 4.8 μm, respectively, whereas according to (4.3-6) these fluxes are strongly increasing functions of r_{80} throughout the stated range of validity of the formulation, implying that the dominant contributions to the fluxes of SSA surface area and volume may result from larger SSA particles, and thus that estimates of these fluxes calculated from this parameterization may be greatly underestimated.

Measurements of the total rate of SSA production from a laboratory breaking wave in seawater at 16 °C in a tank of the same dimensions as that of *Monahan et al.* [1980] (i.e., 3 m × 0.6 m with mean water depth 0.66 m) were made by *Monahan* and coworkers [*Cipriano et al.*, 1987] using a CN counter. These investigators reported that approximately 7% of the SSA particles produced had r_{80} greater than 1 μm and estimated that at least 20% of the particles produced, and possibly more, had r_{80} greater than 0.036 μm. Thus the vast majority of the SSA particles produced would have had $r_{80} < 0.036$ μm. On the assumption that the lower limit of SSA production is $r_{80} = 0.01$ μm, the values of f_{wc} for the size ranges 0.01 μm $< r_{80} < 0.036$ μm and 0.036 μm $< r_{80} < 1$ μm can be calculated; these values are plotted in Fig. 32 at the geometric means of these ranges ($r_{80} = 0.019$ μm and $r_{80} = 0.19$ μm, respectively), where the ranges in r_{80} are indicated by horizontal bars.

4.3.4.3. Factors affecting SSA production from laboratory whitecaps.

The production of SSA particles is affected by any factor that changes the size distribution of the flux of bubbles bursting at the surface and by any factor that influences the bubble bursting process itself. Experiments involving laboratory whitecaps permits systematic investigation of such factors thought to be controlling, allowing determination of their influence on SSA production. Principal factors that have been suggested in this regard are the presence and amount of surface-active materials, the temperature of the water, and its gas saturation (§2.5.1). The results of investigations of the effects of these factors are summarized here.

As with many aspects of drop and bubble behavior (§2.3.4; §4.3.2.3; §4.3.3.3), it is not unexpected that surface-active materials can affect SSA production from laboratory breaking waves, by changing the size distribution of bubble concentration, bubble rise velocities, the energy available for drop production, and the way bubbles interact with each other at the surface, to name but a few possibilities. However, little systematic investigation of the effects of surface-active materials on SSA production from laboratory whitecaps has been reported, although anomalous results are often attributed to the presence of such materials. Measurements of *Monahan et al.* [1980] indicated that the presence of a small amount of oleic acid could reduce SSA production from laboratory breaking waves by up to a factor of 2. Other instances in which results are attributed to surface-active materials are noted below. The extent to which effects of such substances are important in ocean conditions is examined further in §4.3.4.4.

Several studies have investigated the effect of temperature on SSA production from laboratory whitecaps. From measurements of SSA production from a discrete whitecap formed by a laboratory breaking wave, *Bowyer* [1984] concluded that the number of particles with $r_{80} > 2.5$ μm produced per breaking wave event increased with increasing temperature, but that the number of particles with $r_{80} > 0.33$ μm was relatively independent of temperature. Two figures were presented in a report by *Bowyer* [1984], one showing an increase with temperature in production of SSA particles with $r_{80} > 2.5$ μm per breaking wave event by a factor of ~3 between 0 °C and 30 °C (albeit with much scatter—typically a factor of ~5 or more at any given temperature), and the other showing virtually no change in production of SSA particles with $r_{80} > 0.325$ μm for temperatures from 5 °C to 30 °C, but a large range of values (more than a factor of 5) at temperatures near 0 °C. These results were attributed by *Monahan* [1986] to an increase in the number of small bubbles at lower temperatures (§4.4.1.4), resulting in an increase in the number of jet drops, which would comprise the majority of SSA particles with $r_{80} > 2.5$ μm. A brief abstract [*Monahan et al.*, 1984a] reported that the number of drops produced by a laboratory whitecap increased markedly with increasing water temperature because of a shift in the bubble spectrum to smaller bubbles with increasing temperature with resultant shift in the spectrum of drops produced, but results were not quantified and no data were presented.

From a series of experiments involving seawater over the temperature range 7.7 °C to 23.4 °C, *Woolf et al.* [1987] reported that the temperature dependence of SSA production from a laboratory breaking wave varied both with particle size and among different seawater samples. They reported that the production of SSA particles with $r_{80} \gtrsim 2.5$ μm usually increased with increasing water temperature, but that this was not always the situation, and the example they presented (which they claimed was "exceptional" and "atypical" in this regard) exhibited no noticeable increase with increasing water temperature. They also reported that the production of SSA particles with $r_{80} \gtrsim 0.25$ μm exhibited a complex temperature dependence which varied widely between different water samples, sometimes showing a strong decrease with increasing temperature (their results demonstrated a decrease by a factor of 6 between ~7 °C and 15 °C in the number of drops with $r_{80} > 0.75$ μm), which they attributed to the presence of a film of surface-active substance. Based on factors such as time of production of the drops after the breaking of the wave and their electric charge (§4.3.1) the investigators concluded that the majority of SSA particles produced with $r_{80} \lesssim 2$ μm were film drops, that the majority of larger drops were jet drops, and further that jet drop production from a laboratory breaking wave increased with increasing water temperature whereas film drop production exhibited little temperature dependence. These results are consistent with those of *Monahan* [1986] discussed above.

Based on similar experiments, *Woolf and Monahan* [1988] reported that the production of what they characterized as small drops was erratic (which was attributed to the presence of "slicks" on the surface, which, though always present, were presumably modified somehow by temperature), although in general small drop production was much higher at lower temperatures. These investigators also reported that SSA production varied widely between different water samples, and noted that the temperature dependence of drop production in at least one water sample changed during the course of experiments, concluding that the presence of surface-active materials, which would be difficult to characterize, could exert a strong influence on the resulting SSA size distribution.

Similar experiments reported by *Bowyer et al.* [1990] yielded an increase of approximately 2% per °C in the number of SSA particles with $r_{80} \gtrsim 1$ μm produced per breaking wave event over the temperature range 0-30 °C, whereas the number of smaller particles appeared to remain constant or decrease with increasing temperature. These results are consistent with the hypothesis that the drops with $r_{80} \gtrsim 1$-2 μm are mostly jet drops and the smaller ones mostly film drops, as discussed above. These investigators also attributed some of their anomalous results to the presence of a surface-active film. Additionally, they reported that small changes in the undisturbed water level of the two colliding waves had a substantial impact on the initial whitecap area and on SSA production, although they did not quantify this dependence, nor did they state whether the total SSA production was directly proportional to the initial whitecap area.

The effect of gas saturation (§2.5.1) of the seawater on SSA production from a discrete laboratory whitecap in seawater was investigated by *Monahan* and coworkers [*Stramska et al.*, 1990], who reported a size-dependent increase in SSA production per breaking wave event with increasing gas saturation, with an increase in gas saturation from 100% to 125% resulting in an increase in the pro-

duction of SSA particles with $r > 2.5$ μm by a factor of ~3 and an increase in the production of SSA particles with $r > 0.25$ μm by ~25% (presumably the radii are at ambient RH, which was not stated). Gas saturation, reported and measured as oxygen saturation, was varied from 90-130% during the investigation, mainly by varying the temperature of the water over the range 12.9-23.5 °C. The SSA production data exhibited a large amount of scatter; for example, the range in the production of drops with $r > 2.5$ μm per breaking wave event for a given oxygen saturation was at least a factor of 2 and for some values of the oxygen saturation was a factor of 7. The production of drops with $r > 0.25$ μm per breaking wave event for a given oxygen saturation exhibited much less variability, typically ranging from ~20% up to a factor of 2 about the mean value.

These results were interpreted as indicating that the larger particles (i.e., $r_{80} \gtrsim 2.5$ μm) were mostly jet drops which arose from smaller bubbles (which would be more affected by the saturation) and that most of the smaller particles were film drops which were produced by larger bubbles largely unaffected by the state of saturation. The investigators proposed inclusion of a factor dependent on the oxygen saturation (but independent of SSA particle size) in the SSA production flux formulation such as that given by (4.3-4), which according to them represented jet drop production.

That gas saturation could play a role in bubble growth and persistence (and thus affect the size distribution of the bubbles and of the resultant drops) was proposed initially by *Ramsey* [1962a,b]. This effect was seen also in the numerical study by *Thorpe et al.* [1992] and in the laboratory investigation of *Asher and Farley* [1995], who reported an increase in the concentrations of bubbles with $R_{bub} < 0.1$ mm (and hence a resultant change in the size-dependent SSA production flux) with increased gas saturations. Gas saturation can exert a large effect on the behavior of small bubbles, which have small rise velocities (§4.4.2). Sufficiently small bubbles ($R_{bub} \lesssim 0.1$ mm) may dissolve completely before reaching the surface under normal saturation conditions (§4.4.2.2), but the extent of dissolution of bubbles of these sizes would decrease with increasing gas saturation. Bubbles of the size required to produce film drops ($R_{bub} > 1$ mm) have terminal velocities of several tens of centimeters per second (§4.4.2.1) and rise to the surface so quickly that they are little affected by gas saturation (§4.4.2.1).

It would appear that effects of gas saturation are of little importance to oceanic SSA production in most situations. As gas saturation in the investigation of *Stramska et al.* [1990] was varied by changing the temperature, the change in total gas saturation would have been roughly the same as the change in oxygen saturation because of similarity in the temperature dependencies of the solubilities of nitrogen, oxygen, and argon, each of which decreases with increasing temperature (§2.5.1). Thus, an increase the temperature of the water sample would have resulted in an increase in the saturation of each of these gases by roughly the same amount (before equilibration occurred). Change in oxygen saturation in the ocean, however, is generally caused by biological activity (§2.5.1), which has virtually no effect on nitrogen or argon saturation. Thus, the change in oceanic total gas saturation resulting from biological activity would be a factor of three less than the change in oxygen saturation (§2.5.1). The range of oxygen saturations typically found in the ocean surface waters is rather narrow (§2.5.1) and saturations above 110% are not common [*Najjar and Keeling*, 1997]. Thus even for a relatively large oxygen supersaturation of 15%, corresponding to a total gas saturation of about one-third this value (i.e., 5%) under oceanic conditions, the expressions presented by *Stramska et al.* yield an increase in the production of SSA particles with $r > 0.25$ μm of only 16%, and of SSA particles with $r > 2.5$ μm of only 26%, far below the scatter in the data. Furthermore, under situations in which waves are breaking any saturation conditions of the surface waters with respect to the major atmospheric gases that differ appreciably from equilibrium would diminish with time because of exchange of gases between the ocean surface waters and entrained bubbles.

There are additional concerns with the conclusions of *Stramska et al.* [1990]. The investigators based their contention that the effects were due to gas saturation and not temperature on two previous studies which they claimed had demonstrated that film drop production typically exhibited little or no dependence on water temperature and thus that the measured increase in the number of drops with $r > 0.25$ μm (which they assumed to be mostly film drops) must be due to gas saturation. One of these studies was unpublished, and the other one, that of *Woolf et al.* [1987] discussed above, reported that "small drop production showed a complex dependence on temperature, which varied widely between different water samples." Additionally, an increase in temperature, which would accompany an increase in oxygen saturation, would result in a decrease in the viscosity of the seawater (§2.5.1) and thus an increase in the rise velocities of small bubbles (§4.4.2.1), allowing more of them to reach the surface before they dissolve. Temperature might also have an effect on the spectrum of bubbles produced by a breaking wave under otherwise the same conditions, and several laboratory investigations (discussed in §4.4.1.4) have demonstrated a disproportionate increase in the concentration of smaller bubbles compared to larger bubbles with increasing temperature. For these reasons it is difficult to escape the conclusion that the results of *Stramska et al.* are due largely to effects of temperature on the bubble spectrum rather than effects of gas saturation.

4.3.4.4. Considerations regarding interpretation of measurements of SSA production from laboratory whitecaps. Beyond the several concerns with laboratory experiments on SSA production from simulated whitecaps already noted, a key concern is the fidelity with which such experiments represent actual wave-breaking processes in the ocean, especially the small areas (0.0003-0.35 m^2) and depths (0.04-0.7 m) of the laboratory whitecaps and the means of their production. Determinations of $f_{wc}(r_{80})$ obtained using different techniques to generate laboratory whitecaps shown in Fig. 32 exhibit differences of up to two orders of magnitude for the same value of r_{80}, indicating that the value of f_{wc} depends on the particular experimental technique, and calling into question the extent to which any of the methods used can accurately simulate oceanic whitecaps.

Laboratory whitecaps, at least those studied thus far, are orders of magnitude smaller, and consequently much less powerful, than actual whitecaps typical of the open ocean. Bubble plumes from the laboratory breaking waves are not entrained nearly so deeply as those from oceanic breaking waves (for which reported depths are typically at least several meters, and may be more than 10 m; §4.4), and the spectrum of bubbles arriving and breaking at the surface, and the spectrum of drops produced, may be expected to differ considerably between laboratory and oceanic whitecaps. A key potential advantage of studying SSA production from laboratory whitecaps is that the various methods employed to generate these whitecaps provide opportunities to investigate the dependence of SSA production on variables such as the strength of the laboratory waves, tank depth, and the like. Such studies might alleviate concerns about extrapolating to oceanic conditions (if, for instance, breaking waves of two very different strengths yielded nearly the same size-dependent production flux), and might additionally provide support to the assumptions (for which there is little evidence) that production is directly proportional to whitecap coverage and that τ_{wc} is independent of whitecap size; alternatively they may demonstrate that these assumptions are not satisfied. However, no systematic examination of the dependence of SSA production on the size of the breaking wave (or whitecap) appears to have been reported and even if such measurements had been made, concerns would still remain regarding extrapolating such results to much larger oceanic whitecaps. The results of *Bowyer et al.* [1990] noted in §4.3.4.3 are especially pertinent here, that the initial whitecap area and the SSA production are very sensitive to the height of the water parcels that collided to form the laboratory whitecap, implying that the SSA production flux per whitecap area $f_{wc}(r_{80})$ may vary from whitecap to whitecap.

An additional concern is that in none of the laboratory whitecap simulations was the whitecap formed by the action of the wind, as is the situation for oceanic whitecaps formed by breaking waves. In addition to any influences of the different production methods of laboratory whitecaps on the drop production process, the absence of wind precludes accurate simulation of the upward entrainment of SSA particles and restricts these experiments at best to determination of the interfacial SSA production flux. However, the large sedimentation velocities of larger SSA particles, noted by *Cipriano and Blanchard* [1981] in their continuous laboratory whitecap, and the resultant inhomogeneity in the vertical distribution of the SSA particles produced by discrete laboratory whitecaps (§4.3.4.2), yielding differences of nearly an order of magnitude in estimates of the production flux over whitecap area f_{wc} at $r_{80} \approx 8$ μm (depending on the assumptions made regarding this vertical distribution), result in many or most of the larger SSA particles produced not having been sampled. The upper limit of the range of r_{80} for the parameterization of f_{wc} presented by *Monahan et al.* [1983b, 1986] was 8 μm, and data presented by *Cipriano and Blanchard* [1981] did not extend above $r_{80} = 14$ μm; such a limited range further diminishes the utility of this approach for determining the interfacial SSA production flux, for which larger SSA particles are typically most important.

There are further concerns with the means of producing laboratory whitecaps and the extent to which they simulate oceanic whitecaps. For SSA production from a laboratory whitecap to accurately simulate that from an oceanic whitecap, it is necessary that it exhibit the same size-dependent flux of bubbles bursting at the surface as that characterizing oceanic whitecaps. *Woolf* [1985] argued that the size distribution of bubbles beneath a continuous laboratory whitecap forms an steady state distribution with respect to two opposing processes: coalescence and breakup under collision, and breakup under drag forces (for large bubbles), and thus that this size distribution would differ from that beneath an oceanic breaking wave, which is not characterized by a steady state distribution. Presumably the rate and size distribution of bubbles bursting at the surface would also differ between the two, but no oceanic measurements of this flux appear to have been reported. The concentration of bubbles reported by *Cipriano and Blanchard* [1981] beneath the continuous whitecap in their laboratory study was generally one to two orders of magnitude greater than bubble concentrations reported from field measurements (they are, however, comparable to bubble concentrations measured directly under a breaking wave; §4.4.1.2). *Cipriano and Blanchard* argued, however, that the field measurements represented background bubble populations and had similar size distributions, and suggested, based on analogy of the independence of the shape (i.e., size dependence) of raindrop size distributions with rainfall rate [*Blanchard and Spencer*, 1970], that the laboratory bubble size distributions represented a steady

state or background population in the ocean, and thus that the continuous whitecap was a valid simulation of the whitecap formed by an oceanic breaking wave [this was discussed in *Monahan*, 1982; *Cipriano and Blanchard*, 1982; and *Cipriano et al.*, 1983]. One concern with this approach is that the bubble spectrum in such laboratory whitecaps (compared to oceanic whitecaps) may be biased toward large bubbles which rapidly rise to the surface; this was noted also by *Haines and Johnson* [1995], and is discussed further in §5.3.2. *Monahan* [1982] argued that laboratory whitecaps formed by pouring water (as were those of *Cipriano and Blanchard* [1981] and *Cipriano et al.* [1983]) do not accurately simulate typical breaking waves in the open ocean, which are predominantly spilling breakers as opposed to plunging breakers, the categorization being from *Mason* [1952 (cited in *Longuet-Higgins and Turner*, 1974, who also described a third type of breaking wave, a surging breaker, that exists at steep coastlines; *Galvin*, 1968, and *Cokelet*, 1977, described a fourth type, a collapsing breaker, that also occurs only in shallow water)]. In plunging breakers the wave crest curls forward and plunges deeply into the slope of the wave at some distance from the crest, whereas in spilling breakers the broken water develops more gently from an instability at the sharp crest and forms a quasi-steady whitecap on the forward slope [*Longuet-Higgins and Turner*, 1974]. Subsequently *Monahan et al.* [1982; see also *Monahan*, 1986] argued that the discrete breaking waves formed by the collision of two parcels of water produced a plunging breaker, and that these types of breaking waves provide the dominant contribution to the whitecap coverage of the oceans. As noted by *Bortkovskii* [1987, p. 23], the collision of two waves in the open ocean is an uncommon event and thus the laboratory colliding wave likewise would not seem to accurately simulate typical oceanic wave breaking behavior.

Regarding experiments with glass frits, the extent to which the bubble spectrum (and hence drop spectrum) produced by forcing air through a frit accurately simulates that of an oceanic whitecap would likewise appear subject to question, as the mechanism of bubble formation at a frit is entirely different from bubble production by a breaking wave. It might, for example, be expected that the bubble production flux, and hence the SSA production flux, as well as the dependencies of these fluxes on temperature and other factors, would depend strongly on the frit pore size and the air flow rate. *Cipriano and Blanchard* [1983] reported experiments in which the SSA production flux resulting from forcing air through a frit varied by up to a factor of 30 depending on the location of the frit in the water, concluding that neither the size-dependent SSA production flux nor the bubble production fluxes resulting from a frit-type bubbler are representative of those for oceanic conditions because of the arbitrary dependence of production on frit pore size and air flow rate. In the experiments reported by *Mårtensson et al.* [2003] air volume flow rates of 10, 60, and 120 mL min^{-1} were used, but results were presented only for 10 mL min^{-1}, which was stated to have resulted in a bubble concentration that was closest to that observed in oceanic whitecaps (it was also stated that the bubbles produced by the frit were "in the same size range as breaking waves in the real sea but only partly in the same spectra"); however, the shape of the size distribution of bubble concentration (Fig. 2 of that reference) appears very different from that for oceanic bubbles (§4.4.1). A greater SSA production rate was observed for a 13 ml min^{-1} flow rate than for a 10 ml min^{-1} flow rate; this was attributed to larger whitecap area, but data were not presented to permit comparison of the fluxes and the SSA production rates resulting from the two flow rates exhibited size dependencies that depended on temperature and in some situations was lower for the higher flow rate. As with discrete laboratory whitecap experiments, systematic examination of the dependence of SSA production flux on experimental factors such as frit pore size, air flow rate, and other quantities that could possibly affect the spectrum of drops produced might have alleviated concerns about extrapolating to oceanic conditions, or alternatively might have demonstrated the dependence of the SSA production flux on these factors, but no such results were presented.

The effect of surface-active materials on drop production (§2.3.2; §4.2.3.3; §4.3.3.3) may also be expected to differ between laboratory whitecaps and the open ocean. Several experiments involving laboratory whitecaps used artificial seawater [*Monahan and Zietlow*, 1969; *Mårtensson et al.*, 2003]. The applicability of results of such experiments to seawater was questioned by *Scott* [1975a], who argued that the possible inclusion of contaminants in the salt from which the artificial seawater was made might affect bubble behavior and thus lifetime, thereby also affecting drop production.

Several investigators have reached conclusions regarding the effects of surface-active materials on SSA production from large numbers of bubbles bursting collectively at the surface, though as with many consequences of surface-active materials, results are not consistent. *Blanchard* [1963] determined that although the presence of an organic surface-active film may eliminate film drop production for single bubbles, it does not necessarily do so for film drop production from clusters of bubbles. Monolayer films can increase the production of drops by up to threefold in natural seawater for clusters of bubbles according to *Garrett* [1968], though he observed no increase in filtered seawater, concluding that this result was due to differences in the residence time of the bubbles; much variability from one sample to the next was also reported. *Paterson and Spillane* [1969] reported that although small amounts of a surface film can greatly reduce the number of film drops produced by individual seawater bubbles with $R_{bub} = 0.9$ mm (as noted in §4.3.3.2), the effect on clusters of bubbles was not as great. They attributed this result to the creation of new (and clean) water/air surface associated with production of foam (clusters); *Cipriano and Blanchard* [1981] argued that this occurred also in their laboratory whitecap, and *Mason* [1971, p. 78], *Blanchard and Syzdek* [1974a], and *Blanchard* [1983] proposed similar arguments. According to these arguments, the large numbers of bubbles rising after a breaking wave, resulting in upwelling and outflow of water at the surface, produce a region momentarily free of organic films, allowing later bubbles to experience a clean surface on which to burst (although it is not known what fraction of the total number of bubbles remain to experience this clean surface). That bubbling can remove surface-active materials from water or seawater has long been known [e.g., *Blanchard*, 1963, 1964; *Cloke et al.*, 1991; §2.5.2], and several investigators have suggested bubbling as a means for cleaning water of surface-active material [e.g., *Baier*, 1972; *MacIntyre*, 1974b; *Scott*, 1975b]. *Morelli et al.* [1974] reported that surface-active materials have little effect on film drop production from large numbers of bursting bubbles in seawater, but result in a large reduction in jet drop production. *Blanchard and Hoffman* [1978] argued that although organic materials can play a large role in jet drop dynamics, their significance for SSA production may be minimal because of the dependence of the effects on bubble size and the scarcity of bubbles of the sizes for which the largest effects occur. *Monahan et al.* [1979 (cited in *Monahan et al.*, 1983a)] reported that an organic film could reduce drop production (presumably from a simulated laboratory whitecap) by a factor of two or more. As noted in §4.3.2.3 and §4.3.3.3, the residence time of a bubble on the surface is a measure of the cleanliness of the surface and also affects the bursting. *Blanchard* [1983] reported that of several thousand observations of bubbles bursting at sea, only about 1% of the bubbles burst immediately upon arrival at the surface, indicating a clean surface. This mix of observations suggests that application of results from individual bubbles to oceanic conditions requires careful scrutiny for possible influences of surface-active substances.

As with many other aspects of SSA behavior, the effect of surface-active substances on SSA production from laboratory whitecaps has proven difficult to investigate and to quantify, and although these effects are not well characterized, they may be substantial. As noted by *Woolf and Monahan* [1988], "Existing estimates of marine aerosol production [*Monahan*, 1986] are reliant on the principle that the aerosol production from single simulated whitecaps are [sic] representative of the aerosol production from all oceanic whitecaps (with appropriate scaling for the size of the whitecaps). The experiments described in the foregoing sections make it clear that the interference of SAM [surface-active materials] in aerosol production makes this principle untenable, at least in the broadest usage."

4.3.5. Sea Salt Aerosol Production Estimated from Surf Zone Production

A variant of the approach using laboratory whitecaps to determine the size-dependent SSA production flux over whitecap area $f_{wc}(r_{80})$ uses measurements of size-dependent SSA concentrations downwind of an active surf zone (§3.2). The premise of this approach is that production of SSA particles in a surf zone is sufficiently similar to that by whitecaps in the open ocean that the SSA production flux in the surf zone can be equated to the SSA production flux over whitecap area in the open ocean. Application of this method typically assumes a uniformly productive surf zone having "white" area of extent from the coast L in the along-wind direction, with SSA particles produced by the surf zone being uniformly mixed throughout a plume of height H and advected toward the shore with wind speed U. The size-dependent SSA production flux over the surf zone is determined by measurements of the increase (compared to background) in size-dependent SSA number concentration $\Delta n(r_{80})$ measured at the shore during conditions of onshore wind, calculated according to (3.2-4) as $f_{wc}(r_{80}) = \Delta n(r_{80})HU/L$. The few such measurements of f_{wc} that have been reported are examined here and compared with results of laboratory measurements.

Measured values of $\Delta n(r_{80})$ for $r_{80} = 1\text{-}16$ μm resulting from breaking surf were reported by *Woodcock et al.* [1963] at a coastal location in Hawaii during conditions of a 4 m s^{-1} wind blowing onshore at an angle to the beach line of 45°. These data were used by *Monahan* [1968] to calculate f_{wc} on the assumptions (citing *Blanchard* [1964], personal communication) of a surf zone width of 75 m (presumably this distance referred to the extent in the along-wind direction) and a spray plume 5 m high; values of $f_{wc}(r_{80})$ calculated from the data presented in Table 2 of *Woodcock et al.* [1963] using these values for U, H, and L are shown in Fig. 32. However, the units in which f_{wc} were presented by *Monahan* have led to confusion and ambiguity (discussed in §5.2.1.1).

This approach was used also by *Blanchard* [1969] to calculate the total SSA number production flux over white area based on the change in total number concentration of SSA particles with $r_{80} > 0.2$ μm of ~100 cm^{-3} resulting from a surf zone in Hawaii using the values $L = 50$ m, $H = 5$ m, and $U = 4$ m s^{-1}. The value of f_{wc}, calculated on the assumption that all the particles were in the r_{80} range 0.2-1 μm, is shown in Fig. 32 plotted at the geometric mean of this size interval.

Measurements of SSA production from two coastal sites in California reported by *de Leeuw et al.* [2000] were used

to determine the size-dependent SSA production flux over whitecap area $f_{wc}(r_{80})$ on the assumption that concentrations of SSA particles decreased logarithmically with height above the surface over typical plume heights of 20 m or more. Surf zone widths of 30 m and 100 m were reported, although not all values were stated. Values of $f_{wc}(r_{80})$ were presented for r_{80} from 0.4-9 μm in four wind speed intervals, 0-2, 2-4, 4-6, and ≥ 6 m s^{-1}. Although these values exhibited a general increase with increasing wind speed, the differences in f_{wc} among the different wind speed intervals for a given r_{80} were generally less than the range of values in any one wind speed interval, which was typically a factor of ⨉ 2 about the central value. These investigators proposed an expression of the form $f_{wc}(r_{80}) \propto r_{80}^{-0.65}$ for $r_{80} = 0.4\text{-}5$ μm, with magnitude increasing with wind speed as $\exp[0.23U_{10}/(\text{m s}^{-1})]$ for U_{10} from 0 to 9 m s^{-1}; over this wind speed range the increase in the magnitude is a factor of 8 (a factor of 10^6 is required for the expression presented by these investigators, Eq. 4 of that reference, to yield the values shown in Fig. 7a of that reference). This formulation is shown in Fig. 32 for $U_{10} = 5$ m s^{-1} with a multiplicative uncertainty of ⨉ 2.

The main concerns with this variant of the whitecap method are possible dissimilarity of wave breaking and evolution of the resultant bubbles (and thus interfacial SSA production) and of wind conditions (and thus upward entrainment of SSA particles) between the surf zone and the open ocean. Wave breaking at coasts is caused primarily by the interaction of the waves with the ocean floor whereas wave breaking over the oceans is caused primarily by the action of the wind; hence differences in wave breaking between shallow coastal waters and the open ocean are expected. Dependence of wave breaking at coasts on properties and topography of the sea floor and of the nearby coast would also be expected [*de Leeuw et al.*, 2000]. For these reasons wave breaking and resultant SSA production would also be expected to vary from one location to another. Evolution of the bubble plumes resulting from breaking waves would be expected to differ greatly between the open ocean, where such plumes may extend to more than 10 m (§4.4) and coastal areas, where the plumes are constrained by the shallow bottom. Sea salt aerosol concentrations and production fluxes near shore have also been reported to depend on the wind direction [*Zieliński and Zieliński*, 1996; *de Leeuw et al.*, 2000]. At coasts SSA particles can be formed by mechanical disintegration (that is, by water drops rebounding from the collision of the wave on rocks or the beach) in addition to the bubble bursting and spume drop mechanisms. Production of SSA in the surf zone may also be affected by tides. Because of increased concentration of natural and anthropogenic organic materials at the coast, and possibly different formation mechanisms, chemical composition of SSA particles produced in a coastal zone may differ from those produced in the open ocean [*Duce and Woodcock*, 1971; *McKay et al.*, 1994], with possible influence of surface-active materials on SSA production, as discussed elsewhere in this review (e.g., §2.3.2; §4.2.3.3; §4.3.3.3; §4.3.4.4). Because of the influence of the nearby land on the mean and turbulent components of the wind speed, the dispersal of the SSA particles once formed might also differ; this difference would be expected to depend on shoreline topography and thus would also vary with location. Based on these considerations, it is likely that measurements of SSA concentrations and production fluxes inferred from measurements of SSA production from a surf zone would be specific to a given location, and hence the extent to which any measurement could be generalized to other locations might be questioned. These considerations also suggest that local influences might confound attempts to determine f_{wc} by this approach, and further that f_{wc} at coastal locations may differ substantially from that in the open ocean in both magnitude and size dependence.

Other concerns with this approach pertain to sampling and sizing SSA particles. For a wind speed of 5 m s^{-1} the time required for an SSA particle to travel a distance of 50 m, typical of the distance from the middle of the surf zone to the measurement device, is 10 s, which according to (2.5-6) is approximately equal to the characteristic equilibration time with respect to RH $\tau_{98,80}$ for an SSA particle with $r_{80} = 10$ μm. Thus it is unlikely that larger SSA particles would have attained radii close to their equilibrium values by the time of measurement, resulting in possible overestimation of values of r_{80} for these particles (if detected optically, for instance). The extent of such overestimation would vary with distance traveled over the surf zone as well as with wind speed, RH, and r_{80}. Additionally, it might be expected that the RH in the plume resulting from the surf zone production would differ from (and probably be greater than) that outside the plume, possibly resulting in further error in determining the value of r_{80} if this effect is not taken into account. Some loss of larger SSA particles would result from gravitational sedimentation, but it is unlikely that this would have been an important factor in the above investigations, which did not extend above $r_{80} \approx 16$ μm.

A further concern with this method is the assumption, which is inherent to the whitecap method generally, that the SSA production flux over white area is the same in both magnitude and size dependence, for all white areas. The measurements of *de Leeuw et al.* [2000] are pertinent here. Although according to the formulation presented by these investigators the reported shapes of the size distribution $f_{wc}(r_{80})$ are invariant with wind speed, consistent with the assumption of the whitecap method, the dependence of the magnitude of f_{wc} on wind speed (increasing by a factor of 8 as U_{10} increases from 0 to 9 m s^{-1}) raises questions specifically regarding the applicability of values of f_{wc} determined by the surf zone method to calculations of SSA production flux in the open ocean and more generally regarding the assumption that f_{wc} is a universal property of white area.

The several estimates of $f_{wc}(r_{80})$ in Fig. 32—those based on SSA production from continuous and discrete laboratory whitecaps and those based on SSA production from a surf zone—vary by well over an order of magnitude for the same r_{80}. To quantify this uncertainty in knowledge of $f_{wc}(r_{80})$, and to arrive at an expression for $f_{wc}(r_{80})$ that can be used (together with this uncertainty) in calculations to infer the interfacial SSA production flux $f_{int}(r_{80})$ by the whitecap method, a solid line is shown in Fig. 32 together with an accompanying shaded gray region encompassing the majority of the measurements of f_{wc}, from laboratory (§4.3.4) and surf zone studies over the radius range $r_{80} = 0.1$-10 μm. The shaded region is below the estimates of *Cipriano and Blanchard* [1981] for interfacial production fluxes of jet and film drops over whitecap area calculated from measured size distributions of bubble concentration under their laboratory breaking wave. However, those estimates may be too high for the reasons discussed in §4.3.4.1.

The width of the shaded region in Fig. 32 denoting uncertainty in f_{wc} is a factor of $\overset{\times}{\div} 7$ about the central value given by the solid line shown for r_{80} from 0.1 μm to 10 μm. The central value of $f_{wc}(r_{80})$ is constant, independent of r_{80}, for $r_{80} = 0.1$-1 μm and it decreases as $r_{80}^{-1.7}$ for $r_{80} = 1$-10 μm. This uncertainty band is pertinent also to f_{int} at $W = 0.01$ (right axis) but not in general to estimates of f_{int} determined by the whitecap method, which may also contain uncertainty in W (§5.3.1). It is not known to what extent the uncertainty in f_{wc} reflects differences in production mechanisms in the several studies as opposed to errors in measurement or assumptions in interpretation of the measurements.

4.3.6. Investigations with Wind/Wave Tunnels

Experiments in wind/wave tunnels have been performed to study production mechanisms of drops formed by the action of the wind on water surfaces and to measure near-surface along-wind fluxes and interfacial production fluxes of these drops. Key studies, summarized in Table 18, are examined in this section; other studies not intended to simulate oceanic drop formation [e.g., *Mestayer and Lefauconnier*, 1988; *Rouault et al.*, 1991; *Edson and Fairall*, 1994; *Edson et al.*, 1996] in which plumes of bubbles were produced by aeration devices immersed in a tank filled with freshwater are not examined. Most investigations have involved tunnels that are at most a few tens of meters long with widths, water depths, and maximum heights above the water of 0.5 m to 1 m. The laboratory wind speed U_{lab}

Table 18. Laboratory Investigations of Drop Production in Wind/Wave Tunnels

Reference	Tunnel dimensions/m				Medium[a]	Fetch/m	Measurement Height/cm
	Length	Width	Water depth	Air height			
Moore & Mason, 1954[c,d]	6	0.076	0.15	0.23	SW	5	17
Okuda & Hayami, 1959[d]	14.7	0.75	0.43	0.57	FW	7.2	< 35
Toba, 1961a,b	21.6	0.75	0.5	0.52	FW	4.5-13.6	8-25
Mangarella et al., 1973[d]	35	0.91	0.965	0.965	FW		
Wu, 1973[j]	14	1.5	1.2	0.35	FW	7.5	4.2-23.7
Lai & Shemdin, 1974	45.7	1.83	0.915	1.015	FW, saltwater	24.4	10-18
Wang & Street, 1978a,b[n]	35	0.91	0.965	0.965	FW	3-12	~5-~30
Ling et al., 1978	10	0.6	0.6				
Koga, 1981[d]	20	0.6	0.6	0.6	FW	5-16	
Koga & Toba, 1981	20	0.6	0.6	0.6	FW	16	8-26
Anguelova et al., 1999	37	1	0.75	0.55	FW		6-21.5

[a] FW denotes freshwater; SW denotes seawater.

[b] Values of r_{80} for freshwater drops are calculated as $r_{80} = 0.5r_{form}$.

[c] The wind speed range was stated as equivalent to a U_{10} range of 0-25 m s^{-1}.

[d] No size-dependent measurements were reported.

[e] Air bubbles first appeared in the water at 6.4 m s^{-1} and at wind speeds greater than 18 m s^{-1} there was a marked increase in the number of very large spray (i.e., spume) drops, but it was not clear if these wind speeds referred to those in the tunnel or to the oceanic equivalents.

[f] Water drops began to splash from the water surface when the laboratory wind speed exceeded 11-14 m s^{-1}, corresponding roughly to U_{10} of 18-23 m s^{-1}. The total along-wind mass flux increased rapidly when the wind speed exceeded a critical value that depended on the wave form, especially the wave steepness, and the fetch. The investigators indicated that this critical velocity would be different from its oceanic equivalent.

[g] Measurements were reported only for laboratory wind speeds of 5.1-12.1 m s^{-1}.

[h] The critical laboratory wind speed for the onset of production of drops from bubbles formed by the merging of capillary waves near the crests of gravity waves was reported to be 7.5 m s^{-1} for fetch 13 m (stated as corresponding to $U_{10} = 13$ m s^{-1}), increasing with decreasing fetch.

[i] Commencement of spray occurred between laboratory wind speeds of 11 and 14 m s^{-1}.

[j] According to *Wu* [1993], the "Droplet concentration" presented in Fig. 10 of *Wu* [1973] should have been the vertical integral of the along-wind flux and thus the stated units are incorrect; additionally, the label on the logarithmic scale on the figure presenting the values should have been a factor of ten greater.

Table 18. (Continued)

Reference	U_{lab}/(m s^{-1}) Range	Critical U_{lab}/(m s^{-1})	Detection Method	r_{80}/μm Range[b]	Wave Height/cm
Moore & Mason, 1954		6.4, 18[e]	glass slides	> 0.44	
Okuda & Hayami, 1959	2-14	11-14[f]	filter paper		< 10
Toba, 1961a,b	0.5-12.1[g]	7.5[h]	glass slides, filter paper	~5-75	< 5
Mangarella et al., 1973	3.4-14.5	11.1-14.5[i]			
Wu, 1973	2.5-13.4	9.5, 11.5[k]	laser and photo-transistor	35-100	< 5
Lai & Shemdin, 1974	10-18[l]		hot-film anemometer	17.5-175	≤ 20[m]
Wang & Street, 1978a,b	12.5-15	12.5[o]	electrostatic capacitance wire	6.25-110	5-10
Ling et al., 1978	4				
Koga, 1981	14-16	12-16[p]	photographs	200-750	< 10
Koga & Toba, 1981	9-16[q]	13-14[r]	filter paper, glass slides, photographs	>15	< 10
Anguelova et al., 1999	11.4-14.1	9[s]	video	~200-3000	~6

[k] The critical laboratory wind speed for wave breaking was reported to be 9.5 m s^{-1}; at $U_{lab} > 11.5$ m s^{-1} whitecapping was observed at every wave crest.

[l] Size-dependent along-wind fluxes were reported only for laboratory wind speeds of 15-18 m s^{-1}.

[m] Waves with heights up to 20 cm were generated mechanically.

[n] As noted by *Wu* [1982a], the critical value of the dimensionless parameter $u_*\sigma/\nu_w$ for inception of spray reported by *Wang and Street* [1978a,b] as 0.3 should have been 300.

[o] Spray particles were observed when the laboratory wind speed exceeded 12.5 m s^{-1}.

[p] At a laboratory wind speed of 16 m s^{-1} wave breaking with bubble entrainment was rarely observed at a fetch of 5 m but was common at a fetch of 16 m, at which fetch the critical wind speed for the occurrence of wave breaking and for the appearance of small projections on the water surface was $U_{lab} = 12$ m s^{-1}.

[q] Size-dependent along-wind fluxes were reported only for laboratory wind speeds of 14-16 m s^{-1}.

[r] Wave breaking for fetch 16 m occurred for laboratory wind speeds greater than 13 m s^{-1}, and wave breaking with bubble entrainment occurred at nearly every wave crest occurred for laboratory wind speeds greater than 14 m s^{-1}.

[s] Spume drop production was reported to have commenced at a laboratory wind speed of 9 m s^{-1}.

reported in these investigations generally refers to the average wind speed over the tunnel or to the wind speed toward the top of the tunnel above the water surface. So-called critical wind speeds have commonly been reported (Table 18), but these have referred to a variety of phenomena (e.g., onset of bubble or drop production, marked increase in along-wind number or mass flux, wave breaking) and in several instances have depended on the fetch. In most of the experiments drop formation resulted from the action of the wind on the water which created "waves" of height typically a few centimeters, but in some experiments waves were generated mechanically. Along-wind fluxes were typically measured by collecting drops by impaction, or by counting and possibly sizing them using optical techniques. Some of the investigations that are examined have provided insight into the mechanisms by which drops are produced, but for the most part they have provided little information that can be used to obtain quantitative estimates of the interfacial SSA production flux over the ocean, having yielded at best relative values of along-wind fluxes at different drop sizes. Several wind/wave tunnel investigations have referred to the term "sea spray," which in general appears to refer to SSA drops near the surface of the ocean independent of the particular formation mechanism, although in some instances investigators have used the term to denote only spume drops.

Many of the wind/wave tunnel investigations have involved freshwater. Implicit in application of the results of these investigations to SSA production is that the formation mechanism of large drops by processes simulated by these wind/wave tunnels is the same for freshwater and seawater. This is likely to be so for spume drops (which are not produced by the bubble-bursting mechanism), in view of the small differences between the two media in the factors that would be expected to be important—density, surface tension, and viscosity (§2.5.1), whereas for drops produced by bubble bursting substantial differences may arise because of differences in the bubble spectra between freshwater and seawater which result from differences in the electrolytic nature of the two media (§4.4.1.4). For freshwater of course the drops are not SSA particles, but they are referred to as such here, with the value of r_{80} used to denote their sizes taken to be half the formation radius (§2.1.1). The investigations have generally been restricted to drops sufficiently large that measured radii are very close to the radii at formation. For example, the decrease in the radius of an SSA particle with $r_{80} > 10$ μm formed at the seawater surface and thus initially in equilibrium with RH 98%, or in the radius of a freshwater drop with formation radius greater than 20 μm, would be less than ~1 μm during the first second after such a particle is instantaneously injected into an external air flow of 10 m s^{-1} and 80% RH, even taking into account ventilation

effects (increased evaporation due to the relative motion of the air; §2.5.4); the decrease in radius for an SSA particle with $r_{80} > 100$ μm under these same conditions would be less than ~0.25 μm. Hence such drops formed in wind/wave tunnels experience little change in radius between formation and measurement, and r_{80} is accurately taken to be half the measured radius.

In principle, results of wind/wave tunnel measurements that yield the size-dependent along-wind SSA flux $j_x(r_{80}, z)$ as a function of height z above the water surface could be used to determine the interfacial SSA production flux according to the along-wind flux method (§3.6) using (3.6-2). This approach would require knowledge of the vertically integrated along-wind flux, $\int j_x(r_{80}, z)\, dz$, over heights from the surface (i.e., $z = 0$) to as high as necessary to include the dominant contribution to this integral (which may be the full height of the tunnel). However, several investigations presented insufficient information to calculate this quantity. Such an approach would also require knowledge of the mean horizontal distance traveled by SSA particles of given r_{80} produced at the surface $\bar{L}_p(r_{80})$, but determinations of this quantity have not been reported. Over the open ocean as well as in laboratory experiments such as these, it would be expected that $\bar{L}_p(r_{80})$ is strongly dependent on SSA particle size. As noted in §3.6, if the fetch (i.e., the length of the wind tunnel section over which the wind-water contact occurs) is much less than $\bar{L}_p(r_{80})$, then the fetch should be used instead of this quantity in (3.6-2), but satisfaction of this condition is likewise expected to be strongly dependent on particle size, and for sufficiently large particles this condition will probably not be satisfied. Consequently, any uncertainty in $\bar{L}_p(r_{80})$ is reflected in the inference of the interfacial SSA production flux from measurements of $j_x(r_{80})$ and its vertical integral.

In practice, nearly all determinations of the interfacial SSA production flux using the along-wind flux method (§5.7) have relied on only the relative size dependence of the along-wind flux measured from these experiments, often only at a few heights, with the absolute magnitude of the interfacial SSA production flux being adduced by other arguments. However, such a procedure would result in an incorrect size dependence for the interfacial SSA production flux because of the strong dependence of $\bar{L}_p(r_{80})$ on particle size. For these reasons, and because of concerns with extrapolating results of measurements from wind/wave tunnels to the open ocean (examined below in §4.3.6.2), laboratory measurements of the along-wind flux are presented in Fig. 25 in the representation $j_x(r_{80}) \equiv dJ_x(r_{80})/d\log r_{80}$ but with arbitrary ordinate scale, implying that only the shapes of the size dependence of the along-wind fluxes (that is, the relative magnitudes of the fluxes of drops of different sizes and not the absolute magnitudes) are meaningful. Size-dependent along-wind fluxes reported by *Wu et al.* [1984] from oceanic near-surface flux measurements displayed on this figure, also with arbitrary ordinate, were discussed in §4.2.

Some investigations have reported measurements of the mean horizontal velocities of drops, which are required to relate along-wind fluxes of SSA particles to their number concentrations (§2.1.6). Sufficiently large drops require appreciable times for their horizontal velocities to approach that of the surrounding air (§2.6.4), and may fall back to the surface before attaining an appreciable fraction of this velocity. An additional complication is that the horizontal and vertical motions of such drops are not independent (§2.6.4); hence accurate modeling of these velocities would require knowledge of the flow conditions near the surface, which might be difficult to obtain under conditions for which large drops are produced, especially over the open ocean.

4.3.6.1. Sea salt aerosol production in wind/wave tunnels. Several wind/wave tunnel investigations [*Moore and Mason*, 1954; *Okuda and Hayami*, 1959; *Mangarella et al.*, 1973] presented only total along-wind fluxes of number (J_x) or mass, or a threshold laboratory wind speed for some phenomenon, but not size-dependent measurements, and as such cannot be used to determine the interfacial SSA production flux according to the along-wind flux method. In view of concerns over relating these laboratory wind speed thresholds to oceanic equivalents (§4.3.6.2) and the dependence of the threshold wind speeds on fetch (Table 18), these investigations are not considered further. Other investigations [*Toba*, 1961a,b; *Koga*, 1981; *Anguelova et al.*, 1999] focused primarily on the mechanisms of drop production; depending on the experimental arrangement one or another mechanism was dominant. These investigations, all of which employed photography, have yielded both qualitative insight into these mechanisms and quantitative results, and are discussed briefly here.

Toba [1961a,b] concluded that the drops he measured were produced by the bubble mechanism (i.e., that they were jet and film drops) from bubbles formed by the merging of capillary waves near the crests of gravity waves, for which surface tension would be an important factor. Size distributions of the along-wind flux of drops from 8 cm to 25 cm above the water surface were measured. These measurements, together with the assumption that all drops had attained a horizontal component of velocity equal to that of the local wind speed at the height of measurement, were used to calculate number concentrations for three ranges of r_{80}: 5.5-7.5 µm, 11.75-18.25 µm, and 37.5-62.5 µm at U_{lab} = 12.1 m s^{-1}. These size distributions of concentration were fitted to exponentially decreasing functions of height based

on the assumption that only turbulent diffusivity and gravity acted on the particles, using values of the diffusivity and wind speed that were independent of height. For the lowest two size ranges the calculated concentrations were fairly close to measured values, but for the largest size range the calculated concentrations were roughly an order of magnitude less than measured values. These measurements cannot be used to calculated the integral of the along-wind flux over height z, which is required to determine the interfacial production flux according to the along-wind flux method, as it appears that the dominant contribution to the along-wind flux would have been in the region $z < 8$ cm, below the height of the lowest measurement. Additionally, vertical distributions of bubble concentrations down to 12 cm below the surface were measured, from which the flux of bubbles to the surface was obtained as the product of the integral of bubble concentration over depth and the fraction of wave crests of main gravity waves that entrain bubbles, divided by the period of the main wave (as the mean residence time of the bubbles in the water was less than the time between wave crests from which they are formed, the terminal velocities of the bubbles do not occur in the relation). From these several results, *Toba* obtained an estimate for the interfacial SSA production flux, essentially combining the whitecap method, bubble method, steady state dry deposition method, and along-wind flux method. This approach is examined further in §5.5.1.1.

Direct production of spume drops in freshwater at laboratory wind speeds of 14, 15, and 16 m s^{-1} was observed by *Koga* [1981], who used a multi-color photographic technique to resolve drop motion, allowing determination of the trajectories and speeds of drops with $r_{80} > 200$ µm and their precursor surface projections (smaller drops could not be detected on the photographs). Similar photographs of drop formation were also presented in *Koga and Toba* [1981]. The mechanism by which drops were formed was discussed in §2.2; briefly, during high winds small bell-shaped projections roughly 1 cm across and half as high form on the surface and become unstable because of a Kelvin-Helmholtz instability caused by suction due to air flow over the wave crest, subsequently breaking off to form drops. Most drops were produced on the leading slope near the wave crest with initial speeds typically somewhat larger than the speed of the projection from which they were formed. Production of drops with r_{80} up to 750 µm was observed. No size distributions of along-wind fluxes or of concentrations were reported.

The interfacial production flux of spume drops with r_{80} from ~200 µm to 3000 µm was measured by *Anguelova et al.* [1999] in a wind/wave tunnel filled with freshwater using video recordings at four laboratory wind speeds having values at 28 cm above the mean water level of 11.4 m s^{-1} to

14.1 m s^{-1}. A strong correlation was noted between the frequency of wave breaking and drop production flux. Size distributions of the interfacial production flux of drops were reported in 18 intervals of r_{80}. These results, along with a model that was proposed to relate laboratory drop production to oceanic production, were used to estimate the interfacial oceanic SSA production flux; this approach is examined in §5.8. Occasional breakup or coalescence of drops was reported; such phenomena are also not unexpected, as drops with $r_{80} = 3000$ μm have geometric radii of 6 mm, and raindrops of this size and smaller are known to become unstable and break up (§2.6.3). Although coalescence or breakup would change the size distribution of drops, the consequence of these on reported size distributions is probably minor.

Several wind/wave tunnel investigations [*Wu*, 1973; *Lai and Shemdin*, 1974; *Wang and Street*, 1978a,b; *Ling et al.*, 1980; *Koga and Toba*, 1981] presented measurements of the size-dependent along-wind flux as a function of height and laboratory wind speed. Different dependencies of size-dependent or total along-wind fluxes on height and on laboratory wind speed or wind friction velocity were obtained by different investigators. Although results of some of these measurements, along with field measurements of *Wu et al.* [1984], have been applied to determination of interfacial SSA production flux (§5.7.1) by the along-wind flux method (§3.6), only the shapes of the size-dependent along-wind fluxes have been used (even though, as noted above, the shape of the size-dependent along-wind flux and the shape of the interfacial SSA production flux may differ greatly because of the dependence of $\bar{L}_p(r_{80})$ on r_{80}). Thus, the proposed dependencies of the along-wind fluxes on height and wind speed are not examined here. In most of these investigations the range of wind speeds was quite limited (Table 18), calling into question parameterizations of the dependence of the along-wind flux (size-dependent or total) on wind speed. Some of these investigations [*Wu*, 1973; *Lai and Shemdin*, 1974; *Wang and Street*, 1978a,b] attempted to parameterize the near-surface along-wind flux in terms of dimensionless groups formed from quantities such as the mean wave height, characteristic drop size, or characteristic roughness length resulting from extrapolation of the logarithmic velocity profile; the wind friction velocity; and the kinematic viscosity of air or water, surface tension of the water-air interface, and acceleration due to gravity. However, these parameterizations have not proven to be of general validity and in most instances rest on rather weak physical foundations; consequently they are not discussed further here. In several of these investigations containing compilations and comparisons of previous measurements some of the data sets were incorrectly presented (e.g., the data sets of *Preobrazhenskii* [1973], for which difficulties in interpretation were noted in §4.1.4.2, and of *Monahan* [1968], discussed in §4.2), calling into question the conclusions that were reached.

Measurements of size-dependent along-wind fluxes of freshwater drops in a wind/wave tunnel were reported by *Wu* [1973]. Size distributions of the frequency of occurrence (presumably proportional to $dJ_x(r_{80})/dr_{80}$) of drops with 35 μm $\lesssim r_{80} < 100$ μm were presented for three laboratory wind speeds ranging from ~12 m s^{-1} to ~13.5 m s^{-1} at several heights, from which *Wu* concluded that these size distributions were roughly independent of height, varying approximately as r_{80}^{-4} for $r_{80} < 50$ μm, and as r_{80}^{-8} for $r_{80} > 50$ μm (corresponding to size distributions expressed as $j_x(r_{80}) \equiv dJ_x(r_{80})/d\log r_{80}$ being proportional to r_{80}^{-3} and r_{80}^{-7} respectively for these size ranges). These size distributions, expressed in the representation $dJ_x(r_{80})/dr_{80}$, exhibited peaks at r_{80} in the range 35 μm to 50 μm, the value of which increased somewhat over the narrow range of wind speeds (mean r_{80} values of 41.5 μm, 47.75 μm, and 52 μm were stated for specified values of U_{lab} of 11.72, 12.62, and 13.43 m s^{-1}, respectively). These size distributions would appear to be better fitted by lognormal distributions, which represent the data over the entire size range better than power laws. In some instances it appears that the sample rate was only a few particles per minute or less for certain combinations of wind speed and height, leading to concerns about uncertainties for the 5 min sampling period used (counting statistics were not discussed). Frequencies of occurrence in the representation $j_x(r_{80})$ of drops at $z = 15.2$ cm for these laboratory wind speeds [*Wu*, 1973, Table 8] are presented in Fig. 25, as are *Wu's* power law fits.

A concern with quantitative interpretation of the data of *Wu* [1973] is confusion over the quantity reported and the units. A quantity denoted as "concentration," and specified in the text in one instance as "the number of droplets *n* per unit area (square meters) of the mean water surface and per unit time"—which is an interfacial production flux—and in another instance as the number "of droplets *n* per unit area of the water surface," was presented in Fig. 10 of *Wu* [1973] as a function of (laboratory) wind speed with units "Number/m^2/sec." Subsequently, *Wu* [1993] reported that this "concentration" should have referred to the total along-wind flux integrated over height (i.e., the total number of drops per unit time passing through a width of an area perpendicular to the mean water surface, integrated over height), that the units had been incorrectly given in the figure (they should have been m^{-1} s^{-1}), and that the logarithmic ordinate scale on the figure was in error by one decade, resulting in values that were too low by an order of magnitude. These errors have led to confusion and have resulted in the misinterpretation of these data by others (§5.7).

Measurements of size-dependent along-wind fluxes of both freshwater and saltwater (presumably a sodium chloride solution) drops in a wind/wave tunnel were reported by *Lai and Shemdin* [1974] for three values of U_{lab} from 15 m s^{-1} to 18 m s^{-1}, both with and without mechanically generated waves of several heights (up to 15 cm) and wave frequencies; reported values of r_{80} ranged from ~17.5 μm to 175 μm. The dependencies of the total along-wind flux J_x on height and wind friction velocity were reported to be similar for fresh-water and seawater drops. The total along-wind flux was less in the presence of a mechanically generated 5-cm wave height than in the absence of mechanically generated waves (implying decreased drop production), but for wave height greater than 5 cm the total along-wind flux increased with increasing wave height, and for a 15 cm wave height this flux was roughly a factor of 5 greater than it was without mechanically generated waves. Size distributions of the along-wind fluxes of saltwater drops at height $z = 13$ cm (Fig. 7 of that reference) are shown in Fig. 25 in the repre-sentation $j_x(r_{80})$, but with arbitrary magnitude. Size distri-butions of along-wind fluxes of saltwater drops for $U_{lab} = 16.7$ m s^{-1} at heights 13 cm, 15.5 cm, and 18 cm were also presented (Fig. 8 of that reference); however, the size dis-tributions for 13 cm at $U_{lab} = 16.7$ m s^{-1} differ markedly between the two figures. Size distributions of along-wind fluxes of saltwater drops at any given height were fitted by *Lai and Shemdin* to the form $j_x(r_{80}) \propto r_{80}^{-1}$, but the data show considerable scatter and are not well represented by this relation. Size distributions of along-wind fluxes of fresh-water drops (Fig. 11 of that reference) are also presented in Fig. 25 (likewise with arbitrary magnitude), as is a fit of the form $j_x(r_{80}) \propto r_{80}^{-2}$ proposed by these investigators. It appears that these size distributions manifest large uncertain-ties resulting from low numbers of particles sampled; the increase in the size distribution of the freshwater drops at the largest sizes for 13 cm at $U_{lab} = 18$ m s^{-1} arises in part from conversion of the fluxes from the representation $dJ_x(r_{80})/dr_{80}$ to $j_x(r_{80}) \equiv dJ_x(r_{80})/d\log r_{80}$, but probably mostly from count-ing statistics (similar behavior was evident in size distri-butions of SSA concentration; §4.1.4.3).

The size distributions of the along-wind fluxes of fresh-water and saltwater drops exhibited different behaviors, which the investigators attributed to the differences in sur-face tension, bubble spectra, and other physical and electri-cal potential properties (it was stated that measurements of surface tension were made, but such measurements were not reported nor discussed). The difference in surface tension between seawater (or saltwater) and freshwater is too slight to expect that it could make an appreciable difference in the resultant size-distribution of drops produced (§2.5.1); how-ever, as noted above, differences in the bubble spectra of freshwater and seawater can occur because of differences in the electrolytic nature of the two media (§4.4.1.4). For this reason, differences in the along-wind flux (and hence in the

production flux) of drops between saltwater and freshwater observed by *Lai and Shemdin* under nominally the same conditions suggest that most of the drops, especially those with $r_{80} \lesssim 75$ μm, were formed from the bubble-bursting mechanism and hence were probably jet drops, as opposed to spume drops torn from wave crests. Such differences also call into question the extent to which results from other labor-atory investigations for which freshwater was used can be applied to SSA production.

Along-wind fluxes of freshwater drops with r_{80} from ~6 μm to ~100 μm were reported by *Wang and Street* [1978a,b] for four laboratory wind speeds from 12 m s^{-1} to 15 m s^{-1}. These investigators concluded that the inception of spray occurred at a critical Reynolds number determined as the product of the root-mean-squared wave height and the wind friction velocity divided by the kinematic viscosity of air (as noted by *Wu* [1982a], the stated Reynolds numbers were a factor of 10^3 too low; additionally, the stated Reynolds numbers based on the drop boundary layer thickness parameter were a factor of 10^4 too low). Size distributions of the along-wind flux for a height of 10.2 cm at laboratory wind speeds of 12.8 m s^{-1} and 14.9 m s^{-1} [*Wang and Street*, 1978a, Fig. 8] are shown in Fig. 25 in the representation $j_x(r_{80})$ with arbitrary magni-tude, as are size distributions of the frequency of occurrence of drops for a height of 10.2 cm at $U_{lab} = 14.5$ m s^{-1} presented in *Wang and Street* [1978b, Fig. 6]. These data, which appear to have been presented in the representation $dJ_x(r_{80})/d\log r_{80}$, are presumably the same as those in Fig. 8 of *Wang and Street* [1978a] for which the velocity was stated as 14.9 m s^{-1}, although these two data sets differ somewhat (Fig. 25).

Measurements of the along-wind flux of freshwater drops were reported by *Koga and Toba* [1981] for six ranges of r_{80} from ~15 μm to ~450 μm at heights from 9 cm to 26 cm above the mean water surface for laboratory wind speeds of 14, 15, and 16 m s^{-1}. The vertical dependencies of the along-wind fluxes for the various size ranges and wind speeds appeared similar. Examples of these fluxes in the represen-tation $j_x(r_{80})$, again with arbitrary amplitude, are presented in Fig. 25. These size distributions were used to calculated number concentrations and their vertical distribution, although there are serious concerns with these calculations, as discussed below (§4.3.6.2).

One final study that should be noted is that of *Ling et al.* [1978], which examined the production of drops from what were characterized as "hydraulic jumps" of various energy levels created by an air blower developing a wind speed of 4 m s^{-1} in an open tunnel. Little description of the experi-ment was provided, such as how the hydraulic jumps were created, the dimensions of the jumps and the resultant labo-ratory whitecaps, the measurement technique, and whether the liquid was seawater or freshwater. Ratios of the produc-tion rates of drops with r_{80} from 1.5 μm to 100 μm to that at

$r_{80} = 25$ μm were presented graphically, although it is not clear if these rates (presumably proportional to fluxes) referred to rates per unit interval of radius or per unit logarithmic interval of radius (based on *Ling et al.* [1980] it was probably the former). In subsequent papers [*Ling et al.*, 1980; *Ling*, 1993] ratios were presented of the fluxes of drops in size ranges of r_{80} centered at 2.5, 10, 20, 35, and 75 μm (the values listed for the first size range were slightly different in the two papers) to the flux of drops with r_{80} in the size range 5-15 μm, but not the fluxes themselves, limiting confidence in any quantitative interpretation of these results. These investigations are discussed further in §5.7.1.

Comparison of the shapes of the size-dependent along-wind fluxes from different investigations (Fig. 25) demonstrates that there is considerable variation in these shapes from study to study, with the along-wind fluxes measured in the laboratory study of *Wu* [1973] and in the field study of *Wu et al.* [1984] exhibiting a much stronger decrease with r_{80} than those of other studies—a factor of 300 to 1000 between $r_{80} = 50$ μm and $r_{80} = 100$ μm compared to a factor of ~3 for the other investigations. The reasons for such differences between investigations are not known, and differences may have arisen both from different mechanisms of production and from different experimental conditions. It cannot be confidently stated how accurately any of the laboratory shape functions represents the size dependence of the along-wind flux over the ocean. Additionally, it is questionable how accurately the shape functions resulting from the field measurements of *Wu et al.* [1984], for which U_{10} did not exceed 8 m s^{-1}, represent the size dependence of the along-wind flux under conditions for which spume drop production would be expected to be important, as noted above (§4.2). To these concerns must be added uncertainties in converting the size dependence of the along-wind flux distribution to the size distribution of the interfacial SSA production flux arising from the size dependence of $\bar{L}_p(r_{80})$, which is also expected to differ from experiment to experiment.

Several investigations also reported measurements of the horizontal component of the drop velocities. The mean horizontal velocity components of the drops measured by *Wu* [1973] at five heights was slightly less than the mean wind speed at the several heights, increasing slightly with height. For example, for $U_{lab} = 12.62$ m s^{-1} the mean horizontal component of drop velocity increased from near 8 m s^{-1} at 5 cm to near 10 m s^{-1} at 15 cm. However, the mean horizontal component of drop velocity was calculated as an average over all drops and thus would be strongly weighted by drops in the range $r_{80} = 40$-50 μm, which comprised the majority of the drops measured (Fig. 25). As the heights for which these velocities were measured were within the range for which the vertical dependence of the wind speed was described by a logarithmic profile that would have increased with increasing height comparably to the reported increase in drop speed, a comparison of the mean horizontal velocity component of the drops to the mean wind speed at the same height would have been more relevant to considerations of velocity entrainment. *Koga* [1981] reported that drops had been accelerated to roughly half the local wind speed (i.e., they would have been traveling ~6 m s^{-1}) before colliding with the surface; however, no size dependence of drop speed was reported. Horizontal velocity components of drops measured by *Anguelova et al.* [1999] ranged from 1.5 m s^{-1} to 5.4 m s^{-1}, who reported that there was no evident dependence on drop size. This is not unexpected, as drops would have been detected at various times after they had left the surface and at various heights above the surface. Hence these investigations have provided little quantitative information on mean horizontal velocities attained by drops or on mean horizontal distances they traveled, or on the dependencies of these quantities on drop size.

4.3.6.2. Considerations regarding interpretation of wind/wave tunnel experiments. As drops produced in wind/wave tunnels result from the action of the wind on the water, it might be expected that the production mechanisms and the interfacial fluxes of these drops would be similar to those over the ocean. However, there are a variety of concerns regarding the applicability of the results of these experiments to the ocean, regarding extrapolation of laboratory wind speeds to oceanic equivalents, the ability of wind/wave tunnels to accurately simulate oceanic production, and differences between different laboratory experiments, including which drop formation mechanisms operate.

A key issue in interpreting results of these experiments and relating them to oceanic SSA production is the correspondence between a laboratory wind speed U_{lab} and its oceanic equivalent characterized by U_{10}. In several instances threshold laboratory wind speeds were reported above which a certain phenomenon would occur (Table 18), but in the absence of a relation between U_{lab} and U_{10}, the extent to which this information can be used to obtain meaningful estimates of SSA production over the open ocean is limited. The quantities U_{lab} and U_{10} certainly cannot be equated. It has sometimes been assumed that the wind friction velocity u_* can be used to relate these two quantities, on the assumption that the wind friction velocity in the laboratory corresponds to the same wind friction velocity over the ocean, but this has not been established. Moreover, it would seem unlikely that a single parameter such as u_* could accurately characterize all aspects of the wave spectrum and properties relevant to wave breaking and drop production over such differing scales, especially the occurrence, crest length, and heights of breaking waves; hence the ability of such a parameter to relate oceanic and laboratory data is doubtful, as noted by several investigators [*Wu*, 1973; *Hwang et al.*, 1990; *Wu and Hwang*, 1991; *Wu*, 2000b]. In some instances laboratory wind speeds were converted to oceanic equivalents by extrapolation of the logarithmic velocity profile in the laboratory to 10 m, but such a procedure implies that the roughness length (§2.4) is the same for the laboratory and the ocean, an assumption for which little basis has been provided.

A second key issue is the ability of laboratory experiments to accurately simulate the complexities and scales of air flow over the ocean, the production and behavior of waves and wave breaking, bubble injection into the sea, and white-cap formation, as noted by several investigators [*MacIntyre*, 1974b; *Jones and Kenney*, 1977; *Thorpe*, 1984a; *Bortkovskii*, 1987, pp. 37, 83; *Apel*, 1994; *Wu*, 2000b]. Key concerns are the inherent small size and limited fetches of wind/wave tunnels; entry conditions and wall effects; confinement of wind and waves; the representation of the heights, breaking properties, ages (i.e., state of development), and spectra of waves, especially long waves and large breaking waves, which are necessary to model whitecaps over the ocean and to produce spume drops; representation of the vertical transport of bubbles in the water (most of the investigations considered were performed in tunnels in which the water depth was less than 1 m); and matching the level and structure of turbulence and gustiness of the wind. Additionally, the distribution of SSA particles in the along-wind direction is likely to be much more uniform in a wave tank than over the ocean [*Wu*, 1979a].

Perhaps the greatest concern is that wave heights in wind/wave tunnel measurements, typically 5-10 cm (Table 18), are one to as much as two orders of magnitude less than what would be expected in oceanic conditions, and hence that wave breaking and associated phenomena such as drop production cannot be expected to be quantitatively or even qualitatively similar to those in the open ocean. For instance, *Wu* [1973] discussed the occurrence of spray when "waves break violently," and reported that whitecapping was observed at every wave crest when the tunnel wind speed was greater than 11.5 m s^{-1}, but as measurements were reported at 4 cm above the water surface, it can be concluded that the wave heights did not exceed this value and that "violent" breaking as referred to these laboratory experiments probably did not correspond to what is common with oceanic waves.

In addition to these concerns, differences between experiments in different laboratories on the dependencies on fetch and laboratory wind speed, which in many instances appeared to depend on the specific experimental design employed, make it difficult to argue for one data set over another. The only wind/wave tunnel experiment from

Table 18 reporting size-dependent along-wind fluxes in a medium other than freshwater was that of *Lai and Shemdin* [1974], who reported differences in shapes of the size distributions between along-wind fluxes of saltwater and freshwater drops. This result calls into question the applicability to oceanic drop production of other wind/wave tunnel experiments using freshwater. Perhaps more importantly, the mechanisms for drop production appeared to differ among the different experiments, with resultant uncertainty as to which if any of the mechanisms would correspond to mechanisms operating in the ocean; knowledge of the mechanism is important if laboratory measurements are used to infer oceanic fluxes because only for situations in which the formation mechanism is the same can extrapolation be justified. Indeed, considerable controversy has arisen concerning the drop formation mechanism. These concerns are perhaps best conveyed by briefly summarizing the characterizations of the several investigators of the drop production mechanisms occurring in their studies and the conflicting interpretations given by investigators of their own work and the studies of others.

Moore and Mason [1954] restricted their measurements to conditions for which there was not appreciable "spray produced by shearing off of the wave crests"; thus their results apparently did not include spume drops. *Toba* [1961b] hypothesized that most of the drops in his laboratory experiments were film drops and jet drops (which he stated as corresponding to the peaks at r_{80} of 5 μm and 15 μm, respectively, in the representation $dF_{int}(r_{80})/dr_{80}$ resulting from the bursting of bubbles formed by merging of wave tops, and he hypothesized further that this mechanism was not operable until U_{10} exceeded ~13 m s^{-1}. *Toba* also concluded that the occasional drops with $r_{80} > 250$ μm were produced by splashing of water from wave crests. *Wu* and coworkers [1979a, 1981c, 1982a; *Wu et al.*, 1984; *Wu*, 1990c; *Taylor and Wu*, 1992] have repeatedly argued that most spray observed in laboratory investigations, including those of *Wu* [1973], *Lai and Shemdin* [1974], and *Wang and Street* [1978a,b], consisted of jet drops, although in his 1974 paper, *Wu* suggested that the drops in the experiment of *Wu* [1973] were produced by both mechanical tearing of wave crests and bursting bubbles. Later *Wu* [1993] concluded, based on differences in the shapes of the size distributions of the along-wind fluxes, that in the laboratory measurements of *Wu* [1973] the spray was caused by wave tearing (i.e., the drops were spume drops), whereas in the laboratory experiments of *Lai and Shemdin* [1974] and of *Wang and Street* [1978a,b] the spray was caused by bubble bursting. *Lai and Shemdin* [1974] stated that the exact mechanism of drop formation was uncertain, that is, that the drops in their study could have been jet drops or spume drops. *Wang and Street* [1978a] suggested the drops in their experiments and those of *Lai and Shemdin* [1974] were bubble-produced drops, and that most of those of *Wu* [1973] were also bubble-produced drops, although some of the drops observed by *Wu* at the highest wind speeds may have been spume drops. In contrast, *Monahan et al.* [1983b, 1986] based their spume drop formulation (§5.7) on laboratory measurements of *Wu* [1973] and of *Lai and Shemdin* [1974], and *Andreas* [1990b, 1992, 1994a, 1998] and *Andreas et al.* [1995] likewise concluded that the drops produced in the experiments of *Wu* [1973] and *Lai and Shemdin* [1974] were spume drops. Such an array of conclusions makes it difficult to assign a particular production mechanism to a given experiment with any confidence, and hence to extrapolate wind/wave tunnel results to oceanic SSA production.

Several investigators have calculated and reported number concentrations from along-wind fluxes based on assumptions regarding the horizontal velocity components of the drops. Because of the time required for the horizontal velocity of a drop of a given size to attain a given fraction of the velocity of the surrounding air, calculation of the number concentration from the along-wind flux using the mean horizontal wind speed at a given height will result in underestimation of the concentration, the amount by which depends on particle size. Despite this situation, various assumptions have been made for the mean horizontal velocity component of drops of a given size: the tunnel wind speed [*Moore and Mason*, 1954], the wind speed at an arbitrary height of 10 cm above the water in the tunnel [*Toba*, 1961b], the local wind speed at the height of the drop [*Lai and Shemdin*, 1974], and 75% of the local wind speed at the height of the drop [*Wu*, 1974; *Wang and Street*, 1978a], this latter assumption based on the mean drop velocities measured by *Wu* [1973] at various heights (despite concerns with this result such as those presented above). None of these estimates takes into account the dependence of the mean drop velocity on size. *Koga and Toba* [1981] assumed that the horizontal velocity component of a drop at a given height was equal to the average of the horizontal components of the drop at this height on the way up and on the way down during free-fall motion. However, the velocities used were erroneous because the investigators did not account for relative motion of drops in the along-wind and vertical directions (that is, they incorrectly assumed that the horizontal and vertical motions are independent; §2.6.4) and they did not account for drag in the vertical direction, which would have played a key role in controlling the drop motion. These investigators presented number concentrations which they fitted to power laws with different exponents in different size ranges, the difference being attributed to different production and dispersion mechanisms for smaller and larger drops. However, in view of the concerns noted, little confidence can be placed in any of the reported number concentrations.

An assessment of the situation by *Mestayer and Lefauconnier* [1988] offers a rather pessimistic view of the ability of such laboratory studies to simulate drop production in the ocean: "The quantity and spectral distribution of the released droplets depends on the wave geometry and dynamics, the entrainment of the droplets depends on the air flow dynamics in the viscous sublayer and in the wave troughs, and the droplet behavior depends on the temperature and RH close to the surface. So if by 'spray production' we mean 'quantity and spectral distribution of sprays entrained above the waves,' it is futile to try to simulate marine spray production in the laboratory." While this assessment is perhaps too harsh, the ability of laboratory waves of such small dimensions to accurately simulate SSA production in the ocean would certainly seem questionable to anyone who has been on the deck of a ship at sea when waves are breaking at heights of several meters and been thoroughly drenched with spray.

4.3.7. Summary of Drop Production

Laboratory investigations of drop production from individual bursting bubbles, from laboratory whitecaps, and in wind/wave tunnels have been examined in the above sections. These investigations, together with field measurements of SSA concentration at heights near 10 m (§4.1) and near the sea surface (§4.2), have been responsible for virtually all present knowledge of oceanic SSA production, such as the size ranges in which different mechanisms operate, the relative importance of the different mechanisms in a given size range, and the dependencies of the numbers and sizes of drops produced on factors thought to be controlling. While in some instances conclusions can be reached concerning the types of drops that predominate in different size ranges, there is much that remains unknown in many aspects of SSA production, and there has been considerable controversy over several issues, some of which is examined here. Additionally there are concerns regarding the extent to which results from each of these types of experiments can be extrapolated to oceanic SSA production. These concerns and conclusions are reviewed here, and aspects in which little consensus has been reached regarding SSA production are noted.

Extrapolating results of laboratory experiments of drop production from bubbles bursting individually at a calm, flat liquid surface (§4.3.2; §4.3.3) to oceanic conditions requires the assumption that oceanic bubbles burst independently and yield the same number and sizes of drops as do their laboratory counterparts. Under actual conditions of drop production in the ocean, however, when large numbers of bubbles of a wide range of sizes formed by a breaking wave rise to the surface and burst, the surface is neither calm nor flat but vigorously churning. The bubbles do not behave independently but interact, colliding with one another, sometimes forming clusters or a layer of foam, and perhaps coalescing. Screening may occur whereby drops produced by bubbles below clusters may be prevented from reaching the atmosphere. Coalescence (although expected to be of minor importance in seawater; §4.4.1) changes the bubble size distribution and the amount of film area available for film drop production. It is expected that the bursting process, the energy available for drop production, and the partitioning of this energy, differ between bubbles that are in contact with other bubbles and those that burst individually. All of these considerations, together with the strong dependence (§4.3.3.3) of film drop production on the bubble residence time at the surface (and thus on the height of the bubble with respect to the surface when it bursts), provoke serious questions about the extrapolation of results from bubbles bursting individually to conditions after a wave breaks when large numbers of bubbles are jostling and sharing film faces at a disturbed surface before bursting.

Extrapolating results of experiments with laboratory whitecaps (§4.3.4) to oceanic conditions requires the assumption that the laboratory whitecaps accurately simulate oceanic whitecaps. As noted in §4.3.4.4 there are concerns with this assumption, arising mainly from the greatly different physical scales but also from the methods by which the laboratory whitecaps were produced. An additional concern is the assumption that any whitecap, regardless of its magnitude or mechanism of formation, can be characterized by a single quantity—its area (continuous laboratory whitecap) or its initial area (discrete laboratory whitecap), and that for any given r_{80} the SSA production flux (continuous laboratory whitecap) or the total SSA production resulting from such a whitecap (discrete laboratory whitecap) is directly proportional to this quantity.

Concerns with extrapolation of results of experiments in wind/wave tunnels have also been noted (§4.3.6.2), the basic concern being the assumption that wave breaking and drop production from laboratory waves with heights of a few centimeters accurately simulate those in the open ocean. An additional concern is the method by which wind speeds from the laboratory should be extrapolated to oceanic conditions. Although such experiments, when taken individually, permit a variety of conclusions, they do not provide a consistent picture when taken as a whole. For example, little consensus has been reached on even the mechanism of drop production in many of the experiments (§4.3.6.2).

Despite the above concerns regarding extrapolating results of laboratory experiments to the ocean, some general conclusions can be reached concerning the relative importance of SSA production mechanisms to SSA concentrations as a function of particle size over the ocean. However, many key quantities remain unknown. These considerations are examined here.

Bubble-produced jet and film drops can account for all small ($r_{80} \lesssim 1$ μm) and medium (1 μm $\lesssim r_{80} \lesssim 25$ μm) SSA particles present in the marine atmosphere. Upon consideration of measurements of size distributions of SSA concentration from multiple sources (Figs. 22a-f), it is difficult to discern an identifiable spume contribution to concentrations of SSA particles with $r_{80} \lesssim 25$ μm at heights near 10 m at $U_{10} \lesssim 15$ m s^{-1} (§4.1.4.3). The measurements of de Leeuw [1986a,b, 1987, 1990a] noted in §4.2 are especially pertinent in this regard, as these measurements did not suggest any evidence of spume drop production for $r_{80} \lesssim 15$ μm at heights as low as a meter or so above the sea surface at wind speeds up to 14 m s^{-1} (§4.1.4.3). Spume drop production has been documented in laboratory studies of drop production in wind/wave tunnels (§4.3.6.1), but several of the studies did not extend to sizes much below $r_{80} = 25$, and even in such experiments it was difficult to identify different distributions that could be identified in spume production vs. from bursting bubbles; additionally the types of the drops in many of the experiments were not well established (§4.3.6.2).

Based on considerations of jet drop production in the ocean (§4.3.2.4) and on the results of laboratory whitecap experiments (§4.3.4.3), it can confidently be concluded that few small SSA particles ($r_{80} \lesssim 1$ μm) present in the marine atmosphere are jet drops, as the ability of jet drops of these sizes to be produced is limited by constraints on bubble size and on the energy of drop ejection (§4.3.2.4). Hence most small SSA particles in the marine atmosphere are film drops. Formation of small SSA particles from secondary bubbles formed by the collision of spume drops with the surface has been hypothesized (§4.7.2), but little is known about this process and it is not likely to provide a major contribution to SSA production in this size range.

Whereas jet drop production appears fairly well understood, as least concerning the number, sizes, and ejection heights of such drops, film drop production has many unresolved issues. Over most of the range $r_{80} = 0.1$-1 μm, which provides the dominant contribution to atmospheric SSA particle concentrations, size distributions of film drops from individual bubbles are lacking. The number and sizes of film drops produced by a bursting bubble demonstrate much more variability, both among various investigations and within a single investigation, than do comparable quantities for jet drops (Figs. 26 and 27 vs. Figs. 30 and 29), frustrating attempts to quantify the dependence of N_{film} and the sizes of the film drops on R_{bub} and other controlling factors such as temperature. Additionally, these attempts are limited by the relatively small number of investigations that have been performed to determine these dependencies (Table 17), the variety of different patterns of film drops that have been observed, and different experimental techniques (which detect film drops of different sizes) that have been employed. If, as argued in §4.3.3.3, the bubble bursting process is affected by the experimental technique, then the ability to draw conclusions on the number, sizes, and heights of the film drops produced, and to apply these conclusions to the oceanic production of SSA production, would be seriously compromised. A major factor that would appear to call into question the ability of results from individual bubbles bursting in laboratory experiments to be extrapolated to oceanic conditions is the fact that the great majority of bubbles bursting after wave breaking in the oceans do so not in isolation but in close proximity to other bubbles with resulting uncertain effects of clustering, screening, coalescence, and other multiple-bubble processes, and their dependencies on temperature and other factors. To all of these considerations must be added the potentially large role played by surface-active materials in all aspects of drop production by bursting bubbles.

With increasing SSA particle size it can be concluded that jet drops comprise an increasingly greater fraction of the SSA particles present; however, no consensus has been reached on the value of r_{80} above which production of jet drops exceeds that of film drops. Based on the location of a minimum in the size distributions of SSA concentration from field measurements, *Toba* [1965a,b] concluded that this critical value of r_{80} occurs between 4.4 µm and 6.5 µm. In contrast, based on differences in size-dependent SSA concentrations measured over cold waters near Alaska and over warm waters near Hawaii, *Woodcock* [1972] concluded that this critical value of r_{80} occurred between 0.2 µm and 0.5 µm (*Woodcock and Duce* [1972] placed this value near $r_{80} \approx 0.25$ µm). Other estimates of this critical value of r_{80}, based primarily on laboratory results, are 0.8 µm [*Wu*, 1992a], 1.5 µm [*Bowyer*, 1984; *Monahan*, 1986], 2 µm [*Woolf et al.*, 1987], 2.5 µm [*Mestayer and Lefauconnier*, 1988], and 3.75 µm [*Cipriano and Blanchard*, 1981]. Implicit in these arguments is that this critical size is independent of wind speed and other metoeorological or environmental conditions.

Although jet drops provide the majority of SSA particles over much of the size range comprising medium SSA particles (1 µm $\lesssim r_{80} \lesssim$ 25 µm), large gaps in the understanding of jet drop production remain. The number, sizes, and heights of jet drops as a function of R_{bub} reported by different investigators agree fairly well over the size ranges which have been examined (§4.3.2.1; §4.3.2.2; §4.3.2.3), but relatively few data exist for seawater drops with $r_{80} < 25$ µm. Also, few data exist for temperatures below 20 °C, and the interesting temperature dependence of jet drop production reported by *Spiel* [1997a] is not fully characterized or understood.

Other issues of SSA production by the bubble mechanism remain unsolved. For instance, several investigators have attempted to determine whether more film drops or jet drops are produced at the ocean surface, but no consensus has been reached on this point. Measurements of SSA concentration show small drops ($r_{80} < 1$ µm), presumed to be film drops, to be much more numerous than larger drops (§4.1.4.2), but as the residence times of these smaller drops is much longer than those of larger drops, the greater concentrations may not reflect greater production. *Toba* [1961b], based on laboratory measurements of size distributions of bubble concentration and of the number of jet and film drops produced per bubble burst, concluded that many more film drops than jet drops are produced. In contrast, *Blanchard and Woodcock* [1980], based on similar arguments, estimated that 75 times as many jet drops as film drops are produced, though they conceded that this conclusion was based on estimates of film drop production that were low. Several laboratory investigations of SSA production from simulated whitecaps discussed above [e.g., *Cipriano and Blanchard*, 1981; *Cipriano et al.*, 1987; *Woolf et al.*, 1987] concluded that roughly an order of magnitude more film drops than jet drops are produced and that most of these film drops are very small ($r_{80} \lesssim 0.1$ µm). On the other hand, *Wu* [1989a], based on oceanic measurements of size distributions of bubble concentration and laboratory measurements of the number of jet and film drops produced per bubble burst, concluded that the number of jet drops produced was overwhelmingly greater than the number of film drops produced. This assertion prompted arguments from *Woolf* [1990] and *Blanchard and Syzdek* [1990a] that *Wu*, by omitting the peak in film drop production reported by *Blanchard and Syzdek* [1988] at $R_{bub} = 1$-1.25 µm (§4.3.3.2), had greatly underestimated the number of film drops produced (this is discussed further in *Wu* [1990a,b]). *Resch and Afeti* [1991], based on arguments similar to those of *Wu* [1989a], also concluded that the oceanic production flux of jet drops is greater than that of film drops.

Although the question of relative production of jet drops and film drops has been posed without regard to the application to which this information is to be used, this application is of key importance to the interpretation of experiments. For instance, film drops that are ejected at low angles and large jet drops (i.e., those with $r_{80} \gtrsim 25$ µm), both of which fall immediately back into the sea, contribute little if any to SSA concentration at any appreciable height. Likewise, small film drops ($r_{80} < 1$ µm), although they may dominate number concentrations, contribute little to atmosphere-ocean transfer of moisture and momentum, or to mass concentrations in the marine boundary layer. Thus it would seem that questions such as the total number of drops produced of each type without regard to size would be of little practical interest in the absence of a particular application.

With increasing particle size above $r_{80} \gtrsim 25$ µm spume drop production is thought to become increasingly more important; however, this mechanism is also poorly characterized. Virtually nothing is known concerning the numbers and sizes of spume drops produced at the ocean surface for a given set of conditions, or on the dependencies of these quantities on controlling factors. It can be concluded that this mechanism produces mainly large SSA particles, but neither the size nor the wind speed at which oceanic production of these drops commences can be stated with any confidence or with any basis in data. As noted in §4.3.6.2, little consensus has been reached concerning even the mechanism of drop production in laboratory wind/wave tanks. Likewise, little consensus has been reached concerning whether large SSA particles ($r_{80} \gtrsim 25$ µm) above the ocean are predominantly jet drops or spume drops. *Monahan* [1968] assumed that the drops he measured, with values of r_{80} up to ~300 µm, were jet drops (§5.9), although based on the sizes and heights of these drops at least some of them were probably spume drops (§4.2). *Preobrazhenskii* [1973] concluded that spume drops occurred with r_{80} down to 7.5-12.5 µm, based on the premise that the drops he detected near wave tops had recently been formed. As noted above in §4.3.6, *Wu* and coworkers [1979a, 1981c, 1982a; *Wu et al.*, 1984; *Wu*, 1990c; *Taylor and Wu*, 1992] have repeatedly argued that most spray over the open ocean [*Monahan*, 1968; *Preobrazhenskii*, 1973; *Wu et al.*, 1984] consists of jet drops, especially for wind speeds less than 15 m s^{-1}. *Wu et al.* [1984] argued that spume production would be important only at wind speeds U_{10} greater than 15 m s^{-1}, but later *Wu* [1990c] and *Taylor and Wu* [1992] proposed that spume drops were the source of the maximum in the number concentration of SSA particles at a height near 1 m above the mean sea surface that had been reported by *de Leeuw* [1986b, 1987] at wind speeds as low as $U_{10} = 7.5$ m s^{-1} (§4.2), concluding that at $U_{10} = 13$ m s^{-1} about a third of the drops, down to $r_{80} \approx 5$ µm, were spume drops. Later, however, *Wu* [1993] reversed his arguments and concluded that in the oceanic measurements of *Wu et al.* [1984] the spray was caused by wave tearing (i.e., the drops were spume drops), based on the temporal correlation of the fluctuations in measured along-wind flux with the passing of wave crests (§4.2). Based on comparison of size distributions of number concentration reported by *de Leeuw* [1986b] and by *Smith et al.* [1993], *Wu* [2000b] argued that spume drop production started to occur at U_{10} near 15 m s^{-1} for r_{80} as low as 5 µm. *Andreas* [1990b, 1992, 1994a, 1998] and *Andreas et al.* [1995] likewise concluded that the drops measured by *Wu et al.* [1984] were spume drops, and further that drops with $r_{80} > 10$ µm were predominantly spume drops, presumably at all wind speeds. Based on concentration measurements over the ocean, *Smith et al.* [1993] hypothesized that the two

modes in their formulation of $f_{\text{eff}}(r_{80})$ centered about $r_{80} = 2.5$ µm and $r_{80} = 10.7$ µm (near the r_{80} values presumed by *Toba* [1961b] to correspond to film drops and jet drops in the representation $dF_{\text{int}}(r_{80})/dr_{80}$) corresponded to jet drops and spume drops, respectively, implying that spume drop production is important for drops as small as $r_{80} \approx 7$ µm and for wind speeds as low as $U_{10} = 5$ m s^{-1}.

Extremely large SSA particles, those with r_{80} greater than several hundred micrometers, are almost certainly spume drops, as noted by several investigators, as jet drops of these sizes are not produced (§4.3.2.2) and no film drops this large have been reported (§4.3.3.1). *Koga* [1981], for instance, based on laboratory experiments using freshwater, concluded that most of the drops with $r_{80} > 125$ µm he observed were formed by the spume mechanism. However, few quantitative measurements of drops of these sizes have been reported; most along-wind flux measurements have been limited to drops with $r_{80} \lesssim 100$ µm, and only the laboratory investigations of *Koga and Toba* [1981] and of *Anguelova et al.* [1999] extended much above this range. Concerns with extrapolation of laboratory results to the open ocean preclude quantitative criteria in terms of sizes and wind speeds denoting onset of conditions for which production of spume drops becomes more prevalent than that of jet drops.

In summary, bubble bursting is the main mechanism for production of small and medium SSA particles under most oceanic conditions, and hence this mechanism is the most important with regard to the effective SSA production flux. Most small SSA particles ($r_{80} \lesssim 1$ µm) are film drops, and with increasing drop size over the range comprising medium SSA particles (1 µm $\lesssim r_{80} \lesssim 25$ µm) an increasing fraction of the particles are jet drops. Both the jet drop mechanism and the spume drop mechanism contribute to the production of large SSA particles, but whereas it can be confidently concluded that most SSA particles with r_{80} greater than several hundred micrometers are spume drops, for many drop sizes little confidence can be placed in quantitative assessment of the relative importance of these mechanisms.

4.4. BUBBLES IN THE OCEANS

Bubbles in the oceans are the source of the vast majority of small and medium SSA particles, and of many of the large SSA particles also. In principle, if the flux of bubbles arriving and bursting at the ocean surface were known, then the size-dependent production flux of jet and film drops could be obtained using the bubble method (§3.4) from knowledge of the number and sizes of drops of these types (§4.3.2; §4.3.3) produced per bursting bubble, under the assumption that all bubbles burst independently (although spume drop production would have to be determined by other methods). Jet drops in the size range contributing to the effective production flux (i.e., $r_{80} \lesssim 25$ μm) are produced primarily from bubbles with $R_{bub} \lesssim 0.5$ mm (Fig. 27), and film drops are produced mainly from bubbles with $R_{bub} \gtrsim 0.5\text{-}1$ mm (Fig. 30). However, the flux of bursting bubbles at the surface under oceanic conditions does not appear to have been measured directly, although in principle this flux might be measured by photographing the bubbles that collect under a transparent material of known area resting on the sea surface (similar to a bubble trap discussed below). In the absence of such measurements, this flux has been approximated as the product of the size-dependent bubble concentration at the surface $n_{bub}(R_{bub}, 0)$ and the size-dependent terminal velocity of bubbles in still water $v_{bub}(R_{bub})$, where $n_{bub}(R_{bub}, 0)$ has typically been determined by extrapolation of measurements of the size-dependent bubble concentration $n_{bub}(R_{bub}, z)$ fitted to some function of depth z to the surface ($z = 0$). Alternatively, in principle the flux of bubbles to the surface might be calculated from the flux or concentration of bubbles at the depth of measurement taking into account the change in radius of bubbles due to decreased hydrostatic pressure as the bubble rises and gas exchange with the surrounding water. Both approaches require knowledge of size distributions of bubble concentration in the ocean, bubble rise speeds, and gas-exchange properties of bubbles to evaluate the interfacial production flux of bubble-produced drops by the bubble method (§3.4). Size distributions of bubble concentrations at various depths have been determined by both direct and indirect methods, and bubble rise velocities and gas-exchange properties have been measured under laboratory conditions.

The concentration of bubbles of a given volume-equivalent radius R_{bub} (§4.3.1) at a given depth is the result of several processes: initial bubble formation resulting from entrainment of air associated with breaking of waves, coalescence and breakup of bubbles, gas exchange and dissolution, vertical ascent, organized flow of seawater on several scales, and turbulent mixing. Any factor that affects any of these processes will affect bubble concentrations. As the wind blows across the water surface it creates waves which may eventually break. Upon breaking such waves entrain large quantities of air into the ocean surface waters which result in large numbers of bubbles. Laboratory experiments on freshwater breaking waves by *Lamarre and Melville* [1991] indicated that up to 40% of the energy in a wave is lost by breaking and that nearly half of this energy is expended in entraining air which becomes the bubble plume. The volume fraction of air, known as the void fraction, can be up to several tens of percent immediately under a breaking wave, although because of the rapid ascent of large bubbles to the ocean surface this void fraction decreases rapidly with time. Typical background values of the void fraction in the upper meter of the ocean surface waters are

$\sim 10^{-7}$ to $\sim 10^{-6}$, increasing by several orders of magnitude during passage of a bubble plume from a breaking wave [*Lamarre and Melville*, 1991, 1992, 1994a; *Vagle and Farmer*, 1992; *Deane*, 1997; *Farmer et al.*, 1998; *Stokes and Deane*, 1999; *Deane and Stokes*, 1999; *Terrill and Melville*, 2000; *Terrill et al.*, 2001; *Bowyer*, 2001; *Deane and Stokes*, 2002; *Stokes et al.*, 2002].

Bubble clouds resulting from breaking waves can be injected to depths of several meters or more. *Kanwisher* [1963] reported that under some conditions bubbles under a breaking wave in the ocean were acoustically detected as deep as 20 m and that considerable sound reflection existed for as long as 20 s afterwards. Depths of oceanic bubble clouds reported by *Thorpe* [1982] generally increased with increasing wind speed and were typically 1-3 m, and sometimes as much as 8 m, for wind speeds of 10 m s^{-1}. Similar results were reported by *Thorpe* [1986], with bubble clouds extending down to more than 12 m at wind speeds up to 14 m s^{-1}. Average depths of bubble penetration in the sea reported by *Crawford and Farmer* [1987] for wind speeds 8-12 m s^{-1} were around 4 m, with maximum depths near 10 m. Bubble clouds reaching more than 10 m deep for wind speeds around 12 m s^{-1} were reported by *Farmer et al.* [1993]. Field measurements employing acoustic scattering have demonstrated that at low wind speeds bubble clouds are patchy but that with increasing wind speed waves break more frequently than bubble clouds dissipate, resulting in a nearly continuous layer of bubble cloud [*Thorpe*, 1982; *Thorpe et al.*, 1982; *Thorpe and Hall*, 1983].

Although an analogy between bubbles in the ocean and SSA particles in the atmosphere may seem apt [*Bohren*, 1987, Chap. 1], there are major differences. Sea salt aerosol particles are essentially conserved in the absence of precipitation for time scales much greater than those characterizing passage of breaking waves (such as the time interval between whitecaps, which is of order seconds to minutes) or of times of measurements (also typically seconds to minutes). Consequently temporal and spatial inhomogeneities in SSA concentrations are expected to be relatively small; times characterizing the change of such concentrations are of order hours to days for a wide range of particle sizes (§2.9.6). Additionally, because of turbulence and the relatively slow sedimentation velocities of most SSA particles that attain heights of 10 m (terminal velocities of SSA particles with radii up to 25 μm at 80% RH are less than 10 cm s^{-1}; Eq. 2.6-8), SSA concentrations at heights near 10 m are expected to vary little with height over a wide range of particle sizes (§2.9.4), and little horizontally. In contrast, bubble lifetimes are of order seconds to minutes, generally resulting in patchy distributions with characteristic times of around a minute and characteristic depths of several meters [*Thorpe*, 1982, 1986; *Farmer and*

Vagle, 1989; *Dahl and Jessup*, 1995], although becoming increasingly continuous horizontally with increasing wind speed as noted above. Bubbles with $R_{bub} \gtrsim 0.5$ mm rapidly rise to the surface; for instance, a bubble with $R_{bub} = 0.5$ mm will rise a distance of 1 m in ~3 s in still water at 20 °C (§4.4.2.1). Although smaller bubbles are dispersed by turbulence and may be carried to greater depths, many of them dissolve because of the combined effects of hydrostatic pressure and surface tension (§4.4.2). Thus extremely large temporal and spatial variation is expected for bubble concentrations over times of seconds to minutes and over differences in depth of a meter or so, especially immediately after formation of a whitecap by a breaking wave, when the concentration of bubbles of all sizes in the ocean surface waters markedly increases.

The intermittent nature of the bubble field is illustrated in Plate 1 of *Terrill et al.* [2001], for example, in which time series of size distributions of bubble concentration and of void fraction show spikes in these quantities several orders of magnitude greater than background values over times of order tens of seconds. Likewise, Plate 1 of *Stramski and Tegowski* [2001], which presents time series of bubble concentrations as a function of depth, illustrates the sporadic occurrence of dense bubble clouds entrained to depths more than 5 m every few minutes, between which a nearly uniform bubble cloud extends to depths of ~3 m, but with the water below being nearly free of bubbles. Because of the rapid temporal evolution of the size distribution of bubble concentration, the magnitude and shape of the size-resolved concentrations depend strongly on the time after wave breaking, with the concentrations being greatest at the time of or soon after wave breaking. The rapid change in the shape of size-dependent bubble concentrations was noted by *Deane and Stokes* [2002], who described processes occurring during the temporal evolution of a bubble plume after wave breaking, and is illustrated in Fig. 4 of that reference, which presents measurements of size distributions of bubble concentration taken 1.5 s apart immediately after a breaking wave event in seawater in a laboratory wave tunnel 0.5 m deep, showing changes in concentrations of bubbles with $R_{bub} \gtrsim 1$ mm of more than two orders of magnitude. Similar results are illustrated in Fig. 5 of *Bezzabotnov et al.* [1986] for oceanic bubble concentrations. Likewise, Fig. 10 of *Vagle and Farmer* [1992], which displays the temporal behavior of the size distribution of bubble concentration during the passage of a bubble plume past the location of measurement, shows great temporal variability in the shape and magnitude of the size distribution, especially for larger bubbles. Because of these considerations and as a consequence of inevitable differences from wave to wave, large differences can be expected between size distributions of bubble concentration arising from a single breaking wave as a function of time

and depth, between size distributions of bubble concentrations from different waves, and between individual size distributions and averages over large numbers of such distributions or over long sampling times. This inherent variability makes it difficult to present a quantitative description of size distributions of bubble concentration.

Recognition of the distinction between size distributions of bubble concentration measured immediately subsequent to breaking of a wave and averages of these distributions over times of minutes is essential to application of the bubble method. If a given measurement is dominated by an individual breaking wave event, then use of such a size distribution of bubble concentration to estimate the interfacial SSA production flux would require a scaling factor such as the fraction of time a given location experiences such an event, or, equivalently, the whitecap fraction W (§4.5). Alternatively, measurements of size distributions of bubble concentration at an appreciable depth or at an appreciable time after wave breaking will greatly underestimate concentrations of large bubbles and thus cannot be expected to accurately describe the flux of such bubbles to the surface. These considerations raise inherent questions as to the extent to which reported size distributions of bubble concentration at a given depth, whether of individual measurements or averages of several measurements, can be used to infer the flux of bubbles bursting at the sea surface.

In addition to the time following wave breaking, other factors can affect bubble concentrations at a given location. Langmuir circulation [*Langmuir*, 1938; *Thorpe*, 1984c, 1985; *Crawford and Farmer*, 1987; *Zedel and Farmer*, 1991], in which the surface layers of two adjacent masses of water flow toward each other and form windrows—alternating vortices aligned in the direction of the wind with characteristic lengths of a few tens of meters, may cause bubbles to collect at and under these windrows. Likewise freely floating instruments that are used to measure bubble concentrations may yield results that are not representative of most of the ocean surface waters for the same reason [*Thorpe*, 1984b]. Additionally, there is evidence of low background concentrations of bubbles at low wind speed indicative of alternative bubble production mechanisms, such as biological activity; the importance of such bubbles to SSA production depends on the relative magnitude of this background population to that of bubbles formed by breaking waves.

Bubble concentrations in the ocean and their size dependence have been investigated with regard to a number of other topics of geophysical interest in addition to their role in producing SSA particles. Bubbles exchange gas with the surrounding water and affect the saturation of the upper ocean with regard to atmospheric gases, particularly trace gases, and affect the exchange of these gases between the atmosphere and the ocean[1]. Subsurface bubbles scatter light and thus affect ocean color and influence remote sensing of ocean color and/or of atmospheric aerosols [*Zhang et al.*, 1998; *Flatau et al.*, 2000; *Stramski and Tegowski*, 2001; *Terrill et al.*, 2001; *Zhang et al.*, 2002; *Yan et al.*, 2002]. Bubbles are efficient scatterers and absorbers of underwater sound, and as such they have been used as tracers for near-surface currents and breaking waves [*Thorpe et al.*, 1982; *Thorpe*, 1982; *Thorpe and Hall*, 1983; *Thorpe*, 1986; *Farmer and Vagle*, 1989]. Additionally, bubbles are a source of ocean noise and affect the transmission of underwater sound[2]; consequently, bubbles can also interfere with acoustic surveys for fish abundance [*Dalen and Løvik*, 1981], for example. *Lamarre and Melville* [1994b] reported that a newly created bubble plume at a depth of 0.25 m reduced the speed of sound from ~1500 m s^{-1}, characteristic of bubble-free ocean surface water, to ~700 m s^{-1}, and that in a bubble plume 25 s old the speed was ~1100 m s^{-1}. Bubbles transport surface-active materials and other substances to the sea surface (§2.5.2; §4.3.4.4), influencing the surface composition and properties; these materials affect the bubble bursting process (§4.3.2.3; §4.3.3.3) and may consequently be injected into the atmosphere along with SSA particles in amounts that greatly exceed their bulk concentrations (§2.5.2).

The importance of surface-active substances to various aspects of drop production and behavior has been noted throughout this review (§2.3.4; §4.3.2.3; §4.3.3.3; §4.3.4.3; §4.3.4.4), and the effects of these materials on bubble behavior in the ocean and the laboratory is no less pronounced. Nearly every laboratory investigation of bubble motion and gas exchange calls attention to the potentially large effects of surface-active materials, and it is difficult to determine the importance of these materials to bubbles under breaking waves, when large numbers of bubbles are rising to the surface and scavenging surface-active substances in various amounts depending on the size, depth, and time after the breaking event.

In this section, oceanic measurements of size distributions of bubble concentration are examined and the factors that affect these concentrations are examined. Pertinent properties of individual bubbles such as terminal velocity in still water and rate of exchange of gases with the surrounding water, and the dependence of these properties on bubble size and other factors, are also reviewed.

4.4.1. Size Distributions of Bubble Concentration in the Ocean

4.4.1.1. Measurement methods.
Several methods have been used to measure size distributions of bubble concentration in the ocean, the primary ones being bubble traps, photographic techniques, and acoustic techniques.

Measurements using bubble traps [*Blanchard and Woodcock*, 1957; *Kolovayev*, 1976] count and size the bubbles that have risen to a transparent surface at the top of the trap, from which concentrations are calculated from size-dependent rise velocities. This technique necessarily favors large bubbles, which rise rapidly, while giving smaller bubbles time to dissolve or to coalesce, thus underestimating their numbers [*Johnson and Cooke*, 1979; *Johnson*, 1986; *Walsh and Mulhearn*, 1987]. However, as this approach determines bubble fluxes rather than concentrations, results from this method may be more applicable to calculating the flux of bubbles that burst at the surface, the ultimate goal of such measurements for considerations of SSA production, rather than concentration measurements themselves. In principle this approach could be applied to directly determine the size distribution of the bubble flux at the surface, for example by video recording, sizing, and counting bubbles that arrive at the surface in a known area or perhaps by counting bubbles that collect under a transparent material of known area. However, measuring and sizing the large numbers of bubbles rising to and bursting at the surface under conditions for which the flux is desired (i.e., after a breaking wave) might prove challenging.

Photographic methods [*Johnson and Cooke*, 1979; *Walsh and Mulhearn*, 1987; *Loewen et al.*, 1996; *Stokes and Deane*, 1999; *Deane and Stokes*, 1999; *Bowyer*, 2001; *Deane and Stokes*, 2002; *Stokes et al.*, 2002; *de Leeuw and Cohen*, 2002] determine bubble concentrations from photographs or video pictures of a region of known volume in the ocean. Some of these studies used results of theoretical investigations of light scattering by air bubbles in water, such as those by *Davis* [1955], *Kingsbury and Marston* [1981], *Marston et al.* [1982], and *Kokhanovsky* [2003]. In situations for which a series of photographs were taken there was often large variability from one photograph to the next, leading to concerns over the meaningfulness of mean values of concentration obtained from such photographs. *Walsh*

---

[1] *Kanwisher* [1963]; *Atkinson* [1973]; *Thorpe* [1982]; *Merlivat and Memery* [1983]; *Broecker and Siems* [1984]; *Thorpe* [1984a,b]; *Memery and Merlivat* [1985]; *Thorpe* [1986]; *Woolf and Thorpe* [1991]; *Wallace and Wirick* [1992]; *Farmer et al.* [1992]; *Woolf* [1993]; *Keeling* [1993]; *Asher et al.* [1996]; *Woolf* [1997].

[2] e.g., *Strasberg* [1956]; *Wenz* [1962]; *Novarini and Bruno* [1982]; *Farmer and Lemon* [1984]; *Wille and Geyer* [1984]; *Kerman* [1988a]; *Crowther* [1988]; *Prosperetti* [1988]; *Medwin and Beaky* [1989]; *Hall* [1989]; *Medwin and Daniel* [1990]; *Longuet-Higgins* [1990b]; *Loewen and Melville* [1991]; *McDaniel* [1993]; *Lamarre and Melville* [1994a,b]; *Terrill and Melville* [1997]; *Hwang and Teague* [2000].

and Mulhearn [1987], for example, presented an example in which eight out of 32 photographs in a series taken 32 s apart contained no bubbles, eight contained two bubbles, one contained 47 bubbles, and one contained 66 bubbles (the high values occurred when bubble clouds passed the measurement location), with an average of 7 bubbles per photograph but with large standard deviation owing to the small number of photographs with large numbers of bubbles. Photographic methods suffer from inaccurate determination of depth of field, small sample volume (resulting in few bubbles counted), inability to distinguish bubbles from other particles (or to detect bubbles that have adsorbed substances on their surfaces), screening by larger bubbles, and lack of resolution of smaller bubbles ($R_{bub} \lesssim 0.06$-0.2 mm, depending on the system employed), often resulting in underestimating concentrations of these bubbles [*Kingsbury and Marston*, 1981; *MacIntyre*, 1986; *Walsh and Mulhearn*, 1987; *Medwin and Breitz*, 1989; *Breitz and Medwin*, 1989].

Acoustic methods [*Medwin*, 1970, 1977a; *Medwin and Breitz*, 1989; *Breitz and Medwin*, 1989; *Vagle and Farmer*, 1992, 1998; *Farmer et al.*, 1998; *Terrill and Melville*, 2000; *Terrill et al.*, 2001] are indirect methods that employ scattering or attenuation of sound at various frequencies to infer bubble concentrations at the sizes at which these frequencies are resonant (which also depend on depth). The resonant acoustic scattering cross section of a bubble is several orders of magnitude greater than its geometric (and thus optical) cross section, and its resonant acoustic absorption cross section is many times larger still; in addition, the acoustic scattering cross section is 10^{10}-10^{11} times greater than that of a rigid sphere, or of other common marine bodies, of the same radius, which scatter in the fourth-power Rayleigh regime [*Medwin*, 1970, 1977a; *Medwin and Breitz*, 1989; *Breitz and Medwin*, 1989]. The measured acoustic signal is inverted to provide a size distribution of bubble concentration under assumptions about the contributions of off-resonant bubbles [e.g., *Commander and Moritz*, 1989; *Commander and McDonald*, 1991]. Earlier methods employing this technique [*Medwin*, 1970] were based on the assumption that only bubbles at or near resonance contribute to scattering or absorption. Later modifications [*Medwin*, 1977a] included the contributions from off-resonant bubbles, but this contribution to the signal increased the extinction cross section only by a factor of less than two [*Medwin*, 1977b]. The assumptions that only bubbles at or near resonance contributed in any appreciable manner was demonstrated by *Commander and Moritz* [1989] to lead to an overestimate of bubbles with $R_{bub} \lesssim 0.05$ mm by up to several orders of magnitude, the amount increasing with decreasing R_{bub}, as well as to a slight overestimation for larger radii. This overestimation in both size ranges depends upon the shape of the size distribution, making inversion of the acoustic signal to arrive at a size distribution a difficult problem, especially when only a few frequencies are used. Later techniques have employed broadband acoustic pulses [e.g., *Terrill and Melville*, 2000; *Terrill et al.*, 2001]. Other concerns with acoustic methods include differences in results determined using scattering and attenuation, the possibility of the measuring devices themselves interacting with the acoustic signal and/or producing bubbles, limited dynamic ranges of measurement devices, uncertainties in determining sampling volume, unknown values for absorption and scattering coefficients of bubbles that have substances adsorbed onto their surfaces (which may apply to most bubbles in the ocean), and false signals from organisms containing air bubbles (such as bubbles used for buoyancy), or other particles containing gas which would also be efficient scatterers of sound [*Medwin*, 1970, 1977a; *MacIntyre*, 1986; *Vagle and Farmer*, 1992; *Dahl and Jessup*, 1995].

4.4.1.2. Measurements. Measurements of size-dependent bubble concentrations in the ocean, presented in Fig. 33a-c in the representation $n_{bub}(R_{bub}) \equiv dN_{bub}(R_{bub})/d\log R_{bub}$, are summarized in Table 19. As depth and wind speed are thought to be two of the key factors that affect bubble concentrations, and are the main two quantities by which size distributions of bubble concentration in the ocean are classified and parameterized, the measurements are shown in three separate figures classified by these quantities. Figs. 33a and 33b present size distributions of bubble concentration measured at wind speeds less than 10 m s^{-1} and at wind speeds greater than or equal to 10 m s^{-1}, respectively, at depths less than or equal to 1.5 m. Fig. 33c presents size distributions of bubble concentration at depths greater than 1.5 m. The axes on all three graphs are the same to permit comparison, and the ordinate as well as the abcissa has a logarithmic scales to encompass the large range of the data and to permit comparison of the shapes of the distributions. These measurements comprise the majority of size distributions of oceanic bubble concentration that have been reported and are the basis of nearly all estimates of the production flux of SSA particles from bursting bubbles using the bubble method (§3.4).

Several other reported size distributions of bubble concentrations not presented in the figures are briefly noted here. Size distributions of bubbles on the sea surface determined from four photographs taken by *Bortkovskiy and Timanovskiy* [1982] just as a wave was breaking at wind speeds of 6.5-7.2 m s^{-1} would not necessarily be representative of those that subsequently rose to the surface and burst to form drops, nor would similar measurements reported by *Bezzabotnov et al.* [1986]. Size distributions of bubbles measured in surf foam [*Rayzer and Sharkov*, 1980; *Podzimek*, 1984] are not pertinent to bubbles in the open

a

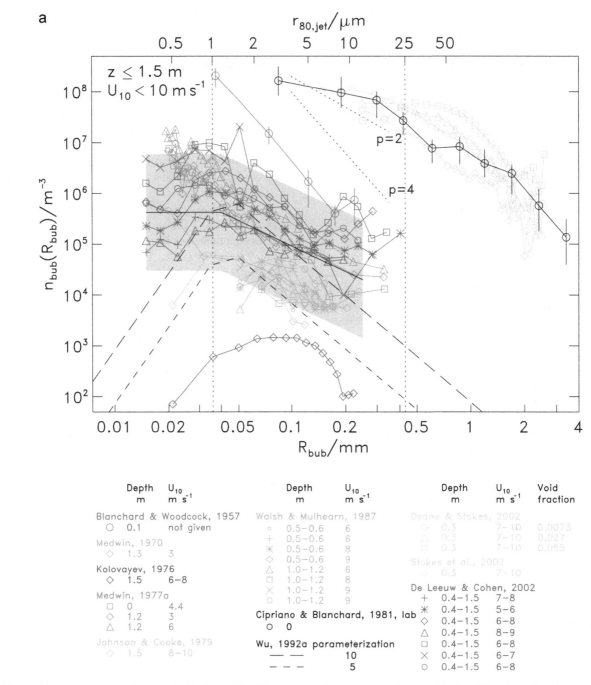

Depth m	U_{10} m s^{-1}		Depth m	U_{10} m s^{-1}		Depth m	U_{10} m s^{-1}	Void fraction
Blanchard & Woodcock, 1957			Walsh & Mulhearn, 1987			Deane & Stokes, 2002		
○ 0.1	not given		○ 0.5–0.6	6		◇ 0.3	7–10	0.0073
Medwin, 1970			+ 0.5–0.6	6		△ 0.3	7–10	0.027
◇ 1.3	3		✳ 0.5–0.6	8		□ 0.3	7–10	0.065
Kolovayev, 1976			◇ 0.5–0.6	9		Stokes et al., 2002		
◇ 1.5	6–8		△ 1.0–1.2	6		○ 0.3	7–10	
Medwin, 1977a			□ 1.0–1.2	8		De Leeuw & Cohen, 2002		
□ 0	4.4		× 1.0–1.2	9		+ 0.4–1.5	7–8	
◇ 1.2	3		○ 1.0–1.2	9		✳ 0.4–1.5	5–6	
△ 1.2	6		Cipriano & Blanchard, 1981, lab			◇ 0.4–1.5	6–8	
Johnson & Cooke, 1979			○ 0			△ 0.4–1.5	8–9	
◇ 1.5	8–10		Wu, 1992a parameterization			□ 0.4–1.5	6–8	
			—— ——	10		× 0.4–1.5	6–7	
			– – –	5		○ 0.4–1.5	6–8	

Figure 33. Measured oceanic size distributions of bubble concentration, represented as $n_{bub}(R_{bub}) \equiv dN_{bub}(R_{bub})/d\log R_{bub}$ as a function of equivalent-volume bubble radius R_{bub} classified by depth z and wind speed at 10 m above the sea surface U_{10}: a) $z \leq 1.5$ m, $U_{10} < 10$ m s^{-1}; b) $z \leq 1.5$ m, $U_{10} \geq 10$ m s^{-1}; c) $z > 1.5$ m. All graphs are presented on the same axes to facilitate comparison; colors are maintained for measurements from the same study appearing in more than one graph, with symbols indicating measurement conditions (summarized in Table 19). All measurements are for the open ocean except surf zone measurements of *Blanchard and Woodcock* [1957] and laboratory measurements of *Cipriano and Blancard* [1981] shown for comparison. Measurements shown represent all reported oceanic measurements in the indicated study except for *Medwin* [1970], for which only daytime attenuation measurements are shown (nighttime attenuation measurements, acoustic scattering results, and photographic determinations are not shown), and *Walsh and Mulhearn* [1987], for which distributions containing only two values of R_{bub} not shown. Measurements of

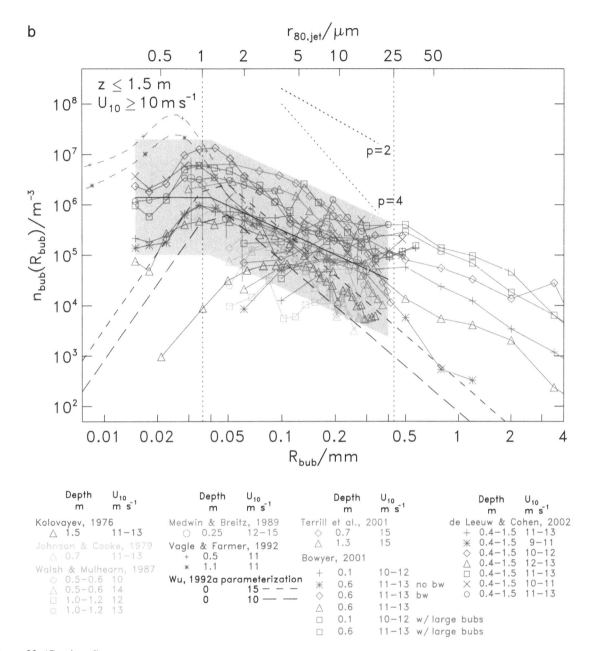

Figure 33. (Continued).

Bowyer [2001] denoted "no bw" and "bw" refer to those excluding and including visible breaking wave events, respectively; those denoted "w / large bubbles" refer to video frames containing bubbles with $R_{bub} > 0.5$ mm, providing a rough indicator of the presence of a breaking wave (as the same symbols are used for these two size distributions of bubble concentration in the original reference, it was not possible to determine which corresponded to which set of conditions). Shaded bands drawn to encompass the bulk of the data denote a factor of $\times 14$ about solid black lines; central line is constant for $R_{bub} = 0.015$-0.04 mm and proportional to $r_{80}^{-1.6}$ for $R_{bub} > 0.04$ mm. Shaded band in b) is a factor of 3.3 higher than in a). Also shown is parameterization of *Wu* [1992a] at U_{10} values of 5, 10, and 15 m s^{-1}. Dotted lines representing power laws of the form $n_{bub}(R_{bub}) \propto R_{bub}^{-p}$ with $p = 2$ and $p = 4$ are shown to permit comparison with measured distributions. Upper axis shows values of r_{80} based on (4.3-3); vertical dotted lines at r_{80} values of 1 μm and 25 μm demarcate small, medium, and large SSA particles (§2.1.2).

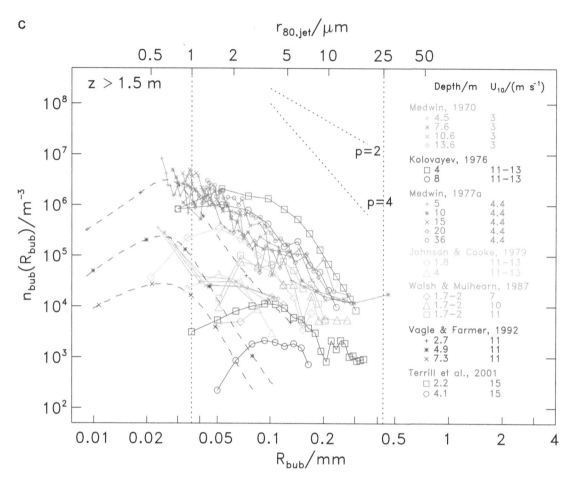

Figure 33. (Continued).

ocean, nor are measurements taken in the surf zone [*Deane*, 1997; *Deane and Stokes*, 1999; *Stokes and Deane*, 1999; *Terrill and Melville*, 2000], typically in water of depths of 1 m or less. However, measurements taken in the surf zone by *Blanchard and Woodcock* [1957] are shown to illustrate the enhancement of concentrations relative to the open ocean and because these data have been used by subsequent investigators to estimate drop production. Size distributions reported by *Walsh and Mulhearn* [1987] containing only two radius intervals are not shown, nor are data of *O'Hern et al.* [1988], for which concentrations were reported for no more than three intervals of R_{bub} (additionally the latter measurements were at depths and under wind conditions for which such bubbles were most probably not formed by the action of the wind on the waves; for the most part such bubbles would not survive long enough to rise to the surface and burst). Measurements reported by *Ling and Pao* [1988] are also not presented, having been smoothed and with the original data not having been reported. Measurements of *Su et al.* [1988], also not shown, indicated little dependence on depth

down to 15 m; the investigators attributed this to the possible influence of wave-platform interactions on the measurements or leakage from the equipment. Measurements of *Vagle and Farmer* [1998], described as individual representative size distributions but for which accompanying data such as wind speed were not provided, are not shown, nor are individual examples presented by *Farmer et al.* [1998].

Many concerns noted in §4.1.1 and §4.1.4.1 regarding measurements and reporting of size distributions of SSA concentrations pertain to bubble concentrations as well. Key among these are the extent to which any measurement or set of measurements is representative of others and can be generalized to other locations and conditions; the amount of variability, both temporal and spatial, that can be expected for measurements taken under nominally the same conditions (such as depth and wind speed); differences due to experimental techniques; data averaging, binning, and sampling statistics; and conversion from one representation to another. Some of the reported size distributions of bubble concentration are individual examples, others were termed "typical," whereas

Table 19. Oceanic Measurements of Size Distributions of Bubble Concentration

Reference	Measurement Technique	R_{bub}/mm Range	Location, Ocean Depth	Measurement Conditions	T_{sea}/°C
Blanchard & Woodcock, 1957[a]	bubble trap $9 \times 6 \times 2$ cm^3	< 0.250	5 m from shore, ~1 m	0.1 m deep, after breaking wave, no wind speed stated, 2 s sample time	21
Medwin, 1970	excess acoustic attenuation and scattering	0.015-0.3	coastal, 9 m, 20 m	5 depths: 1.3-13.6 m, U_{10} < 6 m s^{-1}	13
Kolovayev, 1976	bubble trap 60 cm long, 20 cm diameter	0.015-0.35	open ocean, 3000 m	4 conditions: 1.5 m depth for U_{10} = 6-8 m s^{-1}; 1.5 m, 4 m, 8 m for U_{10} = 11-13 m s^{-1}, each an average over several determinations	14-17
Medwin, 1977a[b]	excess acoustic attenuation	0.015-0.3	coastal, 40 m	depths to 36 m, U_{10} < 6 m s^{-1}, average of 50 determinations over 10 min	not stated
Johnson & Cooke, 1979	photographic	0.017-0.3	coastal, 20-30 m	4 conditions: 1.5 m depth for U_{10} = 8-10 m s^{-1}; 0.7 m, 1.8 m, 4 m for U_{10} = 11-13 m s^{-1}, each an average from many photographs over minutes to hours	2-3
Walsh & Mulhearn, 1987	photographic	0.05-0.306	most open ocean, 300-4800 m; some coastal, 120 m	17 conditions: depths 0.5-0.6 m, 1-1.2 m, 1.7-2 m for U_{10} < 14 m s^{-1}, each an average from more than 32 photographs over 15 min	15-22.4
Medwin & Breitz, 1989[c]; *Breitz & Medwin, 1989*	broadband acoustic extinction, backscattering, and dispersion	0.03-0.24	coastal, 500 m	0.25 m under breaking wave for U_{10} = 12 m s^{-1}, 2.5 min average	not stated
Vagle & Farmer, 1992[d]	multi-frequency (6) acoustic backscattering	0.008-0.13	coastal	5 depths: 0.5-7.3 m, U_{10} = 11 m s^{-1}, 5 min averages	not stated
Bowyer, 2001	photographic	0.6-6.0	coastal	2 conditions: 0.1 m, U_{10} = 10-12 m s^{-1}; 0.6 m, U_{10} = 11-13 m s^{-1}	not stated
Terrill et al., 2001[e]	broadband acoustic extinction and dispersion	0.03-0.4	coastal	4 depths: 0.7-4.1 m, U_{10} = 15 m s^{-1}, taken within 30 s	not stated
Deane & Stokes, 2002; Stokes et al., 2002[f]	photographic	0.15-2.5	coastal	4 size distributions at 0.3 m below breaking wave for U_{10} = 7-10 m s^{-1}	13.2
de Leeuw & Cohen, 2002	photographic	15-575	open ocean	7 pairs of data sets at depths 0.4-1.5 m, each at low (U_{10} = 5-9 m s^{-1}) and high(U_{10} = 9-13 m s^{-1}) wind speeds, 15-20 min averages	11-20

Size distributions of bubble concentration are presented in Fig. 33.

[a] Vertical lines on data of *Blanchard and Woodcock* [1957] show range of values from six separate measurements.

[b] Data for U_{10} = 4.4 m s^{-1} were taken under windrows.

[c] Vertical lines on data connect maximum and minimum values. Data are from Fig. 7 of that reference; values from Fig. 6 of that reference are a factor of 2 too high [*Wu and Hwang*, 1991].

[d] Asterisks denote radii of bubbles resonant with the acoustic frequency used to probe the size distribution; intermediate values were obtained from inversion as described in text.

[e] Data were taken inside a bubble cloud under a breaking wave and were stated as typical.

[f] Data from *Deane and Stokes* [2002], stated to be typical bubble size spectra, were analyzed from multiple images from three separate breaking wave events. Data of *Stokes et al.* [2002] are from a 0.5 s average of 11 breaking wave events.

others are averages over several measurements or over extended periods of time (Table 19). For most reported size distributions estimates of the uncertainties were not presented. Additionally, differences in location (i.e., coastal vs. open ocean) may contribute to differences in size distributions of bubble concentration, although the extent of any such contribution is not known. The several methods used to detect and size bubbles introduce further uncertainties and contribute to the variability in reported size distributions of bubble concentration. As with size distributions of SSA concentration, clear trends can sometimes be observed in size distributions of bubble concentration within a given set of measurements, which are typically taken under similar conditions (wind speed, sea temperature, and time-averaging conditions) using a single technique, but when data from different investigators are compared, such trends are not so apparent.

Relatively few size distributions of oceanic bubble concentration have been reported, and approximately one-third of those presented in Fig. 33 were reported in the last few years [Bowyer, 2001; Terrill et al., 2001; Deane and Stokes, 2002; Stokes et al., 2002; de Leeuw and Cohen, 2002]. These latter investigations generally reported bubble concentrations that were an order of magnitude or more greater than earlier ones [Kolovayev, 1976; Johnson and Cooke, 1979; Walsh and Mulhearn, 1987] upon which much of the literature pertaining to SSA production has been based. Values of R_{bub} in Figs. 33a-c range from near 0.01 mm to 6.0 mm, and values of $n_{bub}(R_{bub})$ vary by up to six orders of magnitude for the same R_{bub}. The key reason for this variability is probably differences in the time subsequent to wave breaking (for size distributions of bubble concentration that are not temporal averages), but differences in wind speed, depth, water temperature, and other factors, and differences in measurement techniques may also contribute.

Most reported measurements of $n_{bub}(R_{bub})$ are for R_{bub} in the range 0.02-0.35 mm. Bubbles of these sizes are too small to produce film drops (Fig. 30), but produce jet drops with r_{80} from ~0.5 μm to ~20 μm according to (4.3-3). Larger bubbles rise rapidly to the surface (§4.4.2.1) and, as noted above, concentrations of these bubbles change rapidly with time after breaking of a wave. Much attention has been given to the shape of the bubble concentration size distributions, specifically the locations of any maxima, but as noted in §2.1.4, little importance can be attached to the location of such a maximum (if it exists), as this location depends on the representation used for the size distribution of bubble concentration. Additionally, any such location would be expected to vary with time subsequent to wave breaking.

The majority of the bubbles reported by all investigators have $R_{bub} < 0.1$ mm, but results from different researchers exhibit different trends for $n_{bub}(R_{bub})$ in this size range: measurements of Kolovayev [1976], Johnson and Cooke [1979], Walsh and Mulhearn [1987], Terrill et al. [2001], and de Leeuw and Cohen [2002] indicate that $n_{bub}(R_{bub})$ attains a maximum at R_{bub} in the range 0.05-0.1 mm, whereas values of n_{bub} reported by Medwin [1970, 1977a] and Medwin and Breitz [1989; also Breitz and Medwin, 1989] continue to increase with decreasing R_{bub}, down to the lowest value measured. Size distributions of bubble concentration reported by Vagle and Farmer [1992] exhibit a maximum in $n_{bub}(R_{bub})$ at $R_{bub} = 0.020$-0.025 mm, though the investigators reported reduced confidence in their reported bubble size distribution for $R_{bub} \lesssim 0.016$ mm. As discussed above, optical methods are expected to underestimate, and acoustic measurements are expected to overestimate, number concentrations of small bubbles ($R_{bub} \lesssim 0.05$ mm), but these effects will not resolve the discrepancy noted above. Walsh and Mulhearn [1987] stated that the peaks in their data were probably not characteristic of the actual distributions, but suggested that such peaks were due to limitations of measuring bubbles with $R_{bub} < 0.050$ mm; these investigators suggested also that the peaks found by Kolovayev [1976] may have been due to the measurement techniques used, which undercounted bubbles with $R_{bub} < 0.085$ mm. They also reported that B. D. Johnson had communicated to them that the peaks reported by Johnson and Cooke [1979] may have not been real but instead were possibly caused by limitations of the measurement system employed. There is thus little consensus on the shape of the size distribution of bubble concentration for $R_{bub} \lesssim 0.05$ mm, despite much discussion [Johnson and Cooke, 1979; Wu, 1981a; MacIntyre, 1986; Walsh and Mulhearn, 1987; Medwin and Breitz, 1989; Commander and Moritz, 1989].

For $R_{bub} > 0.1$ mm, the reported size distributions uniformly decrease for all measurements (except those of Deane and Stokes [2002] and those of Stokes et al. [2002]), but with fluctuations at the largest sizes (for $R_{bub} > 0.2$ mm), where counting statistics play a large role and there is considerable noise in the data.

As noted above, much of the variability in bubble concentrations is attributable to differences in time after wave breaking. The measurements of Blanchard and Woodcock [1957], taken a few seconds after the passage of a breaking wave, yielded much greater concentrations than most of other measurements. However, as wind speeds were not reported, and as the measurements were made in shallow water (~0.1 m deep) very close to shore, the results are probably not representative of bubble populations in the open ocean. The measurements of Medwin and Breitz [1989; also Breitz and Medwin, 1989], taken under breaking waves, also yielded greater concentrations than most other measurements, but these were made at a coastal location at only a single wind speed, so again the extent to which they are representative of other locations and situations is uncertain.

Likewise, measurements of *Bowyer* [2001] which included visible breaking wave events (denoted "bw" in Fig. 33b) are 1-2 orders of magnitude greater than those which did not include such events, with the difference increasing with increasing R_{bub}. Size distributions of bubble concentration that included only video frames in which bubbles larger than $R_{bub} = 0.5$ mm were observed (denoted "w / large bubs" in Fig. 33b), indicating the likelihood of a breaking wave, are 1-3 orders of magnitude greater than those included all frames. Several of the measurements in Table 19 were indicated to be averages over several minutes and would thus be expected to be appreciably lower than those for a short time interval that includes a wave breaking event. Additionally, because of the intermittent nature of wave breaking, a single event could strongly affect such measurements. Residual (i.e., background) bubble concentrations were reported by *Medwin and Breitz* [1989] to be one to two orders of magnitude lower than concentrations at the peak of bubble production under a breaking wave. *Vagle and Farmer* [1992] cautioned that the size distributions of bubble concentration they reported, which were 5 min averages, should not be taken as representative of longer-time averages. The measurements of *Deane and Stokes* [2002] and of *Stokes et al.* [2002], which were limited to the time of wave breaking, are 2 to 4 orders of magnitude greater in the size range of overlap than measurements from other investigations, several of which were long-time averages (Table 19); however, if the fraction of the ocean surface under breaking waves were taken into account, this difference would be greatly reduced.

It would be expected that higher winds, which would produce more and larger breaking waves, would result in greater bubble concentrations. Generally this expectation has been shown to be valid. Mean void fractions at 0.7 m reported by *Terrill et al.* [2001] remained near 10^{-6} for wind speeds up to ~6 m s^{-1} and increased with increasing wind speed to greater than 10^{-4} for wind speeds of 17 m s^{-1}. Most of the size distributions of bubble concentration from individual investigations presented in Fig. 33 also exhibit an increase with increasing wind speed, although not in all instances [e.g., some of the size distributions presented by *Walsh and Mulhearn*, 1987]. Size distributions of bubble concentration at wind speeds less than 10 m s^{-1} (Fig. 33a) are generally less than those at higher wind speeds (Fig. 33b), although this difference is much less than the variability among the data sets on any one graph. Shaded gray regions presented on Figs. 33a and 33b have been drawn to encompass the

majority of the data (other than the concentrations reported by *Deane and Stokes* [2002] and *Stokes et al.* [2002] for large bubbles) that represent bubbles that are likely to burst at the surface and produce SSA particles. These shaded regions are more than two orders of magnitude wide and represent a multiplicative uncertainty factor of ⪅14 about the central values represented by solid lines. In contrast, the central values differ from each other by only a factor of 3.

Bubble concentrations are expected to decrease with increasing depth, and measurements taken simultaneously a different depths clearly exhibit such a decrease [e.g., *Vagle and Farmer*, 1992, Fig. 11; *Farmer et al.*, 1998, Fig. 15; *Vagle and Farmer*, 1998, Fig. 3]. However, size distributions of bubble concentrations over the entire range of R_{bub} (0.015-0.5 mm) reported by *Medwin* [1970, 1977a] at $U_{10} < 6$ m s^{-1}, mostly below wind speeds at which whitecapping is expected (§4.5.1), exhibited little dependence on depth down to 36 m (Fig. 33c), in contrast to most other measurements; large numbers of bubbles with $R_{bub} = 0.010$-0.015 mm at depths to over 30 m under conditions of calm winds were reported also by *O'Hern et al.* [1988]. Differences in the shape and magnitude of size distributions of bubble concentration between night and day were also reported by *Medwin* (suggestive of biological origin), and in some situations (e.g., under windrows) the greatest bubble concentrations occurred at the greatest depths. These measurements also exhibited an increase in $n_{bub}(R_{bub})$ with decreasing R_{bub} over the full range of R_{bub} measured. The presence of small bubbles that persist much longer than expected has long been known, and explanations are required not only for the source of these bubbles but also for this persistence; because of the immense hydrostatic pressures encountered at depths to which they have been measured, the bubbles would be expected to dissolve in a matter of seconds (§4.4.2.2). Various hypotheses for the production and persistence of such bubbles have included coating by substances that alter the gas-exchange properties of the bubbles and allow them to persist without dissolving, attachment to particles or inclusion of bubbles in organisms, and production by biological processes, aerosol particles passing through the sea surface, or cosmic rays[1]. Based on the measurement conditions it can be concluded that the overwhelmingly vast majority of these bubbles were not formed by the action of the wind on waves, and the presence of bubbles under these conditions is indicative of low-level bubble production by mechanisms other than wind-driven breaking waves, the most plausible explanation being that

---

[1] *Fox and Herzfeld* [1954]; *Lieberman* [1957]; *Turner* [1961]; *Ramsey* [1962b]; *Gavrilov* [1969]; *Medwin* [1977a]; *MacIntyre* [1978]; *Yount* [1979]; *Cipriano and Blanchard* [1981]; *Wu* [1981a]; *Mulhearn* [1981]; *Johnson and Cooke* [1981a]; *Glazman* [1983]; *Woolf* [1985]; *MacIntyre* [1986]; *Johnson* [1986]; *Johnson and Wangersky* [1987].

most of the small bubbles are biological in origin and are perhaps stabilized by organic coatings.

4.4.1.3. Parameterizations. The decrease in $n_{bub}(R_{bub})$ with increasing R_{bub} has often been parameterized[1] by a power law relationship in some range of R_{bub}, and several theoretical arguments concerning the size distributions of bubble concentration [*Kerman*, 1986; *Kerman*, 1988b; *Longuet-Higgins*, 1992; *Baldy*, 1993; *Garrett et al.*, 2000; *Deane and Stokes*, 2002] also have attempted to justify a power law relationship. Here the expressions reported by various investigations have been converted to and are reported in the representation $n_{bub}(R_{bub}) \equiv dN_{bub}(R_{bub})/d\log R_{bub} \propto R_{bub}^{-p}$ with the number of significant figures in the exponent p retained. A wide range of values of p in one or another range of R_{bub} have been reported from field measurements of size distributions of bubble concentration, and likewise from laboratory measurements. Values from 2.5-3.6 for field measurements and 0.5-2 for laboratory experiments were reported by *Monahan* [1982]. Values from 1.87-4.99 for oceanic bubbles with $R_{bub} > 0.1$ mm were reported by *Walsh and Mulhearn* [1987]. Values from 3-6, increasing with increasing depth, were reported by *Vagle and Farmer* [1992] for oceanic bubbles with $R_{bub} > 0.02$ mm. Values ranging from 0.5 to 3.7 were reported by *Haines and Johnson* [1995] for measurements of other investigations (some of which were laboratory experiments covering larger size ranges). Values from 0.8-4 were reported by *de Leeuw and Cohen* [2002] from field measurements. *Deane and Stokes* [2002] reported values of p of 0.5 and 2.3 for R_{bub} less than and greater than 1 mm, respectively shortly after a laboratory breaking wave event, and values of p of 1.3 and 5.0 for bubbles of these sizes 1.5 s later. From measurements under oceanic breaking waves taken at intervals of 1 s, these investigators reported values of p of 0.8, 1.5, and 1.9 for $R_{bub} \lesssim 1$ mm, and 3.9, 4.3, and 4.5 for $R_{bub} \gtrsim 1$ mm. For comparison, power laws with $p = 2$ and $p = 4$ are shown in Fig. 33 by dotted lines for $R_{bub} = 0.1$-0.35 mm, and the decrease in the central line of the gray shaded region in Figs. 33a and 33b is a power law with $p \approx 1.6$. For increasingly large R_{bub} the value of the exponent p must exceed 3 in order that the void fraction, evaluated as $\int (4\pi/3) R_{bub}^3 n_{bub}(R_{bub}) d\log R_{bub}$, not diverge, and further constraints on p are necessary to avoid divergence in calculated gas-exchange properties, depending on assumed rise speeds and exchange coefficients [*Keeling*, 1993].

As noted above, size distributions of bubble concentration depend strongly on time after wave breaking, so it would be expected that parameterizations of instantaneous concentrations measured shortly after breaking of a wave and resultant formation of the bubble plume would differ from those of time-average concentrations. Larger bubbles, because of their large terminal velocities (more than 25 cm s^{-1} for $R_{bub} \gtrsim 1$ mm; §4.4.2.1), rapidly rise to the surface, and these bubbles are depleted within several seconds after a breaking wave. Additionally, the range of R_{bub} reported in the several data sets upon which many of these parameterizations were based is rather small. In only two of the four size distributions of bubble concentration reported by *Johnson and Cooke* [1979] were bubbles with $R_{bub} > 0.12$ mm observed, and in only one were there bubbles with $R_{bub} > 0.17$ mm; in only two of the four size distributions of bubble concentration reported by *Kolovayev* [1976] were bubbles with $R_{bub} > 0.22$ mm observed; and the majority of the size distributions of bubble concentration of *Walsh and Mulhearn* [1987] did not extend above $R_{bub} = 0.2$ mm. As noted above, there appeared to be much noise at the largest values of R_{bub} for most of these size distributions. Because of the large range of values of p reported from different investigations and the small range of R_{bub} from which these values were determined, typically 0.1-0.35 mm (half an order of magnitude), it is difficult to place much confidence in any given value of p, let alone attribute physical meaning to such a value.

The magnitude of the bubble concentration has commonly been suggested as being a power law in the wind speed or wind friction velocity, with exponent ranging from 1 to more than 4. The same range of exponents has been proposed for the dependence of ocean noise on wind speed or wind friction velocity [e.g., *Farmer and Lemon*, 1984; *Kerman*, 1984b], although measurements of the backscattering cross section (which exhibits the same wind speed dependence as the bubble concentration) reported by *Dahl and Jessup* [1995] indicate an exponent greater than 7 (though with relatively large error bounds). However, over the fairly narrow range of wind speeds for which size distributions of bubble concentrations have been measured, ~5 m s^{-1} to ~15 m s^{-1} (half an order of magnitude), such parameterizations are of limited utility. For example, for an increase in U_{10} from 10 m s^{-1} to 15 m s^{-1}, a power law with exponent 3 yields an increase by a factor barely greater than three, compared to the spread of more than two orders of magnitude exhibited

---

[1] *Blanchard and Woodcock* [1957]; *Medwin* [1970, 1977a]; *Johnson and Cooke* [1979]; *Monahan* [1982]; *Novarini and Bruno* [1982]; *Farmer and Lemon* [1984]; *Walsh and Mulhearn* [1987]; *Crowther* [1988]; *Su et al.* [1988]; *Hall* [1989]; *Medwin and Breitz* [1989]; *Breitz and Medwin* [1989]; *Wu* [1992b]; *Vagle and Farmer* [1992]; *Dahl and Jessup* [1995]; *Woolf* [1997]; *Bowyer* [2001]; *Deane and Stokes* [2002]; *de Leeuw and Cohen* [2002]; *Farmer et al.* [2002].

by the data in this wind speed range presented in Fig. 33b. Similarly, for an increase in U_{10} from 6 m s^{-1} to 10 m s^{-1}, such a wind speed dependence yields an increase of less than a factor of 5, compared to the more than two orders of magnitude spread exhibited by the data in this wind speed range presented in Fig. 33a.

Several attempts have been made to model the size distributions of bubble concentration by a shape that is independent of wind speed and depth, but with the overall amplitude depending on these quantities[1]; a similar but slightly more complicated expression was proposed by *Hall* [1989]. However, such models, although convenient, fail to adequately represent much of the data. Most investigations covered only a small range of depths, wind speeds, or seawater temperatures, and while such parameterizations may rather accurately describe measurements of a given investigation, it is evident that no such parameterization would exhibit much skill in comparison to the bulk of the measurements. Additionally, although the data of any one investigation may exhibit trends of increasing bubble concentration with decreasing depth and with increasing wind speed, such as those of *Kolovayev* [1976] and *Johnson and Cooke* [1979], each of these studies involved only four pairs of wind speed and depth. The only study for which bubble concentrations were reported for more than a few sets of depths and wind speeds was that of *Walsh and Mulhearn* [1987], and these data showed much variability and did not always exhibit increasing bubble concentrations with increasing wind speeds and decreasing depths. A further consideration is that any depth dependence based on concentration measurements will reflect the effects of the production of bubbles and their injection by breaking waves and subsequent loss by dissolution and rising of larger bubbles, and can therefore not be assumed to be representative of the size distribution of the bubbles that survive to the surface to burst. Indeed, the assumption of the invariance of the size distribution with depth is not supported by the data. Both *Kolovayev* [1976] and *Johnson and Cooke* [1979] reported narrower size distributions at greater depths, which they attributed to the rise of the large bubbles to the surface and the dissolution of the smaller ones. Likewise *Baldy* [1988] reported that the shape of the size distribution of

freshwater bubbles in a laboratory wave tank depended on wind speed and depth, with concentrations of larger bubbles showing a greater increase with increasing wind speed than those of smaller bubbles. Thus, because of the large differences between different investigations, little confidence can be placed in the ability of the parameterizations presented to quantify the dependence of oceanic size-dependent bubble concentrations on factors such as wind speed and depth.

Parameterizations of the depth dependence of size distributions of bubble concentration have been extrapolated to yield concentrations at the surface for use in the bubble method of determining SSA production (§5.5). However, such extrapolations are questionable because of the rapid evolution of the shape of the bubble concentration with time, and consequently the difference in the shape of the size distribution of bubble concentration between measurement depths and the surface. For this reason, measurements at depths greater than 1.5 m (Fig. 33c) are expected to be much less pertinent to the flux of bubbles that burst at the surface than measurements at depths less than or equal to 1.5 m (Figs. 33a and 33b). A model examining the evolution of bubbles between 1 m depth and the surface is presented in §4.4.2.2.

4.4.1.4. Laboratory studies of bubble production and concentrations. In view of the wide range of results for $n_{bub}(R_{bub})$ from ocean measurements, for the many reasons noted, it is useful to consider results from laboratory studies which, while not reproducing the dynamics of bubble production from breaking ocean waves, may nevertheless allow dependence on controlling factors to be identified and characterized. For example, some of the variability in reported size distributions of bubble concentration may have resulted from differences in water temperature, although systematic field investigations on this topic have not been reported. Thus, laboratory investigations of the effect of temperature on bubble formation, coalescence, and breakup might yield pertinent insight.

Laboratory investigations of bubble entrainment and concentrations have been reported for saltwater or seawater[2], freshwater[3], and for both freshwater and saltwater[4]. The size distribution of bubble concentration reported by *Cipriano*

---

[1] *Cipriano and Blanchard* [1981]; *Wu* [1981a]; *Monahan* [1982]; *Novarini and Bruno* [1982]; *Cipriano and Blanchard* [1982]; *Cipriano et al.* [1983]; *Wu* [1986b]; *Kerman* [1986]; *Crawford and Farmer* [1987]; *Wu* [1988a]; *Crowther* [1988]; *Kerman* [1988b]; *Thorpe et al.* [1992]; *Wu* [1992a,b, 1994b]; *Woolf* [1997]; *Farmer et al.* [2002].

[2] *Glotov et al.* [1962]; *Cipriano and Blanchard* [1981]; *Cipriano et al.* [1983]; *Johnson* [1986]; *Asher and Farley* [1995]; *Haines and Johnson* [1995]; *Asher et al.* [1997]; *Terrill and Melville* [2000]; *Deane and Stokes* [2002].

[3] *Toba* [1961b]; *Toba et al.* [1975]; *Koga* [1982]; *Broecker and Siems* [1984]; *Baldy and Bourguel* [1985]; *Resch* [1986]; *Baldy* [1988]; *Medwin and Daniel* [1990]; *Hwang et al.* [1990, 1991]; *Lamarre and Melville* [1991]; *de Leeuw and Leifer* [2002]; *Leifer and de Leeuw* [2002]; *Komori and Misumi* [2002].

[4] *Monahan and Zietlow* [1969]; *Bowyer* [1992]; *Carey et al.* [1993a,b]; *Cartmill and Su* [1993]; *Loewen et al.* [1996].

and Blanchard [1981] from a laboratory experiment with a continuous breaking wave (§4.3.4.1) is shown in Figs. 33a as it has been used by these investigators to determine the interfacial SSA production flux based on the bubble method (§5.5.1); this size distribution is similar in shape and magnitude to distributions obtained from field measurements of *Deane and Stokes* [2002] and *Stokes et al.* [2002]. Laboratory measurements of the SSA production flux over whitecap area have also been used to infer size distributions of the near-surface bubble concentration and the flux of bubbles bursting at the surface [*Monahan*, 1988a,b; *Monahan and Woolf*, 1988].

Although such laboratory studies may be useful in elucidating the roles of various factors affecting bubble populations and behavior in a systematic manner, there are concerns over applying results of such laboratory studies to ocean conditions. Key among these are the bubble production mechanisms and the inability of laboratory experiments to accurately simulate oceanic breaking waves and conditions in the ocean under which bubble concentrations evolve. For example, several of the studies involved bubbles produced by pouring water into a tank, and there is little reason to believe that the resulting size distribution of bubble concentration resembles that from a breaking wave. Concerns regarding the ability of laboratory breaking waves to accurately simulate oceanic breaking waves and bubble populations in the oceans were noted in §4.3.4.4, and concerns over the applicability of results of laboratory experiments in wind/wave tunnels to the ocean were noted in §4.3.6.2. The large difference in magnitude between laboratory and oceanic waves and the shallow depths employed in most laboratory investigations—less than 1 m, raise questions regarding the extent to which such experiments can be applied to the ocean, where velocity fields under a breaking wave may be very different and bubbles are entrained to depths of several meters or more. For example, the size distribution of bubbles produced by breaking events in laboratory wind/wave tunnels has been reported to differ substantially from that beneath whitecaps in moderate seas (U_{10} = 7-10 m s^{-1}), with concentrations of bubbles with R_{bub} < 0.3 mm being at least an order of magnitude less than in oceanic whitecaps, and with the difference between laboratory and oceanic results increasing with decreasing R_{bub} [*Stokes et al.*, 2002]. Thus, although laboratory investigations might yield insight into processes that affect size distributions of bubble concentration, such as coalescence, breakup, and dissolution,

they do not accurately simulate other processes such as the initial bubble formation process and turbulent and large scale mixing. For these reasons laboratory results cannot be expected to yield size distributions of bubble concentration that can be extrapolated to the ocean.

The use of freshwater in laboratory studies of bubble production and behavior may result in further differences compared to the ocean. Although no systematic studies of the effects of salinity on bubble concentrations from breaking waves have been reported, numerous experiments have shown differences in the size distributions of bubbles formed by various methods in freshwater and those in seawater or NaCl or other salt solutions under presumably the same conditions, resulting primarily from differences in coalescence and/or breakup of bubbles in the different media[1], with the result that seawater contains more smaller bubbles ($R_{bub} \lesssim 1$ mm) than does freshwater, typically by an order of magnitude. Comparable differences have been reported between bubble concentrations in freshwater and seawater environments [*Thorpe*, 1982]. Additionally, *Slauenwhite and Johnson* [1999] reported that many more smaller bubbles are produced by the breakup of a larger bubble in seawater than in freshwater and argued that this may also explain some of the difference in bubble concentration size distributions between seawater and freshwater. In contrast, little difference between bubble size distributions in freshwater and NaCl solutions was observed by *Loewen et al.* [1996] for $R_{bub} \gtrsim 0.8$ mm. Based on the latter result, *Wu* [2000a] suggested that the actual process by which bubbles are formed during wave breaking is more important to the bubble concentration size distribution than bubble coalescence or breakup, and that the difference between size distributions of bubble concentration in seawater and freshwater is due to a greater amount of air being entrained during breaking waves in seawater; this suggestion was questioned by *Monahan* [2001] who called attention to several experiments that demonstrated large differences between freshwater and seawater or saltwater [see also *Wu*, 2002a]. However, as *Loewen et al.* had presented size distributions only of bubbles with $R_{bub} \gtrsim 0.8$ mm, their results do not contradict the large number of studies that have found major differences between freshwater and seawater or saltwater with regard to size distributions of concentrations of bubbles with R_{bub} smaller than 0.8 mm.

Although the reasons for the differences between size distributions of bubble concentration in freshwater and

---

[1] *Foulk and Miller* [1931]; *Foulk* [1932]; *Davis* [1940]; *Miyake and Abe* [1948]; *Blanchard and Woodcock* [1957]; *Blanchard* [1963, p. 83]; *Marrucci and Nicodemo* [1967]; *Monahan and Zietlow* [1969]; *Lessard and Zieminski* [1971]; *Zieminski and Whittemore* [1971]; *Scott* [1975a,b]; *Keitel and Onken* [1982]; *Exton et al.* [1986]; *Pounder* [1986]; *Shatkay and Ronen* 1992; *Bowyer* [1992]; *Craig et al.* [1993a,b]; *Carey et al.* [1993a,b]; *Cartmill and Su* [1993]; *Kolaini et al.* [1994]; *Haines and Johnson* [1995]; *Hofmeier et al.* [1995]; *Weissenborn and Pugh* [1995]; *Asher et al.* [1997]; *Slauenwhite and Johnson* [1999].

seawater or saltwater are by no means well understood, it would seem difficult to support hypotheses that attribute such differences to the very small differences in surface tension or viscosity between seawater and freshwater (§2.5.1). It appears that the primary cause of the difference is the difference in the electrolytic nature of the two media, perhaps involving a barrier to coalescence due to properties of the air-water interface. The large differences in bubble concentration size distributions, and by extension bubble surface area concentrations, between the two media call into question the ability of laboratory experiments in freshwater to accurately simulate seawater with regard not only to drop production but also to gas transfer and noise production. Additionally, bubble breakup, with resultant change in bubble size distributions, differs greatly between NaCl solutions and seawater [*Slautenwhite and Johnson*, 1999], apparently because of the presence of Mg^{2+} and SO_4^{2-}, even in the low concentrations in which they occur in seawater (Table 6). Such a result calls into question the ability of NaCl solutions to accurately simulate seawater with regard to bubble concentrations and thus drop production.

In many electrolyte solutions a transition value of the concentration has been reported above which bubble coalescence is much reduced, resulting in an increased number of smaller bubbles in solutions with concentrations above this transition value than in freshwater; this transition has typically been determined to occur at ionic strength near 0.1-0.2 [*Marrucci and Nicodemo*, 1967; *Lessard and Zieminski*, 1971; *Drogaris and Weiland*, 1983; *Craig et al.*, 1993a,b; *Carey et al.*, 1993a,b], much below the seawater value of around 0.7. An ionic strength of 0.15 corresponds to seawater of salinity 7.5, close to that of the Baltic Sea, suggesting that the salinity of the Baltic is near the location of this transition value and thus that under otherwise similar conditions, size distributions of bubble concentration and hence SSA production from the Baltic might differ greatly from most other oceanic locations.

Temperature also can potentially play a large role in oceanic bubble concentrations and their behavior, as viscosity, solubility, and gas-exchange properties are strongly temperature-dependent (§2.5.1), with the kinematic viscosity of seawater varying by nearly a factor of two and the solubilities of the major atmospheric gases decreasing by ~30% over the temperature range 0-20 °C (in contrast, the density and surface tension change little, decreasing by only ~1% and ~5%, respectively, over this temperature range). However, no systematic study of the effects of temperature on seawater bubble concentrations has been reported, and there have been conflicting results from the few investigations on this topic. An increase in coalescence of laboratory seawater bubbles with increasing temperature was reported by *Lessard and Sieminski* [1971]; this would lead to decreasing bubble concentrations. In contrast to these results, measurements

involving laboratory-generated bubbles in seawater reported by *Exton et al.* [1986] and *Pounder* [1986] indicated a decrease in mean bubble size and an increase in bubble concentration with increasing temperature. Laboratory measurements reported by *Craig et al.* [1993a] indicated a slight decrease in coalescence of bubbles in an NaCl solution with increasing temperature. The number of laboratory generated seawater bubbles with $R_{bub} < 0.07$ mm produced by *Asher and Farley* [1995] at 15 °C was less than the number at 0 °C by an amount that increased as R_{bub} decreased to 0.035 mm. Measurements of *Slauenwhite and Johnson* [1999] demonstrated that the number of bubbles produced from breakup of larger bubbles in seawater decreased with increasing temperature, resulting in a decrease in the total bubble concentration by more than 30% as temperature increased from 3 °C to 20 °C. Results of a computer model incorporating the effects of temperature, saturation of the major atmospheric gases, particle concentrations, and turbulence [*Thorpe et al.*, 1992] indicated that bubble concentrations at a depth of 4 m should decrease with increasing temperature by about 7% per °C, although the model did not include any possible effects of temperature on the production of bubbles. The applicability of any of these results to bubble formation by breaking waves in the ocean is questionable because of the methods by which the bubbles were formed, and given the lack of consensus no conclusions can confidently be drawn from these data; however, for the reasons noted, the possibility remains open that temperature can play a role in determining the size distribution of bubble concentration resulting from an oceanic breaking wave.

Gas saturation (§2.5.1) may also affect oceanic bubble concentrations in some size ranges. According to the model of *Thorpe et al.* [1992] noted above, a several-fold increase in the concentration of bubbles at 4 m should occur for every 10% increase in saturation, with the concentration of very small bubbles being even greater. An increase in the concentration of laboratory-generated seawater bubbles with $R_{bub} < 0.1$ mm with increasing gas saturation was reported by *Asher and Farley* [1995], although there was little effect on concentrations of larger bubbles.

Surface-active substances may also affect bubble concentrations, although there do not appear to be any systematic investigations of such affects. According to the model of *Thorpe et al.* [1992] noted above suspended particles may considerably enhance the number of smaller bubbles by providing an organic coating that changes the bubble surface properties, impeding gas transfer and resulting in greater stability with respect to dissolution; this effect may account for some of the differences between results of studies in coastal regions and those in the open ocean.

4.4.1.5. Summary. Despite the appearance of a large number of studies examining size distributions of bubble concentration in the ocean, the number of such measurements is actually rather small. Perhaps not surprisingly, these is large variation in reported results. This variation appears to be due in large part to the conditions of measurement—under, or immediately after, a breaking wave versus in quiescent water (which can result in differences of orders of magnitude), versus attempts to obtain a time-averaged result of such a highly variably quantity. These considerations, together with the relatively few conditions under which concentrations have been measured and the resultant large variability, preclude drawing little more than qualitative conclusions on the size distributions of bubble concentration in the ocean and their dependence on depth, wind speed, and temperature. The shape of the size distribution of bubble concentration generally exhibits a decrease with increase in R_{bub} over the range 0.1-0.35 mm, but there are few measurements of concentrations of larger bubbles (especially of the sizes necessary for film drop production; §4.3.3), and the results for smaller bubbles are inconsistent and inconclusive. Laboratory studies, while permitting investigation of mechanisms of bubble formation and evolution and the dependencies of these phenomena on controlling factors under well defined conditions, do not adequately simulate the conditions and processes occurring in the ocean to provide values of size distributions of bubble concentration that might be used to obtain meaningful estimates of SSA production over the ocean. An important consistent finding of such studies is that behavior of bubbles in freshwater generally differs greatly from that of bubbles in seawater. A further unresolved issue is the inference of size distributions of bubbles that burst at the surface from size distributions of bubbles measured at various depths in the ocean. For all these reasons it seems unlikely, at least at present, that bubble concentrations and fluxes determined from such studies can be used with confidence to calculate SSA production in the ocean to accuracy of better than one or two orders of magnitude.

4.4.2. Behavior of Individual Bubbles

Although it is the size distribution of bubble flux to the surface that determines the size distribution of the SSA particles produced, the commonly measured property of bubbles (§4.4.1) is the size distribution of bubble concentration, necessitating calculation of the flux from the concentration in order to determine the interfacial SSA production flux by the bubble method (§3.4). This calculation requires knowledge of the motion of an air bubble in seawater under the influence of gravity (buoyancy).

Additionally, this calculation requires taking into account the evolution of a bubble from a given depth to the surface, during which time the bubble can change its radius by exchanging gases with the surrounding water and may completely dissolve. Hence application of the bubble method requires knowledge of the terminal velocity of a bubble of a given size $v_{bub}(R_{bub})$ in still water and of size-dependent gas-exchange properties of bubbles. These topics are examined in this section.

4.4.2.1. Kinematics. The kinematics of a gas bubble in a liquid is considerably more complex than that of a solid sphere for several reasons: the conditions at the interface are different, surface tension may play a role, the bubble can deform and/or oscillate in shape, internal motion may occur, and the bubble may exchange gas with the surrounding fluid, resulting in a change in its size. Additionally, the bubble may break up or it may coalesce with another bubble. Furthermore, surface-active materials can exert a strong influence on all of these phenomena. The general case of a drop (bubble) of one fluid moving in another exhibits an extremely wide variety of behavior, many aspects of which have been successfully explained [*Haberman and Morton*, 1953; *Harper*, 1972; *Grace*, 1973; *Wallis*, 1974; *Grace et al.*, 1976; *Clift et al.*, 1978; *Grace*, 1983; *Churchill*, 1988, Chap. 17; *Karamanev*, 1994; *Nguyen*, 1998]. Results are often presented in terms of several dimensionless parameters describing properties of the fluid and the relative importance of viscous, buoyancy, inertial, and surface-tension forces, the most important parameter being the Reynolds number $Re = 2v_{bub}R_{bub}/v_f$, (here v_{bub} is the terminal velocity of the bubble and v_f is the kinematic viscosity of the surrounding fluid, i.e., water or seawater in the present context). Unfortunately, these dimensionless numbers are often used and named differently by different investigators, sometimes resulting in errors by subsequent investigators. Additionally, some investigators have used the volume-equivalent radius R_{bub} to describe the particle shape, whereas other have used alternative measures of bubble size, such as the cross-sectional area in the direction of motion, in some instances reporting better correlations of data [*Calderbank and Lochiel*, 1964; *Raymond and Zieminski*, 1971; *Miyahara, and Takahashi*, 1985; *Karamanev*, 1994; *Nguyen*, 1998]. Results presented in this review are given in terms of the volume-equivalent radius R_{bub}.

Several theoretical[1] and numerical [e.g., *Brabston and Keller*, 1975; *Ryskin and Leal*, 1984] investigations concerning the motion of a gas bubble in a liquid have been reported. However, in some instances boundary conditions

---

[1] *Miyagi* [1925]; *Davies and Taylor* [1950]; *Saffman* [1956]; *Hartunian and Sears* [1957]; *Moore* [1959, 1963, 1965]; *Levich* [1962, Chap. VIII]; *Chao* [1962]; *Taylor and Acrivos* [1964]; *Winnikow and Chao* [1966]; *Collins* [1966]; *Mendelson* [1967]; *Harper and Moore* [1968].

were not accurately treated, and in others approximations for the shape of the bubble (such as symmetry in the direction of motion or about an axis, or constancy of shape) or of the flow field (such as potential flow, axisymmetric flow, or steady flow) were made that might not be valid for actual bubbles. Several of the investigations, because of the range of validity, are of rather limited applicability for air bubbles in water. For example, the range of validity of the treatment of *Moore* [1963] for air bubbles in water at 20 °C was given by $Re^{1/3} \gg 1$ and $Re \ll 130$. If the requirement $Re^{1/3} \gg 1$ can be considered to be met by $Re^{1/3} \geq 3$, then the range of validity can be expressed as $27 \leq Re \ll 130$. In terms of R_{bub} this expression is equivalent to 0.25 mm $\leq R_{bub} \ll 0.45$ mm; that is, a range of R_{bub} less than a factor of two. In other instances the range of validity, as well as the magnitudes of physical quantities such as R_{bub} and v_{bub}, are not immediately apparent as results have often been expressed in terms of the several dimensionless groups.

Several investigations have reported incorrect conclusions. As noted by *Harper* [1972] in a review on this topic, "Further developments have been remarkable for the number of errors perpetrated: most writers on the subject seem to have made at least one in published work." This tendency has persisted. For example, *Thorpe* [1982] used a drag coefficient C_d from *Moore* [1963], but used a different definition for this quantity, resulting in an equation for v_{bub} that is too low by a factor of two for small R_{bub} (the value 18 in the denominator in Eq. 11 of *Thorpe* [1982] should be 9, and the factor 0.091 should be 0.182). The equation for the rise velocity of bubbles of *Wu* [1992a] was presented incorrectly, as noted by *Andreas et al.* [1995]; both terms on the right hand side of Eq. 12 of *Wu* [1992a] should contain the factor ρ_w/ρ_a (instead of the factor ρ_a/ρ_w in the first term). Additionally, this equation did not account for the added mass of an accelerating bubble—the amount of the fluid it "drags" with it, resulting in an effectively increased mass which is described by an added mass coefficient β, the ratio of the volume of the entrained fluid to that of the bubble [*Garrettson*, 1973]; experimentally, $\beta \approx 2$ for air bubbles in water [*Karanfilian and Kotas*, 1978]. The drag coefficient given by *Keeling* [1993, p. 246] was also presented incorrectly; the factor $(1 + 0.566Re^{0.5})$ in the denominator of the expression for C_d should have been $(1 + 0.0566Re^{0.5})$ in the numerator [*Keeling*, personal communication, 1999]. These

incorrect expressions have not infrequently been propagated. For instance, the expression for v_{bub} presented by *Thorpe* [1982] has been used by numerous subsequent investigators in models of gas exchange between the atmosphere and the ocean (§4.4.2.2).

Numerous experimental investigations on the motion of gas bubbles in water (including distilled water or tap water) or seawater have been also reported[1]. Results from these investigations are by no means consistent, often differing greatly. For instance, sizes for the onset of certain behavior varied from one investigation to another, as did the behaviors themselves. Results in distilled water often differed from those in tap water or seawater (though sometimes not), results from filtered tap water were in some instances different from those in unfiltered tap water, and experiments in so-called "hyper clean" water yielded different results from those in distilled water. Some of these differences have been attributed to the presence of small particles or other substances in the water, and nearly every investigation listed above since that of *Bond* [1927] has noted the extremely large influence of surface-active materials on the behavior and motion of bubbles. Indeed, of all aspects of SSA production, bubble behavior and motion are probably the most strongly influenced by surface-active materials.

A bubble can undergo internal circulation which reduces the relative motion between its surface and the surrounding fluid, resulting in reduced drag and thus a greater terminal velocity than that of a solid sphere under the same conditions. Bubbles that exhibit this behavior are referred to as fluid spheres and are termed "clean" bubbles (discussed further below). However, many experimental investigations have reported that small air bubbles in water (or seawater) often behave like solid spheres (that is, as if they have no internal motion), and that larger air bubbles in water (or seawater) often exhibit behavior intermediate to that of a fluid sphere and that of a solid sphere. A bubble whose terminal velocity is the same as that of a solid sphere of the same radius and density is commonly referred to in the literature as "dirty," as such behavior has been attributed to the presence of surface-active substances in the water that are collected by the bubble on its upper surface and swept toward the back of the bubble as it rises, causing a surface tension gradient along the bubble and resulting in formation of a stagnant cap. Such a cap affects the motion of the

---

[1] *Allen* [1900]; *Miyagi* [1925]; *Bond* [1927]; *Bond and Newton* [1928]; *O'Brien and Gosline* [1935]; *van Krevlin and Hortijzer* [1950]; *Garner* [1950]; *Datta et al.* [1950]; *Coppock and Meiklejohn* [1951]; *Haberman and Morton* [1953]; *Peebles and Garner* [1953]; *Garner and Hammerton* [1954]; *Uno and Kintner* [1956]; *Saffman* [1956]; *Houghton et al.* [1957]; *Lieberman* [1957]; *Hartunian and Sears* [1957]; *Calderbank and Lochiel* [1964]; *Redfield and Houghton* [1965]; *Calderbank et al.* [1970]; *Marks* [1973]; *Bachhuber and Sanford* [1974]; *Motarjemi and Jameson* [1978]; *Detwiler and Blanchard* [1978]; *Tedesco and Blanchard* [1979]; *Detsch* [1991]; *Duineveld* [1995]; *Leifer et al.* [2000]; *Patro et al.* [2002].

bubble, inhibiting internal circulation, and influencing the motion of the surrounding fluid near the back of the bubble. As much of the bubble surface may be relatively free of surface-active material, part of the air within the bubble may continue to circulate (in which situation the motion will be intermediate between that of a fluid sphere and that of a solid sphere) and a kinematically dirty bubble may act like a clean bubble with regard to gas exchange; bubbles have been modeled in this manner by *Keeling* [1993].

Whether or not a bubble behaves as if it were clean or dirty depends on impurities in the water and the distance the bubble has traveled (or the time it has spent) in the water [*MacIntyre*, 1974b; *Bachhuber and Sanford*, 1974; *Detwiler and Blanchard*, 1978]; this in turn depends upon the bubble's initial depth and rise speed, the latter of which depends also on bubble size and water temperature. *Bachhuber and Sanford* [1974] and *Detwiler and Blanchard* [1978] demonstrated that bubbles initially (i.e., immediately after being introduced into the water) behave like fluid

spheres (i.e., as if they were clean), but that with increasing time in the water they behave more like solid spheres (i.e., as if they were dirty). The critical age at which this transition occurs generally differs from that beyond which no further change in the jet drop ejection heights occurs [*Detwiler and Blanchard*, 1978], another consequence of the accumulation of surface-active materials onto the surface of the bubble (§4.3.3.3).

The motion of a bubble in water or seawater depends primarily on its size, although this motion can be strongly affected by surface-active materials, as noted above. Additionally, depending on bubble size, temperature may affect bubble rise speeds because of its controlling influence on the kinematic viscosity of water (or seawater), which decreases by nearly a factor of two over the temperature range 0 °C to 20 °C (§2.5.1). The terminal velocity of a bubble v_{bub} with given equivalent-volume radius R_{bub} in still water is shown in Fig. 34, and the shape of a bubble in different ranges of R_{bub} is indicated. Also shown are the time

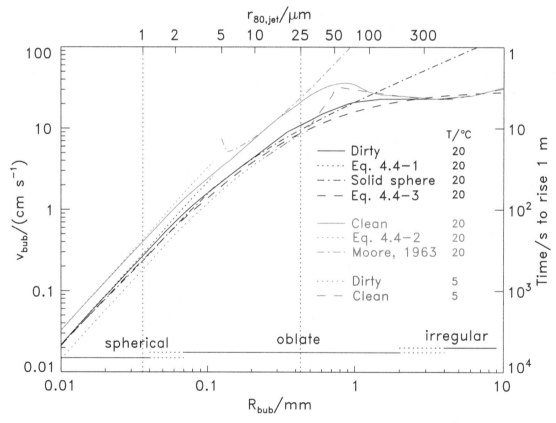

Figure 34. Terminal velocity of a bubble in still water, v_{bub}, as a function of R_{bub}, at 20 °C and at 5 °C, primarily from measurements of *Haberman and Morton* [1953] and *Duineveld* [1995]. Right axis shows time for 1 m rise in still water. Upper axis shows values of r_{80} based on (4.3-3); vertical dotted lines at r_{80} values of 1 μm and 25 μm demarcate small, medium, and large SSA particles (§2.1.2). Also shown for 20 °C are limiting cases of small solid spheres (Eq. 4.4-1) and small fluid spheres (Eq. 4.4-2), approximation (4.3-3) valid over entire range of R_{bub}, values for solid sphere with the same R_{bub}, and theoretical results of *Moore* [1963].

for a bubble to rise a distance of 1 m in still water and the sizes of jet drops produced for a given R_{bub} according to (4.3-3). Bubbles of all sizes attain their terminal velocities very rapidly, requiring less than a few tenths of a second to attain 90% of these terminal velocities when initially at rest in still water (not including the change in terminal velocity with time caused by uptake of surface-active materials discussed above), during which time they will have traveled no more than a few centimeters. Thus for nearly all considerations of bubbles in the oceans pertinent to SSA production these times and distances can be neglected and bubbles can be accurately assumed to rise at their terminal velocities with respect to the surrounding water.

Bubbles with $R_{bub} \lesssim 0.06$ mm are essentially spherical in shape and in still water they rise in straight vertical paths. The terminal velocity of a bubble in this size range is determined primarily by viscosity; hence this velocity depends strongly on temperature. For a sufficiently small bubble whose motion has been affected by surface-active materials to the extent that it behaves as if it were a solid sphere because of suppression of internal motion by accumulated surface-active materials, the terminal velocity can be obtained from Stokes' Law (§2.6.2):

$$v_{bub}(R_{bub}) = \frac{2g}{9v_w} R_{bub}^2, \qquad (4.4\text{-}1)$$

where v_w is the kinematic viscosity of water and g is the acceleration due to gravity. This equation yields $v_{bub}/(\text{cm s}^{-1}) \approx [R_{bub}/(0.07 \text{ mm})]^2$ at 20 °C (Fig. 34) and $v_{bub}/(\text{cm s}^{-1}) \approx [R_{bub}/(0.09 \text{ mm})]^2$ at 0 °C; these expressions are accurate to ~10% for $R_{bub} \lesssim 0.05$ mm and $R_{bub} \lesssim 0.075$ mm, respectively. For a sufficiently small bubble that acts as if it were a fluid sphere, that is, it undergoes internal circulation and its motion is not influenced by surface-active materials, an expression for the terminal velocity can be obtained [Hadamard, 1911; Rybczynski, 1911] which for a given acceleration force has the same functional dependence on R_{bub} as (4.4-1) for the Stokes velocity of a solid sphere with the same radius and density, but which is greater by a factor that depends only the ratio of the dynamic viscosities of the two fluids. For air bubbles in water, this factor is 3/2, and the terminal velocity is given by

$$v_{bub}(R_{bub}) = \frac{g}{3v_w} R_{bub}^2. \qquad (4.4\text{-}2)$$

This expression yields $v_{bub}/(\text{cm s}^{-1}) \approx [R_{bub}/(0.055 \text{ mm})]^2$ at 20 °C (Fig. 34) and $v_{bub}/(\text{cm s}^{-1}) \approx [R_{bub}/(0.075 \text{ mm})]^2$ at 0 °C. Most experimental investigations have observed

terminal velocities of small bubbles intermediate to these two limiting cases.

Determination of v_{bub} for bubbles sufficiently large that their motion is no longer Stokesian requires a relation between the drag coefficient C_d, which for air bubbles in water is given by $C_d = (8/3)(gR_{bub}/v_{bub}^2)$, and the Reynolds number Re, analogous to gravitational deposition of liquid drops in air (§2.6.2). In this size range $v_{bub}(R_{bub})$ increases less than quadratically with R_{bub}. For dirty bubbles, the approximation $v_{bub}/(\text{cm s}^{-1}) \approx 25(R_{bub}/\text{mm}) - 0.7$ is accurate to ~5% for $R_{bub} = 0.05\text{-}0.75$ mm at 20 °C; likewise, the approximation $v_{bub}/(\text{cm s}^{-1}) \approx 21(R_{bub}/\text{mm}) - 1.0$ is accurate to ~5% for $R_{bub} = 0.1\text{-}1$ mm at 0 °C. For clean bubbles few data have been reported, but it is expected that the terminal velocity remains higher than that of a dirty bubble by roughly a factor 3/2, consistent with the theoretical results of Moore [1963], which yield terminal velocities that approximate measured values for air bubbles in water with R_{bub} in the range 0.25-0.45 mm at 20 °C, although they depart appreciably outside of this range.

As R_{bub} increases from 0.6 mm to 2.5 mm, surface tension becomes increasingly important in determining the shape and terminal velocity of a bubble; thus the effect of temperature, which has relatively little effect on surface tension (§2.5.1), becomes increasingly less important in determining v_{bub}. With increasing R_{bub}, the shape of a bubble increasingly distorts from spherical and becomes more oblate, being increasingly flattened in the direction of motion until the radius in the direction perpendicular to motion is nearly twice that in the direction of motion at $R_{bub} \approx 2.5$ mm. Fore-aft symmetry no longer holds for a bubble in this size range. The motion of such a bubble becomes complex, and the bubble may oscillate in both path and shape, with the motion increasingly departing from straight vertical ascent as the bubble rises with a rocking, zigzag, or helical motion (caused by shedding of vortices from the rear of the bubble). With increasing R_{bub} the terminal velocity of a dirty bubble increases to 20-25 cm s^{-1} near $R_{bub} \approx 1$ mm at 20 °C and remains near this value for R_{bub} up to several millimeters. In contrast, with increasing R_{bub} the terminal velocity of a clean bubble increases until it attains a maximum of ~35 cm s^{-1} near $R_{bub} \approx 0.75$ mm at 20 °C (and near $R_{bub} \approx 0.85$ mm at 0 °C), then decreases slightly to just below 25 cm s^{-1} at $R_{bub} \approx 2.5$ mm.

Bubbles with $R_{bub} \gtrsim 2.5$ mm become very irregular in shape, pulsating and rising with a rocking or helical motion up to $R_{bub} \approx 5\text{-}9$ mm. The cleanliness of a bubble has virtually no effect on its motion in this size range, as indicated by the merging of the lines in Fig. 34 denoting velocities of clean and dirty bubbles. Likewise, temperature has little effect on v_{bub} in this size range. The near constancy of v_{bub} over the approximate range of R_{bub} from 1 mm to 5 mm can be attributed partly to the flattening of the bubble, which results in an increase in drag for the same volume-equivalent radius.

Bubbles with $R_{bub} \gtrsim 5\text{-}9$ mm form spherical caps and rise in relatively straight paths, and their velocities increase roughly proportional to $R_{bub}^{1/2}$ (and thus vary by less than 25% over this range of sizes). The velocity of such bubbles is independent of the properties of the liquid such as its density, viscosity, or surface tension. These bubbles may become unstable and break into smaller bubbles.

In several instances approximations for v_{bub} have been extrapolated far beyond their range of validity, leading to erroneous results. For example, *Deane* [1997] extrapolated *Thorpe's* [1982] expression for the terminal velocity of a clean bubble to arrive at a values of $v_{bub} = 140$ cm s^{-1} for $R_{bub} = 6$ mm, compared to measured values of ~25 cm s^{-1} (Fig. 34); likewise, v_{bub} was assumed by *Garrett et al.* [2000] to be given by Stokes' Law for bubbles as large as $R_{bub} = 0.5$ mm, at which size this approximation overestimates the rise velocity at 20 °C by more than a factor of 4 (Fig. 34).

A simple expression for v_{bub} accurate to ~20% for dirty bubbles over the entire range of R_{bub} up to 10 mm at 20 °C is given by

$$\frac{v_{bub}}{\text{cm s}^{-1}} = \frac{33\left(\dfrac{R_{bub}}{\text{mm}}\right)^2}{\left(\dfrac{R_{bub}}{\text{mm}} + 0.37\right)^2}; \qquad (4.4\text{-}3)$$

this expression is shown in Fig. 34. Similar expressions can also be obtained for clean bubbles at 20 °C or for dirty bubbles or clean bubbles at other temperatures.

There are several implications of the above results. The dependency of v_{bub} on surface-active materials results in uncertainty because of incomplete knowledge of the amount and nature of surface-active substances in the ocean surface waters and their effects on bubble motion, in addition to concerns such as whether or not a rising bubble will encounter such substances or whether they will have been scavenged by previously-rising bubbles. Further uncertainty results if the temperature is not known. These effects vary with R_{bub}, but the overall uncertainty in v_{bub}, which must be taken into account in estimation of the interfacial SSA production flux according to the bubble method (§5.5), can be estimated from Fig. 34 to be a factor of $\lesssim 2$. Additionally, at a given temperature the value of R_{bub} corresponding to a given velocity may differ between clean and dirty bubbles by up to 50% for bubbles with $R_{bub} \lesssim 0.25$ mm and up to nearly a factor of two for larger bubbles (and by a greater amount if temperature is

not known); consequently, use of bubble rise velocity to determine bubble size can result in substantial error.

As noted above, bubbles with $R_{bub} \gtrsim 1$ mm, whether clean or dirty, have terminal velocities near or slightly exceeding 25 cm s^{-1}, for the most part independent of temperature and unaffected by the presence of surface active materials. Such bubbles require less than a few seconds to rise to the surface from a depth of 1 m in still water; consequently measurements of size distributions of bubble concentration in the ocean will not include the majority of bubbles in this size range unless they are made within a few seconds after a breaking wave. Additionally, if the brightness that identifies whitecaps is the result of large bubbles which rise rapidly to the surface, then whitecap persistence would not be strongly affected by temperature, contrary to what has often been suggested by some investigators (§4.5.3.1). Bubbles in this size range produce film drops; hence the production of these drops would be expected to occur soon after a breaking wave, whereas jet drop production would occur later, as demonstrated by laboratory experiments of *Woolf et al.* [1987] discussed in §4.3.4.3.

Bubbles sufficiently small to produce jet drops with $r_{80} \lesssim 1$ μm (and thus having $R_{bub} \lesssim 0.04$ mm) would require long times to reach the surface (several minutes for bubbles starting at depths greater than 1 m) during which they may alter their size by gas exchange with the surrounding water and dissolve completely. This is discussed in the next section.

4.4.2.2. Time-evolution. As noted in §4.4.1, concerns over the application of the bubble method using measurements of bubble concentrations at various depths in the ocean arise as a consequence of the influence of exchange of gas between a bubble and the surrounding water on the size and thus the rise speed of the bubble and possible complete dissolution of the bubble before it reaches the surface. Gas exchange depends on not only the size of the bubble, but cleanliness of the interface (which in turn depends on the time the bubble has spent in the water and the amount of surface-active materials present), the gas pressure difference between the interior of the bubble and the surrounding water (which in turn depends on the depth, the saturation of the water with respect to the major atmospheric gases—nitrogen and oxygen, the surface tension, and the size of the bubble), the relative velocity between the bubble and the surrounding water (which affects the shape and thickness of the diffusive layer adjacent to the bubble wall), and the diffusivities of the gases in water. Some of these factors depend also on temperature. Several experimental investigations of gas exchange between bubbles and surrounding water or seawater have been reported[1]; in some instances

[1] *Pattle* [1950]; *Coppock and Meiklejohn* [1951]; *Wyman et al.* [1952]; *Lieberman* [1957]; *Manley* [1960]; *Deindoerfer and Humphrey* [1961]; *Calderbank and Lochiel* [1964]; *Redfield and Houghton* [1965]; *Zieminski and Raymond* [1968]; *Calderbank et al.* [1970]; *Lessard and Zieminski* [1971]; *Motarjemi and Jameson* [1978]; *Goncharov et al.* [1984]; *Detsch* [1990]; *Harris and Detsch* [1991].

differences have been observed in the exchange coefficients of bubbles in freshwater versus seawater. As noted above, bubbles have been labeled "clean" or "dirty" in reference to gas exchange as well as to their motion, although typically dissolution of air bubbles in water or seawater has been found to exhibit behavior intermediate to that of clean or dirty bubbles. In nearly all studies the effects of surface-active materials on the gas-exchange properties were noted. Surface-active materials typically decrease the rate of gas exchange apparently by presenting an additional barrier to interfacial mass transfer of the gas molecules, and they can thereby allow small bubbles to remain stable against complete dissolution [*Johnson and Cooke*, 1981a; *Ramsey*, 1962b], as noted above (§4.4.1.2). Surface-active substances affect the kinematic behavior of a bubble differently from its gas-exchange behavior, and as noted above (§4.4.2.1) the time required for a bubble to become dirty with respect to gas exchange is in general greater than the time required for the bubble to accumulate sufficient surface-active materials to appreciably affect its motion.

The motion of a bubble and its rate of gas exchange with the surrounding fluid may be strongly coupled. Any change in size in the bubble resulting from exchange of gas with the surrounding water affects the motion of the bubble through the dependence of v_{bub} on R_{bub}. Likewise, the motion of a bubble can affect its rate of gas exchange, and as a bubble rises it experiences decreasing hydrostatic pressure, resulting in a change in gas saturation between the interior of the bubble and the surrounding water. Coupled differential equations describing the time evolution of the radius and depth of a single bubble, and which include the effects of surface tension, hydrostatic pressure, and gas exchange, have been presented by *LeBlond* [1969a,b]; these equations relate the rate of change of the bubble radius and its rate of gas exchange to rate of change of hydrostatic pressure or depth (i.e., the rise velocity). Similar equations for the evolution of the size distribution of bubble concentration were presented by *Garrettson* [1973]. Several studies, some of which have investigated the role of bubbles in air-sea gas exchange, have used such equations [*Wyman et al.*, 1952; *Blanchard and Woodcock*, 1957; *Garrettson*, 1973; *Thorpe*, 1982; *Merlivat and Memery*, 1983; *Thorpe*, 1984a; *Memery and Merlivat*, 1985; *Woolf and Thorpe*, 1991; *Thorpe et al.*, 1992; *Woolf*, 1993; *Keeling*, 1993]. In view of the complexities noted above, simplifying assumptions have typically been made in the solution of these equations, and usually only one or two expressions are used to describe the motion of a bubble (usually considered dirty) and its gas-exchange properties over the entire size range. Because of interdependence of kinematics and gas exchange, the behavior of a bubble under given conditions of temperature, saturation, and the like exhibits a rather complex dependence on the initial depth and size of the bubble.

This dependence is illustrated in Fig. 35, in which the evolution of bubbles of various radii has been calculated with a model similar to those of *Wyman et al.* [1952] and *Blanchard and Woodcock* [1957]. Here the bubble starts from a specified initial depth of 4 m in still water at 20 °C that is fully (100%) saturated with air, and is assumed to behave kinematically as a dirty bubble with terminal velocity given by that of a solid sphere of the same radius and density.

A common feature of models such as this is the existence of a critical radius $R_{bub,c}$ for a given depth; bubbles with $R_{bub} < R_{bub,c}$ dissolve completely before reaching the surface because of their slow rise velocities and the combined effects of surface tension and hydrostatic pressure, with only bubbles with $R_{bub} > R_{bub,c}$ surviving to reach the surface.

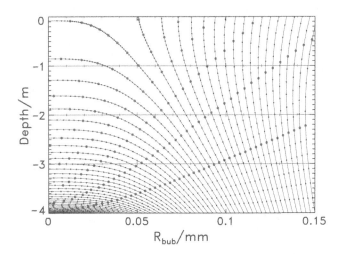

Figure 35. Evolution of size and depth of air bubble in water or seawater as a function of time evaluated with model similar to that of *Wyman et al.* [1952]. Curves denote radius-depth trajectory in still water of bubbles initially at indicated radius and depth of 4 m; small ticks on trajectory curves denote intervals of 10 s and larger ticks denote intervals of 1 min. As bubbles rise their radius may initially decrease as air, under hydrostatic and surface-tension pressure, is transferred into the surrounding water, assumed saturated in air at the surface pressure of 1 atm (1.01 bar). Smaller bubbles continue to decrease in size as they rise and completely dissolve ($R_{bub} = 0$) before reaching the surface; larger bubbles attain a minimum radius but then grow under reduced hydrostatic pressure before arriving at the surface. For bubbles initially at a depth of 4 m, the critical value of R_{bub} below which bubbles dissolve before reaching the surface is $R_{bub,c} \approx 0.115$ mm. The figure may also be used to determine the critical bubble radius for other initial depths; for example, the critical radius for an initial depth of 1 m is ~0.05 mm. Bubbles were assumed to instantaneously attain their terminal velocities and to rise as rigid spheres with no internal circulation ("dirty" bubbles). Calculations were made for 20 °C with gas exchange coefficient $4 \cdot 10^{-9}$ mol cm^{-2} s^{-1} atm^{-1}.

According to the model described above, $R_{bub,c} \approx 0.115$ mm for an initial depth of 4 m (Fig. 35). This critical radius decreases with decreasing depth; for an initial depth of 1 m, $R_{bub,c} \approx 0.05$ mm, the radius to which a bubble with $R_{bub} \approx 0.115$ mm initially at 4 m has evolved. According to (4.3-3), a bubble with $R_{bub} = 0.05$ mm produces jet drops with $r_{80} \approx 1.5$ μm, and smaller jet drops are produced by still smaller bubbles. As the fate of such bubbles is dissolution rather than persistence to the surface and bursting, few jet drops of $r_{80} \lesssim 1$ μm are expected to be produced in the ocean (§4.3.2.4). Bubbles with $R_{bub} > R_{bub,c}$ either initially shrink and later grow as they encounter lower hydrostatic pressure near the surface, or else they grow for all times. Bubbles that are more than slightly larger than the critical size rise to the surface sufficiently rapidly that they experience little change in size (Fig. 35).

The critical radius for a given depth depends on the cleanliness of the bubble, (both with regard to its motion and to its gas-exchange properties), the temperature of the water (which affects viscosity and diffusivity), the saturation of the water with respect to nitrogen and oxygen (other gases occur in such low concentrations as to be unimportant), and of course any downward velocity of the surrounding water (which might occur after wave breaking or under a windrow or above a downwelling portion of a Langmuir cell). Several investigations have examined this topic under a variety of assumptions and with somewhat differing models, and not surprisingly have obtained different estimates for this critical radius. The critical radius for a bubble in 0 °C seawater starting from a depth of 0.3 m was 0.05 mm according to the model of *Blanchard and Woodcock* [1980]; for bubbles entrained at the surface in a uniform downward flow of 3 cm s^{-1} (to account for organized flow from wave breaking) a critical radius of 0.14 mm resulted from the model of *Thorpe* [1982], even under conditions of 10% supersaturation; a critical radius of 0.04 mm for a depth of only 10 cm was obtained by *Merlivat and Memery* [1983] using a model of a bubble near the air-water interface under a surface wave; and a critical radius of around 0.07 mm for a starting depth of 1 m at 0 °C resulted from the model of *Johnson* [1986].

Because of dependencies such as these, it is difficult to model the behavior of a bubble population in anything more than a qualitative sense unless accurate values are known for a variety of factors and environmental conditions, some of which, such as the cleanliness of the bubble and the velocity field under a breaking wave, would be difficult to determine.

Most models contain the assumption that bubbles rise through still water, or at least through water that is not turbulent, and the effects of turbulence (which would occur under a breaking wave) on the rise of bubbles and their gas exchange are not known [*Keeling*, 1993]. Nonetheless, models usefully illustrate key factors influencing the behavior of bubbles, and they clearly demonstrate that the size distribution of bubble concentration can be greatly modified during the evolution of a bubble population with time. The fact that the critical radius for depths at which most bubble populations have been measured occurs at a value equal to or exceeding the size of bubbles that contribute most to bubble concentrations in the ocean raises serious questions concerning the possibility of using measurements of bubble size distributions taken at depth to infer distributions at the surface, where the bubbles burst and form drops.

4.5. OCEANIC WHITECAPS

Whitecaps are remnants of breaking waves, formed when large numbers of bubbles produced by a wave rise to the surface. The bright white area of a whitecap arises from multiple light scattering by elements of sizes comparable to or greater than the wavelength of visible light; for whitecaps these elements are clusters of proximate, closely packed bubbles at or near the surface. It is the bursting of these bubbles that is the major process responsible for SSA production, and thus it is expected that the production of SSA particles increases with increasing fraction of the ocean surface covered by whitecaps. This expectation is the basis for the whitecap method (§3.2), which makes the stronger assumption that the size-dependent interfacial SSA production flux is directly proportional to the whitecap ratio W, the fractional area of the sea surface covered by white area, and infers this production flux from the product of the measured SSA production from laboratory whitecaps and W. Hence W is an important quantity for SSA production. Additionally, the discrete whitecap method (§3.2) assumes that the decay of the area of an individual whitecap is an exponentially decreasing function of time and requires knowledge of this characteristic decay time τ_{wc}. Hence properties of individual whitecaps and their temporal behavior is also important for considerations of SSA production.

There is an extensive body of research on whitecaps, as they have been investigated for their importance to a variety of geophysical processes in addition to SSA production, including the albedo of Earth[1]; gas exchange between the

---

[1] *Payne* [1972]; *Gordon and Jacobs* [1977]; *Koepke* [1984]; *Stabeno and Monahan* [1986]; *Monahan and O'Muircheartaigh* [1987]; *Frouin et al.* [2001].

oceans and the atmosphere[1]; air-sea exchange of heat and moisture [*Bortkovskii*, 1987; *Andreas and Monahan*, 2000]; remote sensing of the oceans and atmosphere[2] noise production in the ocean[3]; and transport and transformation of organic matter [*Monahan and Dam*, 2001]. Additionally, theoretical investigations of wave breaking and subsequent whitecap formation have been reported[4], as have theoretical and field investigations on the frequency of wave breaking and the fraction of waves that break[5]; however, these investigations have not proved fruitful in allowing determination of W and τ_{wc} from easily measurable quantities such as wind speed. Consequently, analyses pertinent to SSA production have been based almost entirely on W determined by observation at sea and τ_{wc} determined for oceanic or laboratory whitecaps. Data on oceanic whitecaps pertinent to SSA production are examined in this section, specifically, measured oceanic whitecap ratios as a function of controlling factors, and the temporal behavior of individual oceanic whitecaps.

4.5.1. Measurements

The whitecap ratio W has been reported as a function of wind speed and other meteorological quantities by several investigations, as summarized in Table 20. Key studies are that of *Toba and Chaen* [1973, hereinafter TC], and those of *Monahan* and coworkers in several large-scale studies (BOMEX+, JASIN, STREX, MIZEX83, MIZEX84, HEXPILOT, and HEXMAX); additionally the arguments and data presented in *Bortkovskii* [1987] are especially pertinent. Tables listing whitecap ratio, wind speed, air and sea temperature, dew point, significant wave period, and number concentration of SSA particles with $r_{80} > 8$ μm for

observation intervals of typical duration 20-30 min were presented by *Toba and Chaen*. Whitecap ratio and standard deviation, wind speed, and air and sea temperature for observation intervals of typical duration 30-60 min were reported by *Monahan* and coworkers. Whitecap ratio and wind speed, classified by sea temperature, were presented in graphical form only by *Bortkovskii* [1987, Chap. 2].

Whitecap ratios have been determined from photographs of the sea surface taken from aircraft, towers, and ships, and from video tapes taken from ships. Typically, photographic determination of W has consisted of measuring the amount of white area from enlarged photographs by cutting and weighing, or by counting squares, with the results of several photographs averaged for a given observation interval of duration several minutes [*Blanchard*, 1963, p. 160; *Kanwisher*, 1963; *Monahan*, 1969; *Monahan and Zietlow*, 1969; *Monahan*, 1971; *Toba and Chaen*, 1973; *Bondur and Sharkov*, 1982; *Bortkovskii*, 1987, pp. 51-59]; photographs have also been analyzed by densitometers [*Nordberg et al.*, 1971; *Ross and Cardone*, 1974]. A typical method for determination of W from video tapes has been to use an automatic area analyzer for each ten second period, independently by at least two investigators, for a five minute observation interval, from which the mean and standard deviation were calculated [*Monahan et al.*, 1988]. More recent determinations have involved digitizing photographs or videos and determining the fraction of pixels that exceed a given threshold for brightness.

Difficulties in determining W have been discussed by several investigators [*Monahan*, 1969, 1970; *Williams*, 1970; *Blanchard*, 1971a; *Monahan*, 1971; *Toba and Chaen*, 1973; *Kondo et al.*, 1973; *Ross and Cardone*, 1974; *Bondur and Sharkov*, 1982; *Monahan*, 1982; *Cipriano and Blanchard*, 1982; *Koepke*, 1984; *Bortkovskii*, 1987; *Stramska and*

---

[1] *Redfield* [1948]; *Kanwisher* [1963]; *Atkinson* [1973]; *Thorpe* [1982, 1984a]; *Monahan and Spillane* [1984]; *Kerman* [1984a]; *Memery and Merlivat* [1985]; *Thorpe* [1986]; *Mason et al.* [1988]; *Torgerson et al.* [1989]; *Woolf and Thorpe* [1991]; *Wallace and Wirick* [1992]; *Asher et al.* [1992]; *Woolf* [1993]; *Keeling* [1993]; *Farmer et al.* [1993]; *Erickson* [1993]; *Asher et al.* [1996]; *Woolf* [1997]; *Asher and Wanninkhof* [1998]; *Kettle and Andreae* [2000]; *Monahan* [2002].

[2] *Cox and Munk* [1954]; *Williams* [1969]; *Nordberg et al.* [1971]; *Stogryn* [1972]; *Ross and Cardone* [1974]; *Tang* [1974]; *Webster et al.* [1976]; *Gordon and Jacobs* [1977]; *Quenzel and Kaestner* [1980]; *Monahan et al.* [1981]; *Koepke and Quenzel* [1981]; *Whitlock et al.* [1982]; *Zheng et al.* [1983a]; *Wentz* [1983]; *Koepke* [1986a]; *Griggs* [1986]; *Monahan and O'Muirchaertaigh* [1986]; *Melville et al.* [1988]; *Jessup et al.* [1990]; *Jessup et al.* [1991a,b]; *Estep and Arnone* [1994]; *Gordon and Wang* [1994]; *Wu* [1995a]; *Frouin et al.* [1996]; *Jessup et al.* [1997]; *Gordon* [1997]; *Wagener et al.* [1997]; *Barber and Wu* [1997]; *Moore et al.* [1998, 2000]; *Terrill et al.* [2001]; *Nicolas et al.* [2001]; *Kahn et al.* [2001].

[3] *Glotov et al.* [1962]; *Perrone* [1970]; *Wilson* [1980, 1983]; *Kerman et al.* [1983]; *Kerman* [1984b]; *Farmer and Vagle* [1989]; *Medwin and Beaky* [1989]; *Loewen and Melville* [1991]; *Carey et al.* [1993a,b]; *Kennedy* [1993]; *Hollett* [1994]; *Oguz* [1994]; *Deane* [1997].

[4] *Unna* [1941]; *Longuet-Higgins and Turner* [1974]; *Banner and Phillips* [1974]; *Cokelet* [1977]; *Snyder and Kennedy* [1983]; *Kennedy and Snyder* [1983]; *Ochi and Tsai* [1983]; *Phillips* [1985]; *Huang et al.* [1986]; *Srokosz* [1986]; *Bortkovskii* [1987]; *Glazman and Weichman* [1989]; *Kerman and Bernier* [1994]; *Kraan et al.* [1996]; *Xu et al.* [2000]; *Zhao and Toba* [2001].

[5] *Donelan et al.* [1972]; *Thorpe and Humphries* [1980]; *Longuet-Higgins and Smith* [1983]; *Weissman et al.* [1984]; *Thorpe* [1986]; *Srokosz* [1986]; *Holthuijsen and Herbers* [1986]; *Xu et al.* [1986]; *Thorpe* [1993]; *Ding and Farmer* [1994]; *Gemmrich and Farmer* [1999]; *Xu et al.* [2000]; *Banner et al.* [2000]; *Melville and Matusov* [2002]; *Banner et al.* [2002].

Table 20. Measurements of Oceanic Whitecap Coverage

Study[a]	Reference	Location	Technique	Observation Intervals (nonzero W)	Maximum $U_{10}/(\text{m s}^{-1})$	$T_{sea}/\,°C$
BOMEX+	Monahan, 1971	Atlantic Ocean (mostly)	photos (432) from shoreline, towers, ship	70 (54)	17.4	17.4-30.55
TC	Toba & Chaen, 1973	East China Sea	photos (164) from ship	41 (36)	16.6	20.9-29.0
JASIN[b]	Monahan et al., 1981a	NE Atlantic Ocean	photos (599) from ship	55 (50)	15.3	12.5-14.0
STREX	Doyle, 1984	Gulf of Alaska	photos (779) from ship	85 (84)	17.2	5.11-11.11
MIZEX83	Monahan et al., 1984b	Greenland Sea/ Arctic Sea	photos (322) from ship	43 (21)	14	−1.4-14.4
MIZEX83	Higgins et al., 1985	Greenland Sea/ Arctic Sea	videos from ship	47 (29)	14.3	−1.8-13.0
MIZEX84	Doyle & Higgins, 1985	Greenland Sea/ Arctic Sea	photos from ship	56 (38)	13.1	−1.17-4.38
MIZEX84	Doyle & Higgins, 1985	Greenland Sea/ Arctic Sea	videos from ship	88 (65)	16.4	−0.64-13.1
HEXPILOT	Higgins et al., 1985	North Sea	videos from tower	28 (27)	14.1	10.2-11.7
HEXMAX	Monahan et al., 1988	North Sea	videos from tower	159 (155)	29.9	9.8-14.6
	Blanchard, 1963[c]	Carribean	photos (5) from aircraft	5 (5)	20	not stated
	Nordberg et al., 1971	Atlantic Ocean, North Sea	photos from aircraft	6 (4)	25[d]	2-10
	Kondo et al., 1973	Japan coast	photos from tower	14 (13)	14	14.8-26.0
	Ross & Cardone, 1974	Atlantic Ocean (mostly)	photos (427) from aircraft	13 (13)	24.7	not stated
	Martsinkevich & Melent'ev, 1975[f]	not stated	photos from aircraft	6 data	25	not stated
	Bondur & Sharkov, 1982	Black and Barents Seas	photos from aircraft	5 (3) wind speed intervals[g]	10.5	not stated
	Bortkovskii, 1987[h]	various locations, covering most of the world oceans	photos (>3000) from ship	> 109	23	≤ 28
	Monahan & Wilson, 1993[i]	Gulf of Alaska	videos from ship	~65 (~65)[j]	17	not stated
	Asher & Wanninkhof, 1998[k]	North Atlantic	videos from ship	31	17	5.5
	Xu et al., 2000	Bohai Bay, China coast	photos from tower	9 (9)[g]	19.8	15.5-16.9
GasEx-98	Asher et al., 2002	North Atlantic	videos from ship	~62 (~62)[j]	16.3	not stated
	Stramska & Petelski, 2003	Polar North Atlantic	photos	~67 (~67)[j]	13.5	2-13

[a] The following acronyms are used:
BOMEX+: Barbados Oceanographic and Meteorological Experiment, and other data.
TC: used in text to denote the data set of *Toba and Chaen* [1973].
JASIN: Joint Air-Sea Interaction.
STREX: Storm Transfer Response Experiment.

Table 20. (Continued)

MIZEX: Marginal Ice Zone Experiment.
HEXPILOT: Humidity Exchange Over the Sea Pilot Experiment.
HEXMAX: Humidity Exchange Over the Sea Main Experiment.
GasEx-98: Gas Exchange 1998.
[b] Data used here are from *Monahan et al.* [1981a]; a classification of these data by different intervals is presented in *Monahan et al.* [1984b].
[c] From photographs taken from U.S. Weather Squadron Two over the Caribbean in June-October, 1952 [*Blanchard*, 1963, Fig. 31].
[d] The uncertainty in U_{10} was estimated to be around 3 m s^{-1}.
[e] Wind speeds refer to those a height of 19.5 m, which will typically be greater than U_{10} by ~1 m s^{-1} under stable conditions (although *Ross and Cardone* [1974] estimated that a wind speed of 18 m s^{-1} at 19.5 m corresponds to $U_{10} \approx 15$ m s^{-1}).
[f] From Fig. 2.5 of *Bortkovskii* [1987].
[g] Data from different observation intervals (the total number of which was not specified) with nearly the same wind speed were averaged.
[h] Data are from several investigations presented in *Bortkovskii* [1987, Fig. 2.5]. It is not clear if zero values are shown in his figure. Some of these same data, although classified by wind speed interval, were presented in *Bortkovskii and Novak* [1993], which additionally presented the number of measurements from the several studies; however, the numbers do not correspond exactly to the data in Fig. 2.5 or to the values in Table 2.1 of *Bortkovskii* [1987].
[i] From Fig. 10 of *Hanson and Phillips* [1999].
[j] Estimated from graphs in original investigation.
[k] Data were specifically chosen from times of relatively high wind speeds.

Petelski, 2003]. Key difficulties are bias due to the oblique angle at which a photograph was taken (with resultant geometric effects both on area and on whiteness threshold) and shadowing by large waves (for photographs or videos taken from ships or towers), inability of the measurement system to resolve all white areas (of importance for photographs taken from aircraft), lack of independence of photographs taken at closely spaced intervals because of the persistence of whitecaps, lack of contrast because of sea color and variable sky reflectance due to variations with illuminations and cloud, determination of a brightness threshold for what constitutes a whitecap, and small-number statistics arising from the stochastic nature of whitecapping (resulting in many photographs or video frames containing no white area). There does not appear to be consensus whether photographs from aircraft are more accurate than those from lower heights (such as those from ships). Whitecap ratios determined by photographs typically are greater than those determined by video by one to two orders of magnitude, as discussed below. Sampling strategies generally have not been discussed, and there is a possibility of bias due to taking pictures only when whitecaps were present, although in the investigations of *Monahan* and coworkers mentioned above, and possibly in others, photographs were taken at fixed intervals or otherwise without regard to the instantaneous presence or absence of whitecaps [*E. Monahan*, personal communication, 2004]. Photographs and videos are necessarily restricted to daytime and to conditions where visibility is not substantially impaired, leading to possible bias toward clear, non-storm conditions. Additionally, when photographs are made from aircraft, uncertainty may arise in determining surface wind speeds and values of other factors thought to be controlling [*Nordberg et al.*, 1971; *Bortkovskii*, 1987, p. 62].

Subjectivity in determining whitecap ratio is manifested by differences in the value of W determined from the same photograph by different analysts. Differences of 15-40% were reported in early studies [*Monahan*, 1969; *Williams*, 1970; *Monahan*, 1971; *Nordberg et al.*, 1971]. A study in which photographs were analyzed by two different observers and by four different methods resulted in differences of more than a factor of two 30% of the time when using a standard projection method, with differences among the several methods often being much greater [*Nolan*, 1988b]. This subjectivity is even more pronounced when W has been determined from video tapes: the value of W determined by an automatic area analyzer was reported to change by a factor of over 10 for two different gain settings, with the result being insensitive to gain only when the gain was so high that W was near zero [*Nolan*, 1988a].

Some investigators have distinguished between various forms of white area, although there is ambiguity and inconsistency over the terms used to described the various structures. *Nordberg et al.* [1971] distinguished between "whitecaps" and "foam streaks"; *Ross and Cardone* [1974] distinguished "actively forming whitecaps and large new foam patches" from "thin foam and foam streaks"; *Bondur and Sharkov* [1982] distinguished "crests", which they termed "dynamic foam", from "strip-like, or patchy structures", which they termed "static foam"; and *Koepke* [1984, 1986a,b] distinguished "whitecaps (foam patches)" and "foam streaks." Several investigators [*Nordberg et al.*, 1971; *Ross and Cardone*, 1974; *Bondur and Sharkov*, 1982] reported the fractional coverages of different types of white area separately. Although these investigators explicitly noted the subjectivity of these categorizations (*Ross and Cardone* [1974], for example, stated that their photographs, after being scanned into displays containing 32 different levels of brightness, were "subjectively divided" into categories of actively forming whitecaps, foam streaks, and foam-free water), they treated these structures not as if they were different stages of whitecap evolution, but rather as different objects. For instance, lifetimes and reflectances of both whitecaps (foam patches) and foam streaks were reported by *Koepke* [1984, 1986b]. Oceanic coverage of "whitecaps" and "foam patches" (frequently referred to by other researchers as foam streaks) were also separately reported by *Bortkovskii* [1987], who presented expressions relating the two quantities and proposed a theory describing the formation, lifetime, and behavior of the two types of structures, which he considered to be successive stages of the same whitecap. A conceptual model describing stages of whitecap development proposed by *Monahan* and coworkers [*Monahan et al.*, 1988; *Monahan and Woolf*, 1988; and *Monahan and Woolf*, 1989] is examined in more detail below.

The irregular nature of whitecapping, noted by several investigators, results in large temporal and spatial variability, increasing the difficulty in determining W and the uncertainty of any determination, in addition to difficulty in determining the dependence of W on controlling factors. The area fraction of white regions in a photograph or video can vary by over an order of magnitude over times of seconds, as illustrated in time series of video signals or reflectivities of the ocean surface [*Jessup et al.*, 1990; *Massouh and le Calve*, 1999; *Moore et al.*, 2000]; thus the size of the region examined, or the duration of the measurement, will affect determination of values of W and their corresponding uncertainties. As W increases, whitecaps become larger and more frequent, but for very low values of W, the spatial and temporal scales necessary to obtain meaningful statistics may be so great as to require averaging over inhomogeneous meteorological and/or environmental conditions.

Whitecap data are commonly reported as a mean and standard deviation of W for each observation interval, but with the standard deviation rather inconsistently representing the combined effects of subjectivity, small-number statistics, and/or variation in W during the observation interval. Often reported standard deviations for W exceeded the mean values. Standard deviations of W reported by *Toba and Chaen* [1973] were 2 to 4 times the mean values, for the JASIN photographic data set the standard deviation equaled or exceeded the mean value in more than 80% of the nonzero observation intervals, and for the MIZEX83 and MIZEX84 photo data sets the standard deviation equaled or exceeded the mean value in half of the nonzero observation intervals. For values of W determined from video data sets of MIZEX83, MIZEX84, HEXPILOT, and HEXMAX, the standard deviation equaled or exceeded the mean value in 93%, 95%, 96%, and 80% of the nonzero observation

intervals, respectively. Such large standard deviations limit confidence in these determinations and raise concern that the standard deviation might not be the appropriate statistic to characterize the uncertainty of the data. In situations for which whitecap coverage is so low that few photographs contain white area, the assumption of Poisson distribution would perhaps be better than that of a Gaussian distribution. Clearly the distributions of the uncertainties are highly skewed, as W cannot of course be less than zero.

Questions may also be raised regarding the representativeness of measurements. As most studies have been conducted at a single location and season, the ranges of meteorological variables such as sea temperature and wind speed are necessarily limited. The ranges of these quantities presented in Table 20 denote the extremes, with the majority of the data generally extending over considerably smaller ranges. For example, the sea temperature was less than 4 °C in 88% of the observation intervals in the MIZEX84 video data set, and was below 10 °C in all but one observation interval. Likewise, the sea temperature was in the range 28.0-29.0 °C in over 60% of the observation intervals in the BOMEX+ data set. This situation inevitably leads to concern whether a given investigation can be considered representative of other locations and seasons and can be generalized to the oceans as a whole. Additionally, attempts to discern the dependence of W on presumed controlling factors are limited by the inability to ascribe differences in W to differences in such factors as U_{10} and T_{sea} and not to other factors such as fetch or location specific to the given study.

4.5.2. Dependence on Wind Speed

Measurements of W have typically been classified by wind speed, U_{10}, based on the implicit assumption that this is the dominant factor controlling W. Data from investigations summarized in Table 20 are presented as a function of U_{10} in Fig. 36; a logarithmic ordinate scale is used to encompass the wide range of values of W. For studies in which investigators distinguished between various forms of white area, the figure indicates the total fraction of the sea surface covered by white. Each data point represents a single observation interval, or, for the investigations of *Bondur and Sharkov* [1982] and *Xu et al.* [2000], averages over a given wind speed interval (it was not specified whether the data of *Martsinkevich and Melent'ev* [1975], which were obtained from Fig. 2.5 of *Bortkovskii* [1987], corresponded to observation intervals or to mean values for a given wind speed interval). Of the approximately 1000 data shown in Fig. 36, the value of W was reported as zero for more than 100 of the measurements, with values of zero reported even for wind speeds as great as 12 m s^{-1}. It is likely that there are more zero values in the data sets shown in *Bortkovskii*

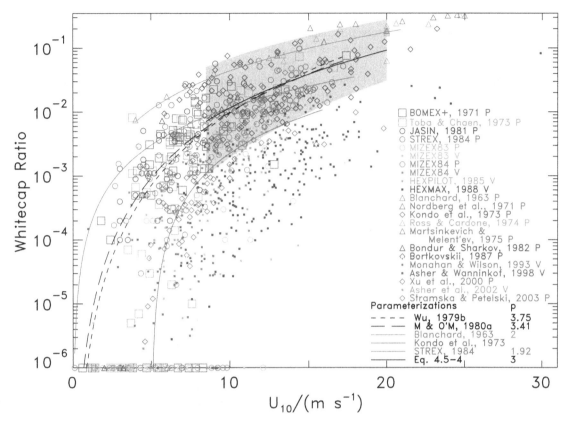

Figure 36. Measurements of whitecap ratio W as a function of the wind speed at 10 m above the sea surface, U_{10}, from investigations summarized in Table 20. P denotes photographic data sets; V denotes video data sets. Each data point represents a single observation interval, or, for data of *Bondur and Sharkov* [1982] and *Xu et al.* [2000], averages over several observation intervals for a given set of conditions. For investigations that distinguished between different types of white area, values presented are the sum of the several types. *Nordberg et al.* [1971] reported "whitecaps" and "foam streaks"; *Ross and Cardone* [1974] reported "actively forming white-caps and large new foam patches" and "thin foam and foam streaks"; *Bondur and Sharkov* [1982] reported "crests", which they termed "dynamic foam", and "strip-like, or patchy structures", which they termed "static foam." Data of *Bortkovskii* [1987] from his Fig. 2.5 represent sums of combined foam and whitecap coverage; data from his Fig. 2.6, showing only whitecap coverage, are not used. Values $W = 0$ are indicated by symbols on the lower axis. Uncertainties in individual values of W, though often comparable or greater than mean values, are not shown. Open symbols denote values of W determined by photographs; triangles denote values determined by photographs taken from aircraft. Small filled squares denote values of W determined by video. Wind speeds plotted for the BOMEX+ data are the midpoint of the range listed. Wind speeds for data of *Ross and Cardone* [1974] and *Bondur and Sharkov* [1982] are at 19.5 m. Solid line given by (4.5-4) represents a cubic wind speed dependence ($p = 3$) over the range of U_{10} from 8.5 m s^{-1} to 20 m s^{-1}; shaded gray band encompasses the bulk of the photographic data over the wind speed range shown, with width about central solid line decreasing from a factor of $\lessgtr$ 7 at $U_{10} = 8.5$ m s^{-1} to $\lessgtr$ 3 at $U_{10} = 20$ m s^{-1}. Several parameterizations are also presented, and the value of the exponent p is listed for power law relations of the form $W = aU_{10}^p$; the relation given by *Kondo et al.* [1973] was not a power law. The parameterization of *Wu* [1979b] is (4.5-1). M & O'M refers to the parameterization of *Monahan and O'Muirchaertaigh* [1980a] (Eq. 4.5-2). The STREX parameterization was that presented by *Monahan and O'Muirchaertaigh* [1986].

[1987][1]. Standard deviations reported for the individual data points are not presented in Fig. 36; as noted above they often exceeded the mean values and on the logarithmic vertical scale would thus extend beyond the lower axis. Additionally, uncertainties in U_{10} are not shown (and were typically not reported); these are expected to be especially important at low wind speeds (§2.3.1). Other whitecap data that have been reported but are not shown in Fig. 36 are those of

---

[1] The data presented in Fig. 36, from Fig. 2.5 of *Bortkovskii* [1987], were from various studies; *Bortkovskii and Novak* [1993] presented some of what were indicated to be the same data, though already combined by wind speed interval, but the number of measurements for the various studies listed in that reference does not correspond exactly to either the data shown in Fig. 2.5 nor to the values listed in Table 2.1 of *Bortkovskii* [1987].

Monahan [1969] taken in freshwater lakes consisting of 292 photographs of whitecaps comprising 20 observation intervals; two data of *Williams* [1969]; data from *Snyder et al.* [1983] obtained from photographs taken only when waves were breaking; three data of *Bezzabotnov et al.* [1986]; data from *Monahan and Monahan* [1986] that were based on 1500 visual observations at a lighthouse near Bremerhaven, Germany (discussed in §4.5.3.5); video data of *Kraan et al.* [1996] for 44 observation intervals for which wind speed was not reported; and sea surface reflectance data of *Moore et al.* [2000].

Values of W determined from the video data sets (indicated in Fig. 36 by small filled squares) are on the whole considerably lower (by one to two orders of magnitude) than photographic (film) data sets and exhibit much more scatter. In an attempt to account for this difference, *Monahan* and coworkers proposed a conceptual model for the evolution of bubble plumes and whitecaps that allowed the classification of whitecaps into Stage A (young) and Stage B (older) whitecaps [*Monahan et al.*, 1988; *Monahan and Woolf*, 1988, 1989]. According to this classification, Stage A whitecaps, with an albedo of 0.5 or greater, last only a few seconds, after which they decay into Stage B whitecaps, with lower albedo, which the investigators considered to be of greater importance for SSA production. These investigators postulated that videos record only Stage A whitecaps, whereas photographs detect both Stage A and Stage B whitecaps, with whitecap ratios W_A and W_B respectively. Many of the video data were collected in the early to mid 1980s; more recent technology may permit adequate recording of the full range of reflectance values of the different white areas [*Kraan et al.*, 1996], allowing videos to determine both Stage A and Stage B whitecaps, but this has yet to be demonstrated. The distinction between the various white areas had not been noted earlier by analysts of the video data, but as indicated above, alternative distinctions had previously been made between "foam strips" or "streaks", and "crests" or "patches" [*Blanchard*, 1971a; *Nordberg et al.*, 1971; *Ross and Cardone*, 1974; *Bondur and Sharkov*, 1982; *Koepke*, 1984, 1986a,b; *Bortkovskii*, 1987].

The ratio W_A/W_B reported by *Monahan et al.* [1988] for various wind speeds falls within estimated bounds of 0.001-0.2, from which they concluded that W_A values obtained from video tapes "were consistent with the present working model of bubble plumes," although the reported range leaves ample room for consistency. W_A and W_B were also found to exhibit different wind-speed dependencies [*Monahan et al.*, 1988; *Monahan and Woolf*, 1989]. Such a result would imply that the pattern of wave breaking is not similar at all wind speeds—hardly a surprising finding—and would thus indicate that the sizes of individual whitecaps vary with the wind speed, with stronger winds producing larger whitecaps with lifetimes that depend on their size. Such a pattern of behavior would violate one of the basic assumptions of the whitecap method (§3.2) required to relate W to SSA production, namely that the decay of area of individual whitecaps is characterized by a single exponential time constant which is the same for all whitecaps, leading to further concerns over the validity of extrapolating lifetimes determined from a laboratory studies to oceanic whitecaps.

Other investigators who classified white areas into similar categories likewise reported different wind-speed dependencies for the two structures, although with wind-speed dependencies that differed greatly from those reported by *Monahan et al.* [1988]. *Nordberg et al.* [1971] reported that as the wind speed increased from 13 m s^{-1} to 25 m s^{-1} the fraction of the sea surface covered by whitecaps was nearly independent of wind speed, but that the ratio of the area covered by streaks to that covered by whitecaps increased from near unity to around five. Roughly similar results were reported by *Ross and Cardone* [1974], although they reported that no area was covered by streaks for wind speeds of 10 m s^{-1}, the lowest wind speed of their investigation. *Bondur and Sharkov* [1982] reported that both the fraction of the sea surface covered by striplike formations and the fraction covered by crests increased with wind speed, with the ratio of the two decreasing from around 50 to around 20 as the wind speed increased from 5 m s^{-1} to 10.5 m s^{-1}. Fractional coverage of foam and whitecaps and fractional coverage of whitecaps only were presented by *Bortkovskii* [1987, pp. 60-61; Figs. 2.5 and 2.6], who additionally presented expressions for these quantities as functions of wind speed for three temperature ranges, with the fractional area of foam coverage ranging from five to ten times that of whitecap coverage.

The categorization of whitecaps into early, bright, Stage A and later, less bright, Stage B has important implications for inferring SSA production by the whitecap method. Certainly the first several seconds after the formation of a whitecap is the most important time for production of SSA particles with $r_{80} \lesssim 1$ μm, which are produced by large bubbles which quickly rise to the surface and burst. If, however, Stage B whitecaps are more important than Stage A whitecaps for SSA production, as has been argued by some investigators, then the interfacial SSA production flux would be more closely related to W_B instead of W_A. If this were the situation, then the use of video data would seem ill-advised for determining whitecap ratios pertinent to SSA production, except perhaps for SSA particles with $r_{80} \lesssim 1$ μm. Given the difficulty in determining Stage A whitecaps by videotape (with values of W_A depending strongly on the gain setting on the video tape analyzer), the large uncertainties in individual values of W_A, the fact that W_A (determined by videos) is so much less than W_B (determined by photographs), and the lack of any physically based or reproducible methods of relating W_A to W_B, subsequent discussion here is restricted to values of W determined by photographs.

Examination of the dependence of W on U_{10} (Fig. 36) demonstrates that overall W increases with increasing wind speed, albeit with considerable scatter in the data, even with the exclusion of video data. At wind speeds below ~3-5 m s^{-1} a substantial fraction of reported values of W are zero. There does not appear to be a minimum wind speed for the onset of whitecapping, and several measurements of nonzero values of W have been reported at U_{10} below ~3-5 m s^{-1}, but whitecapping is not common at such low wind speeds. For $U_{10} \lesssim 8$ m s^{-1} values of W exhibit a large amount of scatter at a given wind speed (over three orders of magnitude, not including zero values) and little systematic trend with wind speed (Fig. 36); consequently it would be difficult to justify any quantitative relation between W and U_{10} in this wind speed range. For $U_{10} \gtrsim 8$ m s^{-1} there are relatively few zero values of W, and W generally increases with increasing wind speed, although at any given wind speed the data still vary by more than an order of magnitude. Few data have been reported at wind speeds above 15 m s^{-1}; of the photographic investigations described by acronyms in Table 20, only the STREX data set contains more than a single observation interval with wind speed greater than 15 m s^{-1}, and the other only data at these wind speeds are some from *Bortkovskii* [1987], from *Xu et al.* [2000], and older ones from aircraft photographs [*Nordberg et al.*, 1971; *Ross and Cardone*, 1974]. Values of W determined from photographs taken from aircraft (indicated by triangles in Fig. 36) are, for the same wind speed, roughly an order of magnitude greater than most of the other photographic determinations. No consensus has been reached on the relative accuracy of determination of W from aircraft photographs compared to determination from photographs taken at lower heights; however, as noted above, such a large discrepancy suggests that the photographs taken from lower heights, typically at oblique angles to the sea surface, might considerably underestimate W. At wind speeds greater than 20 m s^{-1} the few values of W appear more consistent than at lower wind speeds, but this may be due in part to the small number of data at these wind speeds. However, as whitecapping becomes more prevalent at greater wind speeds, small-number statistics are less of a problem, as most photographs would be expected to contain an appreciable fraction of white area. Additionally, wind speed is expected to become a more dominant controlling factor at high wind speeds, resulting in less variability due to factors such as atmospheric stability or sea temperature.

The large amount of scatter in values of W at a given wind speed and the large uncertainties in individual data points raise concerns over the confidence that can be placed in any quantitative relation between W and U_{10}, especially for $U_{10} \lesssim 8$ m s^{-1}, and consequently in the numerical results of any such fit or functional relationship. Nonetheless, it may still be possible to determine a relationship that describes the trend followed by the majority of the data provided the wide range of uncertainty contained in such a relationship is kept in mind. Several such expressions have been proposed, most of which consist of a single functional form over the entire range of wind speeds. Power law expressions, i.e., $W = aU_{10}^p$ (equivalent to $\ln W = \ln a + p \ln U_{10}$) are the most common, although more recent investigations have fitted $W^{1/3}$ to a linear function of wind speed, resulting in an expression of the form $W = b(U_{10} - c)^3$. Several expressions relating W to U_{10} based on photographic determination of W are shown in Fig. 36. The large spread among the several expressions, not only in the magnitudes but also in the wind-speed dependencies, which is comparable to the spread in the data, is indicative of uncertainty that accompanies any such relation. The expression proposed by *Blanchard* [1963, Fig. 31] to five photographs taken from aircraft and that proposed by *Kanwisher* [1963] to six photographs taken from aircraft, both power laws with $p = 2$, are roughly consistent with other aircraft data [*Williams*, 1970; *Nordberg et al.*, 1971; *Ross and Cardone*, 1974; *Martsinkevich and Melent'ev*, 1975 (from *Bortkovskii*, 1987, Fig. 2.5)] and yield values of W much greater than most determinations from photographs not taken from aircraft. In contrast, the expression presented by *Kondo et al.* [1973, Fig. 5] for their data, roughly a quadratic dependence of W on U_{10}, yields values of W considerably lower than most other data sets, which *Kondo et al.* suggested might have been caused by overestimation of W from other investigations because of the photographic technique used. However, as data of *Kondo et al.* were obtained in coastal waters, the lower values might not reflect values typical of the open ocean because of differences in wind properties and wave development. The majority of the photographic determinations of W in Fig. 36 are bounded above and below by the parameterizations of *Blanchard* [1963, Fig. 31] and of *Kondo et al.* [1973]. The expression given by *O'Muirchaertaigh and Monahan* [1986] describing the STREX photographic data set (nearly identical to that given by *Doyle* [1984] to the same data; not shown), a power law with $p = 1.92$, is intermediate; at 10 m s^{-1} this expression is roughly a factor of 4 lower than those of *Blanchard* and of *Kanwisher* and a factor of 5 greater than that of *Kondo et al.*

Two power law expressions relating W and U_{10} that have been widely used in parameterizations of SSA production are those of *Wu* [1979b] and of *Monahan and O'Muirchaertaigh* [1980], both to the combined BOMEX+ and TC data sets. *Wu* [1979b] omitted TC data for which $W < 10^{-5}$, some of the BOMEX+ data for which $W < 10^{-4}$

(which he stated were below the resolution of the measurements), and (without explanation) some of the TC data for higher values of W, and fitted the resulting data to a power law for which the value of p was taken to be 3.75 (for physical reasons discussed below):

$$W = 1.7 \cdot 10^{-6} U_{10}^{3.75} \approx \left(\frac{U_{10}}{35 \text{ m s}^{-1}} \right)^{3.75}. \quad (4.5\text{-}1)$$

Monahan and O'Muirchaertaigh [1980] omitted data for which $W = 0$ and fitted the remaining data to a power law, resulting in $p = 3.41$:

$$W = 3.84 \cdot 10^{-6} U_{10}^{3.41} \approx \left(\frac{U_{10}}{39 \text{ m s}^{-1}} \right)^{3.41} \quad (4.5\text{-}2)$$

(several alternative expressions with different exponents were also presented by *Monahan and O'Muirchaertaigh*, but it is this expression that was used subsequently by *Monahan* and coworkers [*Monahan et al.*, 1982; *Monahan*, 1986; *Monahan and Woolf*, 1988] as the basis for estimating SSA production). The two expressions (4.5-1) and (4.5-2) yield values of W that are quite similar (Fig. 36), differing by less than 20% for wind speeds from 6 m s^{-1} (below which little dependence of W on U_{10} is apparent) to more than 17 m s^{-1} (the maximum wind speed in either data set). In view of the scatter in the data, the protracted discussion [*Monahan and O'Muirchaertaigh*, 1980; *Wu*, 1982b; *Monahan and O'Muirchaertaigh*, 1982; *Monahan and Woolf*, 1989; *Wu*, 1989b] over which expression better represents the data hardly seems justified. The implied precision in each of the expressions would likewise seem difficult to justify, as fewer than half of the reported values of W from the TC and BOMEX+ data sets are within a factor of two of either expression, even if reported values of zero are excluded. Furthermore, the extent to which these expressions accurately represent the data as a whole is questionable. Although both expressions are very similar to that presented by *Monahan et al.* [1983a] for the JASIN data set, the expression subsequently proposed for the STREX data by *O'Muirchaertaigh and Monahan* [1986] exhibits a weaker dependence on wind speed, and the expression presented by *Doyle and Higgins* [1985] for the MIZEX84 photographic data set is roughly a factor of three lower than the *Wu* [1979b] or *Monahan and O'Muirchaertaigh* [1980] fits. All values of W in the MIZEX83 photographic data set fall below both the *Wu* [1979b] and *Monahan and O'Muirchaertaigh* [1980] parameterizations, as do nearly all the MIZEX84 photographic data and all of the video data.

Bortkovskii [1987, pp. 64-65] explicitly stated that neither of these expressions exhibits the dependence on wind speed that characterized his data. For all these reasons the use of either expression to infer whitecap ratio from measured values of U_{10} to estimate SSA production by the whitecap method would result in substantial uncertainty, and possible bias, over the entire U_{10} range, and especially for $U_{10} \lesssim 8.5$ m s^{-1}.

The physical basis for the exponent p equal to 3.75 in (4.5-1) proposed by *Wu* [1979b, 1988b] rests on the hypothesis that under steady state conditions W is proportional to the flux of energy to the ocean from the wind under the assumption that the pattern of wave breaking is similar at all wind speeds (a similar hypothesis had earlier been advanced by *Ross and Cardone* [1974]). This energy flux $\dot{E}$ is the product of the wind stress, by definition proportional to u_*^2 (§2.3.2), and the wind-induced drift velocity, typically assumed to be proportional to u_* [*Wu*, 1975]; hence $\dot{E}$ is proportional to the cube of the wind friction velocity [*Longuet-Higgins*, 1969; *Toba*, 1972]: $\dot{E} \propto u_*^3$. The assumption that the wind stress coefficient is directly proportional to the square root of the wind speed $C_D \propto U_{10}^{1/2}$ [*Wu*, 1969a] yields the 3.75 power law in wind speed proposed by *Wu* [1979b, 1988]:

$$W \propto \dot{E} \propto u_*^3 = (C_D U_{10}^2)^{\frac{3}{2}}$$
$$\propto \left(U_{10}^{\frac{1}{2}} U_{10}^2 \right)^{\frac{3}{2}} = U_{10}^{3.75}. \quad (4.5\text{-}3)$$

Several concerns may be noted with this argument and with the value 3.75 of the exponent p. The hypothesis of proportionality between W and the energy flux from the wind has not been established. The two quantities may be related and are probably well correlated, but any such relation between these quantities involves not only the fraction of the energy flux that goes into wave breaking but also the relation between W and the energy expended in wave breaking, neither of which is necessarily independent of the wind speed or the extent of wave breaking. The distinction between the fraction of the wind stress that goes into wave formation and the fraction that goes into the drift current had earlier been made by *Toba* [1972], and *Woolf* [1985] noted that the majority of the energy imparted by the wind to the sea may go into breaking waves, as proposed by *Banner and Melville* [1976], and not to the drift velocity. Likewise, *Makin et al.* [1995] concluded that for high winds ($U_{10} > 15$ m s^{-1}) the majority of the stress results from waves and not from viscous drag. *Hanson and Phillips* [1999] reported that whitecap ratios (determined from video recordings) were best fitted by the 1.5 power of energy flux, arguing that the exponent 1.5 (instead of 1) implied that not all of the energy flux goes into whitecap formation. Furthermore, *Wu's* argument rests on the assumption that the pattern of wave breaking is similar at all wind speeds, whereas waves do not break at sufficiently low wind speeds, and they break with increasingly more vigor as the wind speed increases. In other words, as the sea becomes more disturbed by the action of the wind, the whitecaps generally increase in size as well as number.

As noted above, the assumption that $p = 3.75$ rests also on the proportionality of the wind stress coefficient and the square root of the wind speed, $C_D \propto U_{10}^{1/2}$. This square-root dependence was based on an expression reported by *Wu* [1969a]. Later *Wu* [1980] proposed a linear relationship between C_D and U_{10}; this would result in a relationship between W and U_{10} that is not a power law. In one paper [*Wu*, 1988b] both expressions were used. When the wind stress coefficients from numerous experiments were fitted to power laws in the wind speed by *Wu* [1980], exponents between 0.08 and 0.74 were obtained; these values would result in exponents for the wind-speed dependence of W ranging from 3.1 to 4.1. As noted above (§2.3.2), numerous different formulations relating C_D and U_{10} have been proposed, but the scatter in the data is such as to limit the confidence that can be placed in any specific formulation or exponent. Similarly, the proportionality of the wind-induced drift velocity and the wind friction velocity was based on laboratory data [*Wu*, 1975], and although the data were fitted to a proportional relation, the data appear to be equally well fitted by a linear function in wind speed or by some power near unity of either the wind speed or the wind friction velocity, resulting in an exponent other than 3.75. *Bortkovskii* [1987, pp. 73-74] argued that although the energy flux from the wind may have been equal to that given by *Wu*, the amount of whitecapping depends on other factors that reduce the exponent p to around 1.5, although he fitted W to a linear function of U_{10}. He noted that this argument applies to whitecaps only and not to foam (which typically has a much higher coverage) and noted further that *Wu* did not distinguish between the two types of white area.

A further concern with any proposed parameterization relating W and U_{10} deals with the methods used to fit the data. Despite the identity of the functional relations $\ln W = \ln a + p \ln U_{10}$ and $W = aU_{10}^p$, the choice of the quantity being minimized in the fit can make a great difference in the resultant fitting expression because of the wide range of values of W and the large uncertainties in the individual data (*cf.* §4.1.2.2 with regard to the dependence of SSA mass concentration M on U_{10}). Typically in fits of expressions such as these the data have been weighted equally instead of by a factor that depends on their associated uncertainties, as quantified by, perhaps, the standard deviation. Both

Wu [1979b] and *Monahan and O'Muirchaertaigh* [1980] excluded zero values of W from the data being fit, but despite the inconvenience of such zero values when fitting to the form $\ln W = \ln a + p \ln U_{10}$ (or when plotting on a logarithmic scale), the omission of these data would hardly seem justified. Fitting the data to the expression $\ln W = \ln a + p \ln U_{10}$ weights each datum by its ratio to the fitted value, resulting in low values of W being as important to the fit as high values. Additionally, this approach implicitly assumes the data are lognormally distributed and thus should be geometrically averaged, but there is no basis for this assumption, and if the uncertainties arise mainly from small-number statistics, then arithmetic averaging rather than geometric averaging would be required to determine W from independent photographs. Alternatively, fitting the data in a least squares sense to the expression $W = aU_{10}^p$ results in the fit being dominated by the largest values of W (or U_{10}). For example, if the uncertainty in W is proportional to the mean value, then for $p = 3.5$ the value of W corresponding to $U_{10} = 15$ m s^{-1} would be nearly fifty times greater than that for $U_{10} = 5$ m s^{-1}, and its contribution to the sum of squared errors (the quantity to be minimized) would be more than 2000 times greater. For considerations of SSA production, it could be argued that high values of W should be weighted more than low values of W; whether W is equal to 10^{-5} or 10^{-6}, for instance, makes little difference in the existing SSA concentration, whereas the SSA production corresponding to $W = 0.1$ will probably dominate the SSA concentration. Restricting the fit to W greater than a certain minimum value can also affect the result considerably. For example, fitting only those data of *Toba and Chaen* [1973] with $W > 10^{-4}$ to the expression $\ln W = \ln a + p \ln U_{10}$ with equal weighting yields an exponent $p = 3.3$, whereas fitting all nonzero values yields $p = 4.7$. Likewise, an exponent $p = 3.6$ was obtained by *Hanson and Phillips* [1999] to the video data of *Monahan and Wilson* [1993 (cited in *Hanson and Phillips*, 1999)] for which $W > 5 \cdot 10^{-5}$, whereas for the complete data set the value $p = 5.2$ was obtained.

Numerical values of parameterizations of W in terms of U_{10} also contain appreciable uncertainties. For example, the root-mean-squared error resulting from the power law fit of the BOMEX+ and TC data to $W = aU_{10}^p$ is less than 1% greater than its minimum value for that of *Monahan and O'Muirchaertaigh* [1980] given above ($p = 3.41$) for any value of p in the range 3.2-4.0, and less than 5% greater than its minimum value for values of p down to 2.8 (the minimum root-mean-squared error is equal to the value of W given by that fit for $U_{10} = 9$ m s^{-1}, implying that the few large values of W dominate the fit). Typical standard errors for p from other fits were also near 0.4 [*Monahan et al.*, 1983a; *O'Muirchaertaigh and Monahan*, 1986; *Monahan and O'Muirchaertaigh*, 1986], and values of p from power law fits to individual data sets ranged from less than 2 to more than 5 [*Monahan and O'Muirchaertaigh*, 1980; *O'Muirchaertaigh and Monahan*, 1986; *Monahan and Monahan*, 1986; *Hanson and Phillips*, 1999]. Such results further diminish confidence in specific numerical values in relations between W and U_{10}, and in such parameterizations in general.

The spread of the measured values of W about any relation between W and U_{10} suggests the utility, as was done above for the dependencies of M and N on U_{10}, of identifying a central value for such a relation and a range that encompasses the bulk of the measurements as an indication of the dependence of W on U_{10} and its variability. This central value and variability range are shown in Fig. 36 as a shaded gray region that encompasses the majority of the photographic data (open symbols) for the wind speed range $U_{10} = 8.5$-20 m s^{-1}; on the logarithmic plot an equal uncertainty above and below a central value corresponds to a multiplicative uncertainty. Below this wind speed there is too much scatter in the data to argue for any discernible relation between U_{10} and W, and above this wind speed there are too few data to support such a relation. The central solid line denotes a cubic dependence of W on U_{10} given approximately by

$$W = 12 \cdot 10^{-6} \, U_{10}^3 \approx \left(\frac{U_{10}}{44 \text{ m s}^{-1}} \right)^3 \qquad (4.5\text{-}4)$$

At $U_{10} = 8.5$ m s^{-1} the shaded gray region spans a factor of 50, narrowing to a factor of slightly less than 10 at $U_{10} = 20$ m s^{-1}; that is, the width of this region about central solid line decreases from a factor of $\lessgtr 7$ at $U_{10} = 8.5$ m s^{-1} to $\lessgtr 3$ at $U_{10} = 20$ m s^{-1} (at $U_{10} = 10$ m s^{-1} the width is approximately a factor of $\lessgtr 6$). This width is substantially greater than the increase in the central value of W between $U_{10} = 8.5$ m s^{-1} and $U_{10} = 17$ m s^{-1} of less than an order of magnitude. Thus, while there is no question that W exhibits a general increase with increasing U_{10}, it is clear, as reflected in the width of the variability band, that there is much variance in W that is due to factors other than wind speed. This variance directly contributes to uncertainty in estimates of SSA production by the whitecap method when W is inferred from U_{10}, as is common in modeling studies.

A relation between wind speed and whitecap ratio would in principle allow inference of U_{10} from W determined by remote (satellite) sensing. Expressions relating U_{10} to W have been determined by inverting power law fits of the form $W = aU_{10}^p$ [e.g., *Wu*, 1979b] or by fitting U_{10} to a power law in W (or $\ln U_{10}$ to a linear function of $\ln W$) [*Monahan and O'Muirchaertaigh*, 1981; *Monahan et al.*, 1981]. However,

as noted by *Monahan and O'Muirchaertaigh* [1981, 1986], these procedures will result in different exponents for W. In any event, from Fig. 35 it would appear that the uncertainties resulting from either approach would be so great as to call into question any such parameterization.

4.5.3. Dependence on Factors Other Than Wind Speed

The large scatter in W when parameterized by U_{10} alone indicates that factors other than wind speed play a key role in determining whitecap ratios. One such potentially important factor is sea temperature T_{sea}, for which there is a physical basis for a dependence through its rather large effect on the kinematic viscosity of seawater ν_{sw}. Other factors that have been suggested are atmospheric stability, wind friction velocity, sea state, fetch, salinity, saturation of the surface waters with respect to the major atmospheric gases, and surface-active materials. However, many of these factors vary little in most situations, and quantities describing them often have not been reported. Dimensionless groups involving wind speed, friction velocity, and air and seawater viscosity have also been investigated. Some of the attempts to determine and describe the dependence of W on these additional factors, and to thereby reduce the scatter in the data, are examined here.

4.5.3.1. Seawater temperature. Several investigators have examined for differences in the dependence of W on U_{10} when data are classified by sea temperature. *Kondo et al.* [1973] classified W by T_{sea}, but their data showed no apparent dependence on T_{sea} over the range 15-26 °C. Values of W from the BOMEX+, TC, JASIN, and STREX photographic data sets (none of which were obtained at values of T_{sea} less than 5 °C), sorted by T_{sea}, were fitted to power laws in U_{10} by *Doyle* [1984] and *Spillane et al.* [1986], who reported an increase in exponent p with increasing T_{sea}, but no corresponding trend in W. *Spillane et al.* noted that the classification by T_{sea} closely matched the breakdown by experiment (only the STREX data set contained values of T_{sea} less than 12.5 °C, for instance), making it difficult to eliminate other effects, such as wind duration, which may have been different in the different studies, the several investigations having been conducted at different locations. Values of W from each of these same data sets and from the MIZEX83 photographic data set (which contained 11 intervals for which $T_{\text{sea}} < 5$ °C), sorted by investigation, were fitted to power laws in U_{10} by *Monahan and O'Muirchaertaigh* [1986], who likewise reported an increase in p with the mean value of T_{sea} from each study, although they did not report an increase in W with increasing T_{sea}. This increase in p was attributed not to sea temperatures *per se*, but rather to the fact that investigations at lower mean sea temperatures had been conducted at higher latitudes, which typically did not exhibit fully developed wave spectra because of shorter duration of high wind events at these latitudes; however, no evidence was presented to support this argument. Fractional areas covered by whitecaps and foam, W_{wc} and W_f, respectively, were fitted by *Bortkovskii* [1987, pp. 60-61] to functions of U_{10} in three different temperature ranges, with W_{wc} increasing linearly with wind speed, and W_f, approximately a factor of 5-10 greater than W_{wc}, generally exhibiting a stronger increase with increasing wind speed. Despite the absence of any trend in the parameters resulting from fits classified by T_{sea}, *Bortkovskii* concluded that the sum of W_{wc} and W_f increased greatly with increasing T_{sea}, being roughly a factor of 7 greater at high temperatures (~27 °C) than at low temperatures (< 3 °C) over the wind speed range of overlap of these data (~9 14.5 m s^{-1}), with results at intermediate temperatures (3-15 °C) being comparable to those at low temperature for $U_{10} \lesssim 14$ m s^{-1}, but becoming greater at higher wind speeds. Such a strong temperature dependence was not apparent in the data sets considered by the previously mentioned investigators, but these did not include the MIZEX84 photographic data set, for which T_{sea} was less than 5 °C for all observation intervals. Data from the BOMEX+, TC, JASIN, STREX, and MIZEX84 photographic data sets, categorized into intervals of u_*, were fitted to power laws in U_{10} and u_* with exponents 3.75 and 3, respectively, by *Wu* [1988b], who argued that any dependence of W on T_{sea} should reside in the coefficient a (the exponents being determined by physical reasons, as discussed in §4.5.2). *Wu* reported no apparent systematic trends in W with T_{sea}, although values near $T_{sea} = 0$ °C (mainly from the MIZEX84 data) appeared to be lower than those from the other studies [see also *Wu*, 1992b]. *Stramska and Petelski* [2003] reported no systematic influence of T_{sea} on W at a given wind speed for temperatures 2-13 °C.

Possible temperature dependence of W on T_{sea} was investigated (Fig. 37) by plotting values of W from investigations summarized in Table 20 that were determined by photographs against U_{10} as in Fig. 36, but classified into three temperature ranges (following the classification of *Bortkovskii* [1987]): cold ($T_{sea} \lesssim 4$ °C), intermediate (4 °C $\lesssim T_{sea} < 17$ °C), and hot (17 °C $< T_{sea}$). The difference in the nonzero values of W for low temperatures ($T_{sea} \lesssim 4$ °C) and those for high temperatures ($T_{sea} > 17$ °C) is quite apparent, with those values at high temperatures being nearly an order of magnitude greater. Although there are still many zero values of W and much scatter for $U_{10} \lesssim 8$ m s^{-1}, as discussed above, the scatter in the nonzero values of W within each of these two temperature ranges for a given wind speed is appreciably less than for the data collectively, with the low temperature data being mostly encompassed a swath with width a factor of 3-5, and the high temperature data by a swath with width a factor of 7-10. Although such a finding is encouraging, any encouragement is tempered by the fact that at most only a few percent of the ocean surface [*Alexander and Mobley*, 1976; *Shea et al.*, 1992], consisting of a narrow zone north of Antarctica and some regions of high latitude in the Northern Hemisphere, experience temperatures corresponding to those in the low temperature category and that some of these regions contain appreciable ice and have limited fetch; thus the process of wave breaking and whitecap formation in these regions may differ from other locations. For intermediate temperatures, there is only a slight reduction in the scatter compared to the data as a whole, and there is considerable overlap between values of W in this temperature range and those in the other two ranges. The large difference noted by *Bortkovskii* [1987] between the values of W at intermediate temperatures and those at high temperatures is not apparent in Fig. 37, nor in the analyses discussed above for the BOMEX+, TC, JASIN, STREX, and MIZEX83 photographic data sets.

Most of the arguments that have been proposed concerning the effect of T_{sea} on W have been based on the strong temperature dependence of the kinetic viscosity of seawater v_{sw} noted above, which decreases by nearly a factor of two between 0 °C and 20 °C (§2.5.1). It has been argued [*Monahan and O'Muirchaertaigh*, 1980; *Monahan et al.*, 1983a] that the rise times of bubbles are longer in colder waters, resulting in longer whitecap lifetimes and hence greater values of W at a given wind speed. However, as noted in §4.4.2.1, the dependence of bubble terminal velocity on temperature decreases with increasing R_{bub}, and for $R_{bub} \gtrsim 1$ mm, the size range that probably contributes the most to the visible white area during whitecapping, the terminal velocity is almost independent of temperature. Early experiments of *Miyake and Abe* [1948], which consisted of shaking a small quantity (about 100 cm^3) of seawater in a jar and measuring persistence of foam have been adduced in support of the contention that whitecapping should persist longer in colder waters, resulting in greater values of W. These investigators reported little effect of temperature on the amount of foam produced, but a large effect on the persistence of the foamy layer, with this lifetime decreasing by about 4% per °C increase in sea temperature (however, both the example presented and Fig. 2 of that reference showed a decrease of close to 8% per °C for temperatures up to 20 °C). These results, if applicable to the open oceans, imply that W would be greater at high latitudes (with lower sea temperatures) than at lower latitudes at a given wind speed [*Blanchard*, 1971a]. However, it would seem difficult to justify these arguments on the basis of laboratory experiments; the decay of foam in a jar [*Miyake and Abe*, 1948] is a very different process from the breaking of a wave, and it is entirely possible that these results were

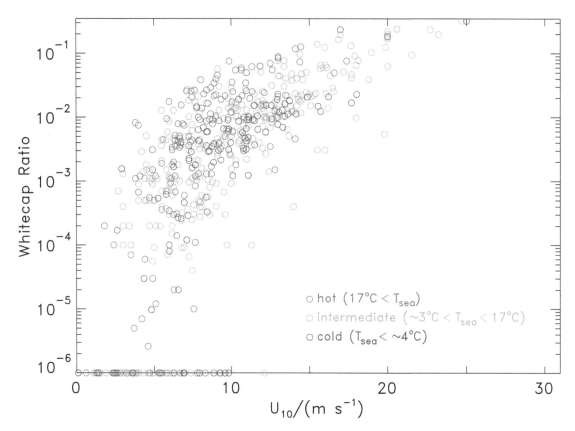

Figure 37. Photographic determinations of whitecap ratio W as a function of the wind speed at 10 m above the sea surface, U_{10}, categorized by sea surface temperature T_{sea} into three temperature ranges: high, intermediate, and low, roughly following the classification of *Bortkovskii* [1987], with changes as noted (values of T_{sea} for individual data were not given in that references). The high temperature category consists of the entire BOMEX+ and TC data sets and those data of *Kondo et al.* [1973] and *Bortkovskii* [1987] for which $T_{sea} > 17$ °C. The data of *Blanchard* [1963] were included in this category although T_{sea} was not reported, based on location and date (they were from photographs taken in the Caribbean in October). The intermediate temperature category consists of the entire JASIN and STREX data sets, those data from the MIZEX83 data set for which $T_{sea} > 3$ °C ($W = 0$ for those data for which $T_{sea} < 3$ °C), four data of *Nordberg et al.* [1971] for which $T_{sea} = 9$-10 °C, three data of *Kondo et al.* [1973] for which $T_{sea} = 14.8$-16.7 °C, those data of *Bortkovskii* [1987] for which 3 °C $< T_{sea} < 15$ °C, all data of *Xu et al.* [2000], and data of *Stramska and Petelski* [2003] for which $T_{sea} > 5.0$ °C. The data of *Ross and Cardone* [1974] were included in this category although T_{sea} was not reported, based on location and date. The low temperature category consists of two data of *Nordberg et al.* [1971] for which $T_{sea} = 2$ °C and 4 °C, the entire MIZEX84 data set (for which $T_{sea} < 4.38$ °C), data of *Bortkovskii* [1987] for which $T_{sea} < 3$ °C, and data of *Stramska and Petelski* for which $T_{sea} < 5.0$ °C. Data of *Martsinkevich and Melent'ev* [1975, from *Bortkovskii*, 1987, Fig. 2.5] and *Bondur and Sharkov* [1982] are not shown as information that would allow T_{sea} to be estimated was not reported. Values $W = 0$ are indicated by symbols on the lower axis.

influenced by surface-active substances specific to the experiment. Similar experiments involving other liquids [*Leike*, 2002] indicated no temperature dependence. The opposite tendency, greater whitecapping at higher temperatures, has also been proposed [*Monahan*, 1986; *Monahan and O'Muircheartaigh*, 1986] based on a brief report by *Pounder* [1986] of a laboratory investigation on the effects of temperature on the number and size distribution of bubbles. The hypothesis put forward was that the shift in the shape of the size distribution of bubble concentration to smaller bubbles with increasing temperatures, and the

consequent lower terminal velocities, more than compensates for the effects of decreased viscosity, resulting in increased lifetimes, and hence increased values of W, in warmer waters than in colder ones. However, the method by which bubbles were produced in the investigation of *Pounder* [1986] was not described, and temperature may affect bubble formation in the ocean differently. Again, there is a likely influence of surface-active substances on these results. Additionally, as noted in §4.4.1.4, differing results for the effect of temperature on the bubble size distribution have been reported. Another proposal [*Monahan and*

O'Muirchaertaigh, 1986] is that W increases with increasing T_{sea} because of lower seawater viscosity, resulting in less viscous dissipation in waves and hence more energy available for wave breaking. According to *Bortkovskii* [1987, p. 64], damping of short gravity-capillary waves also decreases with increasing temperature, affecting the roughness of the sea surface and hence possibly the breaking behavior of waves.

Field measurements that would aid in determining the temperature dependence are lacking; as noted in §4.4.1, no systematic investigations of the effects of sea temperature on bubble concentration size distributions in the oceans have been reported. Attempts to determine the physical basis for a dependence of W on temperature are hampered by the difficulty in isolating the proposed effects that vary simultaneously with T_{sea} from each other (e.g., bubble terminal velocities and the frequency of wave breaking) and from other factors that do not directly depend on T_{sea} (e.g., the saturation of the surface waters, turbulence and downward entrainment of bubbles, or the presence of surface-active materials). Thus, although there is some indication that W is generally greater at high sea temperatures than at low sea temperatures, given the large amount of scatter in the data in any temperature category, comparable to the difference in W in the several categories, and the differing results of *Bortkovskii* and other investigators at intermediate temperatures, it has proven difficult to quantify the dependence of W on sea temperature, and little can be concluded about the physical basis for this dependence.

4.5.3.2. Atmospheric stability. Atmospheric stability (§2.3.2) is thought to affect whitecapping through its influence on air flow over the sea and consequently the interaction of the wind and waves. Atmospheric stability has frequently been parameterized in terms of $\Delta T = T_{sea} - T_{air}$, the sea temperature minus the air temperature; a positive ΔT representing an unstable atmosphere and a negative ΔT, stable. However, in most situations over the ocean ΔT is quite small ($\lesssim$1-2 °C), implying atmospheric stability is close to neutral. Consequently there are few data and a very limited range of ΔT upon which to base conclusions concerning the dependence of whitecapping on atmospheric stability. Not unsurprisingly, contradictory results have been obtained.

Monahan [1969] concluded that freshwater whitecapping was greater in unstable than in stable conditions, but this conclusion was based on only 4 stable and 4 unstable observation intervals for which U_{10} was greater than 8 m s^{-1}, and these data do not appear to exhibit systematic trends with wind speed; consequently, elucidating any effects of stability from these data would be difficult. *Monahan* [1971] reported that no conclusions concerning effects of atmospheric stability could be drawn from the BOMEX+ data because of

the small temperature differences encountered—only a single observation interval for which the magnitude of ΔT was greater than 3 °C. In contrast, *Wu* [1979b] concluded that whitecap ratios determined from both the BOMEX+ and TC data sets were greater by a factor of two in stable conditions than in unstable conditions. However, his relation $W \propto (C_D U_{10}^2)^{3/2}$ (§4.5.2) implies that instability, resulting in greater wind stress (i.e., greater C_D) for the same wind speed, should result in greater values of W. To resolve this discrepancy, *Wu* argued that wind speed affects only the occurrence of wave breaking, whereas W is influenced by the persistence of whitecaps as well, and that this increase in persistence (of unknown cause) exceeds the effect of lower whitecap production in situations of stability. Based on analysis of the same data, *Kerman* [1984a] concluded that W is four to five times greater under moderately stable conditions than under moderately unstable conditions for the same wind friction velocity u_*, although he noted that the apparent variation would be less with greater wind speed (because of greater wind stress in unstable than stable conditions at the same wind speed). Later investigations concluded that W increases with increasing instability for a given value of U_{10}. *Monahan and O'Muirchaertaigh* [1986], based on analysis of photographic data, concluded that W increases exponentially with ΔT by nearly 10% per °C (implying greater values of W under unstable than stable conditions for the same wind speed), although improvement in the fit of W vs. U_{10} was not demonstrated. *Monahan et al.* [1988], from analysis of video data (Stage A whitecaps), concluded that W_A increases exponentially with ΔT by about 20% per °C. Similarly, *Wu* [1988b], based on the hypothesis [*Wu*, 1986a] that C_D varies exponentially with $(\Delta T/U_{10})^{5/3}$, concluded that increasingly unstable conditions would result in greater values of W for the same wind speed. However, the data upon which these conclusions are based are limited mainly to values of ΔT in the range -2 °C $< \Delta T < 3$ °C (*cf.* Fig. 3 of *Monahan and O'Muirchaertaigh* [1986]). *Stramska and Petelski* [2003] reported that there was no obvious relation between ΔT and W at the same value of U_{10} for their data, which contained values of ΔT ranging from -2 °C to 5.5 °C. Although no conclusions can be reached on this question, it might be noted that the magnitudes of the proposed effects of atmospheric stability on W are quite small relative to the spread of the data indicated in Fig. 36: for $\Delta T = 3$ °C, the formula of *Monahan and O'Muirchaertaigh* [1986] yields an increase in W of less than 30%, and that of *Wu* [1988b] at $U_{10} = 10$ m s^{-1} yields an increase in W of less than 15%; by comparison, the side of one of the large squares representing the BOMEX+ or TC data in Fig. 36 corresponds to a change in W of 30%.

4.5.3.3. Wind friction velocity. Other investigators have classified W by the wind friction velocity, u_* (often determined from U_{10} and ΔT, and thus including effects of atmospheric stability), under the assumption that u_*, rather than U_{10}, is the principal controlling variable for W [*Wu*, 1988b]. However, plots of W vs. u_* still exhibit much scatter [*Monahan and Woolf*, 1986; *Wu*, 1988b; *Monahan and Lu*, 1990]. Fitting $W^{1/3}$ vs. u_* was less successful (resulted in a lower correlation coefficient) than fitting $W^{1/3}$ vs. U_{10} for five data sets considered by *Monahan and Lu* [1990], and *Bortkovskii and Novak* [1993] reported virtually no difference in how well their data, and the data of other researchers, could be fitted to u_* as compared to U_{10}.

A dimensionless quantity that includes the effects of both atmospheric stability and temperature is $u_*/(g\nu_{sw})^{1/3}$, where g is the acceleration due to gravity. When values of $W_A^{1/3}$ determined from video tapes and values of $W_B^{1/3}$ determined by photographs were fitted by *Monahan and Lu* [1990] against this quantity, virtually the same correlation coefficients resulted as when they were fitted against $U_{10}/(g\nu_{sw})^{1/3}$, and no evidence was provided that either parameterization was better than one based on U_{10} alone. *Bortkovskii and Novak* [1993] also reported roughly the same correlation coefficients when fitting W against $u_*/(g\nu_{sw})^{1/3}$ as when fitting W to either U_{10} or u_* individually.

4.5.3.4. Wave phase velocity. Several investigators have proposed that W is influenced by the state of development, or age, of waves, which has been typically parameterized by the phase velocity, c_s, (or equivalently, the period or wavelength) of the significant (or dominant) wave (roughly, the wave near the peak in the energy spectrum); unfortunately, this quantity has rarely been reported together with whitecap ratios. *Toba and Chaen* [1973] concluded from their data that the whitecap ratio is more dependent on the period of the significant wave than on wind speed alone, and based on dimensional arguments of *Toba* [1972], these investigators proposed that W is proportional to the 1.5 power of the dimensionless group $u_* c_s^2/(g\nu_a)$. As this parameterization depends on the kinematic viscosity of air ν_a it does not include the effect of sea temperature; additionally ν_a exhibits a much weaker dependence on temperature than ν_{sw}. When nonzero values of W from the data of *Toba and Chaen* [1973] are fitted to a power law in this dimensionless parameter (using a least squares fit of the logarithms of these quantities), the characteristic departure of these data from the fitting function is reduced from that from wind speed alone from a factor of 5.5 to a factor of 3.6. However, when only those data for which $W > 10^{-4}$ are fitted to a power law of this dimensionless parameter (yielding an exponent of the 1.1), the characteristic departure from the fitting function is reduced only slightly, from a factor of 2.4 to a factor of 2.1, implying little improvement over fitting W to a function of U_{10} alone. Later *Toba and Koga* [1986; see also *Zhao and

Toba, 2001], based on similar arguments, concluded that W is directly proportional to this dimensionless group, $u_*^2 c_s/(g\nu_a)$; fitting W in this manner resulted in a reduction of the characteristic departure from the reported fit from a factor of 5.5 to 4.4 for these same data. Although the fit is somewhat improved, the scatter in the data is still quite large, greater than an order of magnitude. Additionally, as noted in §2.3.5, c_s is highly correlated with U_{10}, so that each of the above fits essentially reduces to a power law relation between W and either U_{10} or u_* with an exponent in the range 3-4.5. *Bortkovskii* [1987, pp. 77-79] argued that for fixed wind speed, W decreases with increasing c_s, and presented data supporting this result, although he did not quantify the dependence. *Bortkovskii and Novak* [1993] fitted the fraction of the ocean surface covered by whitecaps W_{wc} to a linear function of the two dimensionless groups, $u_*/(g\nu_{sw})^{1/3}$ and $u_*^2 c_s/(g\nu_{sw})$, and the fraction of the ocean surface covered by foam streaks W_f to a power law in these quantities, but reported only slight improvement over fitting these quantities to similar functions of U_{10} or u_* alone. *Kraan et al.* [1996] fitted the fractional area covered by whitecaps (and not the resulting foam) determined by video cameras to a power law in c_s/u_* (which they argued is the appropriate variable to describe whitecap coverage), resulting in an exponent near 2, but the fit is poor (linear correlation coefficient $r_{corr} \approx 0.2$). *Stramska and Petelski* [2003] reported greater whitecap coverage in the open ocean under fully developed seas than under undeveloped seas, and concluded that wind duration was one of the most important factors affecting W. The measure of the state of development they employed was the difference in the significant wave height from that expected for a given U_{10}, with wave heights lower than expected indicative of an undeveloped sea. As lower wave heights would yield less wave breaking, such a result is not unexpected.

4.5.3.5. Fetch. Fetch, the distance over water that the wind has blown, may affect W by affecting the sea state and the characteristics of the wave spectrum. *Ross and Cardone* [1974] concluded that W would increase with increasing fetch under otherwise similar conditions, but their data do not support this conclusion. *Wu* [1975] concluded from field data and laboratory data that the wind stress coefficient C_D decreases with increasing fetch [*Wu*, 1969c], as does the wind-induced surface drift velocity; according to the arguments of *Wu* presented above leading to (4.5-3), these results imply that W would decrease with increasing fetch, opposite to the conclusion to that reached by *Ross and Cardone*. *Monahan and Monahan* [1986] analyzed data collected visually by lighthouse observers and concluded that in the situation of "extreme limited fetch" W was less than might otherwise be expected at lower wind speeds, but was comparable to the infinite fetch data at wind speeds near 15 m s^{-1}; however, the large uncertainties in the data and the method of calculating W (visual estimates from different observers, who had a "tendency to exaggerate") make it difficult to place much confidence in these results. In any event few oceanic situations would be characterized by "extreme limited fetch," and observations at lighthouses can hardly be considered representative of the open ocean. The few data of *Xu et al.* [2000], four values of W with fetch 60 km, two with fetch 120 km, and three with fetch 170 km, appear to exhibit a decrease in W with increasing fetch at a given wind speed. Reported values of W were appreciably lower than most others reported at similar wind speeds, possibly indicating a systematic bias. These results compared well to a proposed expression relating W to fetch, but this expression has the value of W at a given wind speed decreasing to zero with increasing fetch, and hence would imply that $W = 0$ for most open ocean whitecaps.

4.5.3.6. Salinity. No studies have been reported on the influence of salinity on whitecap ratio over the range of salinities normally encountered in the oceans (33-37), but based on the arguments presented throughout this review (§2.5.1), any such influence would be expected to be extremely small. Some arguments have been presented concerning differences between whitecap ratios in freshwater and seawater which, although not pertinent to SSA production, are briefly summarized here. The freshwater whitecap data upon which these arguments have been based are those measured by *Monahan* [1969] in the Great Lakes for which in most instances the fetch was a few tens of kilometers or less. These data fall below the fits proposed by *Monahan and O'Muirchaertaigh* [1980] and by *Wu* [1979b] to the combined BOMEX+ and TC data. These data are few in number; there were 20 observation intervals, but the data were presented in graphical form [*Monahan*, 1969, Fig. 2] and for only about half of these, for which the whitecap ratio was greater than about $4 \cdot 10^{-4}$, could the values be determined from the figure. Additionally, these data do not exhibit an easily quantifiable trend; for the four highest wind speeds, values of W exhibited a consistent large decrease with increasing wind speed. *Monahan* [1971], arguing that there should be no difference in the rate of whitecap production or in the initial areas of individual whitecaps between freshwater and seawater under the same nominal conditions, concluded that seawater whitecap coverage is expected to be 51% greater than for freshwater because of the difference in the persistence of seawater bubble plumes determined from measurements (§4.3.4.2) of the decay of laboratory whitecaps [*Monahan and Zietlow*, 1969]. However, he stated that measurements of freshwater whitecapping could not be compared with those over the oceans because

of differences in fetch [see also *Monahan*, 1970]. Based on these same measurements of freshwater whitecap ratios, *Wu* [2000a] concluded that freshwater whitecapping is seven times less extensive than for the oceans at the same wind speed. *Monahan* [2001] argued that the oceanic whitecap ratio would be greater than that for freshwater under the same conditions because of the longer lifetimes of seawater whitecaps as a consequence of the difference in shapes of the size distributions of bubble concentration (§4.4.1.4), with freshwater bubbles being on average larger and thus rising to the surface more rapidly. Based on the few data available it would seem difficult to place much confidence in any quantitative comparison between whitecapping over the ocean and that over freshwater.

4.5.3.7. Surface-active substances. The presence of surface-active substances can affect the process of whitecapping and the persistence of whitecaps by increasing the damping of small waves and by changing the sea state or roughness length, surface tension of the seawater-air interface, and the lifetime of foam on the ocean surface (§2.3.4), but systematic examination of these effects in the ocean would be very difficult. Observations reported by *Ariel et al.* [1979; described in *Bortkovskii*, 1987, pp. 81-82] indicated that for a given wind speed, the area covered by whitecaps was smaller, and the area covered by foam was greater, in a coastal zone than in the open ocean, the differences being greater at $U_{10} = 5$ m s^{-1} than at 10 m s^{-1}. This situation was attributed to the presence of foam-stabilizing surface-active substances, with the area covered by foam decreasing with increasing wave breaking associated with increasing wind speed, the wave breaking resulting in bubbles that transport substances to the surface which apparently reduce the stability of the foam. However, an increase in whitecap lifetime (i.e., greater persistence of whitecapping) due to surface-active substances, resulting in greater values of W for the same U_{10}, would not necessarily lead to greater SSA production.

4.5.3.8. Summary. Although factors other than wind speed can be expected to exert an influence on W, investigation of these effects has led to inconclusive or contradictory results. As seen when whitecap ratios are parameterized by U_{10} alone, although a fit to a particular data set might be fairly successful, such a fit generally does not provide a good description of data from other studies, obtained in different locations under different conditions. Although W appears to be greater at high temperatures than at cold temperatures, it is difficult to quantify this dependence given the large scatter in the data and the differing results of *Bortkovskii* and other investigators at intermediate temperatures. Results for other possible factors are less certain, and the effects are typically much less than the uncertainties of individual data sets, and far too small to appreciably reduce the scatter in W for a given wind speed. Additionally, it is difficult to attribute an observed effect to a given factor, as typically there has been little variability of environmental and meteorological conditions within any given study but considerable differences from study to study. Other quantities, such as variables describing sea state (the phase velocity or period of the significant wave, wind duration, and the like) or surface films have not been widely reported in conjunction with measurements of W, and few studies have been conducted on the effects of these quantities on W. For these reasons little confidence can be placed even in qualitative association of W with these quantities and all the more in parameterizations of such relations.

4.5.4. Properties of Individual Whitecaps

Application of the whitecap method to determine the size-dependent interfacial SSA production flux assumes that the area of each whitecap, independent of its initial size, decreases exponentially with the same characteristic time τ_{wc}. However, little systematic investigation of lifetimes and sizes of oceanic whitecaps has been reported, and reported results vary greatly. This section reviews the pertinent available observations.

Sizes of oceanic whitecaps have been reported by several investigators. Bubble trails up to 800 m long during winter gales in the North Atlantic were reported by *Woodcock et al.* [1963]. Maximum areas of individual whitecaps of up to 40 m^2 at $U_{10} = 14$ m s^{-1} were reported by *Kondo et al.* [1973], although mean areas were generally smaller: 1.4 m^2 and 2.2 m^2 for wind speeds of 5-11 m s^{-1} and 11-14 m s^{-1}, respectively. The distribution of areas of individual crests and strip-like structures for each of three wind speeds (5.7, 9.5, and 10.5 m s^{-1}) were parameterized by *Bondur and Sharkov* [1982] as gamma distributions, with the mean areas of individual crests and foam strips increasing from ~0.5 m^2 to ~1.2 m^2 and from ~20 m^2 to ~30 m^2, respectively as wind speed increased over this range. Although individual strips having areas up to nearly 100 m^2 were reported for each of the wind speeds, typical areas were around 10 m^2. Mean areas of oceanic whitecaps in shallow coastal water at low wind speeds (typically around 6 m s^{-1}) of 0.2-0.5 m^2 were reported by *Snyder et al.* [1983]. Whitecap areas of tens of square meters were reported by *Bortkovskii* [1987], with typical whitecap lengths being 1-2 m along the forward slopes of breaking waves, greater at higher temperatures and only slightly increasing as U_{10} increases from 5 m s^{-1} to 20 m s^{-1}. The extents of the whitecaps across the slopes were roughly two to three times these values, resulting in typical areas of 5-10 m^2.

Lifetimes of oceanic whitecaps have also been reported by several investigators. Individual foam patches formed at wind speeds of $U_{10} = 10$ m s^{-1} were reported by *Woodcock* [1953] to remain clearly visible from aircraft for more than two minutes. Lifetimes of oceanic whitecaps at a coastal site in shallow water reported by *Kondo et al.* [1973] ranged from 0.3-3 s, with mean lifetimes of 1.3 s and 1.6 s for wind speeds of 5-11 m s^{-1} and 11-14 m s^{-1}, respectively. Characteristic lifetimes for "crests" and "foam strips" were reported by *Bondur and Sharkov* [1982] as seconds and minutes, respectively. Mean lifetimes of oceanic whitecaps in shallow coastal water at low wind speeds (typically around 6 m s^{-1}) of around 0.15-0.2 s were reported by *Snyder et al.* [1983], who additionally reported that the area of the whitecap grew for roughly the first half of its lifetime and then decreased, often rather abruptly; such behavior was also reproduced in a statistical model of wave breaking [*Snyder and Kennedy*, 1983]. From an analysis of photographs of 13 individual whitecaps at winds of around 8 m s^{-1} and 6 individual foam streaks at winds of around 15 m s^{-1}, at temperatures near 15 °C, *Koepke* [1984, 1986b] reported that whitecaps (foam patches) had areas that increased and reflectances that decreased (from around 40% to below 10%, averaging about 20%) for the first ten seconds after formation, whereas foam streaks, with lifetimes of more than ten seconds, had areas that exhibited only a slight increase with time, and reflectances (generally around 10%) that exhibited only a slight decrease with time. Lifetimes of foam patches and the durations of wave crest breaking reported by *Bortkovskii* [1987, p. 75, Fig. 2.12] and *Bortkovskii and Novak* [1993] were 15-40 s and 3-5 s, respectively, both increasing nearly linearly with wind speed, with lifetimes of foam patches decreasing with increasing temperature, and durations of wave crest breaking exhibiting no apparent temperature dependence. Areas of nine oceanic whitecaps formed at wind speeds between 10 m s^{-1} and 15 m s^{-1} analyzed (using an automatic area analyzer) from video tapes by *Nolan* [1988a] reached a maximum after 5-10 s, after which they decayed. This maximum was a broad function of time, often lasting several seconds. The decay was fitted to an exponential with an average characteristic time of 4.3 s, and an average effective lifetime (the definition of which was not clearly stated) of 10 s; no measure of variance was reported. The results were extremely sensitive to the gain of the analyzer; for a slight increase in gain the decay constant was approximately halved.

Infrared measurements of skin temperature following breaking waves [*Jessup et al.*, 1997] have indicated that the time required for decay of the deviation of the mean associated with whitecaps depends on the strength of the breaking wave, with larger whitecaps requiring longer times. Such an observation would seem to contradict the assumption that whitecap lifetimes are independent of their size.

Based on these measurements it is difficult to determine a characteristic size or time of a typical whitecap, or the dependence of the sizes and times on wind speed. However, these observations certainly suggest that the decay time of whitecaps can vary substantially from the constant ~3 s assumed in application of the whitecap method, certainly by a factor of $\gtrless$ 2, and perhaps more.

4.5.5. Summary of Data on Whitecaps

Despite the large body of work on whitecaps, much uncertainty remains concerning whitecap ratio, properties of individual whitecaps, the dependencies of these quantities on controlling factors, and the relation of whitecaps to SSA production. While all evidence indicates that whitecaps are the major source of SSA particles, that SSA production increases with increasing W, and that W increases with increasing wind speed, the arbitrary nature of what constitutes a whitecap, difficulties in determining W, the large spatial and temporal variability of whitecapping on scales of measurements, and the large variation in W between different studies under seemingly similar conditions result in quite large uncertainties in values of W for a given set of conditions, and all the more so in proposed dependencies of W on wind speed and other factors. It is uncertain whether determinations of W from photographs from aircraft are more accurate than those from lower heights, which yield considerably lower values. Video determinations yield considerably lower values of W than do photographs; the use of such video data, at least as reported so far, seems ill-advised for a variety of reasons (however, this approach might be revisited with newer digital technology and new analysis techniques for discerning whitecaps).

While it is clear that W generally increases with increasing U_{10} above ~8 m s^{-1}, the ability to describe the dependence of W on wind speed and other controlling factors is quite limited, and although some individual data sets appear to exhibit clear trends with some of these other factors, there is little consistency in the data as a whole. Fitted quantities have often been presented with far greater precision than can be justified by the reported variability in individual data sets, and all the more when multiple data sets are examined. Although classifying the dependence of W on U_{10} by T_{sea} reduces the scatter in the data somewhat, the temperature dependence is not accurately known; there is no consensus for intermediate temperatures, and only a small fraction of the ocean surface waters experiences temperatures at the low end of the temperature range of the measurements, which provides much of the evidence for the observed temperature dependence. Attempts to reduce the scatter of the data by including dependence on other factors, such as atmospheric stability and fetch have not been successful;

often there are not sufficient data upon which to draw conclusions, and when there are, the magnitudes of the proposed effects are generally too small to appreciably decrease the variance in calculated values of W.

The assumption that the area of a whitecap decreases exponentially with a lifetime that is independent of its size, which is inherent to the application of the whitecap method, is not supported by data either from laboratory experiments or field measurements. Likewise, the dependence of whitecap lifetime on wind speed and temperature, if any, is not well characterized.

In summary, although the relation of SSA production to whitecaps is qualitatively understood, quantitative relation of whitecap area to wind speed and other potentially controlling factors, which are required for estimation of the SSA production flux according to the whitecap method, remains highly uncertain.

4.6. DRY DEPOSITION

The dry deposition concept was introduced in §2.1.6 and discussed further in §2.7.2, where the dry deposition velocity of an SSA particle of given r_{80}, $v_d(r_{80})$, was introduced and a dry deposition model was described. According to this model, the lower atmosphere is conceptually divided into two layers, a thin (~0.1-1 mm) viscous sublayer adjacent to the sea surface, and above that a surface layer extending several tens of meters. For each layer the contributions of various mechanisms to the dry deposition of SSA particles are characterized by transfer velocities. An expression (Eq. 2.7-7) for the dry deposition velocity was obtained in terms of these transfer velocities by matching the downward flux of SSA particles through the two layers under the assumption that the dry deposition flux of SSA particles of a given size is independent of height. This approach has served as the basis of most estimates for $v_d(r_{80})$.

The steady state dry deposition method for determining the effective SSA production flux (§3.1) is based on the assumption that this production flux for SSA particles of a given r_{80} is equal to the downward dry deposition flux of these particles, calculated as the product of the dry deposition velocity and the number concentration at a height near 10 m. This method, because it pertains to the effective SSA production flux, is necessarily restricted to small and medium SSA particles (i.e., $r_{80} \lesssim 25$ μm). As emphasized throughout this review, large SSA particles do not contribute appreciably to the effective SSA production flux; they will not have equilibrated with respect to the ambient RH during their residence time in the atmosphere (§2.9.1) and their concentrations are expected to exhibit large vertical gradients near 10 m (§2.9.4). Hence only SSA particles with $r_{80} \lesssim 25$ μm need be considered in the context of dry deposition.

Numerous reviews of dry deposition of particles have been presented [*Slinn*, 1974, 1977a; *Slinn et al.*, 1978; *McMahon and Denison*, 1979; *Sehmel*, 1980; *Slinn and Slinn*, 1981; *Slinn*, 1983a,b; *Chamberlain*, 1983b; *van Aalst*, 1985; *Nicholson*, 1988; *Wiman et al.*, 1990; *Ruijgrok et al.*, 1995; *Wesely and Hicks*, 2000], and many models have been proposed. Here a specific model [*Slinn and Slinn*, 1980, 1981] for the dry deposition of SSA particles over the ocean is examined in some detail. This model has been widely used and many later models have been based upon it. This model leads to an explicit formulation of size-dependent transfer velocities and through these to an expression for the size-dependent dry deposition velocity as a function of meteorological parameters, most importantly wind speed. The corresponding size-dependent atmospheric residence time against removal by dry deposition is presented; knowledge of this residence time permits determination of the size range of SSA particles for which the steady state dry deposition method can be used to infer the SSA production flux, based on the conditions specified in §3.1. Modifications to this model and alternative models, including different parameterizations of the various transfer velocities and other transfer mechanisms, are also presented and dry deposition velocities resulting from these modifications and alternative models are compared to the *Slinn and Slinn* dry deposition velocity. Measurements of dry deposition and the extent to which such measurements can be used to evaluate performance of these models are also discussed. Finally, dry deposition models in general and their ability to accurately simulate physical processes in the atmosphere are examined.

4.6.1. Dry Deposition Formulation of Slinn and Slinn

The model of *Slinn and Slinn* [1980, 1981], which incorporates many features described earlier [*Sverdrup*, 1937; *Johnstone et al.*, 1944; *Eriksson*, 1959; *Toba*, 1965a,b; *Sehmel and Hodgson*, 1974; *Slinn*, 1974, 1977a; *Slinn et al.*, 1978], accounts not only for vertical transport of SSA particles but also for increased radii of particles near the sea surface due to the higher RH there. This model employs parameterizations different from those used previously for the several transfer velocities, as described in *Slinn* [1983a,b]. In their original formulation, *Slinn and Slinn* [1980, 1981] had assumed that the particles in the surface layer were dry, but *Fairall and Larsen* [1984] argued that 80% RH is typical for this layer; this certainly seems a much better first approximation, and hence it is used here (as it is throughout this review). Relative humidities of 99% and 100% were considered by *Slinn and Slinn* for the viscous sublayer, however near the ocean surface RH would be limited to near 98% because of the vapor pressure lowering

of water over seawater (§2.5.3) due to the salt content [*Sverdrup*, 1937; *Eriksson*, 1959; *Toba*, 1965b; *Williams*, 1982b; *Sievering*, 1984; *Fairall and Larsen*, 1984]; thus this value also is used here. Because of the various modifications that have been proposed, references to "the model of Slinn and Slinn" remain ambiguous unless the relative humidities of the two layers, as well as the formulation of the various transfer velocities, are explicitly given. The model of *Slinn and Slinn* [1980, 1981], with these choices for the RH values of the viscous sublayer and surface layer, is examined below.

Calculation of the dry deposition velocity, v_d (Eq. 2.7-7), requires evaluation of five different transfer velocities: $v_{V,Stk}$, v_{imp}, and v_{diff}—the gravitational sedimentation velocity, impaction velocity, and Brownian diffusion velocity, respectively, of a particle in the viscous sublayer, and $v_{S,Stk}$ and v_{turb}—the gravitational sedimentation velocity and turbulent diffusion velocity, respectively, of a particle in the surface layer (§2.7.2).

The gravitational sedimentation velocity (§2.6.2) of an SSA particle in each layer is represented by the Stokes velocity, (2.6-7), where the density and radius (and hence the Cunningham correction C) are assumed to be those of a drop in equilibrium at the relative humidity of this layer. Under the assumption that SSA particles entering the viscous sublayer equilibrate instantaneously with the RH of this layer, 98% (examined below), the radius of a particle in the viscous sublayer is twice that in the surface layer (§2.5.3), and over most of the size range of SSA particles of interest the Stokes velocity in the viscous sublayer is slightly more than three times greater than the Stokes velocity in the surface layer (Fig. 7).

Impaction, caused by the inability of an SSA particles to respond rapidly to non-uniform flow near the sea surface, is parameterized in terms of the dimensionless Stokes number St:

$$St = \frac{2}{9}\frac{\rho_p}{\rho_a}\frac{r^2 C C_D U_{10}^2}{v_a^2} \qquad (4.6\text{-}1)$$

where ρ_a and v_a are the density and kinematic viscosity of air, respectively, C_D is the wind stress coefficient (§2.3.2), and r, ρ_p, and C are for a drop in equilibrium at the relative humidity of this viscous sublayer. For C taken as unity (which may be done with little loss in accuracy because of the size range of particles for which impaction plays an appreciable role), St is given approximately by

$$St \approx 0.004\left(\frac{r_{80}}{\mu m}\right)^2 \left(\frac{U_{10}}{m\,s^{-1}}\right)^2 \qquad (4.6\text{-}2)$$

To estimate transport across the viscous sublayer, *Slinn and Slinn* rejected earlier "free-flight" theories [e.g., *Friedlander and Johnstone*, 1957; *Owen*, 1960; *C. Davies*, 1966] in which particles carried by turbulent eddies to the top of the viscous sublayer continue their motion until reaching the sea surface, and adopted a theory proposed by *Owen* [1969], based on experiments of *Kline et al.* [1967], in which particles are transported to the sea surface at the leading edge of turbulent eddies that sporadically burst from this surface. The contribution of impaction to the transfer velocity in the viscous sublayer, v_{imp}, is that given by *Slinn* [1974], based data of *Liu and Agarwal* [1974] for the deposition rate of particles from turbulent flow to the smooth surface of a vertical pipe:

$$v_{imp} = \left(\frac{C_D U_{10}}{\kappa}\right) \times 10^{-\left(\frac{3}{St}\right)} \qquad (4.6\text{-}3)$$

where $\kappa = 0.40$ is the von Karman constant. For particle sizes and wind speeds for which $St < 1$ impaction rarely occurs; rather the particles follow streamlines at the front of the turbulent eddy. For larger particles or higher wind speeds such that St is considerably greater than unity, the particles, because of their inertia, do not follow steamlines but are impacted into the sea surface. For increasingly large St, v_{imp} becomes independent of St and depends only upon U_{10}.

For very small particles transport across the viscous sublayer occurs primarily through Brownian (molecular) diffusion. The rate of this diffusion is parameterized in terms of the (dimensionless) Schmidt number, Sc, defined by $Sc = v_a/D_B$, where v_a is the kinematic viscosity of air and D_B is the Brownian diffusivity of a particle, given by (2.6-2). Numerically the Schmidt number for an SSA particle of given r_{80} is given approximately by $Sc \approx 3 \cdot 10^6 (r_{80}/\mu m)/C$. The contribution of Brownian diffusion to the transfer velocity across this layer is given by

$$v_{diff} = \frac{C_D U_{10}}{\kappa} Sc^p \qquad (4.6\text{-}4)$$

with $p = -1/2$. The $Sc^{-1/2}$ dependence follows from the assumption that the surface of the water is not fixed, but slips (i.e., moves with the wind). At fixed wind speed, v_{diff} decreases with increasing r_{80} (in contrast to $v_{V,Stk}$ and v_{imp}).

The turbulent transfer velocity,

$$v_{turb} = \frac{C_D U_{10}}{1-\kappa}, \qquad (4.6\text{-}5)$$

is based on the Reynolds analogy, namely that rate of mass transfer in the lower atmosphere is the same as the rate of momentum transfer. This implies that all particles, independent of size, are transferred at the same rate by the action of turbulent eddies in the surface layer.

Under the assumption that the several transfer mechanisms in each layer act in parallel and thus that the corresponding transfer velocities in each layer are additive, the transfer velocities through the surface layer and viscous sublayer are given respectively by

$$v_S = \frac{C_D U_{10}}{1-\kappa} + \frac{2}{9}\frac{\rho_p}{\rho_a}\frac{g}{v_a} r^2 C \qquad (4.6\text{-}6)$$

and

$$v_V = \frac{C_D U_{10}}{\kappa}\left(10^{-\left(\frac{3}{St}\right)} + Sc^{-\left(\frac{1}{2}\right)}\right) + \frac{2}{9}\frac{\rho_p}{\rho_a}\frac{g}{v_a} r^2 C, \quad (4.6\text{-}7)$$

where in each expression the values of ρ_p, r, and C appropriate for the RH of that layer must be used. These velocities contain a very slight temperature dependence through the kinematic viscosity, diffusivity, and density of air, but this can be neglected with little sacrifice in accuracy.

4.6.2. Size-Dependent Dry Deposition Velocities and Their Implications

Knowledge of the five transfer velocities specified above allows dry deposition velocities to be calculated from (2.7-7) by matching the fluxes through the two layers. Dry deposition velocities calculated in this manner, under the assumptions of a constant wind stress coefficient $C_D = 0.0013$ (the value used by *Slinn and Slinn* [1980]), an air temperature of 20 °C, and RH in the upper and lower layers of 80% and 98% respectively, are shown in Fig. 38 as a function of r_{80} for several wind speeds. Also shown are v_V and v_S, the transfer velocities through the viscous sublayer and the surface layer, respectively, and $v_{V,Stk}$, $v_{S,Stk}$, v_{diff}, and v_{imp}, the transfer velocities describing gravitational sedimentation through these two layers, Brownian diffusion, and impaction, respectively.

The importance of mechanisms controlling dry deposition varies greatly in different ranges of r_{80}, and v_d more or less asymptotically approaches the transfer velocity of one or another of these mechanisms as this mechanism becomes the limiting factor controlling dry deposition. For a given wind speed, v_d exhibits a strong dependence on r_{80}, increasing by up to three orders of magnitude as r_{80} increases from 0.1 μm to 25 μm. Likewise, for a given r_{80}, v_d is an increasing

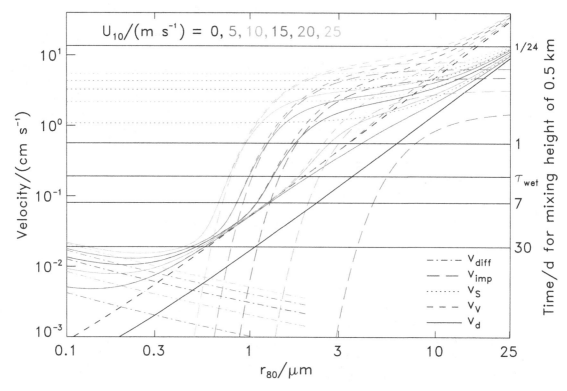

Figure 38. Dry deposition velocity and transfer velocities calculated from the model of *Slinn and Slinn* [1980, 1981] as a function of r_{80} for several values of the wind speed at 10 m above the sea surface U_{10}. Right axis shows values of characteristic dry deposition time τ_{dry} for an assumed mixing height of 0.5 km; horizontal lines denoting 1 hr, 1 d, 1 week, and 1 month are shown, as is line denoting value of τ_{wet} assumed to be 3 d. Characteristic dry deposition times τ_{dry} will be overestimated for SSA particles with $r_{80} \gtrsim 15$ μm for which mixing heights are less than 0.5 km. The quantities v_{diff} and v_{imp} are the transfer velocities due to Brownian and impaction, respectively; v_S and v_V are the transfer velocities through the surface layer and the viscous sublayer, respectively, and v_d is the dry deposition velocity.

function of wind speed, although the strength of this dependence varies greatly with r_{80}, from almost no increase at $r_{80} = 0.5$ μm, to an increase of nearly an order of magnitude at $r_{80} = 1.5$ μm, to again almost no increase at $r_{80} = 25$ μm, as U_{10} increases from 5 m s^{-1} to 25 m s^{-1}. The dependence on r_{80} at different wind speeds is qualitatively similar, although the values of r_{80} for which different effects become important change somewhat, and for some wind speeds not all mechanisms come into play. For instance, at $U_{10} = 5$ m s^{-1} neither impaction or turbulent diffusion is an important mechanism for dry deposition; consequently, neither v_{imp} nor v_{turb} contributes appreciably to v_d for this wind speed. The dependence of v_d on r_{80}, and the importance of various mechanisms controlling dry deposition, are examined below for $U_{10} = 10$ m s^{-1}. At this wind speed, dry deposition of particles with $r_{80} \lesssim 3$ μm is controlled almost exclusively by transfer through the viscous sublayer, whereas for $r_{80} \gtrsim 10$ μm dry deposition is governed largely by transfer through the surface layer. For intermediate sizes, the transfer

velocities through both the viscous sublayer and the surface layer are important in determining v_d.

For r_{80} near 0.1 μm, the dominant contribution to v_V, and hence to v_d, is from Brownian diffusion, and the slight decrease of v_{diff} with increasing r_{80} leads to a similar decrease in v_d. As r_{80} increases above 0.1 μm, the contribution of gravitational sedimentation becomes increasingly important and exceeds that of Brownian diffusion at $r_{80} = 0.3$ μm. The broad minimum in v_d of approximately 0.01 cm s^{-1} near $r_{80} = 0.2$ μm results from the additive contributions of these two mechanisms.

As r_{80} increases to 1 μm, the dominant contribution to v_d is provided by gravitational sedimentation in the viscous sublayer (where the geometric radius is twice the value of r_{80}), but as r_{80} increases above 1 μm, the contribution from impaction becomes increasingly important. The strong dependence of v_{imp} on particle size is reflected by the large increase in v_d (a factor of three) as r_{80} increases from 2 μm to 3 μm; at higher wind speeds this effect is even more pronounced. Additionally, the strong dependence of v_{imp} on wind speed results in an increase in v_d by more than an order of magnitude as U_{10} increases from 10 m s^{-1} to 25 m s^{-1} for an SSA particle with $r_{80} = 1$ μm.

The transfer velocities through the viscous sublayer and the surface layer, v_V and v_S, respectively, are equal at $r_{80} \approx 4$ μm, and for greater r_{80}, up to ~10 μm, the transfer velocities through both layers contribute appreciably to v_d. Both impaction and gravitational sedimentation contribute appreciably to v_V, and turbulent diffusion provides the major contribution to v_S (the turbulent transfer velocity v_{turb}, Eq. 4.6-4, is independent of particle size, as indicated by the dotted lines in Fig. 38 being horizontal in this region). As r_{80} increases from 4 μm to 10 μm, the dry deposition velocity increases slowly, barely doubling from 1.2 cm s^{-1} to 2.8 cm s^{-1}.

As r_{80} increases above 10 μm, v_d is controlled almost exclusively by transfer through the surface layer. The contribution from gravitational sedimentation becomes increasingly important, and it is equal to that from turbulent diffusion at $r_{80} = 12$ μm. As the gravitational sedimentation velocity increases quadratically with r_{80}, $v_{S,Stk}$ rapidly exceeds v_{turb} with further increase in r_{80}, and gravitational sedimentation increasingly becomes the dominant removal mechanism. Thus v_d becomes nearly equal to $v_{S,Stk}$, and hence nearly independent of wind speed, as illustrated by the approach of the various solid lines to the solid black line representing $v_{S,Stk}$.

The characteristic lifetime against removal by dry deposition of an SSA particle with given r_{80} (§2.7.2.3), $\tau_{dry}(r_{80})$, is evaluated from $v_d(r_{80})$ using (2.9-7). These lifetimes are shown on the right axis of Fig. 38 for an assumed mixing height of 0.5 km, with horizontal lines indicating one month, one week, one day, and one hour. At $U_{10} = 10$ m s^{-1} the corresponding values of r_{80} are roughly equal to 0.5, 1, 3, and 30 μm, respectively; however, for r_{80} greater than ~15 μm, the expected mixing heights are much less than 0.5 km (§2.9.5), resulting in correspondingly shorter characteristic removal times than the values given. Also shown is the characteristic time for removal by wet deposition τ_{wet} (§2.7.1), assumed equal to 3 d, for which the corresponding value of r_{80} is ~2 μm. Examination of $\tau_{dry}(r_{80})$ permits evaluation of the satisfaction of the conditions required for the application of the steady state dry deposition method and of the validity of some of the assumptions of specific dry deposition models (§3.1).

Sea salt aerosol particles with $r_{80} \lesssim 1$ μm have such long atmospheric residence times against removal by dry deposition—more than a week for $U_{10} < 20$ m s^{-1}—that it is almost certain that dry deposition will not be the dominant removal mechanism; the condition $\tau_{dry} < \tau_{wet}$ will rarely hold, and in nearly all situations wet deposition is the main removal mechanism for these particles (§2.7.1). Furthermore, the assumption

that steady state conditions hold over such long times will rarely if ever be satisfied. Likewise, the distance such a particle would have traveled during its estimated lifetime against removal by dry deposition is many thousands of kilometers—far too large to expect any correlation between local conditions and those under which the particle was produced.

Sea salt aerosol particles with r_{80} between ~1 μm and ~3 μm have estimated lifetimes against removal by dry deposition from more than a week to less than a day, decreasing with increasing particle size and with increasing wind speed, during which they would be transported from several hundred to several thousand kilometers. As for smaller SSA particles, the assumptions that wind speeds and other conditions remain constant during this time, and that local conditions reflect those under which the particles were produced, are unlikely to be satisfied for particles in this size range. The high sensitivity of the dry deposition velocity to wind speed presents difficulties in interpretation in situations of non-steady winds, and because of the highly nonlinear nature of the problem, it would seem unlikely that any meaningful "average" wind speed or dry deposition velocity could be defined in such situations. Additionally, the high sensitivity of the dry deposition velocity (and thus the lifetime) to particle size, due to the sensitivity of v_{imp} to particle size in this size range, makes the assumption that particles equilibrate to the RH of the viscous sublayer especially critical; this is discussed below in §4.6.3.

Sea salt aerosol particles with r_{80} from ~3 μm to 25 μm have mean lifetimes against removal by dry deposition ranging from nearly a day to much less than an hour (if the mixing height being less than 0.5 km for SSA particles in the upper portion of this size range is taken into account), decreasing with increasing particle size and with increasing wind speed. With increasing particle size in this size range the likelihood thus increases that removal of SSA particles is due entirely to dry deposition. Likewise with increasing particle size, the requirement that conditions remain constant during the time the particles spend in the atmosphere and that current local conditions relate to those under which the particles were produced, are increasingly likely to be satisfied. Therefore the production of SSA particles in this size range can often be balanced by removal through dry deposition, justifying application of the steady state dry deposition method for calculating the effective SSA production flux.

4.6.3. Alternative Formulations and Models

An assessment of the confidence that can be ascribed to modeled dry deposition velocities from a particular formulation, to the use of dry deposition models in general, and to their use in obtaining SSA production fluxes in particular, can be gained from quantifying the uncertainties in the calculated dry deposition velocities. One method of estimating these uncertainties is to consider differences in dry deposition velocities resulting from formulations different from that of *Slinn and Slinn* [1980, 1981] described in §4.6.1. Hence some of the various formulations and different dry deposition models that have been proposed are briefly reviewed with the aim of assessing the spread in dry deposition velocities at a given wind speed; this uncertainty translates directly into the effective SSA production flux inferred by the steady state dry deposition method.

Numerous dry deposition models pertinent to SSA particles over the oceans have been proposed, many of which are similar to the formulation of *Slinn and Slinn* [1980, 1981]. Some include effects of atmospheric stability, some use different formulations for one or another of the several transfer velocities, and some include additional transfer mechanisms. Most of these modifications retain the basic two-layer structure of the original model, with the notable exception of a model proposed by *Fairall and Larsen* [1984; this model was also discussed in *Davidson et al.*, 1982 and *Fairall et al.*, 1982, 1984], which included an additional layer above the surface layer, termed the mixed layer, extending from the top of the surface layer to the top of the marine boundary layer, to incorporate the effects of entrainment of the free troposphere at the top of the marine boundary layer (§3.1). Some of these alternative models are examined in this section. The effects of atmospheric stability on dry deposition are also examined, as are phoretic effects and alternative expressions for the various transfer velocities. Additionally, the extent to which SSA particles equilibrate to the RH of the viscous sublayer, one of the key differences in the model of *Slinn and Slinn* and previous models, is examined. Alternative models that include effects of broken surfaces (that is, regions of the ocean covered by bubbles and/or whitecaps) are also reviewed. Dry deposition velocities resulting from several modifications to the model of *Slinn and Slinn* and from alternative models are compared, and the size range of SSA particles for which these velocities differ appreciably from those obtained from the model of *Slinn and Slinn* is identified.

The effects of atmospheric stability are manifested in different values of the wind stress coefficient C_D (§2.3.2) for the same wind speed U_{10}, resulting in changes in v_{imp}, v_{diff}, and v_{turb} through their dependencies on C_D, and even greater changes in v_{imp} through the additional dependence of St upon C_D (Eqs. 4.6-3, 4.6-4, 4.6-5). Numerical investigations have demonstrated that atmospheric stability typically has a relatively small effect on dry deposition velocities (a few tens of percent at most) except at low wind speeds and for large stable temperature gradients [*Sehmel and Hodgson*, 1974; *Williams*, 1982b; *Rojas et al.*, 1993; *Lo et al.*, 1999], in which situation the estimated dry deposition velocities are even lower than those shown in Fig. 38. However, these effects are generally unimportant, as under such conditions little SSA production occurs. The possibility that atmospheric stability might play a larger role in transport through its effects on the enhancement or suppression of turbulent bursts from the surface than through its effects on the wind stress coefficient and consequent enhanced (or diminished) vertical transport was suggested by *Slinn* [1983b], but the influence of such turbulent bursts either on dry deposition or on SSA production would be difficult to quantify.

Formulations of v_{imp} different from that in (4.6-3) have been proposed by several investigators. Fits to the data of *Liu and Agarwal* [1974] different from that presented by *Slinn and Slinn* [1980] have been used [*Haynie*, 1985; *Giorgi*, 1986a,b, 1988], and *Slinn* [1983a] suggested that a dependence of v_{imp} on St that is smoother than that represented by (4.6-3) may be more appropriate, such as that proposed in *Slinn* [1982] (which was incorrectly presented, as noted in *Slinn* [1983a]) or that in *Slinn* [1983a]. Expressions for v_{imp} with the same dependence on St as (4.6-3) but with coefficients different from that originally proposed by *Slinn* [1974], i.e., $C_D U_{10}/\kappa$, have also been proposed [e.g., *Giorgi*, 1986a,b, 1988; *Hummelshøj et al.*, 1992; *Seinfeld and Pandis*, 1998, p. 965]. In some instances these coefficients differ greatly; that proposed by *Seinfeld and Pandis* [1998, p. 965], for example, is more than an order of magnitude greater than that proposed by *Slinn and Slinn* [1980, 1981].

A variety of expressions for v_{diff} that differ from (4.6-4) have been proposed, both in the functional dependence on Sc and in the coefficient. *Slinn and Slinn* [1981] and *Slinn* [1983a,b] concluded that there is no firm basis to decide whether the exponent p of Sc in the expression for v_{diff}, (4.6-4), is -1, $-2/3$, -0.6, or $-1/2$, each of which has a theoretical (or quasi-theoretical) basis, and further that mechanisms corresponding to each of these dependencies probably operate. They suggested that the correct expression may consist of a sum of terms, each corresponding to one of the above dependencies. There is a similar lack of unique theoretical basis for the coefficient of Sc^p in (4.6-4). The quantity $C_D U_{10}/\beta$ was proposed by *Slinn* [1974], who suggested using $\beta = 0.4$, the same value as the von Karman constant κ, although he stated that this choice was "of no obvious significance" and "convenient," and later referred to β as a "fudge factor" [*Slinn*, 1983a]. Other values have been proposed [e.g., *Giorgi*, 1986a,b, 1988; *Hummelshøj et al.*, 1992; *Seinfeld and Pandis*, 1998, p. 965], sometimes differing greatly, and a formulation in which the exponent p of the Schmidt number is $-1/2$ and the coefficient depends on the Reynolds number based on the roughness length of the surface was presented by *Hummelshøj et al.* [1992]. The effects of such changes are important only for those SSA particles for which v_{diff} is the controlling factor in v_d, that is, particles with $r_{80} \lesssim 0.3$ μm. Because Sc is quite large in this size range, the differences among the different formulations are considerable; at $r_{80} = 0.3$ μm, $Sc \approx 10^6$ and the values of $Sc^{-1/2}$, $Sc^{-0.6}$, $Sc^{-2/3}$, and Sc^{-1} are in the ratio 1000:250:100:1 (differences are greater for smaller particles). The effect of such alternative formulations on the application of the dry deposition method to infer SSA production, however, is irrelevant, as use of a formulation with a dependence of Sc being a power law with exponent -0.6, $-2/3$, or -1 instead of $-1/2$ as in (4.6-4) would result in v_d being even less for particles with $r_{80} \lesssim 0.3$ μm, for which values of τ_{dry} are more than a month—far too long for satisfaction of the conditions required for the steady state dry deposition method.

Alternative expressions for v_{turb} have also been proposed by several investigators. The formulation given above for v_{turb} (Eq. 4.6-5)—the heuristic nature of which was stressed by *Slinn* [1983a]—was chosen by the Reynolds analogy of the equivalence of mass transfer and momentum transfer [*Slinn and Slinn*, 1981]. However, this analogy may fail to hold over the ocean because pressure forces and form drag on waves can provide a mechanism for momentum transport with no corresponding analog for mass transport. Several investigators have suggested that analogies with heat transfer may be more apt; under this assumption C_D in the expression for v_{turb} would be replaced by C_H, the bulk transfer coefficient for sensible heat. Most formulations for C_H show it to be roughly independent of wind speed with a value of around 0.001; thus it is nearly equal to C_D at wind speeds of 10 m s^{-1} and decreases to about half the value of C_D when U_{10} reaches 20 m s^{-1} [*Liu et al.*, 1979; *Blanc*, 1985; *Smith*, 1988]. The effects on v_d resulting from use of C_H instead of C_D would be expected only for particles for which v_{turb} plays a substantial role in determining v_d (i.e., 4 μm $\lesssim r_{80} \lesssim 15$ μm) and then only for high wind speeds (i.e., $U_{10} \gtrsim 15$ μm) where C_H and C_D differ appreciably. A much more complicated expression for v_{turb} than that of *Slinn and Slinn* [1980] was proposed by *Giorgi* [1986a,b; 1988], but the two expressions never differ by more than 25%. The expression proposed for v_{turb} by *Seinfeld and Pandis* [1998, p. 963, 972] is ~40% less than that of *Slinn and Slinn* [1981].

Phoretic effects (§2.6.1) resulting from gradients of water vapor concentration and temperature, of which only Stefan flow is thought to be important in most circumstances, can also influence transport of particles in the viscous sublayer (§2.6.1). For an RH of 98% in the viscous sublayer and 80% in the surface layer there is net evaporation of water from the sea surface, resulting in a Stefan flow given by (2.6-1) that acts to transport particles away from the surface. The rate of this transport can often exceed that through the viscous sublayer toward the surface due to Brownian diffusivity alone. *Sehmel and Sutter* [1974] concluded from wind tunnel experiments that phoretic effects could be important for particles

with $r_{80} \lesssim 0.5$ μm. However, phoretic effects are not often included in dry deposition models. These effects would vary considerably both temporally and spatially because of variability in gradients of RH and temperature, and hence quantities such as evaporation rate in addition to wind speed would be required to parameterize dry deposition velocities for small SSA particles. As discussed earlier, however, for such particles the estimated lifetimes against dry deposition are already so long that the conditions required for the steady state dry deposition method are not satisfied.

Whether or not the radius of a depositing SSA particle initially in the surface layer (80% RH) equilibrates to the higher RH (98%) of the viscous sublayer before reaching the surface can have a large effect on the dry deposition velocity of such a particle because of the dependencies of the Stokes velocity and the impaction velocity on the particle's geometric radius (the effect of the decrease in density at the higher RH is minor). *Eriksson* [1959] concluded that the particles as small as $r_{80} = 0.5$ μm did not equilibrate before they reached the surface. *Slinn and Slinn* [1980, 1981] included the assumption that particles instantaneously equilibrate as one of major differences with previous models, arguing that for particles with r_{80} less than several micrometers, the growth is sufficiently rapid to justify this assumption, whereas for larger particles the dry deposition velocity is controlled by the transfer through the surface layer. As noted above, *Slinn and Slinn* assumed particles in the surface layer were initially dry, and they considered RH 99% and 100% in the viscous sublayer. Clearly, the relative humidities of the two layers will play a role in determining v_d, as they will affect the equilibrium particle radii and thus the various transfer velocities and the time spent in the viscous sublayer, in addition to the characteristic equilibration times. Also, as discussed above, because RH in the marine boundary layer is almost always greater than the efflorescence value, ~45%, SSA particles will almost certainly be solution drops rather than dry particles (§2.5.5).

The effect of equilibration on dry deposition of initially dry sodium chloride particles in the size range $r_{80} = 2.5$-8.4 μm was investigated experimentally by *Jenkin* [1984], who measured the deposition rate of these particles to smooth wet and dry filter paper in a wind tunnel. He reported that the increase in v_d compared to that expected from gravitational sedimentation alone was very small, from which he concluded that particles with $r_{80} > 2$ μm did not have time to fully equilibrate and suggested an effective RH of around 90% for the viscous sublayer. However, there are several reasons why these results may not be applicable to SSA particles over the ocean. Most SSA particles would be solution drops when they reach the viscous sublayer, not dry particles, as in these experiments. In addition, *Jenkin* argued that time available for particle growth was the lifetime of one of the turbulent streaks observed by *Kline et al.* [1967]. This assumption would imply that the primary mechanism for transport through the viscous sublayer is impaction and not gravitational sedimentation; however, the two wind friction velocities considered, $u_* = 5$ cm s^{-1} and 12 cm s^{-1}, corresponding roughly to $U_{10} = 1.5$ m s^{-1} and 3 m s^{-1}, are too low for impaction to be an important transfer mechanism (Fig. 38). Finally, the extent to which conditions in the wind tunnel reflected conditions typical of the surface of the ocean is questionable. Several investigators have concluded that roughness elements on a surrogate surface too small to disturb the flow in the viscous boundary layer may nonetheless efficiently remove particles by interception [*Wells and Chamberlain*, 1967; *Browne*, 1974; *Chamberlain et al.*, 1984], thus increasing the deposition velocity, and *Jenkin* [1984] suggested that this effect might have accounted for some of his results.

The effect of equilibration on the dry deposition of both initially dry and wet ammonium sulfate particles to a water surface (at which the RH was assumed to be 99% or 100%) was modeled by *Zufall et al.* [1998b] for a wind speed of 2 m s^{-1}. These investigators concluded that particles with r_{80} (evaluated as twice the dry radius) greater than ~0.1 μm do not equilibrate before reaching the surface and that hygroscopic growth plays little role in determining dry deposition velocities of liquid drops. However, the results from this investigation are of little use in deciding whether or not equilibrium with respect to RH affects dry deposition velocities under realistic conditions over the oceans for several reasons. The wind speeds they considered are too low for impaction to occur (Fig. 38); the RH profile was not specified and the equation presented for the flux of moisture is not valid near the surface; the stated times for particles with initial diameters of 0.01, 0.1, and 1 μm to reach their equilibrium radii with respect to 99% RH—of order seconds, minutes, and hours, respectively—are two to three orders of magnitude greater than the estimate given above (§2.5.4) for equilibration to 98% RH values and than other numerical [*Ferron*, 1977; *Andreas*, 1989, 1990a] and experimental [*Woodcock et al.*, 1981] studies (the difference between RH growth of ammonium sulfate particles and SSA particles is expected to be small); the effects of latent heat in the effective diffusivity of water vapor (§2.5.4) was omitted in the equation used for drop growth (implying that the equilibration times should be greater yet); and the accommodation coefficient used was apparently that of SO_2 on water instead of water vapor on ammonium sulfate solution. For all these reasons, the results of *Zufall et al.* must be discounted.

An assessment of whether the radius of an SSA particle equilibrates to the RH of the viscous sublayer can be gained from consideration of the time such a particle spends in this layer (of thickness δ) relative to the characteristic time of equilibration for an SSA particle initially at 80% RH instantaneously exposed to 98% RH, $\tau_{80,98}$ (§2.5.4). The requirement for equilibration is that the time spent by the particle in the viscous sublayer be several times greater than the characteristic time $\tau_{80,98}$ (for which the value given by Eq. 2.5-7 for these conditions may be used): $\delta/v_V \succ \tau_{80,98}$ (where the symbol $\succ$ was introduced in §2.1.5.2 to denote "considerably greater than"). As $v_V > v_{V,Stk}$, the requirement $\delta/v_{V,Stk} \succ \tau_{80,98}$ is necessary (though not sufficient). Substituting for $\tau_{80,98}$ and $v_{V,Stk}$, and taking $r_{98} = 2r_{80}$ (the particle radius for equilibrium at 98% RH) yields $r_{80}/\mu m \lesssim 1.5 \, (\delta/mm)^{1/4}$. For $\delta = 1$ mm, typical of the values suggested for the thickness of the viscous boundary layer, this condition requires $r_{80} < 1.5$ μm. This condition is not very sensitive to the value of δ; for example, for $\delta = 0.1$ mm the condition requires $r_{80} \lesssim 1$ μm. Thus it may be confidently concluded that larger particles do not spend sufficient time in the viscous sublayer to equilibrate; consequently, for particles with r_{80} greater than ~1-2 μm up to ~5-7 μm (above which dry deposition velocities are mainly controlled by transfer through the surface layer), dry deposition velocities based on the assumption of RH equilibration in the viscous sublayer (as assumed in the model of *Slinn and Slinn* discussed above and most variants of this model) will be overestimated, and the corresponding atmospheric residence times underestimated. This conclusion holds especially for higher wind speeds, for which v_{imp}, which exhibits a strong increase with increasing particle size, is important in determining v_d. As the above argument considered only the effects of gravitational sedimentation on transfer through the viscous sublayer, it overestimates the time spent by a particle in that layer. Impaction, which drives particles into the surface from the top of the viscous sublayer, would also contribute. This mechanism operates mainly at the leading edge of eddies; hence the actual velocity experienced by an individual SSA particle that is transferred through the viscous sublayer by this mechanism will be substantially greater than v_{imp}, which refers to the average impaction velocity over the sea surface. This impaction process results in a further decrease in the time available for equilibration, and strengthens the conclusion that dry deposition velocities for SSA particles should be evaluated at their equilibrium radius characteristic of the surface layer for r_{80} at least down to 2 μm.

An additional concern with the dry deposition models examined thus far is that they do not accurately represent conditions near the ocean surface. For instance, drops may be ejected completely through the viscous sublayer, making the modeling of a surface source even more difficult (§2.1.6), and other removal mechanisms may come into play: particles may be removed through scavenging by impaction and coagulation on spray and waves [*Toba*, 1965a,b; *Sehmel and Sutter*, 1974; *Slinn et al.*, 1978; *Slinn and Slinn*, 1980, 1981;

Sievering, 1981; *Williams*, 1982b; *Slinn*, 1983a,b; *Miller*, 1987; *Hummelshøj et al.*, 1992], or in dissolving bubbles formed from air engulfed by breaking waves [*Slinn et al.*, 1978; *Slinn*, 1983b]. Based on such arguments, *Toba* [1965b] even went so far as to conclude the absence of a diffusion (viscous) sublayer, leaving gravity and turbulence as the only operable dry deposition mechanisms. Furthermore, the existence of a broken surface (due to whitecapping or bubbles bursting at the surface), or of roughness elements extending through the viscous sublayer, would also allow transport to the ocean directly from the surface layer [*Slinn et al.*, 1978], providing a path for deposition that avoids transfer through the viscous sublayer, and possibly resulting in a nearly constant dry deposition velocity for particles with $r_{80} \lesssim 1$ μm [*Sievering*, 1981].

These concerns led *Williams* [1982b] to extend the formulation of *Slinn and Slinn* [1980] by assuming two parallel paths by which particles can be deposited to the ocean: one the same as that of *Slinn and Slinn* [1980] to a smooth surface, and another to a broken surface, for which the expression given by *Wu* [1979b] for the whitecap ratio, (4.5-1), was used to describe the fraction of the ocean surface that was broken. In addition, *Williams* included a horizontal transfer between the smooth and broken areas, representing an interaction effect, allowing the concentrations at the top of the viscous sublayer over smooth and broken areas to differ (although physically this would require quite rapid horizontal transport of particles as breaking waves were formed or decayed). To describe scavenging by spray, *Williams* [1982b] considered arbitrary broken-surface transfer velocities from 1 cm s^{-1} to 1000 cm s^{-1} (to which the horizontal transfer velocities were set equal), resulting in dry deposition velocities for particles with $r_{80} \lesssim 1$ μm that are greater than the values obtained from the model of *Slinn and Slinn* [1980] by up to two orders of magnitude; for particles with $r_{80} \gtrsim 10$ μm the broken-surface transfer velocity has little effect on the dry deposition velocity. This model was criticized by *Slinn* [1983a,b] because it did not include any size dependence of the broken-surface transfer coefficient, and he argued that for a realistic collision efficiency the differences between the two models would be less.

Another model in which transfer could occur directly from the surface layer to a broken surface, above which there is no viscous sublayer, was proposed by *Hummelshøj et al.* [1992]. This model is similar to that of *Williams* [1982b] but includes also the possibility of scavenging of depositing SSA particles by seawater drops near the surface. This model employs constant values for the scavenging efficiency, drop size and height, and the number of drops per unit area of broken surface. A modification of the model of *Williams* [1982b] was also proposed by *Lo et al.* [1999], who considered RH values in the two layers different

from those considered by *Williams* (who took particles to be dry in the surface layer) and included effects of atmospheric stability. Experiments in an wind-wave tunnel experiment by *Larsen et al.* [1995] indicated that bubbling and spray processes have no appreciable influence on the dry deposition velocity of (non-hygroscopic) MgO particles with radii from 0.1 to 0.65 μm.

Estimates of dry deposition velocities resulting from modifications to the model of *Slinn and Slinn* and from four other models for dry deposition to the ocean are compared in Fig. 39 for wind speed $U_{10} = 10$ m s^{-1}; also shown are the gravitational sedimentation velocity of an SSA particle in the surface layer (80% RH), and the transfer velocity through the surface layer according to the model of *Slinn and Slinn* described above. The differences at other wind speeds are similar. The ratios of the dry deposition velocities resulting from these modifications and other models to those obtained from the model of *Slinn and Slinn* (as described above) are shown in the lower panel of Fig. 39. The modifications are:

1. Using a value of C_D twice as great as that assumed above; this will cover situations of atmospheric instability, the use of C_H instead of C_D in the formulation of v_{turb} (implying that mass transfer corresponds to heat transfer rather than momentum transfer), and the increase in C_D with increasing wind speed,

2. Using a value of C_D half as great as that assumed above; this will cover situations of greater atmospheric stability,

3. Using $p = -2/3$ instead of $p = -1/2$ as exponent of Schmidt number Sc in (4.6-4) for v_{diff},

4. Assuming that no equilibration with respect to RH occurs in the viscous sublayer, and

5. Including a Stefan flow of 0.01 cm s^{-1} (§2.6.1) opposing dry deposition (although according to *Carstens and Martin* [1982] phoretic effects and Brownian diffusion are not additive, for purposes of comparison they are treated as such to show the possible importance of phoretic effects in calculating dry deposition velocities of smaller particles).

The four other models are those of *Williams* [1982b], *Giorgi* [1986a,b], *Hummelshøj et al.* [1992], and *Hoppel et al.* [2002]. The model of *Williams* is similar to that of *Slinn and Slinn* [1980] but provides a possibility of transport to the ocean through a broken surface in addition to the pathway through the viscous sublayer over a smooth surface; a broken-surface transfer velocity and a horizontal transfer velocity each of 100 cm s^{-1} are assumed for this model. The model of *Giorgi* [1986a,b] employs parameterizations for C_D, v_{diff}, v_{imp}, and v_{turb} different from those in the model of *Slinn and Slinn* [1980]. The model of *Hummelshøj et al.* [1992], which contains a formulation for v_{diff} and coefficients for v_{imp} and v_{turb}

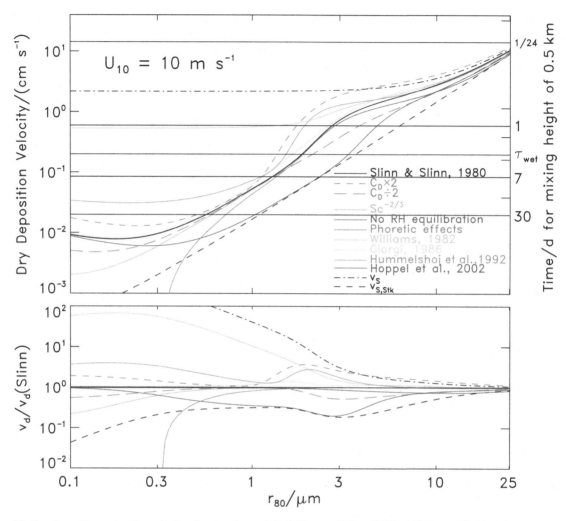

Figure 39. Dry deposition velocities calculated using the model of *Slinn and Slinn* [1980, 1981] specified in §4.6.1, modifications of this model, and other models (described in §4.6.3) as a function of r_{80} for $U_{10} = 10$ m s^{-1}. Upper panel illustrates values of dry deposition velocity, with right axis showing values of characteristic dry deposition time τ_{dry} for an assumed mixing height of 0.5 km; horizontal lines denoting 1 hr, 1 d, 1 week, and 1 month are shown, as is line denoting value of τ_{wet} assumed to be 3 d. Characteristic dry deposition times τ_{dry} will be overestimated for SSA particles with $r_{80} \gtrsim 15$ μm for which mixing heights are less than 0.5 km. Lower panel illustrates ratio of dry deposition velocities to those calculated using model of *Slinn and Slinn* [1980, 1981], with lines denoting 0.5 and 2. $C_D \times 2$ and $C_D \div 2$ denote calculations for which a wind stress coefficient a factor of 2 greater (i.e., 0.0026) and a factor of 2 lower (i.e., 0.00065) were used. $Sc^{-2/3}$ denotes calculation for which v_{diff} was taken as proportional to $Sc^{-2/3}$ instead of $Sc^{-1/2}$. "No RH equilibration" denotes calculation for which the SSA particle radius was taken to be r_{80} in both the surface layer and the viscous sublayer. "Phoretic effects" denotes calculation for which a Stefan flow of 0.01 cm s^{-1} opposing dry deposition was assumed. The quantities v_S and $v_{S,Stk}$ are the transfer velocity through the surface layer and the gravitational sedimentation velocity (i.e., Stokes velocity) in the surface layer, respectively.

different from those of any of the other models considered, and which also allows for direct transfer from the surface layer to a broken surface, was used with several changes: in Eq. 4 of that reference the quantity $v_{g,d}$ in the denominator was changed to $v_{g,w}$ (as required from the previous equations in that reference), and in Eq. 10 of that reference the Stokes number, *St*, which was defined using $v_{g,d}$, was evaluated using $v_{g,w}$ instead—this is required to match Fig. 3 of that reference,

and is consistent with the assumption made there of equilibration to the RH of the viscous sublayer. The model of *Hoppel et al.* [2002] yields a formulation for the dry deposition velocity in which a differential equation rather than a difference equation is used to describe transfer through the surface layer. This formulation is calculated using $30v_d/u_*$ as the height of the viscous sublayer, the value used by *Hoppel et al.* (results are very insensitive to this choice), and the same transfer velocities as those defined in §4.6.1. As above, relative humidities of 80% and 98% are used for the surface layer and viscous sublayer, respectively, for these four other models.

For most of the modifications and other models differences from the model of *Slinn and Slinn* [1980] as used above are less than a factor of three (Fig. 39), with larger differences being for the most part restricted to small SSA particles ($r_{80} \lesssim 1\ \mu m$). In the present context of estimating SSA production fluxes these differences are unimportant for these small particles because of their extremely large residence times; a reduction in residence time from a month to a week does little to justify the assumption of using local conditions to describe the production of these particles or to satisfy the condition that removal be through dry deposition and not wet deposition. Only the model of *Williams* [1982b] yields lifetimes against dry deposition less than a week for SSA particles in this size range, and, as discussed above, for such long times not only will dry deposition likely not be the dominant removal mechanism, but the assumption of steady state cannot be supported. Furthermore, most of the proposed changes reduce the estimated dry deposition velocities in this size range, resulting in even greater lifetimes against dry deposition; only the case of doubled C_D, the model of *Williams* [1982b], and the model of *Hummelshøj et al.* [1992] result in increased dry deposition velocities. These models yield dry deposition velocities that differ from that of the model of *Slinn and Slinn* [1980] by over an order of magnitude for most of this size range, but the resulting (nearly size-independent) lifetimes against dry deposition are nearly a day for the model of *Williams* to nearly a month for the other two models. Although it is possible that the mechanisms described by these other models operate over the ocean and are important in some circumstances, the uncertainties in such models, and the many simplifying assumptions required for their use (including the rather arbitrary values chosen for the broken surface transfer velocities), limit confidence in numerical results obtained from their use.

For particles with $r_{80} \gtrsim 3\ \mu m$, the only alternative model for which changes in the dry deposition velocity of more than a factor of two results is that which incorporates the assumption that SSA particles do not equilibrate to the RH of the viscous sublayer. As shown above, SSA particles in this size range are not expected to equilibrate before passing through this layer. Because of the increased impaction velocity that would result from increased particle size, the assumption of equilibration for SSA particles with r_{80} near 3 μm would result in an overestimation of dry deposition velocities by several fold. This overestimation would in turn result in an underestimation of residence times, and an increase in the lower size of SSA particles for which the conditions required for the implementation of dry deposition models are satisfied.

For SSA particles with $r_{80} \gtrsim 10\ \mu m$, the several alternative formulations result in little change in the estimated dry deposition velocities. For these particles, v_d is well approximated by the transfer velocity through the surface layer, v_S. In other words, only for those particles sufficiently large that the conditions for the steady state dry deposition method of inferring SSA production are satisfied do the several models agree, but in this size range the dry deposition velocities are nearly equal to v_S, and the details of the transfer through the viscous sublayer are relatively unimportant (this is equivalent to the conclusion reached by *Toba* [1965b] that only gravitational sedimentation and eddy diffusion contributed to the motion of SSA particles in this size range; §4.6.5). Thus the basic conclusion reached above remains relatively unchanged: that for SSA particles with $r_{80} \lesssim 3\ \mu m$ the lifetimes against dry deposition are so great that dry deposition is not the controlling removal process, local balance between dry deposition and production rarely occurs, and conditions experienced by the particles during their residence time in the atmosphere do not remain constant. For these reasons dry deposition models cannot be used to infer SSA production for particles in this size range. On the other hand, for particles in the size range 5 μm $\lesssim r_{80} \lesssim$ 25 μm, the steady state dry deposition method appears well suited to determining the effective SSA production flux under situations for which it can be demonstrated that conditions discussed above are satisfied.

For SSA particles of the size for which the steady state dry deposition method can be applied under appropriate conditions ($r_{80} \gtrsim 5\ \mu m$ for wind speeds greater than ~5 m s^{-1}) dry deposition is controlled mostly through transfer through the surface layer. Based on this premise, an alternative model is proposed for calculation of the effective SSA production flux by this method (§5.2.1.5) in which the dry deposition velocity is taken as equal to the transfer velocity through the surface layer, $v_d(r_{80}) \approx v_S(r_{80}) = v_{S,Stk}(r_{80}) + v_{turb}(r_{80})$:

$$\frac{v_d}{cm\ s^{-1}} \approx \frac{U_{10}}{4.6\ m\ s^{-1}} + \left(\frac{r_{80}}{8.5\ \mu m}\right)^2. \qquad (4.6\text{-}8)$$

This approach simplifies the analysis and leads to greater physical insight with little loss in accuracy for this size range. A multiplicative uncertainty of $\stackrel{\times}{\div} 2$ is chosen to quantify the uncertainty associated with this approximation; although this value certainly depends on r_{80}, this choice is expected to be representative for the size range for which the dry deposition method can be applied.

4.6.4. Dry Deposition Measurements

Measurements of dry deposition SSA fluxes over the ocean, from which dry deposition velocities could be obtained from (2.7-2), have not been reported; such measurements would of course be complicated by the simultaneous presence of a surface source, in addition to the experimental difficulties in measuring the flux to the ocean surface. Experiments to infer dry deposition velocities of substances other than SSA particles have been performed, including those involving wind tunnels, measurements of deposition fluxes on surrogate surfaces, measurements of the decrease in concentration with space or time, and measurement of fluxes by micrometeorological techniques. However, the ability of such experiments to determine size-dependent SSA deposition fluxes is limited. Sea salt aerosol particles are hygroscopic, and as the dependence of their size on RH is an important aspect of the dry deposition model considered above, SSA dry deposition velocities differ from those of specific chemical species or radioactive non-SSA particles. Measurements over land cannot be used to determine v_d because of differences in surface conditions and vertical profiles of wind speed and relative humidity, precluding for example inference of the SSA dry deposition flux to the ocean from measurements of the decrease in SSA concentration or deposition flux with distance from the coast. Additionally, measurements of the dry deposition velocity of SSA particles over the ocean are confounded by the fact that under actual conditions, the net flux, which is the measured quantity, is not equal to the gross flux, which is the desired quantity. Under steady state conditions, of course, the net flux is zero, otherwise production would not be balanced by dry deposition. More generally the existence of the two opposing fluxes renders questionable any relation of a measured flux to either the dry deposition flux or the surface production flux, as discussed in §2.1.6. In this section some of the pertinent studies are examined.

Wind tunnel experiments of particle deposition to water have been made for few wind speeds and particles sizes [Möller and Schumann, 1970; Sehmel and Sutter, 1974; Larsen et al., 1995], but the conditions used in these studies are not realistic of those over the ocean [Slinn and Slinn, 1981]. Seawater was not used in these experiments, and substantial differences between freshwater and seawater may exist because of the different RH values above their surfaces [Williams, 1982b; Sievering, 1984], especially the much greater growth over freshwater as RH approaches 100%. Additionally, as noted in §4.3.6.2, little confidence can be placed in extrapolating wind tunnel measurements to actual conditions in the marine environment, especially with regard to vertical transport of SSA particles from 10 m.

Similarly, the flux of SSA particles to surrogate surfaces is expected to differ from the dry deposition flux to the ocean because of differences in surface properties and flow near the surface [van Aalst, 1985; Nicholson, 1988; Ruijgrok et al., 1995; Wesely and Hicks, 2000], especially

if wave properties and scavenging by spray are important. Investigations of dry deposition fluxes of substances other than sea salt have reported differences depending on the surrogate surface [Dasch, 1983; Sickles et al., 1983; Davidson et al., 1985]. As conditions near the ocean surface are expected to control the dry deposition velocities for SSA particles with $r_{80} \lesssim 3$ μm, and to play a role for SSA particles up to $r_{80} \approx 10$ μm, the use of surrogate surfaces is unlikely to provide reliable deposition data for these particles. Settle and Patterson [1982] concluded that dry deposition to the undulating surface of the ocean could not be simulated using a fixed, rigid horizontal surface, and Arimoto et al. [1985] emphasized that the use of flat rimless plastic plates to collect dry deposition samples did not mimic the characteristics of natural water surfaces. As noted by Fairall and Larsen [1984], because the dry deposition flux at a given height above the sea surface is the net downward flux passing through that surface and not the total downward flux, it should be determined not by measuring the amount of a substance deposited on upward facing horizontal surface, but by the difference between the amounts deposited on an upward and a downward facing horizontal surface. In any event, the flow in the vicinity of such surfaces cannot be expected to mimic that near the surface of the ocean. The dry deposition velocity may be increased by interception of particles by roughness elements in the viscous boundary layer on a surrogate surface that are too small to disturb the flow, as discussed above [Wells and Chamberlain, 1967; Browne, 1974; Jenkin, 1984; Chamberlain et al., 1984]. Although small pans of seawater may model RH near the ocean surface, they cannot adequately reproduce the motion of the ocean surface nor the airflow near the surface, neither can they yield any information on the dependence of v_d on r_{80} for SSA particles. Based on these considerations, it must be concluded that dry deposition of SSA particles to surrogate surfaces cannot provide a basis for accurate determination of dry deposition velocities.

An alternative approach to determine dry deposition velocities is by measuring the decrease in concentration with distance or time. Box models relating the measured decrease in concentration along the wind direction with fetch have been used to estimate dry deposition velocities of smoke, nitrate, and sulfate and sulfur dioxide over the North Sea by Lodge [1978], and of phosphate and total mass over Lake Michigan by Delumyea and Petel [1979], for example. Meaningful results from such box models require extremely accurate concentrations measurements; for a mixing layer height of 1 km and a fetch of 100 km, the uncertainty in the calculated dry deposition velocity will be 15 times the uncertainty in the concentration measurements [Williams, 1982a]. Based on the scatter in reported measurements of

SSA concentrations (§4.1.4.2) such an uncertainty would be prohibitively large. Additionally, an assumption is required that particles do not flow into or out of the box; such models would be even more difficult to apply to situations in which SSA particles are produced as well as removed at the sea surface. Changes in the height of the marine boundary layer with time or space may overwhelm the estimated dry deposition velocities of particles that are uniformly mixed over this height [Slinn, 1983a,b; Slinn et al., 1983]; as noted in §3.1, for $U_{10} = 10$ m s^{-1}, a mixing layer height gradient of only 1 m per km, equivalent to an entrainment velocity of 1 cm s^{-1}, would have a larger effect than that of dry deposition on the concentration of particles with $v_d \lesssim 1$ cm s^{-1} ($r_{80} \lesssim 3$ µm). In principle it might be possible to account for dilution resulting from an increase in height of the marine boundary layer by considering dilution of a conservative tracer gas [Horst et al., 1983; Sehmel, 1983], but no such measurements over the ocean have been reported, and in practice there would be inevitable concerns, for instance, over the initial vertical and horizontal distributions of such a tracer.

The size-dependent rate of removal of marine aerosol particles by dry deposition was inferred by M. Smith et al. [1991] from measurements of concentrations at a coastal site over a 90 h period during which air was of maritime origin during the first 48 h and again during the last 17 h. Particle concentrations were measured in seven size intervals from 0.09 µm < r_{amb} < 23.5 µm; as the RH was consistently near 70% throughout this time, ambient radii would have differed little from r_{80} values. As little or no whitecapping was visible beyond the surf zone, the investigators assumed that no SSA production was occurring during this period and that the primary process affecting particle concentrations was dry deposition; the rate of decrease of particle concentration following a decrease in wind speed was examined in §2.9.6. Concentrations for each of the size ranges were fitted by Smith et al. [1991] to exponentially decreasing functions of time, and two characteristic decay times were obtained for each particle size range: one using only the data for the first 48 hour period, and one using data for the entire period for which the air was maritime. Characteristic decay times for $r_{80} \gtrsim 3$ µm were mainly 30-50 h, with little dependence on size (although for the radius ranges 0.09-0.125 µm and 1-2 µm characteristic times were weeks to months or even negative, implying an increase in concentration over the time period). The investigators concluded that these results were consistent with values calculated under the assumption that dry deposition is controlled by gravitational sedimentation and turbulent diffusion (and not transfer through the viscous sublayer). However, several concerns arise with the interpretation of these measurements by Smith et al. [1991], their

ascription of the characteristic times to dry deposition, and the application of these results to infer SSA production. First, absent evidence to the contrary, it seems unlikely that the majority of the smaller particles ($r_{80} \lesssim 1$ µm) were sea salt (§4.1.3.2). Second, for the reported wind speed of 5 m s^{-1} the assumption that dry deposition through the viscous sublayer is not controlling restricts the results to particles with $r_{80} \gtrsim 5$-10 µm (Fig. 38), for which estimated characteristic times would be ~1 d or less, decreasing rapidly with increasing particle size, entirely inconsistent with the results presented. More importantly in the present context, these measurements were made during conditions under which little or no production occurred. Consequently, knowledge of the dry deposition velocities for such wind speeds, even if they could be accurately determined, would be of little utility in inferring SSA production by the steady state dry deposition method.

A few determinations of particle flux over water have been made by micrometeorogical measurements [Sievering, 1981; Wesely et al., 1982; Sievering et al., 1982; Schmidt et al., 1983; Katsaros et al., 1987; Nilsson and Rannik, 2001; Nilsson et al., 2001], but these investigations have provided little information on size-dependent dry deposition velocities of SSA particles over the ocean. Vertical concentration gradients of aerosol particles over Lake Michigan were measured by Sievering [1981] at wind speeds from ~2-8 m s^{-1}, resulting in a mean of 0.28 for the ratio of the dry deposition velocities to the momentum transfer velocity for particles over the size range $r_{amb} = 0.055$-1.5 µm, this ratio increasing only slightly with increasing particle size. For a momentum transfer velocity of around 1 cm s^{-1} (corresponding to $U_{10} \approx 8$ m s^{-1}), this ratio yields a mean dry deposition velocity of 0.3 cm s^{-1}, much greater than expected for SSA particles in this size range (Fig. 38). The composition of the aerosol particles was not reported, but as the lake is freshwater, the particles were certainly not SSA particles and possibly not hygroscopic. The other studies noted above involved eddy correlation measurements, but poor wind speed and particle size resolution, large uncertainties, and possible artifacts caused by RH gradients over the ocean [Wesely et al., 1982; Sievering et al., 1982; Schmidt et al., 1983] limit the ability of the results of these studies to provide a check on dry deposition models. Additionally, as discussed earlier, micrometeorological methods yield the net flux of particles, not the gross flux, and only if it can be ascertained that the dry deposition flux is overwhelmingly larger than the surface production flux can this net flux be equated to the dry deposition flux.

Many studies of deposited material over the ocean have reported mass-weighted, or "overall", dry deposition velocities of particles, evaluated as the ratio of the mass deposition flux to the average mass concentration at a given height, typically near 10 m [*Silker*, 1974; *Settle and Patterson*, 1982; *McDonald et al.*, 1982; *Arimoto et al.*, 1985; *Arimoto and Duce*, 1986; *Erickson and Duce*, 1988; *Milford and Davidson*, 1987; *Dulac et al.*, 1989; *Steiger et al.*, 1989; *Baeyens et al.*, 1990; *Rojas et al.*, 1993; *Zufall et al.*, 1998a]. Although the total dry mass deposited may be of interest in some instances (as for radioactive materials), in view of the strong dependence of v_d on particle size, such mass-weighted dry deposition velocities yield little or no information on the validity of dry deposition models for considerations of SSA, the relative importance of the various mechanisms, or the dependence of dry deposition velocities on particle size, all of which are essential for inferring size-resolved SSA production.

For SSA, the mass-weighted dry deposition velocity increases so rapidly with increasing r_{80} that in most situations the mass-weighted dry deposition flux will be dominated by particles with $r_{80} \gtrsim 10$ μm, for which the dominant removal mechanism is gravitational sedimentation, with virtually no contribution from particles with $r_{80} \lesssim 1$ μm. For example, for the canonical SSA size distribution of concentration (§2.1.5; §4.1.4.4) at $U_{10} = 10$ m s^{-1}, the peak of the mass-weighted dry deposition flux is expected to occur at a value of r_{80} greater than 25 μm—beyond the range for which the approximation for this size distribution is valid, and beyond the range for which the dry deposition model is valid, as SSA particles of this size will not have equilibrated to their equilibrium radii with respect to the ambient RH of the surface layer (§2.9.1). For this canonical SSA size distribution of concentration, the mass-weighted dry deposition flux of SSA particles with r_{80} between 10 μm and 11 μm (that is, the flux of mass to the surface resulting from dry deposition of SSA particles in this size range) is nearly twice the total mass deposition flux of all SSA particles with $r_{80} < 3$ μm. On the basis of such considerations it seems fair to say that any claim that a certain mass-averaged dry deposition velocity implies that a measurement or model is "consistent with the model of *Slinn and Slinn* [1980]" implies at best that the dry deposition of the largest particles is accurately described. As the deposition of the largest particles is controlled almost entirely by gravitational sedimentation, described by Stokes' Law, such a claim cannot be taken as lending support to the more general model at lower r_{80}.

In summary, the few measurements of dry deposition for sea salt particles over the ocean are insufficient and inadequate to test the particle size- and wind-speed dependencies obtained from models, or to determine size-resolved dry deposition fluxes. Mass-averaged dry deposition velocities yield little to no information on the size dependence of dry deposition velocities. Similarly, the applicability of studies of deposition in wind tunnels, deposition to surrogate surfaces, and deposition of other substances is questionable for testing the models discussed above, for which hygroscopic growth and details of the air flow in the viscous boundary layer are important components of the model.

4.6.5. Considerations Regarding Interpretation of Dry Deposition Models

As emphasized by *Slinn and Slinn* [1980, 1981], because of the simplifications and approximations made, and the fact that many of the arguments involved dimensional analysis and order-of-magnitude estimates, the model they proposed was intended to be viewed only as an attempt to heuristically describe the dependence of the dry deposition velocity on the mechanisms that operate [see also *Slinn*, 1983a,b]. Also, because of the extreme sensitivity of the results of the model to wind speed and particle size (Fig. 38), high accuracy was not to be expected. *Slinn and Slinn* [1981] were very explicit in this regard, stating "it is probably unprofitable to pursue layer models to the degree that has already been described in the literature." Later *Slinn* [1983a] referred to the model as "poor," adding "it's all guesswork." The same considerations apply also to alternative models (§4.6.3), whose foundations do not appear to rest on any firmer ground than that of *Slinn and Slinn*. However, these concerns have not always been heeded, and in many investigations dry deposition models appear to have been accepted as accurately describing particle deposition, with implied uncertainties (based on the number of significant digits or on reported sensitivities to atmospheric stability and other factors) of a percent or better. In the absence of measurements that could verify any of these models, or decide on one over another (§4.6.4), the uncertainty in dry deposition models for SSA is undoubtedly much greater—an order of magnitude or more for SSA particles with $r_{80} \lesssim 1$ μm and a factor of several fold for larger particles, depending on their size (a multiplicative uncertainty of $\nleqq 2$ for $r_{80} \gtrsim 5$ μm was estimated for the dry deposition velocity given by Eq. 4.6-8). It would be expected that with increasing particle size the models become more accurate, but as the dominant contribution to dry deposition for these larger particles is gravitational settling, a more complicated model becomes unnecessary.

There are several concerns associated with dry deposition models and their use in determining the effective SSA production flux. Implicit in all such models is the assumption of steady state conditions, in essence restricting the applicability of such models to constant wind speed over the atmospheric residence time of the SSA particles present at the time of measurement. However, wind speed over the ocean rarely stays constant for extended periods of time. Some investigators [*Sievering*, 1984; *Arimoto and Duce*, 1986] have suggested that non-steady state dry deposition may be more important than steady state deposition; that is, that most of the deposition may occur during changing conditions, specifically wind speeds, perhaps of because of the prevalence of such conditions or because physical processes that occur during unsteady flow, such as turbulent eddies bursting from the surface, resulting in greater dry deposition. Additionally, it is highly unlikely that during non-steady state conditions SSA particle transfer can be adequately described by the concentration at only a single height [*Sehmel and Hodgson*, 1974]. For situations in which the wind speed changes appreciably over the mean atmospheric residence time of SSA particles of a given size the nonlinearity in the various dry deposition models with respect to wind speed precludes averaging the dry deposition velocity (or flux) over wind speed, as noted above (§4.6.2). No models to infer the effective SSA production flux that account for varying wind speeds have been proposed.

There also remain difficult issues associated with the concept of dry deposition in the presence of a surface source, which changes the boundary condition at the lower surface—the term n_0 in (2.7-3) is no longer zero, and the gross vertical flux of SSA particles is no longer equal to the net flux. This situation necessitates a conceptual (and somewhat artificial) distinction between the upward flux and the downward flux of SSA particles. Additionally, the premise that the downward dry deposition flux is independent of height, necessary for the derivation of v_d (Eq. 2.7-7) through flux matching (§2.7.2), is tenuous. Taken literally, this premise requires an increase in the number concentration with increasing height; that is, $n_{ref} > n_\delta$, as demonstrated by (2.7-8), taking into account that $v_V \equiv v_{V,Stk} + v_{diff} + v_{imp} > v_{S,Stk}$ because $v_{V,Stk} > v_{S,Stk}$. This increase with height of the SSA number concentration results from the artificial distinction between the upward flux and the downward flux inherent in the dry deposition concept and necessary for flux matching.

Several investigators have attempted to infer the interfacial SSA production flux from measurements of SSA concentrations at heights near 10 m. This approach consists of determining the corresponding steady state concentration of SSA particles of a given size at a lower height z_1 near the sea surface, $n(z_1)$, and makes the assumption that such particles are removed to the surface from this height only by gravitational sedimentation, with terminal velocity $v_{Stk,1}$ corresponding to an SSA particle in equilibrium with the ambient RH at z_1. Thus a balance is assumed between SSA production and removal not at the sea surface but at some height z_1, with SSA particles in effect being introduced into and removed from the marine atmosphere at this height. In this situation the interfacial SSA dry deposition flux would be equal to $n(z_1)v_{Stk,1}$, which under the assumption of steady state (i.e., local balance between SSA production and removal) would equal the interfacial SSA

production flux f_{int} of SSA particles of the given size. In essence, this approach infers the interfacial SSA production flux using a dry deposition velocity of the form $v_d(r_{80}, z_1, 10 \text{ m})$ defined in §2.7.2.1, where z_1 is chosen to represent conditions at the sea surface ($z = 0$). Equivalently, this approach can be viewed as an attempt to determine the ratio of the effective SSA production flux to the interfacial SSA production flux, Ψ_f. As $f_{eff} = n(10 \text{ m})v_d(10 \text{ m})$, this ratio would be given by

$$\Psi_f = \left(\frac{v_d(10 \text{ m})}{v_{Stk,1}}\right)\left(\frac{n(10 \text{ m})}{n(z_1)}\right). \qquad (4.6\text{-}9)$$

The quantity $v_d(10 \text{ m})$ is not necessarily the dry deposition velocity presented in §4.6.1 based on dry deposition models but depends on the choice made for what mechanisms contribute to the dry deposition flux; if only gravitational sedimentation is assumed to remove particles, then v_d would be equal to the gravitational sedimentation velocity at 10 m, $v_{Stk,10 \text{ m}}$. If it is further assumed that RH is the same at z_1 and at 10 m, then $v_{Stk,10 \text{ m}} = v_{Stk,1}$ and Ψ_f would be equal to the ratio of the steady state SSA concentrations at the two heights, $n(10 \text{ m})/n(z_1)$, the same result obtained by heuristic arguments in §2.9.4 and for which values were presented in Fig. 13 for $z_1 = 0.1$ m and $z_1 = 0.01$ m.

This approach has been used by some investigators [e.g., *Eriksson*, 1959; *Blanchard*, 1963] who assumed that the concentrations were independent of height, and thus did not distinguish between the interfacial and effective SSA production fluxes; that is, they implicitly assumed that $\Psi_f(r_{80}) = 1$ for all values of r_{80} considered. These investigations are discussed below (§5.2.1.1).

Toba and coworkers [*Toba*, 1965b; *Chaen*, 1973; *Iida et al.*, 1992] and *Hoppel et al.* [2002] used this approach to infer the interfacial SSA production flux from measurements of SSA concentrations at heights of several meters or more above the sea surface. *Toba* [1965b] extrapolated SSA concentrations from 10 m to yield values at the sea surface under the assumption that only gravity and eddy diffusion control the motion of SSA particles in the lowest 10 m over the ocean, taking into account the change in RH and thus particle size (and gravitational terminal velocity) with height under the assumption that SSA particles instantaneously equilibrate to the local RH. *Toba* assumed that eddy diffusion operates down to the sea surface and is described by $D_{eddy} = \kappa u_* (z + z_0)$, where κ is the von Karman constant (0.40) and z_0 was taken as $4.5 \cdot 10^{-4}$ m, independent of wind speed (*cf.* Eq. 2.4-2); the role of the laminar sublayer was neglected because drops were assumed to be ejected some distance above the sea surface and because spray and bubbles would scavenge drops very close to the surface. The dry deposition velocity at height z_0 was taken to be the Stokes velocity at 98% RH. Thus the quantity Ψ_f was given by the product of the ratio of the steady state SSA concentrations $n(10 \text{ m})/n(z_0)$ and the ratio of the Stokes velocities at 80% RH and 98% RH, $v_{\text{Stk},80}/v_{\text{Stk},98}$. The concentration ratio $n(10 \text{ m})/n(z_0)$ obtained by *Toba* differs from that given by (2.9-4), which was derived under the assumption that the RH was 80% over this range of heights; at $U_{10} = 10 \text{ m s}^{-1}$ the ratio of the concentration at 10 m to that at $z_0 = 4.5 \cdot 10^{-4}$ m given by the procedure used by *Toba* [1965b] for $r_{80} = (5, 10, 15)$ µm is approximately (0.8, 0.25, 0.05) whereas this ratio calculated using (2.9-4) is approximately (0.8, 0.35, 0.1). The ratio $v_{\text{Stk},80}/v_{\text{Stk},98}$ is approximately 0.33, independent of r_{80} over the range of interest (Fig. 7; §2.6.2). Thus the assumptions of *Toba* would imply that the quantity $\Psi_f(r_{80})$ approaches a limiting value of ~0.33 with decreasing particle size and that in this limit the effective SSA production flux remains lower than the interfacial SSA production flux by that factor, contrary to the expectation that the two fluxes should be roughly the same for sufficiently small SSA particles. *Toba's* estimate for Ψ_f is shown in Fig. 13 as a function of r_{80} for several wind speeds.

Similar analyses have been reported by other investigators. *Chaen* [1973] considered the dependence of RH on height, but noted that the assumption of a constant RH of 95%, independent of height, led to very similar concentration ratios as did the more complicated analysis of *Toba*, and calculated concentration ratios based on this assumption. The value of z_1 was chosen as $0.225 H_s$, where H_s is the significant wave height (the height of the highest 1/3 of the waves), for which an expression was presented in terms of the period of the significant wave (§2.3.5). Similar assumptions were made by *Iida et al.* [1992], who chose z_1 as $0.635 H_s$ based on experimental laboratory results of *Koga and Toba* [1981]. At $U_{10} \approx 10 \text{ m s}^{-1}$, based on the data of *Toba and Chaen* [1973] (which were used by *Chaen* and by *Iida et al.*), these values of z_1 correspond approximately to 0.2 m and 0.6 m, respectively. *Hoppel et al.* [2002] discussed the choice of z_1, arguing that it should be greater the height at which SSA particles were injected into the marine atmosphere (and possibly greater than the wave height), and for use in numerical models it should be some value characterizing the lowest level, such as a midpoint. These investigators assumed a constant RH of 80% and considered values of z_1 of 1 m and 10 m.

There are several concerns with this approach. A primary difficulty is the choice of the height z_1 from which SSA particles are transferred to the sea surface by gravitational sedimentation. The steady state SSA concentration cannot be extrapolated directly to the sea surface using an eddy diffusivity given by $D_{\text{eddy}} = \kappa u_* z$ (Eq. 2.4-2), because the value of D_{eddy} and the denominator in Eq. 2.9-4 would be equal to zero at $z = 0$ (this concern could be alleviated by defining an eddy diffusivity by $D_{\text{eddy}} = \kappa u_*(z + z_0)$, as done by *Toba* [1965b], but this is equivalent to assuming $z_1 = z_0$). The steady state SSA concentration ratio $n(10 \text{ m})/n(z_1)$ depends strongly on the choice of z_1 for sizes for which the effective and interfacial SSA production fluxes differ appreciably. For example, at $U_{10} = 10 \text{ m s}^{-1}$ values of the steady state SSA concentration ratio $n(10 \text{ m})/n(z_1)$ for $r_{80} = (5, 10, 15, 20, 25)$ µm calculated by (2.9-4) for $z_1 = 4.5 \cdot 10^{-4}$ m (the value used by *Toba* [1965b]) and $z_1 = 0.1$ m (one of the choices made in §2.9.4 used to evaluate Ψ_f) are approximately (0.8, 0.4, 0.1, 0.02, 0.002) and (0.9, 0.6, 0.4, 0.2, 0.06), respectively.

There are other difficulties with this approach. Very near the sea surface the height becomes a poorly defined quantity because of deviations of smoothness on a variety of scales, wave breaking, bubbles, and the like. Additionally, the wind flow in close proximity to the sea surface is strongly affected by waves which are dynamically coupled to the wind. Consequently, the idealization of the sea surface as flat and calm with wind speed varying logarithmically with height above it (Eq. 2.4-1), and the resultant dependence of the steady state SSA concentration on height as described by (2.9-4), become increasingly questionable in this region.

The assumption that steady state conditions have been attained, necessary to extrapolate SSA concentrations from 10 m to a lower height, restricts this approach to SSA particles with $r_{80} \gtrsim 5$ µm. A further concern is the explicit assumption that removal to the sea surface occurs only by gravitational sedimentation, thus restricting this approach to SSA particles sufficiently large that their transfer to the sea surface is not affected by the viscous sublayer; that is, to this same size range, $r_{80} \gtrsim 5$ µm. However, even for SSA particles of these sizes this assumption is questionable, as particles that reach the height z_1 from above are not necessarily removed to the sea surface but may remain in the atmosphere and cycle through this height time and time again. The assumption that dry deposition occurs only by gravitational sedimentation implies that eddy diffusion (turbulence) acts only to more particles upwards, whereas it acts to mix particles, and thus moves them both upward and downward. Finally, for a given choice of z_1 not equal to zero, SSA particles that do not attain this height are not included in what is determined to be the interfacial SSA production flux; thus such an approach in fact does not calculate the interfacial SSA production flux but rather the effective SSA production flux at height z_1.

Other approaches to dealing with the presence of a surface source have been proposed. *Slinn* [1983a,b] suggested that the surface source could be treated as a volumetric gain term, but did not provide an explicit formulation. *Fairall and Larsen* [1984] suggested that the surface source could be treated as an equivalent concentration at the surface of the ocean, equal in value to the interfacial flux divided by v_{diff}, i.e., $n_0 = f_{\text{int}}/v_{\text{diff}}$. This approach would result in a dry deposition velocity that is no longer equal to the ratio of the downward flux at the reference height to the concentration there, but which contains a term involving the interfacial flux. Additionally, these investigators defined an "effective source" at the 10 m reference height—the flux of particles through a horizontal plane 10 m above the surface—that is related to the interfacial flux by a factor that depends on wind speed and particle size. This factor, equal to the quantity $\Psi_f(r_{80})$ defined by (2.1-27), was given by these investigators as

$$\Psi_f(r_{80}) = \frac{v_{\text{turb}}}{v_V(r_{80}) + v_{\text{turb}}} \qquad (4.6\text{-}10)$$

(v_{turb} is independent of r_{80} and depends only on U_{10}) and is shown in Fig. 13 as a function of r_{80} for several wind speeds, calculated using the expressions presented in §4.6.2 for the transfer velocities. The quantity plotted is the reciprocal of that presented in Fig. 5 of *Fairall and Larsen* [1984], which displays the ratio $f_{\text{int}}/f_{\text{eff}}$ vs. U_{10} for several SSA particle sizes. However, the values in Fig. 13 calculated from (4.6-10) differ somewhat from those in *Fairall and Larsen* [1984], which contained an error in the expression for the impaction velocity, $(10^{-3})/St$ instead of $10^{(-3/St)}$, that evidently was used in preparing Figs. 5 and 7 of that reference (additionally, the exponent of C_D in Eq. 22b of that reference, stated as $-1/2$, should be $+1/2$). The corrected values of $\Psi_f(r_{80})$ are lower over the r_{80} range 1-10 µm by an amount that increases with increasing wind speed, from nearly 30% for $U_{10} = 10$ m s^{-1} to slightly more than 50% for $U_{10} = 20$ m s^{-1}, and they are greater for $r_{80} < 1$ µm by an amount that increases with decreasing r_{80} and decreasing wind speed; for $U_{10} = 10$ m s^{-1} the difference is less than 20% for $r_{80} > 0.1$ µm.

The flux ratio $\Psi_f(r_{80})$ given by (4.6-10) is qualitatively similar to the ratios of steady state SSA concentrations $n(10 \text{ m})/n(0.1 \text{ m})$ and $n(10 \text{ m})/n(0.01 \text{ m})$ obtained from (2.9-4), also shown in Fig. 13, but it exhibits some differences. This flux ratio approaches unity for decreasing r_{80} near 1 µm, but it decreases with increasing wind speed for $U_{10} > 10$ m s^{-1} at this size, contrary to physical expectation. For $r_{80} \approx 3$ µm, values of $\Psi_f(r_{80})$ calculated from (4.6-10) range from ~0.4 to ~0.7; because SSA particles in this size range are well mixed and the role of gravity in their behavior is small, values closer to unity would be expected. For $r_{80} \approx 6$ µm, $\Psi_f(r_{80})$ is almost independent of wind speed (and approximately equal to 0.37), but this quantity exhibits different wind speed dependencies for r_{80} less than and greater than this value. The replacement of the surface source by an equivalent concentration at the top of the viscous sublayer, as assumed by this approach, implies that

the newly produced particles attain this height only by diffusion, whereas in fact many particles are ejected through the viscous sublayer to greater heights. These concerns raise questions as to the extent to which this approach accurately describes the ratio of the effective and interfacial SSA production fluxes.

In summary, the several approaches to inferring the interfacial SSA production flux from SSA concentration measurements at heights near 10 m, or equivalently to determining the size-dependent flux ratio $\Psi_f(r_{80})$, for the most part yield qualitatively similar results. However, none of these attempts is satisfactory. Each of the estimates for Ψ_f (except for that of *Toba* [1965b]) yields values near unity for small SSA particles, as expected, and each exhibits a somewhat similar dependence on r_{80} to that given by the ratio of SSA concentrations at 10 m and at a lower height (§2.9.4) for larger SSA particles, but little confidence can be placed in quantitative results for SSA particles of the sizes for which Ψ_f differs appreciably from unity, i.e., $r_{80} \gtrsim 5$-10 µm. Estimates for $\Psi_f(r_{80})$ are discussed further in §5.1.2.

4.7. MINOR SEA SALT AEROSOL PRODUCTION MECHANISMS

Under most circumstances the primary mechanism by which SSA particles are produced is bursting of bubbles formed by breaking waves (§2.2), giving rise to jet drops (§4.3.2) and film drops (§4.3.3). At high wind speeds spume drops are formed by tearing of wave crests, a phenomenon that is readily observed in high seas. Other mechanisms for drop production have also been suggested. Some of these pertain to the bubble mechanism but from bubble sources other than breaking waves, whereas others involve entirely different mechanisms, such as wave breaking at coasts and impaction of hydrometeors (raindrops, hail, or SSA particles) on the surface of the ocean. Additionally, it has been hypothesized that fracture of existing SSA particles can result in the production of smaller SSA particles in the atmosphere. This process, while not adding mass, would alter the size distribution of SSA concentration and would thereby effectively result in a production flux of smaller SSA particles. Under most circumstances the production of SSA particles from these other mechanisms is likely to be insignificant except perhaps locally, and then only under certain circumstances. In this section these other mechanisms are examined and the extent to which they might be important as a source of SSA is estimated.

4.7.1. Coastal SSA Production

Waves breaking at coasts introduce large numbers of SSA particles into the atmosphere. The plume consisting of SSA particles formed by an active surf zone is readily visible, and SSA particles are present in concentrations that are often one or two orders of magnitude greater than background SSA concentrations coming from the open ocean [e.g., *Woodcock et al.*, 1963; *Blanchard*, 1969; *McKay et al.*, 1994; *de Leeuw et al.*, 2000]. A size-dependent SSA production flux formulation for the surf zone proposed by *de Leeuw et al.* [2000] was discussed in §4.3.5. However, as noted in §4.3.5, SSA production from coastal regions is expected to differ considerably from that over the open ocean because of differences in the wave-breaking mechanisms and possibly in SSA particle formation mechanisms as well, and it is expected also to depend strongly on factors such as sea-bottom and coastal topography, local wind direction, and the like that are specific to a given location, making it difficult to generalize results from one location to the other. Although surf-zone SSA production may be very important locally, it is unlikely to be a major source of SSA particles globally as may be seen from the following back-of-the-envelope calculation. Under the assumptions that wave breaking at the shore creates a white area 100 m in extent that is equally productive with the same amount of white area over the open ocean, and that the global mean whitecap ratio $W = 0.01$ (that is, on average 1% of the ocean is covered with white area at any one time), the production in the surf zone per unit length of coast is equal to that from an area of the ocean surface of width 10 km from the shore. Such an area is negligible on a global scale.

4.7.2. Impaction of Drops on the Ocean Surface

The impaction of hydrometeors (raindrops, hail, or SSA particles) on the ocean surface can produce SSA particles directly from the impact itself or indirectly, either from bubbles entrained by the drop or from bubbles or SSA particles produced by the splashed drops resulting from the original impact [*Blanchard and Woodcock*, 1957]. Impaction of individual drops on a liquid surface has been investigated experimentally[1], numerically [*Harlow and Shannon*, 1967a,b; *Nystuen*, 1986; *Oguz and Prosperetti*, 1990, 1991], and analytically [*Hsiao et al.*, 1988; *Le Méhauté*, 1988; *Longuet-Higgins*, 1990a]. Most investigations have involved impaction of a drop on a still liquid

surface. Several types of behavior have been reported in these experiments. The drop may coalesce into the liquid with little or no splashing, producing a vortex ring, or it may rebound or splash, forming a jet that then produces drops [*Carroll and Mesler*, 1981b; *Rodriguez and Mesler*, 1985]. For a drop of radius r impacting a calm surface of the same liquid with velocity v, a relation for the critical value of the Weber number, $We = (2r\rho/\sigma)^{1/2}v$, where ρ and σ are the density and surface tension of the liquid, respectively, was proposed by *Hsiao et al.* [1988]: for $We > 8$ a jet is produced whereas for $We < 8$ a vortex ring is produced. The value of We for water or seawater is given approximately by $We \approx 0.0016(r/\mu m)^{1/2}[v/(cm\ s^{-1})]$, for which the critical value $We = 8$ occurs for a drop with $r \approx 350\ \mu m$ moving at its gravitational terminal velocity. However, the extent to which many of these investigations are applicable to the impaction of raindrops or SSA particles on the ocean surface under actual conditions is questionable. In many of the experiments only vertical impaction on a calm surface was investigated, but experiments reported by *Siscoe and Levin* [1971] involving the impact of a drop on a wavy water surface resulted in a wide range of splash types which differed from those for a smooth surface. In some experiments dye was added to the drop to aid in visualization, but this may have resulted in changes in the surface tension [*Sainsbury and Cheeseman*, 1950; *Carroll and Mesler*, 1981b] and thus in the resultant behavior, as indicated in experiments of gas-liquid mixtures reported by [*Anderson and Quinn*, 1970]. Additionally, in many experiments the drops had not attained their terminal velocities; whereas a drop with $r = 500\ \mu m$ falling from rest in still air attains (90, 95, 99)% of its gravitational terminal velocity in (1.5, 2.5, 4) m, a drop with $r \gtrsim 1500\ \mu m$ requires approximately (5.5, 7, 12) m to attain these same fractions of its terminal velocity (§2.6.4).

Several investigators have discussed the production of SSA particles resulting from the direct impact of a drop on a water surface. *Stuhlman* [1932], for instance, referred to drops produced by "splashes from the descending shower" of previously produced jet drops, and the well-known photographs by *Edgerton* [*Edgerton and Killian*, 1979, pp. iii, 21] of the "crown" formed in milk after impact shows drops being ejected from the spikes on the crown (similar photographs are presented in *Kilgore and Day* [1963] and

---

[1] *Rogers* [1858 (cited in *Carroll and Mesler*, 1981b)]; *Deacon* [1871 (cited in *Rodriguez and Mesler*, 1985)]; *Reynolds* [1874 (in *Reynolds*, 1900, pp. 86-88)]; *Thompson and Newall* [1885 (cited in *Chapman and Critchlow*, 1967)]; *Raman and Dey* [1920]; *Jones* [1920]; *Barnaby* [1949]; *Ashton and O'Sullivan* [1949]; *Sainsbury and Cheeseman* [1950]; *Blanchard and Woodcock* [1957]; *Franz* [1959]; *Kilgore and Day* [1963]; *Worthington* [1963]; *Jayaratne and Mason* [1964]; *Engel* [1966]; *Hobbs and Kezweeny* [1967]; *Chapman and Critchlow* [1967]; *Siscoe and Levin* [1971]; *Macklin and Metaxas* [1976]; *Carroll and Mesler* [1981a]; *Rodriguez and Mesler* [1985, 1988]; *Prosperetti et al.* [1989]; *Pumphrey et al.* [1989]; *Pumphrey and Elmore* [1990]; *Medwin et al.* [1990, 1992]; *Nystuen and Medwin* [1995]; *Craeye et al.* [1999].

Hobbs and Kezweeny [1967]). The conditions under which such splash drops are formed, the number and sizes of these drops, and the dependence of these quantities on the size, impact velocity, and impact angle of the parent drop, are not well quantified. *Kientzler et al.* [1954] reported that large drops can rebound with a much reduced diameter upon striking a water surface, and that not all drops impacting a water surface will coalesce and thus form a cavity which can result in jet drop production. Laboratory experiments of *Blanchard and Woodcock* [1957] in which drops falling at their terminal velocities impacted calm seawater surfaces demonstrated that nearly 900 splash drops with $r_{80} > 25$ μm could result from the impact of a single raindrop with $r = 2.5$ mm, whereas an average of less than one splash drop with $r_{80} > 25$ μm resulted from the impact of a drop with $r < 0.5$ mm; for both size ranges the vast majority of the splash drops had $r_{80} < 100$ μm. These investigators further reported that an estimated several hundred SSA particles with $r_{80} < 12.5$ μm were produced by the impact of a raindrop with $r = 2.35$ mm. Whether or not drops bounce or coalesce with the surface depends on their size, impact angle, and impact velocity, according to an experimental investigation by *Jayaratne and Mason* [1964]. *Hobbs and Kezweeny* [1967] reported that the number of splash drops formed by the impact of a drop with $r = 1.5$ mm was a linear function of the distance fallen by the parent drop, but this dependence on distance implies that the drop would not have attained its terminal velocity upon impact; thus the pertinence of these results to SSA production by raindrops would be limited.

Sea salt aerosol particles can also be produced indirectly from bubbles entrained by raindrops or by splash drops produced by the original impact. Laboratory experiments of *Blanchard and Woodcock* [1957] in which freshwater drops with radii 0.2, 1.1, 1.5, and 2.35 mm falling at their terminal velocities impacted a calm seawater surface indicated that the number of bubbles produced per impact increased rapidly with increasing drop size, with the impaction of drops with $r = 0.2$ mm producing 2-3 bubbles with $R_{bub} = 0.025$ mm, and the impaction of drops with $r = 2.35$ mm producing 200-400 bubbles, most of which had $R_{bub} < 0.025$ mm although a few had $R_{bub} \approx 0.5$ mm (a bubble with $R_{bub} = 0.025$ mm is expected to produce jet drops with $r_{80} < 1$ μm according to Eq. 4.3-3). Laboratory experiments of *Franz* [1959] in which freshwater drops with radii 1.4, 2.4, 2.9, and 3.5 mm impacted a freshwater surface at velocities less than their terminal velocities indicated that bubble size increased with increasing drop size, although not every impact resulted in bubble entrainment. *Franz* concluded that the bubble formation process is very sensitive to minor variations in the condition of the surface, the shape of the drop, and the angle of impact.

Much of the work on bubble entrainment by raindrops has been motivated by investigations of the effect of rainfall on underwater noise and the discovery of a peak in underwater acoustic signal during rainfall near 15 kHz, which broadens with increasing wind speed, although its intensity does not correlate well with total rainfall rate [*Scrimger*, 1985; *Nystuen*, 1986; *Scrimger et al.*, 1987; *Nystuen and Farmer*, 1987; *Scrimger et al.*, 1989]. This peak has been attributed to radial oscillations of bubbles entrained by raindrops [*Leighton and Walton*, 1987; *Prosperetti et al.*, 1989; *Pumphrey et al.*, 1989], the resonant frequency of a bubble being inversely proportional to its radius [*Minnaert*, 1933]. Subsequent laboratory investigations of the effects of drop size and velocity on bubble formation during normal impact on still freshwater surfaces [*Prosperetti et al.*, 1989; *Pumphrey et al.*, 1989; *Medwin et al.*, 1990] have led to results quite different from those reported by *Blanchard and Woodcock* [1957]. A range of combinations of radii and velocities of drops that always entrained bubbles, termed Type I bubbles, was determined, which if restricted to drops that impact the water surface at their terminal velocities result in only a small range of drop radii, from 0.4 mm to 0.55 mm, that always entrain bubbles (this narrow radius range was the reason for the high values, 95% and 99%, specified as the fraction of the terminal velocities attained by falling drops in the estimates presented in §2.6.4). Typically each impact entrains a single bubble with $R_{bub} \approx 0.2$ mm by the closure of the tip of the crater formed by the splash; according to (4.3-3) bubbles of this size are expected to produce jet drops with $r_{80} \approx 9$ μm. The fraction of drops in this size range entraining bubbles when impacting a freshwater surface at their terminal velocity decreases sharply as the impact angle increases from vertical, from 100% at 0° to 10% at 20°, and the bubble size decreases slightly with increasing angle [*Medwin et al.*, 1990]. These findings explain the behavior of the 15 kHz acoustic signal. As raindrops with $r = 0.4$-0.55 mm are always present in rain but contribute little to the total rainfall rate when larger drops are present, this signal will always be present during rain with light winds, but its amplitude will be nearly independent of total rainfall rate. Furthermore, a drop with $r = 0.5$ mm, which has a terminal velocity of around 4 m s^{-1} (§2.6.3), requires a horizontal wind speed of less than 1.5 m s^{-1} to give it sufficient horizontal velocity that it impacts the surface at an angle of 20° to the vertical; this explains why the width of the strong peak in the signal broadens with increasing wind speed.

Drops with $r > 1.1$ mm impacting a freshwater or (artificial) seawater surface vertically at their terminal velocities have been found to occasionally entrain bubbles, termed Type II bubbles, by the turbulent jet of water at the bottom of the crater created by the impact; these are the bubbles reported by *Franz* [1959]. With increasing radius the percentage of instances in which entrainment occurs increases from zero at $r = 1.1$ mm to over 50% for $r = 1.7$ mm, and remains between 50% and 65% for r up to 2.3 mm [*Medwin et al.*, 1992]. Typically each impact results in entrainment of only a single bubble, with R_{bub} roughly proportional to the volume of the drop, increasing from 0.3 mm to 1.5 mm as the drop radius increases from 1.3 mm to 2.4 mm (the jet drops produced by these bubbles are expected to have $r_{80} \approx$ 16-130 μm according to Eq. 4.3-3), although smaller bubbles can also be entrained. These results were confirmed by *Nystuen and Medwin* [1995], who reported that 50-60% of the impacts from raindrops with radii 1.5, 1.7, and 2.35 mm entrained Type II bubbles upon normal impact with a still freshwater surface, with an average of around one bubble per splash. *Medwin et al.* [1992] speculated that Type II bubbles may be entrained a greater percentage of the time during impaction at oblique angles, although *Nystuen and Medwin* [1995] reported that the number of Type II bubbles formed from the impact of raindrops with $r = 1.7$ mm was reduced by 35% when the water surface was rough.

Bubbles can also be entrained from the impact of splash drops and from the impact of jet and film drops formed from bubbles entrained by the original impact of a raindrop. Laboratory investigations by *Nystuen and Medwin* [1995] of bubbles (which they termed Type III bubbles) formed by the impact of splash drops resulting from the impact of raindrops traveling at their terminal velocity determined that an average of around one such bubble was produced by the splash drops resulting from raindrops with radii 1.5 mm and 1.7 mm, and an average of around 1.6 bubbles was produced from the splash drops resulting from a raindrop with radius 2.35 mm. Based on the frequencies with which they resonate, it was determined that most of these bubbles have $R_{bub} =$ 0.12-1 mm (and according to Eq. 4.3-3 are thus expected to produce jet drops with $r_{80} = 5$-75 μm). As several splash drops are created by the impact of a raindrop, these results imply that most reentry splashes do not entrain bubbles. When the water surface was disturbed in an attempt to simulate waves, each impact of a raindrop with $r = 1.7$ mm resulted on average in the entrainment of only about 0.2 Type III bubbles by the resultant splash drops, although splash drops ejected from the measurement tank (over 40 cm horizontal distance) were observed in almost half of the impacts.

Because of the high variability of the experimental results and because of the limited extent to which the impact of drops on calm surfaces at normal incidence is representative of typical conditions over the ocean, it is difficult to determine the implications of the above results on the oceanic production of SSA particles by rainfall. *Blanchard and Woodcock* [1957] concluded, based on their results (described above), that the total number of bubbles resulting from the impact of a raindrop with $r = 2.5$ mm and its resultant splash drops could exceed 2000 [this was also discussed in *Blanchard*, 1967, p. 59]. Based on the sizes of the bubbles reported, most of the SSA particles produced would be jet drops with $r_{80} < 1$ μm, and few of these drops would be ejected to sufficient heights to be entrained upward into the atmosphere (§4.3.2.4). *Blanchard and Woodcock*, although noting that the production of bubbles by splash drops depended on the velocity and angle with which the drops impact the surface, assumed that the rate of production of these bubbles would be the same as for raindrops at normal incidence; however, any reduction in the production with increase in impact angle from the vertical would result in numbers of splash drops lower than this estimate. Later investigations (discussed above) involving acoustic measurements [*Prosperetti et al.*, 1989; *Pumphrey et al.*, 1989; *Medwin et al.*, 1990; *Medwin et al.*, 1992; *Nystuen and Medwin*, 1995] reported many fewer bubbles entrained from falling drops and from splash drops. Based on these measurements, an appreciable increase would be expected in the production and near-surface concentration of SSA particles with $r_{80} =$ 9 μm formed from the abundant Type I bubbles with $R_{bub} \approx 0.2$ mm during periods of rainfall at very low wind speeds (less than a few m s^{-1}), although such conditions are unlikely over the open ocean and would result in little upward entrainment of these particles. Additionally, the effective SSA production flux and resulting increase in atmospheric concentration of SSA particles of this size would probably be appreciably less in the presence of rain events than in their absence because of scavenging of SSA particles by the falling raindrops, which is expected to provide an efficient removal mechanism for SSA particles of this size [*Dana and Hales*, 1976; *Slinn*, 1977b; §2.7.1].

There are other factors that make it difficult to extrapolate laboratory results to the open ocean. Rain can affect SSA production in several ways (§2.3.3) by lowering the salinity and possibly the temperature of the surface waters and by affecting the stress on the ocean and its surface properties, specifically the damping and breaking of waves, surface roughness, and the removal of surface films, but these effects are not well quantified. Experiments described in *Charters* [1960] indicated that the splash drop resulting from impaction contained nearly all of the liquid in the impacting drop. If this result applies also to the impaction of raindrops on the ocean, then the SSA particles resulting from such an impact would contain much less sea salt for a given radius at production than would drops resulting from the bubble-bursting process.

Few field measurements of raindrop production of aerosols have been reported. Measured total SSA mass concentrations and size distributions of SSA concentration at various heights above the sea were reported by *Marks* [1990] for dry and rainy conditions (although rainfall rates were not reported) at wind speeds greater than 7 m s^{-1}. At wind speeds less than 10 m s^{-1} slightly lower SSA mass concentrations were reported at height 4.5 m for rainy conditions than dry conditions; this was attributed to scavenging by rain or to a change in the sea surface during rainfall resulting in decreased SSA production. At higher wind speeds (up to 15 m s^{-1}) a much greater increase in the SSA mass concentration at 4.5 m with increasing wind speed was reported for rainy conditions relative to dry conditions, and a less pronounced increase in SSA mass concentration at greater heights, roughly similar for rainy and dry conditions. It was suggested that these differences could be due to increased production of SSA particles by raindrops and the subsequent removal of these particles through scavenging by raindrops. However, the individual SSA mass concentrations contained much scatter and did not appear to exhibit any noticeable difference between rainy and dry conditions at any height for the wind speeds considered. Furthermore, the method by which the "average" concentration profiles were determined was not specified, and the individual concentration profiles presented by *Marks* did not appear representative of those discussed. *Marks* also reported that size distributions at 12 m above the sea determined from impactor measurements indicated a more rapid increase with increasing wind speed in the mass concentration of SSA particles with radii (at ambient RH, which was not reported) from 0.245 μm to 1.5 μm under rainy conditions than under dry conditions, which he attributed to scavenging by rainfall and possibly smaller equilibrium radii of SSA drops due to dilution of the surface waters by rain. An alternative explanation might be that the size distribution of SSA concentration was modified during the high RH accompanying rainy conditions because sampling by an impactor characterizes particles by their ambient radius.

An abrupt increase in measured concentrations of marine aerosol particles was observed by *M. Smith et al.* [1991] for all size ranges from 0.09 μm < r_{amb} < 23.5 μm at a coastal site during a brief period of heavy precipitation at low wind speeds (< 7 m s^{-1}) when little or no whitecapping was visible beyond the surf zone (these measurements were discussed also in §4.6.4). The investigators suggested that this increase may have been caused by increased SSA production due to rain, but as the wind speeds had recently increased from ~1 m s^{-1} to nearly 7 m s^{-1}, it is also possible that SSA particles could have been produced by conventional mechanisms or that the increase in particle concentration was due to transport to the measurement site of SSA particles that had been produced earlier.

In summary, although production of SSA particles by rainfall may be important on local scales [*Blanchard and Woodcock*, 1957; *Blanchard*, 1967, p. 59], it is probably not an important source globally compared to production by bursting bubbles in whitecaps, both because of the relatively small amount of production (due in part to the limited fraction of the ocean covered by rain) and because of probable scavenging by the raindrops of the SSA particles produced [*Blanchard*, 1963, p. 159; *Blanchard*, 1967, p. 60; *Miller*, 1987]. Likewise, SSA particles can be formed by hail by both direct impaction and from subsequent bubble entrainment, but as hail occurs much less frequently than rain, these processes are likewise probably not important sources of SSA production globally.

4.7.3. Miscellaneous Mechanisms

Other mechanisms of drop production pertinent to SSA have been suggested, many of which have involved bubble production by mechanisms other than those discussed above, such as snow falling on the sea surface. *Woodcock* [1955] reported that snowflakes falling upon and melting into seawater introduced bubbles with R_{bub} from 0.005 mm to 0.05 mm at a rate of about 25 cm^{-2} s^{-1}; such bubble flux could result in a jet drop flux of roughly 10^6 m^{-2} s^{-1} on the assumption of 5 jet drops per bubble. Up to several hundred bubbles, the majority of which had R_{bub} < 0.025 mm, were observed by *Blanchard and Woodcock* [1957] to be entrained by an individual snowflake falling on a calm seawater surface in the laboratory. Acoustic measurements of the underwater noise produced by snow falling on a freshwater body during low wind conditions exhibited a continuously increasing amplitude with frequency up to 50 kHz, indicating the presence of bubbles with R_{bub} < 0.06 mm [*Scrimger*, 1985; *Scrimger et al.*, 1987], although similar measurements at sea under high winds exhibited much less of an increase in amplitude with frequency from 40 kHz to 80 kHz (indicating bubbles with radii less than 0.075 mm, down to at least 0.0375 mm), with higher signal for greater snowfall [*McConnell et al.*, 1992]. It was suggested that the difference was caused by the greater wind speed rather than the salinity. In laboratory experiments [*Crum et al.*, 1999], snowflakes falling on a freshwater surface produced acoustic signals up to at least 100 kHz, indicating the presence of bubbles with R_{bub} < 0.03 mm, although only around 10% of the snowflakes produced detectible signals (it was not possible to determine the number of bubbles produced for each snowflake). The SSA production resulting from such bubbles would likely be small, as bubbles with R_{bub} < 0.075 mm are expected to produce only jet drops with r_{80} < 2.5 μm.

In an attempt to explain the observed positive vertical fluxes (indicating surface production) of SSA particles with r_{80} < 1 μm over Arctic leads (regions of open water) too small to support whitecaps, *Nilsson et al.* [2001] suggested that the source of these particles might be bubbles entrained by snowflakes or bubbles arising from the melting of wind blown ice crystals [*Drake*, 1968]. Surface production of ions (presumably from bursting bubbles) in Arctic leads with no visible bubble activity that could also be explained by these mechanisms were observed by *Scott and Levin* [1972]. Other sources of bubbles that have been proposed for drop formation in high latitude regions are the action of waves against ice floes [*Stabeno*, 1984] and melting of glacial walls [*Joseberger*, 1980]. The fracture of water drops upon freezing [*Mason and Maybank*, 1960; *Hobbs and Alkezweeny*, 1968; *Dye and Hobbs*, 1968; *Brownscombe and Thordike*, 1968; *Cheng*, 1970, 1971; *Hobbs*, 1971b; *Rosinski et al.*, 1972] has been reported to occur under certain conditions, but evidence for this mechanism in the marine atmosphere appears lacking, and some of the laboratory results have been attributed to initial lack of equilibrium of the drops with respect to temperature and dissolved gases.

Drops can also be produced by the bursting of bubbles formed by the entrainment of air by the merging of slopes of capillary waves [*Toba*, 1961a,b; *Koga*, 1981; *Bortkovskii*, 1987, p. 8; *Longuet-Higgins*, 1988]. The maximum radius of such bubbles is less than 0.2 mm [*Longuet-Higgins*, 1988], and according to (4.3-3) these bubbles would produce only jet drops with $r_{80} < 9$ μm. Entrainment of bubbles into the sea by impaction of the collapsing film of a bursting bubble with the sea surface has been suggested as another source of SSA particles [*Leifer et al.*, 2000a]. Drops can also be produced by bubbles formed from the mixing of different water masses, biological activity, volcanic action, and supersaturation of surface waters as a consequence of changes in temperature or salinity [*Miyake*, 1951; *Ramsay*, 1962a,b; *Blanchard and Woodcock*, 1957], although in most circumstances the number of drops produced from bubbles formed by these mechanisms are probably negligible compared to those produced by bubbles resulting from whitecaps.

Other mechanisms of SSA production not involving bubble bursting have also been proposed. *Spiel* [1998] suggested that drops were formed by the impaction of the toroid resulting from the rolling up of the burst bubble film with the sea surface (§4.33.2). Both "spume" drops and "chop" drops (drops ejected from the crests of choppy waves) were depicted in a sketch in *Monahan et al.* [1983b (and in *Monahan et al.*, 1986, and *Monahan*, 1986)] as being formed by direct production, as opposed to indirect production through bubble bursting. *Woolf et al.* [1987] stated that spume drops are produced from wind-blown wave crests whereas chop drops are produced by wave-wave interactions. A distinction between spume drops, formed by mechanical tearing of wave crests, and splash drops, formed by the vigorous spilling of breaking waves, was made by *Andreas et al.* [1995], who argued that this latter direct mechanism would have the same wind-speed dependence as production by bubbles from breaking waves. The latter

investigators argued that drops formed by this mechanism typically have $r_{80} > 10$ μm, but the mechanism was not clearly defined, and this effect was not considered in a later paper on sea salt aerosol production by *Andreas* [1998]. Little is known about the production mechanisms and sizes of these drops, but it would be expected that they would typically be larger than most bubble-produced drops on account of the surface energy required for their formation.

Other miscellaneous mechanisms have also been suggested. Sea salt aerosol particles can be produced in the spray of certain ships [e.g., *Hooper and James*, 2000]. Sea salt aerosol production from hot lava entering the ocean was investigated by *Woodcock and Spencer* [1961], who concluded that each square meter of lava flowing into the ocean produces roughly the same mass of SSA as is produced by one to six million square meters of the sea surface under normal conditions. Thus this mechanism could be quite important in certain instances and its influence could be widespread, but it is sporadic in time and space and would be difficult to quantify or predict.

Wind-generated particles containing sea salt can also be produced from freshly formed ice with a surface covering of concentrated brine and from so-called frost flowers, fragile crystals of seawater ice that may grow to several centimeters on the surface of young sea ice and which may have different compositions from that of seawater [*Perovich and Richter-Menge*, 1994; *Martin et al.*, 1995, 1996; *Wagenbach et al.*, 1998; *Hall and Woolf*, 1998; *Rankin et al.*, 2000; *Wolff et al.*, 2003]. However, the composition of particles produced from sea ice or unfrozen seawater may differ from that of seawater [*Harvey et al.*, 1991], as progressive freezing of seawater causes different salts to crystallize at different rates [*Nelson and Thompson*, 1954; *Thompson and Nelson*, 1956; *Wellman and Wilson*, 1963; *Herut et al.*, 1990]. In any event, the importance of these mechanisms for SSA production would be limited in geographical extent.

4.7.4. Fracture of Dry Sea Salt Particles

Fracture of sea salt particles upon phase change from a solution drop to a dry particle (i.e., efflorescence; §2.5.5) to yield smaller particles has been proposed by several investigators. This process would not change the SSA mass concentration M but it would change the size distribution of SSA concentration and would increase the number concentration N and thus for example the number of cloud condensation nuclei [*DeFelice and Cheng*, 1998]. As evaporation to crystallization requires that the RH be below the efflorescence value for seawater, ~45%, not the deliquescence RH, ~75% (§2.5.5), this mechanism would be expected to occur (if at all) only at large heights where such low RH values might occur.

The possibility that SSA drops may fracture upon efflorescence, producing secondary particles, was first proposed by *Bennett* [1940; cited in *Dessens, 1949*]. Subsequently *Dessens* [1946a] reported that seawater drops collected on a spider web may explode violently when dried and that when this occurs, the largest of the daughter nuclei is pure NaCl, not a mixture of salts. However, he reported that fracture upon dessication of drops suspended by a spider web occurred infrequently, and he attributed this fracture to the influence of the web [*Dessens, 1946c*]; drops of a mixed solution of NaCl-CaCl$_2$ on a spider web were also observed to crystallize separately by *El Golli et al.* [1977]. To explain the difference in composition between salts in precipitation and those in seawater, *Sugawara et al.* [1949; see also *Sugawara, 1965*] hypothesized that during evaporation seawater drops undergo partial crystallization and form at least two different types of particles: one consisting mainly of sodium chloride, and the other consisting mainly of magnesium, calcium, and sulfate. This hypothesis was later extended to assume that this fractionation occurred just after the spray was formed at the sea surface [*Koyama and Sugawara, 1953*]. Several subsequent laboratory investigations have been performed to determine whether or not seawater drops fracture upon dessication, resulting in secondary production of smaller particles. Results of these investigations, summarized in Table 21, are examined here.

A considerable increase in number of particles was reported by *Facy* [1951a] for seawater drops falling in a heated column, although no fracture was directly observed and the possibility that the fracture occurred upon impact could not be ruled out. *Facy* stressed that the conditions of the experiment did not model those found in nature. Based on higher counts of condensation nuclei, *Twomey and McMaster* [1955] and *Twomey* [1959] concluded that several hundred smaller particles were produced from an individual larger seawater drop supported on a spider web during dessication,

Table 21. Experiments Examining Fragmentation of Seawater Drops upon Evaporation

Reference	Experiment	Detection Method	$r_{80}/\mu m$	Fragmentation?
Dessens, 1946a, 1949	haze drops on spider web	microscope	10	yes
Pacaud, 1949 (cited in *Facy* [1951a] and *Toye* [1956])	sea spray on spider webs	not stated	not stated	no
Facy, 1951a	seawater drops in heated tube	photography	up to tens of μm	yes
Lodge & Baer, 1954	NaCl drops in heated tube	chloride traces on filters	<1-7.5	no
Twomey & McMaster, 1955	seawater drops on spider webs	cloud observed in expansion chamber	not stated	yes
Toye, 1956	NaCl and seawater drops on spider webs	microscope	~2-~10	no
Junge, 1958, *Junge*, 1963, p. 160	freely suspended seawater drops	not stated	not stated	no
Blanchard & Spencer, 1964	NaCl and seawater drops on spider webs, glass fibers, platinum wired, cotton strings, glass and Lucite surfaces	condensation nucleus counter	1-20	no
Radke & Hegg, 1972	NaCl drops in heated chamber	sodium-detecting flame photometer	0.06-0.4	yes
Iribarne et al., 1977	NaCl drops in an air stream	optical and electron microscopy, condensation nucleus counter	0.04-0.7	no
Cheng, 1988	seawater drops on glass slides	microscope	~2-50	yes
Mitra et al., 1992	drops of NaCl, (NH$_4$)$_2$SO$_4$, mixtures of these two, and artificial and natural seawater on hydrophilic and hydrophobic fibers and glass plates, and in a wind tunnel	microscope, electron microscope, optical counter	0.1-1, 5-20, 40-80, 100-300	no, except drops on hydrophilic fibers

although no fracture was observed directly and the drops were not subjected to RH below 60%, leading to questions by *Blanchard and Spencer* [1964] as to whether or not the drops had completely evaporated (and consequently whether or not the higher counts had actually resulted from fracture of the particles). A decrease in the number of large drops and a sometimes substantial increase in the number of smaller drops resulting from the evaporation of NaCl drops in an air stream was reported by *Radke and Hegg* [1972], although the experiments suffered from high internal losses and highly variable results. These investigators also noted that their experiment did not accurately model atmospheric processes, but concluded that the results supported the idea that fracture of seawater drops occurred upon evaporation to a dry particle. Formation of secondary aerosol particles when seawater drops on glass slides were evaporated, or when seawater drops were evaporated on a hot rock, was reported by *Cheng* [1988], who concluded that the most important mechanism for the generation of a larger number of secondary particles was the bursting of dissolved air bubbles in the drop.

Many laboratory experiments have reported evidence of no fracture of drops upon efflorescence. Sea salt particles captured on spider webs and subjected to many cycles of humidification and dessication were not observed to fracture by *Pacaud* [1949; cited in *Facy*, 1951a, and in *Toye*, 1956], although *Dessens* [1949] reported that *Pacaud* found such fracture to be "very rare"—implying that it may have occurred. No evidence of fracture of NaCl drops upon drying in a heated tube was observed by *Lodge and Baer* [1954], who noted that this process is well known for large crystalline aggregates (1 mm or so in diameter), arising from water being trapped inside an outer crystalline shell [*Ranz and Marshall*, 1952a,b], but concluded that fracture of SSA particles cannot play an important role in meteorological processes. Drops of seawater and NaCl suspended on spider webs subjected to 1000 successive crystallizations and rehumidifications by *Toye* [1956] experienced no appreciable change in size. *Junge* [1958, 1963, p. 160; also in discussion on p. 8 of *Twomey*, 1959] reported that he could not confirm the results of *Twomey and McMaster* [1955], and also that he could not substantiate the production of smaller particles from the fracture of a larger sea salt particle when it was not suspended on a spider web. The experiments of *Twomey and McMaster* [1955] were also repeated by *Blanchard and Spencer* [1964] using drops of NaCl, seawater, and concentrated seawater attached to spider webs, glass fibers, platinum wires, cotton strings, and glass and Lucite surfaces. Evidence of small particle production was observed only when stresses were applied to sea salt particles on a platinum wire long after phase change to a dry particle. No evidence of fracture with unsuspended NaCl drops was observed by *Iribarne et al.* [1977]. Finally, no evidence of secondary production

(i.e., fracture) using solutions of NaCl, $(NH_4)_2SO_4$, mixtures of these two salts, and artificial and natural ocean waters in a wind tunnel was reported by *Mitra et al.* [1992]. These investigators also investigated evaporation of drops suspended on fibers and reported that whereas drops on hydrophilic fibers and on glass slides were pulled apart and produced typically two or three fragments, no violent shattering occurred. They further reported that fracture was never observed for drops on hydrophobic fibers, from which they concluded that surface tension forces were the cause of the fracture reported by others. Additionally, no evidence of fracture or secondary production was reported from numerous experiments involving evaporation and crystallization of drops not attached to supports for both solutions of single salts [*Tang et al.*, 1977; *Tang*, 1980; *Leong*, 1981; *Beard et al.*, 1983; *Richardson and Spahn*, 1984; *Richardson and Hightower*, 1987; *Leong*, 1987a,b; *Cohen et al.*, 1987a,c; *Cheng et al.*, 1988; *Baumgärtner et al.*, 1989; *Tang and Munkelwitz*, 1991, 1993, 1994; *Tang et al.*, 1995] and mixed salts [*Spann and Richardson*, 1985; *Cohen et al.*, 1987b,c; *Chan et al.*, 1992; *Tang and Munkelwitz*, 1994b; *Tang*, 1997; *Tang et al.*, 1997].

Indirect evidence for fracture comes from the composition of individual aerosol particles. Residues of drops contained crystals of pure NaCl, KCl, $CaSO_4$, $MgSO_4$, and other substances were reported by *Parungo et al.* [1986a,b, 1988] in marine aerosol particles analyzed by using X-ray energy spectrometry; these investigators claimed that these results could be explained only by assuming that seawater drops, upon evaporation, experienced fractionary recrystallization and formed loosely attached conglomerates of separate species that would easily shatter in the air. However, the separation of seawater components upon evaporation of a seawater drop on a solid substrate has long been known (§2.5.5). No mechanism was proposed by *Parungo et al.* to explain why or how this shattering would occur once the conglomerate was formed; it is difficult to imagine why an aerosol particle in the atmosphere would be so fragile. These investigators hypothesized that rainfall favored conditions for this to occur; in some situations after rainfall up to 50% of the particles sampled with aerodynamic radii 0.25-2.5 µm contained only Na and Cl. S:Na ratios observed by *Mouri and Okada* [1993] and *Mouri et al.* [1993] from SSA particles in the equatorial Pacific were lower than the seawater value with a high correlation between the ratios Ca:Na, Mg:Na, and K:Na with that of S:Na; similar results for the Southern Ocean were reported by *Mouri et al.* [1997], who attributed them to fractionary recrystallization followed by fracture of seawater drops occurring at cloud tops. Fracture of SSA particles had earlier been considered but discounted by *Oddie* [1960] to account for observed size-dependent differences in elemental composition of such particles. The possibility that the results of *Mouri et al.* [1993] were due to the erroneous assumption that the sea salt particles examined were internally homogeneous was suggested by *Pósfai et al.* [1995]; when analyzing entire particles by electron microscopy, both *McInnes et al.* [1994] and *Pósfai et al.* [1995] reported individual particles with S:Na ratios lower than the seawater value, but observed no variation in Ca:Na, Mg:Na, or K:Na ratios. The possibility that fracture may have occurred upon impact with filters or substrates during sampling, or during handling and preparation, such as drying samples for electron microscopy, must also be considered, although several investigators have argued that there would be a recognizable pattern if this had occurred, and that the absence of such an observed pattern eliminates this possibility. Additionally, mechanisms other than fracture for the formation of substances such as $CaSO_4$ might explain some of the observed elemental ratios. *Andreae et al.* [1986] proposed that $CaSO_4$ could result from the reaction of atmospheric SO_2 with marine biogenic $CaCO_3$ particles, and *Artaxo et al.* [1992] concluded that the seasonal variability of fine mode $CaSO_4$ particles in Antarctica is evidence for a biogenic origin.

In summary, numerous experimental investigations have established absence of fracture of SSA particles upon drying, and most of the experimental evidence reporting such fracture is indirect and inconclusive, and it is difficult to rule out experimental interference, such as fracture during impaction. Furthermore, stresses occurring during the drying of suspended particles have no counterpart in the marine atmosphere; similarly, extreme temperatures and drastic changes in temperature and RH do not reflect conditions experienced by a seawater drop in the atmosphere. As SSA particles remain solution drops until the RH is below the efflorescence value of around 45% (§2.5.5), the range of conditions for which the mechanism of fracture upon evaporation could occur is rather restricted. Based on these considerations it can confidently be concluded that fracture of SSA particles upon drying is not an important mechanism in the marine atmosphere.

4.7.5. Summary

While it can be concluded that SSA particles can be produced by a variety of mechanisms other than those of bubble bursting by breaking waves and wave tearing, and while some of these mechanisms may be locally important, there is little evidence that any of these mechanisms is competitive with the breaking wave-bubble production mechanism or the wave-tearing mechanism on a larger scale (for example, more than several tens of kilometers from coasts), and especially not on a global scale. Perhaps more important in the present context of determining the oceanic SSA production flux is the role of such mechanisms in confounding attempts to infer this production, especially by measurements in coastal locations that are commonly influenced by local surf-zone production.

5. Sea Salt Aerosol Production Fluxes: Estimates and Critical Analysis

Size-dependent interfacial and effective production fluxes of SSA particles, $f_{int}(r_{80})$ and $f_{eff}(r_{80})$, respectively, can be calculated according to the several methods outlined in §3 using the data examined in §4. The steady state dry deposition method, concentration buildup method, micrometeorological methods, and the statistical wet deposition method yield the effective SSA production flux, whereas the whitecap method, bubble method, along-wind flux method, direct observation method, and vertical impaction method yield the interfacial SSA production flux. Each method is restricted in its applicability to limited and different ranges of r_{80}, and each rests on a particular set of assumptions and requirements that must be satisfied for its successful implementation. For most methods relatively few determinations have been made. In this section estimates for these fluxes are examined and compared and the approaches themselves are critically evaluated. Where possible, central values representing in some sense "best" estimates of f_{int} or f_{eff} are presented for each method, along with associated uncertainties to quantify the confidence that can be placed in these estimates. Finally, SSA production fluxes for each of the three size ranges of SSA particles are examined and summarized.

5.1. GENERAL CONSIDERATIONS

5.1.1. Uncertainties in Estimates of SSA Production Fluxes

Estimates for SSA production fluxes have previously been reviewed and compared by other investigators [e.g., *Miller*, 1987; *Andreas et al.*, 1995; *Andreas*, 1998; *Pattison and Belcher*, 1999; *Guelle et al.*, 2001; *Reid et al.*, 2001; *Hoppel et al.*, 2002; *Schulz et al.*, 2004]. However, the data and experimental techniques upon which these estimates were based have rarely been examined in depth, and in some instances the distinction between interfacial and effective fluxes was not clearly made. It has often been assumed that one or another of the formulations provides the correct result; alternatively, consistency of size-dependent concentrations

calculated using a given SSA production flux with observed concentrations has been taken as validation of a particular formulation. However, careful examination of the various flux estimates and the foundations upon which they were based leads to the conclusion that there is considerable uncertainty in these estimates. As elsewhere in this review the intent in this section is to provide estimates of the uncertainties associated with values of SSA production fluxes examined here.

As noted in many places throughout this review, the uncertainties associated with quantities that serve as inputs to calculation of flux estimates are quite large: key quantities such as size-dependent SSA concentrations (§4.1.4; Fig. 22), size-dependent bubble concentrations (§4.4.1; Fig. 33), and whitecap ratios (§4.5.2; Fig. 36) all range over at least an order of magnitude at the same wind speed, and measurements of the SSA production flux over white area f_{wc} (§4.3.4; Fig. 32) vary by at least this much for a given r_{80}. Because of their large magnitudes, uncertainties of these quantities have consistently been reported throughout this review as multiplicative uncertainties (denoted by the symbol $\divideontimes$) rather than as additive uncertainties (denoted by the symbol ±). Such multiplicative uncertainties are displayed as error bars of equal width above and below the central estimate when plotted on a logarithmic scale. Uncertainties in these quantities translate directly into uncertainties in the flux estimates. Additionally such estimates contain uncertainties due to the assumptions in applicability of the methods themselves that are much more difficult to quantify, such as the assumption that steady state conditions hold or that all white ocean areas are equally efficient for SSA production. All these uncertainties contribute to the uncertainty of any estimates of SSA production flux. As well, these uncertainties make it difficult to draw confident conclusions regarding the dependence of SSA production fluxes on controlling meteorological factors such as wind speed.

Because of these considerations it is necessary to examine the approximations and assumptions that have been used in the development of the several SSA production flux formulations, the shortcomings of these formulations, the extent to which they are supported by data, and thus the confidence that can be placed in them. Additionally, this examination is necessary to determine the extent to which any of the estimates provides a credible possibility to represent the effective and/or interfacial SSA production flux(es).

Sea Salt Aerosol Production: Mechanisms, Methods, Measurements, and Models
Geophysical Monograph Series 152
Copyright 2004 by the American Geophysical Union.
10.1029/152GM05

In this section these estimates are critically evaluated based on the available data on size-dependent SSA concentrations in the marine boundary layer (§4.1), jet and film drop production as a function of bubble size (§4.3.2; §4.3.3), laboratory and surf zone SSA production flux simulated whitecaps and in wind/wave tunnels (§4.3.4; §4.3.5; §4.3.6), size dependent bubble concentrations in the ocean surface waters (§4.4.1), and whitecap ratios (§4.5), and their controlling factors.

Additionally, estimates of the uncertainty associated with each SSA production flux estimate are provided, where possible, as summarized in Table 22. For some methods there are too few data to quantify the uncertainty. For several of the methods only a single set of measurements has been reported.

As seen throughout this review when a given quantity is reported from multiple investigations the results often range widely, and thus it would seem difficult to evaluate the uncertainties of formulations that are based on a single or only a few measurements. In other methods for which numerous measurements have been made (e.g., the steady state dry deposition method or the whitecap method), quantification of uncertainty is possible. In these methods the estimate for the flux is often evaluated as the product of several factors, each of which is characterized by an associated uncertainty, with the overall uncertainty in the flux estimate containing contributions from these several uncertainties. The procedure chosen to do this is to treat the logarithms of the

Table 22. Uncertainty Estimates for Sea Salt Aerosol Production Fluxes Determined by the Several Methods

Method	Flux[a]	Particle Size[b]	Required Measurement	q_i	Figure No.	Multiplicative Uncertainty[c] υ	Multiplicative Uncertainty[c] Υ
Steady state dry deposition	E	M					4
			Size-dependent number concentration	$n(r_{80})$	22	3	
			Size-dependent dry deposition velocity	$v_d(r_{80})$	38, 39	2[d]	
Whitecap	I	S, M					9-15
			Size-dependent production flux over white area	$f_{wc}(r_{80})$	32	7	
			Oceanic whitecap ratio	W	36	3-7	
Concentration buildup	E	S, M	Size-dependent number concentration as function of along-wind distance and height	$n(r_{80}, x, z)$			?[e]
Bubble[f]	I	M					17
			Number of drops per bubble	$N_{jet}(R_{bub})$	26	2	
			Size distribution of bubble concentration	$n_{bub}(R_{bub})$	33	14	
			Bubble rise velocity	$v_{bub}(R_{bub})$	34	2	
Micrometeorological	E	S					?
			Size-dependent number concentration				
			Size-dependent vertical flux at ~10 m				
Along-wind flux	I	L					?
			Size-dependent along-wind flux near sea surface	$j_x(r_{80})$	25		
			Size-dependent mean horizontal distance traveled	$\bar{L}_p(r_{80})$			
Direct observation	I	L	Count from video				?
Vertical impaction	I	M, L	Size-dependent vertical flux near sea surface				?
Statistical wet deposition	E	S					5
			Size dependent number concentration	$n(r_{80})$	22	4	
			Mean time between precipitation events	τ_{wet}		2	
			Height of marine boundary layer	H_{mbl}		2	

[a] E denotes effective SSA production flux; I denotes interfacial SSA production flux.

[b] Particle size refers to that for which the method is most applicable; S, M, and L denote small, medium, and large SSA particles, respectively (§2.1.2).

[c] Multiplicative uncertainty in quantity q is expressed as $q \overset{\times}{\div} \upsilon$. The uncertainty Υ associated with a quantity Q evaluated as the product $Q = \Pi q_i$ of several factors q_i, each with an associated multiplicative uncertainty υ_i, is estimated as $\Upsilon = \exp\{[\Sigma(\ln \upsilon_i)^2]^{1/2}\}$.

[d] Although the multiplicative uncertainty for v_d certainly varies with r_{80}, the value $\overset{\times}{\div} 2$ is chosen here; §4.6.3.

[e] The symbol ? denotes too few measurements by indicated method to allow inference of uncertainty.

[f] Limited to jet drops (§5.5.2).

respective multiplicative uncertainties in the several factors as if they could be combined in the typical method of treating additive uncertainties (appropriate if the uncertainties in the several factors are lognormally distributed [*Aitchison and Brown*, 1957]). The multiplicative uncertainty Υ in a quantity Q that is the product of several factors q_i (i.e., $Q = \Pi q_i$) each characterized by a multiplicative uncertainty υ_i, is given by $\Upsilon = \exp\{[\Sigma \ (\ln \upsilon_i)^2]^{1/2}\}$, i.e., $\delta(\ln Q) = \{\Sigma[\delta(\ln q_i)]^2\}^{1/2}$. For example, the product of two quantities, one characterized by a multiplicative uncertainty of $\gtrless 5$ and one by $\gtrless 10$, is characterized by multiplicative uncertainty of $\gtrless 17$. This procedure provides an indication of the uncertainties of given estimates of SSA formulations examined below.

5.1.2. Relation Between Interfacial and Effective SSA Production Fluxes

Although it is generally recognized that the principal mechanisms for SSA production are bursting of bubbles from whitecaps formed by breaking waves and at higher wind speeds tearing of drops from wave crests, equally important to the effective SSA production flux is the upward entrainment of newly formed SSA particles. The separation of the production and the entrainment mechanisms is not always straightforward, as both are dependent to great extent on the wind speed, and to some extent on atmospheric stability and other factors. As discussed in §2.1.6, the effective SSA production flux differs from the interfacial SSA production flux because not all particles produced at the surface attain the height of 10 m, a somewhat arbitrary but practical and rather insensitive criterion for an SSA particle having been introduced into the marine atmosphere. The distinction between these two fluxes is important, as the difference between them can be a substantial fraction of the interfacial SSA production flux, depending on particle size. However, although recognized at least as far back as *Toba* [1965a,b, 1966], this distinction has not always been made by subsequent investigators, with resultant ambiguity in presentation and comparison of fluxes. The ratio of the effective SSA production flux to the interfacial SSA production flux, Ψ_f (§2.1.6), is required to relate these two fluxes and compare flux estimates obtained by different methods. As it is assumed that SSA particles that attain a height of 10 m remain in the atmosphere sufficiently long to participate in atmospheric processes of interest and thus contribute to the effective SSA production flux, the quantity Ψ_f, is equal to the probability that an SSA particle produced at the sea surface attains a height of 10 m.

Based on considerations of entrainment and transport of SSA particles in the lower marine boundary layer and on examination of their removal mechanisms (§2.9.4, §4.6.5), the qualitative behavior of Ψ_f was determined as a function of particle size and wind speed. Over a wide range of wind speeds, Ψ_f is near unity for $r_{80} \lesssim 3 \ \mu m$; as the primary removal mechanism for SSA particles in this size range is expected to be wet deposition, such particles, once produced, remain in the atmosphere for extended periods of time (i.e., days) and become well mixed over the entire marine boundary layer. Thus, no distinction between interfacial and effective SSA production fluxes is necessary in this size range. For a given wind speed, Ψ_f decreases with increasing particle size, as gravity plays an increasingly greater role in the fate of SSA particles, and a smaller fraction of particles produced at the sea surface survives to 10 m. As r_{80} approaches 25 μm, Ψ_f decreases to a value that is much less than unity, as has been noted throughout this review. Additionally, with increasing r_{80} near 25 μm several other factors become important. The SSA concentration becomes so strongly dependent on height above the surface that the effective SSA production flux at a specific height is not a well defined nor a useful concept. Failure of SSA particles in this size range to have equilibrated to the ambient RH (§2.9.1) results in even less vertical mixing. The assumption that the eddy diffusivity of such particles is given by (2.4-2) is also less likely to be valid with increasing particle size, but instead it would be expected that D_{eddy} would decrease with increasing particle size as large SSA particles, because of their inertia, may "miss" smaller eddies (§2.6.4). As time scales for vertical mixing of SSA particles of these sizes may be greater than mean lifetimes of such particles, the assumption of a steady state vertical concentration profile is questionable. Moreover, the motion of large SSA particles is inherently nonlinear, and the horizontal and vertical velocity components are coupled (§2.6.3), implying that 1-D models might not provide accurate results for the motion and vertical mixing of such particles. In addition to all these concerns, measurement uncertainties and the low concentrations of large SSA particles more than a few meters above the sea surface, in addition to large vertical concentration gradients of such particles, would render attempts to infer the interfacial production flux of such particles from concentration measurements at heights near 10 m highly uncertain. For these reasons also the effective SSA production flux is restricted to small and medium SSA particles ($r_{80} \lesssim 25 \ \mu m$).

Several quantitative estimates of $\Psi_f(r_{80})$ have been presented (Fig. 13). In §2.9.4 it was proposed that the dependencies of $\Psi_f(r_{80})$ on r_{80} and U_{10} should be similar to those of the steady state SSA concentration ratios $n(10 \ m)/n(0.1 \ m)$ and $n(10 \ m)/n(0.01 \ m)$, the lower heights somewhat arbitrarily chosen but typical of those to which jet drops are ejected and thus expected to yield concentration values characteristic of those near the sea surface. Although this estimate is qualitatively valid, it might be argued that it underestimates Ψ_f. Sea salt aerosol particles for which the concentration ratio is less than unity may still have long atmospheric residence times during which they will be repeatedly mixed by turbulence above the height of 10 m and thus contribute to the effective SSA production flux; that is, the probability that any individual SSA particle of the given size will attain a height of 10 m during its atmospheric residence time (and thus by definition contribute to the effective SSA production flux) is greater than the ratio of the probabilities of finding such a particle at 10 m compared to a lower height. Alternative estimates have been presented (§4.6.5). Those of *Toba and coworkers* [*Toba*, 1965b; *Chaen*, 1973; *Iida et al.*, 1992] and that of *Hoppel et al.* [2002] are based on the ratio of steady state concentrations at 10 m and at a lower height, with the assumption that only gravity acts to remove SSA particles from the lower height to the sea surface. The assumptions of steady state and of removal being solely due to gravity restrict these estimates to SSA particles with r_{80} greater than several micrometers. The choice of the lower height can make a large difference in quantitative results for larger SSA particles. However, this choice is arbitrary, and the assumption that there is an absorbing surface at the given height has no physical basis, as SSA particles are not removed from the atmosphere at this height. This approach essentially relates the effective SSA production flux at 10 m to the effective SSA production flux at a lower height and makes the implicit assumption that the effective SSA production flux at this lower height is equal to the interfacial SSA production flux; however, SSA particles that do not attain this lower height will not be included in what constitutes the interfacial SSA production flux (*cf.* §2.1.6). Additionally, under conditions for which waves are breaking and SSA particles are being produced, the concept of a horizontal smooth surface above which height can be accurately determined and/or defined is tenuous. Another estimate was presented by *Fairall and Larsen* [1984] based on a modification of the dry deposition model of *Slinn and Slinn* [1980, 1981] (§4.6.5). This estimate yields values of $\Psi_f(r_{80})$ that are roughly a factor of two lower than the ratio $n(10 \text{ m})/n(0.1 \text{ m})$ for r_{80} from ~3 μm to ~15 μm; this difference, although substantial, is relatively minor in comparison to the uncertainties in the flux estimates presented below. However, some aspects of the behavior of this estimate seem unphysical, such as a decrease in Ψ_f with increasing wind speed in some size ranges.

In summary, the formulations that have been presented for Ψ_f for the most part yield qualitatively similar dependencies on particle size and wind speed. The value of Ψ_f is near unity for SSA particles with r_{80} less than several micrometers, decreasing with increasing particle size to much less than unity for $r_{80} \approx 25$ μm, above which the quantity is essentially zero. In the size range for which it differs appreciably from unity and from zero, i.e., 5 μm $\lesssim r_{80} \lesssim 25$ μm it is expected to increase with increasing wind speed for a given r_{80}, but more quantitative results that would allow comparison of effective SSA production fluxes and interfacial SSA production fluxes cannot be justified.

5.1.3. Factors Affecting SSA Production

Sea salt aerosol production, as with many other quantities discussed in this review—SSA mass and number concentrations, SSA concentration size distributions, bubble concentration size distributions, and whitecap coverage, for example—has most often been parameterized in terms of the local wind speed, under the implicit presumption that local wind speed is the dominant factor controlling the production of SSA particles at the sea surface and their subsequent entrainment upward. Although classification of these other quantities by wind speed has resulted in much scatter and often low correlations with local wind speed, this situation can in some instances easily be explained by the long atmospheric residence times of SSA particles (especially small SSA particles), resulting in conditions under which such particles were produced being not well characterized by conditions (typically only wind speed) at the location at which they are measured. As production of SSA particles is inherently a local phenomenon, it might be expected to be largely free of such concerns. Additionally, the major role that wind speed plays in forming breaking waves and spume drops, and in transporting SSA particles upward, lends physical support for this presumption. On the other hand, the large scatter in values of whitecap ratio at a given wind speed (Fig. 36) raises concerns that wind speed alone is insufficient to fully characterize production. Additionally, some local conditions such as sea state, as parameterized by mean wave height or significant wavelength, for example, are to some extent dependent on past conditions, and these quantities are typically not reported with measurements pertinent to SSA production (but as noted in §2.3, quantities parameterizing sea state are often highly correlated with local wind speed, and under steady state conditions in the open ocean it would be expected that these quantities would be determined solely, or at least mainly, by local wind speed). The frequency of occurrence of steady state vs. non-steady state conditions of sea state would be difficult to quantify, and additionally non-steady state effects would require detailed knowledge of the dependence of SSA production on sea state; as seen in the comparisons of the various flux estimates below, this knowledge is not available. Additionally, there may be a dependency of effects of these factors on SSA particle size, as small SSA particles ($r_{80} \lesssim 1$ μm), being mostly film drops (§4.3.4.4), are produced by larger bubbles (§4.3.3) which may have production dependencies on sea state that differ from those governing smaller bubbles which produce most jet drops, which in turn comprise the majority of medium SSA particles (§4.3.2; §4.3.4.4).

Several additional factors, although typically not included in SSA production formulations, are thought to affect SSA production. The air temperature, along with the sea temperature, determines the atmospheric stability, which is thought not only to affect wave breaking and the spectrum of ocean waves, but also the vertical entrainment of SSA particles once produced, and thus the effective SSA production flux. Additionally, the sea temperature affects bubble concentrations (§4.4.1.4), rise speeds (§4.4.2.1), and bursting behavior (*cf.* the results of *Spiel* [1994a,b, 1995, 1997a] concerning the sizes of jet drops as a function of temperature; §4.3.2.2), and possibly whitecap coverage (Fig. 37), and thus both the interfacial and effective SSA production fluxes. However, too little is known to draw any conclusions about dependencies of these factors on SSA production over the ocean. The wind friction velocity has been argued to be the primary controlling factor for many aspects of air-sea interaction, but in practice it is usually determined from wind speed and the air-sea temperature difference. These meteorological quantities (wind speed and sea and air temperatures) are easily determined (or readily available) and thus well suited to parameterize SSA flux formulations, especially for use in models for which forecast or archived data are available. In practice, however, no explicit temperature dependence is included in most models. The presence, amount, and type of surface active materials have been shown to have a strong affect on many aspects of drop production, but these factors are extremely difficult to determine or predict and have not been included in any formulations that have been proposed. Some formulations, such as those of *Chaen* [1973] and *Iida et al.* [1992] involve other quantities such as the peak frequency of the wind waves, but these quantities would be more difficult to include in models than parameterizations involving more easily obtained meteorological quantities and have not been widely reported together with measurements pertaining to SSA production.

5.2. STEADY STATE DRY DEPOSITION METHOD

The steady state dry deposition method (§3.1) calculates the size-dependent effective SSA production flux $f_{eff}(r_{80})$ from field measurements of $n(r_{80})$, the size-dependent SSA number concentration at heights near 10 m (§4.1.4), and modeled values of $v_d(r_{80})$, the size-dependent dry deposition velocity (§4.6), as $f_{eff}(r_{80}) = n(r_{80})v_d(r_{80})$, under the explicit assumption that local SSA production and dry deposition are equal and opposite. Measurements of $n(r_{80})$ were presented in Fig. 22, and modeled values of $v_d(r_{80})$ were presented in Fig. 38. As this method is based on SSA concentrations measured at heights typically near 10 m above the sea surface, it necessarily yields an effective SSA production flux (although some investigators have attempted to infer an interfacial SSA production flux from such measurements using

additional assumptions that in essence prescribe the quantity $\Psi_f(r_{80})$; §4.6.5; §5.1.2), and as such it is restricted to small and medium SSA particles, i.e., $r_{80} \lesssim 25$ μm. Based on considerations presented in §4.6 regarding the ability of the dry deposition method to satisfy the conditions set forth in §3.1, the size range of SSA particles for which this method can, under appropriate circumstances, be expected to yield accurate results is additionally restricted to $r_{80} \gtrsim$ 3-5 μm. This method has been one of the most commonly used ones of inferring the SSA production flux.

5.2.1. Estimates of the Effective SSA Production Flux

This section review estimates for the size-dependent effective SSA production flux $f_{eff}(r_{80})$ based on the steady state dry deposition method. Several of these estimates are quite dated, but nonetheless have continued to be modified and cited. An estimate for the effective SSA production flux based on the canonical SSA size distribution of concentration (§2.1.5; §4.1.4.4) is also presented. The several estimates at wind speeds near 10 m s^{-1} are compared in Fig. 40 in the representation $f_{eff}(r_{80}) \equiv dF_{eff}(r_{80})/d\log r_{80}$. Finally, the applicability of the dry deposition method for determination of the effective SSA production flux is examined.

5.2.1.1. Early estimates. An early estimate of the SSA production flux based in part on the steady state dry deposition method was presented by *Moore and Mason* [1954], shown in Fig. 40 for $U_{10} = 10$ m s^{-1}. In Fig. 1 of that reference size-dependent number concentrations of SSA particles with r_{80} from 0.25 μm to 20 μm were presented based on measurements made in the North Atlantic, but the wind speed was not stated and data were not shown (additionally, the shape and magnitude of the size distribution of SSA concentration presented on that figure is inconsistent with relations stated in the text, which were used to calculate the estimate presented in Fig. 40). The production flux of SSA particles with r_{80} greater than a critical value (which was stated to increase from 2 μm to 20 μm as the wind speed increased from 6 m s^{-1} to 15 m s^{-1}) was assumed to be balanced by removal through gravitational sedimentation, as described by the Stokes velocity (Eq. 2.6-7), whereas the production flux of smaller particles was assumed to be greater than their dry deposition flux and was taken as proportional to their concentration, with the factor of proportionality dependent on properties of the airflow but independent of particle size. Hence the dry deposition velocity was constant for SSA particles with r_{80} less than the critical value and was equal to the Stokes velocity for those with r_{80} greater than the critical value. Although this estimate of the production flux is surprisingly similar to later estimates (Fig. 40), it must be considered of historical interest only.

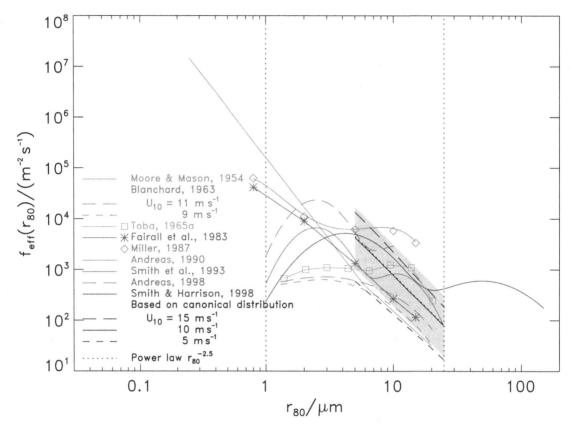

Figure 40. Estimates for the size-dependent effective SSA production flux $f_{eff}(r_{80})$ based on the dry deposition method calculated for $U_{10} = 10$ m s^{-1} unless otherwise noted. Estimate shown for *Miller* [1987] is geometric mean of estimates for $U_{10} = 9$ m s^{-1} and $U_{10} = 11$ m s^{-1}. Curve for *Andreas* [1990b] is fit to *Miller* [1987]. Curve for *Andreas* [1998] is 3.5 times fit of *Smith et al.* [1993] over range shown. Estimates based on canonical distribution were calculated for $U_{10} = 5$, 10, and 15 m s^{-1} from the canonical SSA size distribution of concentration presented in §4.1.4.4 and dry deposition velocity given by (4.6-8); power law $r_{80}^{-2.5}$ is shown as an approximation to $f_{eff}(r_{80})$ for $U_{10} = 10$ m s^{-1}. Vertical dotted lines at r_{80} values of 1 μm and 25 μm demarcate small, medium, and large SSA particles (§2.1.2). Shaded gray band about central solid line denotes uncertainty of $\overset{\times}{\div}$ 4 associated with uncertainties in canonical SSA size distribution of concentration and dry deposition velocity.

Several investigators have used the steady state dry deposition method to estimate the effective SSA production flux using primarily size distributions of SSA number concentration reported by *Woodcock* [1953, Fig. 1] from measurements at cloud base (at least 600 m above sea level) near Hawaii. Size distributions were reported for r_{80} ranging from near 1 μm to just over 20 μm, at five values of U_{10} up to 15 m s^{-1} (this figure also contained a single size distribution for U_{10} near 35 m s^{-1} based on measurements taken from a lighthouse in Florida during a single occurrence of this wind speed); these measurements were discussed in §4.1.2.1 and §4.1.4.2. As the estimates based on these size distributions of SSA concentration involved extrapolation over large vertical distances and used simplified models to obtain dry deposition velocities, they are also mostly of historical interest, but as several later formulations have been based, at least in part, on these estimates, they warrant discussion.

Eriksson [1959], noting that the size-dependent SSA concentrations reported by *Woodcock* [1957] at heights of at least 150 m near Hawaii and those from unpublished data of *Woodcock* at heights of at least 30-150 m near Florida and in the Caribbean exhibited little dependence on height up to 500 m, explicitly assumed that the concentrations in the laminar boundary layer were the same as those at the lowest heights for which they had been reported by *Woodcock* (30 m). Based on arguments presented in *Sverdrup* [1951] regarding details of the air flow near the surface of the ocean, *Eriksson* calculated the height of the laminar boundary layer to be 1.1 mm (for $U_8 = 6$ m s^{-1}) and RH to be 91.4% at the top of this layer. The dry deposition velocity was taken to be the Stokes velocity (Eq. 2.6-7) at 91.4% RH, under assumption that SSA particles would experience little growth as they fall through this layer; this Stokes velocity is about 40% greater than that at 80% RH (Fig. 7; §2.6.2).

Thus the dry deposition flux of SSA particles with r_{80} from around 1 μm to 20 μm was obtained as the product of their number concentrations and their Stokes velocities at 91.4% RH. These fluxes were presented for measurements of *Woodcock* at three locations [*Eriksson*, 1959, Fig. 3.13], but wind speeds were not reported and the wind-speed dependence of this dry deposition flux was not discussed; according to *Blanchard* [1963, p. 128], these unpublished data of *Woodcock* were obtained at a wind speed of ~5 m s^{-1}. *Eriksson* argued that sea salt mass is removed by dry deposition and wet deposition at roughly the same rate, and thus that the production flux of sea salt mass[1] is roughly twice the dry deposition flux of sea salt mass, but without knowledge of the size dependence of the wet deposition flux it is not possible to apply a correction factor to obtain the size dependence of the SSA production flux; consequently no estimate from *Eriksson* is presented in Fig. 40.

Blanchard [1963, pp. 127-132], noting the unavailability of data on the vertical distribution of SSA particles over a wide range of wind speeds, took the size-dependent concentrations of SSA particles at the top of the laminar boundary layer to be equal to those reported by *Woodcock* [1953] from measurements at cloud base (at least 600 m above sea level) near Hawaii. Following *Eriksson* [1959], *Blanchard* [1963] took the dry deposition velocity of an SSA particle to be its Stokes velocities at 91.4% RH, and obtained [*Blanchard*, 1963, Fig. 17] the dry deposition flux of jet drops (which he assumed comprised most of the concentrations measured by *Woodcock*) with r_{80} from ~1 μm to 20 μm for six wind speeds from 5 m s^{-1} to 15 m s^{-1}. According to these estimates, the value of r_{80} at which $dF(r_{80})/d\log r_{80}$ attains a maximum increases with increasing U_{10} from ~2.5 μm at 5 m s^{-1} to ~3.5 μm at 15 m s^{-1}. These estimates for the production flux are presented in Fig. 40 for $U_{10} = 9$ m s^{-1} and $U_{10} = 11$ m s^{-1}. *Blanchard* noted that although the total SSA production flux F_{eff} appeared to be directly proportional to the wind speed, the increase in the production of large drops with increasing wind speed would result in the total SSA dry mass deposition flux being directly proportional to the cube of the wind speed. He also explicitly noted that this method could not be used for SSA particles sufficiently small that their residence times were too long for the assumption of steady state to be satisfied. No explicit criterion for this size was presented, but based on discussion in *Blanchard* [1963, p. 132] it would be between $r_{80} \approx 2.5$ μm and $r_{80} \approx 10$ μm.

Blanchard [1963, p. 181], assuming that wet deposition would remove 1.9 times as many SSA particles as does dry deposition over the oceans and that dry deposition to the continents would provide a contribution of 10% of the total deposition over the oceans (both of these factors taken as independent of size), concluded that the production flux of jet drops would be equal to 3.2 times their dry deposition flux. However, the assumptions that the values 1.9 and 10%, and thus the factor of 3.2, are independent of particle size must be considered highly questionable. Additionally, such assumptions violate the assumption that dry deposition and SSA production are in local balance, which is the premise of this method. Moreover, concentrations of sufficiently large SSA particles would be expected to approach their steady state values well before such particles were removed by wet deposition. Consequently, although such arguments may provide rough estimates of the total production of SSA number on a global scale, for example, it would be difficult to justify their application to determine size-dependent quantities such as the size-dependent effective SSA production flux.

The flux estimates of *Blanchard* have been widely quoted by subsequent investigators, although sometimes incorrectly. The time-dependent vertical profile of SSA concentrations over the ocean was modeled by *Gathman* [1982], who fitted *Blanchard's* dry deposition flux estimates to a lognormal distribution whose parameters depend on wind speed, reflecting the increase in the peak of the size distribution of SSA production flux with increasing wind speed. The fitted distributions [Fig. 2 of *Gathman*, 1982] appear to closely match the shape of the dry deposition flux estimates of *Blanchard* [1963, Fig. 17], but the argument in the fit was erroneously taken as the diameter of a particle in equilibrium at 91.4% RH in the computer code in *Gathman* [1982], not the formation diameter. Although *Blanchard* [1963], following *Eriksson* [1959], assumed that the dry deposition velocity of an SSA particle when it reached the sea surface was the same as that of an SSA particle in equilibrium at 91.4% RH, it is apparent from his text that he meant the diameter designating the size of a particle to refer to the formation diameter, which is nearly 60% greater. The magnitudes of the fits to the fluxes presented in *Gathman* [1982] were a factor of ~12.5 lower than the dry deposition flux estimates of *Blanchard*; the factor of 12.5 appears to be restored in a modified version of the *Gathman* fit presented in *Andreas et al.* [1995] and displayed graphically in *Andreas* [1994a] and *Andreas* [1998], but in each of these presentations the argument was still assumed to refer to an SSA particle in equilibrium at 91.4% RH.

There is further confusion in the subsequent literature over the interpretation of the units of the flux estimates presented by *Blanchard* [1963, p. 130 and Fig. 17] as "the number of drops of sea water cm^{-2} s^{-1} per 0.2 log$D$ interval, where D is the drop diameter" (here "log" refers to log$_{10}$). Based on

---

[1] The symbol m, used by *Eriksson* [1959] to denote mass concentration in the present context, was also used in that reference to denote mass flux in reference to the bubble method.

Fig. 18 of *Blanchard* [1963] it can be inferred that the flux in these units is equal to $(dF(r)/d\log r) \times 0.2$ (note that the quantities $dF(r)/d\log r$ and $dF(D)/d\log D$ are equal). Flux estimates in these same units were presented by *Monahan* [1968, Fig. 8], but it appears, based on his conversion of the flux estimates of *Toba* [1961b], that he erroneously interpreted the flux reported by *Blanchard* to be $(dF/d\log r)/0.2$ rather than $(dF/d\log r) \times 0.2$. The flux estimates presented by *Monahan* in these units were subsequently interpreted by *Bortkovskii* [1987, pp. 38-41, Table 1.6] to be equal to $(dF/d\log r) \times 0.2$, which yields a flux that is a factor of 25 greater than what *Monahan* intended. This confusion may also account for part of the factor of 12.5 discrepancy in the formulation of *Gathman* [1982].

The estimates of *Eriksson* [1959] and *Blanchard* [1963] are subject to additional concern as SSA concentrations at low heights were assumed to be equal to those at heights up to hundreds of meters above the sea surface. However, the ratio of SSA concentrations at two heights is expected to depend on particle size and wind speed, as discussed in §2.9.3. For example, at $U_{10} = 10$ m s^{-1} in a neutrally stable atmosphere, the steady state concentration ratio $n(600 \text{ m})/n(10 \text{ m})$ for SSA particles with $r_{80} = (5, 10, 15, 20)$ μm is approximately $(0.9, 0.7, 0.4, 0.2)$, and under these same conditions the concentration ratio $n(600 \text{ m})/n(0.1 \text{ m})$ is approximately $(0.8, 0.4, 0.15, 0.04)$. The estimates of *Eriksson* and *Blanchard* are in fair agreement with one another and with other estimates based on the steady state dry deposition method (Fig 40), but these estimates must be considered mainly of historical interest.

5.2.1.2. Estimates of Toba and coworkers. Estimates of both the effective (10 m) and the interfacial SSA production fluxes were presented by *Toba* [1965a b, 1966], again based on SSA concentration size distributions of *Woodcock* [1953]. Concentrations at 10 m were obtained by *Toba* [1965a] from extrapolation of the measured size-dependent concentrations of *Woodcock* from an assumed height of 700 m based on an exponential decrease with height. The constant characterizing the exponential dependence was determined by the particle size and the eddy diffusivity D_{eddy}, for which a value of 10 m^2 s^{-1}, independent of height and of wind speed, was assumed; according to (2.4-2) this value corresponds to the eddy diffusivity at a height of 70 m for $U_{10} = 10$ m s^{-1}. The dry deposition velocity was taken to be equal to the Stokes velocity (Eq. 2.6-7) at 80% RH (that being a value typical of the marine boundary layer). Thus *Toba* [1965a, Fig. 12; the captions to Figs. 10 and 12 of that reference apparently were switched] obtained estimates for the effective SSA production flux (i.e., that at 10 m, which *Toba* referred to as the top of the boundary layer) for r_{80} from ~1.5 μm to 20 μm at six wind speeds up to 35 m s^{-1}. The estimate shown in Fig. 40 is that for wind force 5, corresponding to a U_{10} range of 8-11 m s^{-1}.

Based on an apparent "trough" in this flux (which *Toba* represented as $dF(m_{dry})/d\log m_{dry} \equiv 1/3 \times dF(r_{80})/d\log r_{80}$) at r_{80} in the vicinity of 4.5-6.5 μm, *Toba* concluded that SSA particles larger than this size were mostly jet drops and that smaller particles were mostly film drops. However, this trough is hardly pronounced. Furthermore, *Toba* argued that only SSA particles with $r_{80} \gtrsim 4$ μm have atmospheric residence times sufficiently small (less than a few days) that their concentrations are controlled by local production; such a conclusion constitutes further early recognition that the steady state dry deposition method cannot be applied to determine the production flux of smaller SSA particles.

Interfacial SSA production fluxes were obtained by *Toba* [1965b] by the further extrapolating these 10 m SSA concentrations to a height of $4.5 \cdot 10^{-4}$ m, which *Toba* assumed would yield values characteristic of those at the sea surface. This procedure was described in §4.6.5; briefly, it was assumed that only gravity and eddy diffusion control the motion of SSA particles in the lowest 10 m over the ocean, and the dependence of particle size (and thus gravitational sedimentation velocity) on height, z, due to the dependence of particle size on RH, which also varied with height, was taken into account (particles were assumed to have instantaneously equilibrated with the local RH). The resultant calculated concentrations exhibited a pronounced minimum near $r_{80} = 6.5$ μm, which (as noted above) *Toba* argued divided the jet drops and film drops. The dry deposition velocity was taken to be the Stokes velocity at 98% RH, and the interfacial SSA production of SSA particles with r_{80} from ~1.5 to 20 μm was obtained [*Toba*, 1965b, Fig. 10] again for six wind speeds up to 35 m s^{-1}. The increase in the interfacial SSA production flux with increasing particle size led *Toba* [1965b] to conclude that bubble production (as would be reflected in bubble concentrations) also increases with increasing R_{bub} up to 0.75 mm, but this conclusion is not supported by reported bubble size distributions of concentration (Fig. 33), even if it were assumed that v_{bub} varies as R_{bub}^2 instead of more slowly (§4.4.2).

A similar approach was attempted by *Chaen* [1973], who measured size distributions of SSA concentration at 6 m above the sea surface during cruises in the Indian Ocean and the East China Sea for several Beaufort wind forces (§2.4); average values of these size distributions are shown in Fig. 22, and all values reported for Beaufort force 5 are shown in Fig. 23. *Chaen* estimated the interfacial SSA production fluxes by extrapolating SSA concentrations to surface values at height z_1 which was related to the significant wave height as discussed in §4.6.5 using the analysis of *Toba* [1965a,b]. *Iida et al.* [1992] later presented estimates of the interfacial SSA production flux for several Beaufort wind forces based on selected values of *Chaen's* concentration measurements extrapolated to a different height also related to the significant wave height (§4.6.5). The concentration measurements upon which these estimates were based are similar to those of other investigations (§4.1.4.2), but as noted above, attempts to determine interfacial SSA production fluxes from effective SSA production fluxes are questionable because of the choice of the height to which concentrations should be extrapolated.

5.2.1.3. Estimates of Fairall and coworkers. An estimate of the effective SSA production flux based on a steady state dry deposition method that incorporated the effects of entrainment of the free troposphere into the marine boundary layer was presented by *Fairall et al.* [1982, 1983]. The investigators referred to the flux derived in this manner as a surface flux, but as it involved concentration measurements at heights of at least several meters, it refers in the present terminology to an effective SSA production flux rather than an interfacial SSA production flux. Size distributions of SSA concentration for r_{80} = 0.8, 2, 5, 10, and 15 μm were measured during the CEWCOM-78 (Cooperative Experiment for West Coast Oceanography and Meteorology) field study (Fig. 22c). The dry deposition velocity was assumed to be $v_d = v_{Stk} + v_{ent}$, where v_{Stk} is the Stokes velocity of a particle at 80% RH and v_{ent} is the entrainment velocity (§2.4) for which a value of 0.35 cm s^{-1} was used, that being an average for the study period (a similar formulation incorporating the effects of entrainment with dry deposition was presented by *Hoppel et al.* [2002]). From these values the effective SSA production flux was obtained, which was stated to be applicable to U_{10} = 9 m s^{-1}, the mean value over the study period (wind speed ranged from 7.4 m s^{-1} to 10.3 m s^{-1}). *Fairall et al.* argued that because the dry deposition velocity exhibits little dependence on wind speed, the ratio of the effective SSA production fluxes at two wind speeds would be directly proportional to the ratio of SSA concentrations at those wind speeds. Based on the CEWCOM-78 concentrations and associated effective SSA production flux at 9 m s^{-1}, concentration measurements reported from the JASIN (Joint Air-Sea Interaction Experiment) field study (also presented in Fig. 22) were used to determine the effective SSA production flux at wind speeds U_{10} of 6, 9, 11, 13, 15, and 18 m s^{-1} for the values of r_{80} listed above (there are some minor discrepancies between the JASIN concentrations listed in Table 2 of *Fairall et al.*, 1983, and those shown in Fig. 6 of that reference).

According to this treatment the number concentrations of larger SSA particles (those with $v_{Stk} \succ v_{ent}$[1]) and those of smaller SSA particles (those with $v_{ent} \prec v_{Stk}$) would be controlled by different mechanisms; the critical radius at which $v_{ent} = v_{Stk}$ is $r_{80} \approx 5$ μm for the above choice of v_{ent}.

Larger SSA particles would be removed primarily by gravitational sedimentation, whereas the concentration of smaller SSA particles would be diminished at a rate independent of particle size, but dependent on characteristics of the air flow (through the dependence of v_{ent} on these characteristics) because of dilution due to increasing marine boundary layer height caused by entrainment (this argument is similar to that put forth by *Moore and Mason* [1954] discussed in 5.2.1.1). The inclusion of v_{ent} in the expression for v_d implies that entrainment equally affects all SSA particles, independently of size, and thus that SSA particles of all sizes are uniformly mixed throughout the marine boundary layer. Although this assumption is questionable for larger SSA particles, it leads to no serious difficulties, as dilution due to entrainment does not appreciably affect v_d for SSA particles sufficiently large that are not uniformly mixed throughout this layer. One concern with this approach is that the implicit assumption that the value of v_{ent} corresponding to each of the measured JASIN concentrations was the same as that for the CEWCOM-78 data (the statement that v_d is nearly independent of wind speed states merely that the v_{ent} does not strongly depend on the local wind speed and does not imply that it remains constant from one time and location to another), but no evidence to support this assumption was presented. There are several other concerns with these estimates. The values for the effective production flux listed in Table 3 of *Fairall et al.* [1983] are not consistent with the concentrations listed in Table 2 of that reference for the CEWCOM-78 data, in part because the expression presented for v_{Stk} (Eq. 24 in *Fairall et al.* [1982]; Eq. 8a and Fig. 5 in *Fairall et al.* [1983]) is incorrect and yields values that are too large by a factor of two (although there are other discrepancies among the results that cannot be accounted for by this error). The error in Stokes velocity would result in an overestimation of the effective SSA production flux by a factor that depends on particle size, varying from near unity for small particles, for which $v_{ent} \succ v_{Stk}$, to near 2 for large particles, for which $v_{ent} \prec v_{Stk}$. The effective SSA production flux based on the JASIN concentrations would also be overestimated by this same size-dependent factor. The estimate for the effective SSA production flux shown in Fig. 40 was calculated as the geometric means of the JASIN concentrations at 9 m s^{-1} and those at 11 m s^{-1} reported in Table 2 of *Fairall et al.* [1983], using v_{ent} = 0.35 cm s^{-1} and the correct values for v_{Stk}.

Effective SSA production fluxes based on CEWCOM-78 and JASIN concentration data were also reported by *Fairall and Larsen* [1984], but these differed from those reported in *Fairall et al.* [1982, 1983], often by more than a factor of three, and the shapes of the spectra differed also. *Fairall and*

---

[1] The symbols $\succ$ and $\prec$ were introduced in §2.1.5.2 to denote "considerably greater than" and "considerably less than", respectively.

Larsen distinguished between the surface source strength (i.e., the interfacial production flux) and the effective source strength (which was referred to as the surface flux in *Fairall et al.* [1982] and *Fairall et al.* [1983]), and presented a relation between the two (§4.6.5) based on the dry deposition model of *Slinn and Slinn* [1980], which they used to determine the interfacial SSA production flux at $U_{10} = 8.5$ m s^{-1} using the CEWCOM-78 data. As noted in §4.6.5 this relation contained an error in the expression for v_{imp}.

Estimates for the interfacial SSA production flux were presented by *Miller* [1987] and *Miller and Fairall* [1988] based on data from four different sources: effective SSA production flux estimates of *Toba* [1965a] (discussed above), averaged SSA concentrations from the JASIN and STREX (Storm Transfer and Response Experiment) field studies, SSA concentrations given by the original Navy Aerosol Model [*Gathman*, 1983a,b] (discussed in §4.1.4.4), and interfacial SSA production flux estimates given by the formulation of *Monahan et al.* [1983b, 1986] (discussed below in §5.3.1). The effective SSA production flux of *Toba* [1965a], obtained from the extrapolation of concentration measurements of *Woodcock* [1953] from ~700 m to 10 m under the assumption of an exponential dependence on height, was transformed by *Miller* to an interfacial production flux using the relation for Ψ_f given by *Fairall and Larsen* [1984] discussed above and in §4.6.5. Effective SSA production fluxes were obtained from the average of the SSA concentrations from the JASIN and STREX field studies using the CEWCOM-78 concentrations and fluxes, as described above, and from the concentrations predicted by the original Navy Aerosol Model; these were also transformed to interfacial production fluxes using the relation of *Fairall and Larsen* [1984]. These four interfacial SSA production fluxes were presented for $r_{80} = 0.8, 2, 5, 10,$ and 15 μm at six values of U_{10}: 6, 9, 11, 13, 15, and 18 m s^{-1} (presumably data of *Toba* were interpolated and/or extrapolated). The flux of *Toba* for $r_{80} = 0.8$ μm and the fluxes of *Monahan et al.* for $r_{80} = 5, 10,$ and 15 μm were excluded because they did not follow the trend of the other data (however, as *Toba* did not present fluxes for $r_{80} < 1.5$ μm and the formulation of *Monahan et al.* is limited to $r_{80} < 8$ μm, §4.3.4.2, the corresponding data in these size ranges should have been omitted for these reasons). The remaining fluxes, which differed by up to an order of magnitude, were arithmetically averaged to obtain what the investigators characterized as a "consensus" interfacial SSA production flux for the particle sizes and wind speeds listed above [*Miller*, 1987, Fig. 2.13 and Table 2.17; *Miller and Fairall*, 1988, Fig. 2]. The investigators noted that this flux exhibits a shape nearly independent of wind speed and has a magnitude with wind speed dependence nearly directly proportional to $[U_{10}/(\text{m s}^{-1}) + 5]^{3.41}$. The values of f_{eff} shown in Fig. 40 are the geometric means of the consensus interfacial

fluxes at 9 m s^{-1} and 11 m s^{-1} after being transformed from an interfacial SSA production flux using the relation of *Fairall and Larsen* [1984], taking into account the errors in that relation noted above. This estimate exhibits a size dependence that differs considerably from most other estimates obtained using the steady state dry deposition method, with values of f_{eff} for SSA particles with $r_{80} = 10$ μm and 15 μm that are roughly an order of magnitude greater than most others.

Several concerns must be raised with this estimate. In the conversion of *Toba's* flux estimates from the representation $dF/d\log m_{dry}$ to dF/dr_{80} a factor of ln(10) was omitted (in Eq. A1 of *Miller* [1987] *log* was apparently interpreted to refer to *ln* instead of log_{10}), leading to an overestimation by *Miller* of these fluxes by this factor. As noted in §4.1.4.4, the original Navy Aerosol Model does not accurately represent SSA size distributions of concentration, especially for $r_{80} \gtrsim 5$ μm, yielding results that are more than an order of magnitude greater than most other reported concentrations (Fig. 22). Additionally, because of the incorrect relation used for the Stokes velocity in the analysis of *Fairall et al.* [1982, 1983] discussed above, the interfacial SSA production fluxes derived from the JASIN/STREX and from the original Navy Aerosol Model are too high by an additional factor of up to 2, depending on particle size. An additional concern with this estimate and that of *Fairall et al.* discussed above, is that it is highly unlikely that concentrations of SSA particles with $r_{80} = 0.8$ μm and 2 μm, two of the five radii for which SSA production fluxes were determined, would have attained steady state conditions; thus application of the dry deposition method to SSA particles of these sizes is questionable.

The consensus flux of *Miller and Fairall* has served as the basis of several reformulations presented by *Andreas* [1990b, 1992] and *Andreas et al.* [1995]. This consensus flux was fitted by *Andreas* [1990b] to an expression equivalent to

$$\frac{f_{int}(r_{80})}{\text{m}^{-2}\text{s}^{-1}} = \left(\frac{U_{10}}{\text{m s}^{-1}}\right)^{2.22}$$
$$\times 10^{(2.4447 - 1.6784L - 2.4581L^2 + 7.7635L^3 - 3.9667L^4)}, \quad (5.2\text{-}1)$$

where $L = \log_{10}(r_{80}/\mu\text{m})$; the expression originally presented in that reference resulted in fluxes a factor of 8 too low because of misinterpretation of the radius convention used by *Miller* [E. Andreas, personal communication, 1999]. This corrected expression, which was indicated to be applicable for 0.8 μm < r_{80} < 15 μm (−0.1 < L < 1.2) and U_{10} < 20 m s^{-1}, matches the results of *Miller* [1987] to within around 30%; given the number of parameters in the fit and the number of significant digits greater agreement might be expected (although such precision would hardly seem warranted in light of the considerable differences in the individual fluxes that were averaged to obtain this consensus flux). The wind-speed dependence was fitted to match that proposed by *Miller* [1987] and *Miller and Fairall* [1988] discussed above; this would result in the shape of the consensus inter-facial SSA production flux being independent of wind speed. The effective SSA production flux obtained from this expression at $U_{10} = 10$ m s^{-1} is shown in Fig. 40, after being transformed from an interfacial production flux using the relation of *Fairall and Larsen* [1984], taking into account the errors in that relation.

An alternative expression for the interfacial SSA produc-tion flux was presented by *Andreas* [1992] in which the logarithm of the production flux was fitted to a 4th order polynomial in $\log_{10} r_{80}$ with coefficients (each given to 4 significant digits) chosen at each wind speed so that the fitted fluxes were equal to the ones tabulated by *Miller* [1987] at each of the sizes and wind speeds for which fluxes were reported. This formulation allows the size dependence of the production flux to vary with wind speed, but it must be interpolated to yield estimates at wind speeds other than specific values. This was presumably the approach taken by *Andreas et al.* [1995], who graphically presented the result-ing interfacial SSA production flux for $U_{10} = 10$ m s^{-1} and compared it with other formulations, concluding that this formulation was the best and was accurate to probably 50%. However, because of the errors in the various estimates used to obtain the consensus interfacial production flux reported in *Miller* [1987] and the large uncertainties in and differ-ences among these diverse estimates (which differed by up to an order of magnitude), such an accuracy would hardly seem justified.

5.2.1.4. Estimates of Smith and coworkers.

A widely used estimate for the effective SSA production flux was pre-sented by *Smith et al.* [1993]. Size distributions of marine aerosol concentrations of particles with r_{amb} in the range 0.09-23.5 µm were obtained from over 700 hours of mea-surements from a 10 m tower located 14 m above mean sea level at South Uist, an island off the coast of Scotland (the site and some of the measurements are described in *Exton et al.* [1985], *Smith et al.* [1989], and *M. Smith et al.* [1991]. Only data from maritime air masses were considered in an effort to eliminate any continental influences, and efforts were made to ensure that SSA production in the surf zone did not affect the results, although the possibility cannot be discounted that whitecap formation, and hence SSA produc-tion, at this site would not be representative of that over the open ocean. The mean RH for the data set was 77%, with hourly averages rarely exceeding 90%, so the results would apply to radii at 80% RH. As most of the measurements were taken under non-precipitating conditions, especially those at low winds, locally occurring wet deposition would have played at most a minor role in controlling SSA con-centrations. During the course of the measurements instances of both very low and extremely high wind speeds and a variety of meteorological conditions were encoun-tered. The averages of the size-dependent measured number concentrations over wind speed intervals of 1 m s^{-1} were assumed to represent steady state values for those wind speeds, and it was argued that such averaging would elimi-nate the need to consider secondary effects such as dilution by entrainment of air into the free troposphere, atmospheric stability, mixing layer heights, and the like, and further that the variability in number concentration would be greater than the variation due to the variability in these other quanti-ties. These concentrations were discussed in §4.1.4.2 and presented in Fig. 22. Dry deposition velocities used to calcu-late effective SSA production fluxes from these size distri-butions of SSA concentration were calculated from a model based on that of *Slinn and Slinn* [1981], as described in §4.6.2.

The resulting effective SSA production flux was presented as the sum of two lognormal functions, equivalent to:

$$\frac{f_{eff}(r_{80})}{\text{m}^{-2}\text{s}^{-1}} = A_1 \exp\left\{-\frac{1}{2}\left[\frac{\ln(r_{80}/2.5\,\mu\text{m})}{\ln 1.49}\right]^2\right\}$$
$$+ A_2 \exp\left\{-\frac{1}{2}\left[\frac{\ln(r_{80}/10.7\,\mu\text{m})}{\ln 1.48}\right]^2\right\}, \quad (5.2\text{-}2)$$

where the wind speed dependence is contained in the two coefficients $A_1 = \exp\{0.156[U_{10}/(\text{ms}^{-1})] + 7.25\}$ and $A_2 = \exp\{2.21[U_{10}/(\text{m s}^{-1})]^{1/2} - 0.28\}$. The effective SSA production flux according to this formulation at $U_{10} = 10$ m s^{-1} is shown in Fig. 40. The products of the aver-age SSA concentrations and the dry deposition velocities are generally within a factor of two of the flux estimates based on this formulation. The investigators suggested that the mode centered at $r_{80} = 2.5$ µm consisted of drops resulting from bubble bursting and that the mode centered at centered at $r_{80} = 10.7$ µm consisted of spume drops. The range of appli-cability of this formulation was stated as 1 µm $\leq r_{80} \leq 25$ µm and $U_{10} < 34$ m s^{-1}, although as noted in §4.1.2.1 the high winds experienced during these measurements were from a single occurrence. Particles with $r_{80} < 1$ µm were not con-sidered because the investigation was concerned with SSA fluxes relevant to moisture and energy exchange at the ocean surface, to which such particles would contribute little. Results for these sizes would also not be valid because of the possible inclusion of non-SSA particles resulting from the use of optical detectors which do not distinguish particle type (this would lead to an overestimate of fluxes of SSA particles) and uncertainties in inference of particle radius (Mie ambiguities) arising from the employment of optical detectors to size the particles, in addition to concerns arising from the extremely long atmospheric residence times of such particles (weeks to months in the absence of precipitation), which would additionally invalidate the results for $r_{80} \lesssim$ 3-5 μm (§4.6.3). *Smith et al.* cautioned against extrapolation of this formulation to r_{80} greater than 25 μm and noted that although the effective SSA production flux appears to be well represented by the sum of two lognormal modes (with a possibility of another mode at larger sizes), the size dependence for particles with $r_{80} > 20$ μm was not well established and measurements of the concentrations of these particles exhibited considerable scatter because of small-number statistics.

The wind-speed dependence of the effective SSA production flux according to this formulation is manifested only in the coefficients A_1 and A_2 (implying that the location of the peaks are assumed to be independent of wind speed), which were fitted, as indicated, to functions of wind speed. If only the data for wind speeds 5-25 m s^{-1} are considered, then the coefficients can be better fitted (in the sense of lower root-mean-squared deviation) by $A_1 = 31[U_{10}/(\text{m s}^{-1})]^{1.8}$ and $A_2 = 0.0058[U_{10}/(\text{m s}^{-1})]^{3.9}$; this is not to suggest a different fit of their coefficients, but merely to demonstrate that these coefficients can be well represented by a power law in the wind speed. There is no marked reduction in the effective production flux at wind speeds below that at which white-capping might typically be expected to commence (~5 m s^{-1} in the open ocean; Fig. 36), and f_{eff} calculated by the expression of *Smith et al.* does not go to zero at zero wind speed; in fact this model yields a production flux for drops with r_{80} as large as 5 μm at $U_{10} = 0$ that is approximately 20% of that at $U_{10} = 10$ m s^{-1}. This of course results from the fact that SSA concentrations do not go to zero at low wind speeds, implying a lack of steady state conditions and thereby establishing the failure of local and temporal balance between removal through dry deposition and production. The ratio of the magnitude of the second mode to that of the first increases with increasing wind speed from 0.034 at $U_{10} = $ 5 m s^{-1} to near 0.25 at $U_{10} = 15$ m s^{-1}, and over this same wind speed range the value of r_{80} at which the contribution from the second mode exceeds that of the first decreases from 7.5 μm to 6 μm.

A key concern with the formulation of *Smith et al.* is that concentrations of SSA particles with $r_{80} \lesssim 3$-5 μm would not have attained their steady state values (with respect to dry deposition); this concern would not be alleviated even by averaging over long times and different meteorological conditions. Concentrations of SSA particles of these sizes would be below their steady state values, as the characteristic times to attain these values are much longer than the mean residence times of these particles against removal by

wet deposition, resulting in underestimation of SSA concentrations and of SSA production fluxes in this size range. Indeed this flux estimate exhibits a rapid decrease with decreasing r_{80} below ~3 μm. Additionally, as the mean times such particles have spent in the atmosphere would have been quite long, the assumption of local balance, that is, that the local wind speed characterizes that under which these particles were formed, is unlikely to have been satisfied. An additional concern with this estimate is that it may be specific to location. Although surf zone influences appear to have been eliminated, measured SSA mass concentrations at this location, even at the high end of the range, where the method is applicable, were greater than most other reported values (§4.1.2.1; Fig. 22) by a factor of several fold.

A modification of this formulation was proposed by *Andreas* [1998], who concluded that the expression of *Smith et al.* [1993], when multiplied by a factor of 3.5 (as discussed presently), provides the most accurate representation of the production flux of SSA particles with r_{80} up to 10 μm, (above which he proposed a different representation, discussed below); the expression of *Andreas* is also shown in Fig. 40 for $U_{10} = 10$ m s^{-1}. *Andreas* converted the wind speeds from 14 m, the height above mean sea level of the tower from which *Smith et al.* measured concentrations, to 10 m values, but this results in so little change that it hardly seems necessary in light of other uncertainties and of the discussion in §2.3.1 on wind speed determinations (for example, $U_{10} = 10.0$ m s^{-1} corresponds to $U_{14} \approx 10.3$ m s^{-1}; §2.4). Although *Andreas* did not distinguish between the interfacial and effective SSA production fluxes, this estimate was clearly meant to refer to f_{eff}. *Andreas* based his proposed formulation on the contention that the formulation of *Monahan et al.* [1983b, 1986] (§4.3.4.2; §5.3.1) was the most accurate for the production flux of bubble-produced drops and that the formulation of *Smith et al.* [1993] exhibited a similar size dependence to that of *Monahan et al.*, but resulted in a volume flux over the range $r_{80} = 2$-7.5 μm that was roughly 3.5 times lower for $U_{10} = 5$–20 m s^{-1}. Several reasons were suggested to explain the factor of 3.5 discrepancy between the two formulations. As the concentrations of *Smith et al.* [1993] were measured at a tower at an appreciable distance from the tide line (from 10 m to 270 m according to *Exton et al.* [1985]), two effects could have contributed to a bias in these measurements, resulting in an underestimation of the production: evaporation and gravitational settling. According to *Andreas*, the size dependence of the time required for the change of size due to evaporation would explain the smaller fluxes resulting from the *Smith et al.* formulation compared to those of *Monahan et al.* for $r_{80} < 2$ μm; the geometric radius intervals containing the smallest particles would contain a reduced number of particles because particles initially in those intervals would have shifted into smaller ones, but these intervals would not have been replenished by larger particles which require longer times for evaporation and which would not have attained their equilibrium radii before reaching the detectors. Additionally, *Andreas* argued, the greater gravitational settling velocities of larger particles compared to smaller ones, resulting in larger particles being preferentially removed before reaching the sampling location, would explain why the flux of the *Smith et al.* formulation decreases with increasing radius for particles with $r_{80} > 15$ μm instead of increasing like several other formulations (the formulation of *Smith et al.* has a peak near $r_{80} = 15$ μm when presented as a flux of SSA volume in the representation dF^V/dr_{80}, which was the representation considered by *Andreas*).

Neither of these arguments holds up under scrutiny, as noted also by *Reid et al.* [2001]. The argument concerning the failure of some particles to attain their equilibrium radii can in no way account for the difference of nearly two orders of magnitude in the flux estimates of *Monahan et al.* and *Smith et al.* at $r_{80} = 1$ μm and is wholly inconsistent with the characteristic times presented above (Eq. 2.5-6; §2.5.4) or with those in *Andreas* [1989, 1990a] for drops to achieve their equilibrium radii—less than one minute for drops as large as $r_{80} = 25$ μm. As the drops measured by *Smith et al.* required times of seconds to tens of seconds to reach the measuring devices from the shoreline, and as most drops were probably produced at some distance from the shoreline, it can confidently be concluded that the vast majority of these drops would have attained radii within a few percent of their equilibrium values by the time of measurement. If it were a question of drops not attaining their equilibrium radii, then evaluation of the flux estimate of *Smith et al.* at radii that are reduced by a factor of not greater than two (the maximum factor under the assumption that SSA particles experience no change in their radii) should cause the discrepancy to disappear; it does not. Additionally, as noted by *Reid et al.* [2001], this argument does not explain the difference in the number of particles produced according to the two estimates. Likewise, the argument on gravitational settling does not account for any appreciable difference in the concentrations of SSA particles with $r_{80} \lesssim 25$ μm, which have gravitational sedimentation velocities less than 10 cm s^{-1} when in equilibrium at 80% RH (§2.6.2), as during their transit time to the measuring device from the shoreline, of order tens of seconds or less, such drops will fall less than a few meters.

An additional concern is the arbitrary scaling upward by the factor of 3.5. In addition to such scaling being unjustified in general, in the present instance, although not noted by *Andreas*, it is an attempt to relate an effective SSA production flux to an interfacial SSA production flux, and, as discussed in numerous places throughout this review, the relation between these fluxes is expected to depend strongly

on particle size (and also on wind speed), and it is thus highly unlikely that a constant factor (for example, 3.5), would adequately describe this relation over any appreciable size range. The similarity in shape between the flux estimates of *Monahan et al.* and *Smith et al.* in the range $r_{80} = 2$-7.5 μm resulted from both having a peak in this range when presented as volume fluxes in the form $dF^V(r_{80})/dr_{80}$; that of *Monahan et al.* near 2.6 μm and that of *Smith et al.* near 3.4 μm, but these peaks have very different widths, and comparison of (4.3-5) and (5.2-2) demonstrates that the two expressions are quite different when expressed in terms of number fluxes represented as $dF(r_{80})/d\log r_{80}$. Additionally, as noted above, the flux estimate of *Smith et al.* is not likely to be accurate for $r_{80} \lesssim 3$-5 μm because the conditions required for the dry deposition method would not have been satisfied.

Another expression for the 10 m production flux of SSA particles was proposed by *Smith and Harrison* [1998] based on concentrations of SSA particles with radii from 1 μm to 150 μm (presumably the radius variable refers to the geometric radius, not the equilibrium radii) measured at wind speeds up to about 20 m s^{-1} during a cruise in the eastern North Atlantic. No data were reported, and no discussion was presented of the number of data or of the methods by which they were obtained, other than that optical particle counters were used. The model used for dry deposition velocities was not stated, but was presumably the same as that used by *Smith et al.* [1993]. The flux estimated from the expression of *Smith and Harrison* was also be expressed as the sum of two lognormals in terms of r_{80}:

$$\frac{f_{\text{eff}}(r_{80})}{\text{m}^{-2}\text{s}^{-1}} = A_1 \exp\left\{-\frac{1}{2}\left[\frac{\ln(r_{80}/4.2\ \mu\text{m})}{\ln 1.78}\right]^2\right\} + A_2 \exp\left\{-\frac{1}{2}\left[\frac{\ln(r_{80}/50\ \mu\text{m})}{\ln 2.03}\right]^2\right\},$$

where $A_1 = 1.7[U_{10}/(\text{m s}^{-1})]^{3.5}$, and $A_2 = 0.6[U_{10}/(\text{m s}^{-1})]^3$; an estimate based on this formulation is shown in Fig. 40 for $U_{10} = 10$ m s^{-1}. The maxima of the modes occur at values of r_{80} well greater than those of the formulation of *Smith et al.* [1993], and at a given wind speed the size dependencies of the two formulations differ considerably. At $U_{10} = 10$ m s^{-1} the ratio of the value of f_{eff} for $r_{80} = (1, 5, 10, 15, 25)$ μm given by this formulation to that given by *Smith et al.* (Eq. 5.2-2) is approximately (0.4, 3, 2, 1, 5).

It is difficult to place credence in this formulation for several reasons, not the least of which is the lack of presentation of the details of the measurements, the attendant conditions, and the data. Issues associated with

RH equilibration were not addressed, but as SSA particles with $r_{80} \gtrsim 25$ µm would have retained radii near their formation values (§2.9.1), an abrupt decrease in the effective SSA production flux at 10 m (i.e., f_{eff}) would almost certainly be expected because of the strong decrease in Ψ_f at r_{80} near 25 µm; no such decrease is evident in the formulation. Further, concentrations of SSA particles with $r_{80} \gtrsim 25$ µm would be expected to be extremely low at 10 m, resulting in large uncertainties because of poor sampling statistics (§4.1.4.3). The investigators stated that based on vertical profiles of aerosol particle loadings (which were not presented) the fluxes at the 10 m level were estimated to be between 0.25 and 0.8 times those near the surface (size range and lower height not specified). This factor, essentially equal to the quantity Ψ_f, would be expected to be strongly dependent on particle size (§2.1.6; Fig. 13), and, for SSA particles with $r_{80} > 25$ µm, it would be exceedingly low.

5.2.1.5. Estimate based on the canonical SSA size distribution of concentration. The steady state dry deposition method can be applied to any size distribution of SSA concentration that is assumed to represent steady state conditions. *Hoppel et al.* [2002], for instance, used the so-called film drop mode and jet drop mode of the formulation for the size-dependent SSA concentration proposed by *O'Dowd et al.* [1997] discussed in §4.1.4.4, under the assumption that dry deposition acts only by gravitational sedimentation, to arrive at an estimate for effective SSA production flux at 10 m, and by extrapolation of the SSA concentrations to 1 m under the assumption of steady state conditions (as discussed in §4.6.5), the effective SSA production flux at 1 m. Effects of entrainment of the free troposphere were also considered. Additionally, these investigators, noting that concentrations of small SSA particles would not have attained steady state values at the time of measurement, assumed that the lifetimes of such particles in the marine atmosphere would be limited by wet deposition, for which a value of $\tau_{wet} = 3$ d was assumed (*cf.* §2.7.1). By presenting an expression to determine the effective SSA production flux from size distributions of SSA concentration that applies to both larger SSA particles (those with r_{80} greater than ~ 2 µm) and smaller SSA particles, these investigators essentially combined the steady state dry deposition method and the statistical wet deposition method (§3.9; §5.10).

The effective SSA production flux f_{eff} could similarly be calculated from other measurements of SSA concentration, but here it is calculated from the canonical SSA size distribution of concentration presented in §4.1.4.4 and the dry deposition velocity given by (4.6-8) as the transfer velocity through the surface layer, $v_{turb} + v_{S,Stk}$ (§4.6.3). Such an approach also allows estimation of the uncertainty in the

estimate of f_{eff}. This estimate is shown in Fig. 40 for $U_{10} = 10$ m s^{-1} over the r_{80} range 3-25 µm, along with the uncertainty of a factor of $\lesssim 4$ shown as a gray shaded region. This estimate of uncertainty results from an uncertainty of a factor of $\lesssim 3$ associated with the canonical SSA size distribution of concentration (§4.1.4.4) and an uncertainty of a factor of $\lesssim 2$ associated with the dry deposition velocity used (§4.6.3). Also shown are the values of f_{eff} for $U_{10} = 5$ m s^{-1} and $U_{10} = 15$ m s^{-1}, without associated uncertainties.

This estimate of $f_{eff}(r_{80})$ with its associated uncertainty encompasses the majority of previous estimates based on the steady state dry deposition method within the range of validity, despite the simplicity of the estimate and concerns noted with these other estimates. A strength of this estimate of the effective SSA production flux is that it is based on the canonical SSA size distribution of concentration, which is obtained from large numbers of measured data sets, instead of being based on a limited number of data sets, typically taken at a single location and under similar meteorological and environmental conditions (§4.1.4.4), and as such is relatively insensitive to secondary factors such as sea temperature. The accompanying uncertainty of a factor of approximately $\lesssim 4$ indicates the accuracy with which f_{eff} can be specified, implying that there is little to be gained by arguing over details of the size dependence or the magnitude of f_{eff} that are contained within this uncertainty.

According to this formulation the size dependence of f_{eff} is nearly proportional to $r_{80}^{-2.5}$ at $U_{10} = 10$ m s^{-1}, and this slope (i.e., the exponent 2.5) is almost independent of wind speed over the range $U_{10} = 5\text{-}20$ m s^{-1}; the size dependence of the canonical SSA size distribution is independent of wind speed (§4.1.4.4), and the size-dependent transfer velocity through the surface layer (Fig. 38) depends only slightly on wind speed. An approximation for $f_{eff}(r_{80})$ that closely agrees with the one obtained in this manner (within ~20% over most of this range of U_{10} and r_{80}) is given by

$$\frac{f_{eff}(r_{80})}{\text{m}^{-2}\text{s}^{-1}} \approx 800 \left(\frac{U_{10}}{\text{m s}^{-1}}\right)^{2.5} \left(\frac{r_{80}}{\text{µm}}\right)^{-2.5}; \quad (5.2\text{-}3)$$

this expression at $U_{10} = 10$ m s^{-1} is also shown in Fig. 40.

5.2.2. Discussion

The steady state dry deposition method has been used by several investigators to infer the effective SSA production flux. The method is easy to implement, requiring as inputs only measured size-dependent SSA concentrations (§4.1.4) and modeled dry deposition velocities (§4.6). The several conditions required for its successful implementation were specified §3.1 in terms of inequalities relating characteristic times for atmospheric processes and the characteristic time for SSA particles of a given r_{80} against removal through dry deposition, $\tau_{dry} = H_{mix}/v_d$, where H_{mix} is the characteristic mixing height for SSA particles of this size (§2.9.6). These conditions are that entrainment of the free troposphere be negligible (this condition can be relaxed if entrainment is accounted for, as it was by *Fairall et al.* [1982, 1983] and *Hoppel et al.* [2002]), that meteorological conditions remain stationary over the mean atmospheric residence times of SSA particles of a given size against dry deposition (and hence that dry deposition and not wet deposition be the dominant factor governing removal of SSA particles of this size), that there be no appreciable horizontal gradients, and that wet deposition not occur during the measurement period. These conditions ensure that the production flux of SSA particles of a given size is balanced by the dry deposition flux of these particles at the time and location of measurement, and thus that local conditions, which govern the modeled dry deposition velocities, are similar to those under which the particles were produced. The satisfaction of these conditions was examined in §3.1 and in §4.6.2, allowing evaluation of the applicability of the steady state dry deposition method to determine the effective production flux of SSA particles of a given size.

Based on these considerations the size range of SSA particles for which the steady state dry deposition method can be accurately applied was determined to be $r_{80} \gtrsim 3\text{-}5 \ \mu\text{m}$ under most conditions; similar conclusions have been reached by other investigators [e.g., *Blanchard*, 1963, p. 132; *Toba*, 1965a; *Hoppel et al.*, 2002]. Smaller SSA particles, which are expected to be mixed throughout the marine boundary layer (Fig. 12), have such low estimated dry deposition velocities (Fig. 38) and such large corresponding values of τ_{dry} (nearly a day or more), that rarely if ever would the conditions necessary for application of this method be satisfied. Dry deposition is not expected to play a major role in the removal of such small particles, and as the meteorological conditions under which these particles are produced would not be expected to correlate with those under which concentrations are measured, and under which the particles are removed by dry deposition, the assumption of local balance between production and removal through dry deposition would rarely be justified. Not only is it unlikely that conditions experienced by particles would remain constant over such a time, but additionally it is highly unlikely that conditions such as wind speed would have remained constant over such a great distance, necessary for current local conditions to be similar to those responsible for the production of the particles.

The question of whether the requirement of local balance between SSA production and dry deposition is fulfilled may to some extent be assessed by consideration of the dependence of SSA number concentrations on current local conditions, specifically wind speed. If local balance existed, then the size-dependent SSA number concentration, being the ratio (Eq. 2.7-2) of the size-dependent SSA production flux and the size-dependent dry deposition velocity, both of which are expected to be determined by current local conditions, would also be determined solely by current local conditions and would therefore be expected to exhibit a strong dependence on the wind speed at the time and location of measurement. The fact that SSA concentrations generally exhibit only weak dependence on local conditions (e.g., §4.1.3) suggests that this local balance requirement is not necessarily satisfied, especially for smaller particles.

Several formulations of mechanisms that affect the dry deposition of particles of these sizes have been proposed, as have various modifications to the model of *Slinn and Slinn* [1980] to include effects such as drop scavenging, diffusiophoresis, and entrainment (§4.6.3), but in those few situations for which the resultant reductions in estimated lifetimes are substantial, the importance of such effects, and their appropriate formulations are unknown, and the uncertainties so large that little confidence can be placed in the results. Experimental verification of local balance from evaporation rates is hardly possible; as noted earlier (§1.1), estimates for SSA production correspond to a daily removal of a layer of water from the ocean surface of order micrometers thick—far too small a difference to be detected. Based on these considerations, local balance can rarely at best be considered to be satisfied for a wide range of particle sizes. Hence, SSA production fluxes based on this assumption remain subject to question, especially at small values of r_{80}.

Sea salt aerosol particles with $r_{80} \gtrsim 5 \ \mu\text{m}$ have estimated atmospheric residence times of less than a day under most conditions (Fig. 38; Table 8), decreasing with increasing particle size because of both increasing dry deposition velocity and decreasing mixing heights for particles with $r_{80} \gtrsim 10\text{-}15 \ \mu\text{m}$. Thus the conditions for the steady state dry deposition method are often expected to be satisfied for such particles. Dry deposition is expected to be the dominant removal mechanism for these particles, and steady state conditions are increasingly likely to apply, meaning that local conditions and those under which the particles were produced are expected to be similar. These particles are expected to be nearly uniformly mixed to at least the height of measurement near 10 m and to have achieved their equilibrium radii with respect to the RH of the surface layer at the time of measurement. Because of the restriction of the steady state dry deposition method to SSA particles with $r_{80} \gtrsim 3\text{-}5 \ \mu\text{m}$ several of the concerns with this method become unimportant. These include the extent to which the marine boundary layer can be represented by a two (or three) layer model, lack of understanding of details of the viscous boundary layer near the ocean surface, uncertainties in various transfer velocities, and uncertainties in effects such as spray scavenging and transfer to a broken surface, which limit the amount of confidence that can be placed in the results of models of the dry deposition velocity for smaller SSA particles. The dry deposition velocity of SSA particles in this size range is controlled almost entirely by the transfer through the turbulent surface layer, and the viscous sublayer provides little resistance to their motion; thus with increasing particle size for $r_{80} \gtrsim 5$ μm the approximation given by (4.6-8) for the dry deposition velocity as the velocity through the surface layer is expected to be increasingly accurate.

Application of the steady state dry deposition method using averages of measured SSA concentration measurements taken over a variety of conditions, even when classified by wind speed, implies the assumption that this average represents a steady state value. This assumption has sometimes been stated explicitly [*Toba*, 1965a; *Fairall and Larsen*, 1984; *Smith et al.*, 1993]. For instance, *Fairall and Larsen* [1984] stated that "appropriate ensemble averages of many states of the system" would be reasonably represented by steady state conditions; in other words, the assumption is made that the production flux inferred from dry deposition velocities and measured concentrations will be meaningful in some average sense, even though it might not accurately represent the local production, allowing the steady state dry deposition method to be applied even when the conditions enumerated above are not satisfied. *Smith et al.* [1993] suggested that averaging removes the need to consider secondary effects such as entrainment, and concluded that these effects would be less than the uncertainties in the results. A similar assumption was implicitly made in calculating the estimate of f_{eff} from the canonical SSA size distribution of concentration. If the primary reason a measured SSA concentration would not represent a steady state value is removal by wet deposition, then measurements at different times subsequent to the previous precipitation event of appreciable intensity would yield SSA concentrations lower than their steady state values. In turn, the product $v_{term}(r_{80}) \times n(10$ m$)$ would underestimate the effective SSA production flux, and provide a lower bound for this quantity. The extent to which such estimates underestimate f_{eff} would depend on particle size; for instance, concentrations of SSA particles with $r_{80} \gtrsim 15$ μm, which have expected atmospheric residence times of ~1.5 h or less, are expected to have attained their steady state values for most measurements not taken during precipitation. The extent to which non-steady state conditions are due to precipitation events, as opposed to changes in wind speed, for example, is difficult to determine, but similar arguments would apply to these

situations as well, albeit with lesser consequences for SSA concentrations and hence for calculated SSA production fluxes.

Uncertainties in inferred SSA production fluxes result from uncertainties in measured SSA concentrations and uncertainties in dry deposition velocities. As the canonical SSA size distribution was concluded to represent SSA size distributions of concentration as a function of wind speed as accurately as currently possible, use of this distribution contributes an uncertainty in estimates of f_{eff} by this method of $\lessapprox 3$, and perhaps a bit larger at the very upper end of the size range for which this method can be applied (i.e., $r_{80} \approx 25$ μm). Uncertainties in the dry deposition velocity used was estimated to be $\lessapprox 2$. Thus the factor of $\lessapprox 4$ can be taken as the overall uncertainty of this method. The range of wind speeds for which this method can accurately be applied is bounded below by $U_{10} \approx 5$ m s^{-1}, as little whitecapping and hence little production occurs at these wind speeds (Fig. 36). There are few data at high wind speeds (i.e., $U_{10} \gtrsim 20$ m s^{-1}), and the extent to which conditions at these wind speeds can be considered to represent steady state, or to which any such wind speed occurrence is representative of other such occurrences, is questionable. Additionally, the ocean surface is expected to be far from calm under these circumstances, and effects such as spray scavenging that are thought to be negligible in most situations might come into play.

5.3. WHITECAP METHOD

The whitecap method (§3.2) calculates the size-dependent interfacial production flux of bubble-produced SSA particles from measurements of the size-dependent SSA production flux over whitecap area $f_{wc}(r_{80})$ and measurements of the oceanic whitecap ratio $W(U_{10})$ as $f_{int}(r_{80}, U_{10}) = f_{wc}(r_{80})W(U_{10})$. Intrinsic to this method is the assumption that the shape of the size dependence of the interfacial production flux of bubble-produced drops is independent of wind speed, and that the effect of wind speed, which is contained only in the quantity W, is manifested only in the magnitude of the SSA production flux. Measurements of $f_{wc}(r_{80})$ from laboratory simulated whitecaps and from surf zones were discussed in §4.3.4.2 and §4.3.5, respectively, and presented in Fig. 32. Measurements of W were discussed in §4.5 and presented in Fig. 36. Assumptions specific to the different approaches of the whitecap method, which involve different types of white area in the laboratory or surf zone, were examined in §3.2, the key assumption being that in some sense all white areas are equally productive, and hence that laboratory or surf zone measurements of $f_{wc}(r_{80})$ can be appropriately extrapolated to the ocean by multiplication by W. As noted in §3.2, this method does not provide the contribution from spume drops.

5.3.1. Estimates of the Interfacial SSA Production Flux

Estimates for the interfacial SSA production flux $f_{int}(r_{80})$ based on the whitecap method are examined in this section. These estimates can be categorized into those obtained from determinations of $f_{wc}(r_{80})$ based on continuous laboratory whitecaps (§4.3.4.1), those based on discrete laboratory whitecaps (§4.3.4.2), and those based on surf zone measurements (§4.3.5). Several of these estimates of $f_{int}(r_{80})$ are presented in Fig. 32 (right axis) for an assumed whitecap ratio $W = 0.01$, chosen to represent the value of W at $U_{10} = 10$ m s^{-1}. Values of W at this wind speed evaluated from (4.5-1)—the expression presented by *Wu* [1979b], (4.5-2)—the expression presented by *Monahan and O'Muirchaertaigh* [1980], and (4.5-4)—the central value of the shaded gray region in Fig. 36, are 0.0096, 0.0099, and 0.012, respectively; thus these estimates can be directly compared with estimates of f_{int} presented below based on other methods. While the uncertainty in f_{wc} is reflected in the different determinations shown in Fig. 32 and in the width of the shaded gray region shown, it should be recognized that the uncertainty in f_{int}, which additionally contains uncertainty from measurements of W, is substantially greater.

Several investigations determined $f_{wc}(r_{80})$ from measurements of SSA production from continuous laboratory whitecaps, as discussed in §4.3.4.1. Using a continuous laboratory whitecap of area 0.02 m^2 formed by a waterfall in a tank of depth 0.27 m, *Cipriano and Blanchard* [1981] measured the size-dependent production flux over whitecap area of SSA particles with r_{80} from 0.3 μm to 14 μm in 5 size ranges and also the total production flux over whitecap area (including SSA particles with r_{80} as small as ~0.014 μm), from which they determined that approximately half of the drops had r_{80} less than 0.3 μm. *Cipriano et al.* [1983], also using a continuous whitecap of the same dimensions, measured the SSA production flux over whitecap area using several types of instruments and concluded that the vast majority of the drops produced had $r_{80} \lesssim 0.1$ μm. However, the results exhibited large uncertainties and in several instances unphysical results, for example, cloud condensation nuclei (CCN) production appeared to decrease with decreasing critical supersaturation (*cf.* §2.1.5.1). In using their results to estimate the oceanic production of total SSA particle number, *Cipriano and Blanchard* [1981] assumed 0.03 for a typical whitecap ratio. This value was questioned by *Monahan* [1982], who suggested that 0.01 was more accurate; this estimate was accepted by *Cipriano and Blanchard* [1982]. *Cipriano et al.* [1983] also assumed a global average whitecap ratio of 0.01 to compare the total production flux of SSA particles with measurements of the marine aerosol number concentration, using an approach similar to the statistical

wet deposition method (§3.9). No parameterization of f_{wc} was proposed by these investigators. Estimates of $f_{int}(r_{80})$ based on these measurements calculated using $W = 0.01$ are presented in Fig. 32 (right axis). As in these experiments the white area resulted from a continuously recharging bubble plume, which would be similar to that occurring during the initial stages of whitecap formation, use of this value of the whitecap ratio may overestimate oceanic values of f_{int}; this is discussed further in §5.3.2.

Measurements of SSA production from a continuous laboratory whitecap of area $3 \cdot 10^{-4}$ m^2 produced by forcing air through a frit 0.04 m below the water surface were reported by *Mårtensson et al.* [2003], who presented a formulation of $f_{wc}(r_{80})$ for r_{80} from 0.02 μm to 2.8 μm consisting of different expressions in each of three size ranges, each containing an explicit temperature dependence. For evaluation of f_{int} the wind speed dependence of W given by (4.5-2) was used, the same as that used by *Monahan et al.* and by *Woolf et al.* discussed below. This estimate is also shown in Fig. 32 (right axis). The SSA particles detected in this experiment would appear to be jet drops, as the sizes of bubbles reported were smaller than those thought to produced film drops (§4.3.3); it was suggested in §4.3.4.1 that these jet drops were entrained upward by the air flow through the system and thus would not be expected to occur under oceanic conditions.

Several parameterizations for $f_{int}(r_{80})$ have been proposed by *Monahan* and coworkers based on measurements of SSA production from discrete laboratory whitecaps discussed in §4.3.4.2; as noted there, these estimates refer to jet drop production only. An expression for $f_{int}(r_{80})$ stated to be applicable for $r_{80} = 0.8\text{-}8$ μm was proposed by *Monahan et al.* [1983b, 1986]. This formulation uses values of $f_{wc}(r_{80})$ given by (4.3-4), which was based on measurements by *Monahan et al.* [1982] of the SSA production from a discrete laboratory whitecap with initial white area 0.35 m^2, and values of $W(U_{10})$ given by the parameterization of *Monahan and O'Muirchaertaigh* [1980], (4.5-2), which was based on measurements of *Monahan* [1971] and *Toba and Chaen* [1973] (§4.5.1). This expression is equivalent to

$$\frac{f_{int}(r_{80})}{\text{m}^{-2}\text{s}^{-1}} = 3.15 \left(\frac{U_{10}}{\text{m s}^{-1}} \right)^{3.41}$$
$$\times \left(\frac{r_{80}}{\mu\text{m}} \right)^{-2} \left[1 + 0.057 \left(\frac{r_{80}}{\mu\text{m}} \right)^{1.05} \right]$$
$$\times 10^{1.19\exp\left\{ -2.4\left[0.38 - \log_{10}\left(\frac{r_{80}}{\mu\text{m}} \right) \right]^2 \right\}}. \quad (5.3\text{-}1)$$

and is shown for $U_{10} = 10$ m s^{-1} in Fig. 32 (right axis).

Several variants of the *Monahan et al.* [1983b, 1986] formulation have been proposed subsequently by *Monahan* and coworkers. A similar formulation based on the same data was proposed by *Spillane et al.* [1986] (not shown) which employed a different term to describe the effects of the vertical concentration profile (differing from that of *Monahan et al.* by less than 10%) and a different formulation for W, resulting in values of f_{int} that are twice as high as those of *Monahan et al.* at $U_{10} = 5$ m s^{-1} and around 15% less at $U_{10} = 15$ m s^{-1}. An expression stated to be applicable for $r_{80} = 0.8$-10 μm was proposed by *Woolf et al.* [1988] based on experimental results (which were not reported) from the same laboratory, and is given by the product of $f_{wc}(r_{80})$ given by (4.3-6) and the same value for $W(U_{10})$ as that used in the formulation of *Monahan et al.* [1983b, 1986]. This expression, equivalent to

$$\frac{f_{int}(r_{80})}{\text{m}^{-2}\text{s}^{-1}} = \left(\frac{U_{10}}{\text{m s}^{-1}}\right)^{3.41} 10^{(1.37-0.49L-1.08L^2+0.527L^3)}, \quad (5.3\text{-}2)$$

where $L = \log_{10}(r_{80}/\mu\text{m})$, is also shown in Fig. 32 (right axis) for $U_{10} = 10$ m s^{-1}. This expression yields values of f_{int} that are essentially the same as those of *Monahan et al.* at $r_{80} = 0.8$ μm but nearly an order of magnitude greater at $r_{80} = 8$ μm. A similar expression proposed by *Monahan and Woolf* [1988] (not shown) employed a different value of the characteristic whitecap decay time τ_{wc}, 4.27 s instead of 3.53 s, resulting in values of f_{int} that are lower than those of *Woolf et al.* [1988] by about 20%. A formulation proposed by *Stramska et al.* [1990] (not shown), which incorporated the effects of gas saturation, employed the same expression for $f_{wc}(r_{80})$ and used an expression for $W(U_{10}, \Delta T)$, where ΔT is the sea temperature minus the air temperature, from *Monahan and O'Muircheartaigh* [1986] which for $\Delta T = 0$ varies from approximately 30% greater than that given by (4.5-2) at $U_{10} = 5$ m s^{-1} to 50% less at $U_{10} = 15$ m s^{-1}.

Using a similar experimental arrangement in the same laboratory, *Cipriano et al.* [1987] reported an individual value of f_{wc} for each of two size ranges: $r_{80} = 0.01$-0.036 μm and $r_{80} = 0.036$-1 μm; these values are shown in Fig. 32 (right axis) for $W = 0.01$. In evaluating f_{int}, these investigators assumed a whitecap ratio given by (4.5-2), the same as that used for the formulations of *Monahan et al.* and *Woolf et al.* However, as SSA particles in this size range would be film drops produced by large bubbles (§4.3.7), which rise rapidly to the surface (§4.4.2), the active time for which these bubbles were bursting at the surface would be expected to be considerably less than that characterizing jet drop production, and it would seem that the flux of these particles based on assumption of constant SSA production per unit white cap area might lead to substantial overestimation; similar

considerations were noted above for the estimate of *Cipriano and Blanchard* [1983] based on the continuous laboratory whitecaps.

Several investigations determined f_{wc} based on measurements of the SSA production from an active surf zone, also shown in Fig. 32 (right axis) as f_{int} for an assumed $W = 0.01$. Values of $f_{wc}(r_{80})$ for r_{80} from 1-16 μm were reported by *Woodcock et al.* [1963]; as seen in Fig. 32, these measurements were roughly a factor of 5 higher than most other determinations. The total production flux over white area of SSA particle number for $r_{80} > 0.2$ μm was reported by *Blanchard* [1969]; these estimates were likewise up to an order of magnitude greater than estimates based on laboratory determinations. Values of $f_{wc}(r_{80})$ for r_{80} from 0.4-5 μm based on measurements at two coastal sites were reported by *de Leeuw et al.* [2000], who proposed a formulation for this quantity and its dependence on wind speed.

5.3.2. Discussion

The whitecap method provides a relatively straightforward determination of the interfacial SSA production flux. Additionally, in principle this method allows the processes responsible for SSA production—wave breaking, bubble entrainment, and bubble bursting—to be investigated in the laboratory under controlled conditions. However, the few formulations of f_{int} that have been based on this method range over several orders of magnitude, and there are concerns both with the practical implementation of this method (with regard to the determination both of $f_{wc}(r_{80})$ based on laboratory or surf zone measurements and of W based on field measurements) and with the assumptions upon which it is based. These concerns are examined here.

Issues regarding the extrapolation of laboratory results of SSA production to the open ocean were discussed in §4.3.4.4. Key among these are the large differences in sizes of laboratory and open ocean whitecaps, the extent to which the process of wave breaking in the laboratory accurately simulates that of oceanic waves, and the accuracy with which laboratory experiments can measure the SSA particles produced. As oceanic whitecaps are several orders of magnitude larger and have associated bubble plumes that extend much deeper than those studied thus far in the laboratory (§4.5.4), the extent to which results from such laboratory experiments of SSA production can be applied to the ocean is questionable. Turbulence under a breaking wave and the entrainment of bubbles downward in shallow tanks are very different from the open ocean, where bubble plumes can attain depths of several meters to more than 10 m (§4.4), again one to several orders of magnitude greater than the depths of the laboratory tanks in which simulated whitecaps were produced. It is unlikely that SSA production would be independent of such differences in scale. Additionally, in the laboratory bubbles would rise to the surface much more rapidly than in the ocean, and the differences in sizes of the breaking waves might result in differences in bubble size distributions and hence differences in SSA production.

The extent to which the process of wave breaking in the laboratory accurately simulates that in the ocean has also been questioned (§4.3.4.4). It is an implicit assumption of this method that the laboratory wave breaking process is sufficiently similar to that of oceanic waves that the resulting bubble size distribution, and hence the size distribution of SSA drops produced, is similar. The extent to which this assumption is satisfied is not known. However, while it can be argued that discrete laboratory whitecaps employed by *Monahan* and coworkers (§4.3.4.2) might to some extent represent oceanic whitecaps and could yield results pertinent to jet drop production in the open ocean, and that the continuous laboratory whitecaps employed by *Cipriano and Blanchard* (§4.3.4.1) might represent the initial stages of oceanic whitecap formation and could yield results pertinent to film drop production in the open ocean, it would seem that SSA production from a whitecap formed by forcing air through a glass frit is not applicable to oceanic production for several reasons, as discussed in §4.3.4.4. It is likely that the size dependence of the resulting SSA size distribution is determined by factors specific to the apparatus such as pore size. Additionally, it would seem that most of the drops measured in these experiments were jet drops that would not occur under oceanic conditions and which result from an artifact of the experiment.

Another concern with the whitecap method is the ability of laboratory experiments to accurately determine the SSA production flux over whitecap area. This determination requires that the instruments be capable of detecting, counting, and sizing all drops produced within the stated drop size range. As noted in §4.3.4.2, SSA concentrations resulting from discrete laboratory breaking waves were not well mixed vertically, resulting in differences of up to an order of magnitude in estimated SSA production fluxes of SSA particles with $r_{80} \approx 8$ μm (§4.3.4.2). In the laboratory measurements upon which the formulation of *Monahan et al.* [1983b, 1986] was based there was no mechanism to entrain the particles upward to the measurement device other than the turbulence generated by the breaking wave, and it is therefore probable that many more SSA particles had been produced than were detected. It would be expected that such an experimental approach would result in a decreased efficiency of measurement with increasing SSA particle size. For example, in still air a top jet drop with $r_{80} = 10$ μm would be ejected to a height of ~5 cm (Fig. 28), and as a drop of this size would possess a radius near its formation value of 20 μm for quite some time (according to Eq. 2.5-6

the value of $\tau_{98,80}$ for such a drop would be ~8 s), it would fall back to the surface in less than 1 s. It is highly unlikely that an appreciable fraction of such drops above a laboratory breaking wave would be measured within that time. Likewise, the estimate of f_{wc} from measurements of SSA production from a surf zone by *Woodcock et al.* [1963] were made at a wind speed of 4 m s^{-1}, and under such a low wind speed it is unlikely that larger SSA particles would have been sufficiently entrained upward to survive to the shore to be measured. Thus it is probable that the values of f_{wc} for SSA particles toward the upper radius range of the several formulations of f_{int} according to this method are underestimated; the upper limit of r_{80} is ~15 μm from surf zone experiments (Fig. 32; the values at larger r_{80} were determined by the bubble method, discussed in §5.5.1.2), and other determinations did not extend above $r_{80} \approx 10$ μm. If this were the situation, then the general decrease in most laboratory determinations of f_{wc} for $r_{80} \gtrsim 3$ μm (Fig. 32) may result not from decreased production of SSA particles at the surface (i.e., a decrease in f_{int}) but instead from a decrease in the efficiency with which such particles were detected. This situation would call into question the results of such experiments. A concern also arises that the formulations of *Monahan et al.* and *Woolf et al.*, and the data of *Woodcock et al.*, indicate that the interfacial production flux of SSA mass is an increasing function of r_{80} to these sizes, implying that the majority of the contribution to the total interfacial SSA mass flux might be from SSA particles larger than those measured. The above considerations would further increase the uncertainty in the interfacial production flux of SSA mass according to these formulations.

Determination of W and difficulties associated with this determination were discussed in §4.5.1. For most wind speeds reported values of W range over more than an order of magnitude, and measurements from aircraft typically yield values of W that are much greater (by up to an order of magnitude) than those determined from photographs taken closer to the sea surface. As W is typically much less than unity, a large fraction of the photographs taken for a given set of conditions will contain no white area, resulting in large uncertainties in parameterizations of W.

The whitecap method of determining the interfacial SSA production flux rests on several assumptions (§3.2) which are also subject to question. For the discrete whitecap approach the key assumptions are that all whitecaps, independent of their size or method of production, yield the same production of SSA particles of any given size per unit initial white area, and thus that f_{int} is directly proportional to W. This assumption allows application of laboratory measurements to infer oceanic SSA production. Additionally, this approach assumes that the area of any whitecap, again independent of size and method of production, decreases exponentially with the same characteristic time. The continuous whitecap approach and the surf zone approach assume that all white areas are equally productive for SSA particles of a given size, and thus that f_{wc} is proportional to W. As these assumptions are essential to the applications of these approaches, they are examined here.

Although the assumption that the SSA production per whitecap area $f_{wc}(r_{80})$ is independent of the size of the whitecap is crucial to this method, laboratory experiments that might support this assumption do not appear to have been reported. As noted in §4.3.4.2, *Bowyer et al.* [1990, of which *Monahan* was a co-author], using the same laboratory wave tanks as those used by *Monahan* and coworkers (§4.3.4.2), reported that the initial whitecap area and the resulting SSA production are very sensitive to the height of the water parcels that collided to form the laboratory whitecap, implying that the SSA production flux per whitecap area $f_{wc}(r_{80})$ may indeed vary from whitecap to whitecap.

The basic assumption of the discrete whitecap method that the total production of SSA particles of a given r_{80} during the lifetime of the whitecap (including any bubble bursting and SSA production subsequent to the visible phase of the whitecap) is directly proportional to the initial white area implies that all white areas of the ocean surface are equally efficient producers of SSA. This can be demonstrated by noting that were this assumption satisfied, a whitecap with initial area $2A_0$, by the time its area had decreased to A_0, would have produced one-half of the total number of SSA particles it would produce during its lifetime. By extending this argument it could be concluded that SSA production would occur only during the visible phase of the whitecap, and thus that all white areas of the ocean surface are equally efficient producers of SSA. The fact that SSA production can occur when bubbles rise to the surface after the visible stage of the whitecap has passed would also seem to contradict this assumption. (It might also be noted that the fact that different determinations of f_{wc} shown in Fig. 32 have resulted in differences characterized by multiplicative uncertainty $\gtrless 7$, as estimated in §4.3.5, would also call into question such an assumption).

Several examples can be considered to demonstrate that this assumption is not satisfied, and that not all white areas are equally productive. At one extreme, bubbles can remain on the surface without bursting because of influences of surface-active materials (whose presence can vary greatly on spatial scales both larger and smaller than typical whitecap sizes), allowing regions of the sea surface to remain white with little SSA production (such a situation would also violate the assumption that the visible area of whitecaps decreased exponentially with time). Additionally, drop production may be modified by the thickness of the foam on

the surface, clustering of bubbles, screening by overlying bubbles, and other factors, any of which might result in differing rate and size distribution of SSA production. To be sure, it might be expected that the SSA production flux would be greater in brighter areas, corresponding to a high bubble flux, than in areas that are not as white, such as might occur immediately after a breaking wave when large numbers of bubble rise to the surface and burst. However, there is no way to take any of these factors into account by merely determining the amount of surface area of the ocean whose whiteness exceeds a given threshold.

An additional concern is the arbitrariness in definition of and determination of a whitecap. The whiteness of the ocean after a breaking wave results from scattering of light from multiple scatterers (bubbles surfaces) and does not necessarily correlate with SSA production. Intrinsic to the whitecap method is that an area of the sea surface (or that in a laboratory wave tank) is considered to be white, and no differences in SSA productivity are assumed, provided that bubble flux to the surface is sufficiently great to make an area whiter than an arbitrary threshold (usually determined by the measurement apparatus, or arbitrarily by an analyst, rather than from any physical basis). Thus the quality of being "white" is binary, and shades of white or degrees of whiteness are not distinguished.

The assumption that all whitecaps are intrinsically the same might be compared to the development of the knowledge of clouds. Certainly cloud coverage is important for considerations of atmospheric radiation, the global water cycle, and other concerns, but in addition to this fractional coverage (analogous to whitecap ratio) the thickness, composition, and type of cloud is also important. Perhaps when more understanding of whitecaps is available, it will be seen that, as for clouds, all whitecaps are not the same, and other quantities will allow parameterization of their role in SSA production. Until then, whitecap coverage would seem at best a crude first approximation.

As large bubbles, which produce film drops (Fig. 30), rapidly rise to the surface (Fig. 34), it might be expected that film drop production would occur mainly during the first part of whitecap lifetime, as indicated by the laboratory investigation of *Woolf et al.* [1987]. This study provides strong evidence that for small SSA particles not all white areas are equally productive. As a consequence, the whitecap method would overestimate the interfacial production flux for film drops. These considerations would apply also to determinations of f_{wc} from continuous laboratory whitecaps reported by *Cipriano and Blanchard* [1981] and *Cipriano et al.* [1983]; §5.3.1; §5.5.1.4. Additionally, the procedure of determining whitecap area and lifetime from measurements taken several seconds after the breaking of a wave (as in laboratory studies of breaking waves by

Monahan and coworkers; §4.3.6), would appear to miss much of the SSA production of such particles, as noted in §4.3.4.2.

The hypothesis that SSA production is directly proportional to whitecap ratio W is also not supported by field measurements of SSA concentrations. Measured oceanic number concentrations of SSA particles from the JASIN study with r_{80} near 0.5, 2.5, and 5 µm, and total number concentrations of SSA particles with r_{80} greater than 0.5, 2.6, 4.9, and 6.1 µm, demonstrated no obvious trends when plotted against W [*Monahan et al.*, 1980]; although, to be sure, the interpretation of these finding must be tempered by recognition of the long atmospheric residence times of SSA particles of these sizes (Table 8) and resultant expectation of low correlation with local conditions. Number concentrations of SSA particles with $r_{80} > 7.9$ µm, denoted $N(8+)$, from *Toba and Chaen* [1973] and from the JASIN study were fitted to a power law in W by *Monahan et al.* [1983a], resulting in an exponent around 0.7 (that is, $N(8+) \propto W^p$, with $p = 0.7$), but the significance of this finding is called into question by the large scatter in the data (typically one to two orders of magnitude). Additionally, measured SSA concentrations corresponding to $W > 0.01$ exhibit an even weaker dependence on W. Similarly, the corresponding data from the STREX experiment were fitted by *Doyle* [1984], resulting in $p = 0.71$, but although the exponent was similar to that found by *Monahan et al.* [1983a], the scatter in these data was even more extreme—typically two to three orders of magnitude; the data corresponding to $W > 0.003$ displayed no overall increase in $N(8+)$ with increasing W, and even indicated possible decrease at the highest values of W. Concentrations of SSA particle volume for several particle sizes (data of *Chaen* [1973], from the same study as that of *Toba and Chaen* [1973], and JASIN data) were also plotted against W and fitted to power laws by *Monahan et al.* [1983a], resulting in values of p near 0.08, 0.35, and 1 for r_{80} near 0.3, 5, and 16 µm, respectively. *Doyle* [1984] reported values of p of 0.15, 0.36, and 0.72 for $r_{80} = 2$, 5, and 15 µm, respectively for the corresponding STREX data. Values of p increased with increasing particle size, implying a stronger dependence on W, but again the interpretation is tempered by the fact that the concentrations are indicative not just of local production, but, especially for smaller particles, of production that had occurred days earlier. Only the concentration of SSA volume for $r_{80} = 15\text{-}17$ µm reported by *Monahan et al.* [1983a] arguably show any increase with increasing W for $W > 0.0001$. However, taken as a whole, the scatter in the data was so great (up to four orders of magnitude in concentration for a given whitecap ratio), and the uncertainties in the individual measurements so large, as to raise questions of the interpretation of the relations between SSA concentrations and W.

A further concern deals with the assumption that the areas of all whitecaps decrease exponentially with the same characteristic time (3-4 s), independent of their initial area or means of production. As noted in §4.3.4.2, laboratory measurements of whitecap lifetime and decay do not support this assumption, and this assumed exponential decrease cannot be justified merely by fitting measured areas as a function of time to a line on a semilogarithmic graph with no demonstration of the quality of the fit. This assumption is also not supported by observations of oceanic whitecaps (§4.5.4). Studies of *Snyder et al.* [1983] and *Nolan* [1988a] have indicated a very different temporal dependence (§4.5.4). Results of *Koepke* [1984; 1986b] indicate that whitecap decay occurs not so much by a reduction in area as by a gradual diminution of brightness over the entire area until the whitecap disappears. Reported lifetimes of oceanic whitecaps discussed in §4.5.4 [*Woodcock*, 1953; *Bondur and Sharkov*, 1982; *Bortkovskii*, 1987, Fig. 2.12; *Bortkovskii and Norvak*, 1993] are much greater than the characteristic times determined from laboratory whitecaps and increase with increasing wind speed. Additionally, the wind speed dependence of the ratio of the fractional areas of so-called Stage A and Stage B whitecaps W_A/W_B (according to the model of *Monahan* and coworkers [*Monahan et al.*, 1988; *Monahan and Woolf*, 1988; *Monahan and Woolf*, 1989] examined in §4.5.1), reported by *Monahan et al.* [1988] is inconsistent with the temporal behavior of whitecaps being independent of their size. Because whitecap lifetime is governed by the persistence of arrival of bubbles at the surface following entrainment of air during the wave breaking process, it is expected that larger whitecaps, formed by more energetic breaking waves which entrain bubble plumes deeper, would exhibit longer lifetimes. As discussed in §4.4, the depth to which these bubble plumes are injected increases with wind speed in a nearly linear fashion [*Thorpe*, 1982; 1986] and in the ocean may be 10 m or more for wind speeds 8-12 m s^{-1} [*Thorpe*, 1982; *Thorpe*, 1986; *Crawford and Farme*, 1987; *Farmer et al.*, 1993].

For the same reasons that a larger whitecap might have a longer lifetime than a smaller one, it might also be expected to produce, per unit initial area, more SSA particles during the course of its lifetime, again in violation of a key assumption of the whitecap method. For example, if whitecaps were similar in shape in all three dimensions, then the volume of air entrained by a whitecap—and the number of bubbles available for producing aerosol particles—would be directly proportional not to the initial area but to the 3/2 power of this area. However, as W is the sum of the areas of different sized whitecaps over the ocean, such a situation would not be equivalent to the statement that SSA production would be directly proportional to $W^{3/2}$.

A final consideration is that it has not been established that the visible white area of the sea surface is the appropriate quantity characterizing SSA production. Although such a hypothesis is very plausible, the pertinent quantity might be one related to bubble concentration, such as the volume of the bubble plume (as discussed above) or other quantities related to acoustic determination of bubble concentrations. Such quantities may exhibit different wind speed dependencies from W. For example, the active acoustic coverage reported by *Ding and Farmer* [1994] increased nearly linearly with wind speed, whereas measurements of the acoustic backscattering cross section per unit volume fitted by *Dahl and Jessup* [1995] to a power law in wind speed resulted in an exponent near 7.

In summary, although formulations of the interfacial SSA production flux based on the whitecap method have been widely quoted, these formulations cover a limited range of SSA particle sizes, and over a sizeable fraction of this range it is uncertain to what extent they accurately represent the production flux rather than the measurement efficiency. The assumptions upon which such methods are based are in some instances not satisfied, although the extent to which this affects results based on this method is not known. For each of the several approaches to determining f_{int} using the whitecap method (i.e., those involving continuous laboratory whitecap, discrete laboratory whitecap, and surf zone) only a few measurements have been reported, and it is difficult to estimate an uncertainty in f_{wc} for each. Additionally there were concerns with each of the approaches (§4.3.4.4) that precluded choosing one as unambiguously the best, hence the uncertainty in $f_{wc}(r_{80})$ was estimated based on the range of the several reported values (§4.3.5) as $\stackrel{\times}{\div} 7$, independent of r_{80} (Fig. 32). Reported measurements of W also exhibit a large amount of scatter. The uncertainty in W varies with wind speed, but is approximately $\stackrel{\times}{\div} 6$ at $U_{10} = 10$ m s^{-1} (§4.5.2). Together these two uncertainties yield an uncertainty in $f_{int}(r_{80})$ of approximately $\stackrel{\times}{\div} 14$ at this wind speed (§5.1.1), in addition to the uncertainty pertaining to the assumptions intrinsic to this method.

5.4. CONCENTRATION BUILDUP METHOD

The concentration buildup method (§3.3) calculates the size-dependent effective SSA production flux from measurements of the increase in the size-dependent vertical integral (i.e., column burden) of SSA number concentration with increasing along-wind distance under stationary but nonsteady state conditions. To date this method has been used only by *Reid et al.* [2001], who inferred f_{eff} at 30 m above the sea surface for $r_{80} = 0.3$-8 μm based on aircraft measurements using optical detectors for individual occurrences of offshore winds (i.e., blowing from land to sea) with

speeds of 8 m s^{-1} and 12 m s^{-1} off the east coast of the United States.

Values reported for the effective SSA production flux at 30 m for these two wind speeds, calculated as the averages of the values derived for 5 km advection steps, are shown in Fig. 41 (in Table 2 of that reference the uncertainty for diameter 1.25 μm at 8 m s^{-1}, stated as 0.05, should be 0.5 [*J. Reid*, personal communication, 2003]). The effective SSA production flux at 12 m s^{-1} is roughly three times greater than that at 8 m s^{-1}. The uncertainties are fairly large for the lowest value of r_{80} (0.3 μm), which the investigators attributed to the high concentration of non-SSA particles of these sizes, and for the two largest values (6.25 μm and 8 μm), which they attributed to low counting statistics, particularly for the 12 m s^{-1} data. Greater concentrations of particles with $r_{80} > 1$ μm near the surface were reported for 8 m s^{-1} than for 12 m s^{-1}, although the column burden of concentration was nearly twice as great for 12 m s^{-1} as for 8 m s^{-1}, implying much greater vertical mixing at the higher wind speed.

The investigators concluded that during offshore winds at 8 m s^{-1}, a steady state vertical distribution of SSA particles with $r_{80} > 1$ μm was achieved at a distance of ~35 km from shore (corresponding to ~70 min of transport time), at which point the upward production flux was balanced by dry deposition, but that the distance required for the establishment of steady state conditions for $U_{10} = 12$ m s^{-1} was greater than that over which measurements were taken (50 km). However, it is quite unlikely that steady state conditions had been achieved over the measurement region; according to the analysis in §2.9.6 (taking into account the height of the marine boundary layer being ~1 km instead of an assumed 0.5 km), estimated times for the attainment of steady state with respect to dry deposition at $U_{10} = 10$ m s^{-1} for SSA particles of the sizes for which fluxes were reported (r_{80} up to 8 μm) are nearly a day or more, during which the particles would have been transported more than 500 km (Table 8). Additionally, under steady state conditions such particles would be expected to be mixed over the entire marine boundary layer (§2.9.5); although some vertical mixing was evident, concentrations were not uniform over the height of this layer (~1 km) at the end of the measurement region. Even if concentrations were nearly uniform over the marine boundary layer this would not imply that steady state with respect to dry deposition had been attained, as these processes (vertical mixing and dry deposition) are characterized by different times: τ_{conv} (§2.4) and τ_{dry} (§2.7.2.3; §2.9.6), respectively, which differ by at least an order of magnitude for the sizes of SSA particles under consideration. Thus even after nearly uniform vertical concentration profiles had been achieved concentrations could still continue to increase before dry deposition became a controlling factor (a similar situation was discussed in §2.9.6 with

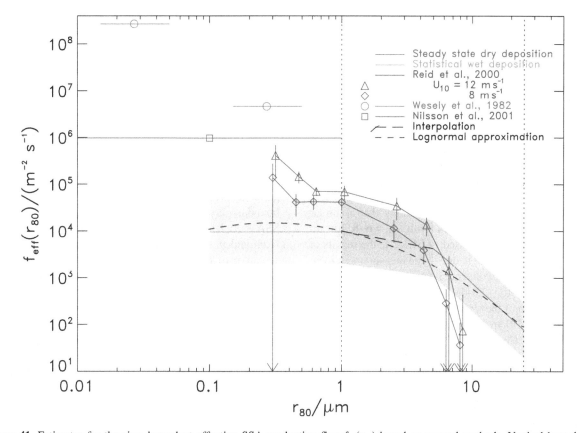

Figure 41. Estimates for the size-dependent effective SSA production flux $f_{\text{eff}}(r_{80})$ based on several methods. Vertical bars denote uncertainties stated by investigators; arrows denote uncertainties that extend further. Horizontal bars denote r_{80} intervals. Estimate based on steady state dry deposition is power law $f_{\text{eff}}(r_{80}) \propto r_{80}^{-2.5}$ for $r_{80} = 5\text{--}25$ µm calculated for $U_{10} = 10$ m s^{-1} (§5.2.1.4); pink shaded band denotes uncertainty of $\overset{\times}{\div}4$. Estimate based on statistical wet deposition method assumes constant $n(r_{80}) = 5$ cm^{-3}, $H_{\text{mbl}} = 0.5$ km, and $\tau_{\text{wet}} = 3$ d (§4.1.4.5); green shaded band denotes uncertainty of $\overset{\times}{\div}5$. Dashed black line and shaded gray region interpolate estimates by steady state dry deposition method and statistical wet deposition method. Dotted line denotes lognormal approximation to estimates of $f_{\text{eff}}(r_{80})$ from statistical wet deposition method, steady state dry deposition method, and interpolation region, with parameters $f_{\text{eff},0} = 1.5 \cdot 10^4$ m^{-2} s^{-1}, $r'_{80} = 0.3$ µm and $\sigma = 4$ at $U_{10} = 10$ m s^{-1}, over r_{80} range 0.1-25 µm. Radii reported by *Reid et al.* [2000] are converted to r_{80} values. Estimates of *Wesely et al.* [1982] are shown at geometric means of limits of size ranges; horizontal bars denote size ranges. Estimate of *Nilsson et al.* [2001] is evaluated from the formula for F_{eff} presented in that reference (Eq. 5.6-1) for $U_{10} = 10$ m s^{-1}, shown at geometric mean of assumed range of $r_{80} = 0.01\text{-}1$ µm; as discussed in §5.6, f_{eff} is relatively insensitive to this choice. Vertical dotted lines at r_{80} values of 1 µm and 25 µm demarcate small, medium, and large SSA particles (§2.1.2).

regard to a numerical model examining the temporal development of the concentration profile by *Stramska* [1987]).

The flux estimates obtained by *Reid et al.* [2001] are greater than those given by the formulation of *Smith et al.* [1993] (Eq. 5.2-2) by roughly a factor of 4 over the size range $r_{80} \approx 2\text{-}5$ µm, and greater by a larger amount for smaller SSA particles. In comparing the results of the two studies, *Reid et al.* (of whom *Smith* was one of the investigators) concluded that the formulation of *Smith et al.* might underestimate fluxes by a factor as large as 4, but attributed this underestimation to too low a value used for the dry deposition velocity rather than to error in or non-representativeness of the SSA concentrations (as had been

suggested by *Andreas* [1998]). Although the dry deposition velocity for SSA particles in this size range is somewhat uncertain (Fig. 39), it is unlikely that it is in error by a factor this large. As noted above (§5.2.1.4), it is probable that the formulation of *Smith et al.* underestimates fluxes for SSA particles of these sizes because of the long atmospheric residence times of such particles, from which it follows that dry deposition is probably not the primary mechanism of their removal and thus that the concentrations upon which *Smith et al.* based their formulation had not attained their steady state values.

Reid et al. suggested that the flux estimates for SSA particles with $r_{80} = 6\text{-}8$ µm they obtained for a height of

30 m above the sea surface (the lowest height at which they measured concentrations) would be lower than effective SSA production fluxes at 10 m (and lower than those reported by *Smith et al.*) because of the lower concentrations at 30 m resulting from a strong vertical concentration gradient over that range of heights such as had been reported by *Blanchard et al.* [1984]. However, it is unlikely that steady state concentrations or the effective SSA production fluxes at these two heights would be very different for particles of these sizes. The concentration gradient reported by *Blanchard et al.* referred to the concentration of SSA mass, which is typically dominated by the largest particles detected (§4.1.4.5) and thus cannot be taken to imply that a similar gradient exists for SSA particles with $r_{80} = 6$–8 μm or smaller. Additionally, as discussed in §4.1.2.1, the results of *Blanchard et al.* were probably due to the measurement location and such a gradient would not be expected over the open ocean. According to (2.9-4), for SSA particles with $r_{80} = 8$ μm the ratio of the steady state concentration at 30 m to that at 10 m $n(30 \text{ m})/n(10 \text{ m})$ would be 0.92 at $U_{10} = 8$ m s^{-1} and 0.95 at $U_{10} = 12$ m s^{-1}. Measurements of number concentrations of marine aerosol particles in several size ranges by *Exton et al.* [1986] at 30 m were more than 75% of those at 10 m (§2.9.4). Hence the difference in heights is expected to make little difference in the effective SSA production flux for the size range considered. *Reid et al.* concluded that "it is likely that no one parameterization or measurement set (including our own) is precise to within an order of magnitude." In view of the limited number of results by the concentration buildup method, it would seem premature to try to account for differences of a factor of 4 between these results and those of the steady state dry deposition method.

The concentration buildup method has several strengths. It directly relates changes in the column burden of the number concentration of SSA particles of a given size to SSA production, and it is free from many of the assumptions required for other methods, such as that of steady state conditions over the atmospheric residence times of the SSA particles. Although stationary over the measurement time is required, this situation occurs much more frequently and can be tested by observation. For situations of off-shore flow at coasts, air mass history and background aerosol concentration can be determined. Additionally, the dry deposition velocity (§4.6) need not be accurately known. For example, *Reid et al.* reported that varying the dry deposition velocity by a factor of 5 had an effect of only 25% on the flux estimates; such a conclusion would imply that conditions were far from steady state over the measurement region (although dry deposition would become increasingly important as steady state conditions were approached). As noted in §3.3, this method requires sufficiently dense measurements in the vertical and along-wind directions to accurately determine both the mean

concentrations and column burdens as a function of distance from shore, and is restricted to limited meteorological and environmental conditions (i.e., offshore winds). The influence of RH on particle size must be taken into account, and corrections must be implemented for contributions from continental aerosol particles and from SSA particles produced in the surf zone, which do not have a counterpart in the open ocean, although *Reid et al.* reported that the latter contributions were minor.

One key concern with this method is the extent to which the wave breaking process (and thus the production of SSA particles at the sea surface) and the marine atmospheric boundary layer development (and thus the entrainment of SSA particles upward) in the near-shore region, with limited fetch and possibly limited sea state development, is representative of that of the open ocean. Additionally, as measurements must be made by aircraft over large horizontal and vertical ranges over limited time during which conditions must not appreciably change, the relatively low concentrations of SSA particles and limited sampling time at each location might result in large uncertainties due to counting statistics. The high concentrations of background aerosol particles in smaller size ranges will increase the difficulty in determining SSA production fluxes because of low signal-to-noise ratios.

The direct nature of the concentration buildup method and its prospect for determining the effective SSA production flux as a function of local conditions suggest its utility for future work. Based on the limited application to date (only that of *Reid et al.* [2001] at two wind speeds), it certainly would seem that this approach is promising. However, many more measurements would be required before the flux estimates obtained by this method can be used with confidence, as for many quantities examined in this review, measurements from more than one investigation often exhibit large differences.

5.5. BUBBLE METHOD

The bubble method (§3.4) calculates the size-dependent interfacial production flux $f_{int}(r_{80})$ of bubble-produced SSA particles (i.e., jet drops and film drops) from values of the size distributions of the near-surface bubble concentration $n_{bub}(R_{bub}, 0) \equiv dN_{bub}(R_{bub}, 0)/d\log R_{bub}$ (typically obtained from extrapolation of measurements at various depths; §4.4.1), size-dependent bubble rise velocities $v_{bub}(R_{bub})$ (§4.4.2.1), and numbers and sizes of the drops produced per bursting bubble as a function of R_{bub} (§4.3.2; §4.3.3), under the assumption that all bubbles burst independently at the sea surface. Size distributions of oceanic bubble concentration were presented in Fig. 33, bubble rise velocities were presented in Fig. 34, the numbers of jet drops and film drops per bubble burst were presented in Fig. 26 and Fig. 30, respectively, and the sizes of the jet drops and film drops produced by individual bursting bubbles were presented in Fig. 27 and Fig. 29, respectively. As noted in §3.4, this method does not include the contribution from spume drops.

5.5.1. Estimates of the Interfacial SSA Production Flux

The bubble method, first suggested by *Eriksson* [1959], has subsequently been used by several investigators to estimate the interfacial SSA production flux of jet and film drops. In this section estimates of f_{int} that have been proposed using this method are reviewed and compared (Fig. 42), and an estimate for f_{int} is presented based on field measurements of size distributions of bubble concentration examined in §4.4.1.2.

5.5.1.1. Early estimates. The production flux of jet drops in the size range $r_{80} = 0.65$-20.5 μm above a breaking wave was estimated by *Eriksson* [1959] using data reported in *Blanchard and Woodcock* [1957] for size-dependent bubble concentrations and bubble rise speeds, the relation between jet drop size and bubble size, and the number of jet drops per bubble and their ejection heights as a function of bubble size. *Eriksson's* results, for which no wind speed dependence was proposed, are shown in Fig. 42, after being modified as described below and multiplied by $W = 0.01$ to allow comparison to other results. As the data for the necessary quantities reported in *Blanchard and Woodcock* [1957] were the only ones available at that time and as they covered only part of the size range of interest, substantial extrapolation was required for some quantities, such as the size

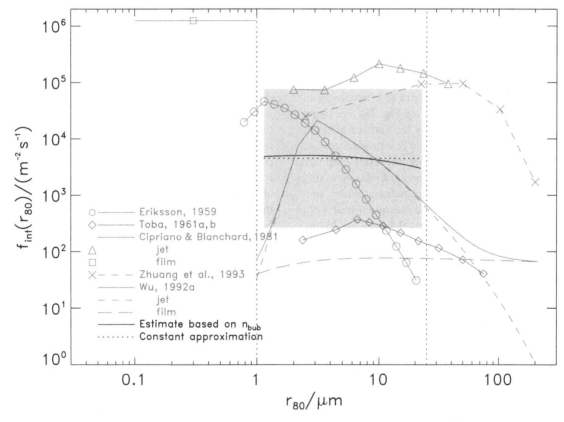

Figure 42. Estimates for the size-dependent interfacial SSA production flux $f_{int}(r_{80})$ at $U_{10} = 10$ m s^{-1} based on the bubble method. Solid black line denotes estimate for jet drop production flux calculated from geometric mean of values of $n_{bub}(R_{bub})$ denoted by central solid lines in Figs. 33a and 33b (to provide the best estimate for $n_{bub}(R_{bub})$ at $U_{10} = 10$ m s^{-1}), using (4.4-1) for $v_{bub}(R_{bub})$ at 20 °C, on the assumption of two jet drops per bubble. Dotted line denotes constant approximation to that estimate; shaded region denotes an uncertainty of $\overset{\times}{\div} 17$ about constant value. Estimate of *Eriksson* [1959] for jet drops is presented as f_{int} evaluated for assumed whitecap ratio $W = 0.01$. Estimate of *Toba* [1961a,b] was evaluated based on extrapolation of wind speed dependence of f_{int} to $U_{10} = 10$ m s^{-1}. Estimates of *Cipriano and Blanchard* [1981] for production flux of jet drops and of film drops from laboratory experiments are presented as f_{int} evaluated for assumed whitecap ratio $W = 0.01$; estimate for production flux of film drops is shown at the geometric mean of assumed range of $r_{80} = 0.1$-1 μm. Estimate of *Wu* [1992a] is evaluated for 20 °C. Vertical dotted lines at r_{80} values of 1 μm and 25 μm demarcate small, medium, and large SSA particles (§2.1.2).

distribution of bubble concentration and the sizes of the jet drops as a function of bubble size. For example, for a drop with $r_{80} = 1$ μm (the data presented in Fig. 2 of *Blanchard and Woodcock* [1957] did not extend below $r_{80} = 2$ μm) the relationship between r_{80} and R_{bub} used by *Eriksson*, equivalent to $r_{80}/$μm $= 37(R_{bub}/$mm$)^{0.83}$, yields $R_{bub} = 0.013$ mm, one-third the value of 0.036 mm given by the 1.3 power law given by (4.3-3); Fig. 27. Additionally, this relation exhibits some discrepancies with the data reported by *Blanchard and Woodcock* for larger drops, yielding a value of R_{bub} over 50% larger than that given by Fig. 2 of *Blanchard and Woodcock* [1957] for a drop with $r_{80} = 20$ μm (there is also an apparent typographic error in Table 3.2 of *Eriksson* [1959]: the value of dD/dw for $\log w = -0.25$ should be 14.2 instead of 19.2).

Eriksson included a factor describing the mean number of jet drops produced per bursting bubble that would contribute to what is referred to in this review as the effective SSA production flux. For small drops this factor would be limited to the number of jet drops per bubble whose ejection heights exceed the height of the laminar sublayer (which he took as 1.1 mm), and was estimated to increase from 0 for $r_{80} \approx 0.64$ μm and 0.2 for $r_{80} \approx 0.78$ μm to a maximum of near 4 for r_{80} in the range 3.5-5.5 μm (*cf.* Fig. 28). For larger drops this factor would be limited by the number of drops produced per bursting bubble that could be entrained upward by turbulence, and would decrease with increasing drop size because of the increasing effect of gravity. *Eriksson*, noting that the choice for this factor was somewhat arbitrary, presented values that decreased from the maximum near 4 for $r_{80} \approx 5.5$ μm to 1.3 for $r_{80} \approx 14$ μm and 0.4 for $r_{80} \approx 20$ μm, implying that only about 10% of the drops of this latter size produced at the sea surface would be entrained upward. As smaller drops that are not ejected above the laminar sublayer would typically not be included in what is considered to be the interfacial SSA production flux, *Eriksson's* estimate for the jet drop production flux for this size range would apply to the interfacial production flux, whereas his estimate for larger sizes would apply to the effective production flux. In this size range (i.e., $r_{80} \gtrsim 3.5$-5.5 μm), the ratio of the factor discussed above to the maximum number of jet drops produced by a bursting bubble would yield an estimate of $\Psi_f(r_{80})$, the ratio of the effective SSA production flux to the interfacial SSA production flux. This estimate, shown in Fig. 13, is remarkably similar to estimates based on steady state concentration ratios for typical wind speeds. The estimate of $f_{int}(r_{80})$ shown in Fig. 42 for *Eriksson* [1959] was calculated from the values presented in Table 3.2 of that reference, after converting his production flux estimate of SSA particles with $r_{80} \gtrsim 3$-5 μm, which are essentially effective production fluxes, to interfacial production fluxes by dividing by the ratio of the mean number of drops produced per

bursting bubble that were entrained further upward to its maximum value of 4.

An estimate for the interfacial SSA production flux was proposed by *Toba* [1961a,b] from laboratory measurements of bubble concentrations and of drop concentrations near the water surface in a wind/wave tunnel (§4.3.6.1). *Toba* concluded that the drops were produced by bubbles formed by the merging of capillary waves near the crests of gravity waves (§4.7.3). Based on an equation relating surface production (which was assumed to be spatially uniform), transport by the mean wind (whose speed was taken to be ~80% of that in the tunnel), turbulent diffusion, and gravitational sedimentation (on the assumption that drops instantaneously attained their gravitational terminal velocities), *Toba* obtained an expression for the concentration of drops, as a function of size, for various heights. The interfacial SSA production flux was calculated from size-dependent bubble fluxes (calculated from measured size-dependent bubble concentrations in the laboratory and modeled bubble rise speeds) together with the number and sizes of jet and film drops produced per bursting bubble. This flux was greater than that required to account for the observed size-dependent concentrations of drops near the surface, from which *Toba* concluded that the efficiency of bubble production was 0.3, independent of drop size; this factor accounts for the reduction in drop production from bubbles bursting collectively from that of bubbles bursting individually resulting from flocking and coalescence. Using this factor along with the calculated interfacial SSA production from the bubble flux (in effect combining the whitecap method, bubble method, steady state dry deposition method, and along-wind flux method), *Toba* arrived at an estimate for the interfacial SSA production flux. Laboratory wind speeds U_{lab} ranged from 5.1 m s^{-1} to 12.1 m s^{-1}, which were stated as corresponding to U_{10} from 8.7 m s^{-1} to 22.8 m s^{-1}, based on extrapolation of the logarithmic height dependence (Eq. 2.4-1). *Toba* concluded that the magnitude of f_{int} varied with wind speed as $\exp\{0.4U_{10}/(\text{m s}^{-1})\}$ for U_{10} from 15 m s^{-1} to 23 m s^{-1}. This estimate for f_{int} is presented in Fig. 42 upon extrapolation of this wind speed dependence down to 10 m s^{-1}.

5.5.1.2. Estimates based on measurements of Cipriano and Blanchard. The production flux of jet drops from a steady state laboratory breaking wave (a continuous laboratory whitecap; §4.3.4.1) was estimated by *Cipriano and Blanchard* [1981] for seven intervals of r_{80} from nearly 1 μm to 50 μm based on measurements of size distributions of bubble concentration with $R_{bub} = 0.025$-4 mm, on the assumption of 5 jet drops per bubble, and using the relation between r_{80} for jet drops and R_{bub} presented in *Blanchard and Woodcock* [1957]. Reported uncertainties in these bubble size distributions were a factor of 2-3. However, it is not apparent from the stated results how the correspondence between r_{80} and R_{bub} was made, as the size interval for the smallest drop sizes corresponds to bubble sizes smaller than those reported. These investigators did not propose a wind speed dependence but assumed a global whitecap ratio of 0.03 to estimate the flux of total SSA particle number over the oceans. Their estimate for the SSA production flux over a laboratory whitecap, multiplied instead by $W = 0.01$ (as discussed in §5.3.1) to yield the oceanic SSA production flux of jet drops, is presented in Fig. 42. These investigators also presented an estimate of the production flux of film drops over laboratory whitecap area using the upper-bound data of *Blanchard* [1963] and *Day* [1964] (Fig. 30), although no sizes of the film drops were given. This estimate of the oceanic SSA production flux of film drops, under the assumption that all the film drops were in the range 0.1-1 μm (*cf.* §4.3.4.1), is presented in Fig. 42 by a single datum at the geometric midpoint of this interval (0.32 μm), again using $W = 0.01$. Use of the upper bound data of film drop production may result in an overestimation by an order of magnitude or more (Fig. 30); *Cipriano et al.* [1983] reported that if instead the average film drop production from these sources is used, the calculated film drop production flux would agree with the production flux of particles in this size range measured from their continuous laboratory whitecap to within a factor of two.

An estimate of the interfacial flux of both jet and film drops was presented graphically by *Zhuang et al.* [1993], based on the production rate of jet drops and the relative production rates of film drops and jet drops from laboratory experiments of *Cipriano and Blanchard* [1981]. This estimate is presented in Fig. 42 after being multiplied by $W = 0.01$, as discussed below. However, there are several concerns with this estimate. No details were presented explaining how the film drop contribution was calculated, and the sizes of the drops do not appear to match (and additionally extended beyond) those reported by *Cipriano and Blanchard* [1981]. The number distribution function used in this investigation, denoted $N(r)$, was not defined; although its units were given as m^{-3}, it appears to represent $dN(r)/dr$ based on some of the figures in that reference, but a different quantity based on others. Furthermore, although a wind speed of $U_{10} = 12$ m s^{-1} was used throughout that reference, it was not clear whether *Zhuang et al.* meant their estimate to refer to an oceanic interfacial SSA production flux f_{int} or a production flux over whitecap area f_{wc}; it appears to more closely match the value of f_{wc} of *Cipriano and Blanchard* and hence is multiplied by $W = 0.01$ before being presented in Fig. 42. Finally, in considering the motion of drops with radii at formation up to 400 μm, the investigators employed a drag coefficient that is accurate only for much smaller

drops and which yields terminal velocities (in still air under the influence of gravity) that for drops with radii 100, 200, and 400 μm are too low by factors of 1.4, 2, and 2.5, respectively (§2.6.3). For these reasons this estimate can be discounted.

5.5.1.3. Estimate of Wu. Another formulation of the SSA production flux using the bubble method was presented by *Wu* [1992a], who incorporated data from several sources. The value of $n_{bub}(R_{bub}, 0)$ was determined using three of the four oceanic size distributions of bubble concentration reported by *Kolovayev* [1976] and the four size distributions reported by *Johnson and Cooke* [1979]; the size distributions reported by *Medwin* [1970; 1977a] were rejected on the grounds that those bubbles were of biological origin and not important for SSA production (§4.4.1). Although the measured bubble concentrations reported by *Kolovayev* [1976] and *Johnson and Cooke* [1979] differed considerably in magnitude (Fig. 33), *Wu* [1981a] argued that they exhibited similar dependence on R_{bub} for $R_{bub} > 0.05$ mm and on this basis normalized the size distributions of bubble concentration at various depths by dividing these distributions by the surface value of the total bubble concentration, $N_{bub}(0)$, which he obtained as described below. This procedure yielded the probability that a bubble at the surface would be in a given size range; the bubble concentration size distribution at the surface, $n_{bub}(R_{bub}, 0)$, was thus calculated as the product of this probability and $N_{bub}(0)$. The probability of occurrence was extended to $R_{bub} < 0.05$ mm based on a few data of *Johnson and Cooke* [1979]; §4.4.1. The resulting expression for the probability for a bubble at the surface being in the size interval $(\log R_{bub}, \log R_{bub} + d\log R_{bub})$ is

$$\frac{dP(R_{bub})}{d \log R_{bub}} = 4.6 \cdot 10^7 \left(\frac{R_{bub}}{mm} \right)^5, \qquad R_{bub} < 0.035 \, mm$$

$$= 6.9 \cdot 10^1 \left(\frac{R_{bub}}{mm} \right), \quad 0.035 \, mm < R_{bub} < 0.05 \, mm$$

$$= 4.3 \cdot 10^{-4} \left(\frac{R_{bub}}{mm} \right)^{-3}, \qquad 0.05 \, mm < R_{bub}$$

(the integral of this probability distribution over all bubble sizes does not yield unity, but 1.16; after dividing the values by this factor, this relation implies that 18% of the bubbles at the surface are in the size range $R_{bub} < 0.035$ mm, 39% in the size range 0.035 mm $< R_{bub} < 0.05$ mm, and 43% in the size range $R_{bub} > 0.05$ mm). Values of the total bubble concentration at the surface, $N_{bub}(0)$, were obtained by *Wu* [1981a] by extrapolating the values of the total bubble concentration $N_{bub}(z)$ determined from data of *Kolovayev*

[1976] for depths z of 1.5 m and 4 m at $U_{10} = 11$–13 m s^{-1} and values of $N_{bub}(z)$ determined from data of *Johnson and Cooke* [1979] for depths of 0.7, 1.8, and 4 m also at $U_{10} = 11$–13 m s^{-1} to the surface under the assumption of an exponential dependence on depth; the surface values obtained in this manner from these two data sets differed by nearly an order of magnitude. The same exponential dependence with depth was used to determine values of $N_{bub}(z)$ from data reported by *Kolovayev* [1976] at a depth of 1.5 m for $U_{10} = 6$–8 m s^{-1} and from data of *Johnson and Cooke* [1979] at a depth of 1.5 m for $U_{10} = 8$–10 m s^{-1}. These four values of $N_{bub}(0)$ were fitted by *Wu* [1981a] to a power law in wind speed with exponent 4.5. Later *Wu* [1988a] forced the exponent to be 3.5, based on acoustic scattering cross section measurements of *Thorpe* [1982], resulting in a total bubble concentration at the surface of $N_{bub}(0)/m^3 = 57[U_{10}/(m\ s^{-1})]^{3.5}$.

Wu [1992a], concluding that all bubbles in the ocean would behave like "dirty" bubbles (that is, as if they were solid spheres; §4.4.2.1), considered two formulations for $v_{bub}(R_{bub})$: the result of *Thorpe* [1982] (which was presented incorrectly; §4.4.2.1), and that calculated using a relation between the drag coefficient C_d and the Reynolds number Re (for which the equation presented by *Wu* for determination of v_{bub}, Eq. 12 of that reference, was stated incorrectly; §4.4.2.1); it appears that the second formulation, but with the correct equation, was used.

The number of jet drops produced per bubble was taken as $N_{jet} = 7\exp\{(-2/3)(R_{bub}/mm)\}$ from *Wu* [1989a], based on a sketch presented by *Blanchard* [1983] (§4.3.2; Fig. 26). The size of each jet drop was taken as $r_{80} = 0.0625R_{bub}$ (equivalent to $R_{bub} = 16r_{80}$), the arithmetic average of the sizes determined by the 10% rule (§4.3.2.2) and the 15% rule. The number of film drops produced per bubble was taken as $N_{film} = 4.75(R_{bub}/mm)^{2.15}$ based on an analysis by *Wu* [1990b] of the data of *Resch et al.* [1986] and of *Blanchard and Syzdek* [1988], in which the peak in the film drop production at R_{bub} near 1 mm (§4.3.3.2) was omitted. The size of each film drop was taken as $r_{80} = 0.011R_{bub}$ (equivalent to $R_{bub} = 91r_{80}$), based on data of *Resch et al.* [1986].

These relations, along with the substitutions $R_{bub} = 16r_{80}$ and $R_{bub} = 91r_{80}$ for jet drops and film drops, respectively, permit calculation of the production flux of jet and film drops in terms of r_{80} by (3.4-3). The resultant interfacial SSA production flux calculated for a temperature of 20 °C using this procedure is shown in Fig 42, as are the separate

contributions from jet drop production and film drop production[1].

According to this formulation, the production flux of jet drops is greater than that of film drops for r_{80} from less than 1 μm to nearly 60 μm, and in the range $r_{80} = 2$–30 μm it is more than an order of magnitude greater. For $r_{80} \approx 10$-60 μm, f_{int} is nearly directly proportional to r_{80}^{-2}. Jet drops of these sizes are from bubbles with R_{bub} in the range 0.16-0.96 mm and film drops of these sizes are from bubbles with R_{bub} in the range 0.9-5.5 mm (no upper limit of applicability on the value of R_{bub}, and hence r_{80}, was specified by *Wu* [1992a], but Fig. 5 of that reference shows values of the SSA production flux for r_{80} to more than 200 μm). The r_{80}^{-2} behavior arises because the probability distribution of bubbles that produce jet drops in this size range varies as R_{bub}^{-3} (and thus as r_{80}^{-3}, according to the relations used above), and v_{bub} varies roughly as R_{bub} (and thus as r_{80}), with the number of jet drops produced per bubble decreasing only slightly with increasing bubble (drop) size; combining these factors yields a production flux of jet drops that varies as r_{80}^{-2}. For film drops with $r_{80} > 1$ μm, which according to the relation used by *Wu* are produced by bubbles with $R_{bub} > 0.09$ mm, the production flux is nearly independent of drop size, increasing only slightly with increasing r_{80}. For bubbles in this size range the probability distribution varies as R_{bub}^{-3} (and thus as r_{80}^{-3}), v_{bub} varies roughly as R_{bub} (and thus roughly as r_{80}, although for larger bubbles v_{bub} increases at a slightly lower rate with increasing R_{bub}), and the number of film drops varies as $R_{bub}^{2.15}$ (and thus $r_{80}^{2.15}$); combining these factors yields a production flux of film drops that is roughly independent of r_{80}.

There are several concerns specific to this formulation (in addition to those of the bubble method in general discussed below). The data upon which the relations were based are scant and in several instances do not cover the necessary size ranges, requiring extrapolation well beyond the range of measurements. The size distribution of bubble concentration used by *Wu* [1992a] was based on few measurements of *Kolovayev* [1976] and *Johnson and Cooke* [1979], which do not seem typical of other measurements (Fig. 33). These measurements contain few values with $R_{bub} > 0.2$ mm (which would produce jet drops with $r_{80} > 12$ μm and film drops with $r_{80} > 2$ μm, according to the relations used in this formulation) and no values with $R_{bub} > 0.35$ mm (which would produce jet drops with $r_{80} > 22$ μm and film drops with $r_{80} > 4$ μm). The relation $n_{bub}(R_{bub}) \propto R_{bub}^{-3}$ for

---

[1] The film drop production flux shown in Fig. 5a of *Wu* [1992a] does not match the values calculated from the relations given in that reference, remaining constant over the range of drop diameters 2-200 μm on that figure instead of decreasing with increasing drop size. Even if the terminal velocity of bubbles with $R_{bub} > 1$ mm is taken to be a constant value near 30 cm s^{-1}, after *Levich* [1962, pp. 451-452], as suggested by *Wu*, the calculated flux does not match that shown in that figure.

$R_{bub} > 0.05$ mm suggested by *Wu* for both the data of *Kolovayev* and of *Johnson and Cooke* is not supported by the data (Fig. 33); the relations determined from Fig. 4 of *Wu* [1981a] are $n_{bub}(R_{bub}) \propto R_{bub}^{-2.5}$ for data of *Kolovayev* and $n_{bub}(R_{bub}) \propto R_{bub}^{-4}$ for data of *Johnson and Cooke*. Furthermore, the fit given by *Wu* [1981a] to the data of *Johnson and Cooke* does not extend beyond $R_{bub} = 0.135$ mm, above which $n_{bub}(R_{bub})$ does not vary as R_{bub}^{-4} but exhibits a much different trend (albeit with much scatter), closer to $R_{bub}^{-0.5}$ (however, the majority of the data with $R_{bub} > 0.135$ mm are from a depth of 8 m, and use of these data to infer surface values is questionable, as discussed below). The extrapolation of the probability distribution to $R_{bub} < 0.05$ mm, corresponding to jet drops with $r_{80} < 3$ μm (where more than half of the probability exists), was based on few data of *Johnson and Cooke* [1979]; as discussed in §4.4.1, the magnitude and size dependence of the concentration of bubbles with $R_{bub} < 0.05$ mm is extremely uncertain. A probability distribution proposed earlier by *Crawford and Farmer* [1987], for which the probability of occurrence of bubbles with $R_{bub} < 0.017$ mm was set equal to zero, was deemed unphysical by *Wu* [1992a]; however, the probability distribution used by *Wu* yields a probability of less than 0.6% for the occurrence of bubbles of this size or smaller, and when normalized to unity, the two distributions are nearly identical. Furthermore, the total bubble volume determined from the bubble probability distribution of *Wu* diverges at large R_{bub} (*cf.* §4.4.1.3). Obviously, the trend cannot continue for larger sizes (recall that no data with $R_{bub} > 0.35$ mm had been used to determine this distribution), although based on figures in *Wu* [1992a], this distribution was apparently assumed to hold for bubbles sizes up to $R_{bub} = 5$ mm.

The wind-speed dependence of the interfacial production flux according to this formulation was based on extrapolation of bubble concentrations to the surface; as noted by *Woolf* [1990] and discussed in §4.4.2.2, this is highly questionable (Fig. 35). The minimum depth in the two data sets fitted by *Wu* was 0.7 m, and even from this depth bubbles with $R_{bub} < 0.05$ mm, which would produce jet drops with $r_{80} < 3$ μm, might dissolve before reaching the surface (Fig. 35)—this would include the peak and the majority of the bubbles (and drops) in this formulation. Therefore the estimate of *Wu* would overestimate the production flux of drops from these bubbles.

Although the data upon which bubble size distributions were based contain few values with $R_{bub} > 0.2$ mm and none with $R_{bub} > 0.35$ mm, the expression for N_{film} was based only on data with $R_{bub} > 0.5$ mm [*Wu*, 1989a, 1990b], and the relation for the size of the film drops was based on measurements of film drops with $r_{80} > 2$-3 μm produced by the bursting of a single bubble of each of four sizes, all with $R_{bub} \geq 2$ mm [*Resch et al.*, 1986]. The assumption of a single size film drop per bubble is not supported by the data (Fig. 29; §4.3.3.1), with too little being known about film drop size distributions to specify a size distribution of film drops produced by a bubble for a given size. Furthermore, bubbles with $R_{bub} \lesssim 0.5$ mm are expected to produce few, if any, film drops (Fig. 30), implying (according to the relation given above) that there would be no production of film drops with $r_{80} \lesssim 5.5$ μm. However, film drops are expected to be the main contributor to the flux of drops with $r_{80} < 1$ μm; thus, the conclusion that the number of film drops produced is less than that of jet drops cannot be justified.

Based on all the above considerations, it must be concluded that little confidence can be placed in the estimate of *Wu* [1992a] for the interfacial production flux of bubble-produced SSA particles using the bubble method.

5.5.1.4. Estimate based on measured size distributions of bubble concentration. A final estimate for the interfacial SSA production flux of jet drops can be made by the bubble method from the estimates for $n_{bub}(R_{bub})$ and their associated uncertainties (Fig. 33) together with bubble rise velocities (Fig. 34) and the number and sizes of jet drops produced as a function of bubble radius (Figs. 26 and 27). Values of $n_{bub}(R_{bub})$ are shown in Figs. 33a and 33b for $U_{10} < 10$ m s^{-1} and for $U_{10} \geq 10$ m s^{-1}, respectively. For each of these wind speed ranges the central value of the shaded region comprising the range of estimates of $n_{bub}(R_{bub})$ was a constant for $R_{bub} < 0.04$ mm and a power law of the form $R_{bub}^{-1.6}$ for $R_{bub} > 0.04$ μm, with an associated multiplicative uncertainty of ⨉14 (§4.4.1.2). Bubbles with $R_{bub} < 0.04$ mm produce jet drops with $r_{80} \lesssim 1.1$ μm according to (4.3-3); however as discussed in §4.3.2.4 there are expected to be few jet drops in this size range (both because many bubbles of these sizes completely dissolve before rising to the surface and because the ejection heights of such jet drops are insufficient to propel them above the viscous boundary layer; Fig. 28). Hence only the power law formulation of $n_{bub}(R_{bub})$ for the size range $R_{bub} > 0.04$ μm need be considered. As the wind speed dependence of $n_{bub}(R_{bub})$ is not well known (§4.4.1.2), the geometric mean of the values of these two wind speed ranges is used to yield an estimate of $n_{bub}(R_{bub})$ at $U_{10} = 10$ m s^{-1}; these estimates differ by a factor of ~3.3, so the geometric mean differs from each by a factor of less than two. Additionally, the same factor of multiplicative uncertainty (⨉14) is kept. The estimate of $n_{bub}(R_{bub})$ at $U_{10} = 10$ m s^{-1} is extended upward to $R_{bub} = 0.4$ mm, at which size (according to Eq. 4.3-3) jet drops with $r_{80} \approx 23$ μm are expected.

The value of v_{bub} is obtained using (4.4-3) for the rise velocities of dirty bubbles at 20 °C (§4.4.2.1); had the bubbles been assumed to act like clean bubbles then the rise velocities and resultant flux would have been up to 1.5-2 times greater (depending on particle size), and for a temperature of 0 °C the values of v_{bub} would have been up to a factor of two lower (Fig. 34). Hence a multiplicative uncertainty of $\underset{\times}{\div} 2$ can be associated with v_{bub}.

Although bubbles in this size range produce 5 or more jet drops (Fig. 26), not all drops will be ejected to sufficient heights that they will be further entrained upward. It is thus assumed that two drops are produced from each bubble (i.e., $N_{jet} = 2$); this assumption likewise contains a multiplicative uncertainty estimated to be $\underset{\times}{\div} 2$. The sizes of the jet drops are related to R_{bub} by the 1.3 power law (Eq. 4.3-3).

The interfacial SSA production flux estimate based on these values is shown in Fig. 42 with its associated uncertainty, calculated to be $\underset{\times}{\div} 17$ (§5.1.1). This estimate yields an interfacial production flux of jet drops that is approximately equal to $4.5 \cdot 10^3$ m^{-2} s^{-1}, nearly independent of r_{80}, over the interval $r_{80} \approx 1\text{-}25$ μm; the decrease in $n_{bub}(R_{bub})$ with increasing R_{bub} (or r_{80}) being nearly compensated for by a corresponding increase in v_{bub} and the 1.3 power relation between R_{bub} and r_{80}. Because of the method by which $n_{bub}(R_{bub})$ was obtained, no wind speed dependence can be inferred from this estimate, and because of the large uncertainty associated with it, little information on the size dependence of this production flux can be determined.

Although this estimate is reasonably consistent with most of the other estimates presented, in view of its large uncertainty this is not surprising. The estimate of *Eriksson* [1959] falls below this uncertainty range for larger drops, but as noted above, this can be accounted for by the relations between R_{bub} and r_{80} he used. Within its range of applicability, this estimate and its associated uncertainty encompass the estimate of *Wu* [1993] also. The estimate of *Cipriano and Blanchard* [1981] based on laboratory experiments lies above this range. The bubble concentration size distribution used by these investigators was measured directly under a laboratory breaking wave, and it appears to be roughly three orders of magnitude greater than the central value in Fig. 33a. For comparison with other estimates on Fig. 42, the estimate of *Cipriano and Blanchard*, which yielded the SSA production flux over whitecap area, was multiplied by an assumed whitecap fraction $W = 0.01$. However, as noted with regard to the determination of the interfacial SSA production flux by these investigators using the whitecap method (§5.3.1) and discussed in §5.3.2, this value of W, which pertains to all white area, might be too high; the experiments of *Cipriano and Blanchard* involved a continuously recharged bubble plume which would correspond to only the initial formation of a whitecap following a breaking wave, which would occur for only part of the life of the white area. These investigators assumed that five drops per bubble were produced, compared to two drops in the estimate presented here; this

difference would also account for some of the difference in the two estimates. Based on the totality of data examined, the present estimate, along with its associated uncertainty, is suggested to be the best that can currently be obtained by the bubble method.

5.5.2. Discussion

The bubble method is appealing in that it provides a physical basis upon which to derive the SSA production flux. However, there are several concerns with the application of this method. Key among these are the extent to which the number and sizes of jet and film drops produced by bubbles at the ocean surface are the same as those determined from bubbles bursting individually at a calm, flat surface in a laboratory (as assumed by this method), and the lack of knowledge of quantities required for the evaluation of the interfacial SSA production flux by this method, currently limiting its use to estimation of only the jet drop production flux.

Concerns with the extent to which drop production from individual bubbles bursting at a calm laboratory surface can be applied to oceanic drop production following a breaking wave, when bubbles are clustered together and burst simultaneously in large numbers, were discussed in §4.3.7. It would be expected that the mean number of drops produced per bubble would be less under oceanic conditions because of screening, in which drops would be prevented from reaching the atmosphere from bubbles above the one bursting, and clustering, which would affect the bursting process, but there is little information upon which to quantify these expectations.

The other main drawback of this method is the uncertainty in, or lack of knowledge of, the quantities required for evaluation of f_{int}. Size distributions of bubble concentration near the ocean surface and their dependence on wind speed were examined in §4.4.1.2, where it was concluded that this quantity was uncertain to a factor $\underset{\times}{\div} 14$, and that although concentrations were generally greater at higher wind speeds, no specific dependence could be justified from the available data. An additional concern is the extrapolation of measured bubble concentrations at depths to their values at the surface. No measurements of the flux of bubbles bursting at the surface have been reported, and this quantity must thus be obtained from extrapolation of measured size distributions of bubble concentrations at depths, $n_{bub}(R_{bub}, z)$. So far most such extrapolations have been made under the assumptions that the water is motionless and that bubbles do not coalesce or otherwise interact, dissolve, or appreciably change size because of pressure differences between the depth at which measurements were made and the surface. This procedure involves evolution of these concentrations with time, and although several models have been proposed (§4.4.2.2; Fig. 35), most of these assume that bubbles rise with their terminal velocities in still water. In reality, the near-surface ocean waters after wave breaking, when these bubbles are formed, would hardly be characterized as still. Small bubbles, i.e., those with $R_{bub} \lesssim 0.05$ mm, which upon bursting would produce jet drops with $r_{80} \lesssim 1$ µm, are likely to dissolve completely before reaching the surface. Bubbles of the size that produce film drops ($R_{bub} \gtrsim 0.5\text{-}1$ mm; Fig. 30) rise to the surface rapidly (Fig. 34), and few measurements of concentrations of bubbles of this size have been reported (Fig. 33). Additionally, there is little information on sizes of film drops with $r_{80} \lesssim 1$ µm (Fig. 29), and measured values of the number of film drops produced per bubble range over two orders of magnitude for the same bubble size (Fig. 30). Consequently, there is too little information to provide an estimate for the interfacial production flux of small SSA particles by this method.

Most measurements of $n_{bub}(R_{bub})$ that have been reported extend only over a rather limited range of bubble sizes (Fig. 33) and do not include bubbles sufficiently large to produce film drops (Fig. 30). As discussed in §4.4.1, (cf. Fig. 33), it would be difficult to argue that the magnitude and size dependence of $n_{bub}(R_{bub})$ is even approximately known. Differences in $n_{bub}(R_{bub})$ account for much of the differences in the estimates of the SSA production flux discussed above (different relations between r_{80} for jet drops and R_{bub} also contributed somewhat to these differences); for the production flux of SSA particles with $r_{80} = 2$ µm, the estimates of *Eriksson* [1959], *Cipriano and Blanchard* [1981], and *Wu* [1992a] are fairly close, but for the production flux of SSA particles with $r_{80} = 10$ µm, the estimate of *Cipriano and Blanchard* is roughly two orders of magnitude greater than that of *Wu*, which in turn is roughly two orders of magnitude greater than that of *Eriksson*. Over most of the bubble size range corresponding to $r_{80} = 2\text{-}10$ µm, the relation for $n_{bub}(R_{bub})$ used by *Eriksson* varied as R_{bub}^{-4}, that used by *Cipriano and Blanchard* [1981] was roughly independent of R_{bub}, and that used by *Wu* varied as R_{bub}^{-3}. Additionally, the dependence of $n_{bub}(R_{bub})$ on wind speed is not well established, and as the wind-speed dependence of the production flux estimated by this method derives exclusively from that of $n_{bub}(R_{bub})$, it too is poorly known. Furthermore, as discussed in §4.4, it is uncertain to what extent most measured size distributions of bubble concentrations pertain to SSA production, and likewise to what extent SSA production results from bubbles located directly beneath a whitecap. Few bubble size distributions have been reported under these conditions, and although turbulence may disperse some of these bubbles before they rise to the surface and burst, the extent to which measured bubble size distributions reflect those bubbles that contribute most to the production

flux of SSA particles is questionable, and undoubtedly depends on R_{bub}; this uncertainty would also be reflected in the size distribution of the SSA production flux. Use of these reported bubble size distributions will underestimate the flux of bubbles bursting at the surface, although because of the uncertainty concerning the numbers of drops produced by clusters of bursting bubbles, it cannot be concluded that the estimates using this method also underestimate the SSA production flux. Based on these considerations, unless more information becomes available on the behavior of large numbers of bubbles bursting collectively, the large scatter in the data on the number and sizes of film drops as a function of R_{bub} is substantially reduced and relations between these quantities are reliably quantified, and accurate values are obtained for bubble concentrations at the surface and their dependence on wind speed over the range of bubble sizes that contribute appreciably to production, the SSA production flux estimated by the bubble method must be considered uncertain to a degree even greater than that reflected by the shaded gray area in Fig. 42.

The terminal velocity of a bubble of a given size (Fig. 34) is arguably well known, containing an uncertainty of perhaps $\lesssim 2$ at most because of incomplete knowledge of the temperature and the state of cleanliness of the bubbles. Likewise, the number (Fig. 26) and sizes (Fig. 27) of jet drops produced per bubble as a function of R_{bub} are arguably fairly well known over a wide range of R_{bub}, although the relation between r_{80} and R_{bub} is poorly known for jet drops with $r_{80} < 5$ µm and few investigations have yielded data for $r_{80} \lesssim 25$ µm (Fig. 27). Because not all jet drops produced are ejected above the viscous boundary layer and entrained further upward, N_{jet}, chosen above as $N_{jet} = 2$, also contains an uncertainty that can be estimated as $\lesssim 2$. However, measurements of the number (Fig. 30) and sizes (Fig. 29) of film drops produced by a bubble of a given R_{bub} vary greatly (the lack of available data on film drop sizes as a function of R_{bub} is the reason the size-dependent estimates of *Eriksson* [1959] and of *Cipriano and Blanchard* [1981] pertain only to jet drops). The quantity that is most limiting to successful implementation of this method, however, would appear to be the bubble size distribution of concentration at the surface.

5.6. MICROMETEOROLOGICAL METHODS

Most micrometeorological measurements yield the net vertical flux of aerosol particles due to turbulent transport at the height of measurement. This measured net flux is added to the dry deposition flux (usually obtained by models) to obtain the net vertical flux, which is the effective production flux. In practice these methods are generally limited to particles with $r_{80} \lesssim 1$ µm, sufficiently small that gravitational sedimentation is negligible compared to the turbulent flux. The primary removal mechanism for SSA particles in this size range is expected to be wet deposition, and their dry deposition velocities are expected to be very low with resultant long times required for concentrations of these particles to achieve their steady state values with respect to dry deposition; consequently, these concentrations are often much less than their steady state values. Under such circumstances the net vertical flux of SSA particles in this size range would be close to the effective SSA production flux. As the SSA concentration approaches its steady state value the measured net flux would decrease and the calculated dry deposition flux would increase until, at steady state, the measured flux would be zero and the method would become the steady state dry deposition method. In general the situation would be expected to be intermediate to these two limiting situations.

Only a few micrometeorological measurements of particle fluxes over the ocean have been reported, all of which employed eddy correlation techniques. These measurements typically had little or no size resolution and no means of distinguishing SSA particles from other types; thus the extent to which they can provide meaningful estimates of the SSA production flux is limited. These measurements are presented here with brief statements of results, but the confidence that can be placed in estimates of the effective SSA production flux from these studies is low.

An early study of particle fluxes over the ocean was that of *Wesely et al.* [1982], who made eddy correlation measurements of fluxes of CO_2 and of aerosol particles over shallow coastal waters near Florida, using a charger [*Wesely et al.*, 1977] for particles with r_{amb} in the range 0.015-0.05 μm and a nephelometer for particles with r_{amb} in the range 0.15-0.5 μm. Based on these measurements the effective SSA production fluxes in the two size ranges were estimated as $f_{eff} \approx 2.7 \cdot 10^8$ m^{-2} s^{-1} and $f_{eff} \approx 4.6 \cdot 10^6$ m^{-2} s^{-1}, respectively, for U_{10} near 10 m s^{-1}. These values are shown in Fig. 41 plotted at the geometric mean of the size interval under the assumption that the ambient RH, which was not stated, was 80% and that all particles were composed of sea salt. However, the lowest size range is almost certainly too small to have contained any SSA particles (§4.1.1.1), and it is likely that many of the particles in the larger size range also were not SSA particles. Under the assumption of uniform mixing over a marine boundary layer height of 0.5 km, these fluxes correspond to a daily increase in the partial number concentrations of particles in these sizes ranges of roughly $2 \cdot 10^4$ cm^{-3} and $4 \cdot 10^2$ cm^{-3}; as such large increases in concentrations of particles in the marine atmosphere are not typically observed, it must be concluded that these results provide little information that can be used to infer the effective SSA production flux.

The primary focus of this investigation was the determination of the vertical flux of CO_2 made at the same time. These fluxes were more than an order of magnitude greater than those determined by other methods, leading *Broecker et al.* [1986] to conclude that the high fluxes were due either to experimental artifacts or to the presence of fluxes internal to the atmospheric boundary layer and not originating at the air-sea interface (e.g., non-uniform upwind conditions)—this latter concern would apply also to particle fluxes. In response to these criticisms, *Wesely* [1986] noted that although advection associated with concentration gradients in the upwind direction could result in an apparent upward flux even when the surface flux was zero (§2.8.3), this would result in a random, not a systematic offset, and that the consistency in his data over a wide range of conditions precluded an explanation of the discrepancy due to these effects. In any event, these considerations call into question these measurements as they pertain to SSA production.

From measurements using a condensation nuclei counter and a scattering spectrometer at the same time and location as the study of *Wesely et al.* [1982], *Sievering et al.* [1982] and *Schmidt et al.* [1983] reported high concentrations of particles with radii less than 0.1 μm which were correlated with wind speed (though over a limited wind speed range) and which were correlated with concentrations of larger particles (radii near 0.5 μm) under conditions favoring predominance of SSA particles—higher wind speeds and longer oceanic fetches; however, there was essentially no correlation of the concentration of these larger particles with wind speed. These small particles, for which concentrations exhibited a large amount of variability (nearly an order of magnitude), contained anthropogenic, sea-derived, and soil-derived components, and the larger mode contained roughly equal numbers of SSA particles and soil-derived particles. The maximum reported upward flux of particles with radii in the range 0.1-0.25 μm (of which only about 16% of the dry mass was sea salt) was $f_{eff} \approx 4.3 \cdot 10^7$ m^{-2} s^{-1}, but the fluxes were highly variable and exhibited little correlation with U_{10} or other ambient meteorological variables (although they were nearly always upward). Eddy correlation measurements of particles with radii in the range 0.4-1.5 μm, which would be expected to be mostly sea salt, also showed little correlation with wind speed. The investigators suggested that the apparent fluxes could possibly be explained by the growth with RH (upwardly moving eddies being typically at higher RH), but the eddy correlation fluxes were not strongly correlated with the water vapor flux, which would have been expected if this were the situation. It was also suggested that the presence of particles that did not have a surface source and thus should have only a downward flux would result in upward fluxes greater than those that were estimated.

Upward fluxes of particles with radii from several tenths of a micrometer to a few micrometers over the North Sea during strong winds were determined by eddy correlation techniques by *Katsaros et al.* [1987], but only ratios of fluxes to concentrations were reported and no wind-speed dependence was presented.

Shipboard eddy correlation measurements of the fluxes of particles with dry diameters greater than 0.01 µm (corresponding to $r_{80} \gtrsim 0.01$ µm for SSA particles) using a condensation nucleus counter over the Arctic Ocean at wind speeds of 4 m s^{-1} to 13.5 m s^{-1} were reported by *Nilsson and Rannik* [2001], and *Nilsson et al.* [2001]. Although the average measured vertical flux over the ocean was near zero for the entire study period, the distribution of fluxes was not normally distributed about this value but consisted of two roughly lognormal distributions of similar magnitudes, one at positive values (upward fluxes) and one at negative values (downward fluxes), each of which was significantly different from zero. The upward fluxes were assumed to result from production at the ocean surface and the negative fluxes from dry deposition, although no reasons were suggested why at a given wind speed only one of these mechanisms would dominate at a given time. The upward fluxes correlated well with wind speed, but there was no such correlation with the concentrations of the number and mass of particles with $r_{80} < 1$ µm or with the hygroscopic growth factor of particles with $r_{80} = 0.165$ µm (SSA particles being characterized by such growth factors being greater than 1.9). Surprisingly, upward fluxes were also reported over pack ice. Number concentrations of marine aerosol particles from this investigation, examined in §4.1.3.2, were determined to be one to two orders of magnitude greater than those typical of SSA particles (Fig. 18), which therefore probably constituted only a small fraction of the total number of particles.

The production flux over the open ocean was determined as the sum of the upward number flux measured by eddy correlation and the mean dry deposition flux calculated by integrating the size-dependent dry deposition velocity over the mean size-dependent concentration of particles with dry diameter greater than 0.01 µm. The dry deposition velocity was calculated using the formulation of *Schack et al.* [1985], which employs expressions for the transfer velocity through the viscous sublayer and that resulting from turbulent diffusion different from those of *Slinn and Slinn* [1980, 1981] described in §4.6.1. The total production flux obtained by this procedure was fitted to an exponential function of wind speed, resulting in

$$\frac{F_{\text{eff}}}{\text{m}^{-2}\text{s}^{-1}} = 1.9 \cdot 10^4 \exp\left\{0.46\left(\frac{U_{10}}{\text{m s}^{-1}}\right)\right\}.$$

The value of f_{eff} at $U_{10} = 10$ m s^{-1} calculated from this expression, under the assumption that the size range is $r_{80} = 0.01\text{-}1$ µm, is presented in Fig. 41 (as noted in §4.3.4.1, the choice of the width of the range does not have a large effect on the values of f_{eff} compared to the range of values reported for this quantity). For a marine boundary layer height of 600 m, typical for the investigation [*Nilsson et al.*, 2001], this value of the flux results in a daily increase in SSA number concentration of nearly 300 cm^{-3}. As this is much greater than measured values of total SSA concentration N (§4.1.3.1) and as loss mechanisms that would compensate for this production are not evident, the pertinence of these measurements to SSA production is questionable.

It was concluded by these investigators, based on linear correlations of the logarithm of the number concentration of particles in a given size range with wind speed, that particles in the mode centered about $r_{80} \approx 0.045$ µm (in the representation dN/dr_{80}), which dominated the number concentrations, were from sources independent of wind speed and that wind-generated particle fluxes had two modes, one centered at $r_{80} \approx 0.1$ µm and one at $r_{80} \approx 0.7$ µm (in the representation dF/dr_{80}). The investigators concluded that the mode centered about $r_{80} \approx 0.1$ µm resulted from film drops or small jet drops from secondary bubbles produced by the impaction of the bubble film with the surface (as proposed by *Leifer et al.* [2000a]; §4.7.3), and that the other mode resulted from jet drops and that this mode could extend to sizes as small as $r_{80} = 0.165$ µm (although as discussed in §4.3.2.4 the ejection heights of these drops would be so small that they would not be expected to be ejected through the viscous sublayer). Several hypotheses were suggested to explain upward fluxes over leads in the ice pack, including melting of ice and snow (§4.7.3).

As noted in (§3.5.5), there are numerous concerns with micrometeorological methods in general, in addition to concerns specific to the above sets of measurements and their use in estimating SSA fluxes over the oceans. Key concerns with the above measurements are the locations of the measurements, the lack of size resolution, and the probable influence of non-SSA particles. The extent to which measurements at a coastal site, or, because of ice packs and other concerns, those in the Arctic, can be extended to other regions is questionable. An additional uncertainty for the shipboard measurements would result from correction taking into account the motion of the sensors. The inability of any of the above flux measurements to discriminate SSA particles from other types would complicate interpretation of these fluxes and could lead to errors in modeling dry deposition fluxes because of unknown composition of other particles and the relatively low abundance of SSA particles, resulting in uncertainties in estimation of dry deposition velocities of these particles necessary to add to measured fluxes to yield the effective SSA production flux (§3.5.5). The majority of the particles measured by *Wesely at al.* [1982] were non-SSA particles. Many of the conclusions reached by *Nilsson et al.* [2001] were based on correlations with wind speed of either measured number concentrations or measured fluxes over a wide range of particle sizes, but the dominant contribution to these quantities would have been from particles with $r_{80} \lesssim 0.1$ μm, and although sea salt was an increasingly common component of the marine aerosol in the Arctic during this study period for particles with $r_{80} > 0.25$ μm [*Bigg and Leck*, 2001], SSA particles were not observed using transmission electron microscopy in smaller particles. Not only does this fact raise questions regarding the applicability of these results to infer the production flux of SSA particles, but it calls into question other conclusions that were reached, such as the existence of a film drop mode at $r_{80} \approx 0.1$ μm.

Complications resulting from particle growth with RH, which can be manifested as an apparent (rather than actual) upward flux [as discussed by *Wesely et al.*, 1982, *Sievering et al.*, 1982 and *Schmidt et al.*, 1983], arise because RH typically decreases with height over the ocean and because the devices that were used to detect particles were sensitive over a fixed range of ambient radii. Thus particles with a given r_{80} at the lower size range of detection (at the RH of measurements) that are in downward moving parcels of air being transported to the measurement device from greater heights, and thus at lower RH, might have smaller sizes than the detection limits, and thus not be detected, whereas the situation would be opposite for upward moving particles with a given r_{80} at the upper size range of detection. If the number concentration were a decreasing function of radius, then more particles would be counted in the upward parcels of air than in the downward parcels of air at the measurement height, mimicking an upward flux of particles. If the number concentration were not uniformly decreasing as a function of radius, as occurred for some of these measurements [*Sievering et al.*, 1982], the situation would be even more complicated.

Finally, size resolution is essential because of the different mechanisms controlling the concentrations, lifetime, and behavior of particles of different types and sizes over the range $r_{80} \approx 0.01$ to $r_{80} \approx 1$ μm, and any conclusions of SSA production fluxes based on measurements of concentrations and fluxes averaged over this size range would be subject to question.

5.7. ALONG-WIND FLUX METHOD

The along-wind flux method (§3.6) determines the interfacial SSA production flux using (3.6-2) from measurements of the integral of the size-dependent along-wind flux $j_x(r_{80}, z)$ over height z together with estimates of the mean distance $\overline{L}_p(r_{80})$ traveled by SSA particles of a given r_{80}. In practice this method is restricted to determination of the production flux of large SSA particles. This method has been applied mainly to measurements from laboratory investigations with wind/wave tunnels, but few such determinations have been made. Most estimates based on this method have assumed that the interfacial SSA production flux exhibits the same shape (i.e., size dependence) as measured along-wind fluxes or concentrations of SSA particles near the water surface, in the laboratory or over the ocean, with the wind speed dependence of the production flux assumed to be the same for all particle sizes, and with the magnitude of the flux determined by some other means, typically by matching that obtained by another method at a given value of r_{80}.

Concerns with applications of this method are that they have not taken into account the size dependence of the mean distance traveled by particles, in essence assuming that this quantity is independent of particle size. Generally such studies have arbitrarily assigned the magnitude of the flux, assuming that the value of the flux determined by another method is accurate for a specific particle size and additionally that the wind speed dependencies of SSA particle production are the same for particles smaller and larger than this size. Failure of these assumptions would result in considerable error not only in the shape of the distribution of the interfacial production flux of large SSA particles, but also in the magnitude of this flux. Despite these concerns, virtually all formulations of the interfacial production flux of large SSA particles, some of which have seen widespread application, are based on the along-wind flux method. These formulations are examined here.

5.7.1. Estimates of the Interfacial SSA Production Flux

Estimates for the interfacial SSA production flux have been presented by *Ling* and coworkers [*Ling and Kao*, 1976; *Ling et al.*, 1978; *Ling et al.*, 1980; *Ling*, 1993] based on measurements (§4.3.6) in a wind tunnel of drop production from a so-called "hydraulic jump" that was presumably meant to simulate a breaking wave [*Ling et al.*, 1978] and on shipboard measurements reported by *Ling* [1993]. *Ling et al.* [1978] presented ratios of the production rates of drops with r_{80} from 1.5 μm to 100 μm to the production rate at $r_{80} = 25$ μm, although it is not clear if the quantities presented (presumably proportional to fluxes) were per unit interval of radius or per unit logarithmic interval of radius (based on *Ling et al.* [1980] it is probably the former). *Ling et al.* [1980] and *Ling* [1993] presented ratios of f_{int} in several size ranges to the interfacial production flux of drops with 5 μm < r_{80} < 15 μm (the ratios were the same in both of these papers except the values for the first size range differed somewhat). *Ling and Kao* [1976] and *Ling et al.* [1978] assumed that the magnitude of f_{int} was directly proportional to the wind speed at 20 times the average wave height, whereas *Ling et al.* [1980] and *Ling* [1993] assumed that the magnitude was directly proportional to U_{10}^2. The estimate for $f_{int}(r_{80})$ presented by *Ling et al.* [1980], shown in Fig. 43, is nearly three orders of magnitude higher than most other estimates for this quantity.

Because of the lack of supporting information and because no plausible reasons were suggested for the assumed wind-speed dependencies, it is difficult to place much credence in this estimate. *Ling* [1993] argued that for U_{10} from 3 m s^{-1} to 19 m s^{-1} the interfacial SSA production flux according to this formulation is consistent with the bubble concentration

size distributions reported by *Ling and Pao* [1988], which he claimed showed a peak in the bubble radius range 0.1 mm to 0.25 mm, under the assumption that each bubble produces 300 film drops; however, *Ling and Pao* did not report measurements of bubbles this large, and bubbles in this size range are too small to produce film drops based on most data shown in Fig. 30.

These investigations concluded that SSA particles contribute appreciably to the moisture and latent heat fluxes to the atmosphere. However, in evaluating the exchange of moisture between the drops and the surrounding atmosphere it was assumed that the drops were pure water drops instead of seawater drops, with no account taken of the effect of the sea salt on this evaporation; additionally the contribution of

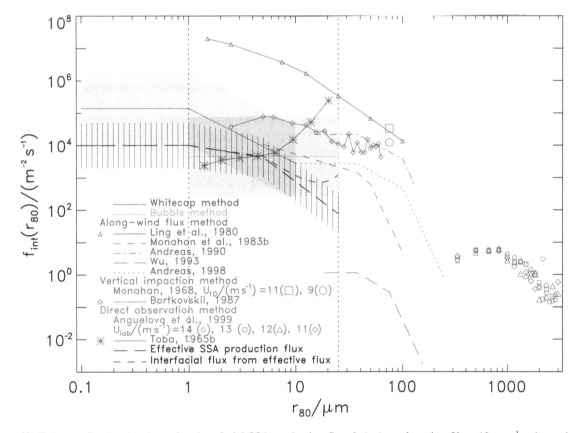

Figure 43. Estimates for the size-dependent interfacial SSA production flux $f_{eff}(r_{80})$, evaluated at $U_{10} = 10$ m s^{-1} unless otherwise noted, based on several methods. Shaded band for whitecap method denotes multiplicative uncertainty of $\overset{\times}{\div}$ 14 about central value from Fig. 32, and that for bubble method denotes multiplicative uncertainty of $\overset{\times}{\div}$ 17 about central value from Fig. 42. Along-wind flux formulation of *Monahan et al.* [1983b, 1986] is modified as described in §5.7.1. Values shown for *Bortkovskii* [1987] are arithmetic means of values reported in Fig. 1.18 of that reference. Values shown for *Toba* [1965b] are based on his extrapolation of measurements from ~700 m above the sea surface, as described in §5.1.2 and §5.10. Estimate of effective SSA production flux is that from Fig. 41, based on steady state dry deposition method for $r_{80} > 5$ μm at $U_{10} = 10$ m s^{-1} and statistical wet deposition method for $r_{80} < 1$ μm (and interpolation region for 1 μm $\leq r_{80} \leq 5$ μm); vertical lines about central value denote multiplicative uncertainty of $\overset{\times}{\div}$ 5 for $r_{80} < 1$ μm and $\overset{\times}{\div}$ 4 for $r_{80} > 5$ μm, with continuous change throughout interpolation region. "Interfacial flux from effective flux" denotes estimate of f_{int} from estimate shown for f_{eff} and value of $\Psi_f(r_{80})$ given by (2.9-4) for $z_1 = 0.1$ m, as discussed in §5.11.2. Vertical dotted lines at r_{80} values of 1 μm and 25 μm demarcate small, medium, and large SSA particles (§2.1.2).

the latent heat of water on the rate of condensation was not taken into account (§2.5.4). The particles were assumed to have instantaneously attained their gravitational terminal velocities and no effect of the mean wind flow on the motion of the particles was considered; as many large particles will not have attained horizontal velocities equal to that of the mean flow and will not have attained their terminal velocities by the time they fall back into the sea (§2.6.3), this assumption is questionable. For all these reasons, in addition to the lack of credible basis for the assumed production flux, these conclusions must be considered suspect.

Most determinations of f_{int} by the along-wind flux method have been based on field measurements of the size-dependent along-wind flux $j_x(r_{80})$ near the sea surface reported by *Wu et al.* [1984] discussed in §4.2, and from laboratory measurements of $j_x(r_{80})$ reported by *Wu* [1973] and *Lai and Shemdin* [1974] discussed in §4.3.6. *Wu et al.* [1984] reported measurements of the probability of occurrence of SSA particles with r_{80} from 10-130 μm at heights from 13 cm to 130 cm above the instantaneous sea surface for $U_{10} = 6$-8 m s^{-1}. *Wu* [1973] presented size distributions of the frequency of occurrence of freshwater drops with r_{80} (calculated as $0.5r_{form}$) from 35-100 μm for three laboratory wind speeds at several heights, and *Lai and Shemdin* [1974] reported measurements of along-wind fluxes of both freshwater and saltwater drops with r_{80} from ~17.5-175 μm for three laboratory wind speeds at several heights. As noted in §4.3.6, the units in which *Wu* [1973] reported his results were incorrectly presented, resulting in confusion by subsequent investigators (discussed further below).

In each of the formulations discussed below it was explicitly assumed by the investigators that the particles to which the formulations applied are spume drops. Additionally, in each formulation it was assumed that the size dependence of the interfacial production flux is independent of wind speed, and that the shape of this size distribution consists of power laws of the form $f_{int}(r_{80}) \propto r_{80}^{-p}$, with different values of p in different ranges of r_{80}.

From observations of increase in the concentration of large SSA particles at heights near 10 m for $U_{10} \gtrsim 9$ m s^{-1}, *Monahan et al.* [1983b, 1986] concluded that spume drop production provides an increasingly important contribution at these wind speeds, and proposed a preliminary expression for the size-dependent interfacial production flux of such drops based on laboratory results of *Wu* [1973] and *Lai and Shemdin* [1974]. However, it appears that *Monahan et al.* made an incorrect assumption regarding the particle size; they fitted size distributions of frequency of occurrence reported by *Wu* [1973], which had been presented in terms of formation radii, down to $r = 10$ μm but combined the resulting spume drop production flux with their formulation of the interfacial production flux of bubble-produced drops

(Eq. 5.3-1), which had been presented in terms of r_{80}. This spume drop production flux, modified here on the assumption that the radii in the original formulation referred to formation radii, is shown in Fig. 43 for $U_{10} = 10$ m s^{-1}. According to this formulation, the wind-speed dependence of the magnitude of f_{int} increases as $\exp\{2.08U_{10}/(\text{m s}^{-1})\}$, and the size dependence of $f_{int}(r_{80})$ consists of power laws of the form r_{80}^{-p} with p equal to 1, 3, and 7 for the ranges $r_{80} = 5$-37.5 μm, $r_{80} = 37.5$-50 μm, and $r_{80} > 50$ μm, respectively. The size dependence of f_{int} for 5 μm $< r_{80} < 37.5$ μm was based on laboratory measurements of *Lai and Shemdin* [1974], who reported along-wind flux values in this size range only at $r_{80} = 18.75$ μm (representing the size range 12.5 μm $\leq r_{80} \leq 25$ μm) and at $r_{80} = 37.5$ μm (representing the size range 25 μm $\leq r_{80} \leq 50$ μm). No justification was given for the extrapolation to lower sizes. The size dependence for $r_{80} \geq 37.5$ μm and the wind-speed dependence were based on laboratory measurements of *Wu* [1973].

Monahan et al. [1983b, 1986] proposed this spume drop production flux formulation along with (5.3-1), which describes the production flux of bubble-produced drops; this implicitly assumes that no jet drops or film drops with $r_{80} > 8$ μm are produced. According to these estimates (after the correction for radius discussed above), the production fluxes of spume drops and of bubble-produced drops with $r_{80} = 8$ μm (the largest size for which Eq 5.3-1 applies) are equal at $U_{10} = 9.7$ m s^{-1}, but the dependence of the spume drop production on wind speed is very much stronger than that of the bubble-produced drops; at $U_{10} = 8$ m s^{-1} the spume term contributes around 5% of the total flux of drops of this size, whereas at $U_{10} = 11$ m s^{-1} it contributes over 90%. Such an abrupt increase in the production of drops of this size at U_{10} near 10 m s^{-1} has not been observed in concentration measurements (Fig. 22), and several investigators [e.g., *Burk*, 1984; *Stramska*, 1987; *Andreas*, 1990b, 1992; *Smith et al.*, 1993] have previously concluded that this spume drop production flux estimate exhibits far too strong of a dependence on wind speed. This was explained [*Wu*, 1993] as being due to mistakes in the presentation of data in *Wu* [1973] noted in §4.3.6 and to misinterpretation of some of the results of that paper (*Monahan et al.* incorrectly took the wind speed in the wind/wave tunnel to be U_{10}). For these reasons the spume drop production formulation of *Monahan et al.* is not considered further.

An expression for the interfacial production flux of spume drops was proposed by *Andreas* [1990b] for which the shape of the size dependence was based on results of *Wu et al.* [1984]. The wind-speed dependence was based on a fit by *Andreas* to the wind speed dependence proposed by *Miller* [1987] and *Miller and Fairall* [1988] (§5.2.1.2), and the magnitude was equated to that given by a fit by *Andreas* (also discussed in §5.2.1.2) to the interfacial production flux estimate of *Miller* [1987] at $r_{80} = 15$ μm. This spume drop interfacial production flux, after having been multiplied by a factor of 8 that was mistakenly omitted in the original expression (because of misinterpretation of the radius convention used by *Miller*; §5.2.1.2), is shown in Fig. 43 for $U_{10} = 10$ m s^{-1}. According to this formulation, the windspeed dependence of the magnitude of f_{int} increases as $U_{10}^{2.22}$, and the shape of $f_{int}(r_{80})$ is described by power laws of the form r_{80}^{-p} with p equal to 0, 1.8, and 7 for the r_{80} ranges 15-37.5 μm, 37.5-100 μm, and > 100 μm, respectively. No upper size limit was given, but the upper size limit of drops measured by *Wu et al.* [1984] was $r_{80} = 130$ μm.

A similar estimate was subsequently proposed by *Andreas* [1992], but with coefficients (again chosen so that the flux matched that given by the fit to *Miller* [1987] at $r_{80} = 15$ μm) presented only at the wind speeds for which the data of *Miller* were presented. Graphs of the interfacial SSA volume flux (presented in terms of $dF^V(r_{98})/dr_{98}$) exhibited maxima near $r_{80} = 100$ μm which Andreas attributed to spume drops, even though this maximum was present at wind speeds as low as 5 m s^{-1}.

Yet another estimate proposed by *Andreas* [1998] had this same size dependence (except with the lower limit on r_{80} extended from 15 μm to 10 μm and the upper limit specified as 250 μm), but with the magnitude chosen so that the production flux is equal to 3.5 times the estimate of f_{eff} of *Smith et al.* [1993] at $r_{80} = 10$ μm (§5.2.1.3). The wind speed dependence of the magnitude of f_{int} according to this formulation is nearly directly proportional to $\exp\{2.21[U_{10}/(\text{m s}^{-1})]^{1/2}\}$ with stated range of applicability as U_{10} up to 32.5 m s^{-1}.

An expression for the interfacial production flux of spume drops was proposed by *Wu* [1993] in which the size dependence was based on laboratory measurements of *Wu* [1973] and field measurements of *Wu et al.* [1984], with the magnitude determined from laboratory investigations of *Wu* [1973], and the wind-speed dependence determined from the along-wind fluxes reported in *Wu* [1973] by matching wind friction values in the laboratory to their equivalent U_{10} values. *Wu* appears to be the only investigator to have appreciated the necessity of integrating the along-wind flux over height and then dividing by the mean distance traveled by drops of a given size, both of which are required for converting along-wind flux to interfacial SSA production flux (§3.6).

Wu [1993] calculated the mean distance traveled by SSA particles of a given size as the product of the mean horizontal velocity and the mean residence time in air for drops of this size. The mean horizontal velocity was assumed to be half of the free-stream wind speed in the tunnel, independent of particle size, and the mean residence time in air was taken to be the mean wave amplitude (~1-2 cm) divided by the vertical component of the velocity of a drop of this size when it hit the surface, which was taken to be 10% of the

terminal velocity of a drop with formation radius 100 μm falling in still air. This procedure resulted in a mean residence time of 0.5-0.9 s and a mean distance traveled of 5-10 m. The value 100 μm was chosen because it was the peak (formation) radius of the size distribution of frequency of occurrence of drops in laboratory experiments reported by *Wu* [1973] (§4.3.6.1), presumably proportional to the along-wind flux in the representation $dJ_x(r_{80})/dr_{80}$. *Wu* [1993] stated that the gravitational terminal velocities of drops with diameter 100 μm would be ~23 cm s^{-1} (which indeed they would), but according to (2.6-11) the gravitational terminal velocities of drops with radii 100 μm, corresponding to the peak radii referred to above, would be nearly three times as large (on p. 18223 of *Wu* [1993] it was incorrectly stated that the size distributions of *Wu* [1973] peaked at "about 100 μm in diameter", although on p. 18225 it was correctly stated that the peak was "located at approximately the radius of 100 μm"). According to the formulation of *Wu* [1993], the wind speed dependence of the magnitude of f_{int} is directly proportional to $\exp\{0.875[U_{10}/(\text{m s}^{-1})]\}$, and the shape of $f_{int}(r_{80})$ is described by power laws of the form r_{80}^{-p} with p equal to 0, 2, and 7 for the ranges $r_{80} = 18.75$-37.5 μm, $r_{80} = 37.5$-75 μm, and $r_{80} > 75$ μm, respectively. The estimate resulting from this formulation at $U_{10} = 10$ m s^{-1} is shown in Fig. 43. In summary, although *Wu* applied a correct expression for calculating the interfacial SSA production flux from the along-wind flux, his assumption that the mean distance traveled by particles was independent of their size (and thus that the size dependence of the interfacial SSA production flux was the same as that of the measured along-wind flux integrated over height) is highly questionable.

5.7.2. Discussion

There are numerous concerns with the formulations that have been proposed for the interfacial SSA production flux based on the along-wind flux method. Key among these are the assumption that the mean distance traveled by SSA particles $\bar{L}_p(r_{80})$ is independent of r_{80}, the determination of the magnitude of the flux, extrapolation of laboratory measurements to the open ocean, and extrapolation of formulations to larger sizes and higher wind speeds than the data upon which they were based.

An implication of the assumption that the shape of the near-surface along-wind flux is the same as that of the interfacial SSA production flux is that the mean horizontal distance traveled by SSA particles of a given size, $\bar{L}_p(r_{80})$, would be independent of r_{80}. However, this is certainly not the situation, as this distance depends strongly on particle size. According to the arguments presented in §2.9.6, the mean distance traveled by SSA particles with $r_{80} = 15$ μm at $U_{10} = 10$ m s^{-1} is 50 km (and the mean residence time is more than 1 h; Table 8), whereas an SSA particle with $r_{80} = 100$ μm would be expected to travel only a few meters before falling back into the sea, especially as the radius of such a particle will remain near its formation value (i.e., 200 μm) during the few seconds it remains airborne. For large SSA particles, which can be assumed to have radii near their formation values during their entire residence time in the atmosphere, it would also be expected that the distances traveled would depend strongly on particle size because the terminal velocities of such particles increase nearly linearly with radius (§2.6.3). There are additional concerns with large SSA particles, as with increasing size it becomes increasingly unlikely that an SSA particle will attain the horizontal velocity component of the surrounding flow, or that it will attain its gravitational terminal velocity before falling back into the surface (§2.6.4). Implications of these concerns for determination of along-wind fluxes were discussed in §4.6.3 where laboratory measurements of this quantity themselves were examined.

The above arguments apply to oceanic situations, and thus pertain to formulations such as those of *Andreas* [1990b, 1992, 1998] and *Wu* [1993] which were based on oceanic measurements of *Wu et al.* [1984]; these formulations were applied to SSA particles with r_{80} from as low as 10-20 μm to well over 100 μm. As noted in §3.6, in determinations of $f_{int}(r_{80})$ from laboratory measurements of the along-wind flux the fetch (that is, the distance over the water from the entrance of the tunnel to the measurement device) should be used instead in (3.6-2) if it is much less than the mean distance traveled by particles (under the assumption that the production is uniform over the surface), in which situation the interfacial SSA production flux would exhibit the same size dependence as the along-wind flux. However, it has not been demonstrated that this situation holds for any laboratory experiments, and for most such experiments (§4.6.3) it is quite unlikely to be so. Even if it were, then the ability of laboratory experiments of such limited fetches to provide results that can be extrapolated with any confidence to the ocean would be questionable.

Another serious concern with most formulations of f_{int} that are based on the along-wind flux method regards the choice of the magnitude of this flux. The magnitudes of the along-wind fluxes measured by *Wu* [1973] and *Wu et al.* [1984] were not presented (only frequencies of occurrence of drops of a given size were presented), and no use has been made of the magnitudes reported by *Lai and Shemdin* [1974] in obtaining these formulations. The assumption typically made that the magnitude of the production flux of spume drops can be obtained by equating it to the production flux determined by another method requires either that this other production flux refers only to spume drops or else that no jet or film drops occur at larger sizes and that the wind speed dependence of the spume drop production is the same as that of jet and film drop production. Neither of these alternatives is tenable. Additionally, there are concerns with the fluxes to which the spume drop flux formulations are matched.

For both the estimates of *Andreas* [1990b] and *Andreas* [1992] the wind-speed dependence of the interfacial production flux of drops with r_{80} greater than 100 μm was determined by that of drops with $r_{80} = 15$ μm based on the formulation of *Miller* [1987]. As discussed in §5.2.1.2, the magnitude of the production flux estimate of *Miller* for drops of this size was based on three sources: *Toba* [1965a], based in turn on measurements of *Woodcock* [1953] taken at 600 m (these would be expected to contain few if any spume drops); JASIN/STREX SSA concentration data; and the original Navy Aerosol Model [*Gathman*, 1983a,b], which was seen to over-predict the concentration of larger drops by an order of magnitude. Similarly, according to the estimate of *Andreas* [1998], the wind-speed dependence of the interfacial production flux of drops with r_{80} as large as 250 μm is determined solely by that of drops with $r_{80} = 10$ μm based on the formulation of *Smith et al.* [1993], with the implicit assumption (discussed above) that the ratio of the interfacial SSA production flux to the 10 m effective SSA production flux is equal to 3.5, independent of r_{80}, for $r_{80} \leq 10$ μm. However, as noted also by others [*Wu*, 1993; *Katsaros and de Leeuw*, 1994; see also *Andreas*, 1994a], the use of measurements of small drops ($r_{80} < 10$-15 μm), which are believed to be produced from bursting bubbles at low wind speeds (6-8 m s^{-1}), to determine the production flux of much larger drops (r_{80} up to 250 μm), which are believed to be produced by wave shearing at wind speeds up to more than 30 m s^{-1}, is highly questionable.

Concerns with applicability of laboratory measurements to oceanic conditions were discussed in §4.3.6.2, not the least of which pertains to the small sizes of the waves in the laboratory (Table 18). The assumption that the process of drop production in a wind/wave tank with wave heights of a few centimeters is the same as that under typical conditions over the open ocean is highly suspect. As noted in §4.3.6.2 little consensus has been reached concerning even the types of drops present in the laboratory experiments; likewise (§4.3.7), little consensus has been reached concerning the types of drops present over the ocean in certain size ranges, including those measured by *Wu et al.* [1984]. In any event, the difference in scale between wave heights in the laboratory experiments and those over the open ocean is so large as to call into question the pertinence of such laboratory results. Some laboratory measurements have reported thresholds for the spume drop production process (§4.3.6), but the correspondence of laboratory wind speeds to oceanic equivalents is uncertain (§4.3.6.2), and the concerns noted above regarding the applicability of laboratory results to the ocean diminish the utility of such thresholds.

The measurements that provide the basis for these formulations are scarce. Only those of *Lai and Shemdin* [1974] extended to values of r_{80} greater than 150 µm, but these measurements exhibited considerable scatter and differed greatly in shape from those of *Wu* [1973] (Fig. 25). Several of the spume drop formulations were extended to values of r_{80} greater (and lower) than the measurements upon which they were based. The assumed wind speed dependencies of the formulations for spume drop production likewise have little basis in measurement and have been extended to wind speeds outside the range of measurements. Most formulations yield spume drop production at wind speeds far below those for which wave tearing is expected, and none exhibits an abrupt increase in spume drop production at any wind speed or yields any indication of a threshold for the onset of spume drop production. For example, a graph showing estimated spume drop production fluxes for U_{10} ranging from 5-18 m s^{-1} was presented by *Wu* [1993, Fig. 5], although 5 m s^{-1} is much lower than values for which evidence of spume drop production has been noted, and the laboratory measurements of *Wu* [1973] were taken at wind friction velocities of 0.6-1.1 m s^{-1}, stated by *Wu* [1993] to be equivalent to U_{10} values of 14-24 m s^{-1}. The only pertinent field measurements are those of *Wu et al.* [1984] at wind speeds 6-8 m s^{-1}, not only below those at which spume drop production is expected, but also too narrow of a range to allow determination of a wind speed dependence. Extrapolation of these results to higher wind speeds is thus highly questionable.

A hypothesis for the wind speed dependence of the interfacial SSA production flux was proposed by *Andreas et al.* [1995], who suggested that the production flux of spume drop surface area (directly proportional to the free energy required to form the drops) would be proportional to the energy flux from the wind. This energy flux is equal to the product of the wind stress and the wind speed at the appropriate height for spume drop formation, which these investigators took to be the mean height of the largest one-third of the waves, $H_{1/3}$. This argument is similar to the one proposed by *Wu* [1979b, 1988b] for the wind-speed dependence of the whitecap ratio (§4.5.2; Eq. 4.5-3), and results in an interfacial production flux of SSA surface area that increases with wind friction velocity u_* somewhat more slowly than u_*^3. *Andreas* [1998] later argued that the surface area production flux was essentially proportional to u_*^3, which is roughly equivalent to U_{10}^p where the value of p is near 4. No data were presented to support this heuristic argument, but it served as the basis for his conclusion that the formulation of *Smith et al.* [1993] provides the best choice with which to match the magnitude of the spume drop production flux proposed by *Andreas* [1998]; as noted in §5.2.1.4., the wind-speed dependence of the amplitude of the second mode of

the estimate of *Smith et al.* [1993] varies approximately as $U_{10}^{3.9}$. However, wind-speed dependence alone (even if it were the correct dependence) is insufficient to determine which of several estimates is the best and can provide no information on the size dependence. Likewise, the argument of *Andreas* provides no information on the size dependence of the spume drop production flux. The hypothesis that for all wind speeds a fixed proportion of the wind contributes to the formation of spume drop surface area implies that spume drop production occurs at all wind speeds (even below those at which waves are breaking). However, if it is the formation of surface area that is important, then the occurrence of even the slightest curvature of the sea surface would overwhelm the contribution of the new surface area resulting from drop production. Additionally, if a drop were transported to a height more than 75 times its formation radius (corresponding to 1.5 cm for a drop with $r_{80} = 100$ µm) then its gravitational potential energy would be greater its surface energy; thus the surface energy of a drop is only a small fraction of its total energy, and there would seem to be little justification for the assumption that this form of energy is somehow related to the energy from the wind.

In conclusion, perhaps the most evident illustration of concern over these spume drop production flux formulations is shown in Fig. 43. Although the size distributions of the estimates of *Andreas* [1990b], *Wu* [1993], and *Andreas* [1998] are all based on the same oceanic measurements [*Wu et al.*, 1984], the magnitudes of these estimates at a given wind speed ($U_{10} = 10$ m s^{-1}) vary by more than 4 orders of magnitude.

5.8. DIRECT OBSERVATION METHOD

Determination of the interfacial SSA production flux by the direct observation method has been reported only by *Anguelova et al.* [1999], who measured the production of freshwater spume drops with r_{80} (presented here as half the formation radius; §2.1.1) from ~200 µm to 3000 µm in a wind/wave tunnel using video recordings taken at 28 cm above the mean water level. Four laboratory wind speeds from 11.4 m s^{-1} to 14.1 m s^{-1} were used, which the investigators reported corresponded to U_{10} of 18.3 m s^{-1} to 25.0 m s^{-1} (based on extrapolation of Eq. 2.4-1). Spume drop production commenced at $U_{lab} \approx 9$ m s^{-1}. Size distributions of the interfacial production flux of drops in 18 intervals of r_{80} up to 3000 µm were reported; a slight bias to larger sizes was noted, and the investigators reported that size distributions for $r_{80} \lesssim 500$ µm (the lowest two size intervals) could not be reliably determined. The total interfacial production flux (which they termed the production rate) of spume drops with $r_{80} \gtrsim 200$ µm for the laboratory situation was fitted to $F_{int} \propto \exp\{0.23 U_{10}/(\text{m s}^{-1})\}$, where

U_{10} was calculated from the laboratory wind speed as described above. To estimate the oceanic SSA production flux, the investigators employed a mean "breaking crest length" [*Phillips*, 1985], defined as the sum of the lengths of all breaking crests per unit area of the sea surface, equivalent to the average width of a whitecap along the wave crest divided by the average area containing one whitecap. Expressions for oceanic values of this quantity were presented based on analysis by *Wu* [1992b] of data of *Bortkovskii* [1987] covering wind speeds U_{10} from approximately 9-19 m s^{-1}. Under the assumption that the ratio of the interfacial SSA production flux to the mean breaking crest length is the same for the laboratory and the ocean, the size-dependent oceanic interfacial SSA production flux was obtained as a function of wind speed. The equivalent oceanic fluxes calculated by this procedure are presented in Fig. 43.

A key concern with this investigation is the extent to which laboratory results of drop production from wave crests that extend only slightly higher than 6 cm can accurately simulate spume drop production over the open ocean. Additionally, according to the expressions presented, laboratory production fluxes and oceanic fluxes would have differed by several orders of magnitude, and such large extrapolation calls into question the accuracy of such results, in addition to concerns regarding the assumption that the correct scaling is the breaking crest length and that this quantity can be accurately determined. No field measurements or other laboratory measurements covering the size range of drops measured by *Anguelova et al.* [1999] have been reported.

5.9. VERTICAL IMPACTION METHOD

The vertical impaction method has seen little use and only two applications have been reported. *Monahan* [1968] calculated the interfacial production flux of SSA particles with $r_{80} = 75$ μm based on measurements of the concentration of SSA particles of this size at 13 cm above the ocean (§4.2), obtaining $f_{int} = 1.2 \cdot 10^4$ m^{-2} s^{-1} for $U_{10} = 11$ m s^{-1} and $f_{int} = 3.25 \cdot 10^4$ m^{-2} s^{-1} for $U_{10} = 16$ m s^{-1}, respectively (Fig. 43). It is difficult to place much confidence in these values; few details of the calculations were reported, and it is not clear how these numerical values were obtained. The fluxes were apparently calculated on the assumption that SSA particles had the same vertical velocities at this height that they would have had if they had been produced by the jet drop mechanism [*Bortkovskii*, 1987, pp. 38-39]; however, at this height and for these wind speeds, it would be expected that some of these drops would be spume drops. Moreover, it is incorrect to assume that the jet drops have the same vertical velocities at that height they would have if ejected into still air, as the vertical and horizontal components of the drop velocity are coupled (§2.6.3). No mention was made of friction drag, but a jet drop of this size would be ejected with an initial velocity of approximately 3.5 m s^{-1} (on the assumption that the initial Reynolds number $Re_0 = 70$; §4.3.2.3). Such a drop would experience a drag force that would be initially nearly 10^4 times the force of gravity; without this drag in still air the drop would attain a height of more than 60 cm. Some confusion has also resulted over the flux units used by *Monahan* [1968], "droplets per cm^2 per sec per 0.2 log$D$," resulting in possible errors of up to a factor of 25 in the interpretation of these fluxes (§5.2.1.1).

Measurements using the vertical impaction method to determine the interfacial production flux of SSA particles with 3 μm $\lesssim r_{80} \lesssim$ 500 μm over the ocean were reported by *Bortkovskii* [1987, pp. 41-46], who detected and later sized drops that impacted a sampling substrate on a device that floated on the sea surface. The measurements were made at wind speeds of 10-20 m s^{-1}, mostly at heights of 15 cm above the instantaneous water surface. Total vertical fluxes of SSA particles at 15 cm were presented which increased with increasing wind speed at a rate greater than linear, from around $F_{int} = 10^4$ m^{-2} s^{-1} at 10 m s^{-1} to approximately 6-8 times this value at 20 m s^{-1}. Two "typical" normalized size distributions of the vertical SSA flux were presented for r_{80} from 3 μm to ~60 μm; these decreased over the range ~6 μm to ~37 μm according to *Bortkovskii* as $f_{int} \propto r_{80}^{-p}$, with $p \approx$ 1-2. The flux shown in Fig. 43 is obtained from the average of these two values using the value of the total number flux at 10 m s^{-1}. *Bortkovskii* attributed a small increase in these distributions (in the representation dF_{int}/dr_{80}) at r_{80} near 30 μm to spume drop production, concluding that most smaller drops are produced by bubble bursting and most larger drops by the spume mechanism; however, when the distributions are displayed in the form $dF_{int}(r_{80})/d\log r_{80}$ the scatter in the data make it difficult to discern any increase (Fig. 43). As the wind speed increased from 10 m s^{-1} to 20 m s^{-1}, the mean value of r_{80} increased roughly linearly from ~5 μm to ~10 μm and the root-mean-cube value of r_{80} increased roughly linearly from ~5 μm to ~20 μm, the greater increase in the root-mean-cube radius with wind speed implying that more larger drops are produced with increasing wind speed.

The main concerns with this method are how well the sampling device can function under oceanic conditions when waves are breaking and the extent to which it can detect drops that are transported nearly horizontally instead of vertically, as would be the situation for spume drops especially and for most drops under high wind conditions.

5.10. STATISTICAL WET DEPOSITION METHOD

The statistical wet deposition method (§3.9) calculates the effective production flux of small SSA particles from measurements of the size-dependent SSA concentration $n(r_{80})$ under the assumptions that these particles are uniformly mixed over the marine boundary layer and that wet deposition is the dominant removal mechanism. Consequences of this latter assumption are that measured concentrations would comprise those particles that had been produced since the last precipitation event and that the size dependence of the effective SSA production flux would be the same as the size dependence of the SSA number concentration. As discussed in §5.2.1.5, a modified version of this method was presented by *Hoppel et al.* [2002], which essentially combines the dry deposition method for SSA particles with $r_{80} \gtrsim 2$ μm and the statistical wet deposition method for smaller particles, allowing determination of the effective SSA production flux from size distributions of SSA concentration over a larger size range than can be applied to either method individually.

An estimate for the effective SSA production flux of small SSA particles ($r_{80} \lesssim 1$ μm) based on the statistical wet deposition method was presented in §4.1.4.5. If H_{mbl} is taken as 0.5 km (§2.4), the mean time since the last precipitation event as $\tau_{wet} = 3$ d (§2.7.1), and a typical value of $n(r_{80})$ characterizing small SSA particles as ~5 cm^{-3} over the range 0.1 μm $\lesssim r_{80} \lesssim 1$ μm, based on the canonical SSA size distribution of concentration at $U_{10} = 10$ m s^{-1} (§4.1.4.5), then $f_{eff}(r_{80})$ is approximately equal to $1 \cdot 10^4$ m^{-2} s^{-1} over this size range. Under these conditions this flux results in a daily increase in SSA number concentration of approximately 2 cm^{-3}. This estimate is shown in Fig. 41 with multiplicative uncertainty $\lessgtr 5$. The multiplicative uncertainty of the canonical SSA size distribution of concentration was estimated as $\lessgtr 3$ (§4.1.4.4), but the few measurements for small SSA particles, mostly at $U_{10} \lesssim 7$ m s^{-1}, are not so consistent as those for other size ranges, and the uncertainty in this size range is probably somewhat higher (Fig. 22). The multiplicative uncertainties in both H_{mbl} and τ_{wet} are estimated as $\lessgtr 2$; these values would result in a multiplicative uncertainty factor in f_{eff} of slightly less than $\lessgtr 5$ (§5.1.1). This method provides no information on the wind speed dependence of f_{eff}; nonetheless this estimate can serve as a rather robust consistency check to eliminate some estimates (e.g., those obtained by micrometeorological methods; §5.6) that would result in unrealistic SSA production fluxes, and it provides information in the size range $r_{80} = 0.1\text{-}1$ μm where few other estimates are available.

5.11. COMPARISON OF FLUX ESTIMATES

In the preceding sections determinations of the size-dependent effective and interfacial SSA production fluxes, $f_{eff}(r_{80})$ and $f_{int}(r_{80})$, respectively, based on the several methods have been presented and examined. Where possible a single estimate with associated uncertainty was chosen for each method. These uncertainties, which were estimated taking into account measurement uncertainties and/or variability of the input data (although not uncertainties associated with the satisfaction of assumptions required for the method), quantify how accurately the pertinent SSA production flux can currently be determined according to each method (Table 22). In this section estimates for the effective SSA production flux or the interfacial SSA production flux based on the several methods are compared with each other. These comparisons permit examination of the extent to which the SSA production fluxes obtained by the several methods can be said to agree to disagree, relative to their uncertainties, and allow identification of best estimates for each of the size ranges characterizing small SSA particles ($r_{80} \lesssim 1$ μm), medium SSA particles (1 μm $\lesssim r_{80} \lesssim 25$ μm), and large SSA particles ($r_{80} \gtrsim 25$ μm), together with the uncertainties in these estimates.

5.11.1. Effective SSA Production Flux

Estimates for the size-dependent effective SSA production flux $f_{eff}(r_{80})$ according to the steady state dry deposition method were presented in §5.2.1 and compared in Fig. 40. The estimate presented in §5.2.1.5 using the canonical SSA size distribution of concentration (§4.1.4.4) and the dry deposition velocity given by (4.6-8) for the transfer velocity through the surface layer, $v_{turb} + v_{S,Stk}$, along with associated uncertainty of a factor of $\lessgtr 4$, was concluded in §5.2 as providing the best estimate currently available based on this method, encompassing the other credible estimates over its range of applicability and providing a quantitative assessment of the confidence that could be placed in this estimate. This estimate, evaluated at $U_{10} = 10$ m s^{-1}, is also presented in Fig. 41, together with all determinations of f_{eff} that have been presented by other methods.

One immediate feature of Fig. 41 is the paucity of estimates of f_{eff} it contains. In addition to the estimate based on the steady state dry deposition method, the only other values are individual determinations of f_{eff} at two wind speeds based on the concentration buildup method (§5.4), three determinations based on micrometeorological methods (§5.6), and the estimate based on the statistical wet deposition method along with its associated uncertainty (§5.10). As noted in (§5.6), the determinations of f_{eff} by micrometeorological methods are almost certainly not pertinent to SSA production, in part because of the large fraction of non-SSA particles that were present. In addition they are inconsistent with measured SSA number concentrations, and thus are not considered further. The two determinations of f_{eff} based on the concentration buildup method are for the most part consistent with the estimate based on the steady state dry deposition method over the limited range of overlap, if the large error bars in these determinations are taken into account. These determinations are roughly an order of magnitude higher than the central value of the estimate based on the statistical wet deposition method over the range of overlap. To what extent, if any, this difference is an artifact of the measurement location (coastal) or the measurement technique of the concentration buildup method is not known. The determinations did not extend below $r_{80} \approx 0.25$ μm, and contained large uncertainty at this size, so whether or not the apparent trend of increasing f_{eff} with decrease in r_{80} extends to lower sizes is also unknown. The determinations from the concentration buildup method, for a marine boundary layer height of 0.5 km, yield a daily increase in the total SSA number concentration of 10-20 cm^{-3}, resulting in values for this quantity toward the high end of those typically observed (§4.1.3.1). Although this method holds much promise, its limited use to date argues against relying too heavily on these determinations.

The remaining estimates, that based on the statistical wet deposition method and that based on the steady state dry deposition method, cover only part of the range for which the effective SSA production flux is desired, $r_{80} = 0.1$-25 μm. In the r_{80} range constituting small SSA particles ($r_{80} \lesssim 1$ μm) only the estimate based on the statistical wet deposition method can be applied. This estimate results in values of total SSA number concentrations that are consistent with measured values (as this was the basis for its determination). Thus, this estimate of f_{eff} and the multiplicative uncertainty of $\times\!\!\!\!\div$ 5 associated with it must be considered the best currently available, although this estimate provides no size resolution and no wind speed dependence (§5.10). Likewise, in the size range 1 μm $\lesssim r_{80} \lesssim 25$ μm, characterizing medium SSA particles, the estimate for f_{eff} based on the steady state dry deposition method using the canonical SSA size distribution of concentration, along with the multiplicative uncertainty of $\times\!\!\!\!\div$ 4, must be considered the best currently available, although this estimate is applicable only to part of this size range, $r_{80} \gtrsim 5$ μm. An interpolation for the r_{80} range 1-5 μm which merges the estimate of f_{eff} based on the statistical wet deposition method and that based on the steady state dry deposition method at $U_{10} = 10$ m s^{-1}, along with an uncertainty that varies with size in this range, is also shown in Fig. 41. The two determinations of f_{eff} based on the concentration buildup method are for the most part consistent with this interpolation. Thus, over the range $r_{80} = 0.1$-25 μm the shaded regions in Fig. 41 are concluded to provide the best currently available estimate of f_{eff}.

The wind speed dependence of the effective SSA production flux is also available by the steady state dry deposition method from considerations of the wind speed dependence as reflected in the canonical SSA size distribution of concentration and dry deposition velocity. An expression for

$f_{\mathrm{eff}}(r_{80})$ resulting from this approach, (5.2-3), was obtained, according to which the value of f_{eff} is directly proportional to $U_{10}{}^{2.5}$ over the range $U_{10} = 5$-20 m s^{-1}. According to this expression, the size dependence of f_{eff} is directly proportional to $r_{80}{}^{-2.5}$, independent of wind speed. The size-dependent effective production flux of SSA mass $f_{\mathrm{eff}}^{\mathrm{M}}(r_{80})$ is therefore directly proportional to $r_{80}{}^{0.5}$, an increasing function of r_{80} up to 25 μm. It would thus seem that the dominant contribution to this flux would be from larger SSA particles. However, as the failure of SSA particles with $r_{80} \gtrsim 25$ μm to equilibrate to the ambient RH (§2.9.1) and other reasons noted in §2.9 result in an abrupt decrease in the effective SSA production flux near this size, this contribution extends only to $r_{80} \approx 25$ μm. The total effective production flux of SSA mass $F_{\mathrm{eff}}^{\mathrm{M}}$ can therefore be evaluated by (2.1-31) for this formulation. With the choice $U_{10} = 10$ m s^{-1} as the wind speed yielding the global mean value of $U_{10}{}^{2.5}$, an estimate for the annual global SSA mass production is obtained as $(5 \times\!\!\!\!\div 4) \cdot 10^{12}$ kg; this estimates encompasses most of the other estimates in Table 1. Based on the uncertainties in the canonical SSA size distribution of concentration and dry deposition velocities, it would be difficult to justify an estimate with narrower uncertainty range.

No wind speed dependence was associated with the estimate for f_{eff} for small SSA particles using the statistical wet deposition method. Small SSA particles are mostly film drops produced by bubble bursting, whereas most medium SSA particles are jet drops, which are also formed by bubble bursting. The sizes of bubbles forming these two types of drops are different, with film drops generally being formed by much larger bubbles. Although the dependence on meteorological conditions (i.e., wind speed) of the formation of larger bubbles which produce mostly film drops may be different from that of smaller bubbles which produce most of the jet drops, it is reasonable to assume that the production flux of small SSA particles would exhibit a wind speed dependence similar to that of medium SSA particles. Thus, a wind speed dependence for f_{eff} of $U_{10}{}^{2.5}$ is assumed for small SSA particles also.

This wind speed dependence is empirical, but provides a good approximation to f_{eff} determined by this method over the wind speed range $U_{10} = 5$-20 m s^{-1}, encompassing wind speeds most often encountered over the oceans (§2.3.1). Although physical arguments have been made for wind speed dependencies of quantities pertinent to SSA such as whitecap ratio W (§4.5.2), these quantities generally exhibit such large uncertainty (e.g., Fig. 36) that little is to be gained by favoring such dependencies over a simple empirical fit. Additionally, given the large uncertainties in determinations of the interfacial and effective SSA production fluxes (Table 22), it would seem difficult to justify one specific dependence over another. Over the wind speed range $U_{10} = 5$-20 m s^{-1}, for example, a wind speed dependence proportional to U_{10}^2 or U_{10}^3 would differ from one proportional to $U_{10}^{2.5}$ by no more than ~40%. Such a difference is well within the uncertainty of $\gtrless 4$ associated with the estimate for f_{eff}, and it would be difficult to discern one dependence over another on the basis of measurements. As stated above, the range of applicability of this formulation is $U_{10} = 5\text{-}20$ m s^{-1}. Below this little SSA production is expected to occur, and there are too few data above this range to confidently draw conclusions. Spume drop production, which very likely exhibits a different wind speed dependence with higher onset threshold, may be an important mechanism of production of medium SSA particles at higher wind speeds.

A single expression that fairly closely matches the estimate of f_{eff} based on the statistical wet deposition method, that based on the steady state dry deposition method at $U_{10} = 10$ s^{-1}, and the interpolation region, over the size range 0.1 μm $\lesssim r_{80} \lesssim 25$ μm, is a lognormal for $dF_{\text{eff}}/d\log r_{80}$ in the representation given by (2.1-9), with $f_{\text{eff},0} = 1.5 \cdot 10^4$ m^{-2} s^{-1}, $r'_{80} = 0.3$ μm, and $\sigma = 4$, shown by the dotted line in Fig. 41. A multiplicative uncertainty $\gtrless 4\text{-}5$ should be associated with this estimate. The magnitude of f_{eff} at other wind speeds over the range $U_{10} \approx 5\text{-}20$ m s^{-1} would likewise be directly proportional to $U_{10}^{2.5}$. This expression readily allows quantities of interest pertinent to the effective SSA production flux.

In summary, based on these arguments it is concluded that the estimates for f_{eff} shown by the shaded regions in Fig. 41, consisting of the estimate based on the statistical wet deposition method with uncertainty $\gtrless 5$ for SSA particles with 0.1 μm $\lesssim r_{80} \lesssim 1$ μm, the estimate based on the steady state dry deposition method with uncertainty $\gtrless 4$ for SSA particles with 5 μm $\lesssim r_{80} \lesssim 25$ μm, and an interpolation region for $r_{80} \approx 1\text{-}5$ μm with comparable uncertainty, along with a wind speed dependence $U_{10}^{2.5}$ for $U_{10} = 5\text{-}20$ m s^{-1}, provides a robust estimate for f_{eff} and its uncertainty. No estimate is provided for the effective production flux of SSA particles with $r_{80} \lesssim 0.1$ μm. Sea salt aerosol particles in this size range have been detected in the marine atmosphere (§4.1.1.1) and production of SSA particles of these sizes has been observed in laboratory experiments (§4.3.3.1; §4.3.4), but the relative abundance of these particles in the marine atmosphere is low, and the concentrations of these particles are low compared to other sizes ranges of SSA particles (§4.1.1.1). As discussed throughout this review, the value of f_{eff} for SSA particles with $r_{80} \gtrsim 25$ μm is taken to be zero, as these particles do not contribute to the effective SSA production flux.

5.11.2. Interfacial SSA Production Flux

Estimates of the interfacial SSA production flux $f_{\text{int}}(r_{80})$ according to the several methods are presented in Fig. 43.

The estimate for $U_{10} = 10$ m s^{-1} based on the whitecap method (§5.3.1) with its associated uncertainty $\gtrless 14$ is shown as a shaded blue region; this estimate was based on the central value of the shaded gray region in Fig. 32 for estimates of the SSA production flux over whitecap area $f_{\text{wc}}(r_{80})$. The estimate based on the bubble method (§5.5.1.4) with its associated uncertainty $\gtrless 17$ is shown as a shaded green region; this estimate was based on the central value of the shaded gray region in Fig. 42 for estimates of f_{int} according to the bubble method. As noted in §5.5.2, this estimate refers to production of jet drops only. Values of f_{int} calculated from formulations of spume drop production at $U_{10} = 10$ m s^{-1} based on the along-wind flux method (§5.7.1), and determinations based on the direct observation method (§5.8) and the vertical impaction method (§5.9), are also shown in Fig. 43. Additionally, the best estimate for the effective SSA production flux determined above, consisting of the estimates using the canonical SSA size distribution of concentration based on the statistical wet deposition method and steady state dry deposition method, and the interpolation region, is shown in Fig. 43, along with its associated uncertainty indicated by vertical lines.

For small SSA particles the only estimate of f_{int} is provided by the whitecap method, but as discussed in §5.3.2, application of this method may result in overestimation of the production flux of such particles. Small SSA particles, which are predominantly film drops, are produced mainly from bubbles with $R_{\text{bub}} \gtrsim 1$ mm (Fig. 30) which rise rapidly to the surface (Fig. 34) and would be expected to burst and produce small SSA particles during the first portion of the whitecap lifetime. For small SSA particles the interfacial SSA production flux and the effective SSA production flux are expected to be nearly identical (as noted throughout this review), but the central value of the estimate of the production flux by the whitecap method is roughly an order of magnitude greater than that obtained by the statistical wet deposition method, albeit with much greater uncertainty. The central value of the estimate for f_{int} based on the whitecap method in this size range would result in a daily increase in SSA number concentration of ~20 cm^{-3} for a marine boundary layer height of 0.5 km. Under the assumption of several days between occurrences of wet deposition, such a flux would result in SSA number concentrations of up to 100 cm^{-3}—much greater than those typically measured (§4.1.3.1). Hence it can be concluded that the estimate of f_{eff} provided by the statistical wet deposition method and its associated uncertainty would be more accurate in this size range. The wind speed dependence of f_{int} according to the whitecap method is given by the wind speed dependence of W, for which power laws of the form $W \propto U_{10}^p$ for a range of p from ~2 to ~4 appear to represent the measured values for U_{10} up to 20 m s^{-1} (Fig. 36) equally well, considering the large amount of scatter in the data. Hence little justification can be made for a wind speed dependence different from that presented above for f_{eff} with $p = 2.5$.

In the range of medium SSA particles estimates for f_{int} have been obtained by the whitecap method (§5.3.1), bubble method (§5.5.1), along-wind flux method (§5.7.1), and vertical impaction method (§5.9). There are too few measurements of *Bortkovskii* [1987] by the vertical impaction method to evaluate its accuracy, but these measurements appear to be consistent with estimates based on the whitecap method and bubble method within the large uncertainties associated with these estimates, although substantially greater than estimates of f_{eff} from the steady state dry deposition method. As noted below, values of the ratio of the effective SSA production flux to the interfacial SSA production flux Ψ_{f} are not sufficiently less from unity to account for this difference. The estimate of *Ling et al.* [1980] can be discounted, as noted in §5.7.1. Estimates by the along-wind flux method by *Monahan et al.* [1983b] and *Andreas* [1990b, 1998] do not provide meaningful information in this size range, as their magnitudes were chosen to match the value of f_{int} obtained by another method at a specified value of r_{80} in this size range; additionally, the size dependencies of these estimates are incorrect, as discussed in §5.7.2. Within the r_{80} range up to ~10 µm, the best estimate of f_{eff} obtained above (§5.11.1) is consistent with both the estimates obtained by the whitecap method and the bubble method. In this size range the value of Ψ_{f} is expected to be between 0.5 and 1 (Fig. 13); thus f_{int} and f_{eff} would have comparable values. As the uncertainty associated with the estimate of f_{eff} is less than that of either the estimate of f_{int} by the whitecap method or that by the bubble method, it provides the best constraint on the SSA production flux in this size range.

For the size range $r_{80} = 10\text{-}25$ µm the only credible estimates of f_{int} are that based on the bubble method and the determinations using the vertical impaction method. However, neither of these estimates provides any indication of the wind speed dependence of f_{int}, or how it might differ from that of f_{eff} in this size range. Additionally, the uncertainty associated with the estimate of f_{int} based on the bubble method is quite large ($\gtrless 17$), and only two determinations using the vertical impaction method were reported. Hence it is difficult to choose a best estimate or to estimate an associated uncertainty. The possibility suggests itself of using the ratio of the effective SSA production flux to the interfacial SSA production flux $\Psi_{\text{f}}(r_{80}) \equiv f_{\text{eff}}(r_{80})/f_{\text{int}}(r_{80})$ together with an estimate of f_{eff} to arrive at an estimate of f_{int}. Although several formulations for Ψ_{f} have been proposed (§5.1.2; Fig. 13), the one considered here is the steady state concentration ratio $n(10 \text{ m})/n(0.1 \text{ m})$ given by (2.9-4) at $U_{10} = 10$ m s^{-1}. This ratio decreases from 0.8 at $r_{80} = 10$ µm to ~0.06 at $r_{80} = 25$ µm (Fig. 13) indicating that f_{eff} becomes increasingly less

than f_{int} with increasing particle size. The value of f_{int} determined from this value of Ψ_{f} and the best estimate for f_{eff} determined above (given by Eq. 5.2-3) is shown in Fig. 43. Another estimate of f_{int} by *Toba* [1965b], determined similarly (§5.1.2), is also shown (although as noted in §5.2.1.2, this estimate was based on size distributions of SSA concentration measured at ~700 m). These estimates differ considerably for $r_{80} \gtrsim 5$ µm, with that of *Toba* becoming increasingly greater as r_{80} approaches 25 µm, in part because of the lower height to which 10 m SSA concentrations were extrapolated, $4.5 \cdot 10^{-4}$ m vs. 0.1 m (§4.6.5). Although the value of f_{int} according to each estimate is an increasing function of r_{80} at 25 µm, the uncertainties at this size both in measurements of $n(r_{80})$ (and thus the estimate for f_{eff}) and in the estimate for Ψ_{f} would limit the confidence that could be placed in this conclusion. It would seem that this approach could not account for the difference noted above between the estimate of f_{eff} by the dry deposition method and the determination of f_{int} by *Bortkovskii* [1987] by the vertical impaction method. As discussed in §5.1.2, the value of Ψ_{f} is strongly dependent on the choice of the lower height to which 10 m concentrations are extrapolated, and little justification can be given for one choice over another, resulting in large uncertainty in estimates of f_{int} from f_{eff} at sizes for which Ψ_{f} differs appreciably from unity. Consequently, it is not possible to constrain estimates of f_{int} in the size range $r_{80} = 10\text{-}25$ µm by this approach.

For large SSA particles estimates have been presented based on the along-wind flux method (§5.7.1), the direct observation method (§5.8), and the vertical impaction method (§5.9). The determinations of f_{int} based on the direct observation method presented by *Anguelova et al.* [1999] were based on laboratory measurements on small waves, and the extent to which these results are applicable to oceanic whitecap production is questionable, especially the method of extrapolating these results to the ocean (§5.8). Little confidence can be placed in the estimates of *Monahan et al.* [1968] by the vertical impaction method, as noted in §5.9, and only two determinations were reported by *Bortkovskii* [1987, Fig. 1.18]. The several other estimates of the interfacial SSA production flux of large SSA particles are based on the along-wind flux method, and as discussed in §5.7.2, these cannot be considered credible. Failure to take into account the size dependence of the mean distance $\overline{L}_{\text{p}}(r_{80})$ traveled by SSA particles results in incorrect size dependence of f_{int}. In the formulations of *Monahan et al.* [1983b] and *Andreas* [1990b, 1998], the magnitude of f_{int} was determined by matching its value to the value of f_{int} (or in some instances f_{eff}) obtained by some other method, an approach for which there is little physical basis. The magnitude of the estimate of f_{int} by *Wu* [1993] was not obtained by this procedure, but this estimate is many orders of magnitude lower than determinations by other methods at $r_{80} = 25$ µm (including determinations of f_{eff}, compared to which it should certainly be greater), and as the magnitude and wind speed dependence of f_{int} were determined from laboratory measurements, concerns with the applicability of results from small (several centimeter) laboratory waves to oceanic spume drop production and with scaling laboratory wind speeds to their oceanic equivalents raise questions over the confidence that can be placed in this estimate. In addition to the above concerns, the field measurements upon which all of these formulations were based occurred at such low wind speeds that it is questionable to what extent spume drops contributed, and the types of drops present in laboratory studies have not been conclusively determined. For these reasons both the shape and the magnitude of the size distribution of the interfacial production flux of large SSA particles, and their dependencies on wind speed, must be considered unknown, and the formulations that have been proposed can be discounted. There appears to be little information that would aid in determining the size dependence, magnitude, or wind-speed dependence of the interfacial production flux for large SSA particles. Although this evaluation might seem overly pessimistic, it is difficult to draw other conclusion based on the data available. Consequently, it is not currently possible to evaluate quantities pertinent to considerations of air-sea transport of mass, momentum, or energy such as the total interfacial flux of SSA mass F_{int}^M (§2.1.7), even to within orders of magnitude. Although it might be possible to bound total fluxes by other constraints, such constraints would provide no indication of the size dependence or wind-speed dependence of the interfacial SSA production flux.

5.11.3. Summary

Although SSA particles with $r_{80} \lesssim 0.1$ µm are certainly present in the marine atmosphere, they generally comprise a small fraction of the total number of marine aerosol particles in that size range and contribute little to concentrations of SSA number. Virtually nothing is known about their production (other than they are almost certainly film drops) or about the dependence of this production on particle size, wind speed, or other controlling factors.

Small SSA particles, $r_{80} \approx 0.1$-1 µm, provide the dominant contribution to SSA number concentration. These particles are mostly film drops produced by the bursting of bubbles. Concentrations of these particles, which have long atmospheric residence times against removal by dry deposition, are controlled primarily by wet deposition; consequently these concentrations rarely if ever attain steady state values, and they exhibit little correlation with local conditions. As wet deposition is expected to remove a large fraction of SSA particles in this size range, the size dependence of the

production flux of SSA particles in this size range is nearly the same as the size dependence of their concentration. However, there have been relatively few measurements of concentrations of SSA particles of these sizes, and this size distribution is not well known. The interfacial and effective production fluxes of these particles are nearly the same (i.e., $\Psi_f \approx 1$), but little is known concerning the size dependence or the wind speed dependence of these production fluxes. The estimate $f_{eff}(r_{80}) \approx 1 \cdot 10^4$ m^{-2} s^{-1} with an uncertainty $\lesssim 5$, nearly independent of size, was presented in §5.10.

Medium SSA particles, $r_{80} \approx 1$-25 µm, provide the dominant contribution to many atmospheric processes of interest. These particles are mainly jet drops produced by bursting bubbles. Concentrations of these particles are controlled by both wet deposition and dry deposition, and there is large variability in atmospheric residence times for SSA particles over this size range, from hours to days. The effective and interfacial production fluxes of these particles are nearly the same in the lower portion of this size range (i.e., 1 µm $\lesssim$ $r_{80} \lesssim 5$ µm), but begin to differ appreciably with further increase in particle size. The best estimate for the effective production flux of medium SSA particles is that presented in §5.11.1 for f_{eff} based on the steady state dry deposition method and the canonical SSA size distribution of concentration. This estimate has an associated uncertainty of $\lesssim 4$, and a wind speed dependence given by $U_{10}^{2.5}$; although numerous formulations could be proposed for this dependence, there is little to justify choice of one over another. For SSA particles in the lower portion of this size range (i.e., $r_{80} \lesssim 10$ µm), the best estimate for the interfacial SSA production flux is also given by the above estimate for f_{eff}, and it expected that the wind speed dependence is likewise that given above. For SSA particles in the upper portion of this size range (i.e., $r_{80} \approx 10$-25 µm) the interfacial SSA production flux becomes increasingly greater than the effective SSA production flux, but little can be stated with confidence about the magnitude or size dependence of f_{int}.

Large SSA particles, $r_{80} \gtrsim 25$ µm, provide the dominant contribution to interfacial fluxes of mass and momentum. These particles are mainly jet drops and spume drops, but little is known about the relative abundance of these types and the dependence of this abundance on size or meteorological conditions. These particles occur mainly in the lowest several meters above the sea surface and have residence times of order seconds, during which their radii change little from their formation values. As SSA particles of these sizes typically do not attain heights of 10 m, they do not contribute to the effective SSA production flux. The formulations that have been proposed for interfacial production fluxes of SSA particles in this size range are known to be incorrect, and virtually nothing is known about this flux or its dependencies on particle size and wind speed.

In summary, for a wide range of particle sizes of interest little can be stated with confidence concerning the effective and interfacial SSA production fluxes. Even when these quantities can be estimated the associated uncertainties are more than half an order of magnitude; consequently, little significance can be attached to numerical values past the first digit in any proposed formulation or numerical result. Likewise, in most situations there is little or no physical basis for expressions or functional forms that have been proposed. The current situation is perhaps best summed up by *Hoppel et al.* [2002], who noted that "all current formulations are crude and must be viewed as little more than order-of-magnitude estimates."

6. Applications and Implications

This review was stimulated by the need to represent radiative and cloud-modifying influences of aerosol particles in chemical transport models and climate models to permit examination of these influences on climate change. Sea salt aerosol is but one category of atmospheric aerosol, a natural aerosol, and a rather simple one, in that most SSA particles undergo little modification in the atmosphere that would affect their concentrations, properties, influences, transport, or fate. Nonetheless SSA particles, by directly scattering radiation (§2.1.5.3) and by modifying cloud properties (§2.1.5.1) exert important influences on climate. Perhaps just as importantly SSA, as the major aerosol species naturally present in the marine boundary layer, affects the evolution, cloud-modifying properties, and climatic influence of other aerosol species, in particular the influences of anthropogenic aerosols, necessitating accurate knowledge of SSA concentrations and properties if these influences are to be accurately described. This section reviews the basis for these climatic influences, examines application of a particular formulation for the SSA production flux that has been widely applied in chemical transport models, and concludes with considerations on improvement of quantitative understanding of SSA production and its dependence on controlling factors.

6.1. THE CLIMATE CONNECTION

In 1987 a highly influential paper was published that proposed a regulatory feedback mechanism for control of climate by marine phytoplankton through the medium of biogenic aerosol affecting cloud microphysical properties, the so-called CLAW hypothesis after the initials of the authors, *Charlson, Lovelock, Andreae, and Warren* [1987]. According to this hypothesis biogenic aerosol particles result from emissions of dimethylsulfide (DMS) by marine phytoplankton. This gaseous trace species is oxidized in the atmosphere to form sulfate and methylsulfonic acid (MSA), both of which are low-vapor-pressure materials which can condense on existing aerosol particles and/or nucleate to

form new particles. These particles serve as cloud condensation nuclei (CCN), increasing the concentration of cloud drops and increasing multiple scattering of sunlight within clouds and thereby increasing cloud albedo. According to the CLAW hypothesis, an increase in the population of phytoplankton would result in an increase in DMS emissions, enhanced concentrations of marine aerosol particles and CCN, enhanced concentrations of cloud drops in marine stratiform clouds, enhanced cloud reflectivity, and decreased insolation at the sea surface. If this decreased insolation resulted in decreased phytoplankton population, then the overall process would constitute a regulatory feedback system for biogenic control of climate.

Publication of this paper stimulated a flurry of research that has hardly abated—as of 2004 the paper has been cited more than 1200 times. An early concern regarding the hypothesis was the assumption in its examination of the sensitivity of the feedback system that concentrations of sea salt particles at cloud height are typically not more than 1 cm^{-3}. *Blanchard and Cipriano* [1987] cited several studies showing number concentrations of SSA particles of 15-20 cm^{-3} (*cf.* §4.1.4.3) Such a higher background concentration would greatly diminish the sensitivity of any perturbation of cloud reflectivity and related effects by incremental cloud drop concentrations.

An immediate consequence of the CLAW paper was the recognition by a broader scientific community of the potential for influence of climate by anthropogenic aerosols in the marine environment. The mechanism on which the CLAW hypothesis rested had been advanced in a series of papers beginning in the 1970s by *Twomey* and coworkers [*Twomey*, 1974, 1977ab; *Twomey et al.*, 1984; *Twomey*, 1991], which had demonstrated that modest increases in cloud drop concentrations by anthropogenic aerosol particles would lead to enhancement of cloud reflectivity, exerting a cooling influence on climate. A key actor here is sulfate, formed in large part by atmospheric oxidation of sulfur dioxide emitted into the atmosphere as a byproduct of combustion of fossil fuel that contains sulfur as an impurity. A survey of sulfate in the marine atmosphere stimulated by the CLAW paper showed, by comparison of the anthropogenically perturbed northern hemisphere to the relatively pristine southern hemisphere, that sulfate concentrations in the remote marine atmosphere were enhanced on account of anthropogenic emissions

Sea Salt Aerosol Production: Mechanisms, Methods, Measurements, and Models
Geophysical Monograph Series 152
Copyright 2004 by the American Geophysical Union.
10.1029/152GM06

[*Schwartz*, 1988b]. Shortly afterwards it was suggested [*Albrecht*, 1989] that increased cloud drop concentrations resulting from anthropogenic aerosols might have a further cooling influence on climate by inhibiting precipitation development from marine clouds, thereby enhancing their lifetimes and their contribution to planetary albedo. These influences have been dramatically demonstrated by enhanced cloud reflectivity in exhaust plumes of ships—so-called ship trails [*Conover*, 1966; *Scorer*, 1987; *Coakley et al.*, 1987; *Radke et al.*, 1989]—and less dramatically, but more importantly in terms of global radiation budgets, in widespread enhancement of albedo of marine stratiform clouds attributed to anthropogenic emissions [e.g. *Albrecht et al.*, 1995a; *Brenguier et al.*, 2000; *Schwartz et al.*, 2002].

To the extent that anthropogenic emissions influence clouds in the marine atmosphere, these emissions, which have been increasing over the industrial period more or less in parallel with emissions of carbon dioxide, may be offsetting some of the "greenhouse" infrared radiative forcing by carbon dioxide over this period. Failure to take into account the cooling influence of anthropogenic aerosols in studies of climate change, as was the situation until the mid 1990's, would have the consequence of decreasing the apparent sensitivity of Earth's climate to radiative forcing by greenhouse gases that might be inferred from examination of climate change over the industrial period. Recognition of the potential climate influence of sulfate and other aerosol species has stimulated much research in the subsequent decade and a half since the CLAW hypothesis was proposed. This research has been reviewed in the most recent assessment report of the Intergovernmental Panel for Climate Change (IPCC) [*Penner et al.*, 2001], and a concomitant assessment of radiative forcing of climate change has identified radiative forcing by anthropogenic aerosols as the greatest uncertainty in secular forcing of climate change over the industrial period [*Ramaswamy et al.*, 2001].

Much of the research on aerosol forcing has been directed to distinguishing the influences of anthropogenic aerosols from those of natural aerosols, of which sea salt aerosol is a dominant contributor, especially in the marine atmosphere. The indirect radiative forcing is especially sensitive to incremental concentrations of aerosol particles. Other factors being held constant, an increase in the number concentration of cloud drops of 30% in marine stratus clouds and the resultant albedo enhancement would result in a global average cooling radiative forcing of 1 W m^{-2} [*Charlson et al.*, 1992]; such a forcing compares to the greenhouse forcing due to enhanced CO_2 concentrations over the industrial period of 1.6 W m^{-2} [*Ramaswamy et al.*, 2001]. In this context it is clear that accurate knowledge of SSA is essential to define the base case in the absence of influences of anthropogenic aerosols necessary to evaluate the forcing over the industrial period as a difference relative to the present situation. Also, as the forcing depends on the ratio of the cloud drop concentration with and without the incremental aerosol [*Charlson et al.*, 1992], it is quite sensitive to cloud drop concentrations taken for the preindustrial case and hence to concentrations of SSA particles. These considerations make it imperative to gain a quantitative description of the properties and processes not only of anthropogenic aerosols but also of natural aerosols, and to be able to represent these processes in chemical transport models (CTMs) and ultimately in climate models. In this respect accurate knowledge of SSA properties and size-dependent and total SSA number concentrations is especially important.

Accurate representation of aerosols in CTMs is a necessary and important first step toward quantifying aerosol influences in climate models. Such quantification is needed to examine aerosol influence on atmospheric radiation, clouds, and precipitation in the context of quantifying the role of anthropogenic aerosols in climate change. Once the accuracy of calculations with a CTM of size distributions of concentration of aerosol species of interest is established, it becomes possible to calculate with known confidence the concentrations, geographical distribution, and properties of the several aerosol species for various emission scenarios—present, preindustrial, and as a function of intermediate time—and thereby determine the forcing as a difference relative to preindustrial. Considerable effort is being directed to development and evaluation of such models [*Penner et al.*, 2001]. In this context SSA is a key aerosol species of interest.

6.2. REPRESENTATION OF SSA PRODUCTION FLUX IN CHEMICAL TRANSPORT MODELS

Perhaps the most important required input to chemical transport models is the source strength of the species being modeled, and it was this requirement for SSA that stimulated the present review, specifically the need for a formulation that accurately represents the size-resolved SSA production flux as a function of meteorological conditions and location. As concentrations of SSA are, to good first approximation, linear in the production flux (*cf.*, §2.8.3), in models as in nature, any multiplicative uncertainty in SSA production flux translates into a roughly equivalent multiplicative uncertainty in modeled SSA concentrations. Inevitably the skill of CTMs must be evaluated by comparison with observations. Consequently a further requirement of production fluxes for chemical transport models is an estimate of the associated uncertainties necessary to make meaningful comparisons with observations. Here the present status of representing production of SSA in CTMs is examined.

The quantity representing production of SSA that is probably most useful as input to CTMs is the size-dependent effective SSA production flux $f_{eff}(r_{80}) \equiv dF_{eff}(r_{80})/d\log r_{80}$ as a function of wind speed and any other controlling variables. The size-dependent effective SSA production flux $f_{eff}(r_{80})$, as defined here for 10 m, is indicated rather than the size-dependent interfacial SSA production flux $f_{int}(r_{80})$, as the effective flux denotes the flux of SSA particles that remain in the atmosphere sufficiently long to require being modeled on regional to global scales treated by CTMs. Virtually all parameterizations of $f_{eff}(r_{80})$ are presented in terms of the wind speed at 10 m; although this quantity is generally not available from meteorological drivers used in chemical transport modeling, it is readily calculable from wind speed at the lowest model level and the temperature profile.

Perhaps the most widely used parameterization for SSA production flux in CTMs is the formulation of interfacial SSA production flux presented by *Monahan et al.* [1983b, 1986] (§5.3.1) for the stated size range of validity r_{80} = 0.8-8 µm, and extensions by subsequent investigators to lower and higher r_{80}. As this formulation yields the interfacial SSA production flux $f_{int}(r_{80})$, in principle it should be corrected by multiplying by $\Psi_f(r_{80})$, the ratio of the size-dependent effective and interfacial SSA production fluxes (§2.1.6), were this quantity known, although in practice typically no correction is made (for most of the indicated size range this correction would be minimal). As the formulation of *Monahan et al.* has served as the basis for evaluation of the skill of CTMs, and has been used to examine radiative and cloud modifying influences of SSA, it is briefly reviewed here.

As described in §5.3.1 the formulation of *Monahan et al.* [1983b, 1986] was obtained by the whitecap method (§3.2) as the product of the size-dependent SSA production flux over whitecap area $f_{wc}(r_{80})$ and the whitecap ratio W. The production flux per whitecap area f_{wc} used by *Monahan et al.* was determined in laboratory experiments with breaking waves from the measured increase in concentration of SSA particles $\Delta n(r_{80})$ for three individual breaking wave events denoted 372, 374, and 387 in Fig. 6 of *Monahan et al.* [1982] (shown also in Fig. 32), all measured on a single day [§4.3.4.2; Fig. 4 of *Monahan et al.*, 1982]. These values of f_{wc} exhibit a scatter about their mean at any value of r_{80} in the range 0.8-8 µm of roughly a factor of 2. The range of applicability of the formulation was based on the criterion that the signal to background ratio in the measurements of Δn exceed 50%. The upper limit of this size range was later extended to 10 µm [*Monahan*, 1986; *Spillane et al.*, 1986]. Subsequent papers by *Monahan* and coworkers [*Woolf et al*, 1988; *Monahan and Woolf*, 1988] presented alternative formulations for $f_{wc}(r_{80})$ which departed from the earlier formulation by up to an order of magnitude at $r_{80} \approx 8$ µm (Fig. 32), but still later papers [e.g., *Monahan and Lu*, 1990, *Monahan*, 1993] reverted to the earlier formulation. The

dependence of $f_{int}(r_{80})$ on wind speed was obtained under the assumption (inherent in the whitecap method) that f_{wc} is independent of wind speed and that the wind-speed dependence is therefore contained entirely in the whitecap ratio W, that is, $f_{int}(r_{80}, U_{10}) = f_{wc}(r_{80})W(U_{10})$. The formulation for $W(U_{10})$ adopted by *Monahan et al.* (Eq. 4.5.2) was based on measurements with U_{10} up to 17 m s^{-1} (§4.5.2).

The SSA production flux formulation of *Monahan* [1983b, 1986] has been used in numerous chemical transport models [e.g., *Pandis et al.*, 1994; *Russell et al.*, 1994; *Gong et al.*, 1997a; *Fitzgerald et al.*, 1998b; *Capaldo et al.*, 1999; *Lohmann et al.*, 1999, 2000; *van den Berg et al.*, 2000; *Pryor and Sørenson*, 2000; *Song and Carmichael*, 2001a; *Grini et al.*, 2002; *Chin et al.*, 2002]. In such studies this formulation has frequently been applied far beyond the stated size range of applicability r_{80} = 0.8-8 µm. For instance, in an investigation of the turbulent transfer and deposition of SSA particles, *Burk* [1984] used this formulation for r_{80} as large as 25 µm and *Stramska* [1987] used it for r_{80} as large as 30 µm. In studies involving marine aerosol production and growth, *Pandis et al.* [1994], *Russell et al.* [1994], and *Capaldo et al.* [1999] used a lognormal expression (similar to Eq. 4.3-4) based on the *Monahan et al.* formulation, which was assumed to apply to all particle sizes, to calculate concentrations of SSA particles that act as cloud condensation nuclei. *Gong et al.* [1997a] used this formulation in a chemical transport model to describe the production of SSA particles with r_{80} as small as 0.06 µm. In studies of chemistry and aerosol dynamics *Fitzgerald et al.* [1998b] and *van den Berg et al.* [2000] used this formulation to describe the production of SSA particles as small as $r_{80} = 0.4$ µm. *Lohmann et al.* [1999, 2000] used a lognormal expression based on this formulation centered at $r_{80} \approx 0.42$ µm (smaller by a factor of 2 than the lower limit of applicability stated by *Monahan et al.*) to model the contribution of SSA particles to the concentration of cloud concentration nuclei. *Song and Carmichael* [2001a] used this formulation to model the evolution of dust and of SSA particles as small as $r_{80} = 0.005$ µm. *Grini et al.* [2002] used this formulation to describe the production of SSA particles with r_{80} down to 0.03 µm. To model the contribution of SSA to aerosol optical depth *Chin et al.* [2002] used this formulation for SSA particles as small as $r_{80} = 0.1$ µm. Finally, the formulation of *Monahan et al.* [1983b, 1986] was extended down to $r_{80} = 0.062$ µm and up to $r_{80} = 16$ µm by *Guelle et al.* [2001], down to $r_{80} = 0.3$ µm and up to $r_{80} = 25$ µm by *Hoppel et al.* [2002], and down to $r_{80} \approx 0.5$ µm and up to $r_{80} = 180$ µm by *Schulz et al.* [2004], for comparison with other SSA production flux formulations.

As noted in §4.3.4.2, extrapolation is quite sensitive to the fitting function; the formulation of *Monahan et al.* [1983b, 1986] for $f_{wc}(r_{80})$ and the lognormal expression (4.3-2) agree to within 10% over the range of the measurements (i.e., $r_{80} = 0.8$-8 μm), but when extrapolated to $r_{80} = 0.03$ μm these two expressions differ by some six orders of magnitude (Fig. 32). Such a difference amply illustrates the perils of extrapolation, and calls into question the results of any of any model based on extrapolation of the *Monahan et al.* formulation below $r_{80} = 0.8$ μm.

The skill of chemical transport models representing SSA on global scales was assessed in a recent intercomparison conducted as part of the IPCC examination of performance of aerosol chemical transport models [*Penner et al.*, 2001]. Six models participated in the comparison, all of which used the *Gong et al.* [1997a,b] extension ($r_{80} = 0.06$-16 μm) of the *Monahan et al.* [1983b, 1986] formulation (Fig. 32). Mass concentration of SSA was calculated in two size categories, radius (RH not specified) less than and greater than 1 μm, but only total SSA mass concentrations were reported (thus these results would be little affected by the extension of the size range below $r_{80} = 0.8$ μm where little mass concentration would reside). Annual average SSA mass concentrations were compared to each other and to measurements of *Prospero* and coworkers at 7 locations in the North and South Pacific (including also Cape Grim, Tasmania). This intercomparison indicated that relative to the annual average SSA mass concentration, 16.1 μg m^{-3} (cf. Fig. 17), the average bias for the six models ranged from -13.5 μg m^{-3} to $+5.3$ μg m^{-3}. As seen in Fig. 44, none of the models exhibited skill in capturing the site-to-site variability.

Several concerns arise with studies such those noted above. A primary concern regards extrapolation of the SSA production flux beyond the region of applicability. The SSA mass production flux given by the *Monahan et al.* [1983b, 1986] formulation increases with increasing r_{80} throughout much of the originally specified range of applicability; if extended beyond this range there is a slight minimum at $r_{80} \approx 10$ μm above which the production flux again increases. Consequently, SSA mass flux and mass concentration are sensitive to the upper limit chosen for the size distribution of the production flux; extrapolation of the SSA mass flux from $r_{80} = 8$ μm to 16 μm would increase the SSA mass production flux by 40%. Much more importantly, as according to the *Monahan et al.* formulation the production flux of SSA particles increases with decreasing r_{80} over the entire range of applicability, the dominant contribution to the SSA number flux in an extrapolation would come from particles with r_{80} less than this stated range. Thus, for example, extrapolation of the number flux from $r_{80} = 0.8$ μm to 0.06 μm would increase the total SSA number production flux by a factor of 18.

A second concern regards the lack of consideration in such studies of the magnitude of uncertainty in the SSA production flux formulation and the consequences of this

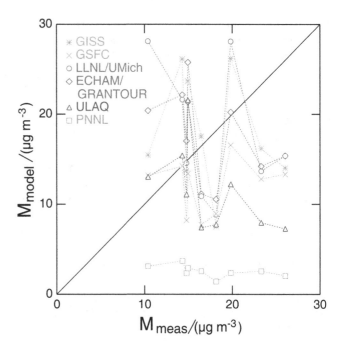

Figure 44. Comparison of annual average SSA mass concentration from six chemical transport models with measured multi-year average concentration at nine sites in the North and South Pacific (including also Cape Grim, Tasmania). Model results are from intercomparison conducted by the Intergovernmental Panel on Climate Change [*Penner et al.*, 2001] in which the models are identified and described. All models use the *Gong et al.* [1997a,b] extension of the *Monahan et al.* [1983b, 1986] formulation of the SSA production flux. Measurements are from *Prospero* and colleagues [*Prospero et al.*, 1985; *D. Savoie*, personal communication, 2003]; the measured value for American Samoa is 18.1 μg m^{-3} [*D. Savoie*, personal communication, 2003], not 5.4 μg m^{-3} as given in Fig. 5.9 of *Penner et al.* [2001]. Solid line denotes one-to-one relation between measured and modeled SSA mass concentrations.

uncertainty in comparison of modeled and observed concentrations. As illustrated in Figs. 41 and 43, which present estimates of f_{eff} and f_{int}, respectively, these uncertainties are substantial. In the *Monahan et al.* formulation of SSA production flux by the whitecap method uncertainty arises from uncertainty in both f_{wc} and W (in addition to uncertainties pertaining to the assumptions of the method itself, which are difficult to quantify). Although the multiplicative uncertainty in values of f_{wc} from the particular few laboratory experiments that gave rise to the *Monahan et al.* [1983b, 1986] formulation is perhaps as small as $\times 2$ (Fig. 32), when the ensemble of determinations of f_{wc} (from laboratory and surf-zone studies) is considered, a substantially greater uncertainty, $\times 7$, must be ascribed to this quantity (§4.3.5). An estimate of the uncertainty in f_{int} requires that the uncertainty in f_{wc} be convolved with the uncertainty in W, which ranges from $\lessgtr$ 7 at $U_{10} = 8.5$ m s^{-1} to $\lessgtr$ 3 at 20 m s^{-1} (§4.5.2). The resulting multiplicative uncertainty in f_{int} therefore ranges from $\lessgtr$ 9 to $\lessgtr$ 16 over this wind speed range (§5.3.1). Although the multiplicative uncertainties of the SSA production flux by other methods are somewhat lower than this value (for instance, $\lessgtr$ 4 by the steady state dry deposition method; §5.2.2; and $\lessgtr$ 5 by the statistical wet deposition method; §5.10), for the size range pertinent to SSA mass and CCN number concentration, $r_{80} \approx 0.1$-25 μm, these uncertainties are still substantial. As multiplicative uncertainty in this production flux translates directly into comparable multiplicative uncertainty in modeled concentrations, in the absence of other measures of the skill of the chemical transport models employed, little can be confidently concluded about whether departures between modeled and observed concentrations of magnitude less than this are to be attributed to the SSA production flux formulations or to the skill of the CTM in modeling concentrations for a specified formulation.

6.3. STATUS AND PROSPECT

While a picture of SSA and the processes responsible for its production and concentration is beginning to emerge, this picture remains quite blurry. Despite substantial research on sea salt aerosol, there remains much work to be done in quantifying the SSA production flux. Uncertainties remain in the first digit, and not infrequently in the power of ten, in description and understanding of many of the processes pertinent to SSA production. Measurements of the size distribution of SSA concentration at a given wind speed show spreads of an order of magnitude at minimum (Fig. 22), and estimates of SSA production flux exhibit even greater uncertainties (Figs. 41 and 43). Such a situation inevitably raises the question of how such uncertainties might be narrowed or indeed whether they can be narrowed.

Here two analogies may be pertinent. Consider first an analogy with near-surface air temperature. Surface air temperature is highly variable at any given location and globally, but the nature and extent of this variation and the reasons for it have not always been known. Seasonal and day-to-day variation in temperature were known to ancients, but further understanding awaited the development of instruments and of written records. As humans traveled they began to compare temperatures at different locations and develop generalizations: it tended to be warmer at low latitudes. Key among advances were systematic measurements by mariners that were diligently recorded in ships' logs. On the basis of such measurements it became possible to develop a highly differentiated picture of temperature as a function of location. With sufficient data it became possible to adduce pertinent statistics and to draw generalizations:

for example, despite temperature ranges that overlap greatly, it is possible to state with some confidence that the average temperature in Frankfurt is 1.1 K higher than in Hamburg. With more and better data further generalizations became possible, for example by type of location (continental or marine) at a given latitude. It then became possible to identify exceptions to these patterns (e.g., Cornwall versus Newfoundland) and to identify the reasons for the exceptions (oceanic currents that transport water from low latitudes to higher latitudes). The picture of surface temperature and the reasons for its spatial and temporal variability were greatly enhanced by widespread and rapid communications that permitted examination of the time dependence of temperature at numerous locations. With advances in the nature and scope of measurements it became possible to develop understanding of the processes responsible for changes in temperature, such as frontal passage. Now satellite-borne instruments determine temperatures with global coverage at high precision and fairly good accuracy, although accurate in-situ measurements remain essential. Ultimately with sufficient understanding—based on measurements and theory—it has become possible to model (forecast) temperatures with considerable skill. This success, in turn, contributes to the confidence of inclusion of the pertinent processes in large scale atmospheric models that are used to understand climate and climate change.

A second analogy might be to the emerging picture of the mixing ratio of carbon dioxide in the atmosphere. With the industrial revolution it became recognized that the rate of combustion of fossil fuels might be sufficiently great that it would appreciably increase the amount of CO_2 in the atmosphere, but major questions remained as to the extent of this increase, in large part because of lack of knowledge of the rate at which any excess CO_2 would be taken up by vegetation or other reservoirs. Early studies [e.g., Callendar, 1938, 1958] suggested, on the basis of a handful of measurements over some decades that perhaps there was a discernible increase in atmospheric CO_2, but it was not until Keeling [1960] began systematic high quality measurements that it became possible to establish the magnitude and rate of this increase, to compare the rate of increase to the rate of fossil fuel combustion, and to discern the modulation of this increase due to the annual cycle of growth and senescence of vegetation in the terrestrial biosphere. Much concerted subsequent research has established the seasonal, latitudinal, and to limited extent, longitudinal dependence of the atmospheric CO_2 mixing ratio, but even with such a highly developed picture many uncertainties remain in key terms of the CO_2 budget and in such major issues as the extent to which afforestation in North America is a sink for CO_2. The answers to these questions await further systematic research.

Like CO_2, SSA is to good approximation a conservative atmospheric tracer, so that consideration of its budget requires knowledge only of sources, sinks, and transport. However, description of SSA is much more complex than CO_2, for two main reasons. First, SSA is not a single constituent, like CO_2, but is more accurately thought of as a suite of constituents, characterized by their sizes and having differing mechanisms and rates of introduction into, transport within, and removal from the atmosphere. The second major complexity is the fact that the rather short lifetimes of SSA particles result in distributions of this material that are highly heterogeneous in space and time, with typical variability an order of magnitude or more. This heterogeneity makes it unreasonable to expect well defined properties of SSA such as size distribution of concentration or total mass concentration characteristic of the global marine boundary layer in the way that the current (i.e., 2004) global atmospheric mixing ratio of CO_2 can be described as $376\ \mu\mathrm{mol}\ \mathrm{mol}^{-1}$ with a variability of about 1%. The SSA that is present at any given time or location is the consequence of processes, intimately related to weather, that have introduced SSA into the air parcel under examination, have transported this air parcel to the location of interest, and have removed SSA particles previously present in this air parcel—that is to say, wind and rain. Mixing ratios of CO_2 are more or less immune to these influences. Although the argument can certainly be made that exactly the same considerations are required to understand the variation of local CO_2 from a global mean, these variations are usually no more than a few percent, whereas for SSA they are an order of magnitude or more.

Given this situation, one might ask whether the necessary understanding can be acquired for SSA that would allow development of a picture of its size-dependent concentrations and processes controlling these concentrations that is much more successful than at present. It is certainly to be hoped that the answer is yes.

How then might progress be made? Despite some half a century of research as summarized in the present review, it would appear that a rather convincing case can be made that the data base of measurements is actually quite sparse. The present review summarizes virtually all of the pertinent measurements. Often for key quantities and for key means of determining the SSA production flux only a handful of measurements are available, some of which are severely flawed, and some of which (laboratory studies) require extrapolations over several orders of magnitude and are thus of questionable applicability to oceanic SSA production. Some key quantities, such as those pertaining to film drop production, are poorly known at best. This situation places the study of SSA more or less at the status of early measurements of the global distribution of temperature, or, by the CO_2 analogy, well before the *Keeling* era. It can thus be

argued that understanding would be substantially enhanced by a greatly expanded data base of quality measurements. Such measurements might in the first instance consist of determinations of the size distribution of SSA concentration. However, such measurements alone would not suffice; they would need to be accompanied by local measurements of pertinent meteorological variables and by systematic examination of the governing meteorology along upwind trajectories. Such an extensive and systematic data set would go a long way to providing both a knowledge base and a measurement data base against which to test models of SSA production, transport, and removal. Some data sets of this nature already exist, such as that collected by *Smith and coworkers* [*Exton et al.*, 1985, 1986; *Smith et al.*, 1989, 1991, 1993], but other such data sets from different locations and using different methods of particle collection and sizing are required.

Additionally, targeted studies would be required of SSA production. This review has distinguished interfacial and effective SSA production fluxes; the former is more amenable to study in the laboratory, and such studies are necessary to understand the pertinent physics under controlled conditions, for example systematic examination of the effects of surface-active substances, temperature, and the like. However, it is the effective flux that governs the production of SSA particles that are important to cloud microphysics (§2.1.5.1), atmospheric chemistry (§2.1.5.2), and radiation transfer (§2.1.5.3). Although to be sure the two fluxes are related (by the flux ratio Ψ_f, which might be modeled as a function of atmospheric conditions), it would seem ultimately that field measurements are required from which the effective SSA production flux can be determined over a rather wide range of conditions so that it is possible to adduce the dependence of this production on controlling factors. This review has identified several approaches to such measurements, some of which have been employed only in a handful of instances, and none of which has seen widespread application. The concentration buildup method (§3.3), for example, has been only recently employed at a single location; this method, if applied to other locations, could possibly reduce uncertainty in $f_{\mathrm{eff}}(r_{80})$. In support of such a measurement-based approach is the availability also of new instrumentation, to which one might imagine the addition of a suite of new measurement techniques that might be developed in the future. The existence of multiple approaches to determination of SSA production flux suggests the further utility of determining this flux by several methods at the same time, as a check on the accuracy and consistency of the several methods.

Only once there is a substantial body of work examining the SSA production flux and its dependence on controlling factors will it be possible to ascertain the extent to which uncertainties in this quantity have been narrowed and whether it would be possible and fruitful to develop parameterizations that can be employed in models. Such parameterizations would need to be stringently evaluated to assess the confidence with which they might be applied. Finally models would need to be developed to represent the life cycle of SSA particles, as a function of particle size, and these models would likewise need to be evaluated by comparison with observation. Such an approach has been the path forward with respect to the distribution and concentrations of other trace atmospheric constituents and more generally in other areas of geophysics, and there is every reason to expect that it would be successful if applied to sea salt aerosol.

Principal Symbols

Symbol	Quantity	Section
a	size-dependent SSA surface area concentration	2.1.4.1, 4.1.4
A	total SSA surface area concentration	2.1.4.1, 4.1.4
c	mean molecular speed	2.1.5.2, 2.5.4
c_s	phase velocity of significant (or dominant) wave	2.3.5, 4.5.3.4
C	concentration	2.1.5.2, 2.4
C	Cunningham slip factor (Cunningham correction)	2.6.1, 2.6.2
C_d	drag coefficient	2.6.3, 2.6.4, 4.4.2.1
C_D	wind stress coefficient	2.3.2, 4.5.2, 4.5.3.2
D	diffusion coefficient (diffusivity)	2.4, 2.1.5.2, 2.6.1
f	size-dependent SSA flux	2.1.6, 2.8, 5
f_{bub}	size-dependent bubble flux	3.4
f_{wc}	size-dependent SSA production flux over white area	3.2, 4.3.4, 5.3
F	flux	
F	total SSA flux	2.1.6, 2.1.7
g_V	size-dependent volumetric gain rate of SSA particle concentration	2.8
h	maximum height attained by a jet drop	4.3.2.3, 4.3.2.4
H	height	
H	Henry's law solubility coefficient	2.1.5.2
j	size-dependent SSA number flux	2.1.6, 2.6.1, 2.8, 3.6
j_w	mass evaporative flux of water	2.4, 2.6.1
J	total SSA number flux	2.1.6, 4.3.6.1
k	rate constant	2.1.5.2, 4.1.4.5
l_V	size-dependent volumetric loss rate of SSA particle concentration	2.8
L_p	horizontal distance traveled by SSA particle produced at the surface	3.6
m	size-dependent SSA dry mass concentration	2.1.4.1, 4.1.4.3, 4.1.4.4
m_{dry}	dry mass of SSA particle	2.1.1, 2.5.3, 4.1.4
M	total SSA dry mass concentration	2.1.4.1, 2.5.3, 4.1
M	molecular weight	
n	size-dependent SSA number concentration	2.1.4, 4.1.4
n_{bub}	size-dependent bubble concentration	3.4, 4.4.1
n_0	amplitude of lognormal mode	2.1.4.2, 4.1.4.4
N	total SSA number concentration	2.1.4, 2.1.5.1, 4.1.3
N_{bub}	total bubble concentration	5.5.1.3
N_{film}	mean number of film drops produced per bubble	4.3.3.3
N_{jet}	mean number of jet drops produced per bubble	4.3.2.1, 4.3.2.4
$p_{w,sat}$	saturation vapor pressure of water	2.4, 2.5.4
q	dimensionless radius r/r_{aq}	2.1.5.2
Q	factor accounting for spatial non-uniformity of the concentration of dissolved gas reacting in a drop	2.1.5.2
Q_s	light-scattering efficiency	2.1.5.3
r	radius	
r_{aq}	penetration distance characterizing the competing effects of diffusion and reaction	2.1.5.2
r_{corr}	linear (Pearson product-moment) correlation coefficient	
r_{dry}	equivalent dry radius of SSA particle	2.1.1, 2.5.3

Symbol	Quantity	Section
r_{eff}	effective radius	2.1.5, 4.1.4.5
r_{meas}	measured radius of SSA particle	2.9.1
r_{mix}	characteristic mixing radius	2.9.3
r_{rh}	radius of SSA particle in equilibrium at fractional relative humidity rh	2.1.1, 2.5.3
r_{s}	radius of small particle (up to a few tens of nanometers)	2.1.5.2, 4.1.4.5
r_{80}	radius of SSA particle in equilibrium with RH 80%, used throughout text as size variable of SSA particles	2.1.1, 2.5.3
r'_{80}	geometric mean of lognormal mode	2.1.4.2, 4.1.4.4
$r'_{80,\mathrm{ent}}$	maximum value of r_{80} for which appreciable entrainment upward from near sea surface can be expected	2.9.2
$r_{80,\mathrm{eq}}$	maximum value of r_{80} for which RH equilibration can be expected	2.9.1
rh	fractional relative humidity	2.1.1
R	total SSA radius concentration	2.1.4.1, 4.1.4.3, 4.1.4.4
R_{bub}	radius of bubble	4.3.1, 4.4
Re	Reynolds number	2.6.3, 4.3.2.3, 4.4.2.1
s	supersaturation ($\equiv rh-1$)	2.1.5.1
Sc	Schmidt number	4.6.1
St	Stokes number	4.6.1
t	time	
T	temperature	
T_{s}	period of significant (or dominant) wave	2.3.5, 4.1.4.4
u	horizontal component of wind velocity	
u	along-wind component of particle velocity	
u_*	wind friction velocity	2.3.2
U_z	mean wind speed at height z	2.3.1
U_{10}	mean wind speed at 10 m above sea surface, used throughout text to characterize wind speed	2.1.5, 2.3.1
v	velocity	
v	cross-wind (y-) component of wind velocity	
v	size-dependent SSA volume concentration	2.1.4.1, 4.1.4.3, 4.1.4.4
V	total SSA volume concentration	2.1.4.1, 2.5.3, 4.1.4
w	vertical wind speed	
w	vertical component of particle velocity	
W	weighting factor for SSA property of interest	2.1.5, 2.1.7
W	oceanic whitecap ratio	3.2, 4.5
x	along-wind direction; distance in along-wind direction	
X	characteristic distance with respect to removal by dry deposition	2.9.6
y	cross-wind direction	
z	vertical direction; height above water surface; depth below sea surface	
α	Ångström exponent	2.1.5.3, 4.1.4.5
α	accommodation coefficient	2.1.5.2, 2.5.4
δ	thickness of viscous sublayer	2.4, 2.7.2.2, 4.6.3
δ	height at which particles are introduced to and removed from the atmosphere	4.6.5
δQ	denotes partial integral of quantity Q	2.1.3
Δ	denotes difference	
$\Delta_{1/2}\log r_{80}$	full width at half-maximum of lognormal mode	2.1.4.2
ζ	ratio r/r_{80}	2.1.1, 2.5.3
v	kinematic viscosity	
ρ	density	
σ	geometric standard deviation of lognormal mode	2.1.4.2, 4.1.4.4
σ	light-scattering coefficient	2.1.5.3, 4.1.4.5
σ_w	root-mean-squared value of the vertical component of the wind speed	2.4
τ	characteristic time	

Symbol	Quantity	Section
τ_{sp}	scattering optical thickness of SSA particles (vertical integral of light-scattering coefficient of SSA particles)	2.1.5.3, 4.1.4.5
τ_u^{-1}	effective first-order rate constant for irreversible uptake	2.1.5.2, 4.1.4.5
Ψ_f	ratio of the size-dependent effective SSA production flux to the size-dependent interfacial SSA production flux	2.1.6, 2.9.4, 4.6.5, 5.1.2

Section numbers indicate section in which quantity is introduced, defined, or extensively discussed.
For size-distributed quantities, lower-case and upper-case symbols typically denote differential and total or integral quantities, respectively (§2.1.3).

Physical Constants

Symbol	Quantity	Temp/°C	Value
$C_{p,a}$	specific heat of air at constant pressure	20	$1.0 \cdot 10^3$ m^2 s^{-2} K^{-1}
$C_{p,sw}$	specific heat of seawater of salinity 35 at constant pressure	20	$4.0 \cdot 10^3$ m^2 s^{-2} K^{-1}
$C_{w,sat}$	saturation mass concentration of water vapor in air	20	$1.7 \cdot 10^{-2}$ kg m^{-3}
		0	$4.9 \cdot 10^{-3}$ kg m^{-3}
D_w	diffusivity of water vapor in air	20	$2.4 \cdot 10^{-5}$ m^2 s^{-1}
$D_{w,eff}$	effective diffusivity of water vapor in air	20	$6.3 \cdot 10^{-6}$ m^2 s^{-1}
		0	$1.2 \cdot 10^{-5}$ m^2 s^{-1}
g	acceleration due to gravity		9.8 m s^{-2}
k	Boltzmann constant		$1.4 \cdot 10^{-23}$ kg m^2 s^{-2} K^{-1}
k_a	thermal conductivity of air	20	$2.5 \cdot 10^{-2}$ kg m^2 s^{-3} K^{-1}
L_w	latent heat of vaporization of water	20	$2.5 \cdot 10^6$ m^2 s^{-2}
M_a	mean molecular weight of air		$2.9 \cdot 10^{-2}$ kg mol^{-1}
M_w	molecular weight of water		$1.8 \cdot 10^{-2}$ kg mol^{-1}
$p_{w,sat}$	saturation vapor pressure of water	20	$2.3 \cdot 10^3$ kg m^{-1} s^{-2}
		0	$6.1 \cdot 10^2$ kg m^{-1} s^{-2}
R	gas constant		8.3 kg m^2 s^{-2} K^{-1} mol^{-1}
κ	von Karman constant		0.40
κ_a	thermal diffusivity of air	20	$2.1 \cdot 10^{-5}$ m^2 s^{-1}
λ	mean free path of air molecules	20	$6.4 \cdot 10^{-8}$ m
ν_a	kinematic viscosity of air	20	$1.5 \cdot 10^{-5}$ m^2 s^{-1}
ν_{sw}	kinematic viscosity of seawater of salinity 35	20	$1.0 \cdot 10^{-6}$ m^2 s^{-1}
		0	$1.8 \cdot 10^{-6}$ m^2 s^{-1}
ρ_a	density of air	20	1.2 kg m^{-3}
ρ_{ss}	density of dry sea salt		$2.2 \cdot 10^3$ kg m^{-3}
ρ_{sw}	density of seawater of salinity 35	20	$1.0 \cdot 10^3$ kg m^{-3}
σ	surface tension of seawater (salinity 35)/air interface	20	$7.4 \cdot 10^{-2}$ kg s^{-2}

Values presented are suitable for evaluation of formulae at 1 atm ($\approx 1.0 \cdot 10^5$ Pa); values which vary substantially with temperature are presented at 0°C and 20°C.

Special Symbols

Symbol	Description	Section
$\prec$	considerably less than	2.1.5.2
$\succ$	considerably greater than	2.1.5.2
$\pm$	plus or minus, denotes additive uncertainty	
$\overset{\times}{\div}$	times or divided by, denotes a multiplicative uncertainty	4.1.2.2, 5.1.1
$\bar{q}$	overbar denotes time average of quantity q	2.8.2
q'	prime denotes fluctuating value of quantity q about time average $\bar{q}$	2.8.2

Principal Subscripts and Acronyms

a	air	sw	seawater
act	activation	S	surface layer
amb	at ambient RH	Stk	Stokes
aq	aqueous	term	terminal
bub	bubble	turb	turbulent
B	Brownian	T	thermal
d	dry deposition	u	uptake
eff	effective	V	viscous sublayer
ent	entrainment	V	volume
form	formation	w	water
g	gas	wc	whitecap
i	interface		
imp	impaction	10	10 m (similarly for other heights)
int	interfacial	80	80% RH (similarly for other RH)
mbl	marine boundary layer		
M	mass	SSA	sea salt aerosol
N	number	RH	relative humidity
p	particle	CCN	cloud condensation nuclei
sp	scattering by particulate matter	CN	condensation nuclei
ss	sea salt	CTM	chemical transport model

List of Figures

List of Tables

References

Abbatt, J.P.D., and G.C.G. Waschewsky, Heterogeneous interactions of HOBr, HNO_3, O_3, NO_2 with deliquescent NaCl aerosols at room temperature, *J. Phys. Chem.*, *102A*, 3719-3725, 1998.

Abraham, F.F., Functional dependence of drag coefficient of a sphere on Reynolds number, *Physics of Fluids*, *13*, 2194-2195, 1970.

Afeti, G.M., and F.J. Resch, Distribution of the liquid aerosol produced from bursting bubbles in sea and distilled water, *Tellus*, *42B*, 378-384, 1990.

Ahmed, A., Fluxes and processes of deposition of atmospheric sea salt to the earth surface, M.S. thesis, University of Rhode Island, Kingston, RI, 1978.

Aitchison, J., and J.A.C. Brown, *The Lognormal Distribution*, 176 pp., Cambridge University Press, Cambridge, 1957.

Aitken, J., On dusts, fogs, and clouds, *Trans. Royal Soc. Edinburgh*, *30*, 337-368, 1881.

Aitken, J., On the effect of oil on a stormy sea, *Proc. Royal Soc. Edinburgh*, *12*, 1883.

Albrecht, B.A., Aerosols, cloud microphysics, and fractional cloudiness, *Science*, *245*, 1227-1230, 1989.

Albrecht, B.A., C.S. Bretherton, D. Johnson, W.H. Schubert, and A.S. Frisch, The Atlantic Stratocumulus Transition Experiment - ASTEX, *Bull. Am. Meteorol. Soc.*, *76* (6), 889-904, 1995a.

Albrecht, B.A., M.P. Jensen, and W.J. Syrett, Marine boundary layer structure and fractional cloudiness, *J. Geophys. Res.*, *100* (D7), 14209-14222, 1995b.

Alcock, R.K., and D.G. Morgan, Investigations of wind and sea state with respect to the Beaufort scale, *Weather*, *33*, 271-277, 1978.

Alexander, R.C., and R.L. Mobley, Monthly average sea-surface temperatures and ice-pack limits on a 1° global grid, *Monthly Weather Rev.*, *104*, 143-148, 1976.

Aliverti, G., and G. Lovera, Sui nuclei di condensazione di origine marittima, *Geofisica Pura e Applicata*, *16*, 133-135, 1950.

Allen, H.S., The motion of a sphere in a viscous fluid, *Philosoph. Mag.*, *7*, 323-338, 519-534, 1900.

Alpers, W., and H. Hühnerfuss, The damping of ocean waves by surface films: A new look at an old problem, *J. Geophys. Res.*, *94* (C5), 6251-6265, 1989.

Alty, T., and C.A. Mackay, The accommodation coefficient and the evaporation coefficient of water, *Proc. Royal Soc. London*, *149A*, 104-116, 1935.

Ambler, H.R., and A.A.J. Bain, Corrosion of metals in the tropics, *J. Applied Chem.*, *5*, 437-467, 1955.

Anderson, J.L., and J.A. Quinn, Bubble columns: flow transitions in the presence of trace contaminants, *Chem. Eng. Sci.*, *25*, 373-380, 1970.

Anderson, J.R., P.R. Buseck, T.L. Patterson, and R. Arimoto, Characterization of the Bermuda tropospheric aerosol by combined individual-particle and bulk-aerosol analysis, *Atmos. Environ.*, *30* (2), 319-338, 1996.

Anderson, R.E., L. Stein, M.L. Moss, and N.H. Gross, Potential infectious hazards of common bacteriological techniques, *Journal of Bacteriology*, *64*, 473-481, 1952.

Andreae, M.O., Marine aerosol chemistry at Cape Grim, Tasmania and Townsville, Queensland, *J. Geophys. Res.*, *87* (C11), 8875-8885, 1982.

Andreae, M.O., The ocean as a source of atmospheric sulfur compounds, in *The Role of Air-Sea Exchange in Geochemical Cycling*, edited by P. Buat-Ménard, pp. 331-362, D. Reidel Publishing Company, Dordrecht, 1986.

Andreae, M.O., Climatic effects of changing atmospheric aerosol levels, in *Future Climates of the World*, edited by A. Henderson-Sellers, pp. 341-392, Elsevier, New York, 1995.

Andreae, M.O., and P.J. Crutzen, Atmospheric aerosols: biogeochmical sources and role in atmospheric chemistry, *Science*, *276*, 1052-1058, 1997.

Andreae, M.O., R.J. Charlson, F. Bruynseels, J. Storms, R. van Grieken, and W. Maenhaut, Internal mixtures of sea salt, silicates, and excess sulfate in marine aerosols, *Science*, *232*, 1620-1623, 1986.

Andreae, M.O., H. Berresheim, T.W. Andreae, M.A. Kritz, T.S. Bates, and J.T. Merrill, Vertical distribution of dimethylsulfide, sulfur dioxide, aerosol ions, and radon over the Northeast Pacific Ocean, *J. Atmos. Chem.*, *6*, 149-173, 1988.

Andreae, M.O., T.W. Andreae, D. Meyerdierks, and C. Thiel, Marine sulfur cycling and the atmospheric aerosol over the springtime North Atlantic, *Chemosphere*, *52*, 1321-1343, 2003.

Andreas, E.L., Thermal and size evolution of sea spray drops, *CRREL Report 89-11*, pp. 37, U.S. Army Cold Regions Research and Engineering Laboratory Report, 1989.

Andreas, E.L., Time constants for the evolution of sea spray droplets, *Tellus*, *42B*, 481-497, 1990a.

Andreas, E.L., Model estimates of the effects of sea spray on air-sea heat fluxes, in *Modelling the Fate and Influence of Marine Spray. Whitecap Report No. 7*, edited by P.G. Mestayer, E.C. Monahan, and P.A. Beetham, pp. 115-127, Marine Sciences Institute, University of Connecticut, Groton, CT, 1990b.

Andreas, E.L., Sea spray and turbulent air-sea heat fluxes, *J. Geophys. Res.*, *97* (C7), 11429-11441, 1992.

Andreas, E.L., Reply, *J. Geophys. Res.*, *99* (C7), 14345-14350, 1994a.

Andreas, E.L., Comments on 'On the contribution of spray droplets to evaporation' by Lutz Hasse, *Boundary-Layer Meteorol.*, *68*, 207-214, 1994b.

Andreas, E.L., The temperature of evaporating sea spray droplets, *J. Atmos. Sci.*, *52* (7), 852-862, 1995.

Andreas, E.L., Reply, *J. Atmos. Sci.*, *52* (11), 1642-1645, 1996.

Andreas, E.L., A new sea spray generation function for wind speeds up to 32 m s^{-1}, *J. Phys. Oceanogr.*, *28*, 2175-2184, 1998.

Andreas, E.L., and K.A. Emanuel, Effects of sea spray on tropical cyclone intensity, *J. Atmos. Sci.*, *58* (24), 3741-3751, 2001.

Andreas, E.L., and E.C. Monahan, The role of whitecap bubbles in air-sea heat and moisture exchange, *J. Phys. Oceanogr.*, *30*, 433-442, 2000.

Andreas, E.L., J.B. Edson, E.C. Monahan, M.P. Rouault, and S.D. Smith, The spray contribution to net evaporation from the sea: a review of recent progress, *Boundary-Layer Meteorol.*, *72*, 3-52, 1995.

Andreas, E.L., M.J. Pattison, and S.E. Belcher, "Production rates of sea-spray droplets" by M.J. Pattison and S.E. Belcher: Clarification and elaboration, *J. Geophys. Res.*, *106* (C4), 7157-7161, 2001.

Andrews, E., and S.M. Larson, Effect of surfactant layers on the size changes of aerosol particles as a function of relative humidity, *Environ. Sci. Technol.*, *27*, 857-865, 1993.

Andsager, K., K.V. Beard, and N.F. Laird, Laboratory measurements of axis ratios for large raindrops, *J. Atmos. Sci.*, *56*, 2673-2683, 1999.

Ångström, A., On the atmospheric transmission of sun radiation and on dust in the air, *Geografiska Annaler*, *11*, 156-166, 1929.

Anguelova, M., R.P. Barber, Jr., and J. Wu, Spume drops produced by the wind tearing of wave crests, *J. Phys. Oceanogr.*, *29*, 1156-1165, 1999.

Anscombe, F.J., Graphs in statistical analysis, *Amer. Stat.*, *27*, 17-21, 1973.

Apel, J.R., An improved model of the ocean surface wave vector spectrum and its effects on radar backscatter, *J. Geophys. Res.*, *99* (C8), 16269-16291, 1994.

Archer, R.J., and V.K. La Mer, The rate of evaporation of water through fatty acid monolayers, *J. Phys. Chem.*, *59*, 200-208, 1954.

Arimoto, R., and R.A. Duce, Dry deposition models and the air/sea exchange of trace elements, *J. Geophys. Res.*, *91* (D2), 2787-2792, 1986.

Arimoto, R., R.A. Duce, B.J. Ray, and C.K. Unni, Atmospheric trace elements at Enewetak Atoll: 2. Transport to the ocean by wet and dry deposition, *J. Geophys. Res.*, *90* (D1), 2391-2408, 1985.

Arimoto, R., R.A. Duce, D.L. Savoie, and J.M. Prospero, Trace elements in aerosol particles from Bermuda and Barbados: concentrations, sources and relationships to aerosol sulfate, *J. Atmos. Chem.*, *14*, 439-457, 1992.

Arons, A.B., and C.F. Kientzler, Vapor pressure of sea-salt solutions, *Trans. Am. Geophys. Union*, *35* (5), 722-728, 1954.

Art, H.W., F.H. Bormann, G.K. Voigt, and G.M. Woodwell, Barrier island forest ecosystems: Role of meteorologic nutrient inputs, *Science*, *184*, 60-62, 1974.

Artaxo, P., M.L. Rabello, W. Maenhaut, and R. van Grieken, Trace elements and individual particle analysis of atmospheric aerosols from tha Antarctic peninsula, *Tellus*, *44B*, 318-334, 1992.

Asher, W.E., and P.J. Farley, Phase-Doppler anemometer measurement of bubble concentrations in laboratory-simulated breaking waves, *J. Geophys. Res.*, *100* (C4), 7045-7056, 1995.

Asher, W., and R. Wanninkhof, Transient tracers and air-sea gas transfer, *J. Geophys. Res.*, *103* (C8), 15939-15958, 1998.

Asher, W.E., P.J. Farley, R. Wanninkhof, E.C. Monahan, and T.S. Bates, Laboratory and field experiments on the correlation of fractional area whitecap coverage with air/sea gas transport, in *Precipitation Scavenging and Atmospheric-Surface Exchange*, edited by S.E. Schwartz, and W.G.N. Slinn, pp. 815-827, Hemisphere Publishing Corporation, Washington, DC, 1992.

Asher, W.E., L.M. Karle, B.J. Higgins, P.J. Farley, E.C. Monahan, and I.S. Leifer, The influence of bubble plumes on air-seawater gas transfer velocities, *J. Geophys. Res.*, *101* (C5), 12027-12041, 1996.

Asher, W.E., L.M. Karle, and B.J. Higgins, On the differences between bubble-mediated air-water transfer in freshwater and seawater, *J. Marine Res.*, *55*, 813-845, 1997.

Asher, W., J. Edson, W. McGillis, R. Wannnkhof, D. Ho, and T. Litchendorf, Fractional area whitecap coverage and air-sea gas transfer velocities measured during GasEx-98, in *Gas Transfer at Water Surfaces*, edited by M.A. Donelan, W.M. Drennan, E.S. Saltzman, and R. Wanninkhof, pp. 383, American Geophysical Union, Washington, DC, 2002.

Ashton, E.W.S., and J.K. O'Sullivan, Effect of rain in calming the sea, *Nature*, *164*, 320-321, 1949.

Atkinson, L.P., Effect of air bubble solution of air-sea gas exchange, *J. Geophys. Res.*, *78* (6), 962-968, 1973.

Aubert, J., Les aérosols marins, vecteurs de microorganismes, *J. Recherches Atmos.*, *8*, 541-554, 1974.

Azbel, D., and A.I. Liapis, Mechanisms of liquid entrainment, in *Handbook of Fluids in Motion*, edited by N.P. Cheremisinoff, and R. Gupta, pp. 1202, Ann Arbor Science, Ann Arbor, MI, 1983.

Bachhuber, C., and C. Sanford, The rise of small bubbles in water, *J. Applied Phys.*, *45* (6), 2567-2569, 1974.

Baeyens, W., F. Dehairs, and H. Dedeurwaerder, Wet and dry deposition fluxes above the North Sea, *Atmos. Environ.*, *24A* (7), 1693-1703, 1990.

Baier, R.E., Organic films on natural waters: Their retrieval, identification, and modes of elimination, *J. Geophys. Res.*, *77* (27), 5062-5075, 1972.

Baier, R.E., D.W. Goupil, S. Perlmutter, and R. King, Dominant chemical composition of sea-surface films, natural slicks, and foams, *J. Recherches Atmos.*, *8*, 571-600, 1974.

Bakan, S., Note on the eddy correlation method for CO_2 flux measurements, *Boundary-Layer Meteorol.*, *14*, 597-600, 1978.

Baker, M.B., and R.J. Charlson, Bistability of CCN concentrations and thermodynamics in the cloud-topped boundary layer, *Nature*, *345*, 142-145, 1990.

Baldy, S., Bubbles in the close vicinity of breaking waves: statistical characteristics of the generation and dispersion mechanism, *J. Geophys. Res.*, *93* (7), 8239-8248, 1988.

Baldy, S., A generation-dispersion model of ambient and transient bubbles in the close vicinity of breaking waves, *J. Geophys. Res.*, *98* (C10), 18277-18293, 1993.

Baldy, S., and M. Bourguel, Measurements of bubbles in a stationary field of breaking waves by a laser-based single-particle scattering technique, *J. Geophys. Res.*, *90* (C1), 1037-1047, 1985.

Banner, M.L., and W.K. Melville, On the separation of air flow over water waves, *J. Fluid Mechanics*, *77* (4), 825-842, 1976.

Banner, M.L., and O.M. Phillips, On the incipient breaking of small scale waves, *J. Fluid Mechanics*, *65*, 647-656, 1974.

Banner, M.L., A.V. Babanin, and I.R. Young, Breaking probability for dominant waves on the sea surface, *J. Phys. Oceanogr.*, *30*, 3145-3160, 2000.

Banner, M.L., J.R. Gemmrich, and D.M. Farmer, Multiscale measurements of ocean wave breaking probability, *J. Phys. Oceanogr.*, *32*, 3364-3375, 2002.

Bao, J.-W., J.M. Wilczak, J.-K. Choi, and L.H. Kantha, Numerical simulations of air-sea interaction under high wind conditions using a coupled model: A study of hurricane development, *Monthly Weather Rev.*, *128*, 2190-2210, 2000.

Barber, R.P.J., and J. Wu, Sea brightness temperature and effects of spray and whitecaps, *J. Geophys. Res.*, *102* (C3), 5823-5827, 1997.

Barbier, M., D. Tusseau, J.C. Marty, and A. Saliot, Sterols in aerosols, surface microlayer and subsurface water in the North-Eastern tropical Atlantic, *Oceanologica Acta*, *4* (1), 77-84, 1981.

Barger, W.R., and W.D. Garrett, Surface active organic material in the marine atmosphere, *J. Geophys. Res.*, *75* (24), 4561-4566, 1970.

Barger, W.R., W.D. Garrett, E.L. Mollo-Christensen, and K.W. Ruggles, Effects of artificial sea slick upon the atmosphere and the ocean, *J. Applied Meteorol.*, *9*, 396-400, 1970.

Barker, D.R., and H. Zeitlin, Metal-ion concentrations in sea-surface microlayer and size-separated atmospheric aerosol samples in Hawaii, *J. Geophys. Res.*, *77* (27), 5076-5086, 1972.

Barnaby, C.F., Effect of rain in calming the sea, *Nature*, *149*, 968, 1949.

Barnhardt, E.A., and J.L. Streete, A method for predicting atmospheric aerosol scattering coefficients in the infrared, *Applied Optics*, *9* (6), 1337-1344, 1970.

Baron, P.A., Calibration and use of the aerodynamic particle sizes (APS 3300), *Aerosol Sci. Technol.*, *5*, 55-67, 1986.

Bartley, D.L., A.B. Martinez, P.A. Baron, D.R. Secker, and E. Hirst, Droplet distortion in accelerating flow, *J. Aerosol Sci.*, *31* (12), 1447-1460, 2000.

Bates, T.S., V.N. Kapustin, P.K. Quinn, D.S. Covert, D.J. Coffman, C. Mari, P.A. Durkee, W.J. de Bruyn, and E.S. Saltzman, Processes controlling the distribution of aerosol particles in the lower marine boundary layer during the First Aerosol Characterization Experiment (ACE 1), *J. Geophys. Res.*, *103* (D13), 16369-16383, 1998.

Batoosingh, E., G.A. Filey, and B. Keshwar, An analysis of experimental methods for producing particulate organic matter in seawater by bubbling, *Deep-Sea Res.*, *16*, 213-219, 1969.

Baumgartner, A., and E. Reichel, *The World Water Balance*, 179 pp., Elsevier Scientific Publishing Company, Amsterdam, 1975.

Baumgartner, F., A. Zanier, and K. Krebs, On the mechanical stability of hollow NaCl-particles, *J. Aerosol Sci.*, *20* (8), 883-886, 1989.

Baylor, E.R., W.H. Sutcliffe, and D.S. Dirschfeld, Adsorption of phosphate onto bubbles, *Deep-Sea Res.*, *9*, 120-124, 1962.

Baylor, E.R., M.B. Peters, and M.B. Baylor, Water-to-air transfer of virus, *Science*, *197*, 763-764, 1977a.

Baylor, E.R., M.B. Baylor, D.C. Blanchard, L.D. Syzdek, and C. Appel, Virus transfer from surf to wind, *Science*, *198*, 575-580, 1977b.

Beard, K.V., Terminal velocity and shape of cloud and precipitation drops aloft, *J. Atmos. Sci.*, *33*, 851-864, 1976.

Beard, K.V., On the acceleration of large water drops to terminal velocity, *J. Applied Meteorol.*, *16*, 1068-1071, 1977.

Beard, K.V., and H.R. Pruppacher, A determination of the terminal velocity and drag of small water drops by means of a wind tunnel, *J. Atmos. Sci.*, *26*, 1066-1072, 1969.

Beard, K.V., H.T.I. Ochs, and K.H. Leong, Particle scavenging by evaporating cloud drops, in *Precipitation Scavenging, Dry Deposition, and Resuspension*, edited by H.R. Pruppacher, R.G. Semonin, and W.G.N. Slinn, pp. 517-527, Elsevier, New York, 1983.

Beck, J.B., Observations on salt storms, and the influence of salt and saline air upon animal and vegetable life, *Am. J. Sci.*, *1*, 388-397, 1819.

Belot, Y., C. Caput, and D. Gauthier, Transfer of americium from sea water to atmosphere by bubble bursting, *Atmos. Environ.*, *16* (6), 1463-1466, 1982.

Bennett, M.G., The condensation of water in the atmosphere, *Quar. J. Royal Meteorol. Soc.*, *60*, 3-14, 1934.

Benner, R., J.D. Pakulaki, M. McCarthy, J.I. Hedges, and P.G. Hatcher, Bulk chemical characteristics of dissolved organic matter in the ocean, *Science*, *255*, 1561-1564, 1992.

Bennett, S.J., J.S. Bridge, and J.L. Best, Fluid and sediment dynamics of upper stage plane beds, *J. Geophys. Res.*, *103* (C1), 1239-1274, 1998.

Benson, B.B., and P.D.M. Parker, Nitrogen/argon and nitrogen isotope ratios in aerobic sea water, *Deep-Sea Res.*, *7*, 237-253, 1961.

Berg, O.H., E. Swietlicki, and R. Krejci, Hygroscopic growth of aerosol particles in the marine boundary layer over the Pacific and Southern Oceans during the First Aerosol Characterization Experiment (ACE 1), *J. Geophys. Res.*, *103* (D13), 16535-16545, 1998.

Bernstein, A.B., Examination of certain terms appearing in Reynolds' equations under unsteady conditions and their implications for micrometeorology, *Quar. J. Royal Meteorol. Soc.*, *92*, 533-542, 1966.

Berry, E.X., and M.R. Pranger, Equations for calculating the terminal velocities of water drops, *J. Applied Meteorol.*, *13*, 108-113, 1974.

Best, A.C., Empirical formulae for the terminal velocity of water drops falling through the atmosphere, *Quar. J. Royal Meteorol. Soc.*, *76*, 302-311, 1950.

Bevington, P.R., *Data Reduction and Error Analysis for the Physical Sciences*, 336 pp., McGraw-Hill Book Company, New York, 1969.

Bezdek, H.F., Size determination of sea water drops, *J. Phys. Chem.*, *75*, 3623-3625, 1971.

Bezdek, H.F., and A.F. Carlucci, Surface concentration of marine bacteria, *Limnol. Oceanogr.*, *17*, 566-569, 1972.

Bezdek, H.F., and A.F. Carlucci, Concentration and removal of liquid microlayers from a seawater surface by bursting bubbles, *Limnol. Oceanogr.*, *19*, 126-132, 1974.

Bezzabotnov, V.S., R.S. Bortkovskiy, and D.F. Timanovskiy, On the structure of the two-phase medium formed when wind waves break, *Atmos. Oceanic Phys.*, *22* (11), 922-928, 1986.

Bigg, E.K., Comparison of aerosol at four baseline atmospheric monitoring stations, *J. Applied Meteorol.*, *19*, 521-533, 1980.

Bigg, E.K., and C. Leck, Properties of the aerosol over the central Arctic Ocean, *J. Geophys. Res.*, *106* (D23), 32101-32109, 2001.

Bigg, E.K., J.L. Brownscombe, and W.J. Thompson, Fog modification with long-chain alcohols, *J. Applied Meteorol.*, *8*, 75-82, 1969.

Bigg, E.K., J.L. Gras, and D.J.C. Mossop, Wind-produced submicron particles in the marine atmosphere, *Atmos. Res.*, *36*, 55-68, 1995.

Bikerman, J.J., Persistence of bubbles on inorganic salt solutions, *J. Applied Chem.*, *18*, 266-269, 1968.

Blanc, T.V., Variation of bulk-derived surface flux, stability, and roughness results due to the use of different transfer coefficient schemes, *J. Phys. Oceanogr.*, *15*, 650-669, 1985.

Blanchard, D.C., Bursting of bubbles at an air-water interface, *Nature*, *173*, 1048, 1954a.

Blanchard, D.C., A simple method for the production of homogeneous water drops down to 1 micron radius, *J. Colloid Sci.*, *9*, 321-328, 1954b.

Blanchard, D.C., Electrified droplets from the bursting of bubbles at an air-sea interface, *Nature*, *175*, 334-336, 1955.

Blanchard, D.C., Electrically charged drops from bubbles in sea water and their meteorological significance, *J. Meteorol.*, *15*, 383-396, 1958.

Blanchard, D.C., Comments on the breakup of raindrops, *J. Atmos. Sci.*, *19*, 119-120, 1962.

Blanchard, D.C., The electrification of the atmosphere by particles from bubbles in the sea, in *Progr. Oceanogr.*, edited by M. Sears, pp. 73-202, Pergamon Press, New York, 1963.

Blanchard, D.C., Sea-to-air transport of surface active material, *Science*, *146*, 396-397, 1964.

Blanchard, D.C., Positive space charge from the sea, *J. Atmos. Sci.*, *23*, 507-515, 1966.

Blanchard, D.C., *From Raindrops to Volcanoes*, 180 pp., Doubleday and Company, Inc., Garden City, NY, 1967.

Blanchard, D.C., The oceanic production rate of cloud nuclei, *J. Recherches Atmos.*, *4* (1), 1-6, 1969.

Blanchard, D.C., Whitecaps at sea, *J. Atmos. Sci.*, *28*, 645, 1971.

Blanchard, D.C., Airborne sea salt sedimentation measurements and a method of reproducing ambient sedimentation rates for the study of its effect on vegetation, *Atmos. Environ.*, *11*, 565-566, 1977.

Blanchard, D.C., Jet drop enrichment of bacteria, virus, and dissolved organic material, *Pure Appl. Geophys.*, *116*, 302-308, 1978.

Blanchard, D.C., The production, distribution, and bacterial enrichment of the sea-salt aerosol, in *Air-Sea Exchange of Gases and Particles*, edited by P.S. Liss, and W.G.N. Slinn, pp. 407-454, D. Reidel Publishing Company, Dordrecht, 1983.

Blanchard, D.C., The oceanic production of atmospheric sea salt, *J. Geophys. Res.*, *90* (C1), 961-963, 1985a.

Blanchard, D.C., Flow of electrical current from world ocean to atmosphere, *J. Geophys. Res.*, *90* (C5), 9147-9148, 1985b.

Blanchard, D.C., The size and height to which jet drops are ejected from bursting bubbles in seawater, *J. Geophys. Res.*, *94* (C8), 10999-11002, 1989a.

Blanchard, D.C., The ejection of drops from the sea and their enrichment with bacteria and other materials: A review, *Estuaries*, *12* (3), 127-137, 1989b.

Blanchard, D.C., Surface-active monolayers, bubbles, and jet drops, *Tellus*, *42B*, 200-205, 1990.

Blanchard, D.C., and R.J. Cipriano, Biological regulation of climate, *Nature*, *330*, 526, 1987.

Blanchard, D.C., and E.J. Hoffman, Control of jet drop dynamics by organic material in seawater, *J. Geophys. Res.*, *83* (C12), 6187-6191, 1978.

Blanchard, D.C., and A.T. Spencer, Condensation nuclei and the crystallization of saline drops, *J. Atmos. Sci.*, *21*, 182-186, 1964.

Blanchard, D.C., and A.T. Spencer, Experiments on the generation of raindrop-size distributions by drop breakup, *J. Atmos. Sci.*, *27*, 101-108, 1970.

Blanchard, D.C., and L. Syzdek, Mechanism for the water-to-air transfer and concentration of bacteria, *Science*, *170*, 626-628, 1970.

Blanchard, D.C., and L.D. Syzdek, Concentration of bacteria in jet drops from bursting bubbles, *J. Geophys. Res.*, *77* (27), 5087-5099, 1972a.

Blanchard, D.C., and L. Syzdek, Variations in Aitken and giant nuclei in marine air, *J. Phys. Oceanogr.*, *2*, 255-262, 1972b.

Blanchard, D.C., and L.D. Syzdek, Importance of bubble scavenging in the water-to-air transfer of organic material and bacteria, *J. Recherches Atmos.*, *8*, 529-540, 1974a.

Blanchard, D.C., and L.D. Syzdek, Bubble tube: Apparatus for determining rate of collection of bacteria by an air bubble rising in water, *Limnol. Oceanogr.*, *19*, 133-138, 1974b.

Blanchard, D.C., and L.D. Syzdek, Electrostatic collection of jet and film drops, *Limnol. Oceanogr.*, *20*, 762-774, 1975.

Blanchard, D.C., and L.D. Syzdek, Seven problems in bubble and jet drop researches, *Limnol. Oceanogr.*, *23*, 389-400, 1978a.

Blanchard, D.C., and L.D. Syzdek, Reply to comment by F. MacIntyre, *Limnol. Oceanogr.*, *23*, 573, 1978b.

Blanchard, D.C., and L.D. Syzdek, Water-to-air transfer and enrichment of bacteria in drops from bursting bubbles, *Applied Environ. Microbiol.*, *43* (5), 1001-1005, 1982.

Blanchard, D.C., and L.D. Syzdek, Film drop production as a function of bubble size, *J. Geophys. Res.*, *93* (C4), 3649-3654, 1988.

Blanchard, D.C., and L.D. Syzdek, Reply, *J. Geophys. Res.*, *95* (C5), 7393, 1990a.

Blanchard, D.C., and L.D. Syzdek, Apparatus to determine the efficiency of transfer of bacteria from a bursting bubble to the jet drops, *Limnol. Oceanogr.*, *35* (1), 136-143, 1990b.

Blanchard, D.C., and A.H. Woodcock, Bubble formation and modification in the sea and its meteorological significance, *Tellus*, *9*, 145-158, 1957.

Blanchard, D.C., and A.H. Woodcock, The production, concentration, and vertical distribution of the sea-salt aerosol, *Ann. NY Acad. Sci.*, *338* (330-347), 1980.

Blanchard, D.C., L.D. Syzdek, and M.E. Weber, Bubble scavenging of bacteria in freshwater quickly produces bacterial enrichment in airborne jet drops, *Limnol. Oceanogr.*, *26* (5), 961-964, 1981.

Blanchard, D.C., A.H. Woodcock, and R.J. Cipriano, The vertical distribution of the concentration of sea salt in the marine atmosphere near Hawaii, *Tellus*, *36B*, 118-125, 1984.

Blifford, I.H., Jr., Tropospheric aerosols, *J. Geophys. Res.*, *75* (15), 3099-3103, 1970.

Blifford, I.H., Jr., and D.A. Gillette, Applications of the lognormal frequency distribution to the chemical composition and size distribution of naturally occurring atmospheric aerosols, *Water Air Soil Pollut.*, *1*, 106-114, 1971.

Bloch, M.R., and W. Luecke, Geochemistry of ocean water bubble spray, *J. Geophys. Res.*, *77* (27), 5100-5105, 1972.

Bloch, M.R., D. Kaplan, V. Kertes, and J. Schnerb, Ion separation in bursting air bubbles: an explanation for the irregular ion ratios in atmospheric precipitations, *Nature*, *209*, 802-803, 1966.

Bodhaine, B.A., and R.F. Pueschel, Flame photometric analysis of the transport of sea salt particles, *J. Geophys. Res.*, *77* (27), 5106-5115, 1972.

Boers, R., and A.K. Betts, Saturation point structure of marine statocumulus clouds, *J. Atmos. Sci.*, *45* (7), 1156-1175, 1988.

Boers, R., P.B. Krummel, S.T. Siems, and G.D. Hess, Thermo-dynamic structure and entrainment of stratocumulus over the Southern Ocean, *J. Geophys. Res.*, *103* (D13), 16637-16650, 1998.

Bohren, C.F., *Clouds in a Glass of Beer*, 195 pp., John Wiley and Sons, Inc., New York, 1987.

Bohren, C.F., and D.R. Huffman, *Absorption and Scattering of Light by Small Particles*, 530 pp., John Wiley and Sons, New York, 1983.

Bohren, D.F., and S.B. Singham, Backscattering by nonspherical particles: A review of methods and suggested new approaches, *J. Geophys. Res.*, *96* (D3), 5269-5277, 1991.

Bolin, B., and H. Rodhe, A note on the concepts of age distribution and transit time in natural reservoirs, *Tellus*, *25*, 58-62, 1973.

Bond, W.N., Bubbles and drops and Stokes' law, *Philosoph. Mag.*, *4*, 889-898, 1927.

Bond, W.N., and D.A. Newton, Bubbles, drops, and Stokes' Law (Paper 2), *Philosoph. Mag.*, *5*, 794-800, 1928.

Bondur, V.G., and E.A. Sharkov, Statistical properties of whitecaps on a rough sea, *Oceanology*, *22* (3), 274-279, 1982.

Bonsang, B., B.C. Nguyen, A. Gaudry, and G. Lambert, Sulfate enrichment in marine aerosols owing to biogenic gaseous sulfur compounds, *J. Geophys. Res.*, *85* (C12), 7410-7416, 1980.

Borisenkov, Y.P., and M.A. Kuznetsov, Parameterization of the interaction between the atmosphere and the ocean under stormy weather conditions as applied to models of general atmospheric circulation, *Atmos. Oceanic Phys.*, *14* (5), 362-368, 1978.

Borisenkov, Y.P., and M.A. Kuznetsov, On the theory of heat and moisture exchange of finely dispersed storm spray with turbulent air, *Atmos. Oceanic Phys.*, *21* (11), 902-906, 1985.

Bortkovskii, R.S., On the mechanism of interaction between the ocean and the atmosphere during a storm, *Fluid Mechanics-Soviet Research*, *2* (2), 87-94, 1973.

Bortkovskii, R.S., *Air-Sea Exchange of Heat and Moisture During Storms*, 194 pp., D. Reidel Publishing Company, Dordrecht, 1987.

Bortkovskii, R.S., and V.A. Novak, Statistical dependencies of sea state characteristics on water temperature and wind-wave age, *J. Marine Sys.*, *4*, 161-169, 1993.

Bortkovskiy, R.S., and D.F. Timanovskiy, On the microstructure of the breaking crests of waves, *Atmos. Oceanic Phys.*, *18* (3), 255-256, 1982.

Boulton-Stone, J.M., and J.R. Blake, Gas bubbles bursting at a free surface, *J. Fluid Mechanics*, *254*, 437-466, 1993.

Boutin, J., and J. Etcheto, Seasat scatterometer versus scanning multichannel microwave radiometer wind speeds: A comparison on a global scale, *J. Geophys. Res.*, *95C* (12), 22275-22288, 1990.

Bowyer, P.A., Aerosol production in the whitecap simulation tank as a function of water temperature (Appendix E), in *Whitecaps and the Marine Atmosphere, Report No. 7*, edited by E.C. Monahan, M.C. Spillane, P.A. Bowyer, M.R. Higgins, and P.J. Stabeno, pp. 95-103, University College, Galway, Ireland, 1984.

Bowyer, P.A., The rise of bubbles in a glass tube and the spectrum of bubbles produced by a splash, *J. Marine Res.*, *50*, 521-543, 1992.

Bowyer, P.A., Video measurements of near-surface bubble spectra, *J. Geophys. Res.*, *106* ((C7)), 14,179-14,190, 2001.

Bowyer, P.A., D.K. Woolf, and E.C. Monahan, Temperature dependence of the charge and aerosol production associated with a breaking wave in a whitecap simulation tank, *J. Geophys. Res.*, *95* (C4), 5313-5319, 1990.

Boyce, S.G., Sources of atmospheric salts, *Science*, *113*, 620-621, 1951.

Boyce, S.G., The salt spray community, *Ecol. Monogr.*, *24* (1), 29-67, 1954.

Boys, C.V., *Soap Bubbles*, 192 pp., Dover Publications, Inc., New York, 1959.

Brabston, D.C., and H.B. Keller, Viscous flows part spherical gas bubbles, *J. Fluid Mechanics*, *69* (1), 179-189, 1975.

Bradley, E.F., P.A. Coppin, and J.S. Godfrey, Measurements of sensible and latent heat flux in the Western Equatorial Pacific Ocean, *J. Geophys. Res.*, *96*, 3375-3389, 1991.

Bradley, R.S., The rate of evaporation of micro-drops in the presence of insoluble monolayers, *J. Colloid Sci.*, *10*, 571-575, 1955.

Braun, C., and U.K. Krieger, Two-dimensional angular light-scattering in aqueous NaCl single aerosol particles during deliquescence and efflorescence, *Optics Express*, *8* (6), 314-321, 2001.

Breitz, N., and H. Medwin, Instrumentation for *in situ* acoustical measurements of bubble spectra under breaking waves, *J. Acoustical Soc. Am.*, *86* (2), 739-743, 1989.

Brenguier, J.-L., H. Pawlowska, L. Schüller, R. Preusker, J. Fischer, and J. Fouquart, Radiative properties of boundary layer clouds: Droplet effective radius versus number concentration, *J. Atmos. Sci.*, *57*, 803-821, 2000.

Bressan, D.J., and F.K. Lepple, Dependence of sea-salt aerosol concentration on various environmental parameters, in *Mapping Strategies in Chemical Oceanography*, edited by A. Zirino, pp. 75-98, American Chemical Society, Washington, DC, 1985.

Bretherton, C.S., P. Austin, and S.T. Siems, Cloudiness and marine boundary layer dynamics in the ASTEX Lagrangian experiments. Part II: Cloudiness, drizzle, surface fluxes, and entrainment, *J. Atmos. Sci.*, *52* (16), 2724-2735, 1995.

Brewer, L., Critical evaluation of typically unreliable high-temperature data, *Bull. Alloy Phase Diagr.*, *9* (2), 99-100, 1988.

Broecker, H.C., and W. Siems, The role of bubbles for gas transfer from water to air at higher windspeeds. Experiments in the wind-wave facility at Hamburg., in *Gas Transfer at Water Surfaces*, edited by W. Brutsaert, and G.H. Jirka, pp. 229-236, D. Reidel Publishing Company, Dordrecht, 1984.

Broecker, H.C., J. Petermann, and W. Siems, The influence of wind on CO_2-exchange in a wind-wave tunnel, including the effects of monolayers, *J. Marine Res.*, *36*, 595-610, 1978.

Broecker, W.S., J.R. Ledwell, T. Takahashi, R. Weiss, L. Merlivat, L. Memery, T.H. Peng, B. Jahne, and K.O. Munnich, Isotopic versus micrometeorologic ocean CO_2 fluxes: a serious conflict, *J. Geophys. Res.*, *91* (C9), 10517-10527, 1986.

Brown, G.S., Estimation of surface wind speeds using satellite-borne radar measurements at normal incidence, *J. Geophys. Res.*, *84* (B8), 3874-3978, 1979.

Browne, L.W.B., Deposition of particles on rough surfaces during turbulent gas-flow in a pipe, *Atmos. Environ.*, *8*, 801-817, 1974.

Brownscombe, J.L., and N.S.C. Thorndyke, Freezing and shattering of water droplets in free fall, *Nature*, *220*, 687-689, 1968.

Brutsaert, W., *Evaporation into the Atmosphere*, 299 pp., D. Reidel Publishing Company, Dordrecht, 1982.

Bruyevich, S.V., and Y.Z. Kulik, Chemical interaction between the ocean and the atmosphere (salt exchange), *Oceanology*, *7*, 279-293, 1967.

Buat-Ménard, P., Particle geochemistry in the atmosphere and the oceans, in *Air-Sea Exchange of Gases and Particles*, edited by P.S. Liss, and W.G.N. Slinn, pp. 455-532, D. Reidel Publishing Company, Dordrecht, 1983.

Bullrich, K., and G. Hänel, Effects of organic aerosol constituents on extinction and absorption coefficients and liquid water contents of fogs and clouds, *Pure Appl. Geophys.*, *116*, 293-301, 1978.

Bunker, A.F., Computations of surface energy flux and annual air-sea interaction cycles of the North Atlantic Ocean, *Monthly Weather Rev.*, *104*, 1122-1140, 1976.

Burger, S.R., and J.W. Bennett, Droplet enrichment factors of pigmented and nonpigmented *Serratio marcescens*: Possible selective function for prodigiosin, *Applied Environ. Microbiol.*, *50* (2), 487-490, 1985.

Burger, S.R., and D.C. Blanchard, The persistence of air bubbles at a seawater surface, *J. Geophys. Res.*, *88* (C12), 7724-7726, 1983.

Burk, S.D., The generation, turbulent transfer and deposition of the sea-salt aerosol, *J. Atmos. Sci.*, *41* (20), 3040-3051, 1984.

Burke, S.P., and W.B. Plummer, Suspension of macroscopic particles in a turbulent gas stream, *Indust. Eng. Chem.*, *20* (17), 1200-1204, 1928.

Buseck, P.R., and M. Pósfai, Airborne minerals and related aerosol particles: Effects on climate and the environment, *Proc. Natl. Acad. Sci., U.S.A.*, *96*, 3372-3379, 1999.

Businger, J.A., Evaluation of the accuracy with which dry deposition can be measured with current micrometeorological techniques, *J. Clim. Applied Meteorol.*, *25*, 1100-1124, 1986.

Businger, J.A., and S.P. Oncley, Flux measurements with conditional sampling, *J. Atmos. Oceanic Technol.*, *7*, 349-352, 1990.

Buzorius, G., Ü. Rannick, J.M. Mäkelä, T. Vesala, and M. Kulmala, Vertical aerosol particle fluxes measured by eddy covariance technique using condensational particle counter, *J. Aerosol Sci.*, *29*, 157-171, 1998.

Buzorius, G., Ü. Rannik, J.M. Mäkelä, P. Keronen, T. Vesala, and M. Kulmala, Vertical aerosol fluxes measured by the eddy covariance method and deposition of nucleation mode particles above a Scots pine forest in Southern Finland, *J. Geophys. Res.*, *105* (D15), 19905-19916, 2000.

Cadenhead, D.A., Monomolecular films at the air-water interface. Some practical applications, *Indust. Eng. Chem.*, *61* (4), 22-28, 1969.

Cadle, R.D., and R.C. Robbins, Kinetics of atmospheric chemical reactions involving aerosols, *Discuss. Faraday Soc.*, *30*, 155-161, 1960.

Caffrey, P., W.A. Hoppel, G. Frick, J. Fitzgerald, N. Shantz, W.R. Leaitch, L. Pasternack, T. Albrechcinski, and J. Ambrusko, Chamber measurements of Cl depletion in cloud-processed sea-salt aerosol, *J. Geophys. Res.*, *106* (D21), 27635-27645, 2001.

Calderbank, P.H., and A.C. Lochiel, Mass transfer coefficients, velocities and shapes of carbon dioxide bubbles in free rise through distilled water, *Chem. Eng. Sci.*, *19*, 485-503, 1964.

Calderbank, P.H., D.S.L. Johnson, and J. Loudon, Mechanics and mass transfer of single bubbles in free rise through some Newtonian and non-Newtonian liquids, *Chem. Eng. Sci.*, *25*, 235-256, 1970.

Caldwell, D.R., and W.P. Elliott, Surface stresses produced by rainfall, *J. Phys. Oceanogr.*, *1*, 145-148, 1971.

Caldwell, D.R., and W.P. Elliott, The effect of rainfall on the wind in the surface layer, *Boundary-Layer Meteorol.*, *3*, 146-151, 1972.

Callendar, G.S., The artificial production of carbon dioxide and its influence on temperature, *Quar. J. Royal Meteorol. Soc.*, *64*, 223-248, 1938.

Callendar, G.S., On the amount of carbon dioxide in the atmosphere, *Tellus*, *10*, 243-248, 1958.

Cambray, R.S., and J.D. Eakins, Pu, [241]Am and ^{137}Cs in soil in West Cumbria and a maritime effect, *Nature*, *300*, 46-48, 1982.

Cammenga, H.K., Evaporation mechanisms in liquids, in *Current Topics in Material Science*, edited by E. Kaldis, pp. 335-446, North-Holland Publishing Company, 1980.

Campuzano-Jost, P., C.D. Clark, H. Maring, D.S. Covert, S. Howell, V. Kapustin, K.A. Clarke, E.S. Saltzman, and A.J. Hynes, Near-real-time measurement of sea-salt aerosol during the SEAS campaign: Comparison of emission-based sodium detection with an aerosol volatility technique, *J. Atmos. Oceanic Technol.*, *20*, 1421-1430, 2003.

Cantrell, W., G. Shaw, and R. Benner, Cloud properties inferred from bimodal aerosol number distributions, *J. Geophys. Res.*, *104* (D22), 27615-27624, 1999.

Capaldo, K.P., P. Kasibhatla, and S.N. Pandis, Is aerosol production within the remote marine boundary layer sufficient to maintain observed concentrations?, *J. Geophys. Res.*, *104* (D3), 3483-3500, 1999.

Cardone, V.J., J.G. Greenwood, and M.A. Cane, On trends in historical marine wind data, *J. Climate*, *3*, 113-127, 1990.

Carey, W.M., J.W. Fitzgerald, E.C. Monahan, and Q. Wang, Measurement of the sound produced by a tipping trough with fresh and salt water, *J. Acoustical Soc. Am.*, *93* (6), 3178-3192, 1993a.

Carey, W.C., J.W. Fitzgerald, E.C. Monahan, and Q. Wang, Erratum: Measurement of the sound produced by a tipping trough with fresh and salt water [*J. Acoust. Soc. Am.* 93, 3178, 1993], *J. Acoustical Soc. Am.*, *94* (5), 3018, 1993b.

Carlucci, A.F., and P.M. Williams, Concentration of bacteria from sea water by bubble scavenging, *J. du Conseil*, *30*, 28-33, 1965.

Carmichael, G.R., M.-S. Hong, H. Ueda, L.-L. Chen, K. Murano, J.K. Park, H. Lee, Y. Kim, C. Kang, and S. Shim, Aerosol composition at Cheju Island, Korea, *J. Geophys. Res.*, *102* (D5), 6047-6061, 1997.

Carrico, C.M., M.J. Rood, and J.A. Ogren, Aerosol light scattering properties at Cape Grim, Tasmania, during the First Aerosol Characterization Experiment (ACE 1), *J. Geophys. Res.*, *103* (D13), 16565-16574, 1998.

Carrier, G.F., On Slow Viscous Flow, *Report NR-062-163*, pp. 35, Brown University, Providence, RI, 1953.

Carroll, K., and R. Mesler, Part II: Bubble entrainment by drop-formed vortex rings, *Am. Inst. Chem. Eng. J.*, *27* (5), 853-856, 1981a.

Carroll, K., and R. Mesler, Splashing liquid drops form vortex rings and not jets at low Froude numbers, *J. Applied Phys.*, *52* (1), 507, 1981b.

Carstens, J.C., and J.J. Martin, A comparison of in-cloud scavenging by Brownian diffusion and thermo and diffusio-phoresis, in *Precipitation Scavenging, Dry Deposition, and Resuspension*, edited by H.R. Pruppacher, R.G. Semonin, and W.G.N. Slinn, pp. 529-540, Elsevier, New York, 1983.

Carter, D.J.T., P.G. Challenor, and M.A. Srokosz, An assessment of Geosat wave height and wind speed measurements, *J. Geophys. Res.*, *97* (C7), 11383-11392, 1992.

Cartmill, J.W., and M.Y. Su, Bubble size distribution under saltwater and freshwater breaking waves, *Dyn. Atmos. Oceans*, *20*, 25-31, 1993.

Castelman, R.A., The resistance to the steady motion of small spheres in fluids, *TN 231*, pp. 12, National Advisory Committee for Aeronautics, Washington, DC, 1926.

Castelman, R.A., Jr., The mechanism of the atomization of liquids, *Bur. Stand. J. Res.*, *6*, 369-376, 1931.

Cattell, F.C.R., and W.D. Scott, Copper in aerosol particles produced by the ocean, *Science*, *202*, 429-430, 1978.

Cauer, H., Some problems of atmospheric chemistry, in *Compendium of Meteorology*, edited by T.F. Malone, pp. 1126-1136, American Meteorological Society, Boston, 1951.

Chabas, A., and R.A. Lefèvre, Chemistry and microscopy of atmospheric particulates at Delos ·(Cyclades-Greece), *Atmos. Environ.*, *34*, 225-238, 2000.

Chabas, A., D. Jeannette, and R.A. Lefèvre, Crystallization and dissolution of airborne sea-salts on weathered marble in a coastal environment at Delos (Cyclades-Greece), *Atmos. Environ.*, *34*, 219-224, 2000.

Chaen, M., Studies on the production of sea-salt particles on the sea surface, *Memoirs of the Faculty of Fisheries, Kagoshima University*, *22* (2), 49-106, 1973.

Chalmers, J.J., and F. Bavarian, Microscopic visualization of insect cell-bubble interactions. II: The bubble film and bubble rupture, *Biotech. Progr.*, *7*, 151-158, 1991.

Chamberlain, A.C., Roughness length of sea, sand, and snow, *Boundary-Layer Meteorol.*, *25*, 405-409, 1983.

Chamberlain, A.C., and R.C. Chadwick, Deposition of airborne radioiodine vapor, *Nucleonics*, *11* (8), 22-25, 1953.

Chamberlain, A.C., J.A. Garland, and A.C. Wells, Transport of gases and particles to surfaces with widely spaced roughness elements, *Boundary-Layer Meteorol.*, *29*, 343-360, 1984.

Chameides, W.L., and A.W. Stelson, Aqueous-phase chemical processes in deliquescent sea-salt aerosols: a mechanism that couples the atmospheric cycles of S and sea salt, *J. Geophys. Res.*, *97* (D18), 20565-20580, 1992.

Chan, C.K., R.C. Flagan, and J.H. Seinfeld, Water activities of $NH_4NO_3/(NH_4)_2SO_4$ solutions, *Atmos. Environ.*, *26A* (9), 1661-1673, 1992.

Chang, D.P.Y., and R.C. Hill, Retardation of aqueous droplet evaporation by air pollutants, *Atmos. Environ.*, *14*, 803-807, 1980.

Chao, B.T., Motion of spherical gas bubbles in a viscous liquid at large Reynolds numbers, *Physics of Fluids*, *5* (1), 69-79, 1962.

Chapman, D.S., and P.R. Critchlow, Formation of vortex rings from falling drops, *J. Fluid Mechanics*, *29* (1), 177-185, 1967.

Chapman, S., Carrier mobility spectra of spray electrified liquids, *Physical Rev.*, *52*, 184-190, 1937.

Chapman, S., Carrier mobility spectra of liquids electrified by bubbling, *Physical Rev.*, *54*, 520-527, 1938a.

Chapman, S., Interpretation of carrier mobility spectra of liquids electrified by bubbling and spraying, *Physical Rev.*, *54*, 528-533, 1938b.

Charlesworth, D.H., and W.R.J. Marshall, Evaporation from drops containing dissolved solids, *Am. Inst. Chem. Eng. J.*, *6*, 9-23, 1960.

Charlson, R.J., D.S. Covert, and T.V. Larson, Observation of the effect of humidity on light scattering by aerosols, in *Hygroscopic Aerosols*, edited by L.H. Ruhnke, and A. Deepak, pp. 35-44, Deepak Publishing, Hampton, VA, 1984.

Charlson, R.J., J.E. Lovelock, M.O. Andreae, and S.G. Warren, Oceanic phytoplankton, atmospheric sulphur, cloud albedo and climate, *Nature*, *326*, 655-661, 1987.

Charlson, R.J., S.E. Schwartz, J.M. Hales, R.D. Cess, J.A. Coakley, Jr., J.E. Hansen, and D.J. Hofmann, Climate forcing by anthropogenic aerosols, *Science*, *255*, 423-430, 1992.

Charnock, H., Air-sea interaction, in *Evolution of Physical Oceanography*, edited by B.A. Warren, and C. Wunsch, pp. 482-503, MIT Press, Cambridge, MA, 1981.

Charters, A.C., High-speed impact, *Scientific Am.*, *203* (4), 128-140, 1960.

Chelton, D.B., and P.J. McCabe, A review of satellite altimeter measurement of sea surface wind speed: with a proposed new algorithm, *J. Geophys. Res.*, *90* (C3), 4707-4720, 1985.

Chelton, D.B., and F.J. Wentz, Further development of an improved altimeter wind speed algorithm, *J. Geophys. Res.*, *91* (C12), 14250-14260, 1986.

Cheng, R.J., Water drop freezing: ejection of microdroplets, *Science*, *170*, 1395-1396, 1970.

Cheng, R.J., Microdroplets and water drop freezing, *Science*, *173*, 849-850, 1971.

Cheng, R.J., The generation of secondary marine aerosols: the crystallization of seawater droplets, in *Atmospheric Aerosols and Nucleation*, edited by P.E. Wagner, and G. Vali, pp. 589-592, Springer-Verlag, Berlin, 1988.

Cheng, R.J., D.C. Blanchard, and R.J. Cipriano, The formation of hollow sea-salt particles from the evaporation of drops of seawater, *Atmos. Res.*, *22*, 15-25, 1988.

Cheng, Y.S., B.T. Chen, and H.C. Yeh, Behaviour of isometric nonspherical aerosol particles in the aerodynamic particle sizer, *J. Aerosol Sci.*, *5*, 701-710, 1990.

Chesselet, R., J. Morelli, and P. Buat-Menard, Variations in ionic ratios between reference sea water and marine aerosols, *J. Geophys. Res.*, *77* (27), 5116-5131, 1972.

Chester, W., and D.R. Breach, On the flow past a sphere at low Reynolds number, *J. Fluid Mechanics*, *47* (4), 751-760, 1969.

Chiapello, I., G. Bergametti, B. Chatenet, F. Dulac, I. Jankowiak, C. Liousse, and E.S. Soares, Contribution of the different aerosol species to the aerosol mass load and optical depth over the northeastern tropical Atlantic, *J. Geophys. Res.*, *104* (D4), 4025-4035, 1999.

Chin, M., P. Ginoux, S. Kinne, O. Torres, B.N. Holben, B.N. Duncan, R.V. Martin, J.A. Logan, A. Higurashi, and T. Nakahima, Tropospheric aerosol optical thickness from the GOCART model and comparisons with satellite and sun photometer measurements, *J. Atmos. Sci.*, *59*, 461-483, 2002.

Chowdhury, K.C.R., and W. Fritz, Sinkversuche mit isometrischen Teilchen in Flüssigkeiten, *Chem. Eng. Sci.*, *11*, 92-98, 1959.

Chuang, P.Y., D.R. Collins, H. Pawlowska, J.R. Snider, H.H. Jonsson, J.L. Brenguier, R.C. Flagan, and J.H. Seinfeld, CCN measruements during ACE-2 and their relationship to cloud microphysical properties, *Tellus*, *52B*, 843-867, 2000.

Churchill, S.W., *Viscous Flows - The Practical Use and Theory*, 602 pp., Butterworths, Boston, 1988.

Cincinelli, A., A.M. Stortini, M. Perugini, L. Checchini, and L. Lepti, Organic pollutants in sea-surface microlayer and aerosol in the coastal environment of Leghorn (Tyrrhenian Sea), *Marine Chem.*, *76*, 77-98, 2001.

Cini, R., P. Desideri, and L. Lepri, Transport of organic compounds across the air/sea interface of artificial and natural marine aerosols, *Analytica Chimica Acta*, *291*, 329-340, 1994.

Cipriano, R.J., and D.C. Blanchard, Bubble and aerosol spectra produced by a laboratory 'breaking wave', *J. Geophys. Res.*, *86* (C9), 8085-8092, 1981.

Cipriano, R.J., and D.C. Blanchard, Reply, *J. Geophys. Res.*, *87* (C8), 5869-5870, 1982.

Cipriano, R.J., D.C. Blanchard, A.W. Hogan, and G.G. Lala, On the production of Aitken nuclei from breaking waves and their role in the atmosphere, *J. Atmos. Sci.*, *40*, 469-479, 1983.

Cipriano, R.J., E.C. Monahan, P.A. Bowyer, and D.K. Woolf, Marine condensation nucleus generation inferred from whitecap

simulation tank results, *J. Geophys. Res.*, *92* (C6), 6569-6576, 1987.

Clapeyron, E., La force motrice de la chaleur, *Journal école polytech. (Paris)*, *14* (23), 153, 1834.

Clark, C.D., P. Campuzano-Jost, D.S. Covert, R.C. Richter, J. Marine, A.J. Hynes, and E.S. Saltzman, Real-time measurement of sodium in single aerosol particles by flame emission: laboratory characterization, *J. Aerosol Sci.*, *32*, 765-778, 2001.

Clarke, A.D., A thermo-optic technique for *in situ* analysis of size-resolved aerosol physicochemistry, *Atmos. Environ.*, *25A* (3/4), 635-644, 1991.

Clarke, A.D., Atmospheric nuclei in the remote free-troposphere, *J. Atmos. Chem.*, *14*, 479-488, 1992.

Clarke, A.D., and J.N. Porter, Pacific marine aerosol 2. Equatorial gradients in chlorophyll, ammonium, and excess sulfate during SAGA 3, *J. Geophys. Res.*, *98* (D9), 16997-17010, 1993.

Clarke, A.D., and V.N. Kapustin, The Shoreline Environmental Aerosol Study (SEAS): A context for marine aerosol measurements influenced by a coastal environment and long-range transport, *J. Atmos. Oceanic Technol.*, *20*, 1351-1361, 2003.

Clarke, A.D., N.C. Ahlquist, and D.S. Covert, The Pacific marine aerosol: evidence for natural acid sulfates, *J. Geophys. Res.*, *92* (D4), 4179-4190, 1987.

Clarke, A.D., J.N. Porter, F.P.J. Valero, and P. Pilewskie, Vertical profiles, aerosol microphysics, and optical closure during the Atlantic Stratocumulus Transition Experiment: Measured and modeled column optical properties, *J. Geophys. Res.*, *101* (D2), 4443-4453, 1996.

Clarke, A.D., J.L. Varner, F. Eisele, R.L. Mauldin, D. Tanner, and M. Litchy, Particle production in the remote marine atmosphere: Cloud outflow and subsidence during ACE-1, *J. Geophys. Res.*, *103* (D13), 16397-16904, 1998.

Clarke, A., V. Kapustin, S. Howell, K.F. Moore, B. Lienert, S. Masonis, T. Anderson, and D. Covert, Sea-salt size distributions from breaking waves: Implications for marine aerosol production and optical extinction measurements during SEAS, *J. Atmos. Oceanic Technol.*, *20*, 1362-1374, 2003.

Clarke, A.G., and M. Radojevic, Oxidation rates of SO_2 in sea-water and sea-salt aerosols, *Atmos. Environ.*, *18* (12), 2761-2767, 1984.

Clausius, R., Ueber die bewegende Kraft der Wärme, *Annalen der Physik und Chemie*, *79*, 368-397, 500-524, 1850.

Clayton, J.L., Salt spray and mineral cycling in two California coastal ecosystems, *Ecology*, *53*, 74-81, 1972.

Clegg, N.A., and R. Toumi, Sensitivity of sulphur dioxide oxidation in sea salt to nitric acid and ammonia gas phase concentrations, *J. Geophys. Res.*, *102* (D19), 23241-23249, 1997.

Clegg, N.A., and R. Toumi, Non-sea-salt-sulphate formation in sea-salt aerosol, *J. Geophys. Res.*, *103* (D23), 31095-31102, 1998.

Clift, R., J.R. Grace, and M.E. Weber, *Bubbles, Drops, and Particles*, 380 pp., Academic Press, New York, 1978.

Cloke, J., W.A. McKay, and P.S. Liss, Laboratory investigation into the effect of marine organic material on the sea-salt aerosol generated by bubble bursting, *Marine Chem.*, *34*, 77-95, 1991.

Coakley, J.A., Jr., R.D. Cess, and F.B. Yurevich, The effect of tropospheric aerosols on the Earth's radiation budget: A parameterization for climate models, *J. Atmos. Sci.*, *40* (1), 116-138, 1983.

Coakley, J.A., Jr., R.L. Bernstein, and P.A. Durkee, Effect of ship-stack effluents on cloud reflectivity, *Science*, *237*, 1020-1022, 1987.

Cohen, M.D., R.C. Flagan, and J.H. Seinfeld, Studies of concentrated electrolyte solutions using the electrodynamic balance. 1. Water activities for single-electrolyte solutions, *J. Phys. Chem.*, *91*, 4563-4574, 1987a.

Cohen, M.D., R.C. Flagan, and J.H. Seinfeld, Studies of concentrated electrolyte solutions using the electrodynamic balance. 2. Water activities for mixed-electrolyte solutions, *J. Phys. Chem.*, *91*, 4575-4582, 1987b.

Cohen, M.D., R.C. Flagan, and J.H. Seinfeld, Studies of concentrated electrolyte solutions using the electrodynamic balance. 3. Solute nucleation, *J. Phys. Chem.*, *91*, 4583-4590, 1987c.

Cokelet, E.D., Breaking waves, *Nature*, *367*, 969-974, 1977.

Cole, I.S., D.A. Paterson, and W.D. Ganther, Holistic model for atmospheric corrosion: Part 1 - Theoretical framework for production, transportation, and deposition of marine salts, *Corr. Eng. Sci. Technol.*, *38* (2), 129-134, 2003a.

Cole, I.S., W.D. Ganther, D.A. Paterson, G.A. King, S.A. Furman, and D. Lau, Holisitc model for atmospheric corrosion: Part 2 - Experimental measurement of deposition of marine salts in a number of long-range studies, *Corr. Eng. Sci. Technol.*, *38* (4), 259-266, 2003b.

Cole, I.S., D.A. Paterson, W.D. Ganther, A. Neufeld, B. Hinton, G. McAdam, M. McGeachie, R. Jeffery, L. Chotimongkol, C. Bhamornsut, N.V. Hue, and S. Purwadaria, Holistic model for atmospheric corrosion: Part 3: Effect of natural and man made landforms on deposition of marine salts in Australia and south-east Asia, *Corr. Eng. Sci. Technol.*, *38* (4), 267-274, 2003c.

Cole, I.S., W.Y. Chan, G.S. Trinidad, and D.A. Paterson, Holistic model for atmospheric corrosion: Part 4 - Geographic information system for predicting airborne salinity, *Corr. Eng. Sci. Technol.*, *39* (1), 89-96, 2004.

Coletti, A., Light scattering by nonspherical particles: A laboratory study, *Aerosol Sci. Technol.*, *3*, 39-52, 1984.

Collins, R., A second approximation for the velocity of a large gas bubble rising in an infinite liquid, *J. Fluid Mechanics*, *25* (3), 469-480, 1966.

Commander, K.W., and R.J. McDonald, Finite-element solution of the inverse problem in bubble swarm acoustics, *J. Acoustical Soc. Am.*, *89* (2), 592-597, 1991.

Commander, K., and E. Moritz, Off-resonance contributions to acoustical bubble spectra, *J. Acoustical Soc. Am.*, *85* (6), 2665-2669, 1989.

Conover, J.H., Anomalous cloud lines, *J. Atmos. Sci.*, *23*, 778-785, 1966.

Coppock, P.D., and G.T. Meiklejohn, The behaviour of gas bubbles in relation to mass transfer, *Trans. Inst. Chem. Eng.*, *29*, 75-86, 1951.

Cornford, S.G., Fall speeds of precipitation elements, *Quar. J. Royal Meteorol. Soc.*, *91*, 91-94, 1965.

Corradini, C., and G. Tonna, Condensation nuclei supersaturation spectrum: analysis of the relationship between the saturation droplet radius and the critical supersaturation in the Laktionov isothermal chamber, *J. Aerosol Sci.*, *10*, 465-469, 1979.

Coste, J.H., and H.L. Wright, The nature of the nucleus in hygroscopic droplets, *Philosoph. Mag.*, *20*, 209-234, 1935.

Covert, D.S., R.J. Charlson, and N.C. Ahlquist, A study of the relationship of chemical composition and humidity to light scattering by aerosols, *J. Applied Meteorol.*, *11*, 968-976, 1972.

Covert, D.S., V.N. Kapustin, P.K. Quinn, and T.S. Bates, New particle formation in the marine boundary layer, *J. Geophys. Res.*, *97* (D18), 20581-20589, 1992.

Covert, D.S., J.L. Gras, A. Wiedensohler, and F. Stratmann, Comparison of directly measured CCN with CCN modeled from the number-size distribution in the marine boundary layer during ACE 1 at Cape Grim, Tasmania, *J. Geophys. Res.*, *103* (D13), 16597-16608, 1998.

Cox, C., and W. Munk, Measurement of the roughness of the sea surface from photographs of the sun's glitter, *J. Optical Soc. Am.*, *44* (11), 838-850, 1954.

Craeye, C., and P. Schlüssel, Rainfall on the sea: surface renewals and wave damping, *Boundary-Layer Meteorol.*, *89*, 349-355, 1998.

Craeye, C., P.W. Sobieski, L.F. Bliven, and A. Guissard, Ring-waves generated by water drops impacting on water surfaces at rest, *IEEE J. Oceanic Eng.*, *24* (3), 323-331, 1999.

Craig, H., and T. Hayward, Oxygen supersaturation in the ocean: biological versus physical contributions, *Science*, *235*, 199-202, 1987.

Craig, V.S.J., B.W. Ninham, and R.M. Pashley, The effect of electrolytes on bubble coalescence in water, *J. Phys. Chem.*, *97*, 10192-10197, 1993a.

Craig, V.S.J., B.W. Ninham, and R.M. Pashley, Effects of electrolytes on bubble coalescence, *Nature*, *364*, 317-319, 1993b.

Craven, C.J., and O.J. Stuhlman, The mechanics of effervescence, *Physical Rev.*, *37*, 1699, 1931.

Crawford, G.B., and D.M. Farmer, On the spatial distribution of ocean bubbles, *J. Geophys. Res.*, *92* (C8), 8231-8243, 1987.

Crawford, J., J. Olson, D. Davis, G. Chen, J. Barrick, R. Shetter, B. Lefer, C. Jordan, B. Anderson, A. Clarke, G. Sachse, D. Blake, H. Singh, S. Sandolm, D. Tan, Y. Kondo, M. Avery, F. Flocke, F.L. Eisele, L. Mauldin, M. Zondlo, W. Brune, H. Harder, M. Martinez, R. Talbot, A. Bandy, and D. Thorton, Clouds and trace gas distributions during TRACE-P, *J. Geophys. Res.*, *108* (D21), 8818, doi:10.129/2002JD003177, 2003.

Crowther, P.A., Bubble noise creation mechanisms, in *Sea Surface Sound*, edited by B.R. Kerman, pp. 131-150, Kluwer Academic Publishers, Dordrecht, 1988.

Crum, L.A., H.C. Pumphrey, R.A. Roy, and A. Prosperetti, The underwater sounds produced by impacting snowflakes, *J. Acoustical Soc. Am.*, *106* (4), 1765-1770, 1999.

Csanady, G.T., Turbulent diffusion of heavy particles in the atmosphere, *J. Atmos. Sci.*, *20*, 201-208, 1963.

Culkin, F., The major constituents of sea water, in *Chemical Oceanography*, edited by J.P. Riley, and G. Skirrow, pp. 121-161, Academic Press, London, 1965.

Cunningham, E., On the velocity of steady fall of spherical particles through fluid medium, *Proc. Royal Soc. London*, *83A*, 357-365, 1910.

Cziczo, D.J., and J.P.D. Abbatt, Infrared observations of the response of NaCl, MgCl$_2$, NH$_4$HSO$_4$, and NH$_4$NO$_3$ aerosols to changes in relative humidity from 298 to 238 K, *J. Phys. Chem.*, *104A*, 2038-2047, 2000.

Cziczo, D.J., J.B. Nowak, J.H. Hu, and J.P.D. Abbatt, Infrared spectroscopy of model tropospheric aerosols as a function of relative humidity: observation of deliquescence and crystallization, *J. Geophys. Res.*, *102* (D15), 18843-18850, 1997.

Dabberdt, W.F., D.H. Lenschow, T.W. Horst, P.R. Zimmerman, S.P. Oncley, and A.C. Delany, Atmospheric-surface exchange measurements, *Science*, *260*, 1472-1481, 1993.

Dahl, P.H., and A.T. Jessup, On bubble clouds produced by breaking waves: An event analysis of ocean acoustic measurements, *J. Geophys. Res.*, *100* (C3), 5007-5020, 1995.

Dahneke, B., Simple kinetic theory of Brownian diffusion in vapors and aerosols, in *Theory of Dispersed Multiphase Flow*, edited by R.E. Meyer, pp. 97-133, Academic Press, New York, 1983.

Dai, D.J., S.J. Peters, and G.E. Ewing, Water adsorption and dissociation on NaCl surfaces, *J. Phys. Chem.*, *99*, 10299-10304, 1995.

Dalen, J., and A. Løvik, The influence of wind-induced bubbles on echo integration surveys, *J. Acoustical Soc. Am.*, *69* (6), 1653-1659, 1981.

Dana, M.T., and J.M. Hales, Statistical aspects of the washout of polydisperse aerosols, *Atmos. Environ.*, *10*, 45-50, 1976.

Danckwerts, P.V., Absorption by simultaneous diffusion and chemical reaction into particles of various shapes and into falling drops, *Trans. Faraday Soc.*, *47*, 1014-1023, 1951.

Daniels, A., Measurements of atmospheric sea salt concentrations in Hawaii using a Tala kite, *Tellus*, *41B*, 196-206, 1989.

Dasch, J.M., A comparison of surrogate surfaces for dry deposition collection, in *Precipitation Scavenging, Dry Deposition, and Resuspension*, edited by H.R. Pruppacher, R.G. Semonin, and W.G.N. Slinn, pp. 883-902, Elsevier, New York, 1983.

Datta, R.L., D.J. Napier, and D.M. Newitt, The properties and behaviour of gas bubbles formed at a circular orifice, *Trans. Inst. Chem. Eng.*, *28*, 14-26, 1950.

Davidson, C.I., S.E. Lindberg, J.A. Schmidt, L.G. Cartwright, and L.R. Landis, Dry deposition of sulfate onto surrogate surfaces, *J. Geophys. Res.*, *90* (D1), 2123-2130, 1985.

Davidson, K.L., and L. Schütz, Observational results on the influence of surface layer stability and inversion entrainment on surface layer marine aerosol number density (1 micrometer), *Opt. Eng.*, *22* (1), 45-49, 1983.

Davidson, K.L., T.M. Houlihan, C.W. Fairall, and G.E. Schacher, Observation of the temperature structure function parameter, C^2_T, over the ocean, *Boundary-Layer Meteorol.*, *15*, 507-523, 1978.

Davidson, K.L., C.W. Fairall, and G.E. Schacher, Mixed layer modeling of aerosols in the marine boundary layer, *Techn. Report NP-563-82-002*, pp. 51, Naval Postgraduate School, Monterey, CA, 1982.

Davidson, K.L., C.W. Fairall, P.J. Boyle, and G.E. Schacher, Verification of an atmospheric mixed-layer model for a coastal region, *J. Clim. Applied Meteorol.*, *23*, 617-636, 1984.

Davies, C.N., Definitive equations for the fluid resistance of spheres, *Proc. Phys. Soc.*, *57*, 259-270, 1945.

Davies, C.N., Deposition from moving aerosols, in *Aerosol Science*, edited by C.N. Davies, pp. 393-446, Academic Press, London, 1966.

Davies, C.N., Size distribution of atmospheric aerosols, *Aerosol Sci.*, *5*, 293-300, 1974.

Davies, J.T., The effect of surface films in dampling eddies at a free surface of a turbulent liquid, *Proc. Royal Soc. London*, *290A*, 515-526, 1966.

Davies, R.M., and G.I. Taylor, The mechanics of large bubbles rising through extended liquids and through liquids in tubes, *Proc. Royal Soc. London*, *200A*, 375-390, 1950.

Davis, G.E., Scattering of light by an air bubble in water, *J. Optical Soc. Am.*, *45* (7), 572-581, 1955.

Davis, R.F., The physical aspect of steam generation at high pressures and the problem of steam contamination, *Proc. Inst. Mech. Eng.*, *149*, 198-216, 1940.

Day, J.A., Production of droplets and salt nuclei by the bursting of air-bubble films, *Quar. J. Royal Meteorol. Soc.*, *90*, 72-78, 1964.

Day, J.A., Bursting air bubbles studied by the time exposure technique, *Nature*, *216*, 1097-1099, 1967.

De Angelis, M., J.P. Steffensen, M.R. Legrand, H. Clausen, and C. Hammer, Primary aerosol (sea salt and dust) deposited in Greenland ice during the last climatic cycle: Comparison with east Antarctic records, *J. Geophys. Res.*, *102* (C12), 26681-26698, 1997.

De Bruyn, W.J., T.S. Bates, J.M. Cainey, and E.S. Saltzman, Shipboard measurements of dimethyl sulfide and SO$_2$ southwest of Tasmania during the First Aerosol Characterization Experiment (ACE-1), *J. Geophys. Res.*, *103* (D13), 16703-16711, 1998.

De Haan, D.O., and B.J. Finlayson-Pitts, Knudsen cell studies of the reaction of gaseous nitric acid with synthetic sea salt at 298 K, *J. Phys. Chem.*, *101A*, 9993-9999, 1997.

de Leeuw, G., Size distributions of giant aerosol particles close above sea level, *J. Aerosol Sci.*, *17* (3), 293-296, 1986a.

de Leeuw, G., Vertical profiles of giant particles close above the sea surface, *Tellus*, *38B*, 51-61, 1986b.

de Leeuw, G., Near-surface particle size distribution profiles over the North Sea, *J. Geophys. Res.*, *92* (C13), 14631-14635, 1987.

de Leeuw, G., Investigations on turbulent fluctuations of particle concentrations and relative humidity in the marine atmospheric surface layer, *J. Geophys. Res.*, *94* (C3), 3261-3269, 1989a.

de Leeuw, G., Modeling of extinction and backscatter profiles in the marine mixed layer, *Applied Optics*, *28* (7), 1356-1359, 1989b.

de Leeuw, G., Profiling of aerosol concentrations, particle size distributions and relative humidity in the atmospheric surface layer over the North Sea, *Tellus*, *42B*, 342-354, 1990a.

de Leeuw, G., Comment on "Vertical distributions of spray droplets near the sea surface: influences of jet drop ejection and surface tearing" by J. Wu, *J. Geophys. Res.*, *95* (C6), 9779-9782, 1990b.

de Leeuw, G., Aerosols near the air-sea interface, *Trends Geophys. Res.*, *2*, 55-70, 1993.

de Leeuw, G., and L.H. Cohen, Bubble size distributions on the North Atlantic and North Sea, in *Gas Transfer at Water Surfaces*, edited by M.A. Donelan, W.M. Drennan, E.S. Saltzman, and R. Wanninkhof, pp. 271-277, American Geophysical Union, Washington, DC, 2002.

de Leeuw, G., and I. Leifer, Bubbles outside the plume during the LUMINY wind-wave experiment, in *Gas Transfer at Water Surfaces*, edited by M.A. Donelan, W.M. Drennan, E.S. Saltzman, and R. Wanninkhof, pp. 295-301, American Geophysical Union, Washington, DC, 2002.

de Leeuw, G., F.P. Neele, M. Hill, M.H. Smith, and E. Vignati, Production of sea spray aerosol in the surf zone, *J. Geophys. Res.*, *105* (D24), 29397-29409, 2000.

Dean, S.W., Classifying atmospheric corrosivity - a challenge for ISO, *Materials Perf.*, *32* (10), 53-58, 1993.

Deane, G.B., Sound generation and air entrainment by breaking waves in the surf zone, *J. Acoustical Soc. Am.*, *102* (5), 2671-2689, 1997.

Deane, G.B., and M.D. Stokes, Air entrainment processes and bubble size distributions in the surf zone, *J. Phys. Oceanogr.*, *29*, 1393-1403, 1999.

Deane, G.B., and M.D. Stokes, Scale dependence of bubble creation mechanisms in breaking waves, *Nature*, *418*, 839-844, 2002.

Debye, P., Der Lichtdruck auf Kugeln von beliebigem Material, *Annalen der Physik*, *30*, 57-136, 1909.

DeCosmo, J., K.B. Katsaros, S.D. Smith, R.J. Anderson, W.A. Oost, K. Bumke, and H. Chadwick, Air-sea exchange of water vapor and sensible heat: The Humidity Exchange Over the Sea (HEXOS) results, *J. Geophys. Res.*, *101* (C5), 12001-12016, 1996.

Defant, A., *Physical Oceanography*, 729 pp., Pergamon Press, Oxford, 1961.

DeFelice, T.P., and R.J. Cheng, On the phenomenon of nuclei enhancement during the evaporative stage of a cloud, *Atmos. Res.*, *47-48*, 15-40, 1998.

Deindoerfer, F.H., and A.E. Humphrey, Mass transfer from individual gas bubbles, *Indust. Eng. Chem.*, *53*, 755-759, 1961.

Deirmendjian, D., Scattering and polarization properties of water clouds and haze in the visible and infrared, *Applied Optics*, *3* (2), 187-196, 1964.

Dekker, H., and G. de Leeuw, Bubble excitation of surface waves and aerosol droplet production: a simple dynamical model, *J. Geophys. Res.*, *98* (C6), 10223-10232, 1993.

DeLuise, J.J., I.H. Blifford, Jr., and J.A. Takamine, Models of tropospheric aerosol size distributions derived from measurements at three locations, *J. Geophys. Res.*, *77* (24), 4529-4538, 1972.

Delumyea, R., and R.L. Petel, Deposition velocity of phosphorus-containing particles over Southern Lake Huron, April-October, 1975, *Atmos. Environ.*, *13*, 287-294, 1979.

Derjaguin, B.V., V.A. Fedoseyev, and L.A. Rosenzweig, Investigation of the adsorption of cetyl alcohol vapor and the effect of this phenomenon on the evaporation of water drops, *J. Colloid Interface Sci.*, *22*, 45-50, 1966.

Desjardins, R.L., Energy budget by an eddy correlation method, *J. Applied Meteorol.*, *16*, 248-250, 1977a.

Desjardins, R.L., Description and evaluation of a sensible heat flux detector, *Boundary-Layer Meteorol.*, *11*, 147-154, 1977b.

Despiau, S., S. Cougnenc, and F. Resch, Concentrations and size distributions of aerosol particles in coastal zone, *J. Aerosol Sci.*, *27* (3), 403-415, 1996.

Dessens, H., Les noyaux de condensation de l'atmosphere, *Comptes Rendus de l'Academies des Sci., Paris*, *223*, 915-917, 1946a.

Dessens, H., Étude d'une particule de brume, *Annales de Géophysique*, *2*, 343-346, 1946b.

Dessens, H., The use of spider's threads in the study of condensation nuclei, *Quar. J. Royal Meteorol. Soc.*, *75*, 23-26, 1949.

Detsch, R.M., Dissolution of 100 to 1000 μm diameter air bubbles in reagent grade water and seawater, *J. Geophys. Res.*, *95* (C6), 9765-9773, 1990.

Detsch, R.M., Small air bubbles in reagent grade water and seawater 1. Rise velocities of 20- to 100-μm-diameter bubbles, *J. Geophys. Res.*, *96* (C5), 8901-8906, 1991.

Detwiler, A., and D.C. Blanchard, Aging and bursting bubbles in trace-contaminated water, *Chem. Eng. Sci.*, *33*, 9-13, 1978.

Deuzé, J.L., M. Herman, P. Goloub, D. Tanré, and A. Marchand, Characterization of aerosols over ocean from POLDER/ADEOS-1, *Geophys. Res. Lttrs*, *26* (10), 1421-1424, 1999.

Ding, L., and D.M. Farmer, Observations of breaking surface wave statistics, *J. Phys. Oceanogr.*, *24*, 1368-1387, 1994.

Dinger, J.E., H.B. Howell, and T.A. Wojciechowski, On the source and composition of cloud nuclei in a subsident air mass over the North Atlantic, *J. Atmos. Sci.*, *27*, 791-797, 1970.

Dobbie, S., J. Li, R. Harvey, and P. Chýlek, Sea-salt optical properties and GCM forcing at solar wavelengths, *Atmos. Res.*, *65*, 211-233, 2003.

Dobson, E., Validation of Geosat altimeter-derived wind speeds and significant wave heights using buoy data, *J. Geophys. Res.*, *92* (C10), 10719-10731, 1987.

Dobson, F.W., The wind blows, the waves come, *Oceanus*, *17*, 29-35, 1974.

DOE, Handbook of methods for the analysis of the various parameters of the carbon dioxide system in sea water, version 2, A. G. Dickson and C. Goyet, eds., ORNL/CDIAC-74, 1994.

Donelan, M., and M. Miyake, Spectra and fluxes in the boundary layer of the trade-wind zone, *J. Atmos. Sci.*, *30*, 444-464, 1973.

Donelan, M., M.S. Longuet-Higgins, and J.S. Turner, Periodicity in whitecaps, *Nature*, *239*, 449-451, 1972.

Donnelly, R.J., and W. Glaberson, Experiments on the capillary instability of a liquid jet, *Proc. Royal Soc. London*, *290A*, 547-556, 1966.

Dorsey, N.E., *Properties of Ordinary Water-Substance*, Hafner Publishing Company, New York, 1940.

Doubliez, L., The drainage and rupture of a non-foaming liquid film formed upon bubble impact with a free surface, *Internatl. J. Multiphase Flow*, *17* (6), 783-803, 1991.

Doyle, D.M., Whitecaps and the Marine Atmosphere, pp. 140, University College, Galway, Ireland, 1984.

Doyle, D., and M. Higgins, Analysis of the MIZEX-84 whitecap results, in *Whitecaps and the Marine Atmosphere, Report No. 8*, edited by E.C. Monahan, P.A. Bowyer, D.M. Doyle, M.R. Higgins, and D.K. Woolf, pp. 54-62, University College, Galway, Ireland, 1985.

Dragcevic, D., M. Vukovic, D. Cukman, and V. Pravdic, Properties of the seawater-air interface. Dynamic surface tension studies, *Limnol. Oceanogr.*, 24 (6), 1022-1030, 1979.

Drake, J.C., Electrification accompanying the melting of ice particles, *Quar. J. Royal Meteorol. Soc.*, 94, 176-191, 1968.

Drogaris, G., and P. Weiland, Studies of coalescence of bubble pairs, *Chemical Engineering Communications*, 23, 11-26, 1983.

Duan, B., C.W. Fairall, and D.W. Thomson, Eddy correlation measurements of the dry deposition of particles in wintertime, *J. Applied Meteorol.*, 27, 642-652, 1988.

Duce, R.A., On the source of gaseous chlorine in the marine atmosphere, *J. Geophys. Res.*, 74 (18), 4597-4599, 1969.

Duce, R.A., and E.J. Hoffman, Chemical fractionation at the air/sea interface, in *Annual Review of Earth and Planetary Sciences*, edited by F.A. Donath, pp. 187-228, 1976.

Duce, R.A., and A.H. Woodcock, Difference in chemical composition of atmospheric sea salt particles produced in the surf zone and on the open sea in Hawaii, *Tellus*, 23, 427-435, 1971.

Duce, R.A., A.H. Woodcock, and J.L. Moyers, Variation of ion ratio with size among particles in tropical oceanic air, *Tellus*, 19, 369-379, 1967.

Duce, R.A., W. Stumm, and J.M. Prospero, Working symposium on sea-air chemistry: summary and recommendations, *J. Geophys. Res.*, 77 (27), 5059-5061, 1972a.

Duce, R.A., J.G. Quinn, C.E. Olney, S.R. Piotriwicz, B.J. Ray, and T.L. Wade, Enrichment of heavy metals and organic compounds in the surface microlayer of Narragansett Bay, Rhode Island, *Science*, 176, 161-163, 1972b.

Duce, R.A., F. MacIntyre, and B. Bonsang, Enrichment of sulfate in maritime aerosols, *Atmos. Environ.*, 16 (8), 2025-2026, 1982.

Duineveld, P.C., The rise velocity and shape of bubbles in pure water at high Reynolds number, *J. Fluid Mechanics*, 292, 325-332, 1995.

Dulac, F., P. Buat-Ménard, U. Ezat, S. Melki, and G. Bergametti, Atmospheric input of trace metals to the western Mediterranean: uncertainties in modelling dry deposition from cascade impactor data, *Tellus*, 41B, 362-378, 1989.

Durbin, W.G., and G.D. White, Measurements of the vertical distribution of atmospheric chloride particles, *Tellus*, 13, 260-275, 1961.

Durkee, P.A., F. Pfeil, E. Frost, and R. Shema, Global analysis of aerosol particle characteristics, *Atmos. Environ.*, 25A (11), 2457-2471, 1991.

Dye, J.E., and P.V. Hobbs, The influence of environmental prameters on the freezing and fragmentation of suspended water drops, *J. Atmos. Sci.*, 25, 82-96, 1968.

Dyer, A.J., Flow distortion by supporting structures, *Boundary-Layer Meteorol.*, 20, 243-251, 1981.

East, T.W.R., and J.S. Marshall, Turbulence in clouds as a factor in precipitation, *Quar. J. Royal Meteorol. Soc.*, 80, 26-47, 1954.

Ebert, M., S. Weinbruch, P. Hoffmann, and H.M. Ortner, Chemical characterization of North Sea aerosol particles, *J. Aerosol Sci.*, 31 (5), 613-632, 2000.

Ebuchi, N., Statistical distribution of wind speeds and directions globally observed by NSCAT, *J. Geophys. Res.*, 104 (C5), 11393-11403, 1999.

Ebuchi, N., H.C. Graber, and M.J. Caruso, Evaluation of wind vectors observed by QuikSCAT/SeaWinds using ocean buoy data, *J. Atmos. Oceanic Technol.*, 19, 2049-2062, 2002.

Edgerton, H.E., and J.R. Killian, Jr., *Moments of Vision: the Stroboscopic Revolution in Photography*, 177 pp., MIT Press, Cambridge, MA, 1979.

Edson, J.B., and C.W. Fairall, Spray droplet modeling 1. Lagrangian model simulation of the turbulent transport of evaporating droplets, *J. Geophys. Res.*, 99 (C12), 25959-25311, 1994.

Edson, J.B., S. Anquentin, P.G. Mestayer, and J.F. Sini, Spray droplet modeling 2. An interactive Eulerian-Lagrangian model of evaporating spray droplets, *J. Geophys. Res.*, 101 (C1), 1279-1293, 1996.

Edson, J.B., A.A. Hinton, K.E. Prada, J.E. Hare, and C.W. Fairall, Direct covariance flux estimates from mobile platforms at sea, *J. Atmos. Oceanic Technol.*, 15, 547-562, 1998.

Edwards, R.S., and S.M. Claxton, The distribution of air-borne salt of marine origin in the Aberystwyth area, *J. Applied Ecol.*, 1, 253-263, 1964.

Einstein, A., Über die von der Molekularkinetischen Theori der Wärme geforderte Bewegung von in ruhenden Flüssigkeiten suspendierten Teilchen, *Annalen der Physik*, 17, 549-560, 1905.

Eisner, H.S., B.W. Quince, and C. Slack, The stabilization of water mists by insoluble monolayers, *Discuss. Faraday Soc.*, 30, 86-95, 1960.

Ekman, V.W., On the influence of the earth's rotation on ocean currents, *Arkiv för Matematik, Astronomi och Fysik*, 2 (11), 1-53, 1905.

El Golli, S., G. Arnaud, J. Bricard, and C. Triener, Evaporation of volatile solvent from saline multi-component droplets carried in a stream of air, *J. Aerosol Sci.*, 8, 39-54, 1977.

Emanuel, K., A similarity hypothesis for air-sea exchange at extreme wind speeds, *J. Atmos. Sci.*, 60, 1420-1428, 2003.

Emerson, S., P.D. Quay, C. Stump, D. Wilbur, and R. Schudlich, Chemical tracers of productivity and respiration in the subtropical ocean, *J. Geophys. Res.*, 100 (C8), 15873-15887, 1995.

Emery, W.J., and J. Meincke, Global water masses: summary and review, *Oceanologica Acta*, 9 (4), 383-391, 1986.

Engel, O.G., Crater depth in fluid impacts, *J. Applied Phys.*, 37 (4), 1798-1808, 1966.

EPA (Environmental Protection Agency), *National primary and secondary ambient air quality standards*, Code of Federal Regulations 40, Part 50, U. S. Government Printing Office, 2002.

Erickson, D.J., III., A stability dependent theory for air-sea gas exchange, *J. Geophys. Res.*, 98 (C5), 8471-8488, 1993.

Erickson, D.J., III., and R.A. Duce, On the global flux of atmospheric sea salt, *J. Geophys. Res.*, 93 (C11), 14079-14088, 1988.

Erickson, D.J., J.T. Merrill, and R.A. Duce, Seasonal estimates of global atmospheric sea-salt distributions, *J. Geophys. Res.*, *91* (D1), 1067-1072, 1986.

Erickson, D.J., III., C. Seuzaret, W.C. Keene, and S.L. Gong, A general circulation model based calculation of HCl and ClNO$_2$ production from sea salt dechlorination: Reactive Chlorine Emissions Inventory, *J. Geophys. Res.*, *104* (D7), 8347-8372, 1999.

Eriksson, E., The yearly circulation of chloride and sulfur in nature; meteorological, geochemical and pedological implications. Part 1, *Tellus*, *11* (4), 375-403, 1959.

Eriksson, E., The yearly circulation of chloride and sulfur in nature; meteorological, geochemical and pedological implications. Part II, *Tellus*, *12*, 63-109, 1960.

Eriksson, E., The yearly circulation of sulfur in nature, *J. Geophys. Res.*, *68* (13), 4001-4008, 1963.

Eriksson, E., Compartment models and reservoir theory, in *Ann. Rev. Ecology Sys.*, pp. 67-84, 1971.

Espenscheid, W.F., M. Kerker, and E. Matijevic, Logarithmic distribution functions for colloidal particles, *J. Phys. Chem.*, *68* (11), 3093-3097, 1964.

Estep, L., and R. Arnone, Effect of whitecaps on determination of chlorophyll concentration from satellite data, *Remote Sens. Environ.*, *50* (3), 328-334, 1994.

Etcheto, J., and L. Banege, Wide-scale validation of Geosat altimeter-derived wind speeds, *J. Geophys. Res.*, *97* (C7), 11393-11409, 1992.

Eugster, H.P., The beginnings of experimental petrology, *Science*, *173*, 481-489, 1971.

Eugster, H.P., C.E. Harvie, and J.H. Weare, Mineral equilibria in a six-component seawater system, Na-K-Mg-Ca-SO$_4$-Cl-H$_2$O, at 25° C, *Geochimica et Cosmochimica Acta*, *44*, 1335-1347, 1980.

Ewing, G., Slicks, surface films and internal waves, *J. Marine Res.*, *9* (3), 161-187, 1950.

Exton, H.J., J. Latham, P.M. Park, S.J. Perry, M.H. Smith, and R.R. Allan, The production and dispersal of marine aerosol, *Quar. J. Royal Meteorol. Soc.*, *111*, 817-837, 1985.

Exton, H.J., J. Latham, P.M. Park, and M.H. Smith, The production and dispersal of maritime aerosol, in *Oceanic Whitecaps and Their Role in Air-Sea Exchange Processes*, edited by E.C. Monahan, and G. MacNiocaill, pp. 175-193, D. Reidel Publishing Company, Dordrecht, 1986.

Facchini, M.C., M. Mircea, S. Fuzzi, and R.J. Charlson, Cloud albedo enhancement by surface-active organic solutes in growing droplets, *Nature*, *401*, 257-259, 1999.

Facchini, M.C., M. Mircea, S. Fuzzi, and R.J. Charlson, Comments on "Influence of soluble surfactant properties on the activation of aerosol particles containing inorganic solute, *J. Atmos. Sci.*, *58*, 1465-1467, 2001.

Facy, L., Embruns et noyaux de condensation, *J. Sci. Meteorologie*, *3*, 62-68, 1951a.

Facy, L., Éclatement des lames minces et noyaux de condensation, *J. Sci. Meteorologie*, *3*, 86-98, 1951b.

Fairall, C.W., Interpretation of eddy-correlation measurements of particulate deposition and aerosol flux, *Atmos. Environ.*, *18* (7), 1329-1337, 1984.

Fairall, C.W., and K.L. Davidson, Dynamics and modeling of aerosols in the marine boundary layer, in *Oceanic Whitecaps and Their Role in Air-Sea Exchange Processes*, edited by E.C. Monahan, and G. MacNiocaill, pp. 195-208, D. Reidel Publishing Company, Dordrecht, 1986.

Fairall, C.W., and S.E. Larsen, Dry deposition, surface production and dynamics of aerosols in the marine boundary layer, *Atmos. Environ.*, *18* (1), 69-77, 1984.

Fairall, C.W., K.L. Davidson, and G.E. Schacher, Humidity effects and sea salt contamination of atmospheric temperature sensors, *J. Applied Meteorol.*, *18*, 1237-1239, 1979.

Fairall, C.W., K.L. Davidson, and G.E. Schacher, Meteorological models for optical properties in the marine boundary layer, *Opt. Eng.*, *21* (5), 847-857, 1982.

Fairall, C.W., K.L. Davidson, and G.E. Schacher, An analysis of the surface production of sea-salt aerosols, *Tellus*, *35B*, 31-39, 1983.

Fairall, C.W., K.L. Davidson, and G.E. Schacher, Application of a mixed-layer model to aerosols in the marine boundary layer, *Tellus*, *36B*, 203-211, 1984.

Fairall, C.W., J.D. Kepert, and G.J. Holland, The effect of sea spray on surface energy transports over the ocean, *Global Atmos. Ocean Sys.*, *2*, 121-142, 1994.

Farmer, D., L. Ding, D. Booth, and M. Lohmann, Wave kinematics at high sea states, *J. Atmos. Oceanic Technol.*, *19*, 225-239, 2002.

Farmer, D.M., and D.D. Lemon, The influence of ambient noise in the ocean at high wind speeds, *J. Phys. Oceanogr.*, *14*, 1762-1778, 1984.

Farmer, D.M., and S. Vagle, Waveguide propagation of ambient sound in the ocean-surface bubble layer, *J. Acoustical Soc. Am.*, *86* (5), 1897-1908, 1989.

Farmer, D.M., C.L. McNeil, and B.D. Johnson, Evidence for the importance of bubbles in increasing air-sea gas flux, *Nature*, *361*, 620-623, 1993.

Farmer, D.M., S. Vagle, and A.D. Booth, A free-flooding acoustic resonator for measurement of bubble size distributions, *J. Atmos. Oceanic Technol.*, *15*, 1132-1146, 1998.

Fasching, J.L., R.A. Courant, R.A. Duce, and S.R. Piotrowicz, A new surface microlayer sampler utilizing the bubble microtome, *J. Recherches Atmos.*, *8*, 649-652, 1974.

Fassina, V., A survey on air pollution and deterioration of stonework in Venice, *Atmos. Environ.*, *12*, 2205-2211, 1978.

Feingold, G., and P.Y. Chuang, Analysis of the influence of film-forming compounds on droplet growth: Implications for cloud microphysical processes and climate, *J. Atmos. Sci.*, *59*, 2006-2018, 2002.

Feingold, G., W.R. Cotton, S.M. Kreidenweis, and J.T. Davis, The impact of giant cloud condensation nuclei on drizzle formation in stratocumulus: Implications for cloud radiative properties, *J. Atmos. Sci.*, *56*, 4100-4117, 1999.

Ferron, G.A., The size of soluble aerosol particles as a function of the humidity of the air. Application to the human respiratory tract, *J. Aerosol Sci.*, *8*, 251-267, 1977.

Ferron, G.A., and S.C. Soderholm, Estimation of the times for evaporation of pure water droplets and for stabilization of salt solution particles, *J. Aerosol Sci.*, *21* (3), 415-429, 1990.

Finlayson-Pitts, B., Reaction of NO_2 with NaCl and atmospheric implications of NOCl formation, *Nature*, *306*, 676-677, 1983.

Finlayson-Pitts, B.J., The tropospheric chemistry of sea salt: A molecular-level view of the chemistry of NaCl and NaBr, *Chem. Rev.*, *103*, 4801-4822, 2003.

Finlayson-Pitts, B.J., and J.C. Hemminger, Physical chemistry of airborne sea salt particles and their components, *J. Phys. Chem.*, *104A*, 11463-11477, 2000.

Finlayson-Pitts, B.J., M.J. Ezell, and J.N.J. Pitts, Formation of chemically active chlorine compounds by reactions of atmospheric NaCl particles with gaseous N_2O_5 and $ClONO_2$, *Nature*, *337*, 241-244, 1989.

Fitzgerald, J.W., Approximation formulas for the equilibrium size of an aerosol particle as a function of its dry size and composition and the ambient relative humidity, *J. Applied Meteorol.*, *14*, 1044-1049, 1975.

Fitzgerald, J.W., On the growth of atmospheric aerosol particles with relative humidity, *Memorandum Report 3847*, pp. 11, Naval Research Laboratory, Washington, DC, 1978.

Fitzgerald, J.W., Marine aerosols: a review, *Atmos. Environ.*, *25A*, 533-545, 1991.

Fitzgerald, J.W., W.A. Hoppel, and F. Gelbard, A one-dimensional sectional model to simulate multicomponent aerosol dynamics in the marine boundary layer 1. Model description, *J. Geophys. Res.*, *103* (D13), 16085-16102, 1998a.

Fitzgerald, J.W., J.J. Marti, W.A. Hoppel, G.M. Frick, and F. Gelbard, A one-dimensional sectional model to simulate multicomponent aerosol dynamics in the marine boundary layer 2. Model application, *J. Geophys. Res.*, *103* (D13), 16103-16117, 1998b.

Fladerer, A., and R. Strey, Growth of homogeneously nucleated water droplets: a quantitative comparison of experiment and theory, *Atmos. Res.*, *65*, 161-187, 2003.

Flamant, C., V. Trouillet, P. Chazettte, and J. Pelon, Wind speed dependence of atmospheric boundary layer optical properties and ocean surface reflectance as observed by airborne backscatter lidar, *J. Geophys. Res.*, *103* (C11), 25137-25158, 1998.

Flatau, P.J., M. Flatau, J.R.V. Zaneveld, and C.D. Mobley, Remote sensing of bubble clouds in sea water, *Quar. J. Royal Meteorol. Soc.*, *126*, 2511-2523, 2000.

Fleming, R.H., and R. Revelle, Physical processes in the ocean, in *Recent Marine Sediments*, edited by P.D. Trask, pp. 48-144, Thomas Murby & Co., London, 1939.

Forch, C., M. Knudsen, and S.P.L. Sørenson, Berichte über die Konstantenbestimmungen zur Aufstellung der hydrographischen Tabellen, *Kgl. Danske Videnskab. Selskabs Skrifter, 7 Raekke, Naturvidensk. Og Mathem. Afd.*, *12*, 1-151, 1902.

Forgan, B.W., Aerosol optical depth, in *Baseline Atmospheric Program (Australia) 1985*, edited by B.W. Forgan, and P.J. Fraser, pp. 56, Department of Science/Bureau of Meteorology and CSIRO/Division of Atmospheric Research. Australia, 1987.

Forrester, F.J., How strong is the wind?, *Weatherwise*, *39*, 147-151, 1986.

Foulk, C.W., A theory of liquid film formation, *Indust. Eng. Chem.*, *21*, 815-817, 1929.

Foulk, C.W., Foaming and priming of boiler water, *Trans. Am. Soc. Mechan. Eng.*, *54*, 105-113, 1932.

Foulk, C.W., and J.N. Miller, Experimental evidence in support of the balanced-layer theory of liquid film formation, *Indust. Eng. Chem.*, *23*, 1283-1288, 1931.

Fournier d'Albe, E.M., Sur les embruns marins, *Bull. de l'Institut Oceanographique*, *995*, 1-7, 1951.

Fox, F., and K.F. Herzfeld, Gas bubbles with organic skin as cavitation nuclei, *J. Acoustical Soc. Am.*, *26* (6), 984-989, 1954.

François, F., W. Maenhaur, J.-L. Colin, R. Losno, M. Schulz, T. Stahlschmidt, L. Spokes, and T. Jickells, Intercomparison of elemental concentrations in total and size-fractionated aerosol samples collected during the Mace Head Experiment, April, 1991, *Atmos. Environ.*, *29* (7), 837-849, 1995.

Frank, E.R., J.P.J. Lodge, and A. Goetz, Experimental sea salt profiles, *J. Geophys. Res.*, *77* (27), 5147-5151, 1972.

Franklin, B., *Writings*, 1605 pp., The Library of America, 1987.

Franz, G.J., Splashes as sources of sound in liquids, *J. Acoustical Soc. Am.*, *31* (8), 1080-1096, 1959.

Freilich, M.H., and P.G. Challenor, A new approach for determining fully empirical altimeter wind speed model functions, *J. Geophys. Res.*, *99* (C2), 25051-25062, 1994.

Freilich, M.H., and R.S. Dunbar, The accuracy of the NSCAT 1 vector winds: Comparisons with the National Data Buoy Center buoys, *J. Geophys. Res.*, *104* (C5), 11231-11246, 1999.

Frew, N.M., J.C. Goldman, M.R. Dennett, and A.S. Johnson, Impact of phytoplankton-generated surfactants on air-sea gas exchange, *J. Geophys. Res.*, *95* (C3), 3337-3352, 1990.

Frey, D.D., and C.J. King, Effects of surfactants on mass transfer during spray drying, *Am. Inst. Chem. Eng. J.*, *32* (3), 437-443, 1986.

Friedlander, S.K., *Smoke, Dust, and Haze*, 407 pp., John Wiley and Sons, New York, 2000.

Friedlander, S.K., and H.F. Johnstone, Deposition of suspended particles from turbulent gas streams, *Indust. Eng. Chem.*, *49*, 1151-1156, 1957.

Friehe, C.A., and K.F. Schmitt, Parameterization of air-sea interface fluxes of sensible heat and moisture by the bulk aerodynamic formulas, *J. Phys. Oceanogr.*, *6*, 801-809, 1976.

Frouin, R., M. Schwindling, and O.-Y. Deschamps, Spectral reflectance of sea foam in the visible and near-infrared: In situ measurements and remote sensing implications, *J. Geophys. Res.*, *101* (C6), 14361-14371, 1996.

Frouin, R., S.F. Iacobellis, and P.-Y. Deschamps, Influence of oceanic whitecaps on the global radiation budget, *Geophys. Res. Lttrs*, *28* (8), 1523-1526, 2001.

Fry, B., E.T. Peltzer, C.S. Hopkinson, Jr., A. Nolin, and L. Redmond, Analysis of marine DOC using a dry combustion method, *Marine Chem.*, *54*, 191-201, 1996.

Fuchs, N.A., *Evaporation and Droplet Growth in Gaseous Media*, 72 pp., Pergamon Press, New York, 1959.

Fuchs, N.A., *The Mechanics of Aerosols*, 408 pp., Dover Publications, Inc., New York, 1964.

Fuchs, N.A., and A.G. Sutugin, High-dispersed aerosols, in *Topics in Current Aerosol Research*, edited by G.M. Hidy, and J.R. Brock, pp. 1-60, Pergamon Press, New York, 1971.

Fukuta, N., and L.A. Walter, Kinetics of hydrometeor growth from a vapor-spherical model, *J. Atmos. Sci.*, *27*, 1160-1172, 1970.

Fung, K.H., I.N. Tang, and H.R. Munkelwitz, Study of condensational growth of water droplets by Mie resonance spectroscopy, *Applied Optics*, *26* (7), 1282-1287, 1987.

Gaddum, J.H., Lognormal distributions, *Nature*, *156*, 463-466, 746-747, 1945.

Gallagher, M.W., B.K. M., J. Duyzer, H. Westrate, T.W. Choularton, and P. Hummelshoj, Measurements of aerosol fluxes to speulder forest using a micrometeorological technique, *Atmos. Environ.*, *31* (3), 359-373, 1997.

Galton, F., The geometric mean, in vital and social statistics, *Proc. Royal Soc. London*, *29*, 365-367, 1879.

Galvin, C.J., Jr., Breaker type classification on three laboratory beaches, *J. Geophys. Res.*, *73* (12), 3651-3659, 1968.

Gard, E.E., M.J. Kleeman, D.S. Gross, L.S. Hughes, J.O. Allen, B.D. Morrical, D.P. Fergenson, T. Dienes, M.E. Galli, R.J. Johnson, G.R. Cass, and K.A. Prather, Direct observation of heterogeneous chemistry in the atmosphere, *Science*, *279*, 1184-1187, 1998.

Garland, J.A., Enrichment of sulphate in maritime aerosols, *Atmos. Environ.*, *15* (5), 787-791, 1981.

Garland, J.A., Author's reply, *Atmos. Environ.*, *16* (8), 2026-2027, 1982.

Garland, J.A., and L.C. Cox, Deposition of small particles to grass, *Atmos. Environ.*, *16* (11), 2699-2702, 1982.

Garner, F.H., Diffusion mechanism in the mixing of fluids, *Trans. Inst. Chem. Eng.*, *28*, 88-96, 1950.

Garner, F.H., and D. Hammerton, Circulation inside gas bubbles, *Chem. Eng. Sci.*, *3* (1), 1-11, 1954.

Garner, F.H., S.R.M. Ellis, and J.A. Lacey, The size distribution and entrainment of droplets, *Trans. Inst. Chem. Eng.*, *32*, 222-235, 1954.

Garratt, J.R., Review of drag coefficients over oceans and continents, *Monthly Weather Rev.*, *105*, 915-929, 1977.

Garratt, J.R., *The Atmospheric Boundary Layer*, 316 pp., Cambridge University Press, Cambridge, 1992.

Garratt, J.R., G.D. Hess, W.L. Physick, and P. Bougeault, The atmospheric boundary layer - advances in knowledge and application, *Boundary-Layer Meteorol.*, *78*, 9-37, 1996.

Garrett, C., M. Li, and D. Farmer, The connection between bubble size spectra and energy dissipation rates in the upper ocean, *J. Phys. Oceanogr.*, *30*, 2163-2171, 2000.

Garrett, W.D., Stabilization of air bubbles at the air-sea interface by surface-active material, *Deep-Sea Res.*, *14*, 662-672, 1967a.

Garrett, W.D., The organic chemical composition of the ocean surface, *Deep-Sea Res.*, *14*, 221-227, 1967b.

Garrett, W.D., Damping of capillary waves at the air-sea interface by oceanic surface-active material, *J. Marine Res.*, *25* (3), 279-291, 1967c.

Garrett, W.D., The influence of monomolecular surface films on the production of condensation nuclei from bubbled sea water, *J. Geophys. Res.*, *73* (16), 5145-5150, 1968.

Garrett, W.D., Retardation of water drop evaporation with monomolecular surface film, *J. Atmos. Sci.*, *28*, 816-819, 1971.

Garrett, W.D., Reply, *J. Atmos. Sci.*, *29*, 786-787, 1972.

Garrett, W.D., and J.D. Bultman, Capillary-wave damping by insoluble organic monolayers, *J. Colloid Sci.*, *18*, 798-801, 1963.

Garrettson, G.A., Bubble transport theory with application to the upper ocean, *J. Fluid Mechanics*, *59*, 187-206, 1973.

Gathman, S.G., A time-dependent oceanic aerosol profile model, *NRL Report 8536*, pp. 35, Naval Research Laboratory, Washington, DC, 1982.

Gathman, S.G., Optical properties of the marine aerosol as predicted by the Navy aerosol model, *Opt. Eng.*, *22* (1), 57-62, 1983a.

Gathman, S.G., Optical properties of the marine aerosol as predicted by a BASIC version of the Navy Aerosol Model, *NRL Memorandum Report 5157*, pp. 31, Naval Research Laboratory, Washington, DC, 1983b.

Gathman, S.G., A preliminary description of NOVAM, the Navy Oceanic Vertical Aerosol Model, *NRL Report 9200*, pp. 22, Naval Research Laboratory, Washington, DC, 1989.

Gathman, S.G., and K.L. Davidson, The Navy Oceanic Vertical Aerosol Model, *Techn. Report 1634*, pp. 112, Naval Command, Control and Ocean Surveillance Center, San Diego, CA, 1993.

Gathman, S.G., and W.A. Hoppel, Surf electrification, *J. Geophys. Res.*, *75* (24), 4525-4529, 1970.

Gathman, S., and E.M. Trent, Space charge over the open ocean, *J. Atmos. Sci.*, *25*, 1075-1079, 1968.

Gathman, S., and E.A. Trent, Reply, *J. Atmos. Sci.*, *26*, 785, 1969.

Gathman, S., and B. Ulfers, On the accuracy of IR extinction predictions made by the Navy Aerosol Model, in *Ninth Conference on Aerospace and Aeronautical Meteorology*, pp. 194-198, American Meteorological Society, Omaha, NE, 1983.

Gautier, A., Quantité maximum de chlorures contenus dans l'air de la mer, *Société Chimique de France: Bulletin*, *21*, 391-392, 1899.

Gavrilov, L.R., On the size distribution of gas bubbles in water, *Soviet Phys. Acoustics*, *15* (1), 22-24, 1969.

Gay, M.J., R.F. Griffiths, J. Latham, and C.P.R. Saunders, The terminal velocities of charged raindrops and cloud droplets falling in strong electric fields, *Quar. J. Royal Meteorol. Soc.*, *100*, 682-687, 1974.

Gebel, M.E., B.J. Finlayson-Pitts, and J.A. Ganske, The uptake of SO_2 on synthetic sea salt and some of its components, *Geophys. Res. Lttrs*, *27* (6), 887-890, 2000.

Geernaert, G.L., Bulk parameterizations for the wind stress and heat fluxes, in *Surface Waves and Fluxes*, edited by G.L. Geernaert, and W.L. Plant, pp. 91-172, Kluwer Academic Publishers, Dordrecht, 1990.

Geernaert, G.L., S.E. Larsen, and F. Hansen, Measurements of the wind stress, heat flux, and turbulence intensity during storm conditions over the North Sea, *J. Geophys. Res.*, *92* (C12), 13127-13139, 1987.

Geernaert, L.L.S., G.L. Geernaert, K. Granby, and W.A.H. Asman, Fluxes of soluble gases in the marine atmosphere surface layer, *Tellus*, *50B*, 111-127, 1998.

Gemmrich, J.R., and D.M. Farmer, Observations of the scale and occurrence of breaking surface waves, *J. Phys. Oceanogr.*, *29*, 2595-2606, 1999.

Genthon, C., Simulations of desert dust and sea-salt aerosols in Antarctica with a general circulation model of the atmosphere, *Tellus*, *44B*, 371-389, 1992a.

Genthon, C., Simulations of the long range transport of desert dust and sea-salt in a general circulation model, in *Precipitation*

Scavenging and Atmosphere-Surface Exchange, edited by S.E. Schwartz, and W.G.N. Slinn, pp. 1783-1794, Hemisphere Publishing Corporation, Washington, DC, 1992b.

Gerber, H.E., Relative-humidity parameterization of the Navy Aerosol Model (NAM), pp. 13, Naval Research Laboratory, Washington, DC, 1985.

Gerber, H., Probability distribution of aerosol backscatter in the lower marine atmosphere at CO_2 wavelengths, *J. Geophys. Res.*, *96* (D3), 5307-5314, 1991.

Gershey, R.M., Characterization of seawater organic matter carried by bubble-generated aerosols, *Limnol. Oceanogr.*, *28* (2), 309-319, 1983a.

Gershey, R.M., A bubble adsorption device for the isolation of surface-active organic material in seawater, *Limnol. Oceanogr.*, *28* (2), 395-400, 1983b.

Ghan, S.J., G. Guzman, and Abdul-Razzak, Competition between sea salt and sulfate particles as cloud condensation nuclei, *J. Atmos. Sci.*, *55*, 3340-3347, 1998.

Giddings, W.P., and M.B. Baker, Sources and effects of monolayers on atmospheric water droplets, *J. Atmos. Sci.*, *34*, 1957-1964, 1977.

Giles, C.H., Franklin's teaspoon of oil, *Chem. Industry*, 1616-1624, 1969.

Gilhousen, D.B., A field evaluation of NDBC moored buoy winds, *J. Atmos. Oceanic Technol.*, *4*, 94-104, 1987.

Gill, P.S., T.E. Graedel, and C.J. Weschler, Organic films of atmospheric aerosol particles, fog droplets, cloud droplets, raindrops, and snowflakes, *Rev. Geophys. Space Phys.*, *21* (4), 903-920, 1983.

Gillette, D.A., I.H.J. Blifford, and C.R. Fenster, Measurements of aerosol size distributions and vertical fluxes of aerosols on land subject to wind erosion, *J. Applied Meteorol.*, *11*, 977-987, 1972.

Ginsberg, T., Aerosol generation by liquid breakup resulting from sparging of molten pools of corium by gases released during core/concrete interactions, *Nuclear Sci. Eng.*, *89*, 36-48, 1985.

Giorgi, F., A particle dry-deposition parameterization scheme for use in tracer transport models, *J. Geophys. Res.*, *91* (D9), 9794-9806, 1986a.

Giorgi, F., Correction to "A Particle Dry-Deposition Parameter Scheme for Use in Tracer Transport Models" by Filippo Giorgi, *J. Geophys. Res.*, *91* (D11), 11915, 1986b.

Giorgi, F., Dry deposition velocities of atmospheric aerosols as inferred by applying a particle dry deposition parameterization to a general circulation model, *Tellus*, *40B*, 23-41, 1988.

Giorgi, F., and W.L. Chameides, Rainout lifetimes of highly soluble aerosols and gases as inferred from simulations with a general circulation model, *J. Geophys. Res.*, *91* (D13), 14367-14376, 1986.

Glass, S.J.J., and M.H. Matteson, Ion enrichment in aerosols dispersed from bursting bubbles in aqueous salt solutions, *Tellus*, *25*, 272-280, 1973.

Glazman, R.E., Effects of adsorbed films of gas bubble radial oscillations, *J. Acoustical Soc. Am.*, *74* (3), 980-986, 1983.

Glazman, R.E., and S.H. Pilorz, Effects of sea maturity on satellite altimeter measurements, *J. Geophys. Res.*, *95* (C3), 2857-2870, 1990.

Glazman, R.E., and P.B. Weichman, Statistical geometry of a small surface patch in a developed sea, *J. Geophys. Res.*, *94* (C4), 4998-5010, 1989.

Glazman, R.E., F.F. Pihos, and J. Ip, Scatterometer wind speed bias induced by the large-scale component of the wave field, *J. Geophys. Res.*, *93* (C2), 1317-1328, 1988.

Glotov, V.P., P.A. Kolobaev, and G.G. Neuimin, Investigation of the scattering of sound by bubbles generated by an artificial wind in sea water and the statistical distribution of bubble sizes, *Soviet Phys. Acoustics*, *7* (4), 341-345, 1962.

Goedde, E.F., and M.C. Yuen, Experiments on liquid jet instability, *J. Fluid Mechanics*, *40* (3), 495-511, 1970.

Goldman, J.C., M.R. Dennett, and N.M. Frew, Surfactant effects on air-sea gas exchange under turbulent conditions, *Deep-Sea Res.*, *35* (12), 1953-1970, 1988.

Goldsmith, P., and F.G. May, Diffusiophoresis and thermophoresis in water vapour systems, in *Aerosol Sci.*, edited by C.N. Davies, pp. 163-194, Academic Press, London, 1966.

Goldsmith, P., H.J. Delafield, and L.C. Cox, The role of diffusiophoresis in the scavenging of radioactive particles from the atmosphere, *Quar. J. Royal Meteorol. Soc.*, *89*, 43-61, 1963.

Goldstein, S., The steady flow of viscous fluid past a fixed spherical obstacle at small Reynolds numbers, *Proc. Royal Soc. London*, *123A*, 225-235, 1929.

Goncharov, V.K., S.N. Kuznetsova, G.G. Neuimin, and N.A. Sorokina, Determination of the diffusion constant of a gas in sea water from the solution of air bubbles in the medium, *Soviet Phys. Acoustics*, *30* (4), 273-275, 1984.

Gong, S.L., L.A. Barrie, and J.-P. Blanchet, Modeling sea-salt aerosols in the atmosphere 1. Model development, *J. Geophys. Res.*, *102* (D3), 3805-3818, 1997a.

Gong, S.L., L.A. Barrie, J.M. Prospero, D.L. Savoie, G.P. Ayers, J. P. Blanchet, and L. Spacek, Modeling sea-salt aerosols in the atmosphere 2. Atmospheric concentrations and fluxes, *J. Geophys. Res.*, *102* (D3), 3819-3830, 1997b.

Gong, S.L., L.A. Barrie, J.-P. Blanchet, and L. Spacek, Modeling size-distributed sea salt aerosols in the atmosphere: an application using Canadian climate models, in *Air Pollution Modeling and Its Application XII*, edited by S.-E. Gryning, and M. Chaumerliac, pp. 337-345, Plenum, New York, 1998.

Gorbunov, B., R. Hamilton, N. Clegg, and R. Toumi, Water nucleation on aerosol particles containing both organic and soluble inorganic substances, *Atmos. Res.*, *47-48*, 271-283, 1998.

Gordon, C.M., E.C. Jones, and R.E. Larson, The vertical distribution of particulate Na and Cl in a marine atmosphere, *J. Geophys. Res.*, *82* (C6), 988-990, 1977.

Gordon, H.R., Atmospheric correction of ocean color imagery in the Earth Observing System era, *J. Geophys. Res.*, *102* (D14), 17081-17106, 1997.

Gordon, H.R., and M.M. Jacobs, Albedo of the ocean-atmosphere system: influence of sea foam, *Applied Optics*, *16* (8), 1977, 1977.

Gordon, H.R., and M. Wang, Influence of oceanic whitecaps on atmospheric correction of ocean-color sensors, *Applied Optics*, *33* (33), 7754-7763, 1994.

Goroch, A., S. Burk, and K.L. Davidson, Stability effects on aerosol size and height distributions, *Tellus*, *32*, 245-250, 1980.

Goroch, A.K., C.W. Fairall, and K.L. Davidson, Modeling wind speed dependence of marine aerosol distribution by a gamma function, *J. Applied Meteorol.*, *21*, 666-671, 1982.

Grace, J.R., Shapes and velocities of bubbles rising in infinite liquids, *Trans. Inst. Chem. Eng.*, *51*, 116-120, 1973.

Grace, J.R., Hydrodynamics of liquid drops in immiscible liquids, in *Handbook of Fluids in Motion*, edited by N.P. Cheremisinoff, and R. Gupta, pp. 1003-1025, Ann Arbor Science, Ann Arbor, MI, 1983.

Grace, J.R., T. Wairegi, and T.H. Nguyen, Shapes and velocities of single drops and bubbles moving freely through immiscible liquids, *Trans. Inst. Chem. Eng.*, *54*, 167-173, 1976.

Graedel, T.E., and W.C. Keene, Tropospheric budget of reactive chlorine, *Global Biogeochem. Cycles*, *9* (1), 47-77, 1995.

Graedel, T.E., and C.J. Weschler, Chemistry within aqueous atmospheric aerosols and raindrops, *Rev. Geophys. Space Phys.*, *19* (4), 505-539, 1981.

Graf, H.G., *Hydraulics of Sediment Transport*, 513 pp., McGraw-Hill Book Company, New York, 1971.

Graham, W.F., S.R. Piotrowicz, and R.A. Duce, The sea as a source of atmospheric phosphorus, *Marine Chem.*, *7*, 325-342, 1979.

Gras, J.L., Condensation nucleus size distribution at Mawson, Antarctica: Seasonal cycle, *Atmos. Environ.*, *27A* (9), 1417-1425, 1993.

Gras, J.L., CN, CCN and particle size in Southern Ocean air at Cape Grim, *Atmos. Res.*, *35*, 233-251, 1995.

Gras, J.L., and G.P. Ayers, Marine aerosol at southern mid-latitudes, *J. Geophys. Res.*, *88* (C15), 10661-10666, 1983.

Gravenhorst, G., Maritime sulfate over the North Atlantic, *Atmos. Environ.*, *12*, 707-713, 1978.

Green, T., and D.F. Houk, The removal of organic surface films by rain, *Limnol. Oceanogr.*, *24*, 966-970, 1979.

Greenberg, J.P., and N. Møller, The prediction of mineral solubilities in natural waters: A chemical equilibrium model for the Na-K-Ca-Cl-SO$_4$-H$_2$O system to high concentration from 0 to 250°C, *Geochimica et Cosmochimica Acta*, *53*, 2503-2518, 1989.

Gregory, P.H., The dispersion of air-borne spores, *Trans. British Mycol. Soc.*, *28*, 26-72, 1945.

Griffiths, W.D., S. Patrick, and A.P. Rood, An aerodynamic particle size analyser tested with spheres, compact particles and fibres having a common settling rate under gravity, *J. Aerosol Sci.*, *15* (4), 491-502, 1984.

Griffiths, W.D., P.J. Iles, and N.P. Vaughan, The behaviour of liquid droplet aerosols in an APS 3300, *J. Aerosol Sci.*, *17* (6), 921-930, 1986.

Griggs, M., Satellite measurements of aerosols over ocean surfaces, in *Oceanic Whitecaps and Their Role in Air-Sea Exchange Processes*, edited by E.C. Monahan, and G. MacNiocaill, pp. 245-250, D. Reidel Publishing Company, Dordrecht, 1986.

Grini, A., G. Myhre, J. Sundet, and I.S.A. Isaksen, Modeling the annual cycle of sea salt in the global 3D model Oslo CTM2: Concentrations, fluxes, and radiative input, *J. Climate*, *15*, 1717-1730, 2002.

Grossman, R.L., B.R. Bean, and W.E. Marlatt, Airborne infrared radiometer investigation of water surface temperature with and without an evaporation-retarding monomolecular layer, *J. Geophys. Res.*, *74* (10), 2471-2476, 1969.

Guazzotti, S.A., K.R. Coffee, and K.A. Prather, Continuous measurements of size-resolved particle chemistry during INDOEX-Intensive Field Phase 99, *J. Geophys. Res.*, *106* (D22), 28607-28627, 2001.

Guelle, W., M. Schulz, and Y. Balkanski, Influence of the source formulation on modeling the atmospheric global distribution of sea salt aerosol, *J. Geophys. Res.*, *106* (D21), 27509-27524, 2001.

Guillaume, A., and N.M. Mognard, A new method for the validation of altimeter-derived sea state parameters with results from wind and wave models, *J. Geophys. Res.*, *97* (C6), 9705-9717, 1992.

Guimbaud, C., F. Arens, L. Gutzwiller, H.W. Gäggeler, and M. Ammann, Uptake of HNO$_3$ to deliquescent sea-salt particles: a study using the short-lived radioactive isotope tracer ^{13}N, *Atmos. Chem. Phys.*, *2*, 249-257, 2002.

Gunn, R., and G.D. Kinzer, The terminal velocity of fall for water droplets in stagnant air, *J. Meteorol.*, *6*, 243-248, 1949.

Gurel, S., S.G. Ward, and R.L. Whitmore, Studies of the viscosity and sedimentation of suspensions, *British J. Applied Phys.*, *6*, 83-87, 1955.

Gysel, M., E. Weingartner, and U. Baltensperger, Hygroscopicity of aerosol particles at low temperatures. 2. Theoretical and experimental hygroscopic properties of laboratory generated aerosols, *Environ. Sci. Technol.*, *36*, 63-68, 2002.

Haberman, W.L., and R.K. Morton, An experimental investigation of the drag and shape of air bubbles rising in various liquids, pp. 55, Navy Department, David W. Taylor Model Basin, Washington, DC, 1953.

Hadamard, M., Mouvement permanent lent d'une sphere liquide et resqueuse dans un liquide visqueux, *Comptes Rendus de l'Academies des Sci., Paris*, *152*, 1735-1738, 1911.

Hahn, L.A., J.J. Stukel, K.H. Leong, and P.K. Hopke, Turbulent deposition of submicron particles on rough walls, *J. Aerosol Sci.*, *16* (1), 81-86, 1985.

Haines, M.A., and B.D. Johnson, Injected bubble populations in seawater and fresh water measured by a photographic method, *J. Geophys. Res.*, *100* (C4), 7057-7068, 1995.

Hall, J.S., and E.W. Wolff, Causes of seasonal and daily variations in aerosol sea-salt concentrations at a coastal Antarctic station, *Atmos. Environ.*, *32* (21), 3669-3677, 1998.

Hall, M.V., A comprehensive model of wind-generated bubbles in the ocean and predictions of the effects on sound propogation at frequencies up to 40 kHz, *J. Acoustical Soc. Am.*, *86* (3), 1103-1117, 1989.

Hämeri, K., A. Laaksonen, M. Väkevä, and T. Suni, Hygroscopic growth of ultrafine sodium chloride particles, *J. Geophys. Res.*, *106* (D18), 20749-20757, 2001.

Hamrud, M., and H. Rodhe, Lagrangian time scales connected with clouds and precipitation, *J. Geophys. Res.*, *91* (D13), 14377-14383, 1986.

Hamrud, M., H. Rohde, and J. Grandell, A numerical comparison between Lagrangian and Eulerian rainfall statistics, *Tellus*, *33*, 235-241, 1981.

Hänel, G., The properties of atmospheric aerosol particles as functions of the relative humidity at thermodynamic equilibrium with the surrounding moist air, *Advances in Geophysics, v. 19*, pp. 73-188, edited by H.E. Landsberg and J. Van Mieghem, Academic Press, New York, 1976.

Hänel, G., The role of aerosol properties during the condensational stage of cloud: a reinvestigation of numerics and microphysics, *Beitr. Phys. Atmosph.*, *60* (3), 321-339, 1987.

Hänel, G., and B. Zankl, Aerosol size and relative humidity: Water uptake by mixtures of salts, *Tellus*, *31*, 478-486, 1979.

Hansen, J.E., and L.D. Travis, Light scattering in planetary atmospheres, *Space Sci. Rev.*, *16*, 527-610, 1974.

Hanson, J.L., and O.M. Phillips, Wind sea growth and dissipation in the open ocean, *J. Phys. Oceanogr.*, *29*, 1633-1648, 1999.

Hardy, J.T., The sea surface microlayer: biology, chemistry, and anthropogenic enrichment, *Progr. Oceanogr.*, *11*, 307-328, 1982.

Hardy, W.J., Chemistry at interfaces, *J. Chem. Soc.*, *127*, 1207-1227, 1925.

Harlow, F.H., and J.P. Shannon, Distortion of a splashing liquid drop, *Science*, *157*, 547-550, 1967a.

Harlow, F.H., and J.P. Shannon, The splash of a liquid drop, *J. Applied Phys.*, *38*, 3855-3866, 1967b.

Harper, J.F., The motion of bubbles and drops through liquids, in *Advances in Applied Mechanics*, edited by C.-S. Yia, pp. 59-129, Academic Press, New York, 1972.

Harper, J.F., and D.W. Moore, The motion of a spherical liquid drop at high Reynolds number, *J. Fluid Mechanics*, *32* (2), 367-391, 1968.

Harris, I.A., and R.M. Detsch, Small air bubbles in reagent grade water and seawater 2. Dissolution of 20- to 500-μm-diameter bubbles at atmospheric pressure, *J. Geophys. Res.*, *96* (C5), 8907-8910, 1991.

Harrison, R.M., and C.A. Pio, Size-differentiated composition of inorganic atmospheric aerosols of both marine and polluted continental origin, *Atmos. Environ.*, *17* (9), 1733-1738, 1983.

Hartunian, R.A., and W.R. Sears, On the instability of small gas bubbles moving uniformly in various liquids, *J. Fluid Mechanics*, *3*, 27-47, 1957.

Harvey, M.J., G.W. Fisher, I.S. Lechner, P. Isaac, N.E. Flower, and A.L. Dick, Summertime aerosol measurements in the Ross Sea region of Antarctica, *Atmos. Environ.*, *25A* (3/4), 569-580, 1991.

Harvie, C.E., J.H. Weare, L.A. Hardie, and H.P. Eugster, Evaporation of seawater: calculated mineral sequences, *Science*, *208*, 498-500, 1980.

Hasse, L., On the contribution of spray droplets to evaporation, *Boundary-Layer Meteorol.*, *61*, 309-313, 1992.

Hasse, L., Reply to Andreas (1994), *Boundary-Layer Meteorol.*, *69*, 335-339, 1994.

Hayami, S., and Y. Toba, Drop production by bursting of air bubble on the sea surface (1) Experiments at still sea water surface, *J. Oceanogr. Soc. Japan*, *14* (4), 145-150, 1958.

Haynie, F.H., Theoretical model of soiling of surfaces by airborne particles, in *Aerosols: Research, Risk Assessment and Control Strategies*, edited by S.D. Lee, T. Schneider, L.D. Grant, and P.J. Verkerk, pp. 951-959, Lewis Publishers, Inc., Williamsburg, VA, 1985.

Haywood, J.M., V. Ramaswamy, and B.J. Soden, Tropospheric aerosol climate forcing in clear-sky satellite observations over the oceans, *Science*, *283*, 1299-1303, 1999.

Hegg, D.A., Particle production in clouds, *Geophys. Res. Lttrs*, *18* (6), 995-998, 1991.

Hegg, D.A., Dependence of marine stratocumulus formation on aerosols, *Geophys. Res. Lttrs*, *26* (10), 1429-1432, 1999.

Hegg, D.A., D.S. Covert, and V.N. Kaputsin, Modeling a case of particle nucleation in the marine boundary layer, *J. Geophys. Res.*, *97* (D9), 9851-9857, 1992.

Hegg, D.A., P.V. Hobbs, R.J. Ferek, and A.P. Waggoner, Measurements of some aerosol properties relevant to radiative forcing on the East Coast of the United States, *J. Applied Meteorol.*, *34*, 2306-2315, 1995.

Hegg, D.A., D.S. Covert, M.J.Rood, and P.V. Hobbs, Measurement of aerosol optical properties in marine air, *J. Geophys. Res.*, *101* (D8), 12893-12903, 1996.

Heintzenberg, J., Properties of the log-normal particle size distribution, *Aerosol Sci. Technol.*, *21*, 46-48, 1994.

Heintzenberg, J., D.C. Cover, and R. van Dingenen, Size distribution and chemical composition of marine aerosols: a compilation and review, *Tellus*, *52B*, 1104-1122, 2000.

Heiss, J.F., and J. Coull, The effect of orientation and shape on the settling velocity of non-isometric particles in a viscous medium, *Chem. Eng. Progr.*, *48*, 133-140, 1952.

Henry, W., Experiments on the quantity of gases absorbed by water, at different temperatures, and under different pressures, *Philosoph. Trans. Royal Soc. London*, *93*, 29-43, 1803a.

Henry, W., Appendix to Mr. William Henry's paper, on the quantity of gases absorbed by water, at different temperatures, and under different pressures, *Philosoph. Trans. Royal Soc. London*, *93*, 274-276, 1803b.

Herdan, G., *Small Particle Statistics*, 418 pp., Butterworths, London, 1960.

Hertz, H., Ueber den Druck des gesättigten Quecksilberdampfes, *Annalen der Physik und Chemie*, *17*, 193-200, 1882.

Herut, B., A. Starinsky, A. Katz, and A. Bein, The role of seawater freezing in the formation of subsurface brines, *Geochimica et Cosmochimica Acta*, *54*, 13-21, 1990.

Hicks, B.B., Reply, *Boundary-Layer Meteorol.*, *10*, 237-240, 1976.

Hicks, B.B., and R.T. McMillen, A simulation of the eddy accumulation method for measuring pollutant fluxes, *J. Clim. Applied Meteorol.*, *23*, 637-643, 1984.

Hidy, G.M., and J.R. Brock, An assessment of the global sources of tropospheric aerosols, in *Proceedings of the Second International Clean Air Congress*, edited by H.M. Englund, and W.T. Beery, pp. 1088-1097, Academic Press, 1971.

Higgins, M.R., D.M. Doyle, A. Duffy, and P. Mohr, Whitecap observations, data, and results from the Hexos 84 (HEXPILOT) experiment, and the final results from the analysis of the MIZEX 83 video results, in *Whitecaps and the Marine Atmosphere, Report No. 8*, edited by E.C. Monahan, P.A. Bowyer, D.M. Doyle, M.R. Higgins, and D.K. Woolf, pp. 117-124, University College, Galway, Ireland, 1985.

Hinds, W.C., *Aerosol Technology*, 483 pp., John Wiley and Sons, New York, 1999.

Hobbs, P.V., Simultaneous airborne measurements of cloud condensation nuclei and sodium-containing particles over the ocean, *Quar. J. Royal Meteorol. Soc.*, *97*, 263-271, 1971a.

Hobbs, P.V., Microdroplets and water drop freezing, *Science*, *173*, 849, 1971b.

Hobbs, P.V., and A.J. Kezweeny, Splashing of a water drop, *Science*, *155*, 1112-1114, 1967.

Hobbs, P.V., and A.J. Alkezweeny, The fragmentation of freezing water droplets in free fall, *J. Atmos. Sci.*, *25*, 881-888, 1968.

Hoffer, T.E., and S.C. Mallen, Evaporation of contaminated and pure water droplets in a wind tunnel, *J. Atmos. Sci.*, *27*, 914-918, 1970.

Hoffman, E.J., and R.A. Duce, The organic carbon content of marine aerosols collected on Bermuda, *J. Geophys. Res.*, *79* (30), 1974.

Hoffman, E.J., and R.A. Duce, Factors influencing the organic carbon content of marine aerosols: a laboratory study, *J. Geophys. Res.*, *81* (21), 3667-3670, 1976.

Hoffman, E.J., and R.A. Duce, Alkali and alkaline earth metal chemistry of marine aerosols generated in the laboratory with natural seawaters, *Atmos. Environ.*, *11*, 367-372, 1977a.

Hoffman, E.J., and R.A. Duce, Organic carbon in marine atmospheric particulate matter: concentration and particle size distribution, *Geophys. Res. Lttrs*, *4* (10), 449-452, 1977b.

Hoffman, E.J., G.L. Hoffman, I.S. Fletcher, and R.A. Duce, Further consideration of alkali and alkaline earth geochemistry of marine aerosols: results of a study of marine aerosols collected on Bermuda, *Atmos. Environ.*, *11*, 373-377, 1977.

Hoffman, E.J., G.L. Hoffman, and R.A. Duce, Particle size dependence of alkali and alkaline earth metal enrichment in marine aerosols from Bermuda, *J. Geophys. Res.*, *85* (C10), 5499-5502, 1980.

Hoffman, G.L., and R.A. Duce, Consideration of the chemical fractionation of alkali and alkaline earth metals in the Hawaiian marine atmosphere, *J. Geophys. Res.*, *77* (27), 5161-5169, 1972.

Hofmeier, U., V.V. Yaminsky, and H.K. Christenson, Observations of solute effects on bubble formation, *J. Colloid Interface Sci.*, *174*, 199-210, 1995.

Högström, U., Review of some basic characteristics of the atmospheric surface layer, *Boundary-Layer Meteorol.*, *78*, 215-246, 1996.

Hollett, R.D., Observations of underwater sound at frequencies below 1500 Hz from breaking waves at sea, *J. Acoustical Soc. Am.*, *95* (1), 165-170, 1994.

Holthuijsen, L.H., and T.H.C. Herbers, Statistics of breaking waves observed as whitecaps in the open sea, *J. Phys. Oceanogr.*, *16*, 290-297, 1986.

Hooper, W.P., and J.E. James, Lidar observations of ship spray plumes, *J. Atmos. Sci.*, *57*, 2649-2655, 2000.

Hooper, W.P., and L.U. Martine, Scanning lidar measurements of surf-zone aerosol generation, *Opt. Eng.*, *38* (2), 250-255, 1999.

Hoppel, W.A., and G.M. Frick, Submicron aerosol size distributions measured over the tropical and South Pacific, *Atmos. Environ.*, *24A* (3), 645-659, 1990.

Hoppel, W.A., J.W. Fitzgerald, G.M. Frick, and R.E. Larson, Aerosol size distributions and optical properties found in the marine boundary layer over the Atlantic Ocean, *J. Geophys. Res.*, *95* (D4), 3659-3686, 1990.

Hoppel, W., L. Pasternack, P. Caffrey, G. Frick, J. Fitzgerald, D. Hegg, S. Gao, J. Ambrusko, and T. Albrechcinski, Sulfur dioxide uptake and oxidation in sea-salt aerosol, *J. Geophys. Res.*, *106* (D21), 27575-27585, 2001.

Hoppel, W.A., G.M. Frick, and J.W. Fitxgerald, Surface source function for sea-salt aerosol and aerosol dry deposition to the ocean surface, *J. Geophys. Res.*, *107* (D19), doi: 10.1029/2001JD002014, 2002.

Horrocks, W.M., Experiments made to determine the conditions under which "specific" bacteria derived from sewage may be present in the air of ventilating pipes, drains, inspection chambers and sewers, *Proc. Royal Soc. London*, *79B*, 255-266, 1907.

Horst, T.W., J.C. Doran, and P.W. Nickola, Dual tracer measurements of plume depletion, in *Precipitation Scavenging, Dry Deposition, and Resuspension*, edited by H.R. Pruppacher, R.G. Scmonin, and W.G.N. Slinn, pp. 1027-1035, Elsevier, New York, 1983.

Horváth, L., E. Mészáros, E. Antal, and A. Simon, On the sulfate, chloride and sodium concentration in maritime air around the Asian continent, *Tellus*, *33*, 382-386, 1981.

Houghton, G., P.D. Ritchie, and J.A. Thomson, Velocity of rise of air bubbles in sea-water, and their types of motion, *Chem. Eng. Sci.*, *7* (111-112), 1957.

Houghton, H.G., The size and size distribution of fog particles, *Physics*, *2*, 467-475, 1932.

Houghton, H.G., Problems connected with the condensation and precipitation processes in the atmosphere, *Bull. Am. Meteorol. Soc.*, *19*, 152-159, 1938.

Houk, D., and T. Green, A note on surface waves due to rain, *J. Geophys. Res.*, *81* (24), 4482-4484, 1976.

Howell, S., A.A.P. Pszenny, P. Quinn, and B. Huebert, A field intercomparison of three cascade impactors, *Aerosol Sci. Technol.*, *29*, 475-492, 1998.

Howell, W.E., The growth of cloud drops in uniformly cooled air, *J. Meteorol.*, *6*, 134-149, 1949.

Hristov, T.S., S.D. Miller, and C.A. Friehe, Dynamical coupling of wind and ocean waves through wave-induced air flow, *Nature*, *422*, 55-58, 2003.

Hsiao, M., S. Lichter, and L.G. Quintero, The critical Weber number for vortex and jet formation for drops impinging on a liquid pool, *Physics of Fluids*, *31* (12), 3560-3562, 1988.

Hsiung, J., Mean surface energy fluxes over the global ocean, *J. Geophys. Res.*, *91* (C9), 10585-10606, 1986.

Hsu, S.A., and T.I. Whelan, Transport of atmospheric sea salt in coastal zone, *Environ. Sci. Technol.*, *10* (3), 281-283, 1976.

Huang, N.E., L.F. Bliven, S.R. Long, and P.S. DeLeonibus, A study of the relationship among wind speed, sea state, and the drag coefficient for a developing wave field, *J. Geophys. Res.*, *91* (C6), 7733-7742, 1986a.

Huang, N.E., L.F. Bliven, S.R. Long, and C.-C. Tung, An analytic model for oceanic whitecap coverage, *J. Phys. Oceanogr.*, *16*, 1597-1604, 1986b.

Hudson, J.G., Cloud condensation nuclei near marine cumulus, *J. Geophys. Res.*, *98* (D2), 2693-2702, 1993.

Hudson, J.G., and S.S. Yum, Cloud condensation nuclei spectra and polluted and clean clouds over the Indian Ocean, *J. Geophys. Res.*, *107* (D19), 8022, doi:10.1029/2001JD000829, 2002.

Huebert, B.J., and A.L. Lazrus, Bulk composition of aerosols in the remote troposphere, *J. Geophys. Res.*, *85* (C12), 7337-7344, 1980.

Huebert, B.J., S. Howell, P. Lai, J.E. Johnson, T.S. Bates, P.K. Quinn, V. Yegorov, A.D. Clarke, and J.N. Porter, Observations of the atmospheric sulfur cycle on SAGA 3, *J. Geophys. Res.*, *98* (D9), 16985-16995, 1993.

Huebert, B.J., D.J. Wylie, L. Zhuang, and J.A. Heath, Production and loss of methanesulfonate and non-sea salt sulfate in the equatorial Pacific marine boundary layer, *Geophys. Res. Lttrs*, *23* (7), 737-740, 1996.

Hughes, H.G., Aerosol extinction coefficient variations with altitude at 3.75 μm in a coastal environment, *J. Applied Meteorol.*, *19*, 803-808, 1980.

Hughes, H.G., Evaluation of the LOWTRAN 6 Navy maritime aerosol model using 8 to 12 μm sky radiances, *Opt. Eng.*, *26* (11), 1155-1160, 1987.

Hughes, H.G., and J.H. Richter, Extinction coefficients calculated from aerosol size distributions measured in a marine environment, *Opt. Eng.*, *19* (4), 616-620, 1980.

Hughes, R.B., and J.F. Stampfer, Jr., Enhanced evaporation of small, freely falling water drops due to surface contamination, *J. Atmos. Sci.*, *28*, 1244-1251, 1971.

Hühnerfuss, H., and W.D. Garrett, Experimental sea slicks: their practical applications and utilization for basic studies of air-sea interaction, *J. Geophys. Res.*, *86* (C1), 439-447, 1981.

Hühnerfuss, H., W. Alfers, W.L. Jones, P.A. Lange, and K. Richter, The damping of ocean surface waves by a monomolecular film measured by wave staffs and microwave radars, *J. Geophys. Res.*, *86* (C1), 429-438, 1981.

Hummelshøj, P., N.O. Jensen, and S.E. Larsen, Particle dry deposition to a sea surface, in *Precipitation Scavenging and Atmospheric-Surface Exchange*, edited by S.E. Schwartz, and W.G.N. Slinn, pp. 829-840, Hemisphere Publishing Corporation, Washington, DC, 1992.

Hunter, K.A., Processes affecting particulate trace metals in the sea surface microlayer, *Marine Chem.*, *9*, 49-70, 1980.

Hwang, P.A., and W.J. Teague, Low-frequency resonant scattering of bubble clouds, *J. Atmos. Oceanic Technol.*, *17*, 847-853, 2000.

Hwang, P.A., Y.-H.L. Hsu, and J. Wu, Air bubbles produced by breaking wind waves: a laboratory study, *J. Phys. Oceanogr.*, *20*, 19-28, 1990.

Hwang, P.A., Y.-K. Poon, and J. Wu, Temperature effects on generation and entrainment of bubbles induced by a water jet, *J. Phys. Oceanogr.*, *21*, 1602-1605, 1991.

Hwang, P.A., W.J. Reague, G.A. Jacobs, and D.W. Wang, A statistical comparison of wind speed, wave height, and wave period derived from satellite altimeters and ocean buoys in the Gulf of Mexico region, *J. Geophys. Res.*, *103* (C5), 10451-10468, 1998.

Iacono, M.J., and D.C. Blanchard, An investigation of vortex rings from bursting bubbles, *Atmos. Res.*, *21*, 139-149, 1987.

Iida, N., Y. Toba, and M. Chaen, A new expression for the production rate of sea water droplets on the sea surface, *J. Oceanogr.*, *48* (4), 439-460, 1992.

Ikegami, M., K. Okada, Y. Zaizen, and Y. Makino, Sea-salt particles in the upper tropical troposphere, *Tellus*, *46B*, 142-151, 1994.

IPCC (Intergovernmental Panel on Climate Change), *Climate Change 2001: The Scientific Basis. Contribution of Working Group I to the Third Assessment Report of the Intergovernmental Panel on Climate Change*, Cambridge University Press, Cambridge, 2001.

Iribarne, J.V., and B.J. Mason, Electrification accompanying the bursting of bubbles in water and dilute aqueous solutions, *Trans. Faraday Soc.*, *63*, 2234-2245, 1967.

Iribarne, J.V., D. Corr, B.Y.H. Liu, and D.Y.H. Pui, On the hypothesis of particle fragmentation during evaporation, *Atmos. Environ.*, *11*, 639-642, 1977.

Isemer, H.-J., and L. Hasse, The scientific Beaufort equivalent scale: effects on wind statistics and climatological air-sea flux estimates in the North Atlantic Ocean, *J. Climate*, *4*, 819-836, 1991.

Isono, K., On sea-salt nuclei in the atmosphere, *Geofisica Pura e Applicata*, *36*, 156-164, 1957.

Jacob, D.J., Group report: What factors influence atmospheric aerosols, How have they changed in the past, and How might they change in the future?, in *Aerosol Forcing of Climate*, edited by R.J. Charlson, and J. Heintzenberg, pp. 183-195, John Wiley and Sons, Chichester, 1995.

Jacobs, W.C., A preliminary report of a study of atmospheric chlorides, *Bull. Am. Meteorol. Soc.*, *17*, 301-303, 1936.

Jacobs, W.C., Preliminary report on a study of atmospheric chlorides, *Monthly Weather Rev.*, *65*, 147-151, 1937.

Jacobs, W.C., A discussion of physical factors governing the distribution of microorganisms in the atmosphere, *J. Marine Res.*, *2*, 218-224, 1939.

Jacobson, M.J., Global direct radiative forcing due to multi-component anthropogenic and natural aerosols, *J. Geophys. Res.*, *106* (D2), 1551-1568, 2001.

Jaenicke, R., Physical properties of atmospheric particulate sulfur compounds, *Atmos. Environ.*, *12*, 161-169, 1978.

Jaenicke, R., Aerosol physics and chemistry, in *Landolt-Börnstein, New Series, Group V, Meteorology*, edited by G. Fischer, pp. 391-457, Springer-Verlag, Berlin, 1988.

Jaenicke, R., and C.N. Davies, The mathematical expression of the size distribution of atmospheric aerosols, *J. Aerosol Sci.*, *7*, 255-259, 1976.

Jaenicke, R., and S. Matthias-Maser, Natural sources of atmospheric aerosol particles, in *Precipitation Scavenging and Atmospheric-Surface Exchange*, edited by S.E. Schwartz, and W.G.N. Slinn, pp. 1617-1639, Hemisphere Publishing Corporation, Washington, DC, 1992.

Jarvis, N.L., The effect of monomolecular films on surface temperature and convective motion at the water/air interface, *J. Colloid Interface Sci.*, *17*, 512-522, 1962.

Jarvis, N.L., Adsorption of surface-active material at the sea-air interface, *Limnol. Oceanogr.*, *12* (2), 213-221, 1967.

Jayaratne, O.W., and B.J. Mason, The coalescence and bouncing of water drops at an air/water interface, *Proc. Royal Soc. London*, *280A*, 545-565, 1964.

Jenkin, M.E., An investigation into the enhancement of deposition of hygroscopic aerosols to wet surfaces in a wind tunnel, *Atmos. Environ.*, *18* (5), 1017-1024, 1984.

Jennings, S.G., and C.D. O'Dowd, Volatility of aerosol at Mace Head, on the west coast of Ireland, *J. Geophys. Res.*, *95* (D9), 13937-13948, 1990.

Jessup, A.T., W.C. Keller, and W.K. Melville, Measurements of sea spikes in microwave backscatter at moderate incidence, *J. Geophys. Res.*, *95* (C6), 9679-9688, 1990.

Jessup, A.T., W.K. Melville, and W.C. Keller, Breaking waves affecting microwave backscatter. 1. Detection and verification, *J. Geophys. Res.*, *96* (C11), 20547-20559, 1991a.

Jessup, A.T., W.K. Melville, and W.C. Keller, Breaking waves affecting microwave backscatter. 2. Dependence on wind and wave conditions, *J. Geophys. Res.*, *96* (C11), 20561-20569, 1991b.

Jessup, A.T., C.J. Zappa, M.R. Loewen, and V. Hesany, Infrared remote sensing of breaking waves, *Nature*, *385*, 52-55, 1997.

Johnson, B., Bubble populations: background and breaking waves, in *Oceanic Whitecaps and Their Role in Air-Sea Exchange Processes*, edited by E.C. Monahan, and G. MacNiocaill, pp. 69-73, D. Reidel Publishing Company, Dordrecht, 1986.

Johnson, B.D., and R.C. Cooke, Bubble populations and spectra in coastal waters: a photographic approach, *J. Geophys. Res.*, *84* (C7), 3761-3766, 1979.

Johnson, B.D., and R.C. Cooke, Generation of stabilized microbubbles in seawater, *Science*, *213*, 209-211, 1981.

Johnson, B.D., and P.J. Wangersky, Microbubbles: stabilization by monolayers of adsorbed particles, *J. Geophys. Res.*, *92* (C13), 14641-14647, 1987.

Johnson, D.B., The role of giant and ultragiant aerosol particles in warm rain initiation, *J. Atmos. Sci.*, *39*, 448-460, 1982.

Johnson, D.L., D. Leith, and P.C. Reist, Drag on non-spherical, orthotropic aerosol particles, *J. Aerosol Sci.*, *18* (1), 87-97, 1987.

Johnstone, H.F., W.E. Winsche, and L.W. Smith, The dispersion and deposition of aerosols, *Chem. Rev.*, *44*, 353-371, 1944.

Jonas, P.R., and B.J. Mason, Systematic charging of water droplets produced by break-up of liquid jets and filaments, *Trans. Faraday Soc.*, *64*, 1971-1982, 1968.

Jones, A., D.L. Roberts, M.J. Woodage, and C.E. Johnson, Indirect sulphate aerosol forcing in a climate model with an interactive sulphur cycle, *J. Geophys. Res.*, *106* (D17), 20293-20310, 2001.

Jones, A.T., The sounds of splashes, *Science*, *52*, 295-296, 1920.

Jones, E.P., and S.D. Smith, The air density correction to eddy flux measurements, *Boundary-Layer Meteorol.*, *15*, 357-360, 1978.

Jones, E.P., T.V. Ward, and H.H. Zwick, A fast response atmospheric CO_2 sensor for eddy correlation flux measurements, *Atmos. Environ.*, *12*, 845-851, 1978.

Jones, I.S.F., and B.C. Kenney, The scaling of velocity fluctuations in the surface mixed layer, *J. Geophys. Res.*, *82* (9), 1392-1396, 1977.

Josberger, E.G., The effect of bubbles released from a melting ice wall on the melt-driven convection in salt water, *J. Phys. Oceanogr.*, *10*, 474-477, 1980.

Junge, C., Die Rolle der Aerosole und der gasförmigen Beimengungen der Luft im Spurenstoffhaushalt der Troposphäre, *Tellus*, *5*, 1-26, 1953.

Junge, C.E., The chemical composition of atmospheric aerosols, I: Measurements at Round Hill Field Station, June-July 1953, *J. Meteorol.*, *11*, 323-333, 1954.

Junge, C., The size distribution and aging of natural aerosols as determined from electrical and optical data on the atmosphere, *J. Meteorol.*, *12*, 13-25, 1955.

Junge, C.E., Recent investigations in air chemistry, *Tellus*, *8*, 127-139, 1956.

Junge, C.E., The vertical distribution of aerosols over the ocean, in *Artificial Stimulation of Rain*, edited by H. Weickmann, and W. Smith, pp. 89-96, Pergammon Press, New York, 1957a.

Junge, C.E., Remarks about the size distribution of natural aerosols, in *Artificial Stimulation of Rain*, edited by H. Weickmann, and W. Smith, pp. 3-17, Pergammon Press, New York, 1957b.

Junge, C.E., Atmospheric chemistry, in *Advances Geophys.*, edited by H.E. Landsberg, and J. van Mieghem, pp. 1-105, Academic Press, New York, 1958.

Junge, C.E., *Air Chemistry and Radioactivity*, 382 pp., Academic Press, New York, 1963.

Junge, C.E., Comments on "Concentration and size distribution measurements of atmospheric aerosols and a test for the theory of self-preserving size distributions", *J. Atmos. Sci.*, *26*, 603-608, 1969.

Junge, C.E., Our knowledge of the physico-chemistry of aerosols in the undisturbed marine environment, *J. Geophys. Res.*, *77* (27), 5183-5200, 1972.

Junge, C.E., and P.E. Gustafson, On the distribution of sea salt over the United States and its removal by precipitation, *Tellus*, *9*, 164-173, 1957.

Junge, C.E., E. Robinson, and F.L. Ludwig, A study of aerosols in Pacific air masses, *J. Applied Meteorol.*, *8*, 340-347, 1969.

Kahn, R., P. Banerjee, D. McDonald, and J. Martonchik, Aerosol properties derived from aircraft multiangle imaging over Monterey Bay, *J. Geophys. Res.*, *106* (D11), 11977-11995, 2001.

Kaimal, J.C., J.C. Wyngaard, D.A. Haugen, O.R. Cote, Y. Izumi, S.J. Caughey, and C.J. Readings, Turbulent structure in the convective boundary layer, *J. Atmos. Sci.*, *33*, 2152-2169, 1976.

Kalogiros, J.A., and Q. Wang, Aerodynamic effects on wind turbulence measurements with research aircraft, *J. Atmos. Oceanic Technol.*, *19*, 1567-1576, 2002.

Kanwisher, J., On the exchange of gases between the atmosphere and the sea, *Deep-Sea Res.*, *10*, 195-207, 1963.

Karamanev, D.G., Rise of gas bubbles in quiescent liquids, *Am. Inst. Chem. Eng. J.*, *40* (8), 1418-1421, 1994.

Karanfilian, S.K., and T.J. Kotas, Drag on a sphere in unsteady motion in a liquid at rest, *J. Fluid Mechanics*, *87* (1), 85-96, 1978.

Karlsson, R., and E. Ljungström, Nitrogen dioxide and sea salt particles - a laboratory study, *J. Aerosol Sci.*, *26*, 39-50, 1995.

Karlsson, R., and E. Ljungström, A laboratory study of the interaction of NH_3 and NO_2 with sea salt particles, *Water Air Soil Pollut.*, *103*, 55-70, 1998.

Kármán, T.v., and M.A. Biot, *Mathematical Methods in Engineering*, 505 pp., McGraw-Hill Book Company, New York, 1940.

Kasper, G., Dynamics and measurements of smokes. I., *Aerosol Sci. Technol.*, *1*, 187-199, 1982.

Kasten, F., Visibility forecast in the phase of precondensation, *Tellus*, *21*, 631-635, 1969.

Katen, P.C., and J.M. Hubbe, An evaluation of optical particle counter measurements of dry deposition of atmospheric aerosol particles, *J. Geophys. Res.*, *90* (D1), 2145-2160, 1985.

Katsaros, K., and K.J.K. Buettner, Influence of rainfall on temperature and salinity of the ocean surface, *J. Applied Meteorol.*, *8*, 15-18, 1969.

Katsaros, K.B., and G. de Leeuw, Comment on "Sea spray and turbulent air-sea heat fluxes" by Edgar L. Andreas, *J. Geophys. Res.*, *99* (C7), 14339-14343, 1994.

Katsaros, K.B., S.D. Smith, and W.A. Oost, HEXOS - Humidity Exchange Over the Sea. A program for research on water-vapor and droplet fluxes from sea to air at moderate to high wind speeds, *Bull. Am. Meteorol. Soc.*, *68*, 466-476, 1987.

Kaufman, J., A. Smirnov, B.N. Holben, and O. Dubovik, Baseline maritime aerosol: methodology to derive the optical thickness and scattering properties, *Geophys. Res. Lttrs*, *28* (17), 3251-3254, 2001.

Keeling, C.D., The concentration and isotopic abundances of carbon dioxide in the atmosphere, *Tellus*, *12*, 200-203, 1960.

Keeling, R.F., On the role of large bubbles in air-sea gas exchange and supersaturation in the ocean, *J. Marine Res.*, *51*, 237-271, 1993.

Keene, W.C., and D.L. Savoie, the pH of deliquesced sea-salt aerosol in polluted marine air, *Geophys. Res. Lttrs*, *25* (12), 2181-2184, 1998.

Keene, W.C., and D.L. Savoie, Correction to "The pH of deliquesced sea-salt aerosol in polluted marine air", *Geophys. Res. Lttrs*, *26* (9), 1315-1316, 1999.

Keene, W.C., A.A.P. Pszenny, J.N. Galloway, and M.E. Hawley, Sea-salt corrections and interpretation of constituent ratios in marine precipitation, *J. Geophys. Res.*, *91* (D6), 6647-6658, 1986.

Keene, W.C., A.A.P. Pszenny, D.J. Jacob, R.A. Duce, J.N. Galloway, J.J. Schultz-Tokos, H. Sievering, and J.F. Boatman, The geochemical cycling of reactive chlorine through the marine troposphere, *Global Biogeochem. Cycles*, *4* (4), 407-430, 1990.

Keene, W.C., R. Sander, A.A.P. Pszenny, R. Vogt, P.J. Crutzen, and J.N. Galloway, Aerosol pH in the marine boundary layer: a review and model evaluation, *J. Aerosol Sci.*, *29* (3), 339-356, 1998.

Keitel, G., and U. Onken, Inhibition of bubble coalescence by solutes in air/water dispersions, *Chem. Eng. Sci.*, *37* (11), 1635-1638, 1982.

Keith, C.H., and A.B. Arons, The growth of sea-salt particles by condensation of atmospheric water vapor, *J. Meteorol.*, *11*, 173-184, 1954.

Kennedy, R.M., Acoustic radiation due to surface wave breaking, *J. Acoustical Soc. Am.*, *94* (4), 2443-2445, 1993.

Kennedy, R.M., and R.L. Snyder, On the formation of whitecaps by a threshold mechanism. Part II: Monte Carlo experiments, *J. Phys. Oceanogr.*, *13*, 1493-1504, 1983.

Kent, E.C., and P.K. Taylor, Choice of a Beaufort equivalent scale, *J. Atmos. Oceanic Technol.*, *14*, 228-242, 1997.

Kepert, J.D., Comments on "The temperature of evaporating sea spray droplets", *J. Atmos. Sci.*, *53* (11), 1634-1641, 1996.

Kerker, M., E. Matijevic, W.F. Espenscheid, W.A. Farone, and S. Kitani, Aerosol studies by light scattering. I. Particle size distribution by polarization ratio method, *J. Colloid Sci.*, *19*, 213-222, 1964.

Kerman, B.R., A model of interfacial gas transfer for a well-roughened sea, *J. Geophys. Res.*, *89* (D1), 1439-1446, 1984a.

Kerman, B.R., Underwater sound generation by breaking wind waves, *J. Acoustical Soc. Am.*, *75* (1), 149-165, 1984b.

Kerman, B.R., Distribution of bubbles near the ocean surface, *Atmosphere-Ocean*, *24* (2), 169-188, 1986.

Kerman, B.R., Sea Surface Sound. Natural Mechanisms of Surface Generated Noise in the Ocean, pp. 639, Kluwer Academic Publishers, Dordrecht, Netherlands, 1988a.

Kerman, B.R., On the distribution of bubbles near the ocean surface, in *Sea Surface Sound*, edited by B.R. Kerman, pp. 185-196, Kluwer Academic Publishers, Dordrecht, 1988b.

Kerman, B.R., and L. Bernier, Multifractal representation of breaking waves on the ocean surface, *J. Geophys. Res.*, *99* (C8), 16179-16196, 1994.

Kerman, B.R., D.L. Evans, D.R. Watts, and D. Halpern, Wind dependence of underwater ambient noise, *Boundary-Layer Meteorol.*, *26*, 105-113, 1983.

Kerminen, V.-M., K. Teinilä, R. Hillamo, and T. Pakkanen, Substitution of chloride in sea-salt particles by inorganic and organic anions, *J. Aerosol Sci.*, *29* (8), 929-942, 1998.

Kerminen, V.-M., K. Teinilä, and R. Hillamo, Chemistry of sea-salt particles in the summer Antarctic atmosphere, *Atmos. Environ.*, *34*, 2817-2825, 2000.

Kessler, E., *On the distribution and continuity of water substance in atmospheric circulations*, 84 pp., American Meteorological Society, Boston, 1969.

Kettle, A.J., and M.O. Andreae, Flux of dimethylsulfide from the oceans: A comparison of updated data sets and flux models, *J. Geophys. Res.*, *105* (D22), 26793-26808, 2000.

Khemani, L.T., G.A. Momin, M.S. Naik, P.S. Prakasa Roa, R. Kumar, and B.V. Ramana Murty, Trace elements and sea salt aerosols over the sea areas around the Indian sub-continent, *Atmos. Environ.*, *19* (2), 277-284, 1985.

Kientzler, C.F., A.B. Arons, D.C. Blanchard, and A.H. Woodcock, Photographic investigation of the projection of droplets by bubbles bursting at a water surface, *Tellus*, *6*, 1-7, 1954.

Kilgore, E.G., and J.A. Day, Energy dissipation of water drops striking an undisturbed water surface, *Tellus*, *15*, 367-369, 1963.

Kim, Y., H. Sievering, and J. Boatman, Volume and surface area size distribution, water mass and model fitting of GCE/CASE/WATOX marine aerosols, *Global Biogeochem. Cycles*, *4* (2), 165-177, 1990.

Kinch, E., The amount of chlorine in rain-water collected at Cirencester, *J. Chem. Soc.*, *51*, 92-94, 1887.

King, W.D., and C.T. Maher, The spatial distribution of salt particles at cloud levels in Central Queensland, *Tellus*, *28*, 11-23, 1976.

Kingsbury, D.L., and P.L. Marston, Mie scattering near the critical angle of bubbles in water, *J. Optical Soc. Am.*, *71* (3), 358-361, 1981.

Kitchener, J.A., and C.F. Cooper, Current concepts in the theory of foaming, *Quar. Rev. Chem. Soc.*, *13*, 71-97, 1959.

Kleefeld, C., C.D. O'Dowd, S. O'Reilly, S.G. Gennings, P. Aalto, E. Becker, G. Kunz, and G. de Leeuw, Relative contribution of submicron and supermicron particles to aerosol light scattering in the marine boundary layer, *J. Geophys. Res.*, *107* (D19), 8103, doi:10.1029/2000JD000262, 2002.

Kline, S.J., W.C. Reynolds, F.A. Schraub, and P.W. Runstadler, The structure of turbulent boundary layers, *J. Fluid Mechanics*, *30*, 741-773, 1967.

Knelman, F., N. Dombrowski, and D.M. Newitt, Mechanism of the bursting of bubbles, *Nature*, *173*, 261, 1954.

Knipping, E.M., and D. Dabdub, Impact of chlorine emissions from sea-salt aerosol on coastal urban ozone, *Environ. Sci. Technol.*, *37*, 275-284, 2003.

Knudsen, M., Die molekulare Wärmeleitung der Gase und der Akkommodationskoeffizient, *Annalen der Physik*, *34* (4), 593-656, 1911.

Knudsen, M., Die maximale Verdampfungsgeschwindigkeit des Quecksilbers, *Annalen der Physik*, *47*, 697-708, 1915.

Knudsen, M., and S. Weber, Luftwiderstand gegen die langsame Bewegung kleiner Kugeln, *Annelen der Physik*, *36*, 981-994, 1911.

Koch, M.K., A. Voßnacke, J. Starflinger, W. Schütz, and H. Unger, Radionuclide re-entrainment at bubbling water pool surfaces, *J. Aerosol Sci.*, *31* (9), 1015-1028, 2000.

Kocmond, W.C., W.D. Garrett, and E.J. Mack, Modification of laboratory fog with organic surface films, *J. Geophys. Res.*, *77* (18), 3221-3231, 1972.

Koepke, P., Effective reflectance of oceanic whitecaps, *Applied Optics*, *23* (11), 1816-1824, 1984.

Koepke, P., Remote sensing signatures of whitecaps, in *Oceanic Whitecaps and Their Role in Air-Sea Exchange Processes*, edited by E.C. Monahan, and G. MacNiocaill, pp. 251-260, D. Reidel Publishing Company, Dordrecht, 1986a.

Koepke, P., Oceanic whitecaps: their effective reflectance, in *Oceanic Whitecaps and Their Role in Air-Sea Exchange Processes*, edited by E.C. Monahan, and G. MacNiocaill, pp. 272-274, D. Reidel Publishing Company, Dordrecht, 1986b.

Koepke, P., and H. Quenzel, Turbidity of the atmosphere determined from satellite: calculation of optimum wavelength, *J. Geophys. Res.*, *86* (C10), 9801-9805, 1981.

Koga, M., Direct production of droplets from breaking wind-waves - its observation by a multi-colored overlapping exposure photographic technique, *Tellus*, *33*, 552-563, 1981.

Koga, M., Bubble entrainment in breaking wind waves, *Tellus*, *34*, 481-489, 1982.

Koga, M., and Y. Toba, Droplet distribution and dispersion processes on breaking wind waves, *Tohoku Geophys. J.*, *28* (1), 1-25, 1981.

Köhler, H., Untersuchungen über die Elemente des Nebels und der Wolken, *Meddelanden Från Statens Meteorologisk-Hydrografiska Anstalt*, *2* (5), 1-73, 1925.

Köhler, H., The nucleus in and the growth of hygroscopic droplets, *Trans. Faraday Soc.*, *32*, 1152-1161, 1936.

Köhler, H., and M. Båth, Quantitative chemical analysis of condensation nuclei from sea water, *Nova Acta Regiae Societatis Scientiarum Upsaliensis, Series IV*, *15* (7), 4-24, 1953.

Kokhanovsky, A.A., Optical properties of bubbles, *J. Optics A: Pure Applied Optics*, *5*, 47-52, 2003.

Kolaini, A.R., L.A. Crum, and R.A. Roy, Bubble production by capillary-gravity waves, *J. Acoustical Soc. Am.*, *95* (4), 1913-1921, 1994.

Kolovayev, P.A., Investigation of the concentration and statistical size distribution of wind-produced bubbles in the near-surface ocean layer, *Oceanology*, *15*, 659-661, 1976.

Komori, S., and R. Misumi, The effects of bubbles on mass transfer across the breaking air-water interface, in *Gas Transfer at Water Surfaces*, edited by M.A. Donelan, W.M. Drennan, E.S. Saltzman, and R. Wanninkhof, pp. 285-290, American Geophysical Union, Washington, DC, 2002.

Kondo, J., Y. Fujinawa, and G. Naito, High-frequency components of ocean waves and their relation to the aerodynamic roughness, *J. Phys. Oceanogr.*, *3*, 197-202, 1973.

Koop, T., A. Kapilashrami, L.T. Molina, and M.J. Molina, Phase transitions of sea salt/water mixtures at low temperatures: Implications for ozone chemistry in the polar marine boundary layer, *J. Geophys. Res.*, *105* (D21), 26393-26402.

Korolev, V.S., S.A. Petrichenko, and V.D. Pudov, Heat and moisture exchange between the ocean and atmosphere in tropical storms Tess and Skip, *Soviet Meteorol. Hydrol.*, 92-94, 1990.

Korzh, V.D., Ocean as a source of atmospheric iodine, *Atmos. Environ.*, *18* (12), 2707-2710, 1984.

Koschmieder, H., Theorie der horizontalen Sichtweite, *Beiträge zur Physik der Freien Atmosphäre*, *12*, 33-53, 1924a.

Koschmieder, H., Theorie der horizontalen Sichtweite II: Kontrast und Sichtweite, *Beiträge zur Physik der Freien Atmosphäre*, *12*, 171-181, 1924b.

Kottler, F., The distribution of particle sizes, *J. Franklin Inst.*, *250*, 339-356, 419-441, 1950.

Kottler, F., The logarithmico-normal distribution of particle sizes: homogeneity and heterogeneity, *J. Phys. Chem.*, *56*, 442-448, 1952.

Kowalski, A.S., Deliquescence induces eddy covariance and estimable dry deposition errors, *Atmos. Environ.*, *35*, 4843-4851, 2001.

Koyama, T., and K. Sugawara, Separation of the components of atmospheric salt and their distribution (continued), *Bull. Chem. Soc. Japan*, *26*, 123-126, 1953.

Kraan, C., W.A. Oost, and P.A.E.M. Janssen, Wave energy dissipation by whitecaps, *J. Atmos. Oceanic Technol.*, *13*, 262-267, 1996.

Krämer, L., U. Pöschl, and R. Niessner, Microstructural rearrangement of sodium chloride condensation aerosol particles on interaction with water vapor, *J. Aerosol Sci.*, *31* (6), 673-685, 2000.

Kramm, G., and R. Dlugi, Modelling of the vertical fluxes of nitric acid, ammonia, and ammonium nitrate, *J. Atmos. Chem.*, *18*, 319-357, 1994.

Kraus, E.B., Wind stress along the sea surface, in *Advances Geophys.*, edited by H.E. Landsberg, and J. van Mieghem, pp. 213-255, Academic Press, New York, 1967.

Kraus, E.B., and R.E. Morrison, Local interactions between the sea and the air at monthly and annual time scales, *Quar. J. Royal Meteorol. Soc.*, *92*, 114-127, 1966.

Kreidenweis, S.M., L.M. McInnes, and F.J. Brechtel, Observations of aerosol volatility and elemental composition at Macquarie Island during the First Aerosol Characterization Experiment (ACE 1), *J. Geophys. Res.*, *103* (D13), 16511-16524, 1998.

Krieger, U.K., and C. Braun, Light-scattering intensity fluctuations in single aerosol particles during deliquescence, *J. Quan. Spectros. Rad. Transf.*, *70*, 545-554, 2001.

Krischke, U., R. Staubes, T. Brauers, M. Gautrois, J. Burkert, D. Stöbener, and W. Jaeschke, Removal of SO_2 from the marine boundary layer over the Atlantic Ocean: A case study on the kinetics of the heterogeneous S(IV) oxidation on marine aerosols, *J. Geophys. Res.*, *105* (D11), 14413-14422, 2000.

Krishna Moorthy, K., S.K. Satheesh, and B.V. Krishna Murthy, Investigations of marine aerosols over the tropical Indian Ocean, *J. Geophys. Res.*, *102* (D15), 18827-18842, 1997.

Kristament, I.S., J.B. Liley, and M.J. Harvey, Aerosol variability in the vertical in the Southwest Pacific, *J. Geophys. Res.*, *98* (D4), 7129-7139, 1993.

Kristensen, L., J. Mann, S.P. Oncley, and J.C. Wyngaard, How close is close enough when measuring scalar fluxes with displaced sensors?, *J. Atmos. Oceanic Technol.*, *14*, 814-821, 1997.

Kritz, M.A., Use of long-lived radon daughters as indicators of exchange between the free troposphere and the marine boundary layer, *J. Geophys. Res.*, *88* (C13), 8569-8573, 1983.

Kritz, M.A., and J. Rancher, Circulation of Na, Cl, and Br in the tropical marine atmosphere, *J. Geophys. Res.*, *85* (C13), 1633-1639, 1980.

Krumbein, W.C., Settling-velocity and flume-behavior of non-spherical particles, *Trans. Am. Geophys. Union*, *23*, 621-633, 1942.

Kulkarni, M.R., B.B. Adiga, R.K. Kapoor, and V.V. Shirvaikar, Sea salt in coastal air and its deposition on porcelain insulators, *J. Applied Meteorol.*, *21*, 350-355, 1982.

Kuo, Y.M., and C.S. Wang, Characteristics of droplets generated by bubble bursting from chromic acid solutions, *J. Aerosol Sci.*, *30* (9), 1171-1179, 1999.

Kuo, Y.-M., and C.-S. Wang, Droplet fractionation of hexavalent chromium from bubbles bursting at liquid surfaces of chromic acid solutions, *J. Aerosol Sci.*, *33*, 297-306, 2002.

Kuroiwa, D., The composition of sea-fog nuclei as identified by electron microscope, *J. Meteorol.*, *13*, 408-410, 1956.

Kuśmierczyk-Michulec, J., M. Schulz, S. Ruellan, O. Krüger, E. Plate, R. Marks, G. de Leeuw, and H. Cachier, Aerosol composition and related optical properties in the marine boundary layer over the Baltic Sea, *J. Aerosol Sci.*, *32*, 933-955, 2001.

La Mer, V., *Retardation of Evaporation by Monolayers: Transport Process*, 277 pp., Academic Press, New York, 1962.

Lai, R.J., and O.H. Shemdin, Laboratory study of the generation of spray over water, *J. Geophys. Res.*, *79* (21), 3055-3063, 1974.

Lamarre, E., and W.K. Melville, Air entrainment and dissipation in breaking waves, *Nature*, *351*, 469-472, 1991.

Lamarre, E., and W.K. Melville, Instrumentation for the measurement of void-fraction in breaking waves: Laboratory and field results, *IEEE J. Oceanic Eng.*, *17* (2), 204-215, 1992.

Lamarre, E., and W.K. Melville, Void-fraction measurements and sound-speed fields in bubble plumes generated by breaking waves, *J. Acoustical Soc. Am.*, *95* (3), 1317-1326, 1994a.

Lamarre, E., and W.K. Melville, Sound-speed measurements near the ocean surface, *J. Acoustical Soc. Am.*, *96* (6), 3605-3616, 1994b.

Lamaud, E., Y. Brunet, A. Labatut, A. Lopez, J. Fontan, and A. Druilhet, The Landes experiment: Biosphere-atmosphere exchanges of ozone and aerosol particles above a pine forest, *J. Geophys. Res.*, *99* (D8), 16511-16521, 1994.

Langer, S., R.S. Pemberton, and B.J. Finlayson-Pitts, Diffuse reflectance infrared studies of the reaction of synthetic sea salt mixtures with NO_2: A key role for hydrates in the kinetics and mechanism, *J. Phys. Chem.*, *101A*, 1277-1286, 1997.

Langmuir, I., The vapor pressure of metallic tungsten, *Physical Rev.*, *2* (5), 329-342, 1913.

Langmuir, I., Surface motion of water induced by wind, *Science*, *87*, 119-123, 1938.

Langmuir, I., The production of rain by a chain reaction in cumulus clouds at temperatures above freezing, *J. Meteorol.*, *5* (5), 175-192, 1948.

Langmuir, I., *The Collected Works of Irving Langmuir*, 451 pp., Pergamon Press, Oxford, 1961.

Langmuir, I.L., and V.J. Schaefer, Rates of evaporation of water through compressed monolayers on water, *J. Franklin Inst.*, *235*, 119-162, 1943.

Lapple, C.E., and C.B. Shepherd, Calculation of particle trajectories, *Indust. Eng. Chem.*, *32*, 605-617, 1940.

Larsen, S.E., J.B. Edson, P. Hummelshøj, N.O. Jensen, G. de Leeuw, and P.G. Mestayer, Dry deposition of particles to ocean surfaces, *Ophelia*, *42*, 193-204, 1995.

Latham, J., and M.H. Smith, Effect on global warming of wind-dependent aerosol generation at the ocean surface, *Nature*, *347*, 372-373, 1990.

Laws, J.O., Measurements of the fall-velocity of water-drops and raindrops, *Trans. Am. Geoph. Union*, *22*, 709-723, 1941.

Le Méhauté, B., Gravity-capillary rings generated by water drops, *J. Fluid Mechanics*, *197*, 415-427, 1988.

Le Méhauté, B., and T. Khangaonkar, Dynamic interaction of intense rain with water waves, *J. Phys. Oceanogr.*, *20*, 1805-1812, 1990.

Leaitch, W.R., C.M. Banic, G.A. Isaac, M.D. Couture, P.S.K. Liu, I. Gultepe, and S.-M. Li, Physical and chemical observations in marine stratus during the 1993 North Atlantic Regional Experiment: Factors controlling cloud droplet number concentrations, *J. Geophys. Res.*, *101* (D22), 29123-29135, 1996.

Leavitt, E., and C.A. Paulson, Statistics of surface layer turbulence over the tropical ocean, *J. Phys. Oceanogr.*, *5*, 143-156, 1975.

LeBlond, P.H., Gas diffusion from ascending gas bubbles, *J. Fluid Mechanics*, *35*, 711-719, 1969a.

LeBlond, P.H., Corrigendum, *J. Fluid Mechanics*, *38*, 861, 1969b.

Lee, C.-T., and W.-C. Hsu, The measurement of liquid water mass associated with collected hygroscopic particles, *J. Aerosol Sci.*, *31* (2), 189-197, 2000.

Lee, I.Y., and M.L. Wesely, Effects of surface wetness on the evolution and vertical transport of submicron particles, *J. Applied Meteorol.*, *28*, 176-184, 1989.

Lee, X., and T.A. Black, Relating eddy correlation sensible heat flux to horizontal sensor separation in the unstable atmospheric surface layer, *J. Geophys. Res.*, *99* (D9), 18545-18553, 1994.

Lee, Y.H., G.T. Tsao, and P.C. Wankat, Hydrodynamic effect of surfactants on gas-liquid oxygen transfer, *Am. Inst. Chem. Eng. J.*, *26* (6), 1008-1012, 1980.

Legrand, M.R., C. Lorius, N.I. Barkov, and V.N. Petrov, Vostok (Antarctica) ice core: atmospheric chemistry changes over the last climatic cycle (160,000 years), *Atmos. Environ.*, *22* (2), 317-333, 1988.

Leifer, I., and G. de Leeuw, Bubble measurements in breaking-wave generated bubble plumes during the LUMINY wind-wave experiment, in *Gas Transfer at Water Surfaces*, edited by M.A. Donelan, W.M. Drennan, E.S. Saltzman, and R. Wanninkhof, pp. 303-309, American Geophysical Union, Washington, DC, 2002.

Leifer, I., G. de Leeuw, and L.H. Cohen, Secondary bubble production from breaking waves: the bubble burst mechanism, *Geophys. Res. Lttrs*, *27* (24), 4077-4080, 2000a.

Leifer, I., R. Patro, and P. Bowyer, A study on the temperature variation of rise velocity for large clean bubbles, *J. Atmos. Oceanic Technol.*, *17*, 1392-1402, 2000b.

Leighton, T.G., and A.J. Walton, An experimental study of the sound emitted from gas bubbles in liquids, *European J. Phys.*, *8*, 98-104, 1987.

Leike, A., Demonstration of the exponential decay law using beer foam, *European J. Phys.*, *23*, 21-26, 2002.

Lemlich, R., Adsubble processes: foam fractionation and bubble fractionation, *J. Geophys. Res.*, *77* (27), 5204-5210, 1972.

Lenschow, D.J., and M.R. Raupach, The attenuation of fluctuations in scalar concentrations through sampling tubes, *J. Geophys. Res.*, *96* (D8), 15259-15268, 1991.

Lenschow, D.H., R. Pearson, Jr., and B.B. Stankov, Measurements of ozone vertical flux to ocean and forest, *J. Geophys. Res.*, *87* (C11), 8833-8837, 1982.

Lenschow, D.H., P.B. Krummel, and S.T. Siems, Measuring entrainment, divergence, and vorticity on the meoscale from aircraft, *J. Atmos. Oceanic Technol.*, *16*, 1384-1400, 1999.

Leong, K.H., Morphology of aerosol particles generated from the evaporation of solution drops, *J. Aerosol Sci.*, *12*, 417-435, 1981.

Leong, K.H., Morphological control of particles generated from the evaporation of solution droplets: theoretical considerations, *J. Aerosol Sci.*, *18* (5), 511-524, 1987a.

Leong, K.H., Morphological control of particles generated from the evaporation of solution droplets: experiment, *J. Aerosol Sci.*, *18* (5), 525-552, 1987b.

Lepple, F.K., D.J. Bressan, J.B. Hoover, and R.E. Larson, Sea salt aerosols, atmospheric radon and meteorological observations in the Western South Atlantic Ocean (February 1981), *NRL Memorandum Report 5154*, pp. 63, Naval Research Laboratory, Washington, DC, 1983.

Lepri, L., P. Desideri, R. Cini, F. Masi, and M.S. van Erk, Transport of organochlorine pesticides across the air/sea interface during the aerosol process, *Analytica Chimica Acta*, *317*, 149-160, 1995.

Lessard, R.R., and S.A. Zieminski, Bubble coalescence and gas transfer in aqueous electrolyte solutions, *Industrial and Engineering Chemistry. Fundamentals*, *10* (2), 260-269, 1971.

Levich, V.G., *Physicochemical Hydrodynamics*, 700 pp., Prentice Hall, Inc., Englewood Cliffs, NJ, 1962.

Levitus, S., Climatological Atlas of the World Ocean, pp. 173, National Oceanic and Atmospheric Administration, Rockville, MD, 1982.

Levitus, S., and A.H. Oort, Global analysis of oceanographic data, *Bull. Am. Meteorol. Soc.*, *58*, 1270-1284, 1977.

Li, X., H. Marine, D. Savoie, K. Voss, and J.M. Prospero, Dominance of mineral dust in aerosol light-scattering in the North Atlantic trade winds, *Nature*, *380*, 416-419, 1996.

Li, Y.Q., P. Davidovits, Q. Shi, J.T. Jayne, C.E. Kolb, and D.R. Worsnop, Mass and thermal accommodation coefficients of $H_2O(g)$ on liquid water as a function of temperature, *J. Phys. Chem.*, *105A* (47), 10627-10634, 2001.

Li, Z., A.L. Williams, and M.J. Rood, Influence of soluble surfactant properties on the activation of aerosol particles containing inorganic solute, *J. Atmos. Sci.*, *55*, 1859-1866, 1998.

Licht, W., and G.S.R. Narasimhamurty, Rate of fall of single liquid droplets, *Am. Inst. Chem. Eng. J.*, *1*, 366-373, 1955.

Lieberman, L., Air bubbles in water, *J. Applied Phys.*, *28* (2), 205-217, 1957.

Lighthill, J., G. Holland, W. Gray, C. Landsea, G. Craig, J. Evans, Y. Kurihara, and C. Guard, Global climate change and tropical cyclones, *Bull. Am. Meteorol. Soc.*, *75* (11), 2147-2157, 1994.

Li-Jones, Z., H.B. Maring, and J.M. Prospero, Effect of relative humidity on light scattering by mineral dust aerosol as measured in the marine boundary layer over the tropical Atlantic Ocean, *J. Geophys. Res.*, *103* (D23), 31113-31121, 1998.

Lilly, D.K., Models of cloud-topped mixed layers under a strong inversion, *Quar. J. Royal Meteorol. Soc.*, *94*, 292-309, 1968.

Limpert, E., W.A. Stahel, and M. Abbt, Log-normal distributions across the sciences: Keys and clues, *BioScience*, *51* (5), 341-352, 2001.

Lin, S.P., and R.D. Reitz, Drop and spray formation from a liquid jet, *Ann. Rev. Fluid Mech.*, *30*, 85-105, 1998.

Lin, W., L.P. Sanford, S.E. Suttles, and R. Valigura, Drag coefficients with fetch-limited wind waves, *J. Phys. Oceanogr.*, *32*, 3058-3074, 2002.

Ling, S.C., Effect of breaking waves on the transport of heat and vapor fluxes from the ocean, *J. Phys. Oceanogr.*, *23*, 2360-2372, 1993.

Ling, S.C., and T.W. Kao, Parameterization of the moisture and heat transfer process over the ocean under whitecap sea states, *J. Phys. Oceanogr.*, *6*, 306-315, 1976.

Ling, S.C., and H.P. Pao, Study of micro-bubbles in the North Sea, in *Sea Surface Sound*, edited by B.R. Kerman, pp. 197-210, Kluwer Academic Publishers, Dordrecht, 1988.

Ling, S.C., A. Saad, and T.W. Kao, Mechanics of multiphase fluxes over the ocean, in *Turbulent Fluxes Through the Sea Surface, Wave Dynamics, and Prediction*, edited by A. Favre, and K. Hasselmann, pp. 185-197, Plenum Press, New York, 1978.

Ling, S.C., T.W. Kao, and A.I. Saad, Microdroplets and transport of moisture from ocean, *J. Eng. Mech. Div.*, *106*, 1327-1339, 1980.

Lion, L.W., and J.O. Leckie, The biogeochemistry of the air-sea interface, *Ann. Rev. Earth Planetary Sci.*, *9*, 449-486, 1981.

Liss, P.S., Chemistry of the sea surface microlayer, in *Chemical Oceanography*, edited by J.P. Riley, and G. Skirrow, pp. 193-243, Academic Press, London, 1975.

Liu, B.Y.H., and J.K. Agarwal, Experimental observation of aerosol deposition in turbulent flow, *Aerosol Sci.*, 5, 145-155, 1974.

Liu, P.C., and D.J. Schwab, A comparison of methods for estimating u_* from given u_z and air-sea temperature differences, *J. Geophys. Res.*, 92 (C6), 6488-6494, 1987.

Liu, W.T., K.B. Katsaros, and J.A. Businger, Bulk parameterization of air-sea exchanges of heat and water vapor including the molecular constraints at the interface, *J. Atmos. Sci.*, 36, 1722-1735, 1979.

Liu, W.T., W. Tang, and P.P. Niiler, Humidity profiles over the ocean, *J. Climate*, 4, 1023-1034, 1991.

Liu, Y., and P.H. Daum, The effect of refractive index on size distributions and light scattering coefficients derived from optical particle counters, *J. Aerosol Sci.*, 31 (8), 945-957, 2000.

Liu, Y., and F. Liu, On the description of aerosol particle size distribution, *Atmos. Res.*, 31, 187-198, 1994.

Lo, A.K.-F., L. Zhang, and H. Sievering, The effect of humidity and state of water surfaces on deposition of aerosol particles onto a water surface, *Atmos. Environ.*, 33, 4727-4737, 1999.

Lodge, J.P., A study of sea-salt particles over Puerto Rico, *J. Meteorol.*, 12, 493-499, 1955.

Lodge, J.P., and F. Baer, An experimental investigation of the shatter of salt particles on crystallization, *J. Meteorol.*, 11, 420-421, 1954.

Lodge, J.P., Jr., An estimate of deposition velocities over water, *Atmos. Environ.*, 12, 973-974, 1978.

Loeb, L.B., *Static Electrification*, 240 pp., Springer-Verlag, Berlin, 1958.

Loewen, M., Inside whitecaps, *Nature*, 418, 830, 2002.

Loewen, M.R., and W.K. Melville, A model of the sound generated by breaking waves, *J. Acoustical Soc. Am.*, 90 (4), 2075-2080, 1991.

Loewen, M.R., M.A. O'Dor, and M.G. Skafel, Bubbles entrained by mechanically generated breaking waves, *J. Geophys. Res.*, 101 (C9), 20759-20769, 1996.

Lohmann, U., J. Feichter, C.C. Chuang, and J.E. Penner, Prediction of the number of cloud droplets in the ECHAM GCM, *J. Geophys. Res.*, 104 (D8), 9169-9198, 1999.

Lohmann, U., J. Feichter, J. Penner, and R. Leaitch, Indirect effect of sulfate and carbonaceous aerosols: a mechanistic treatment, *J. Geophys. Res.*, 105 (D10), 12193-12206, 2000.

Longuet-Higgins, M.S., On wave breaking and the equilibrium spectrum of wind-generated waves, *Proc. Royal Soc. London*, 310A, 151-159, 1969.

Longuet-Higgins, M.S., Limiting forms for capillary-gravity waves, *J. Fluid Mechanics*, 194, 351-375, 1988.

Longuet-Higgins, M.S., An analytical model of sound production by raindrops, *J. Fluid Mechanics*, 214, 395-410, 1990a.

Longuet-Higgins, M.S., Bubble noise spectra, *J. Acoustical Soc. Am.*, 87 (2), 652-661, 1990b.

Longuet-Higgins, M.S., The crushing of air cavities in a liquid, *Proc. Royal Soc. London*, 439A, 611-626, 1992.

Longuet-Higgins, M.S., and N.D. Smith, Measurement of breaking waves by a surface jump meter, *J. Geophys. Res.*, 88 (C14), 9823-9831, 1983.

Longuet-Higgins, M.S., and J.S. Turner, An 'entrainment plume' model of a spilling breaker, *J. Fluid Mechanics*, 63 (1), 1-20, 1974.

Lovett, R.F., Quantitative measurement of airborne sea-salt in the North Atlantic, *Tellus*, 30, 358-364, 1978.

Lowe, J.A., M.H. Smith, B.M. Davison, S.E. Benson, M.K. Hill, C.D. O'Dowd, R.M. Harrison, and C.N. Hewitt, Physico-chemical properties of atmospheric aerosol at South Uist, *Atmos. Environ.*, 30 (22), 3765-3776, 1996.

Lucassen, J., Effect of surface-active material on the dampling of gravity waves: A reappraisal, *J. Colloid Interface Sci.*, 85 (1), 52-58, 1982.

Ludlam, F.H., The production of showers by the coalescence of cloud droplets, *Quar. J. Royal Meteorol. Soc.*, 77, 402-417, 1951.

MacIntyre, F., Bubbles: a boundary-layer "microtome" for micron-thick samples of a liquid surface, *J. Phys. Chem.*, 72 (2), 589-592, 1968.

MacIntyre, F., Geochemical fractionation during mass transfer from sea to air by breaking bubbles, *Tellus*, 22, 451-461, 1970.

MacIntyre, F., Flow patterns in breaking bubbles, *J. Geophys. Res.*, 77 (27), 5211-5228, 1972.

MacIntyre, F., The top millimeter of the ocean, *Scientific Am.*, 230 (5), 62-77, 1974a.

MacIntyre, F., Chemical fractionation and sea-surface microlayer processes, in *The Sea*, edited by E.D. Goldberg, pp. 245-299, John Wiley and Sons, New York, 1974b.

MacIntyre, F., Non-lipid-related possibilities for chemical fractionation in bubble film caps, *J. Recherches Atmos.*, 8, 515-527, 1974c.

MacIntyre, F., Additional problems in bubble and jet drop research, *Limnol. Oceanogr.*, 23, 571-573, 1978.

MacIntyre, F., On reconciling optical and acoustic bubble spectra in the mixed layer, in *Oceanic Whitecaps and Their Role in Air-Sea Exchange Processes*, edited by E.C. Monahan, and G. MacNiocaill, pp. 75-94, D. Reidel Publishing Company, Dordrecht, 1986.

MacIntyre, F., and J.W. Winchester, Phosphate ion enrichment in drops from breaking bubbles, *J. Phys. Chem.*, 73, 2163-2169, 1969.

Macklin, W.C., and G.J. Metaxas, Splashing of drops on liquid layers, *J. Applied Phys.*, 47 (9), 3963-3970, 1976.

MacRitchie, F., Evaporation retarded by monolayers, *Science*, 163, 929-931, 1969.

Makin, V.K., Air-sea exchange of heat in the presence of wind waves and spray, *J. Geophys. Res.*, 103 (C1), 1137-1152, 1998.

Makin, V.K., V.N. Kudryavtsev, and C. Mastenbroek, Drag of the sea surface, *Boundary-Layer Meteorol.*, 73, 159-182, 1995.

Mallinger, W.D., and T.P. Mickelson, Experiments with monomolecular films on the surface of the open sea, *J. Phys. Oceanogr.*, 3, 328-336, 1973.

Mamane, Y., and J. Gottlieb, Nitrate formation on sea-salt and mineral particles - a single particle approach, *Atmos. Environ.*, 26A (9), 1763-1769, 1992.

Mangarella, P.A., A.J. Chambers, R.L. Street, and E.Y. Hsu, Laboratory studies of evaporation and energy transfer through a wavy air-water interface, *J. Phys. Oceanogr.*, *3*, 93-101, 1973.

Manley, D.M.J.P., Change of size of air bubbles in water containing a small dissolved air content, *British J. Applied Phys.*, *11*, 38-42, 1960.

Mansfield, W.W., Influence of monolayers on the natural rate of evaporation of water, *Nature*, *175*, 247-249, 1955.

Mansfield, W.W., Evaporation retardation by monolayers, *Science*, *176*, 944-945, 1972.

Manton, M.J., On the attenuation of sea waves by rain, *Geophys. Fluid Dyn.*, *5*, 249-260, 1973.

Marcelin, R., Échange de matière entre un liquide ou un solide et sa vapeur saturée, *Comptes Rendus de l'Academies des Sci., Paris*, *158*, 1674-1676, 1914.

Mari, C., K. Suhre, T.S. Bates, J.E. Johnson, R. Rosset, A.R. Bandy, F.L. Eisele, R.L.I. Mauldin, and D.C. Thorton, Physico-chemical modeling of the First Aerosol Characterization Experiment (ACE 1) Lagrangian B. 2. DMS emission, transport and oxidation at the mesoscale, *J. Geophys. Res.*, *103* (D13), 16457-16473, 1998.

Maring, H., D.L. Savoie, M.A. Izafuirre, L. Custals, and J.S. Reid, Vertical distributions of dust and sea-salt aerosols over Puerto Rico during PRIDE measured from a light aircraft, *J. Geophys. Res.*, *108* (D19), 8587, doi:10.1029/2002JD002544, 2003.

Marion, G.M., and R.E. Farren, Mineral solubilities in the Na-K-Mg-Ca-Cl-SO₄-H₂O system: A re-evaluation of the sulfate chemistry in the Spencer-Møller-Weare model, *Geochimica et Cosmochimica Acta*, *63* (9), 1305-1318, 1999.

Marks, C.H., Measurements of the terminal velocity of bubbles rising in a chain, *J. Fluids Eng.*, *95*, 17-22, 1973.

Marks, R., Preliminary investigations on the influence of rain on the production, concentration, and vertical distribution of sea salt aerosol, *J. Geophys. Res.*, *95* (C12), 22299-22304, 1990.

Markson, R., J. Sedlacek, and C.W. Fairall, Turbulent transport of electrical charge in the marine boundary layer, *J. Geophys. Res.*, *86* (C12), 12115-12121, 1981.

Marrucci, G., and L. Nicodemo, Coalescence of gas bubbles in aqueous solutions of inorganic electrolytes, *Chem. Eng. Sci.*, *22*, 1257-1265, 1967.

Marston, P.L., D.S. Langley, and D.L. Kingsbury, Light scattering by bubbles in liquids: Mie theory, physical-optics approximations, and experiments, *Applied Sci. Res.*, *38*, 373-383, 1982.

Martens, C.S., J.J. Wesolowski, R.C. Harriss, and R. Kaifer, Chlorine loss from Puerto Rican and San Francisco Bay area marine aerosols, *J. Geophys. Res.*, *78* (36), 8778-8792, 1973.

Mårtensson, E.M., E.D. Nilsson, G. de Leeuw, L.H. Cohen, and H.-C. Hansson, Laboratory simulations and parameterization of the primary marine aerosol production, *J. Geophys. Res.*, *108* (D9), 4297, doi:10.1029/2002JD002263, 2003.

Martin, G.M., D.W. Johnson, and A. Spice, The measurement and parameterization of effective radius of droplets in warm stratocumulus clouds, *J. Atmos. Sci.*, *51* (13), 1823-1842, 1994.

Martin, J.H., and S.E. Fitzwater, Dissolved organic carbon in the Atlantic, Southern and Pacific oceans, *Nature*, *356*, 699-700, 1992.

Martin, J.M., A.J. Thomas, and C. Jeadel, Transfert atmosphérique des radionucléides artificiels de la mer vers le continent, *Oceanologica Acta*, *4* (3), 263-266, 1981.

Martin, S., R. Drucker, and M. Fort, A laboratory study of frost flower growth on the surface of young sea ice, *J. Geophys. Res.*, *100* (C4), 7027-7036, 1995.

Martin, S., Y. Yu, and R. Drucker, The temperature dependence of frost flower growth on laboratory sea ice and the effects of the flowers on infrared observations of the surface, *J. Geophys. Res.*, *101* (C5), 12111-12125, 1996.

Martin, S.T., Phase transitions of aqueous atmospheric particles, *Chem. Rev.*, *100*, 3403-3453, 2000.

Marty, J.C., A. Saliot, P. Buat-Ménard, R. Chesselet, and K.A. Hunter, Relation between the lipid compositions of marine aerosols, the sea surface microlayer, and subsurface water, *J. Geophys. Res.*, *84* (C9), 5707-5716, 1979.

Mason, B.J., Bursting of air bubbles at the surface of sea water, *Nature*, *174*, 470 471, 1954.

Mason, B.J., The oceans as a source of cloud-forming nuclei, *Geofisica Pura e Applicata*, *36*, 148-155, 1957a.

Mason, B.J., The nuclei of atmospheric condensation, *Geofisica Pura e Applicata*, *36*, 9-20, 1957b.

Mason, B.J., *The Physics of Clouds*, 671 pp., Clarendon Press, Oxford, 1971.

Mason, B.J., The role of sea-salt particles as cloud condensation nuclei over the remote oceans, *Quar. J. Royal Meteorol. Soc.*, *127*, 2023-2032, 2001.

Mason, B.J., and F.H. Ludlam, The microphysics of clouds, *Rep. Progr. Phys.*, *14*, 147-195, 1951.

Mason, B.J., and J. Maybank, The fragmentation and electrification of freezing water drops, *Quar. J. Royal Meteorol. Soc.*, *86*, 176-186, 1960.

Mason, R., T. Torgersen, W.F. Fitzgerald, M.P. Dowling, J. Kim, E.C. Monahan, M.B. Wilson, and D.K. Woolf, The role of breaking waves in the control of the gas exchange properties of the sea surface, in *Oceanic Whitecaps and the Fluxes of Droplets from, Bubbles to, and Gases through, the Sea Surface. Whitecap Report No. 4*, edited by E.C. Monahan, R.J. Cipriano, W.F. Fitzgerald, R. Marks, R. Mason, P.F. Nolan, T. Torgersen, M.B. Wilson, and D.K. Woolf, Marine Sciences Institute, University of Connecticut, Groton, CT, 1988.

Massman, W.J., The attenuation of concentration fluctuations in turbulent flow through a tube, *J. Geophys. Res.*, *96* (D8), 15269-15273, 1991.

Massouh, L., and O. le Calve, Measurement of whitecap coverage during F.E.T.C.H. 98 experiment, *J. Aerosol Sci.*, *30*, S177-S178, 1999.

Maxwell, J.C., Illustrations of the dynamical theory of gases, *Philosoph. Mag.*, 1860.

Maxwell, J.C., Diffusion, in *Encyclopedia Brittanica*, pp. 82, 1877.

May, K.R., Comments on "Retardation of water drop evaporation with monomolecular surface films", *J. Atmos. Sci.*, *29*, 784-785, 1972.

Mayewski, P.A., L.D. Meeker, S. Whitlow, M.S. Twickler, M.C. Morrison, R.B. Alley, P. Bloomfield, and K. Taylor, The atmosphere during the Younger Dryas, *Science*, *261*, 195-197, 1993.

Mayewski, P.A., L.D. Meeker, S. Whitlow, M.S. Twickler, M.C. Morrison, P. Bloomfield, G.C. Bond, R.B. Alley, A.J. Gow, P.M. Grootes, D.A. Meese, M. Ram, K.C. Taylor, and W. Wumkes, Changes in atmospheric circulation and ocean ice cover over the North Atlantic during the last 41,000 years, *Science, 263,* 1747-1751, 1994.

Mayewski, P.A., L.D. Meeker, M.S. Twickler, S. Whitlow, Q. Yang, W.B. Lyons, and M. Prentice, Major features and forcing of high-latitude northern hemisphere atmospheric circulation using a 110,000-year-long glaciochemical series, *J. Geophys. Res., 102* (C12), 26345-26366, 1997.

McAlister, D., The law of the geometric mean, *Proc. Royal Soc. London, 29,* 367-376, 1879.

McBain, J.W., and R. DuBois, Further experimental tests of the Gibbs adsorption theorem. The structure of the surface of ordinary solutions, *J. Am. Chem. Soc., 51,* 3534-3549, 1929.

McConnell, S.O., M.P. Schilt, and J.G. Dworski, Ambient noise measurements from 100 Hz to 80 kHz in an Alaskan fjord, *J. Acoustical Soc. Am., 91* (4), 1990-2003, 1992.

McCoy, B.J., Evaporation of water through surfactant layers, *Am. Inst. Chem. Eng. J., 28* (5), 844-847, 1982.

McDaniel, S.T., Sea surface reverberation: A review, *J. Acoustical Soc. Am., 94* (4), 1905-1922, 1993.

McDonald, J.E., Remarks on "Sea salt in a tropical storm", *J. Meteorol., 8,* 362-363, 1951.

McDonald, R.L., C.K. Unni, and R.A. Duce, Estimation of atmospheric sea salt dry deposition: wind speed and particle size dependence, *J. Geophys. Res., 87* (C2), 1246-1250, 1982.

McGann, B.T., and S.G. Jennings, The efficiency with which drizzle and precipitation sized drops collide with aerosol particles, *Atmos. Environ., 25A* (3/4), 791-799, 1991.

McGovern, F.M., A. Krasenbrink, S.G. Gennings, B. Georgi, T.G. Spain, M. Below, and T.C. O'Connor, Mass measurements of aerosol at Mace Head, on the west coast of Ireland, *Atmos. Environ., 28* (7), 1311-1318, 1994.

McGraw, R., P.I. Huang, and S.E. Schwartz, Optical properties of atmospheric aerosols from moments of the particle size distribution, *Geophys. Res. Lttrs, 22,* 2929-2932, 1995.

McInnes, L.M., D.S. Covert, P.K. Quinn, and M.S. Germani, Measurements of chloride depletion and sulfur enrichment in individual sea-salt particles collected from the remote marine boundary layer, *J. Geophys. Res., 99* (D4), 8257-8268, 1994.

McInnes, L.M., P.K. Quinn, D.S. Covert, and T.L. Anderson, Gravimetric analysis, ionic composition, and associated water mass of the marine aerosol, *Atmos. Environ., 30* (6), 869-884, 1996.

McInnes, L., D. Covert, and B. Baker, The number of sea-salt, sulfate, and carbonaceous particles in the marine atmosphere: EM measurements consistent with the ambient size distribution, *Tellus, 49B,* 300-313, 1997.

McInnes, L., M. Bergin, J. Ogren, and S. Schwartz, Apportionment of light scattering and hygroscopic growth to aerosol composition, *Geophys. Res. Lttrs, 25* (4), 513-516, 1998.

McKay, W.A., and N.J. Pattenden, The transfer of radionuclides from sea to land via the air: A review, *J. Environ. Radioactivity, 12,* 49-77, 1990.

McKay, W.A., J.A. Garland, D. Livesley, C.M. Halliwell, and M.I. Walker, The characteristics of the shore-line sea spray aerosol and the landward transfer of radionuclides discharged to coastal sea water, *Atmos. Environ., 28* (20), 3299-3309, 1994.

McMurry, P.H., A review of atmospheric aerosol measurements, *Atmos. Environ., 34,* 1959-1999, 2000.

McNown, J.S., and J. Malaika, Effects of particle shape on settling velocity at low Reynolds numbers, *Trans. Am. Geophys. Union, 31* (1), 74-82, 1950.

Medrow, R.A., and B.T. Chao, Charges on jet drops produced by bursting bubbles, *J. Colloid Interface Sci., 35,* 683-688, 1971.

Medwin, H., In situ acoustic measurements of bubble populations in coastal ocean waters, *J. Geophys. Res., 75* (3), 599-611, 1970.

Medwin, H., In situ acoustic measurements of microbubbles at sea, *J. Geophys. Res., 82* (6), 971-976, 1977a.

Medwin, H., Acoustic determinations of bubble-size spectra, *J. Acoustical Soc. Am., 62* (4), 1041-1042, 1977b.

Medwin, H., and M.M. Beaky, Bubble sources of the Knudsen sea noise spectra, *J. Acoustical Soc. Am., 86* (3), 1124-1130, 1989.

Medwin, H., and N.D. Breitz, Ambient and transient bubble spectral densities in quiescient seas and under spilling breakers, *J. Geophys. Res., 94* (C9), 12751-12759, 1989.

Medwin, H., and A.C. Daniel, Jr., Acoustical measurements of bubble production by spilling breakers, *J. Acoustical Soc. Am., 88* (1), 408-412, 1990.

Medwin, H., A. Kurgan, and J.A. Nystuen, Impact and bubble sound from raindrops at normal and oblique incidence, *J. Acoustical Soc. Am., 88* (1), 413-418, 1990.

Medwin, H., J.A. Nystuen, P.W. Jacobus, L.H. Ostwald, and D.E. Snyder, The anatomy of underwater noise, *J. Acoustical Soc. Am., 92* (3), 1613-1623, 1992.

Meirink, J.F., and V.K. Makin, The impact of sea spray evaporation in a numerical weather prediction model, *J. Atmos. Sci., 58* (23), 3626-3638, 2001.

Meissner, T., D. Smith, and F. Wentz, A 10 year intercomparison between collocated Special Sensor Microwave Imager oceanic surface wind speed retrievals and global analyses, *J. Geophys. Res., 106* (C6), 11731-11742, 2001.

Melville, W.K., and P. Matusov, Distribution of breaking waves at the ocean surface, *Nature, 417,* 58-63, 2002.

Melville, W.K., M.R. Loewen, F.C. Felizardo, A.T. Jessup, and M.J. Buckingham, Acoustic and microwave signatures of breaking waves, *Nature, 336,* 54-56, 1988.

Memery, L., and L. Merlivat, Modelling of gas flux through bubbles at the air-water interface, *Tellus, 37B,* 272-285, 1985.

Mendelson, H.D., The prediction of bubble terminal velocities from wave theory, *Am. Inst. Chem. Eng. J., 13* (2), 250-253, 1967.

Merlivat, L., and L. Memery, Gas exchange across an air-water interface: experimental results and modeling of bubble contribution to transfer, *J. Geophys. Res., 88* (C1), 707-724, 1983.

Merrington, A.C., and E.G. Richardson, The break-up of liquid jets, *Proc. Phys. Soc., 59* (1), 1-13, 1947.

Merry, M., and H.A. Panofsky, Statistics of vertical motion over land and water, *Quar. J. Royal Meteorol. Soc., 102,* 255-260, 1976.

Mestayer, P., and C. Lefauconnier, Spray droplet generation, transport, and evaporation in a wind wave tunnel during the Humidity Over the Sea Experiments in the simulation tunnel, *J. Geophys. Res.*, *93* (C1), 572-586, 1988.

Mestayer, P.G., A.M.J. Van Eijk, G. De Leeuw, and B. Tranchant, Numerical simulation of the dynamics of sea spray over the waves, *J. Geophys. Res.*, *101* (C9), 20771-20797, 1996.

Mészáros, A., and K. Vissy, Concentration, size distribution and chemical nature of atmospheric aerosol particles in remote oceanic areas, *Aerosol Sci.*, *5*, 101-109, 1974.

Mészáros, E., On the atmospheric input of sulfur into the ocean, *Tellus*, *34*, 277-282, 1982.

Metnieks, A.L., The size spectrum of large and giant sea-salt nuclei under maritime conditions, *Geophys. Bull.*, *15*, 1-50, 1958.

Middlebrook, A.M., D.M. Murphy, and D.S. Thomson, Observations of organic material in individual marine particles at Cape Grim during the First Aerosol Characterization Experiment (ACE-1), *J. Geophys. Res.*, *103* (D13), 16475-16483, 1998.

Middleton, W.E.K., *Vision through the Atmosphere*, 250 pp., University of Toronto Press, Toronto, 1952.

Mie, G., Beiträge zur Optik trüber Medien, speziell kolloidaler Metallösungen, *Annalen der Physik*, *25* (2), 377-445, 1908.

Mihara, Y., Frost protection by fog droplets coated with monomolecular films, *Nature*, *212*, 602-603, 1966.

Milford, J.B., and C.I. Davidson, The sizes of particulate sulfate and nitrate in the atmosphere - a review, *J. Air Pollut. Contr. Assoc.*, *37* (2), 125-134, 1987.

Millard, R.C., and G. Seaver, An index of refraction algorithm for seawater over temperature, pressure, salinity, density, and wavelength, *Deep-Sea Res.*, *37* (12), 1909-1926, 1990.

Miller, D.F., D. Lamb, and A.W. Gertler, SO_2 oxidation in cloud drops containing NaCl or sea salt as condensation nuclei, *Atmos. Environ.*, *21* (4), 991-993, 1987.

Miller, M.A., An investigation of aerosol generation in the marine planetary boundary layer, Masters thesis, Pennsylvania State University, 1987.

Miller, M.A., and C.W. Fairall, A new parameterization of spray droplet production by oceanic whitecaps, in *Seventh Congress on Ocean-Atmosphere Interaction*, pp. 174-177, American Meteorological Society, Anaheim, CA, 1988.

Miller, M.A., M.P. Jensen, and E.E. Clothiaux, Diurnal cloud and thermodynamic variations in the stratocumulus transition regime: A case study using in situ and remote sensors, *J. Atmos. Sci.*, *55*, 2294-2310, 1998.

Millero, F.J., Seawater as a multicomponent electrolyte solution, in *The Sea*, edited by E.D. Goldberg, pp. 3-80, John Wiley and Sons, New York, 1974.

Millero, F.J., Thermodynamics of seawater: the PVT properties, *Ocean Sci. Eng.*, *7*, 403-466, 1982.

Millero, F.J., and W.H. Leung, The thermodynamics of seawater at one atmosphere, *Am. J. Sci.*, *276*, 1035-1077, 1976.

Millero, F.J., and A. Poisson, International one-atmosphere equation of state of seawater, *Deep-Sea Res.*, *28A* (6), 625-625, 1981.

Millero, F.J., and A. Poisson, Errata, *Deep-Sea Res.*, *29* (2A), 284, 1982.

Millero, F.J., and M.L. Sohn, *Chemical Oceanography*, 531 pp., CRC Press, Boca Raton, FL, 1992.

Millero, F.J., G. Perron, and J.E. Desnoyers, Heat capacity of seawater solutions from 5° to 35° C and 0.5 to 22‰ chlorinity, *J. Geophys. Res.*, *78* (21), 4499-4507, 1973.

Millikan, R.A., The general law of fall of a small spherical body through a gas, and its bearing upon the nature of molecular reflection from surfaces, *Physical Rev.*, *22* (1), 1-23, 1923.

Mills, A.F., and R.A. Seban, The condensation coefficient of water, *Internatl. J. Heat Mass Transfer*, *10*, 1815-1827, 1967.

Ming, Y., and L.M. Russell, Predicted hygroscopic growth of sea salt aerosol, *J. Geophys. Res.*, *106* (D22), 28259-28274, 2001.

Minnaert, M., On musical air-bubbles and the sounds of running water, *Philosoph. Mag.*, *Series 7*, *16*, 235-248, 1933.

Mitchell, R.L., Permanence of the log-normal distribution, *J. Optical Soc. Am.*, *58* (9), 1267-1272, 1968.

Mitra, S.K., J. Brinkmann, and H.R. Pruppacher, A wind tunnel study on the drop-to-particle conversion, *J. Aerosol Sci.*, *23* (3), 245-256, 1992.

Miyagi, O., The motion of air bubbles rising in water, *Philosoph. Mag.*, *50*, 112-140, 1925.

Miyahara, T., and T. Takahashi, Drag coefficient of a single bubble rising through a quiescent liquid, *International Chemical Engineering*, *25* (1), 146-148, 1985.

Miyake, M., R.W. Stewart, and R.W. Burling, Spectra and cospectra of turbulence over water, *Quar. J. Royal Meteorol. Soc.*, *96*, 138-143, 1970a.

Miyake, M., M. Donelan, G. McBean, C. Paulson, F. Badgley, and E. Leavitt, Comparison of turbulent fluxes over water determined by profile and eddy correlation measurements, *Quar. J. Royal Meteorol. Soc.*, *96*, 132-137, 1970b.

Miyake, Y., The possibility and the allowable limit of formation of air bubbles in the sea, *Papers in Meteor.*, *2*, 95-101, 1951.

Miyake, Y., and T. Abe, A study on the foaming of sea water. Part 1, *J. Marine Res.*, *7*, 67-73, 1948.

Miyake, Y., and M. Koizumi, The measurement of the viscosity coefficient of sea water, *J. Marine Res.*, *7* (2), 63-66, 1948.

Moldanová, J., and E. Ljungström, Sea-salt aerosol chemistry in coastal areas: A model study, *J. Geophys. Res.*, *106* (D1), 1271-1296, 2001.

Möller, D., On the global natural sulphur emission, *Atmos. Environ.*, *18* (1), 29-39, 1984.

Möller, D., The Na/Cl ratio in rainwater and the seasalt chloride cycle, *Tellus*, *42B*, 254-262, 1990.

Möller, U., and G. Schumann, Mechanisms of transport from the atmosphere to the earth's surface, *J. Geophys. Res.*, *75* (15), 3013-3019, 1970.

Mollo-Christensen, E., Upwind distortion due to probe support in boundary-layer observation, *J. Applied Meteorol.*, *18*, 367-370, 1979.

Monahan, E.C., Sea spray as a function of low elevation wind speed, *J. Geophys. Res.*, *73* (4), 1127-1137, 1968.

Monahan, E.C., Fresh water whitecaps, *J. Atmos. Sci.*, *26*, 1026-1029, 1969.

Monahan, E.C., Reply, *J. Atmos. Sci.*, *27*, 1220-1221, 1970.

Monahan, E.C., Oceanic whitecaps, *J. Phys. Oceanogr., 1*, 139-144, 1971.

Monahan, E.C., Comments on "Variations in Aitken and giant nuclei in marine air", *J. Phys. Oceanogr., 3*, 167-168, 1973.

Monahan, E.C., Comment on 'Bubble and aerosol spectra produced by a laboratory "breaking wave"' by R.J. Cipriano and D.C. Blanchard, *J. Geophys. Res., 87* (C8), 5865-5867, 1982.

Monahan, E.C., The ocean as a source for atmospheric particles, in *The Role of Air-Sea Exchange in Geochemical Cycling*, edited by P. Buat-Menard, pp. 129-163, D. Reidel Publishing Company, Dordrecht, 1986.

Monahan, E.C., Whitecap coverage as a remotely monitorable indication of the rate of bubble injection into the oceanic mixed layer, in *Sea Surface Sound*, edited by B.R. Kerman, pp. 85-96, Kluwer Academic Publishers, Dordrecht, 1988a.

Monahan, E.C., Near-surface bubble concentration and oceanic whitecap coverage, in *Seventh Congress on Ocean-Atmosphere Interaction*, pp. 178-181, American Meteorological Society, Anaheim, CA, 1988b.

Monahan, E.C., Occurrence and evolution of acoustically relevant sub-surface bubble plumes and their associated, remotely monitorable, surface whitecaps, in *Natural Physical Sources of Underwater Sound*, edited by B.R. Kerman, pp. 503-517, Kluwer Academic Publishers, Dordrecht, 1993.

Monahan, E.C., Comments on "Bubbles produced by breaking waves in fresh and salt water", *J. Phys. Oceanogr., 31*, 1931-1932, 2001.

Monahan, E.C., Oceanic whitecaps: Sea surface features detectable via satellite that are indicators of the magnitude of the air-sea gas coefficient, *Proc. Indian Acad. Sci. - Earth Planetary Sci., 111* (3), 315-319, 2002.

Monahan, E.C., and H.G. Dam, Bubbles: An estimate of their role in the global oceanic flux of carbon, *J. Geophys. Res., 106* (C5), 9377-9383, 2001.

Monahan, E.C., and M. Lu, Acoustically relevant bubble assemblages and their dependence on meteorological parameters, *IEEE J. Oceanic Eng., 15* (4), 340-349, 1990.

Monahan, E.C., and C.F. Monahan, The influence of fetch on whitecap coverage as deduced from the Alte Weser lightstation obvserver's log, in *Oceanic Whitecaps and Their Role in Air-Sea Exchange Processes*, edited by E.C. Monahan, and G. MacNiocaill, pp. 275-277, D. Reidel Publishing Company, Dordrecht, 1986.

Monahan, E.C., and I. O'Muircheartaigh, Optimal power-law description of oceanic whitecap coverage dependence on wind speed, *J. Phys. Oceanogr., 10*, 2094-2099, 1980.

Monahan, E.C., and I. O'Muircheartaigh, Improved statement of the relationship between surface wind speed and oceanic whitecap coverage as required for the interpretation of satellite data, in *Oceanography from Space*, edited by J.F.R. Gower, pp. 751-755, Plenum Press, New York, 1981.

Monahan, E.C., and I. O'Muircheartaigh, Reply, *J. Phys. Oceanogr., 12*, 751-752, 1982.

Monahan, E.C., and I. O'Muircheartaigh, Whitecaps and the passive remote sensing of the ocean surface, *Int. J. Remote Sensing, 7* (5), 627-642, 1986.

Monahan, E.C., and I. O'Muircheartaigh, Comments on "Albedos and glitter patterns of a wind-roughened sea surface", *J. Phys. Oceanogr., 17* (4), 549-550, 1987.

Monahan, E.C., and M.C. Spillane, The role of oceanic whitecaps in air-sea gas exchange, in *Gas Transfer at Water Surfaces*, edited by W. Brutsaert, and G.H. Jirka, pp. 495-503, D. Reidel Publishing Company, Dordrecht, 1984.

Monahan, E.C., and D.K. Woolf, Oceanic whitecaps and their contribution to air-sea exchange and their influence on the MABL. Whitecap Report No. 1, Marine Sciences Institute, University of Connecticut, Groton, CT, 1986.

Monahan, E.C., and D.K. Woolf, Comprehensive model relating the marine aerosol population of the atmospheric boundary layer to the bubble population of the oceanic mixed layer, in *Third International Colloquium on Drops and Bubbles*, edited by T.G. Wang, pp. 451-457, American Institute of Physics, Monterey, CA, 1988.

Monahan, E.C., and D.K. Woolf, Comments on "Variations in whitecap coverage with wind stress and water temperature", *J. Phys. Oceanogr., 19*, 706-709, 1989.

Monahan, E.C., and C.R. Zietlow, Laboratory comparisons of fresh-water and salt-water whitecaps, *J. Geophys. Res., 74* (28), 6961-6966, 1969.

Monahan, E.C., B.D. O'Regan, and K.L. Davidson, Marine aerosol production from whitecaps, in *The influence of whitecaps on the marine atmosphere. Annual report for the year ending 30 June 1980*, edited by E.C. Monahan, B.D. O'Regan, and D.M. Doyle, University College, Galway, Ireland, 1980.

Monahan, E.C., I. O'Muircheartaigh, and M.P. Fitzgerald, Determination of surface wind speed from remotely measured whitecap coverage, a feasibility assessment, in *Proceedings of an EARSeL-ESA Symposium: Application of Remote Sensing Data on the Continental Shelf*, pp. 103-109, European Space Agency, Voss, Norway, 1981.

Monahan, E.C., K.L. Davidson, and D.E. Spiel, Whitecap aerosol productivity deduced from simulation tank measurements, *J. Geophys. Res., 87* (C11), 8898-8904, 1982.

Monahan, E.C., C.W. Fairall, K.L. Davidson, and P.J. Boyle, Observed inter-relations between 10 m winds, ocean whitecaps and marine aerosols, *Quar. J. Royal Meteorol. Soc., 109*, 379-392, 1983a.

Monahan, E.C., D.E. Spiel, and K.L. Davidson, Model of marine aerosol generation via whitecaps and wave disruption, in *Ninth Conference on Aerospace and Aeronautical Meteorology*, pp. 147-152, American Meteorological Society, Omaha, NE, 1983b.

Monahan, E.C., P.A. Bowyer, and M.C. Spillane, The temperature dependence of whitecap aerosol productivity and the implications for regional sea-air salt fluxes, *Terra Cognita, 4*, 347-348, 1984a.

Monahan, E.C., M.C. Spillane, P.A. Bowyer, M.R. Higgins, and P.J. Stabeno, Whitecaps and the Marine Atmosphere, pp. 103, University College, Galway, Ireland, 1984b.

Monahan, E.C., D.E. Spiel, and K.L. Davidson, A model of marine aerosol generation and wave disruption, in *Oceanic Whitecaps and Their Role in Air-Sea Exchange Processes*, edited by E.C. Monahan, and G. MacNiocaill, pp. 167-174, D. Reidel Publishing Company, Dordrecht, 1986.

Monahan, E.C., M.B. Wilson, and D.K. Woolf, HEXMAX whitecap climatology, in *Oceanic Whitecaps and the Fluxes of Droplets from, Bubbles to, and Gases through, the Sea Surface. Whitecap Report No. 4*, edited by E.C. Monahan, R.J. Cipriano, W.F. Fitzgerald, R. Marks, R. Mason, P.R. Nolan, R. Torgersen, M.B. Wilson, and D.K. Woolf, pp. 20-39, Marine Sciences Institute, University of Connecticut, Groton, CT, 1988.

Monaldo, F., Expected differences between buoy and radar altimeter estimates of wind speed and significant wave height and their implications on buoy-altimeter comparisons, *J. Geophys. Res.*, 93 (C3), 2285-2302, 1988.

Montgomery, R.B., Water characteristics of Atlantic Ocean and of world ocean, *Deep-Sea Res.*, 5, 134-148, 1958.

Moore, C.J., Frequency response corrections for eddy correlation systems, *Boundary-Layer Meteorol.*, 37, 17-35, 1986.

Moore, D.J., Measurements of condensation nuclei over the North Atlantic, *Quar. J. Royal Meteorol. Soc.*, 78, 596-602, 1952.

Moore, D.J., and B.J. Mason, The concentration, size distribution and production rate of large salt nuclei over the oceans, *Quar. J. Royal Meteorol. Soc.*, 80, 583-590, 1954.

Moore, D.W., The rise of a gas bubble in a viscous liquid, *J. Fluid Mechanics*, 6, 113-130, 1959.

Moore, D.W., The boundary layer on a spherical gas bubble, *J. Fluid Mechanics*, 16, 161-176, 1963.

Moore, D.W., The velocity of rise of distorted gas bubbles in a liquid of small viscosity, *J. Fluid Mechanics*, 23 (4), 749-766, 1965.

Moore, K.D., K.J. Voss, and H.R. Gordon, Spectral reflectance of whitecaps: Instrumentation, calibration, and performance in coastal waters, *J. Atmos. Oceanic Technol.*, 15, 496-509, 1998.

Moore, K.D., K.J. Voss, and H.R. Gordon, Spectral reflectance of whitecaps: Their contribution to water-leaving radiance, *J. Geophys. Res.*, 105 (C3), 6493-6499, 2000.

Morcillo, M., B. Chico, E. Otero, and L. Mariaca, Effect of marine aerosol on atmospheric corrosion, *Materials Perf.*, 38 (4), 72-77, 1999.

Morcillo, M., B. Chico, L. Mariaca, and E. Otero, Salinity in marine atmosphere corrosion: its dependence on the wind regime existing in the site, *Corrosion Sci.*, 42, 91-104, 2000.

Morelli, J., P. Buat-Ménard, and R. Chesselet, Production expérimentale d'aérosols a la surface de la mer, *J. Recherches Atmos.*, 8, 961-986, 1974.

Moropoulou, A., P. Theoulakis, and T. Chrysophakis, Correlation between stone weathering and environmental factors in marine atmosphere, *Atmos. Environ.*, 29 (8), 895-903, 1995.

Morris, A.W., and J.P. Riley, The direct gravimetric determination of salinity in sea water, *Deep-Sea Res.*, 11, 899-904, 1964.

Motarjemi, M., and G.J. Jameson, Mass transfer from very small bubbles - the optimal bubble size for aeration, *Chem. Eng. Sci.*, 33, 1415-1423, 1978.

Mouri, H., and K. Okada, Shattering and modification of sea-salt particles in the marine atmosphere, *Geophys. Res. Lttrs*, 20 (1), 49-52, 1993.

Mouri, H., K. Okada, and K. Shigehara, Variation of Mg, S, K and Ca contents in individual sea-salt particles, *Tellus*, 45B, 80-85, 1993.

Mouri, H., K. Okada, and S. Takahashi, Giant sulfur-dominated particles in remote marine boundary layer, *Geophys. Res. Lttrs*, 22 (5), 595-598, 1995.

Mouri, H., I. Nagao, K. Okada, S. Koga, and H. Tanaka, Elemental compositions of individual aerosol particles collected over the Southern Ocean: A case study, *Atmos. Res.*, 43, 183-195, 1997.

Mouri, H., I. Nagao, K. Okada, S. Koga, and H. Tanaka, Individual-particle analyses of coastal Antarctic aerosols, *Tellus*, 51B, 603-611, 1999.

Mozurkewich, M., Aerosol growth and the condensation coefficient of water: a review, *Aerosol Sci. Technol.*, 5, 223-236, 1986.

Mozurkewich, M., Mechanisms for the release of halogens from sea-salt particles by free radical reactions, *J. Geophys. Res.*, 100 (D7), 14199-14207, 1995.

Mugele, R.A., and H.D. Evans, Droplet size distribution in sprays, *Indust. Eng. Chem.*, 43, 1317-1324, 1951.

Mulhearn, P.J., Distribution of microbubbles in coastal waters, *J. Geophys. Res.*, 86 (C7), 6429-6434, 1981.

Munk, W.H., Wind stress on water: an hypothesis, *Quar. J. Royal Meteorol. Soc.*, 81, 320-332, 1955.

Murayama, T., H. Okamoto, N. Kaneyasu, H. Kamataki, and K. Miura, Application of lidar depolarization measurement in the atmospheric boundary layer: Effects of dust and sea-salt particles, *J. Geophys. Res.*, 104 (D24), 31781-31792, 1999.

Murphy, D.M., D.S. Thomson, and A.M. Middlebrook, Bromine, iodine, and chlorine in single aerosol particles at Cape Grim, *Geophys. Res. Lttrs*, 24 (24), 3197-3200, 1997.

Murphy, D.M., J.R. Anderson, P.K. Quinn, L.M. McInnes, F.J. Brechtel, S.M. Kreidenweis, A.M. Middlebrook, M. Pósfai, D.S. Thomson, and P.R. Buseck, Influence of sea-salt on aerosol radiative properties in the Southern Ocean marine boundary layer, *Nature*, 392, 62-65, 1998a.

Murphy, D.M., D.S. Thomson, A.M. Middlebrook, and M.E. Schein, In situ single-particle characterization at Cape Grim, *J. Geophys. Res.*, 103 (D13), 16485-16491, 1998b.

Musgrave, D.L., J. Chou, and W.J. Jenkins, Application of a model of upper-ocean physics for studying seasonal cycles of oxygen, *J. Geophys. Res.*, 93 (C12), 15679-15700, 1988.

Nabavian, K., and L.A. Bromley, Condensation coefficient of water, *Chem. Eng. Sci.*, 18, 651-660, 1963.

Najjar, R.G., and R.F. Keeling, Analysis of the mean annual cycle of the dissolved oxygen anomaly in the World Ocean, *J. Marine Res.*, 55, 117-151, 1997.

Napari, M., H. Noppel, H. Vehkamäki, and M. Kulmala, Parametrization of ternary nucleation rates for H_2SO_4-NH_3-H_2O vapors, *J. Geophys. Res.*, 107 (D19), 10.1029/2002JD002132, 2002.

Neiburger, M., and M.G. Wurtele, On the nature and size of particles in haze, fog, and stratus of the Los Angeles region, *Chem. Rev.*, 44, 321-335, 1949.

Nelis, P.M., D. Branford, and M.H. Unsworth, A model of the transfer of radioactivity from sea to land in sea spray, *Atmos. Environ.*, 28 (20), 3213-3223, 1994.

Nelson, K.H., and T.G. Thompson, Deposition of salts from sea water by frigid concentration, *J. Marine Res.*, 13, 166-182, 1954.

Neubauer, K.R., M.V. Johnston, and A.S. Wexler, Humidity effects on the mass spectra of single aerosol particles, *Atmos. Environ.*, *32* (14/15), 2521-2529, 1998.

Newitt, D.M., Reply to Bursting of bubbles at an air-water interface, *Nature*, *173*, 1048-1049, 1954.

Newitt, D.M., N. Dombrowski, and F.H. Knelman, Liquid entrainment. 1. The mechanism of drop formation from gas or vapour bubbles, *Trans. Inst. Chem. Eng.*, *32*, 244-261, 1954.

Nguyen, A.V., Prediction of bubble terminal velocities in contaminated water, *Am. Inst. Chem. Eng. J.*, *44* (1), 226-230, 1998.

Nguyen, B.C., N. Mihalopoulos, J.P. Putaud, A. Gaudry, W.C. Keene, and J.N. Galloway, Covariations in oceanic dimethyl sulfide, its oxidation products and rain acidity at Amsterdam Island in the Southern Indian Ocean, *J. Atmos. Chem.*, *15*, 39-53, 1992.

Nicholls, S., The dynamics of stratocumulus: aircraft observations and comparisons with a mixed layer model, *Quar. J. Royal Meteorol. Soc.*, *110*, 783-820, 1984.

Nicholson, K.W., The dry deposition of small particles: a review of experimental measurements, *Atmos. Environ.*, *22* (12), 2653-2666, 1988.

Nicolas, J.-M., P.-Y. Deschamps, and R. Frouin, Spectral reflectance of oceanic whitecaps in the visible and near infrared: Aircraft measurements over open ocean, *Geophys. Res. Lttrs*, *28* (23), 4445-4448, 2001.

Nie, K., T.E. Kleindienst, R.R. Arnts, and J.E.I. Sickles, The design and testing of a relaxed eddy accumulation system, *J. Geophys. Res.*, *100* (D6), 11415-11423, 1995.

Nilsson, E.D., and Ü. Rannik, Turbulent aerosol fluxes over the Arctic Ocean. 1. Dry deposition over sea and pack ice, *J. Geophys. Res.*, *106* (D23), 32125-32137, 2001.

Nilsson, E.D., Ü. Rannik, E. Swietlicki, C. Leck, P.P. Aalto, J. Zhou, and M. Norman, Turbulent fluxes over the Arctic Ocean. 2. Wind-driven sources from the sea, *J. Geophys. Res.*, *106* (D23), 32139-32154, 2001.

Nolan, P.F., Decay characteristics of individual whitecaps, in *Oceanic Whitecaps and the Fluxes of Droplets from, Bubbles to, and Gases through, the Sea Surface. Whitecap Report No. 4*, edited by E.C. Monahan, R.J. Cipriano, W.F. Fitzgerald, R. Marks, R. Mason, P.R. Nolan, R. Torgersen, M.B. Wilson, and D.K. Woolf, pp. 41-56, Marine Sciences Institute, University of Connecticut, Groton, CT, 1988a.

Nolan, P.F., Variation of apparent whitecap coverage as a consequence of selecting different techniques for image analysis, in *Oceanic Whitecaps and the Fluxes of Droplets from, Bubbles to, and Gases through, the Sea Surface. Whitecap Report No. 4*, edited by E.C. Monahan, R.J. Cipriano, W.F. Fitzgerald, R. Marks, R. Mason, P.R. Nolan, R. Torgersen, M.B. Wilson, and D.K. Woolf, pp. 137-140, Marine Sciences Institute, University of Connecticut, Groton, CT, 1988b.

Nordberg, W., J. Conaway, D.B. Ross, and T. Wilheit, Measurements of microwave emission from a foam-covered, wind-driven sea, *J. Atmos. Sci.*, *28*, 429-435, 1971.

Novarini, J.C., and D.R. Bruno, Effects of the sub-surface bubble layer on sound propogation, *J. Acoustical Soc. Am.*, *72* (2), 510-514, 1982.

Nukiyama, S., and Y. Tanasawa, Experiments on atomization of liquid by means of air stream, *Trans. Soc. Mech. Eng Japan*, *4* (14), 86-93, 1938.

Nystuen, J.A., Rainfall measurements using underwater ambient noise, *J. Acoustical Soc. Am.*, *79* (4), 972-982, 1986.

Nystuen, J.A., A note on the attenuation of gravity waves by rainfall, *J. Geophys. Res.*, *95* (C10), 18353-18355, 1990.

Nystuen, J.A., and D.M. Farmer, The influence of wind on the underwater sound generated by light rain, *J. Acoustical Soc. Am.*, *82* (1), 270-274, 1987.

Nystuen, J.A., and H. Medwin, Underwater sound produced by rainfall: secondary splashes of aerosols, *J. Acoustical Soc. Am.*, *97* (3), 1606-1613, 1995.

O'Brien, M.P., Review of the theory of turbulent flow and its relation to sediment-transportation, *Trans. Am. Geophys. Un.*, *14*, 487-491, 1933.

O'Brien, M.P., and J.E. Gosline, Velocity of large bubbles in vertical tubes, *Indust. Eng. Chem.*, *27*, 1436-1440, 1935.

Ochi, M.K., and C.-H. Tsai, Prediction of occurrence of breaking waves in deep water, *J. Phys. Oceanogr.*, *13*, 2008-2019, 1983.

Oddie, B.C.V., The variation in composition of sea-salt nuclei with mode of formation, *Quar. J. Royal Meteorol. Soc.*, *86*, 549-551, 1960.

O'Dowd, C.D., Biogenic coastal aerosol production and its influence on aerosol radiative properties, *J. Geophys. Res.*, *106* (D2), 1545-1549, 2001.

O'Dowd, C.D., and M.H. Smith, Physicochemical properties of aerosols over the Northeast Atlantic: evidence for wind-speed-related submicron sea-salt aerosol production, *J. Geophys. Res.*, *98* (D1), 1137-1149, 1993.

O'Dowd, C.D., M.H. Smith, I.E. Consterdine, and J.A. Lowe, Marine aerosol, sea-salt, and the marine sulphur cycle: a short review, *Atmos. Environ.*, *31* (1), 73-80, 1997.

O'Dowd, C.D., J.A. Lowe, and M.H. Smith, Coupling sea-salt and sulphate interactions and its impact on cloud droplet concentration predictions, *Geophys. Res. Lttrs*, *26* (9), 1311-1314, 1999a.

O'Dowd, C.D., J.A. Lowe, M.H. Smith, and A.D. Kaye, The relative importance of non-sea-salt sulphate and sea-salt aerosol to the marine cloud condensation nuclei population: An improved multi-component aerosol-cloud droplet parameterization, *Quar. J. Royal Meteorol. Soc.*, *125*, 1295-1313, 1999b.

O'Dowd, C.D., J.A. Lowe, and M.H. Smith, Observations and modelling of aerosol growth in marine stratocumulus - case study, *Atmos. Environ.*, *33*, 3053-3062, 1999c.

O'Dowd, C.D., E. Becker, and M. Kulmala, Mid-latitude North-Atlantic aerosol characteristics in clean and polluted air, *Atmos. Res.*, *58*, 167-185, 2001.

Ogren, J.A., A systematic approach to *in situ* observations of aerosol properties, in *Aerosol Forcing of Climate*, edited by R.J. Charlson, and J. Heintzenberg, pp. 215-226, John Wiley and Sons, New York, 1995.

Oguz, H.N., A theoretical study of low-frequency oceanic ambient noise, *J. Acoustical Soc. Am.*, *95* (4), 1895-1912, 1994.

Oguz, H.N., and A. Prosperetti, Bubble entrainment by the impact of drops on liquid surfaces, *J. Fluid Mechanics*, *219*, 143-179, 1990.

Oguz, H.N., and A. Prosperetti, Numerical calculation of the underwater noise of rain, *J. Fluid Mechanics*, *228*, 417-442, 1991.

Ohba, R., K. Okabayashi, M. Yamamoto, and M. Tsuru, A method for predicting the content of sea salt particles in the atmosphere, *Atmos. Environ.*, *24A* (4), 925-935, 1990.

O'Hern, T.J., L. d'Agostino, and A.J. Acosta, Comparison of holographic and Coulter counter measurements of cavitation nuclei in the ocean, *J. Fluids Eng.*, *110*, 200-207, 1988.

Okita, T., M. Okuda, K. Murano, T. Itoh, I. Kanazawa, M. Hirota, H. Hara, Y. Hashimoto, S. Tsunogai, S. Ohta, and Y. Ikebe, The characterization and distribution of aerosol and gaseous species in the winter monsoon over the Western Pacific Ocean, *J. Atmos. Chem.*, *4*, 343-358, 1986.

Okuda, S., and S. Hayami, Experiments on evaporation from wavy water surface, *Rec. Oceanogr. Works Japan*, *5* (1), 6-13, 1959.

Olson, R.M., *Essentials of Engineering Fluid Mechanics*, International Textbook Company, Scranton, PA, 1961.

O'Muircheartaigh, I.G., and E.C. Monahan, Statistical aspects of the relationship between oceanic whitecap coverage wind speed and other environmental factors, in *Oceanic Whitecaps and Their Role in Air-Sea Exchange Processes*, edited by E.C. Monahan, and G. MacNiocaill, pp. 125-128, D. Reidel Publishing Company, Dordrecht, 1986.

Oncley, S.P., A.C. Delany, T.W. Horst, and P.P. Tans, Verification of flux measurement using relaxed eddy accumulation, *Atmos. Environ.*, *27A* (15), 2417-2426, 1993.

Oost, W.A., Errors in eddy correlation measurements of momentum fluxes and their correction, *J. Marine Sys.*, *4*, 171-181, 1993.

Oosting, H.J., and W.S. Billings, Factors effecting vegetational zonation on coastal dunes, *Ecology*, *23*, 131-142, 1942.

Oppo, C., S. Bellandi, N. Degli Innocenti, A.M. Stortini, G. Loglio, E. Schiavuta, and R. Cini, Surfactant components of marine organic matter as agents for biogeochemical fractionation and pollutant transport via marine aerosols, *Marine Chem.*, *63*, 235-253, 1999.

Orr, C.J., F.K. Hurd, and W.J. Corbett, Aerosol size and relative humidity, *J. Colloid Sci.*, *13*, 472-482, 1958a.

Orr, C., F.K. Hurd, W. Hendrix, and C. Junge, The behavior of condensation nuclei under changing humidities, *J. Meteorol.*, *15*, 240-242, 1958b.

Oseen, C.W., Über den Gültigkeitsbereich der Stokesschen Widerstandsformel, *Arkiv för Matematik, Astronomi och Fysik*, *9* (16), 1-15, 1913.

Otani, Y., and C.S. Wang, Growth and deposition of saline droplets covered with a monolayer of surfactant, *Aerosol Sci. Technol.*, *3*, 155-166, 1984.

Owen, P.R., Dust deposition from a turbulent air stream, *J. Air Water Pollut.*, *3*, 8-25, 1960.

Owen, P.R., Pneumatic transport, *J. Fluid Mechanics*, *39* (2), 407-432, 1969.

Owens, J.S., Condensation of water from the air upon hygroscopic crystals, *Proc. Royal Soc. London*, *110*, 738-752, 1926.

Owens, J.S., Sea-salt and condensation nuclei, *Quar. J. Royal Meteorol. Soc.*, *66*, 2, 1940.

Pakkanen, T.A., Study of formation of coarse particle nitrate aerosol, *Atmos. Environ.*, *30* (14), 2475-2482, 1996.

Pandis, S.N., L.M. Russell, and J.H. Seinfeld, The relationship between DMS flux and CCN concentration in remote marine regions, *J. Geophys. Res.*, *99* (D8), 16945-16957, 1994.

Papachristodoulou, A., F.R. Foulkes, and J.W. Smith, Bubble characteristics and aerosol formation in electrowinning cells, *J. Applied Electrochem.*, *15*, 581-590, 1985.

Park, P.M., and H.J. Exton, The effect of stability on the concentration of aerosol in the marine atmospheric boundary layer, in *Oceanic Whitecaps and Their Role in Air-Sea Exchange Processes*, edited by E.C. Monahan, and G. MacNiocaill, pp. 277, D. Reidel Publishing Company, Dordrecht, 1986.

Park, P.M., M.H. Smith, and H.J. Exton, The effect of mixing height on maritime aerosol concentrations over the North Atlantic Ocean, *Quar. J. Royal Meteorol. Soc.*, *116*, 461-476, 1990.

Parker, B., and G. Barsom, Biological and chemical significance of surface microlayers in aquatic ecosystems, *BioScience*, *20* (2), 87-93, 1970.

Parungo, F., and C. Nagamoto, Recent investigation on marine aerosols, in *Atmospheric Aerosols and Nucleation*, edited by P.E. Wagner, and G. Vali, pp. 269-272, Springer-Verlag, Berlin, 1988.

Parungo, F.P., C.T. Nagamoto, and J.M. Harris, Temporal and spatial variation of marine aerosols over the Atlantic Ocean, *Atmos. Res.*, *20*, 23-37, 1986a.

Parungo, F.P., C.T. Nagamoto, J. Rosinski, and P.L. Haagenson, A study of marine aerosols over the Pacific Ocean, *J. Atmos. Chem.*, *4*, 199-226, 1986b.

Parungo, F.P., C.T. Nagamoto, R. Madel, J. Rosinski, and P.L. Haagenson, Marine aerosols in Pacific upwelling regions, *J. Aerosol Sci.*, *18* (3), 277-290, 1987.

Paterson, M.P., and R.S. Scorer, The chemistry of sea salt aerosol and its measurement, *Nature*, *254*, 491-495, 1975.

Paterson, M.P., and K.T. Spillane, Surface films and the production of sea-salt aerosol, *Quar. J. Royal Meteorol. Soc.*, *95*, 526-534, 1969.

Patro, R.K., I. Leifer, and P. Bowyer, Better bubble process modeling: Improved bubble hydrodynamics parameterization, in *Gas Transfer at Water Surfaces*, edited by M.A. Donelan, W.M. Drennan, E.S. Saltzman, and R. Wanninkhof, pp. 315-320, American Geophysical Union, Washington, DC, 2002.

Pattenden, N.J., R.S. Cambray, and K. Playford, Trace elements in the sea-surface microlayer, *Geochimica et Cosmochimica Acta*, *45*, 93-100, 1981.

Patterson, E.M., C.S. Kiang, A.C. Delany, A.F. Wartburg, A.C.D. Leslie, and B.J. Huebert, Global measurements of aerosols in remote continental and marine regions: concentrations, size distribution, and optical properties, *J. Geophys. Res.*, *85* (C12), 7361-7376, 1980.

Pattey, E., R.L. Desjardins, and P. Rochette, Accuracy of the relaxed eddy-accumulation technique, evaluated using CO_2 flux measurements, *Boundary-Layer Meteorol.*, *66*, 341-355, 1993.

Pattison, M.J., and S.E. Belcher, Production rates of sea-spray droplets, *J. Geophys. Res.*, *104* (C8), 18379-18407, 1999.

Pattle, R.E., The aeration of liquids Part 1. The solution of gas from rising bubbles, *Trans. Inst. Chem. Eng.*, *28*, 27-31, 1950.

Payne, R.E., Albedo of the sea surface, *J. Atmos. Sci.*, *29*, 959-970, 1972.

Peebles, F.N., and H.J. Garber, Studies on the motion of gas bubbles in liquids, *Chem. Eng. Progr.*, *49*, 88-97, 1953.

Penman, H.L., The water cycle, *Scientific Am.*, *223*, 98-108, 1970.

Penndorf, R., Tables of the refractive index for standard air and the Rayleigh scattering coefficient for the spectral region between 0.2 and 20.0 micrometers and their application to atmospheric optics, *J. Optical Soc. Am.*, *47* (3), 176-182, 1957.

Pennell, W.T., and M.A. LeMone, An experimental study of turbulence structure in the fair-weather trade wind boundary layer, *J. Atmos. Sci.*, *31*, 1308-1323, 1974.

Penner, J.E., M. Andreae, H. Annegarn, L. Barrie, J. Feitchter, D. Hegg, A. Hayaraman, R. Leaitch, D. Murphy, J. Nganga, and G. Pitari, Aerosols, their direct and indirect effects, in *Climate Change 2001: The Scientific Basis. Contribution of Working Group I to the Third Assessment Report of the Intergovernmental Panel on Climate Change*, edited by J.T. Houghton, Y. Ding, D.J. Griggs, M. Noguer, P. van der Linden, Z. Dai, and K. Maskell, pp. 289-348, Cambridge University Press, Cambridge, 2001.

Penner, J.E., S.Y. Zhang, M. Chin, C.C. Chuang, J. Geichter, Y. Feng, I.V. Geogdzhayev, P. Ginoux, M. Herzog, A. Higurashi, D. Koch, C. Land, U. Lohmann, M. Mishchenko, T. Nakahima, G. Pitari, B. Soden, I. Tegen, and L. Stowe, A comparison of model- and satellite-derived aerosol optical depth and reflectivity, *J. Atmos. Sci.*, *59*, 441-460, 2002.

Perovich, D.K., and J.A. Richter-Menge, Surface characteristics of lead ice, *J. Geophys. Res.*, *99* (C8), 16341-16350, 1994.

Perrone, A.J., Ambient-noise-spectrum levels as a function of water depth, *J. Acoustical Soc. Am.*, *48*, 362-370, 1970.

Perry, R.J., A.J. Hunt, and D.R. Huffman, Experimental determinations of Mueller scattering matrices for nonspherical particles, *Applied Optics*, *17* (17), 2700-2710, 1978.

Peterson, E.W., and L. Hasse, Did the Beaufort scale or the wind climate change?, *J. Phys. Oceanogr.*, *17*, 1071-1074, 1987.

Peterson, J.T., and C.E. Junge, Sources of particulate matter in the atmosphere, in *Man's Impact on the Climate*, edited by W.J. Matthews, W.W. Kellogg, and G.D. Robinson, pp. 310-320, M.I.T. Press, Cambridge, Mass., 1971.

Petrenchuk, O.P., On the budget of sea salts and sulfur in the atmosphere, *J. Geophys. Res.*, *85* (C12), 7439-7444, 1980.

Pettyjohn, E.S., and E.B. Christiansen, Effect of particle shape on free-settling rates of isometric particles, *Chem. Eng. Progr.*, *44*, 157-172, 1948.

Phillips, F.C., Oceanic salt deposits, *Quar. Rev. Chem. Soc. London*, *1*, 91-111, 1947.

Phillips, O.M., *The Dynamics of the Upper Ocean*, 261 pp., Cambridge University Press, Cambridge, 1966.

Phillips, O.M., Spectral and statistical properties of the equilibrium range in wind-generated gravity waves, *J. Fluid Mechanics*, *156*, 505-531, 1985.

Pielke, R.A., and T.J. Lee, Influence of sea spray and rainfall on the surface wind profile during conditions of strong wind, *Boundary-Layer Meteorol.*, *55*, 305-308, 1991.

Pierson, W.J., Jr., Examples of, reasons for, and consequences of the poor quality of wind data from ships for the marine boundary layer: implications for remote sensing, *J. Geophys. Res.*, *95* (C8), 13313-13340, 1990.

Pilinis, C., J.H. Seinfeld, and D. Grosjean, Water content of atmospheric aerosols, *Atmos. Environ.*, *23* (7), 1601-1606, 1989.

Pinnick, R.G., and H.J. Auvermann, Response characterics of Knollenberg light-scattering aerosol counters, *J. Aerosol Sci.*, *10*, 55-74, 1979.

Pinnick, R.G., D.E. Carroll, and D.J. Hofmann, Polarized light scattered from monodisperse randomly oriented nonspherical aerosol particles: measurements, *Applied Optics*, *15* (2), 384-393, 1976.

Pio, C.A., and D.A. Lopes, Chlorine loss from marine aerosol in a coastal atmosphere, *J. Geophys. Res.*, *103* (D19), 25263-25272, 1998.

Piotrowicz, S.R., R.A. Duce, J.L. Fasching, and C.P. Weisel, Bursting bubbles and their effect on the sea-to-air transport of Fe, Cu and Zn, *Marine Chem.*, *7*, 307-324, 1979.

Pirjola, L., C. O'Dowd, I.M. Brooks, and M. Kulmala, Can new particle formation occur in the clean marine boundary layer?, *J. Geophys. Res.*, *105* (D21), 26531-26546, 2000.

Pockels, A., Surface tension, *Nature*, *43*, 437-439, 1891.

Podzimek, J., Advances in marine aerosol research, *J. Recherches Atmos.*, *14* (1), 35-61, 1980.

Podzimek, J., Size spectra of bubbles in foam patches and of sea salt nuclei over the ocean, *Tellus*, *36B*, 192-202, 1984.

Podzimek, J., and A.N. Saad, Retardation of condensation nuclei growth by surfactant, *J. Geophys. Res.*, *80* (24), 3386-3392, 1975.

Ponche, J.L., C. George, and P. Mirabel, Mass transfer at the air/water interface: Mass accommodation coefficients of SO_2, HNO_3, NO_2, and NH_3, *J. Atmos. Chem.*, *16*, 1-21, 1993.

Pond, S., G.T. Phelps, J.E. Paquin, G. McBean, and R.W. Stewart, Measurements of the turbulent fluxes of momentum, moisture and sensible heat over the ocean, *J. Atmos. Sci.*, *28*, 901-917, 1971.

Poon, Y.-K., S. Tang, and J. Wu, Interactions between rain and wind waves, *J. Phys. Oceanogr.*, *22*, 976-987, 1992.

Porch, W.M., and D.A. Gillette, A comparison of aerosol and momentum mixing in dust storms using fast-response intruments, *J. Applied Meteorol.*, *16*, 1273-1281, 1977.

Pöschl, U., M. Canagaratns, J.T. Jayne, L.T. Moline, D.R. Worsnop, C.E. Kolb, and M.J. Molina, Mass accommodation coefficient of H_2SO_4 vapor on aqueous sulfuric acid surfaces and gaseous diffusion coefficient of H_2SO_4 in N_2/H_2O, *J. Phys. Chem.*, *102A*, 10082-10089, 1998.

Pósfai, M., J.R. Anderson, P.R. Buseck, T.W. Shattuck, and N.W. Tindale, Constituents of a remote Pacific marine aerosol: A TEM study, *Atmos. Environ.*, *28* (10), 1747-1756, 1994.

Pósfai, M., J.R. Anderson, P.R. Buseck, and H. Sievering, Compositional variations of sea-salt-mode aerosol particles from the North Atlantic, *J. Geophys. Res.*, *100* (D11), 23063-23074, 1995.

Pósfai, M., J. Li, J.R. Anderson, and P.R. Buseck, Aerosol bacteria over the Southern Ocean during ACE-1, *Atmos. Res.*, *66*, 231-240, 2003.

Pounder, C., Sodium chloride and water temperature effects on bubbles, in *Oceanic Whitecaps and Their Role in Air-Sea Exchange Processes*, edited by E.C. Monahan, and G. MacNiocaill, pp. 278, D. Reidel Publishing Company, Dordrecht, 1986.

Preobrazhenskii, L.Y., Estimate of the content of spray-drops in the near-water layer of the atmosphere, *Fluid Mechanics - Soviet Research*, *2* (2), 95-100, 1973.

Prideaux, E.B.R., The deliquescence and drying of ammonium and alkali nitrates and a theory of the absorption of water vapour by mixed salts, *Journal of the Society of Chemistry and Industry*, *39* (13), 182T-185T, 1920.

Princen, H.M., Shape of a fluid drop at a liquid-liquid interface, *J. Colloid Sci.*, *18*, 178-195, 1963.

Princen, H.M., and S.G. Mason, Shape of a fluid drop at a fluid-liquid interface I. Extension and test of a two-phase theory, *J. Colloid Sci.*, *29*, 156-172, 1965.

Prodi, F., G. Santachiara, and F. Oliosi, Characterization of aerosols in marine environments (Mediterranean, Red Sea, and Indian Ocean), *J. Geophys. Res.*, *88* (C15), 10957-10968, 1983.

Prosperetti, A., Bubble-related ambient noise in the ocean, *J. Acoustical Soc. Am.*, *84* (3), 1042-1054, 1988.

Prosperetti, A., L.A. Crum, and H.C. Pumphrey, The underwater noise of rain, *J. Geophys. Res.*, *94* (C3), 3255-3259, 1989.

Prospero, J.M., Mineral and sea salt aerosol concentrations in various ocean regions, *J. Geophys. Res.*, *84* (C2), 725-731, 1979.

Prospero, J.M., Long-term measurements of the transport of African mineral dust to the southeastern United States: Implications for regional air quality, *J. Geophys. Res.*, *194* (D13), 15917-15927, 1999.

Prospero, J.M., The chemical and physical properties of marine aerosols. An introduction., in *Chemistry of Marine Water and Sediments*, edited by A. Gianguzza, E. Pellizzetti, and S. Sammarano, pp. 35-82, Springer-Verlag, Berlin, 2002.

Prospero, J.M., D.L. Savoie, R.T. Nees, R.A. Duce, and J. Merrill, Particulate sulfate and nitrate in the boundary layer over the North Pacific Ocean, *J. Geophys. Res.*, *90* (D6), 10586-10596, 1985.

Proudman, I., Appendix. Modified computation of the drag coefficient of a sphere, *J. Fluid Mechanics*, *37* (4), 759-760, 1969.

Proudman, I., and J.R.A. Pearson, Expansions at small Reynolds numbers for the flow past a sphere and a circular cylinder, *J. Fluid Mechanics*, *2*, 237-262, 1957.

Pruppacher, H.R., and K.V. Beard, A wind tunnel investigation of the internal circulation and shape of water drops falling at terminal velocity in air, *Quar. J. Royal Meteorol. Soc.*, *96*, 247-256, 1970.

Pruppacher, H.R., and R. Jaenicke, The processing of water vapor and aerosols by atmospheric clouds, a global estimate, *Atmos. Res.*, *38*, 283-295, 1995.

Pruppacher, H.R., and J.D. Klett, *Microphysics of Clouds and Precipitation*, 954 pp., Kluwer Academic Publishers, Dordrecht, 1997.

Pryor, S.C., and L.L. Sørenson, Nitric acid-sea salt reactions: implications for nitrogen deposition to water surfaces, *J. Applied Meteorol.*, *39*, 725-731, 2000.

Pryor, S.C., R.J. Barthelmie, L.L.S. Geernaert, T. Ellermann, and K.D. Perry, Speciated particle dry deposition to the sea surface: results from ASEPS '97, *Atmos. Environ.*, *33*, 2045-2058, 1999.

Pueschel, R.F., and C.C. Van Valin, The mixed nature of laboratory-produced aerosols from seawater, *J. Recherches Atmos.*, *8*, 601-610, 1974.

Pueschel, R.F., R.J. Charlson, and N.C. Ahlquist, On the amomalous deliquescence of sea-spray aerosols, *J. Applied Meteorol.*, *8*, 995-998, 1969.

Pueschel, R.F., B.A. Bodhaine, and B.G. Mendonca, The proportion of volatile aerosols on the island of Hawaii, *J. Applied Meteorol.*, *12*, 308-215, 1973.

Pumphrey, H.C., and P.A. Elmore, The entrainment of bubbles by drop impacts, *J. Fluid Mechanics*, *220*, 539-567, 1990.

Pumphrey, H.C., L.A. Crum, and L. Bjørnø, Underwater sound produced by individual drop impacts and rainfall, *J. Acoustical Soc. Am.*, *85* (4), 1518-1526, 1989.

Putaud, J.-P., R. van Dingenen, M. Mangoni, A. Virkkula, F. Raes, J. Maring, J.M. Prospero, E. Swietlicki, O.H. Berg, R. Hillamo, and T. Mäkelä, Chemical mass closure and assessment of the origin of the submicron aerosol in the marine boundary layer and the free troposphere at Tenerife during ACE-2, *Tellus*, *52B* (2), 141-168, 2000.

Qian, S. X., J.B. Snow, H.-M. Tzeng, and R.K. Chang, Lasing droplets: Highlighting the liquid-air interface by laser emission, *Science*, *231*, 486-488, 1986.

Quayle, R.G., Climatic comparisons of estimated and measured winds from ships, *J. Applied Meteorol.*, *19*, 142-156, 1980.

Quenzel, H., Determination of size distributions of atmospheric aerosol particles from spectral solar radiation measurements, *J. Geophys. Res.*, *75* (15), 2915-1921, 1970.

Quenzel, H., and M. Kaestner, Optical properties of the atmosphere: calculated variability and application to satellite remote sensing of phytoplankton, *Applied Optics*, *19* (8), 1338-1344, 1980.

Quinby-Hunt, M.S., L.L. Erskine, and A.J. Hunt, Polarized light scattering by aerosols in the marine atmospheric boundary layer, *Applied Optics*, *36* (21), 5168-5184.

Quinn, J.A., R.A. Steinbrook, and J.L. Anderson, Breaking bubbles and the water-to-air transport of particulate matter, *Chem. Eng. Sci.*, *30*, 1177-1184, 1975.

Quinn, P.K., and D.J. Coffman, Local closure during the First Aerosol Characterization Experiment (ACE 1): Aerosol mass concentration and scattering and backscattering coefficients, *J. Geophys. Res.*, *103* (D13), 16575-16596, 1998.

Quinn, P.K., and D.J. Coffman, Comment on "Contribution of different aerosol species to the global aerosol extinction optical thickness: Estimates from model results" by Tegen *et al.*, *J. Geophys. Res.*, *104* (D4), 4241-4248, 1999.

Quinn, P.K., S.F. Marshall, T.S. Bates, D.S. Covert, and V.N. Kapustin, Comparison of measured and calculated aerosol properties relevant to the direct radiative forcing of tropospheric sulfate aerosol on climate, *J. Geophys. Res.*, *100* (D5), 8977-8991, 1995.

Quinn, P.K., V.N. Kapustin, T.S. Bates, and D.S. Covert, Chemical and optical properties of marine boundary layer aerosol particles of the mid-Pacific in relation to sources and meteorological transport, *J. Geophys. Res.*, *101* (D3), 6931-6951, 1996.

Quinn, P.K., D.J. Coffman, V.N. Kapustin, T.S. Bates, and D.S. Covert, Aerosol optical properties in the marine boundary layer during the First Aerosol Characterization Experiment (ACE 1) and the underlying chemical and physical properties, *J. Geophys. Res.*, *103* (D13), 16547-16563, 1998.

Raabe, O., Particle size analysis utilizing grouped data and the log-normal distribution, *Aerosol Sci.*, *2*, 289-303, 1971.

Radke, L.F., J.A.J. Coakley, and M.D. King, Direct and remote sensing observations of the effects of ships and clouds, *Science*, *246*, 1146-1149, 1989.

Radke, L.F., and D. Hegg, The shattering of saline droplets upon crystallization, *J. Recherches Atmos.*, *6*, 447-455, 1972.

Radke, L.F., and P.V. Hobbs, Measurement of cloud condensation nuclei, light scattering coefficient, sodium-containing particles, and Aitken nuclei in the Olympic Mountains of Washington, *J. Atmos. Sci.*, *26*, 281-288, 1969.

Radke, L.F., P.V. Hobbs, and J.E. Pinnons, Observations of cloud condensation nuclei, sodium-containing particles, ice nuclei and the light-scattering coefficient near Barrow, Alaska, *J. Applied Meteorol.*, *15*, 982-995, 1976.

Raemdonck, H., W. Maenhaut, and M.O. Andreae, Chemistry of marine aerosol over the Tropical and Equatorial Pacific, *J. Geophys. Res.*, *91* (D8), 8623-8636, 1986.

Raes, F., R. Van Dingenen, E. Vignati, J. Wilson, J.-P. Putaud, J.H. Seinfeld, and P. Adams, Formation and cycling of aerosols in the global troposphere, *Atmos. Environ.*, *34*, 4215-4240, 2000.

Rahmstorf, S., Improving the accuracy of wind speed observations from ships, *Deep-Sea Res.*, *36* (8), 1267-1276, 1989.

Raman, C.V., and A. Dey, On the sounds of splashes, *Philosoph. Mag., Series 6, 39*, 145-147, 1920.

Ramaswamy, V., O. Boucher, J. Haigh, D. Hauglustaine, J.M. Haywood, G. Myhre, T. Nakajima, G.Y. Shi, and S. Solomon, Radiative forcing of climate change, in *Climate Change 2001: The Scientific Basis. Contribution of Working Group I to the Third Assessment Report of the Intergovernmental Panel on Climate Change*, edited by J.T. Houghton, Y. Ding, D.J. Griggs, M. Noguer, P. van der Linden, Z. Dai, and K. Maskell, pp. 349-416, Cambridge University Press, Cambridge, 2001.

Ramdas, L.A., Oil contamination as a climatic factor, *Nature*, *119*, 652, 1927.

Ramsey, W.L., Dissolved oxygen in shallow near-shore water and its relation to possible bubble formation, *Limnol. Oceanogr.*, *7*, 453-461, 1962a.

Ramsey, W.L., Bubble growth from dissolved oxygen near the sea surface, *Limnol. Oceanogr.*, *7*, 1-7, 1962b.

Rankama, K., and T.G. Sahama, *Geochemistry*, 912 pp., University of Chicago Press, Chicago, 1950.

Rankin, A.M., V. Auld, and E.W. Wolff, Frost flowers as a source of fractionated sea salt aerosol in the polar regions, *Geophys. Res. Lttrs*, *27* (21), 3469-3472, 2000.

Rannik, Ü., and T. Vesala, Autoregressive filtering versus linear detrending in estimation of fluxes by the eddy covariance method, *Boundary-Layer Meteorol.*, *91*, 259-280, 1999.

Ranz, W.E., Some experiments on the dynamics of liquid films, *J. Applied Phys.*, *30*, 1950-1955, 1959.

Ranz, W.E., and W.R.J. Marshall, Evaporation from drops, Part I, *Chem. Eng. Progr.*, *48*, 141-146, 1952a.

Ranz, W.E., and W.R.J. Marshall, Evaporation from drops, Part II, *Chem. Eng. Progr.*, *48*, 173-180, 1952b.

Raoult, F.-M., Loi générale des tensions de vapeur des dissolvants, *Comptes Rendus de l'Academies des Sci., Paris*, *104*, 1430-1433, 1887.

Rayleigh, L., On the scattering of light by small particles, *Philosoph. Mag.*, *41*, 447-454, 1871.

Rayleigh, L., On the instability of jets, *Proc. London Math. Soc.*, *10*, 4-13, 1879.

Rayleigh, L., Foam, *Proc. Royal Inst.*, *13*, 85-97, 1890.

Rayleigh, L., Some applications of photography, *Nature*, *94*, 249-254, 1891.

Rayleigh, L., *The Theory of Sound*, 504 pp., McMillan and Co., London, 1896.

Rayleigh, L., On the transmission of light through an atmosphere containing small particles in suspension, and on the origin of the blue in the sky, *Philosoph. Mag.*, *47*, 375-384, 1899.

Rayleigh, L., Note as to the application of the principle of dynamic similarity, Report for the Advisory Committee for Aeronautics, pp. 38, 1909-1910.

Raymond, D.R., and S.A. Zieminski, Mass transfer and drag coefficients of bubbles rising in dilute aqueous solutions, *Am. Inst. Chem. Eng. J.*, *17* (1), 57-65, 1971.

Rayzer, V.Y., and Y.A. Sharkov, On the dispersed structure of sea foam, *Atmos. Oceanic Phys.*, *16* (7), 548-550, 1980.

Reader, M.C., and N. McFarlane, Sea-salt aerosol distribution during the Last Glacial Maximum and its implications for mineral dust, *J. Geophys. Res.*, *108* (D18), 4253, doi:10.1029/2002JD002063, 2003.

Redfield, A.C., The exchange of oxygen across the sea surface, *J. Marine Res.*, *7*, 347-361, 1948.

Redfield, J.A., and G. Houghton, Mass transfer and drag coefficients for single bubbles at Reynolds numbers of 0.02-5000, *Chem. Eng. Sci.*, *20*, 131-139, 1965.

Reeburgh, W.S., Some implications of the 1940 redefinition of chlorinity, *Deep-Sea Res.*, *13*, 975-976, 1966.

Reeks, M.W., and G. Skyrme, The dependence of particle deposition velocity on particle inertia in turbulent pipe flow, *J. Aerosol Sci.*, *7*, 485-495, 1976.

Reid, J.S., H.H. Jonsson, M.H. Smith, and A. Smirnov, Evolution of the vertical profile and flux of large sea-salt particles in a coastal zone, *J. Geophys. Res.*, *106* (D11), 12039-12053, 2001.

Reid, J.S., J.J. Jonsson, H.B. Maring, A. Smirnov, D.L. Savoie, S.S. Cliff, E.A. Reid, J.M. Livingston, M.M. Meier, O. Dubovik, and S.-C. Tsay, Comparison of size and morphological measurements of coarse mode dust particles from Africa, *J. Geophys. Res.*, *108* (D19), 8593, doi:10.1029/2002JD002485, 2003a.

Reid, J.S., J.E. Kinney, D.L. Westphal, B.N. Holben, E.J. Welton, S.-C. Tsay, D.P. Eleuterio, J.R. Campbell, S.A. Christopher, P.R. Colarco, H.H. Jonsson, J.M. Livingston, H. Maring, M.M. Meier, P. Pilewskie, J.M. Prospero, E.A. Reid, L.A. Remer, P.B. Russell, D.L. Savoie, A. Smirnov, and D. Tanrè, Analysis of measurements of Saharan dust by airborne and ground-based remote sensing methods during the Puerto Rico Dust Experiment (PRIDE), *J. Geophys. Res.*, *108* (D19), 8586, doi:10.1029/2002JD002493, 2003b.

Reinke, N., A. Voßnacke, W. Schütz, M.K. Koch, and H. Unger, Aerosol generation by bubble collapse at ocean surfaces, in *Air-Surface Exchange of Gases and Particles*, edited by D. Fowler, C. Pitcairn, and J.-W. Erisman, pp. 333-340, Kluwer Academic Publishers, Dordrecht, 2001.

Reitan, C.H., and R.R. Braham, Jr., Observations of salt nuclei over the Midwestern United States, *J. Meteorol.*, *11*, 503-506, 1954.

Resch, F., Oceanic air bubbles as generators of marine aerosols, in *Oceanic Whitecaps and Their Role in Air-Sea Exchange Processes*, edited by E.C. Monahan, and G. MacNiocaill, pp. 101-112, D. Reidel Publishing Company, Dordrecht, 1986.

Resch, F., and G. Afeti, Film drop distributions from bubbles bursting in seawater, *J. Geophys. Res.*, *96* (C6), 10681-10688, 1991.

Resch, F., and G. Afeti, Submicron film drop production by bubbles in seawater, *J. Geophys. Res.*, *97* (C3), 3679-3683, 1992.

Resch, F.J., J.S. Darrozes, and G.M. Afeti, Marine liquid aerosol production from bursting of air bubbles, *J. Geophys. Res.*, *91* (C1), 1019-1029, 1986.

Reynolds, O., An experimental investigation of the circumstances which determine whether the motion of water shall be direct or sinuous and the law of resistance in parallel channels, *Philosoph. Trans. Royal Soc. London*, *174A*, 935-982, 1883.

Reynolds, O., On the dynamical theory of incompressible viscous fluids and the determination of the criterion, *Philosoph. Trans. Royal Soc. London*, *186A*, 123-164, 1894.

Reynolds, O., *Papers on Mechanical and Physical Subjects*, The University Press, Cambridge, 1900.

Richardson, C.B., and R.L. Hightower, Evaporation of ammonium nitrate particles, *Atmos. Environ.*, *21* (4), 972-975, 1987.

Richardson, C.B., and T.D. Snyder, A study of heterogeneous nucleation in aqueous solutions, *Langmuir*, *10*, 2462-2465, 1994.

Richardson, C.B., and J.F. Spann, Measurement of the water cycle in a levitated ammonium sulfate particle, *J. Aerosol Sci.*, *15* (5), 563-571, 1984.

Richardson, C.B., H.-B. Lin, R. McGraw, and I.N. Tang, Growth rate measurements for single suspended droplets using the optical resonance method, *Aerosol Sci. Technol.*, *5*, 103-112, 1986.

Richardson, E.G., The impact of a solid on a liquid surface, *Proc. Phys. Soc.*, *61*, 352-367, 1948.

Rideal, E.K., On the influence of thin surface films on the evaporation of water, *J. Phys. Chem.*, *29*, 1585-1588, 1925.

Robbins, R.C., R.D. Cadle, and D.L. Eckhardt, The conversion of sodium chloride to hydrogen chloride in the atmosphere, *J. Meteorol.*, *16*, 53-56, 1959.

Robinson, G.D., Some current projects for global meteorological observation and experiment, *Quar. J. Royal Meteorol. Soc.*, *93*, 409-418, 1967.

Robinson, R.A., The vapour pressure and osmotic equivalence of sea water, *J. Marine Biol. Assoc. UK*, *33*, 449-454, 1954.

Rodhe, H., Modeling Biogeochemical Cycles, in *Global Biogeochem. Cycles*, edited by S.S. Butcher, R.J. Charlson, G.H. Orians, and G.V. Wolfe, pp. 55-72, Academic Press, London, 1992.

Rodhe, H., and J. Grandell, On the removal time of aerosol particles from the atmosphere by precipitation scavenging, *Tellus*, *24*, 442-454, 1972.

Rodriguez, F., and R. Mesler, Some drops don't splash, *J. Colloid Interface Sci.*, *106* (2), 347-352, 1985.

Rodriguez, F., and R. Mesler, The penetration of drop-formed vortex rings into pools of liquid, *J. Colloid Interface Sci.*, *121* (1), 121-129, 1988.

Rohr, J., and R. Detsch, Low sea-state study of the quieting effect of monomolecular films on the underlying ambient-noise field, *J. Acoustical Soc. Am.*, *92* (1), 365-383, 1992.

Rohr, J., and G. Updegraff, The effect of monomolecular films on low sea state ambient noise, in *Natural Physical Sources of Underwater Sound*, edited by B.R. Kerman, pp. 137-150, Kluwer Academic Publishers, Dordrecht, 1993.

Rohr, J., R. Glass, and B. Catile, Effect of monomolecular films on the underlying ocean ambient-noise field, *J. Acoustical Soc. Am.*, *85* (3), 1148-1157, 1989.

Rojas, C.M., R.E. Van Grieken, and R.W. Laane, Comparison of three dry deposition models applied to field measurements in the Southern Bight of the North Sea, *Atmos. Environ.*, *27A* (3), 363-370, 1993.

Rood, M.J., and A.L. Williams, Reply, *J. Atmos. Sci.*, *58*, 1468-1473, 2001.

Rosenfeld, D., R. Lahav, A. Khain, and M. Pinsky, The role of sea spray in cleansing air pollution over the ocean via cloud processes, *Science*, *297* (6), 1667-1670, 2002.

Rosin, P., and E. Rammler, The laws governing the fineness of powdered coal, *Journal of the Institute of Fuel*, *7*, 29-36, 1933.

Rosinski, J., G. Langer, and C.T. Nagamoto, On the ejection of microdroplets from the surface of a freezing water drop, *J. Applied Meteorol.*, *11*, 405-406, 1972.

Ross, D.B., and V. Cardone, Observations of oceanic whitecaps and their relation to remote measurements of surface wind speed, *J. Geophys. Res.*, *79* (3), 444-452, 1974.

Rossby, C.-G., Current problems in meteorology, in *The Atmosphere and the Sea in Motion. Scientific Contributions to the Rossby Memorial Volume*, edited by B. Bolin, pp. 9-50, The Rockefeller Institute Press, New York, 1959.

Rossi, M.J., Heterogeneous reactions on salts, *Chem. Rev.*, *103*, 4823-4882, 2003.

Roth, B., and K. Okada, On the modification of sea-salt particles in the coastal atmosphere, *Atmos. Environ.*, *32* (9), 1555-1569, 1998.

Rouault, M.P., P.G. Mestayer, and R. Schiestel, A model of evaporating spray droplet dispersion, *J. Geophys. Res.*, *96* (C4), 7181-7200, 1991.

Rouse, H.R., Modern conceptions of the mechanics of turbulence, *Transactions of the American Society of Civil Engineers*, *102*, 463-543, 1937.

Rubel, G.O., and J.W. Gentry, Measurements of the kinetics of solution droplets in the presence of adsorbed monolayers: Determination of water accomodation coefficients, *J. Phys. Chem.*, *88*, 3142-3148, 1984.

Rudolf, R., A. Vrtala, M. Kulmala, T. Vesala, Y. Viisanen, and P.E. Wagner, Experimental study of sticking probabilities for condensation of nitric acid-water vapor mixtures, *J. Aerosol Sci.*, *32*, 913-932, 2001.

Ruggles, K.W., The vertical mean wind profile over the ocean for light to moderate winds, *J. Applied Meteorol.*, *9*, 389-395, 1970.

Ruijgrok, W., C.I. Davidson, and K.W. Nicholson, Dry deposition of particles, *Tellus*, *47B*, 587-601, 1995.

Rumscheidt, F.D., and S.G. Mason, Break-up of stationary liquid threads, *J. Colloid Sci.*, *17*, 260-269, 1962.

Russell, L.M., S.N. Pandis, and J.H. Seinfeld, Aerosol production and growth in the marine boundary layer, *J. Geophys. Res.*, *99* (D10), 20989-21003, 1994.

Rybczynski, W., Über die fortschreitende Bewegung einer flüssigen Kugel in einem zähen Medium, *Bulletin Academie de Sciences de Cracovie (Series A)*, 40-46, 1911.

Ryskin, G., and L.G. Leal, Numerical solution of free-boundary problem in fluid mechanics. Part 2. Buoyancy-driven motion of a gas bubble through a quiescent liquid, *J. Fluid Mechanics*, *148*, 19-35, 1984.

Sabra, A.I., Ibn al-Haytham, *Harvard Magazine*, *106* (1), 54, 2003.

Sadasivan, S., Trace elements in size separated aerosols over sea, *Atmos. Environ.*, *12*, 1677-1683, 1978.

Saffman, P.G., On the rise of small air bubbles in water, *J. Fluid Mechanics*, *1*, 249-275, 1956.

Sageev, G., R.C. Flagan, J.H. Seinfeld, and S. Arnold, Condensation rate of water on aqueous droplets in the transition regime, *J. Colloid Interface Sci.*, *113* (2), 421-429, 1986.

Sainsbury, G.L., and I.C. Cheeseman, Effect of rain in calming the sea, *Nature*, *166*, 79, 1950.

Saint-Louis, R., and E. Pelletier, Sea-to-air flux of contaminants via bubbles bursting. An experimental approach for tributyltin, *Marine Chem.*, *84*, 211-224, 2004.

Sakai, M., Ion distribution at a nonequilibrium gas/liquid interface, *J. Colloid Interface Sci.*, *127* (1), 156-166, 1989.

Salhotra, A.M., E.E. Adams, and D.R.F. Harleman, The alpha, beta, gamma of evaporation from saline water bodies, *Water Resources Res.*, *23* (9), 1769-1774, 1987.

Sander, R., and P.J. Crutzen, Model study indicating halogen activation and ozone destruction in polluted air masses transported to the sea, *J. Geophys. Res.*, *101* (D4), 9121-9138, 1996.

Sander, S.P., R.R. Friedel, A.R. Ravishankara, D.M. Golden, C.E. Kolb, M.J. Jurylo, R.E. Huie, V.L. Orkin, M.J. Molina, G.K. Moortgat, and B.J. Finlayson-Pitts, Chemical Kinetics and Photochemical Data for Use in Atmospheric Studies. Evaluation Number 14, NASA Jet Propulsion Laboratory, California Institute of Technology, Pasadena, CA, 2003.

Sandwell, D.T., and R.W. Agreen, Seasonal variation in wind speed and sea state from global satellite measurements, *J. Geophys. Res.*, *89* (C2), 2041-2051, 1984.

Sartor, J.D., and C.E. Abbott, Prediction and measurement of the accelerated motion of water drops in air, *J. Applied Meteorol.*, *14*, 232-239, 1975.

Satheesh, S.K., Aerosol radiative forcing over tropical Indian Ocean: Modulation by sea-surface winds, *Current Sci.*, *82* (3), 310-316, 2002.

Satheesh, S.K., and D. Lubin, Short wave versus long wave radiative forcing by Indian Ocean aerosols: Role of sea-surface winds, *Geophys. Res. Lttrs*, *30* (13), 1695, doi:10.1029/2003GL017499, 2003.

Satheesh, S.K., V. Ramanathan, X. Li-Jones, J.M. Lobert, I.A. Podgorny, J.M. Prospero, B.N. Holben, and N.G. Lowb, A model for the natural and anthropogenic aerosols over the tropical Indian Ocean derived from the Indian Ocean Experiment data, *J. Geophys. Res.*, *104* (D22), 27421-27440, 1999.

Savoie, D.L., and J.M. Prospero, Aerosol concentration statistics for the Northern Tropical Atlantic, *J. Geophys. Res.*, *82* (37), 5954-5964, 1977.

Savoie, D.L., and J.M. Prospero, Water-soluble potassium, calcium, and magnesium in the aerosols over the Tropical North Atlantic, *J. Geophys. Res.*, *85* (C1), 385-392, 1980.

Savoie, D.L., and J.M. Prospero, Particle size distribution of nitrate and sulfate in the marine atmosphere, *Geophys. Res. Lttrs*, *9* (10), 1207-1210, 1982.

Saxena, P., L. Hildemann, P.H. McMurry, and J.H. Seinfeld, Organics alter hygroscopic behavior of atmospheric particles, *J. Geophys. Res.*, *100* (D9), 18755-18770, 1995.

Schacher, G.E., K.L. Davidson, C.W. Fairall, and D.E. Spiel, Calculation of optical extinction from aerosol spectral data, *Applied Optics*, *20* (22), 3951-3957, 1981.

Schack, C.J., Jr., S.E. Pratsinis, and S.K. Friedlander, A general correlation for deposition of suspended particles from turbulent gases to completely rough surfaces, *Atmos. Environ.*, *19* (6), 953-960, 1985.

Schaefer, V.J., Atmospheric electrical measurements in the Hawaiian Islands, *Geofisica Pura e Applicata*, *34*, 211-220, 1956.

Schiller, L., and A. Naumann, Über die grundlegenden Berechnungen bei der Schwerkraftaufbereitung, *Zeitschrift des Vereines deutscher Ingenieure*, *77* (12), 318-320, 1933.

Schlüssel, P., A.V. Soloviev, and W.J. Emery, Cool and freshwater skin of the ocean during rainfall, *Boundary-Layer Meteorol.*, *82*, 437-472, 1997.

Schmidt, J., J. Eastman, and H. Sievering, Influence of relative humidity and sea salt nuclei on the eddy flux determination of small particle dry deposition over the sea, in *Precipitation Scavenging, Dry Deposition, and Resuspension*, edited by H.R. Pruppacher, R.G. Wemonin, and W.G.N. Slinn, pp. 1233-1242, Elsevier, New York, 1983.

Schmidt, W., Der Massenaustauch in Freier Luft und verwandte Erscheinungen, *Probleme der Kosmichen Physik, Band 7*, 1925.

Schmitt, K.F., C.A. Friehe, and C.H. Gibson, Humidity sensitivity of atmospheric temperature sensors by salt contamination, *J. Phys. Oceanogr.*, *8*, 151-161, 1978.

Schudlich, R., and S. Emerson, Gas supersaturation in the surface ocean: The roles of heat flux, gas exchange, and bubbles, *Deep-Sea Res. II*, *43* (2-3), 569-589, 1996.

Schulz, M., G. de Leeuw, and Y. Balkanski, Sea-salt aerosol source functions and emissions, in *Emissions of Atmospheric Trace Compounds*, edited by C. Granier, P. Artaxo, and C.E. Reeves, pp. 333-359, Kluwer Academic Publishers, Dordrecht, 2004.

Schwartz, S.E., Mass-transport considerations pertinent to aqueous phase reaction of gases in liquid-water clouds, in *Chemistry of Multiphase Atmospheric Systems*, edited by W. Jaeschke, Springer-Verlag, Berlin, 1986.

Schwartz, S.E., Are global cloud albedo and climate controlled by marine phytoplankton?, *Nature*, *336*, 441-445, 1988a.

Schwartz, S.E., Mass-transport limitation to the rate of in-cloud oxidation of SO_2: Re-examination in the light of new data, *Atmos. Environ.*, *22* (11), 2491-2499, 1988b.

Schwartz, S.E., and J.E. Freiberg, Mass-transport limitation to the rate of reaction of gases in liquid droplets: Application to oxidation of SO$_2$ in aqueous solutions, *Atmos. Environ.*, *15* (7), 1129-1144, 1981.

Schwartz, S.E., and A. Slingo, Enhanced shortwave cloud radiative forcing due to anthropogenic aerosols, in *Clouds, Chemistry, and Climate - Proceedings of a NATO Advanced Research Workshop*, edited by P. Crutzen, and V. Ramanathan, pp. 191-236, Springer, Heidelberg, 1996.

Schwartz, S.E., Harshvardhan, and C.M. Benkovitz, Influence of anthropogenic aerosol on cloud optical depth and albedo shown by satellite measurements and chemical transport modeling, *Proc. National Academy of Sciences, U.S.A.*, *99*, 1784-1789, 2002.

Schweitzer, P.H., Mechanism of disintegration of liquid jets, *J. Applied Phys.*, *8*, 513-521, 1937.

Scorer, R.S., Ship trails, *Atmos. Environ.*, *21* (6), 1417-1425, 1987.

Scott, J.C., The influence of surface-active contamination on the initiation of wind waves, *J. Fluid Mechanics*, *56* (3), 591-606, 1972.

Scott, J.C., The role of salt in whitecap persistence, *Deep-Sea Res.*, *22*, 653-657, 1975a.

Scott, J.C., The preparation of water for surface-clean fluid mechanics, *J. Fluid Mechanics*, *69* (2), 339-351, 1975b.

Scott, W.D., and Z. Levin, Open channels in sea ice (leads) as ion sources, *Science*, *177*, 425-426, 1972.

Scrimger, J.A., Underwater noise caused by precipitation, *Nature*, *318*, 647-649, 1985.

Scrimger, J.A., D.J. Evans, G.A. McBean, D.M. Farmer, and B.R. Kerman, Underwater noise due to rain, hail, snow, *J. Acoustical Soc. Am.*, *81* (1), 79-86, 1987.

Scrimger, J.A., D.J. Evans, and W. Yee, Underwater noise due to rain - Open ocean measurements, *J. Acoustical Soc. Am.*, *85* (2), 726-731, 1989.

Sedunov, Y.S., *Physics of Drop Formation in the Atmosphere*, 234 pp., John Wiley and Sons, New York, 1974.

Sehmel, G.A., Particle and gas dry-deposition: a review, *Atmos. Environ.*, *14*, 983-1011, 1980.

Sehmel, G.A., Particle dry deposition measurements with dual tracers in field experiments, in *Precipitation Scavenging, Dry Deposition, and Resuspension*, edited by H.R. Pruppacher, R.G. Semonin, and W.G.N. Slinn, pp. 1013-1025, Elsevier, New York, 1983.

Sehmel, G.A., and W.H. Hodgson, Predicted dry deposition velocities, in *Atmospheric-Surface Exchange of Particles and Gaseous Pollutants*, edited by R.J. Engelman, and G.A. Sehmel, pp. 399-422, Technical Information Center, Office of Public Affairs, Energy Research and Development Administration, Richland, WA, 1974.

Sehmel, G.A, and S.L. Sutter, Particle deposition rates on a water surface as a function of particle diameter and air velocity, *J. Recherches Atmos.*, *8*, 911-920, 1974.

Seinfeld, J.H., and S.N. Pandis, *Atmospheric Chemistry and Physics*, 1326 pp., John Wiley and Sons, New York, 1998.

Sellegri, K., J. Gourdeau, J.-P. Putaud, and S. Despiau, Chemical composition of marine aerosol in a Mediterranean coastal zone during the FETCH experiment, *J. Geophys. Res.*, *106* (D11), 12023-12037, 2001.

Serafini, J.S., Inpingement of water droplets on wedges and double-wedge airfoils at supersonic speeds, National Advisory Committee for Aeronautics, *Report No. 1159*, 1954.

Seto, F.Y.B., and R.A. Duce, A laboratory study of iodine enrichment on atmospheric sea-salt particles produced by bubbles, *J. Geophys. Res.*, *77* (27), 5339-5349, 1972.

Settle, D.M., and C.C. Patterson, Magnitudes and sources of precipitation and dry deposition fluxes of industrial and natural leads to the North Pacific at Enewetak, *J. Geophys. Res.*, *87* (11), 8857-8869, 1982.

Shanks, D., Non-linear transformations of divergent and slowly convergent sequences, *J. Math. Phys.*, *34*, 1-42, 1955.

Shatkay, M., and D. Ronen, Bubble populations and gas exchange in hypersaline solutions: a preliminary study, *J. Geophys. Res.*, *97* (C5), 7361-7372, 1992.

Shaw, R.A., and D. Lamb, Experimental determination of the thermal accommodation and condensation coefficients of water, *J. Chem. Phys.*, *111* (23), 10659-10663, 1999.

Shea, D.J., K.E. Trenberth, and R.W. Reynolds, A global monthly sea surface temperature climatology, *J. Climate*, *5*, 987-1001, 1992.

Shockley, W., On the statistics of individual variations of productivity in research laboratories, *Proc. Inst. Radio Eng.*, *45*, 279-290, 1957.

Shulman, M.L., M.C. Jacobson, R.J. Carlson, R.E. Synovec, and T.E. Young, Dissolution behavior and surface tension effects of organic compounds in nucleating cloud droplets, *Geophys. Res. Lttrs*, *23* (3), 277-280, 1996.

Shulman, M.L., R.J. Charlson, and E.J. Davis, The effects of atmospheric organics on aqueous droplet evolution, *J. Aerosol Sci.*, *28* (5), 737-752, 1997.

Sickles, J.E.I., W.D. Bach, and L.L. Spiller, Comparison of several techniques for determining dry deposition flux, in *Precipitation Scavenging, Dry Deposition, and Resuspension*, edited by H.R. Pruppacher, R.G. Semonin, and W.G.N. Slinn, pp. 979-989, Elsevier, New York, 1983.

Sievering, H., Profile measurements of particle mass transfer at the air-water interface, *Atmos. Environ.*, *15*, 123-129, 1981.

Sievering, H., Profile measurements of particle dry deposition velocity at an air-land interface, *Atmos. Environ.*, *16* (2), 301-306, 1982.

Sievering, H., Small-particle dry deposition on natural waters: modeling uncertainty, *J. Geophys. Res.*, *89* (D6), 9679-9681, 1984.

Sievering, H., Small-particle dry deposition under high wind speed conditions: eddy flux measurements at the Boulder Atmospheric Observatory, *Atmos. Environ.*, *21* (10), 2179-2185, 1987.

Sievering, H., J. Eastman, and J.A. Schmidt, Air-sea particle exchange at a nearshore oceanic site, *J. Geophys. Res.*, *87* (C13), 11027-11037, 1982.

Sievering, H., J. Boatman, L. Gunter, H. Horvarth, D. Wellman, and S. Wilkison, Size distributions of sea-source aerosol particles: a physical explanation of observed nearshore versus open-sea differences, *J. Geophys. Res.*, *92* (D12), 14850-14860, 1987.

Sievering, H., J. Boatman, J. Galloway, W. Keene, Y. Kim, M. Luria, and J. Ray, Heterogeneous sulfur conversion in sea-salt aerosol particles: the role of aerosol water content and size distribution, *Atmos. Environ.*, *25A* (1479-1487), 1991.

Sievering, H., J. Boatman, E. Gorman, Y. Kim, L. Anderson, G. Ennis, M. Luria, and S. Pandis, Removal of sulphur from the marine boundary layer by ozone oxidation in sea-salt aerosols, *Nature*, *360*, 571-573, 1992.

Sievering, H., E. Gorman, T. Ley, A. Pszenny, M. Springer-Young, J. Boatman, Y. Kim, C. Nagamoto, and D. Wellman, Ozone oxidation of sulfur in sea-salt aerosol particles during the Azores Marine Aerosol and Gas Exchange Experiment, *J. Geophys. Res.*, *100* (D11), 23075-23081, 1995.

Sievering, J., B. Lerner, J. Slavich, J. Anderson, M. Posfai, and J. Cainey, O_3 oxidation of SO_2 in sea-salt aerosol water: Size distribution of non-sea-salt sulfate during the First Aerosol Characterization Experiment (ACE 1), *J. Geophys. Res.*, *104* (D17), 21707-21717, 1999.

Sigerson, G., Micro-atmospheric researches, *Proc. Royal Irish Academy, Series 2 (Science)*, *1*, 13-22, 1870.

Silker, W.B., Air to sea transfer of marine aerosol, in *Atmosphere-Surface Exchange of Particulates and Pollutants*, edited by R.J. Engelmann, and G.A. Sehmel, pp. 391-398, Technical Information Center, Office of Public Affairs, Energy Research and Development Administration, Richland, WA, 1974.

Simpson, G.C., Sea-salt and condensation nuclei, *Quar. J. Royal Meteorol. Soc.*, *67*, 163-169, 1941a.

Simpson, G.C., On the formation of cloud and rain, *Quar. J. Royal Meteorol. Soc.*, *67*, 99-134, 1941b.

Simpson, J.J., On the exchange of oxygen and carbon dioxide between ocean and atmosphere in an eastern boundary current, in *Gas Transfer at Water Surfaces*, edited by W. Brutsaert, and G.H. Jirka, pp. 505-514, D. Reidel Publishing Company, Dordrecht, 1984.

Sinclair, D., Stability of aerosols and behavior of aerosol particles, in *Military Problems with Aerosols and Persistent Gases*, edited by W.C. Pierce, pp. 301-313, Washington, DC, 1946.

Siscoe, G.L., and Z. Levin, Water-drop-surface-wave interactions, *J. Geophys. Res.*, *76*, 5112-5116, 1971.

Skartveit, A., Wet scavenging of sea-salts and acid compounds in a rainy, coastal area, *Atmos. Environ.*, *16* (11), 2715-2724, 1982.

Skop, R.A., J.T. Viechnicki, and J.W. Brown, A model for microbubble scavenging of surface-active lipid molecules from seawater, *J. Geophys. Res.*, *99* (C8), 16395-16402, 1994.

Slauenwhite, D.E., and B.D. Johnson, Bubble shattering: Differences in bubble formation in fresh water and seawater, *J. Geophys. Res.*, *104* (C2), 3265-3275, 1999.

Slinn, S.A., and W.G.N. Slinn, Predictions for particle deposition on natural waters, *Atmos. Environ.*, *14*, 1013-1016, 1980.

Slinn, S.A., and W.G.N. Slinn, Modeling of atmospheric particulate deposition to natural waters, in *Atmospheric Pollutants in Natural Waters*, edited by S.J. Eisenreich, pp. 23-53, Ann Arbor Science, Ann Arbor, MI, 1981.

Slinn, W.G.N., Dry deposition and resuspension of aerosol particles - a new look at some old problems, in *Atmospheric-Surface Exchange of Particles and Gaseous Pollutants*, edited by R.J. Engelman, and G.A. Sehmel, pp. 1-40, Technical Information Center, Office of Public Affairs, Energy Research and Development Administration, Richland, WA, 1974.

Slinn, W.G.N., Atmospheric aerosol particles in surface-level air, *Atmos. Environ.*, *9*, 763-764, 1975.

Slinn, W.G.N., Some approximations for the wet and dry removal of particles and gases from the atmosphere, *Water Air Soil Pollut.*, *7*, 513-543, 1977a.

Slinn, W.G.N., Precipitation scavenging: some problems, approximate solutions, and suggestions for future research, in *Precipitation Scavenging (1974)*, edited by R.G. Semonin, and R.W. Beadle, pp. 1-60, Technical Information Center, Energy Research and Development Administration, 1977b.

Slinn, W.G.N., Predictions for particle deposition to vegetative canopies, *Atmos. Environ.*, *16* (7), 1785-1794, 1982.

Slinn, W.G.N., Air-to-sea transfer of particles, in *Air-Sea Exchange of Gases and Particles*, edited by P.S. Liss, and W.G.N. Slinn, pp. 299-405, D. Reidel Publishing Company, Dordrecht, 1983a.

Slinn, W.G.N., A potpoutti of deposition and resuspension questions, in *Precipitation Scavenging, Dry Deposition, and Resuspension*, edited by H.R. Pruppacher, R.G. Semonin, and W.G.N. Slinn, pp. 1361-1416, Elsevier, New York, 1983b.

Slinn, W.G.N., Precipitation scavenging, in *Atmospheric Science and Power Production*, edited by D. Randerson, pp. 466-532, United States Department of Energy, 1984.

Slinn, W.G.N., L. Hasse, B.B. Hicks, A.W. Hogan, D. Lal, P.S. Liss, K.O. Munnich, G.A. Sehmel, and O. Vittori, Some aspects of the transfer of atmospheric trace constituents past the air-sea interface, *Atmos. Environ.*, *12*, 2055-2087, 1978.

Slinn, W.G.N., L.F. Radke, and P.C. Katen, Inland transport, mixing, and dry deposition of sea-salt particles, in *Precipitation Scavenging, Dry Deposition, and Resuspension*, edited by H.R. Pruppacher, R.G. Semonin, and W.G.N. Slinn, pp. 1037-1046, Elsevier, New York, 1983.

SMIC (Report of the Study of Man's Impact on Climate), *Inadvertent Climate Modification*, 308 pp., MIT Press, Cambridge, MA, 1971.

Smirnov, A.V., and K.S. Shifrin, Relationship of optical thickness to humidity of air above the ocean, *Atmos. Oceanic Phys.*, *25* (5), 374-379, 1989.

Smirnov, A., G.N. Holben, Y.J. Kaufman, O. Dubovik, T.F. Eck, I. Slutsker, C. Pietras, and R.N. Halthore, Optical properties of atmospheric aerosol in maritime environments, *J. Atmos. Sci.*, *59*, 501-523, 2002.

Smirnov, A., B.N. Holben, O. Dubovik, R. Frouin, T.F. Eck, and I. Slutsker, Maritime component in aerosol optical models derived from Aerosol Robotic Network data, *J. Geophys. Res.*, *108* (D1), 4033, doi:10.1029/2002JD002701, 2003a.

Smirnov, A., B.N. Holben, T.F. Eck, O. Dubovik, and I. Slutsker, Effect of wind speed on columnar aerosol optical properties at Midway Island, *J. Geophys. Res.*, *108* (D24), 4802, doi:10.1029/2003JD003879, 2003b.

Smirnov, V.I., The rate of quasi-steady growth and evaporation of small drops in a gaseous medium, *Pure Appl. Geophys.*, *86*, 184-194, 1971.

Smith, F.B., and D.J. Carson, Some thoughts on the specification of the boundary-layer relevant to numerical modeling, *Boundary-Layer Meteorol.*, *12*, 307-330, 1977.

Smith, M.H., and N.M. Harrison, The sea spray generation function, *J. Aerosol Sci.*, *29*, S189-S190, 1998.

Smith, M.J., I.E. Consterdine, and P.M. Park, Atmospheric loadings of marine aerosol during a Hebridean cyclone, *Quar. J. Royal Meteorol. Soc.*, *115*, 383-395, 1989.

Smith, M.J., P.M. Park, and I.E. Consterdine, North Atlantic aerosol remote concentrations measured at a Hebridean coastal site, *Atmos. Environ.*, *25A*, 547-555, 1991.

Smith, M.J., P.M. Park, and I.E. Consterdine, Marine aerosol concentrations and estimated fluxes over the sea, *Quar. J. Royal Meteorol. Soc.*, *119*, 809-824, 1993.

Smith, S.D., Thrust-anemometer measurements of wind turbulence, Reynolds stress, and drag coefficient over the sea, *J. Geophys. Res.*, *75*, 6758-6770, 1970.

Smith, S.D., Coefficients for sea surface wind stress, heat flux, and wind profiles as a function of wind speed and temperature, *J. Geophys. Res.*, *93* (C12), 15467-15472, 1988.

Smith, S.D., and E.G. Banke, Variation of the sea surface drag coefficient with wind speed, *Quar. J. Royal Meteorol. Soc.*, *101*, 665-673, 1975.

Smith, S.D., and E.P. Jones, Dry-air boundary conditions for correction of eddy flux measurements, *Boundary-Layer Meteorol.*, *17*, 375-379, 1979.

Smith, S.D., R.J. Anderson, E.P. Jones, R.L. Desjardins, R.M. Moore, O. Hertzman, and B.D. Johnson, A new measurement of CO_2 eddy flux in the nearshore atmospheric surface layer, *J. Geophys. Res.*, *96* (C5), 8881-8887, 1991.

Smith, S.R., M.A. Bourassa, and R.J. Sharp, Establishing more truth in true winds, *J. Atmos. Oceanic Technol.*, *16*, 939-952, 1999.

Snead, C.C., and J.T. Zung, The effects of insoluble films upon the evaporation kinetics of liquid droplets, *J. Colloid Interface Sci.*, *27* (1), 25-31, 1968.

Snyder, R.L., and R.M. Kennedy, On the formation of whitecaps by a threshold mechanism. Part I: Basic formulation, *J. Phys. Oceanogr.*, *13*, 1482-1492, 1983a.

Snyder, R.L., L. Smith, and R.M. Kennedy, On the formation of whitecaps by a threshold mechanism. Part III: Field experiment and comparison with theory, *J. Phys. Oceanogr.*, *13*, 1505-1518, 1983b.

Sollazzo, M.J., L.M. Russell, D. Percival, S. Osborne, R. Wood, and D.W. Johnson, Entrainment rates during ACE-2 Lagrangian experiments calculated from aircraft measurements, *Tellus*, *52B* (2), 335-347, 2000.

Song, C.H., and G.R. Carmichael, The aging process of naturally emitted aerosol (sea-salt and mineral aerosol) during long range transport, *Atmos. Environ.*, *33*, 2203-2218, 1999.

Song, C.H., and G.R. Carmichael, A three-dimensional modeling investigation of the evolution processes of dust and sea-salt particles in east Asia, *J. Geophys. Res.*, *106* (D16), 18131-18154, 2001a.

Song, C.H., and G.R. Carmichael, Gas-particle partitioning of nitric acid modulated by alkaline aerosol, *J. Atmos. Chem.*, *40* (1), 1-22, 2001b.

Spann, J.F., and C.B. Richardson, Measurement of the water cycle in mixed ammonium acid sulfate particles, *Atmos. Environ.*, *19* (5), 819-825, 1985.

Speer, R.E., K.A. Peterson, T.G. Ellestad, and J.L. Durham, Test of a prototype eddy accumulator for measuring atmospheric vertical fluxes of water vapor and particulate sulfate, *J. Geophys. Res.*, *90* (D1), 2119-2122, 1985.

Spencer, R.J., N. Møller, and J.H. Weare, The prediction of mineral solubilities in natural waters: A chemical equilibrium model for the Na-K-Ca-Cl-SO_4-H_2O system at temperatures below 25°C, *Geochimica et Cosmochimica Acta*, *54*, 575-590, 1990.

Spiel, D.E., Acoustic measurements of air bubbles bursting at a water surface: bursting bubbles as Helmholtz resonators, *J. Geophys. Res.*, *97* (C7), 11443-11452, 1992.

Spiel, D.E., The number and size of jet drops produced by air bubbles bursting on a fresh water surface, *J. Geophys. Res.*, *99* (C5), 10289-10296, 1994a.

Spiel, D.E., The sizes of jet drops produced by air bubbles bursting on sea- and fresh-water surfaces, *Tellus*, *46B*, 325-338, 1994b.

Spiel, D.E., On the births of jet drops from bubbles bursting on water surfaces, *J. Geophys. Res.*, *100* (C3), 4995-5006, 1995.

Spiel, D.E., More on the births of jet drops from bubbles bursting on seawater surfaces, *J. Geophys. Res.*, *102* (C3), 5815-5821, 1997a.

Spiel, D.E., A hypothesis concerning the peak in film drop production as a function of bubble size, *J. Geophys. Res.*, *102* (C1), 1153-1161, 1997b.

Spiel, D.E., On the births of film drops from bubbles bursting on seawater surfaces, *J. Geophys. Res.*, *103* (C11), 24907-24918, 1998.

Spillane, M.C., E.C. Monahan, P.A. Bowyer, D.M. Doyle, and P.J. Stabeno, Whitecaps and global fluxes, in *Oceanic Whitecaps and Their Role in Air-Sea Exchange Processes*, edited by E.C. Monahan, and G. MacNiocaill, pp. 209-218, D. Reidel Publishing Company, Dordrecht, 1986.

Spitzer, W.S., and W.J. Jenkins, Rates of vertical mixing, gas exchange and new production: Estimates from seasonal gas cycles in the upper ocean near Bermuda, *J. Marine Res.*, *47*, 169-196, 1989.

Spokes, L.J., S.G. Yeatman, S.E. Cornell, and T.D. Jickells, Nitrogen deposition to the eastern Atlantic Ocean. The importance of south-easterly flow, *Tellus*, *52B* (1), 37-49, 2000.

Springer, T.G., and R.L. Pigford, Influence of surface turbulence and surfactants on gas transport through liquid interfaces, *Industrial and Engineering Chemistry. Fundamentals*, *9* (3), 458-465, 1970.

Srokosz, M.A., On the probability of wave breaking in deep water, *J. Phys. Oceanogr.*, *16*, 382-385, 1986.

Stabeno, P.J., Formation of foam about ice floes, in *Whitecaps and the Marine Atmosphere, Report No. 7*, edited by E.C. Monahan, M.C. Spillane, P.A. Bowyer, M.R. Higgins, and P.J. Stabeno, pp. 44-59, University College, Galway, Ireland, 1984.

Stabeno, P.J., and E.C. Monahan, The influence of whitecaps on the albedo of the sea surface, in *Oceanic Whitecaps and Their Role in Air-Sea Exchange Processes*, edited by E.C. Monahan, and G. MacNiocaill, pp. 261-266, D. Reidel Publishing Company, Dordrecht, 1986.

Starflinger, J., M.K. Koch, U. Brockmeier, W. Schütz, and H. Unger, Determination of aerosol release energies provided by bubbles resting at liquid surfaces by means of the code RESUS, *J. Aerosol Sci.*, *26*, S583-S584, 1995.

Stauffer, C.E., The measurement of surface tension by the pendant drop technique, *J. Phys. Chem.*, *69*, 1933-1938, 1965.

Stefan, R.L., and A.J. Szeri, Surfactant scavenging and surface deposition by rising bubbles, *J. Colloid Interface Sci.*, *212*, 1-13, 1999.

Steiger, M., M. Schulz, M. Schwikowski, K. Naumann, and W. Dannecker, Variability of aerosol size distributions above the North Sea and its implications to dry deposition estimates, *J. Aerosol Sci.*, *20* (8), 1229-1232, 1989.

Stoffelen, A., Toward the true near-surface wind speed: Error modeling and calibration using triple collocation, *J. Geophys. Res.*, *103* (C4), 7755-7766, 1998.

Stogryn, A., The emissivity of sea foam at microwave frequencies, *J. Geophys. Res.*, *77* (9), 1658-1666, 1972.

Stokes, D., G. Deane, S. Vagle, and D. Farmer, Measurements of large bubbles in open-ocean whitecaps, in *Gas Transfer at Water Surfaces*, edited by M.A. Donelan, W.M. Drennan, E.S. Saltzman, and R. Wanninkhof, pp. 279-284, American Geophysical Union, Washington, DC, 2002.

Stokes, G.G., On the effect of the internal friction of fluids on the motion of pendulums, *Trans. Cambridge Philosph. Soc.*, *9*, 8-106, 1850.

Stokes, M.D., and G.B. Deane, A new optical instrument for the study of bubbles at high void fractions within breaking waves, *IEEE J. Oceanic Eng.*, *24* (3), 300-311, 1999.

Stong, C.L., The amateur scientist, *Scientific Am.*, 135-138, 1966.

Stramska, M., Vertical profiles of sea salt aerosol in the atmospheric surface layer: a numerical model, *Acta Geophys. Polonica*, *35* (1), 87-100, 1987.

Stramska, M., and T. Petelski, Observations of oceanic whitecaps in the north polar waters of the Atlantic, *J. Geophys. Res.*, *108* (C3), 3086, doi:10.1029/2002JC001321, 2003.

Stramska, M., R. Marks, and E.C. Monahan, Bubble-mediated aerosol production as a consequence of wave breaking in supersaturated (hyperoxic) seawater, *J. Geophys. Res.*, *95* (C10), 18281-18288, 1990.

Stramski, D., and J. Tegowski, Effects of intermittent entrainment of air bubbles by breaking wind waves on ocean reflectance and underwater light field, *J. Geophys. Res.*, *106* (C12), 31345-31360, 2001.

Strasberg, M., Gas bubbles as sources of sound in liquids, *J. Acoustical Soc. Am.*, *28* (1), 20-26, 1956.

Stringham, G.E., D.B. Simons, and H.P. Guy, The behavior of large particles falling in quiescent liquids, pp. 36, United States Department of the Interior, Geological Survey, Washington, DC, 1969.

Strong, A.E., Comments on "Space charge over the open ocean", *J. Atmos. Sci.*, *26*, 784-785, 1969.

Struthwolf, M., and D.C. Blanchard, The residence time of air bubbles < 400 μm diameter at the surface of distilled water and seawater, *Tellus*, *36B*, 294-299, 1984.

Stuhlman, O.J., The mechanics of effervescence, *J. Applied Phys.*, *2*, 457-466, 1932.

Stull, R.B., *An Introduction to Boundary Layer Meteorology*, 666 pp., Kluwer Academic Publishers, Dordrecht, 1988.

Su, M.-Y., S.-C. Ling, and J. Cartmill, Optical microbubble measurements in the North Sea, in *Sea Surface Sound*, edited by B.R. Kerman, pp. 211-223, Kluwer Academic Publishers, Dordrecht, 1988.

Sugawara, K., Syn-bubble-bursting fractionation of sea salt. Ejection of spray droplets with a salt composition different from that of the main sea water when a foam bursts, in *International Oceanographic Conference*, edited by M. Sears, pp. 875-877, American Association for the Advancement of Science, 1959.

Sugawara, K., Exchange of chemical substances between air and sea, *Oceanogr. Marine Biol. Ann. Rev.*, *3*, 59-77, 1965.

Sugawara, K., S. Oana, and T. Koyama, Separation of the components of atmospheric salt and their distribution, *Bull. Chem. Soc. Japan*, *22*, 47-52, 1949.

Sugimoto, N., I. Matsui, Z. Lui, A. Shimizu, I. Tamamushi, and K. Asai, Observation of aerosols and clouds using a two-wavelength polarization lidar during the Nauru99 experiment, *Sea and Sky*, *76*, 93-98, 2000.

Sumer, B.M., Settlement of solid particles in open-channel flow, *J. Hydraulics Div.*, *103* (1323-1337), 1977.

Sutcliffe, W.H.J., E.R. Baylor, and M.D. W., Sea surface chemistry and Langmuir circulation, *Deep-Sea Res.*, *10*, 233-243, 1963.

Sutton, C., M. O'Brien, and D. Ward, Identifying and enjoying Atlantic Coast seabirds from shore. Part 1, *Birding*, *36* (3), 254-263, 2004.

Suzuki, T., and S. Tsunogai, Daily variation of aerosols of marine and continental origin in the surface air over a small island, Okushiri, in the Japan Sea, *Tellus*, *40B*, 42-49, 1988.

Sverdrup, H.U., On the evaporation from the oceans, *J. Marine Res.*, *1*, 3-14, 1937.

Sverdrup, H.U., The humidity gradient over the sea surface, *J. Meteorol.*, *3*, 1-8, 1946.

Sverdrup, H.U., Evaporation from the oceans, in *Compendium of Meteorology*, edited by T.F. Malone, pp. 1071-1081, American Meteorological Society, Boston, 1951.

Sverdrup, H.U., M.W. Johnson, and R.H. Fleming, *The Oceans. Their Physics, Chemistry, and General Biology*, 1087 pp., Prentice-Hall, New York, 1942.

Swift, C.T., Passive microwave remote sensing of he ocean - a review, *Boundary-Layer Meteorol.*, *18*, 25-54, 1980.

Szekielda, K.-H., S.L. Kupferman, V. Klemas, and D.F. Polis, Elemental enrichment in organic films and foam associated with aquatic frontal systems, *J. Geophys. Res.*, *77* (27), 5278-5282, 1972.

Takemura, T., H. Okamoto, Y. Maruyama, A. Numaguti, A. Higurashi, and T. Nakajima, Global three-dimensional simulation of aerosol optical thickness distribution of various origins, *J. Geophys. Res.*, *105* (D14), 17853-17873, 2000.

Tang, C.C.H., The effects of droplets in the air-sea transition zone on the sea brightness temperature, *J. Phys. Oceanogr.*, *4*, 579-593, 1974.

Tang, I.N., Phase transformation and growth of aerosol particles composed of mixed salts, *J. Aerosol Sci.*, *7*, 361-371, 1976.

Tang, I.N., Deliquescence properties and particle size change of hygroscopic aerosols, in *Generation of Aerosols*, edited by K. Willeke, Ann Arbor Science, Ann Arbor, MI, 1980.

Tang, I.N., Thermodynamic and optical properties of mixed-salt aerosols of atmospheric importance, *J. Geophys. Res.*, *102* (D2), 1883-1893, 1997.

Tang, I.N., and H.R. Munkelwitz, An investigation of solute nucleation in levitated solution droplets, *J. Colloid Interface Sci.*, *98* (2), 430-438, 1984.

Tang, I.N., and H.R. Munkelwitz, Simultaneous determination of refractive index and density of an evaporating aqueous solution droplet, *Aerosol Sci. Technol.*, *15*, 201-207, 1991.

Tang, I.N., and H.R. Munkelwitz, Composition and temperature dependence of the deliquescence properties of hygroscopic aerosols, *Atmos. Environ.*, *27A* (4), 467-473, 1993.

Tang, I.N., and H.R. Munkelwitz, Aerosol phase transformation and growth in the atmosphere, *J. Applied Meteorol.*, *33* (7), 791-796, 1994.

Tang, I.N., H.R. Munkelwitz, and J.G. Davis, Aerosol growth studies - II. Preparation and growth measurements of monodisperse salt aerosols, *J. Aerosol Sci.*, *8*, 149-159, 1977.

Tang, I.N., K.H. Fung, D.G. Imre, and H.R. Munkelwitz, Phase transformation and metastability of hygroscopic microparticles, *Aerosol Sci. Technol.*, *23*, 443-453, 1995.

Tang, I.N., A.C. Tridico, and K.H. Fung, Thermodynamic and optical properties of sea salt aerosols, *J. Geophys. Res.*, *102* (D19), 23269-23275, 1997.

Tang, S., and J. Wu, Suppression of wind-generated ripples by natural films: a laboratory study, *J. Geophys. Res.*, *97* (C4), 5301-5306, 1992.

Taylor, N.J., and J. Wu, Simultaneous measurements of spray and sea salt, *J. Geophys. Res.*, *97* (C5), 7355-7360, 1992.

Taylor, T.D., and A. Acrivos, On the deformation and drag of a falling viscous drop at low Reynolds number, *J. Fluid Mechanics*, *18*, 1964, 1964.

Tedesco, R., and D.C. Blanchard, Dynamics of small bubble motion and bursting in freshwater, *J. Recherches Atmos.*, *13* (3), 215-226, 1979.

Tegen, I., P. Hollrig, M. Chin, I. Fung, D. Jacob, and J. Penner, Contribution of different aerosol species to the global aerosol extinction optical thickness: Estimates from model results, *J. Geophys. Res.*, *102* (D9), 23895-23915, 1997.

Tegen, I., D. Koch, A.A. Lacis, and M. Sato, Trends in tropospheric aerosol loads and corresponding impact on direct radiative forcing between 1950 and 1990: A model study, *J. Geophys. Res.*, *105* (D22), 26971-26989, 2000.

Telford, J.W., P.B. Wagner, and A. Vaziri, The measurement of air motion from aircraft, *J. Applied Meteorol.*, *16*, 156-166, 1977.

ten Brink, H.M., Reactive uptake of HNO_3 and H_2SO_4 in sea-salt (NaCl) particles, *J. Aerosol Sci.*, *29* (1/2), 57-64, 1998.

ten Harkel, M.J., The effects of particle-size distribution and chloride depletion of sea-salt aerosols on estimating atmospheric deposition at a coastal site, *Atmos. Environ.*, *31* (3), 417-427, 1997.

Tennekes, H., The logarithmic velocity wind profile, *J. Atmos. Sci.*, *30*, 234-238, 1973.

Tennekes, H., Similarity relations, scaling laws, and spectral dynamics, in *Atmospheric Turbulence and Air Pollution Modelling*, edited by F.T.M. Nieuwstadt, and H. van Dop, pp. 37-68, D. Reidel Publishing Company, Dordrecht, 1982.

Terrill, E., and W.K. Melville, Sound-speed measurements in the surface-wave layer, *J. Acoustical Soc. Am.*, *102* (5), 2607-2625, 1997.

Terrill, E.J., and W.K. Melville, A broadband acoustic technique for measuring bubble size distributions: laboratory and shallow water measurements, *J. Atmos. Oceanic Technol.*, *17*, 220-239, 2000.

Terrill, E.J., W.K. Melville, and D. Stramski, Bubble entrainment by breaking waves and their influence on optical scattering in the upper ocean, *J. Geophys. Res.*, *106* (C8), 16815-16823, 2001.

Tervahattu, H., K. Hartonen, V.-M. Kerminen, K. Kupiainen, P. Aarnio, T. Koskentalo, A.F. Tuck, and V. Vaida, New evidence of an organic layer on marine aerosols, *J. Geophys. Res.*, *107* (D7), 10.1029/2000JD000282, 2002.

Thompson, T.G., and K.H. Nelson, Concentration of brines and deposition of salts from sea water under frigid conditions, *Am. J. Sci.*, *254*, 227-238, 1956.

Thomson, S.W., On the equilibrium of vapour at a curved surface of liquid, *Philosoph. Mag.*, *4*, 448-452, 1871.

Thorpe, S.A., On the cloud of bubbles formed by a breaking wind-wave in deep water, and their role in air-sea gas transfer, *Philosoph. Trans. Royal Soc. London*, *304A*, 155-210, 1982.

Thorpe, S.A., The role of bubbles produced by breaking waves in super-saturating the near-surface ocean mixing layer with oxygen, *Annales Geophysicae*, *2*, 53-55, 1984a.

Thorpe, S.A., A model of the turbulent diffusion of bubbles below the sea surface, *J. Phys. Oceanogr.*, *14*, 841-854, 1984b.

Thorpe, S. A., The effect of Langmuir circulation on the distribution of submerged bubbles caused by breaking wind waves, *J. Fluid Mech.*, *142*, 151-170, 1984c.

Thorpe, S.A., Small-scale processes in the upper ocean boundary layer, *Nature*, *318*, 519-522, 1985.

Thorpe, S. A., Measurements with an automatically recorded inverted echo sounder; ARIES and the bubble clouds, *J. Phys. Oceanogr.*, *16*, 1462-1478, 1986.

Thorpe, S.A., Energy loss by breaking waves, *J. Phys. Oceanogr.*, *23*, 2498-2502, 1993.

Thorpe, S.A., and A.J. Hall, The characteristics of breaking waves, bubble clouds, and near-surface currents observed using side-scan sonar, *Continental Shelf Res.*, *1* (4), 353-384, 1983.

Thorpe, S.A., and P.N. Humphries, Bubbles and breaking waves, *Nature*, *283*, 463-465, 1980.

Thorpe, S.A., A.R. Stubbs, A.J. Hall, and R.J. Turner, Wave-produced bubbles observed by side-scan sonar, *Nature*, *296*, 636-638, 1982.

Thorpe, S.A., P. Bowyer, and D.K. Woolf, Some factors affecting the size distributions of oceanic bubbles, *J. Phys. Oceanogr.*, *22*, 382-389, 1992.

Toba, Y., Drop production by bursting of air bubbles on the sea surface (II) Theoretical study on the shape of floating bubbles, *J. Oceanogr. Soc. Japan*, *15* (2), 121-130, 1959.

Toba, Y., Drop production by bursting of air bubbles on the sea surface. III Study by use of a wind flume (Short report), *J. Oceanogr. Soc. Japan*, *17* (4), 169-178, 1961a.

Toba, Y., Drop production by bursting of air bubbles on the sea surface (III). Study by use of a wind flume, *Memoirs of the College of Science, University of Kyoto, Series A*, *29* (3), 313-344, 1961b.

Toba, Y., On the giant sea-salt particles in the atmosphere I. General features of the distribution, *Tellus*, *17*, 131-145, 1965a.

Toba, Y., On the giant sea-salt particles in the atmosphere II. Vertical distribution in the 10-m layer over the ocean, *Tellus*, *17*, 365-382, 1965b.

Toba, Y., On the giant sea-salt particles in the atmosphere III. An estimate of the production and distribution over the world ocean, *Tellus*, *18*, 132-145, 1966.

Toba, Y., Local balance in the air-sea boundary processes: 1. On the growth process of wind waves, *J. Oceanogr. Soc. Japan*, *28*, 109-121, 1972.

Toba, Y., and M. Chaen, Quantitative expression of the breaking of wind waves on the sea surface, *Rec. Oceanogr. Works Japan*, *12* (1), 1-11, 1973.

Toba, Y., and M. Koga, A parameter describing overall conditions of wave breaking, whitecapping, sea-spray production and wind stress, in *Oceanic Whitecaps and Their Role in Air-Sea Exchange Processes*, edited by E.C. Monahan, and G. MacNiocaill, pp. 37-47, D. Reidel Publishing Company, Dordrecht, 1986.

Toba, Y., M. Tokuda, K. Okuda, and S. Kawai, Forced convection accompanying wind waves, *J. Oceanogr. Soc. Japan*, *31*, 192-198, 1975.

Tomaides, M., and K.T. Whitby, Generation of aerosols by bursting of single bubbles, in *Fine Particles*, edited by B.Y.H. Liu, pp. 837, Academic Press, New York, 1976.

Tomasi, C., and F. Prodi, Measurements of atmospheric turbidity and vertical mass loading of particulate matter in marine environments (Red Sea, Indian Ocean, and Somalian Coast), *J. Geophys. Res.*, *87* (C2), 1279-1286, 1982.

Torgerson, T., R. Mason, D.K. Woolf, J. Benoit, M.P. Dowling, M.B. Wilson, and E.C. Monahan, The role of breaking waves in the control of the gas exchange properties of the sea surface, in *The Climate and Health Implications of Bubble-Mediated Sea-Air Exchange*, edited by E.C. Monahan, and M.A. Van Patten, pp. 155-162, Connecticut Sea Grant College Program, 1989.

Torobin, L.B., and W.H. Gauvin, Fundamental aspects of solids-gas flow. Part 1. Introductory concepts and idealized sphere motion in viscous regime, *Canadian J. Chem. Eng.*, *37*, 129-141, 1959a.

Torobin, L.B., and W.H. Gauvin, Fundamental aspects of solids-gas flow. Part II. The sphere wave in steady laminar fluids, *Canadian J. Chem. Eng.*, *37*, 167-176, 1959b.

Torobin, L.B., and W.H. Gauvin, Fundamental aspects of solids-gas flow. Part III. Accelerated motion of a particle in a fluid, *Canadian J. Chem. Eng.*, *37*, 224-236, 1959c.

Torobin, L.B., and W.H. Gauvin, Fundamental aspects of solids-gas flow. Part IV. The effects of particle rotation, roughness and shape, *Canadian J. Chem. Eng.*, *38*, 142-153, 1960a.

Torobin, L.B., and W.H. Gauvin, Fundamental aspects of solids-gas flow. Part V. The effects of fluid turbulence on the particle drag coefficient, *Canadian J. Chem. Eng.*, *38*, 189-200, 1960b.

Torobin, L.B., and W.H. Gauvin, Fundamental aspects of solids-gas flow. Part VI. Multiparticle behavior in turbulent fluids, *Canadian J. Chem. Eng.*, *39*, 113-120, 1961.

Toye, M.-J., Fragmentation des noyaux salins par dessication, *Bull. de l'Obseratoire du Puy de Dome*, *4*, 148-151, 1956.

Toyota, K., M. Takahashi, and H. Akimoto, Modeling multiphase halogen chemistry in the marine boundary layer with size-segregated aerosol module: Implications for quasi-size-dependent approach, *Geophys. Res. Lttrs*, *28* (15), 2899-2902, 2001.

Tseng, R.-S., J.T. Viechnicki, R.A. Skop, and J.W. Brown, Sea-to-air transfer of surface-active organic compounds by bursting bubbles, *J. Geophys. Res.*, *97* (C4), 5201-5206, 1992.

Tsimplis, M., and S.A. Thorpe, Wave damping by rain, *Nature*, *342*, 893-895, 1989.

Tsimplis, M.N., The effect of rain in calming the sea, *J. Phys. Oceanogr.*, *22*, 404-412, 1992.

Tsunogai, S., O. Saito, K. Yamada, and S. Nakaya, Chemical composition of oceanic aerosol, *J. Geophys. Res.*, *77* (27), 5283-5292, 1972.

Tufte, E.R., *The Visual Display of Quantitative Information*, 197 pp., Graphics Press, Cheshire, CT, 1983.

Turner, J.S., The salinity of rainfall as a function of drop size, *Quar. J. Royal Meteorol. Soc.*, *81*, 418-429, 1955.

Turner, W.R., Microbubble persistence in fresh water, *J. Acoustical Soc. Am.*, *33* (9), 1223-1233, 1961.

Tusseau, D., M. Barbier, J.-C. Marty, and A. Saliot, Les stérols de l'atmosphère marine, *Comptes Rendus de l'Academies des Sci., Paris, Série C*, *290*, 109-111, 1980.

Twohy, C.H., and J.G. Hudson, Measurements of cloud condensation nuclei spectra within maritime cumulus cloud drops: Implications for mixing processes, *J. Applied Meteorol.*, *34* (4), 815-833, 1995.

Twohy, C.H., P.H. Austin, and R.J. Charlson, Chemical consequences of the initial diffusional growth of cloud droplets: a clean marine case, *Tellus*, *41B*, 51-60, 1989.

Twomey, S., The identification of individual hygroscopic particles in the atmosphere by a phase-transition method, *J. Applied Phys.*, *21* (9), 1099-1102, 1953.

Twomey, S., The composition of hygroscopic particles in the atmosphere, *J. Meteorol.*, *11*, 334-338, 1954.

Twomey, S., The distribution of sea-salt nuclei in air over land, *J. Meteorol.*, *12*, 81-86, 1955.

Twomey, S., Hygroscopic particles in the atmosphere and their identification by a phase-transition method, in *Atmospheric Chemistry of Chlorine and Sulfur Compounds*, edited by J.P. Lodge, Jr., pp. 1-10, American Geophysical Union, 1959.

Twomey, S., On the nature and origin of natural cloud nuclei, *Bull. de l'Obseratoire du Puy de Dome*, *1* (1), 1-19, 1960.

Twomey, S., The composition of cloud nuclei, *J. Atmos. Sci.*, *28*, 377-381, 1971.

Twomey, S.A., Pollution and the planetary albedo, *Atmos. Environ.*, *8*, 1251-1256, 1974.

Twomey, S., *Atmospheric Aerosols*, 302 pp., Elsevier Scientific Publishing Company, Amsterdam, 1977a.

Twomey, S., The influence of pollution on the short-wave albedo of clouds, *J. Atmos. Sci.*, *34*, 1149-1152, 1977b.

Twomey, S., Aerosols, clouds and radiation, *Atmos. Environ.*, *25A* (11), 2435-2442, 1991.

Twomey, S., and K.N. McMaster, The production of condensation nuclei by crystallizating salt particles, *Tellus*, *7*, 458-461, 1955.

Twomey, S.A., M. Piepgrass, and T.L. Wolfe, An assessment of the impact of pollution on global cloud albedo, *Tellus*, *36B*, 356-366, 1984.

Tyler, E., Instability of liquid jets, *Philosoph. Mag.*, *16*, 504-518, 1933.

Tyler, E., and E.G. Richardson, The characteristic curves of liquid jets, *Proc. Phys. Soc. London*, *37*, 297-311, 1925.

Tyler, E., and F. Watkin, Experiments with capillary jets, *Philosoph. Mag.*, *14*, 849-881, 1932.

Uematsu, M., K. Ohta, K. Matsumoto, and I. Uno, Short term variation of marine organic aerosols under the Northwestern Pacific high pressure region in the summer of 1999, *Geochem. J.*, *35*, 49-57, 2001.

Ueno, Y., Studies of salt-solution aerosols - XIII Effect of surface-active substances on the stability of aqueous salt solution aerosols, *Atmos. Environ.*, *10*, 409-413, 1976.

Ueno, Y., and I. Sano, Studies on salt solution aerosols. V. The effect of surface active substance upon the droplet size of an aqueous NaCl solution aerosol, *Bull. Chem. Soc. Japan*, *44*, 908-911, 1971.

UNESCO, Background papers and supporting data on the Practical Salinity Scale 1978, *UNESCO Techn. Papers in Marine Science No. 37*, 1981.

Unna, P.J.H., "White Horses", *Nature*, *148*, 226-227, 1941.

Uno, S., and R.C. Kintner, Effect of wall proximity on the rate of rise of single air bubbles in a quiescent liquid, *Am. Inst. Chem. Eng. J.*, *2* (3), 420-425, 1956.

Vagle, S., and D.M. Farmer, The measurement of bubble-size distributions by acoustical backscatter, *J. Atmos. Oceanic Technol.*, *9*, 630-644, 1992.

Vagle, S., and D.M. Farmer, A comparison of four methods for bubble size and void fraction measurements, *IEEE J. Oceanic Eng.*, *23* (3), 211-222, 1998.

Valencia, M.J., Recycling of pollen from an air-water interface, *Am. J. Sci.*, *265*, 843-847, 1967.

Valenzuela, G.R., and M.B. Laing, Study of Doppler spectra of radar sea echo, *J. Geophys. Res.*, *75* (3), 551-563, 1970.

Van Aalst, R.M., Dry deposition of aerosol particles, in *Aerosols: Research, Risk Assessment and Control Strategies*, edited by S.D. Lee, T. Schneider, L.D. Grant, and P.J. Verkerk, pp. 933-949, Lewis Publishers, Inc., Williamsburg, VA, 1985.

Van de Hulst, H.C., *Light Scattering by Small Particles*, 470 pp., John Wiley and Sons, New York, 1957.

Van den Berg, A., F. Dentener, and J. Lelieveld, Modeling the chemistry of the marine boundary layer: sulphate formation and the role of sea-salt aerosol particles, *J. Geophys. Res.*, *105* (D9), 11671-11698, 2000.

Van Dingenen, R., F. Raes, and N.R. Jensen, Evidence for anthropogenic impact on number concentration and sulfate content of cloud-processed aerosol particles over the North Atlantic, *J. Geophys. Res.*, *100* (D10), 21057-21067, 1995.

Van Doren, J.M., L.R. Watson, P. Davidovits, D.R. Worsnop, M.S. Zahniser, and C.E. Kolb, Temperature dependence of the uptake coefficients of HNO_3, HCl, and N_2O_5 by water droplets, *J. Phys. Chem.*, *94*, 3265-3269, 1990.

Van Dorn, W.G., Wind stress on an artificial pond, *J. Marine Res.*, *12*, 249-276, 1953.

Van Dyke, M., *Perturbation Methods in Fluid Mechanics*, 271 pp., The Parabolic Press, Stanford, CA, 1975.

Van Eijk, A.M.J., B.S. Tranchant, and P.G. Mestayer, SeaCluse: Numerical simulation of evaporating sea spray droplets, *J. Geophys. Res.*, *106* (C2), 2573-2588, 2001.

Van Grieken, R.E., T.B. Johansson, and J.W. Winchester, Trace metal fractionation effects between sea water and aerosols from bubble bursting, *J. Recherches Atmos.*, *8*, 611-621, 1974.

Van Krevelen, D.W., and P.J. Hoftijzer, Studies of gas-bubble formation. Calculation of interfacial area in bubble contactors, *Chem. Eng. Progr.*, *46* (1), 29-35, 1950.

Várhelyi, G., and G. Gravenhorst, Production rate of airborne sea-salt sulfur deduced from chemical analysis of marine aerosols and precipitation, *J. Geophys. Res.*, *88* (C11), 6737-6751, 1983.

Vickers, D., and L. Mahrt, Fetch limited drag coefficients, *Boundary-Layer Meteorol.*, *85*, 53-79, 1997.

Vignati, E., G. de Leeuw, and R. Berkowicz, Modeling coastal aerosol transport and effects of surf-produced aerosols in the marine atmospheric boundary layer, *J. Geophys. Res.*, *106* (D17), 20225-20238, 2001.

Vignati, E., G. de Leeuw, M. Schulz, and E. Plate, Characterization of aerosols at a coastal site near Vindeby (Denmark), *J. Geophys. Res.*, *104* (C2), 3277-3287, 1999.

Villevalde, Y.V., A.V. Smirnov, N.T. O'Neill, S.P. Smyshlyaev, and V.V. Yakovlev, Measurement of aerosol optical depth in the Pacific Ocean and the North Atlantic, *J. Geophys. Res.*, *99* (D10), 20983-20988, 1994.

Vogt, R., and B.J. Finlayson-Pitts, A diffuse reflectance infrared Fourier transform spectroscopic (DRIFTS) study of the surface reaction of NaCl with gaseous NO_2 and HNO_3, *J. Phys. Chem.*, *94*, 3747-3755, 1994.

Vogt, R., P.J. Crutzen, and R. Sander, A mechanism for halogen release from sea-salt aerosol in the remote marine boundary layer, *Nature*, *383*, 327-330, 1996.

Volgin, V.M., O.A. Yershov, A.V. Smirnov, and K.S. Shifrin, Optical depth of aerosol in typical sea areas, *Atmos. Oceanic Phys.*, *24* (10), 772-777, 1988.

von Glasow, R., and P.J. Crutzen, Tropospheric Halogen Chemistry, in *The Atmosphere*, edited by R.F. Keeling, pp. 21-64, Elsevier, Amsterdam, 2004.

von Glasow, R., and R. Sander, Variation of sea salt aerosol pH with relative humidity, *Geophys. Res. Lttrs*, *28* (2), 247-250, 2001.

von Kármán, T., *Aerodynamics*, 203 pp., McGraw-Hill Book Company, New York, 1954.

Vonnegut, B., and R.L. Neubauer, Counting sodium-containing particles in the atmosphere by their spectral emission in a hydrogen flame, *Bull. Am. Meteorol. Soc.*, *34* (4), 163-169, 1953.

Voronov, G.I., and A.S. Gavrilov, Stochastic modeling of water-salt ejection from the ocean surface, *Atmos. Oceanic Phys.*, *25* (3), 233-237, 1989.

Wagenbach, D., F. Ducroz, R. Mulvaney, L. Keck, A. Minikin, M. Legrand, F.S. Hall, and E.W. Wolff, Sea-salt aerosol in coastal Antarctic regions, *J. Geophys. Res.*, *103* (D9), 10961-10974, 1998.

Wagener, R., S. Nemesure, and S.E. Schwartz, Aerosol optical depth over oceans: High space- and time-resolution retrieval and error budget from satellite radiometry, *J. Atmos. Oceanic Technol.*, *14* (3), 577-590, 1997.

Wagner, F.S., Jr., Composition of the dissolved organic compounds in seawater: a review, *Contrib. Marine Sci.*, *14*, 115-153, 1969.

Wagner, P.E., Aerosol growth by condensation, in *Aerosol Microphysics*, edited by W.H. Marlow, pp. 129-178, Springer-Verlag, Berlin, 1982.

Wagner, P.E., and F.G. Pohl, Experimental test of recent droplet growth theories by means of a process controlled expansion cloud chamber, *J. Aerosol Sci.*, *10*, 204-205, 1979.

Waldmann, L., and K.H. Schmitt, Thermophoresis and diffusiophoresis of aerosols, in *Aerosol Sci.*, edited by C.N. Davies, pp. 137-162, Academic Press, London, 1966.

Walker, M.I., W.A. McKay, N.J. Pattenden, and P.S. Liss, Actinide enrichment in marine aerosols, *Nature*, *323*, 141-143, 1986.

Wallace, D.W.R., and C.D. Wirick, Large air-sea gas fluxes associated with breaking waves, *Nature*, *356*, 694-696, 1992.

Wallace, G.T., Jr., and R.A. Duce, Concentration of particulate trace metals and particulate organic carbon in marine surface waters by a bubble flotation mechanism, *Marine Chem.*, *3*, 157-181, 1975.

Wallace, G.T., Jr., and R.A. Duce, Transport of particulate organic matter by bubbles in marine waters, *Limnol. Oceanogr.*, *23* (6), 1155-1167, 1978.

Wallace, G.T., Jr., G.I. Loeb, and D.F. Wilson, On the flotation of particulates in sea water by rising bubbles, *J. Geophys. Res.*, *77* (27), 5293-5301, 1972.

Wallace, W.J., *The Development of the Chlorinity/Salinity Concept in Oceanography*, 227 pp., Elsevier Scientific Publishing Company, Amsterdam, 1974.

Wallis, G., The terminal speed of single drops or bubbles in an infinite medium, *Internatl. J. Multiphase Flow*, *1*, 491-511, 1974.

Walsh, A.L., and P.J. Mulhearn, Photographic measurements of bubble populations from breaking wind waves at sea, *J. Geophys. Res.*, *92* (C13), 14553-14565, 1987.

Wang, C.S., and R.L. Street, Measurements of spray at an air-water interface, *Dyn. Atmos. Oceans*, *2*, 141-152, 1978a.

Wang, C.S., and R.L. Street, Transfers across an air-water interface at high wind speeds: the effect of spray, *J. Geophys. Res.*, *83* (C6), 2959-2969, 1978b.

Wang, P.K., and H.R. Pruppacher, Acceleration to terminal velocity of cloud and raindrops, *J. Applied Meteorol.*, *16*, 275-280, 1977.

Wang, Y., J.D. Kepert, and G.J. Holland, The effect of sea spray evaporation on tropical cyclone boundary layer structure and intensity, *Monthly Weather Rev.*, *129*, 2481-2500, 2001.

Wangersky, P., The organic chemistry of sea water, *American Scientist*, *53*, 358-374, 1965.

Wangersky, P.J., The surface film as a physical environment, *Ann. Rev. Ecology Sys.*, *7*, 161-176, 1976.

Wangwongwatana, S., V. Scarpino, K. Willeke, and P.A. Baron, System for characterizing aerosols from bubbling liquids, *Aerosol Sci. Technol.*, *13*, 297-307, 1990.

Wanninkhof, R., W. Asher, R. Weppernig, H. Chen, P. Schlosser, C. Langdon, and R. Sambrotto, Gas transfer experiments on Georges Bank using two volatile deliberate tracers, *J. Geophys. Res.*, *98* (C11), 20237-20248, 1993.

Warner, J., and W.G. Warne, The effect of surface films in retarding the growth by condensation of cloud nuclei and their use in fog suppression, *J. Applied Meteorol.*, *9*, 639-650, 1970.

Warner, T.B., and J.B. Hoover, A mapping strategy for sea-salt aerosols, in *Mapping Strategies in Chemical Oceanography*, edited by A. Zirino, pp. 57-74, American Chemical Society, Washington, DC, 1985.

Warshay, M., E. Bogusz, M. Johnson, and R.C. Kintner, Ultimate velocity of drops in stationary liquid media, *Canadian J. Chem. Eng.*, *37*, 29-36, 1959.

Weare, B.C., P.T. Strub, and M.D. Samuel, Annual mean surface heat fluxes in the Tropical Pacific Ocean, *J. Phys. Oceanogr.*, *11*, 705-717, 1981.

Webb, E.K., G.I. Pearman, and R. Leuning, Correction of flux measurements for density effects due to heat and water vapour transfer, *Quar. J. Royal Meteorol. Soc.*, *106*, 85-100, 1980.

Weber, M.E., D.C. Blanchard, and L.D. Syzdek, The mechanism of scavenging waterborne bacteria by a rising bubble, *Limnol. Oceanogr.*, *28* (1), 101-105, 1983.

Webster, W.J.J., T.T. Wilheit, and P. Gloersen, Spectral characteristics of the microwave emission from a wind-driven foam-covered sea, *J. Geophys. Res.*, *81* (18), 3095-3099, 1976.

Wei, Y., and J. Wu, In situ measurements of surface tension, wave damping, and wind properties modified by natural films, *J. Geophys. Res.*, *97* (C4), 5307-5313, 1992.

Weibull, W., A statistical distribution of wide applicability, *J. Applied Mech.*, *18*, 293-297, 1951.

Weingartner, E., U. Baltensperger, and J. Burtscher, Growth and structural change of combustion aerosols at high relative humidity, *Environ. Sci. Technol.*, *29*, 2982-2986, 1995.

Weis, D.D., and G.E. Ewing, Water content and morphology of sodium chloride aerosol particles, *J. Geophys. Res.*, *104* (D17), 21275-21285, 1999.

Weisel, C.P., R.A. Duce, J.L. Fasching, and R.W. Heaton, Estimates of transport of trace metals from the ocean to the atmosphere, *J. Geophys. Res.*, *89* (D7), 11607-11618, 1984.

Weiss, R.F., The solubility of nitrogen, oxygen and argon in water and seawater, *Deep-Sea Res.*, *17*, 721-735, 1970.

Weissenborn, P.K., and R.J. Pugh, Surface tension and bubble coalescence phenomena of aqueous solutions of electrolytes, *Langmuir*, *11*, 1422-1426, 1995.

Weissman, D.E., M.A. Bourassa, and J. Tongue, Effect of rain rate and wind magnitude on SeaWinds scatterometer wind speed errors, *J. Atmos. Oceanic Technol.*, *19*, 738-746, 2002.

Weissman, M.A., S.S. Ataktürk, and K.B. Katsaros, Detection of breaking events in a wind-generated wave field, *J. Phys. Oceanogr.*, *14*, 1608-1619, 1984.

Welander, P., On the generation of wind streaks on the sea surface by action of surface film, *Tellus*, *15*, 67-71, 1963.

Weller, R.A., R.E. Payne, W.G. Large, and W. Zenk, Wind measurements from an array of oceanographic moorings and from F/S *Meteor* during JASIN 1978, *J. Geophys. Res.*, *88* (C14), 9689-9705, 1983.

Wellman, H.W., and A.T. Wilson, Salts on sea ice in McMurdo Sound, Antarctica, *Nature*, *200*, 462-463, 1963.

Wells, A.C., and A.C. Chamberlain, Transport of small particles to vertical surfaces, *British J. Applied Phys.*, *18*, 1793-1799, 1967.

Wells, B.W., and I.V. Shunk, Seaside shrubs: wind forms vs. spray forms, *Science*, *85*, 499, 1937.

Wells, W.C., G. Gal, and M.W. Munn, Aerosol distributions in maritime air and predicted scattering coefficients in the infrared, *Applied Optics*, *16* (3), 654-659, 1977.

Wentz, F.J., A model function for ocean microwave brightness temperature, *J. Geophys. Res.*, *88* (C3), 1892-1908, 1983.

Wentz, F.J., and L.A. Mattox, New algorithms for microwave measurements of ocean winds: applications to SEASAT and the special sensor microwave imager, *J. Geophys. Res.*, *91* (C2), 2289-2307, 1986.

Wentz, F.J., and D.K. Smith, A model function for the ocean-normalized radar cross section at 14 GHz derived from NSCAT observations, *J. Geophys. Res.*, *104* (C5), 11499-11514, 1999.

Wentz, F.J., V.J. Cardone, and L.S. Fedor, Intercomparison of wind speeds inferred by the SASS, altimeter, and SMMR, *J. Geophys. Res.*, *87* (C5), 3378-3384, 1982.

Wentz, F.J., S. Peterherych, and L.A. Thomas, A model function for ocean radar cross sections at 14.6 GHz, *J. Geophys. Res.*, *89* (C3), 3689-3704, 1984.

Wenz, G.M., Acoustic ambient noise in the ocean: spectra and sources, *J. Acoustical Soc. Am.*, *34* (12), 1936-1956, 1962.

Wesely, M.L., Response to "Isotopic versus micrometeorologic ocean CO_2 fluxes: a serious conflict" by W. Broecker *et al.*, *J. Geophys. Res.*, *91* (C9), 10533-10535, 1986.

Wesely, M.L., and B.B. Hicks, A review of the current status of knowledge on dry deposition, *Atmos. Environ.*, *34*, 2261-2282, 2000.

Wesely, M.L., B.B. Hicks, W.P. Dannevik, S. Frisella, and R.B. Husar, An eddy-correlation measurement of particulate deposition from the atmosphere, *Atmos. Environ.*, *11*, 561-563, 1977.

Wesely, M.L., D.R. Cook, and R.M. Williams, Field measurement of small ozone fluxes to snow, wet bare soil, and lake water, *Boundary-Layer Meteorol.*, *20*, 459-471, 1981.

Wesely, M.L., D.R. Cook, R.L. Hart, and R.M. Williams, Air-sea exchange of CO_2 and evidence for enhanced upward fluxes, *J. Geophys. Res.*, *87* (C11), 8827-8832, 1982.

Wesely, M.L., D.R. Cook, R.L. Hart, and R.E. Speer, Measurements of parameterization of particulate sulfur dry deposition over grass, *J. Geophys. Res.*, *90* (D1), 2131-2143, 1985.

Wexler, A.S., and J.H. Seinfeld, Second-generation inorganic aerosol model, *Atmos. Environ.*, *25A* (12), 2731-2748, 1991.

Whitby, K.T., The physical characteristics of sulfur aerosols, *Atmos. Environ.*, *12*, 135-159, 1978.

Whitby, K.T., R.B. Husar, and B.Y.H. Liu, The aerosol size distribution of Los Angeles smog, *J. Colloid Interface Sci.*, *39* (1), 177-204, 1972.

Whitlock, C.H., D.S. Bartlett, and E.A. Gurganus, Sea foam reflectance and influence on optimum wavelength for remote sensing of ocean aerosols, *Geophys. Res. Lttrs*, *9* (6), 719-722, 1982.

Wilkerson, J.C., and M.D. Earle, A study of differences between environmental reports by ships in the Voluntary Observing Program and measurements from NOAA buoys, *J. Geophys. Res.*, *95* (C3), 3373-3385, 1990.

Wilkniss, P.E., and D.J. Bressan, Fractionation of the elements F, Cl, Na, and K at the sea-air interface, *J. Geophys. Res.*, *77* (27), 5307-5315, 1972.

Wille, P.C., and D. Geyer, Measurements on the origin of the wind-dependent ambient noise variability in shallow water, *J. Acoustical Soc. Am.*, *75* (1), 173-185, 1984.

Willeke, K., and K.T. Whitby, Atmospheric aerosols: size distribution interpretation, *J. Air Pollut. Contr. Assoc.*, *25* (5), 529-534, 1975.

Williams, D.J., and B.C. Moser, Airborne sea salt sedimentation measurements and a method of reproducing ambient sedimentation rates for the study of its effect on vegetation, *Atmos. Environ.*, *10*, 531-534, 1976.

Williams, G.F., Jr., Microwave radiometry of the ocean and the possibility of marine wind velocity determination from satellite observations, *J. Geophys. Res.*, *74* (18), 4591-4594, 1969.

Williams, G.F., Jr., Comments on "Fresh Water Whitecaps", *J. Atmos. Sci.*, *27*, 1220, 1970.

Williams, P.J.B., Biological and chemical aspects of dissolved organic material in sea water, in *Chemical Oceanography*, edited by J.P. Riley, and G. Skirrow, pp. 301-363, Academic Press, London, 1975.

Williams, R.M., Uncertainties in the use of box models for estimating dry deposition velocity, *Atmos. Environ.*, *16* (11), 2707-2708, 1982a.

Williams, R.M., A model for the dry deposition of particles to natural water surfaces, *Atmos. Environ.*, *16* (8), 1933-1938, 1982b.

Wilson, A.T., Surface of the ocean as a source of air-borne nitrogenous material and other plant nutrients, *Nature*, *184*, 99-101, 1959a.

Wilson, A.T., Organic nitrogen in New Zealand snows, *Nature*, *183*, 318-319, 1959b.

Wilson, J.H., Low-frequency wind-generated noise produced by the impact of spray with the ocean's surface, *J. Acoustical Soc. Am.*, *68* (3), 852-956, 1980.

Wilson, J.H., Wind-generated noise modeling, *J. Acoustical Soc. Am.*, *73* (1), 211-216, 1983.

Wilson, S.R., and B.W. Forgan, Aerosol optical depth at Cape Grim, Tasmania, 1986-1999, *J. Geophys. Res.*, *107* (D8), 10.1029/2001JD000398, 2002.

Wilson, T.R.S., Salinity and the major elements of seawater, in *Chemical Oceanography*, edited by J.P. Riley, and G. Skirrow, pp. 365-413, Academic Press, London, 1975.

Wiman, B.L.B., M.H. Unsworth, S.E. Lindberg, B. Bergkvist, R. Jaenicke, and H.-C. Hansson, Perspectives on aerosol deposition to natural surfaces: interactions between aerosol residence times, removal processes, the biosphere and global environmental change, *J. Aerosol Sci.*, *21* (3), 313-338, 1990.

Winkler, P., The growth of atmospheric aerosol particles with relative humidity, *Physica Scripta*, *37*, 223-230, 1988.

Winkler, P., and C.E. Junge, Comments on "Anomalous deliquescence of sea spray aerosols", *J. Applied Meteorol.*, *10*, 159-163, 1971.

Winnikow, S., and B.T. Chao, Droplet motion in purified systems, *Physics of Fluids*, *9* (1), 50-61, 1966.

Winter, B., and P. Chýlek, Contribution of sea salt aerosol to the planetary clear-sky albedo, *Tellus*, *49B*, 72-79, 1997.

Witter, D.L., and D.B. Chelton, A Geosat attimeter wind speed algorithm and a method for altimeter wind speed algorithm development, *J. Geophys. Res.*, *96* (C5), 8853-8860, 1991.

Wolff, E.W., A.M. Rankin, and R. Röthlisberger, An ice core indicator of Antarctic sea ice production?, *Geophys. Res. Lttrs*, *30* (22), 2158, doi:10.1029/2003GL018454, 2003.

Woodcock, A.H., Note concerning human respiratory irritation associated with high concentrations of plankton and mass mortality of marine organisms, *J. Marine Res.*, *7*, 56-62, 1948.

Woodcock, A.H., Condensation nuclei and precipitation, *J. Meteorol.*, *7*, 161-162, 1950a.

Woodcock, A.H., Sea salt in a tropical storm, *J. Meteorol.*, *7*, 397-401, 1950b.

Woodcock, A.H., Reply, *J. Meteorol.*, *8*, 363, 1951.

Woodcock, A.H., Atmospheric salt particles and raindrops, *J. Meteorol.*, *9*, 200-212, 1952.

Woodcock, A.H., Salt nuclei in marine air as a function of altitude and wind force, *J. Meteorol.*, *10*, 362-371, 1953.

Woodcock, A.H., Bursting bubbles and air pollution, *Sewage Ind. Wastes*, *27*, 1189-1192, 1955.

Woodcock, A.H., Atmospheric sea-salt nuclei data from Project Shower, *Tellus*, *9*, 521-524, 1957.

Woodcock, A.H., The release of latent heat in tropical storms due to the fall-out of sea-salt particles, *Tellus*, *10*, 355-371, 1958.

Woodcock, A.H., Solubles, in *The Sea*, edited by M.N. Hill, pp. 305-311, Interscience, New York, 1962.

Woodcock, A.H., Smaller salt particles in oceanic air and bubble behavior in the sea, *J. Geophys. Res.*, *77* (27), 5316-5321, 1972.

Woodcock, A.H., Marine fog droplets and salt nuclei - Part I, *J. Atmos. Sci.*, *35*, 657-664, 1978.

Woodcock, A.H., and D.C. Blanchard, Tests of the salt-nuclei hypothesis of rain formation, *Tellus*, *7*, 437-448, 1955.

Woodcock, A.H., and R.A. Duce, The "large" salt nuclei hyphothesis of raindrop growth in Hawaii: further measurements and discussion, *Journal de Recherche Atmosphériques*, *6*, 639-649, 1972.

Woodcock, A.H., and M.M. Gifford, Sampling atmospheric sea-salt nuclei over the ocean, *J. Marine Res.*, *8*, 177-197, 1949.

Woodcock, A.H., and A.T. Spencer, An airborne flame photometer and its use in the scanning of marine atmospheres for sea-salt particles, *J. Meteorol.*, *14*, 437-447, 1957.

Woodcock, A.H., and A.T. Spencer, Lava-sea-air contact areas as sources of sea-salt particles in the atmosphere, *J. Geophys. Res.*, *66* (9), 2873-2887, 1961.

Woodcock, A.H., C.F. Kientzler, A.B. Arons, and D.C. Blanchard, Giant condensation nuclei from bursting bubbles, *Nature*, *172*, 1144-1145, 1953.

Woodcock, A.H., D.C. Blanchard, and C.G.H. Rooth, Salt-induced convection and clouds, *J. Atmos. Sci.*, *20*, 159-169, 1963.

Woodcock, A.H., R.A. Duce, and J.L. Moyers, Salt particles and raindrops in Hawaii, *J. Atmos. Sci.*, *28*, 1252-1257, 1971.

Woodcock, A.H., D.C. Blanchard, and J.E. Jiusto, Marine fog droplets and salt nuclei - Part II, *J. Atmos. Sci.*, *38*, 129-140, 1981.

Woolf, D.K., Bubble and aerosol production at the ocean surface, in *Whitecaps and the Marine Atmosphere, Report No. 8*, edited by E.C. Monahan, P.A. Bowyer, D.M. Doyle, M.R. Higgins, and D.K. Woolf, University College, Galway, Ireland, 1985.

Woolf, D.K., Comment on an article by J. Wu, *Tellus*, *42B*, 385-386, 1990.

Woolf, D.K., Bubbles and the air-sea transfer velocity of gases, *Atmosphere-Ocean*, *31* (4), 517-540, 1993.

Woolf, D.K., Bubbles and their role in gas exchange, in *The Sea Surface and Global Change*, edited by P.S. Liss, and R.A. Duce, pp. 173-205, Cambridge University Press, Cambridge, 1997.

Woolf, D.K., and E.C. Monahan, Laboratory investigations of the influence on marine aerosol production of the interaction of oceanic whitecaps and surface-active material, in *Aerosols and Climate*, edited by P.V. Hobbs, and M.P. McCormick, pp. 1-8, Deepak Publishing, 1988.

Woolf, D.K., and S.A. Thorpe, Bubbles and the air-sea exchange of gases in near-saturation conditions, *J. Marine Res.*, *49*, 435-466, 1991.

Woolf, D.K., P.A. Bowyer, and E.C. Monahan, Discriminating between film drops and jet drops produced by a simulated whitecap, *J. Geophys. Res.*, *92* (C5), 5142-5150, 1987.

Woolf, D.K., E.C. Monahan, and D.E. Spiel, Quantification of the marine aerosol produced by whitecaps, in *Seventh Congress on Ocean-Atmosphere Interaction*, pp. 182-185, American Meteorological Society, Anaheim, CA, 1988.

Worthington, A.M., *A Study of Splashes*, 170 pp., Macmillan Company, New York, 1963.

Worthington, L.V., The water masses of the world ocean: some results of a fine-scale census, in *Evolution of Physical Oceanography. Scientific Surveys in Honor of Henry Stommel*, edited by B.A. Warren, and C. Wunsch, pp. 42-69, MIT Press, Cambridge, MA, 1981.

Wright, H.L., Atmospheric opacity: A study of visibility observations in the British Isles, *Quar. J. Royal Meteorol. Soc.*, *65*, 411-442, 1939.

Wright, H.L., The origin of sea-salt nuclei, *Quar. J. Royal Meteorol. Soc.*, *66*, 11-12, 1940a.

Wright, H.L., Sea-salt nuclei, *Quar. J. Royal Meteorol. Soc.*, *66*, 3-12, 1940b.

Wright, H.L., Atmospheric opacity at Valentia, *Quar. J. Royal Meteorol. Soc.*, *66*, 66-77, 209-213, 1940c.

Wu, J., Wind stress and surface roughness at air-sea interface, *J. Geophys. Res.*, *74* (2), 444-455, 1969a.

Wu, J., Correction to paper by Jin Wu, 'Wind stress and surface roughness at air-sea interface', *J. Geophys. Res.*, *74* (13), 3450, 1969b.

Wu, J., Froude number scaling of wind-stress coefficients, *J. Atmos. Sci.*, *26*, 408-413, 1969c.

Wu, J., A note on sea spray and wind-stress discontinuity, *Deep-Sea Res.*, *18*, 1041-1042, 1971a.

Wu, J., Evaporation retardation by monolayers: another mechanism, *Science*, *174*, 283-285, 1971b.

Wu, J., Sea-surface slope and equilibrium wind-wave spectra, *Physics of Fluids*, *15* (5), 741-747, 1972.

Wu, J., Spray in the atmospheric surface layer: laboratory study, *J. Geophys. Res.*, *78* (3), 511-519, 1973.

Wu, J., Evaporation due to spray, *J. Geophys. Res.*, *79* (27), 4107-4109, 1974.

Wu, J., Wind-induced drift currents, *J. Fluid Mechanics*, *68*, 49-70, 1975.

Wu, J., Spray in the atmospheric surface layer: review and analysis of laboratory and oceanic results, *J. Geophys. Res.*, *84* (C4), 1693-1704, 1979a.

Wu, J., Oceanic whitecaps and sea state, *J. Phys. Oceanogr.*, *9*, 1064-1068, 1979b.

Wu, J., Wind-stress coefficients over sea surface near neutral conditions - a revisit, *J. Phys. Oceanogr.*, *10*, 727-740, 1980.

Wu, J., Bubble populations and spectra in near-surface ocean: summary and review of field measurements, *J. Geophys. Res.*, *86* (C1), 457-463, 1981a.

Wu, J., Evidence of sea spray produced by bursting bubbles, *Science*, *212*, 324-326, 1981b.

Wu, J., Sea spray: a further look, *J. Geophys. Res.*, *87* (C11), 8905-8912, 1982a.

Wu, J., Comments on "Optimal power-law description of oceanic whitecap coverage dependence on wind speed", *J. Phys. Oceanogr.*, *12*, 750-751, 1982b.

Wu, J., Wind-stress coefficients over sea surface from breeze to hurricane, *J. Geophys. Res.*, *87* (C12), 9704-9706, 1982c.

Wu, J., Stability parameters and wind-stress coefficients under various atmospheric conditions, *J. Atmos. Oceanic Technol.*, *3*, 333-339, 1986a.

Wu, J., Whitecaps, bubbles, and spray, in *Oceanic Whitecaps and Their Role in Air-Sea Exchange Processes*, edited by E.C. Monahan, and G. MacNiocaill, pp. 113-124, D. Reidel Publishing Company, Dordrecht, 1986b.

Wu, J., Bubbles in the near-surface ocean: a general description, *J. Geophys. Res.*, *93* (C12), 587-590, 1988a.

Wu, J., Variations in whitecap coverage with wind stress and water temperature, *J. Phys. Oceanogr.*, *18*, 1448-1453, 1988b.

Wu, J., Contributions of film and jet drops to marine aerosols produced at the sea surface, *Tellus*, *41B*, 469-473, 1989a.

Wu, J., Reply, *J. Phys. Oceanogr.*, *19*, 710-711, 1989b.

Wu, J., Reply to D. K. Woolf, *Tellus*, *42B*, 387-388, 1990a.

Wu, J., Comment on "Film drop production as a function of bubble size" by D.C. Blanchard and L. D. Syzdek, *J. Geophys. Res.*, *95* (C5), 7389-7391, 1990b.

Wu, J., Vertical distributions of spray droplets near the sea surface: influences of jet drop ejection and surface tearing, *J. Geophys. Res.*, *95* (C6), 9775-9778, 1990c.

Wu, J., On parameterization of sea spray, *J. Geophys. Res.*, *95* (C10), 18269-18279, 1990d.

Wu, J., Bubble flux and marine aerosol spectra under various wind velocities, *J. Geophys. Res.*, *97* (C2), 2327-2333, 1992a.

Wu, J., Individual characteristics of whitecaps and volumetric description of bubbles, *IEEE J. Oceanic Eng.*, *17* (1), 150-158, 1992b.

Wu, J., Production of spume drops by the wind tearing of wave crests: the search for quantification, *J. Geophys. Res.*, *98* (C10), 18221-18227, 1993.

Wu, J., Film drops produced by air bubbles bursting at the surface of seawater, *J. Geophys. Res.*, *99* (C8), 16403-16407, 1994a.

Wu, J., Bubbles in the near-surface ocean: their various structures, *J. Phys. Oceanogr.*, *24*, 1955-1965, 1994b.

Wu, J., Microwave specular reflections from breaking waves - measurements of hurrican winds from space, *J. Phys. Oceanogr.*, *25*, 3231-3236, 1995a.

Wu, J., Sea surface winds - a critical input to oceanic models, but are they accurately measured, *Bull. Am. Meteorol. Soc.*, *76* (1), 13-19, 1995b.

Wu, J., Bubbles produced by breaking waves in fresh and salt waters, *J. Phys. Oceanogr.*, *30*, 1809-1813, 2000a.

Wu, J., Concentration of sea-spray droplets at various wind velocities: separating production through bubble bursting and wind tearing, *J. Phys. Oceanogr.*, *30*, 195-200, 2000b.

Wu, J., Production functions of film drops by bursting bubbles, *J. Phys. Oceanogr.*, *31* (11), 3249-3257, 2001.

Wu, J., Reply, *J. Phys. Oceanogr.*, *32*, 1265-1267, 2002a.

Wu, J., Jet drops produced by bubbles bursting at the surface of seawater, *J. Phys. Oceanogr.*, *32*, 3286-3290, 2002b.

Wu, J., and P.A. Hwang, Comment on "Ambient and transient bubble spectral densities in quiescent seas and under spilling breakers" by H. Medwin and N. D. Breitz, *J. Geophys. Res.*, *96* (C1), 865-866, 1991.

Wu, J., J.J. Murray, and R.J. Lai, Production and distributions of sea spray, *J. Geophys. Res.*, *89* (C5), 8163-8169, 1984.

Wyman, J.J., P.F. Scholander, G.A. Edwards, and L. Irving, On the stability of gas bubbles in sea water, *J. Marine Res.*, *11*, 47-62, 1952.

Wyngaard, J.C., The effects of probe-induced flow distortion on atmospheric turbulence measurements, *J. Applied Meteorol.*, *20*, 784-794, 1981.

Wyngaard, J.C., The effects of probe-induced flow distortion on atmospheric turbulence measurements: extension to scalars, *J. Atmos. Sci.*, *45* (22), 3400-3412, 1988a.

Wyngaard, J.C., Flow-distortion effects on scalar flux measurements in the surface layer: Implications for sensor design, *Boundary-Layer Meteorol.*, *42*, 19-26, 1988b.

Xu, D., P.A. Hwang, and J. Wu, Breaking of wind-generated waves, *J. Phys. Oceanogr.*, *16*, 2172-2178, 1986.

Xu, D., X. Liu, and D. Yu, Probability of wave breaking and whitecap coverage in a fetch-limited sea, *J. Geophys. Res.*, *105* (C6), 14253-14259, 2000.

Yamomoto, Y., N. Mitsuishi, and S. Kadoya, Design and operation of evaporators for radioactive wastes, *Techn. Report 87*, pp. 115, International Atomic Energy Agency, Vienna, 1968.

Yan, B., B. Chen, and K. Stamnes, Role of oceanic air bubbles in atmospheric correction of ocean color imagery, *Applied Optics*, *41* (12), 2202-2212, 2002.

Yang, Z., S. Tang, and J. Wu, An experimental study of rain effects on fine structures of wind waves, *J. Phys. Oceanogr.*, *27*, 419-430, 1997.

Yao, X., M. Fang, and C.K. Chak, The size dependence of chloride depletion in fine and coarse sea-salt particles, *Atmos. Environ.*, *37*, 743-751, 2003.

Yelland, M.J., B.I. Moar, P.K. Taylor, R.W. Pascal, J. Hutchings, and V.C. Cornell, Wind stress measurements from the open ocean corrected for airflow distortions by the ship, *J. Phys. Oceanogr.*, *28*, 1511-1526, 1998.

Yelland, M.J., B.I. Moat, R.W. Pascal, and D.I. Berry, CRD model estimates of the airflow distortion over research ships and the impact on momentum flux measurement, *J. Atmos. Oceanic Technol.*, *19*, 1477-1499, 2002.

Yershov, O.A., A.V. Smirnov, and K.S. Shirfin, Spectral transparency and solar halo in the atmosphere above the ocean, *Atmos. Oceanic Phys.*, *26* (4), 287-292, 1990.

Yoon, Y.J., and P. Brimblecombe, Modelling the contribution of sea salt and dimethyl sulfide derived aerosol to marine CCN, *Atmos. Chem. Phys.*, *2*, 17-30, 2002.

Yoshizumi, K., and K. Asakuno, Characterization of atmospheric aerosols in Chichi of the Ogasawara (Bonin) Islands, *Atmos. Environ.*, *30* (1), 151-155, 1996.

Yount, D.E., Skins of varying permeability: a stabilization mechanism for gas cavitation nuclei, *J. Acoustical Soc. Am.*, *65* (6), 1429-1439, 1979.

Yudine, M.I., Physical considerations on heavy-particle diffusion, *Advances Geophys.*, *6*, 185-191, 1959.

Yum, S.S., and J.G. Hudson, Maritime/continental microphysical contrasts in stratus, *Tellus*, *54B*, 61-73, 2002.

Yum, S.S., J.G. Hudson, and Y. Xie, Comparison of cloud microphysics with cloud condensation nuclei spectra over the summertime Southern Ocean, *J. Geophys. Res.*, *103* (D13), 16625-16636, 1998.

Zedel, L., and D. Farmer, Organized structures in subsurface bubble clouds: Langmuir circulation in the open ocean, *J. Geophys. Res.*, *96* (C5), 8889-8900, 1991.

Zeisse, C.R., NAM6: Batch code for the Navy Aerosol Model, *Techn. Report 1804*, pp. 82, SPAWAE Systems Center, San Diego, 1999.

Zhang, D., and Y. Iwasaka, Chlorine deposition on dust particles in marine atmosphere, *Geophys. Res. Lttrs*, *28* (18), 3613-3616, 2001.

Zhang, X., M. Lewis, and B. Johnson, Influence of bubbles on scattering of light in the ocean, *Applied Optics*, *37* (27), 6525-6536, 1998.

Zhang, X., M. Lewis, M. Lee, B. Johnson, and G. Korotaev, The volume scattering function of natural bubble populations, *Limnol. Oceanogr.*, *47* (5), 1273-1282, 2002.

Zhao, D., and Y. Toba, Dependence of whitecap coverage on wind and wind-wave properties, *J. Oceanogr.*, *57*, 603-616, 2001.

Zheng, Q.A., V. Klemas, G.S. Hayne, and N.E. Huang, The effect of oceanic whitecaps and foams on pulse-limited radar altimeters, *J. Geophys. Res.*, *88* (C4), 2571-2578, 1983a.

Zheng, Q.A., V. Klemas, and Y.-H.L. Hsu, Laboratory measurements of water surface bubble life time, *J. Geophys. Res.*, *88* (1), 701-706, 1983b.

Zhuang, Y., E.P. Lozowski, J.D. Wilson, and G. Bird, Sea spray dispersion over the ocean surface: a numerical simulation, *J. Geophys. Res.*, *98* (C9), 16547-16553, 1993.

Zielinski, A., and T. Zielinski, Aerosol fluxes and their gradients in the marine boundary layer over the coastal zone, *Bull. Polish Acad. Sci., Earth Sciences*, *44* (4), 203-211, 1996.

Zieminski, S.A., and D.R. Raymond, Experimental study of the behavior of single bubbles, *Chem. Eng. Sci.*, *23*, 17-28, 1968.

Zieminski, S.A., and R.C. Whittemore, Behavior of gas bubbles in aqueous electrolyte solutions, *Chem. Eng. Sci.*, *26*, 509-520, 1971.

ZoBell, C.E., and H.M. Mathews, A qualitative study of the bacterial flora of sea and land breezes, *Proc. Natl. Acad. Sci., U.S.A.*, *22*, 567-572, 1936.

Zufall, M.J., C.I. Davidson, P.F. Caffrey, and J.M. Ondov, Airborne concentrations and dry deposition fluxes of particulate species to surrogate surfaces deployed in southern Lake Michigan, *Environ. Sci. Technol.*, *32*, 1623-1628, 1998a.

Zufall, M.J., M.H. Bergin, and C.I. Davidson, Effects of non-equilibrium hygroscopic growth of $(NH_4)_2SO_4$ on dry deposition to water surfaces, *Environ. Sci. Technol.*, *32*, 584-590, 1998b.

Index

About the Authors

Ernie R. Lewis, physicist and oceanographer, received his education at the California Institute of Technology, the von Karman Institute for Fluid Dynamics, and the University of Texas at Austin. His recent work focused on the measurement of physical chemical properties of seawater and precise determination of oceanic carbon dioxide system parameters. He has participated in nine seagoing research cruises at locations as diverse as the South Indian Ocean and the Arctic off Ellesmere Island and has thus tasted more than his share of sea salt. He is a member of the American Geophysical Union. He enjoys nature photography and birding, and hopes to soon attain a life list of 500 birds.

Stephen E. Schwartz was educated as a physical chemist at Harvard and the University of California, Berkeley. As an atmospheric chemist he has focused on cloud and aerosol chemistry and microphysics and examination of the influences of aerosols on climate change. He is a member of the American Chemical Society, the American Physical Society, the American Geophysical Union, the American Meteorological Society, the American Association for Aerosol Research, and the Gesellschaft für Aerosolforschung and a fellow of the American Association for the Advancement of Science. He is the recipient of the 2003 *Atmospheric Environment* Haagen-Smit Award. He enjoys sailing and watching breaking waves (and sea salt aerosol production) on Moriches Bay off the south shore of Long Island.

Printed in the United States
By Bookmasters